Lecture Notes in Computer Science 11802

More information about this series at http://www.springer.com/series/7408

Thomas Schiex · Simon de Givry (Eds.)

Principles and Practice of Constraint Programming

25th International Conference, CP 2019
Stamford, CT, USA, September 30 – October 4, 2019
Proceedings

Editors
Thomas Schiex ⓘ
INRA
Castanet Tolosan, France

Simon de Givry ⓘ
INRA
Castanet Tolosan, France

ISSN 0302-9743 ISSN 1611-3349 (electronic)
Lecture Notes in Computer Science
ISBN 978-3-030-30047-0 ISBN 978-3-030-30048-7 (eBook)
https://doi.org/10.1007/978-3-030-30048-7

LNCS Sublibrary: SL2 – Programming and Software Engineering

This Springer imprint is published by the registered company Springer Nature Switzerland AG
The registered company address is: Gewerbestrasse 11, 6330 Cham, Switzerland

Preface

This volume contains the proceedings of the 25th International Conference on the Principles and Practice of Constraint Programming (CP 2019), which was held in Stamford, Connecticut, USA, during September 30 – October 4, 2019. Detailed information about the conference should be available at http://cp2019.a4cp.org. The CP conference is the annual international conference on constraint programming. It is concerned with all aspects of computing with constraints, including theory, algorithms, environments, languages, models, systems, and applications such as decision making, resource allocation, scheduling, configuration, planning, or automated design. These two facets of CP are represented by the technical and application track of the conference.

The CP community is increasingly keen to ensure CP remains open to interdisciplinary research at the intersection between constraint programming and other directly related fields. The senior Program Committee of the CP 2019 edition, therefore, included people with mixed backgrounds beyond CP – propositional logic and integer linear programming, among others. To reach out beyond these direct 'discrete optimization' connections, the CP 2019 edition continued to offer specialized thematic tracks targeted at the frontier between CP and another specific area. With the current progress and excitement around machine learning technology and the new opportunities that such progress offers for modeling and solving, the CP, Data Science and Machine Learning Track was the most successful thematic track of this edition, attracting even more submissions than the application track. Other thematic tracks were targeted at specific application domains that often raise specific challenges for CP, including the Testing and Verification Track, the Multi-agent and Parallel CP Track, the Computational Sustainability Track and the CP and Life Sciences Track. Each track has a specific sub-committee to ensure that specialist reviewers from the relevant domains vetted papers in their track.

We invited submissions to all tracks and we received 118 submissions (excluding abstracts). The review process of CP 2019 relied on a multitier approach involving one senior Program Committee with Program Committees for all tracks and additional reviewers recruited by Program Committee members. Authors could submit either full papers, with a maximum length of 15 pages without references and abstracts that are not included in these proceedings. All full paper submissions were assigned to a senior Program Committee member and three members of the relevant track Program Committee. Authors were given the opportunity to respond to reviews, generating discussions overseen by the senior Program Committee members and the chairs. Abstracts were directly managed by the chairs. Meetings between the conference chairs and all members of the senior Program Committee were held at the end of June, chaired by the program chairs, where the reviews, author feedback, and discussions were revisited in detail, based on meta-reviews previously prepared by senior Program Committee members. The result of this was that the acceptance rate was 39%. The senior Program Committee and the chairs awarded the Best Conference Paper Prize to Alex Mattenet, Ian Davidson, Siegfried Nijssen, and Pierre Schaus for "Generic Constraint-based Block Modeling using Constraint Programming," the Best Student Paper Prize to Rocsildes Canoy and Tias Guns for "Vehicle Routing by Learning from Historical Solutions," and the Distinguished Student Paper Price to Mohd Hafiz Hasan and Pascal Van Hentenryck for "The Flexible and Real-Time Commute Trip Sharing Problems." The program chairs also invited two papers for direct publication in the *Constraints* journal (Editor-in-Chief Michela Milano). These papers were presented at the conference like any other paper and appeared later in the *Constraints* journal. Awarded papers and nominated papers were also invited to submit an extended version of their paper in the *JAIR* journal (with its own rigorous reviewing process being applied to these extended submissions).

The conference program featured four invited talks by Ian Davidson, Bistra Dilkina, Nina Narodystka, and Phebe Vayanos. These invited talks were selected with the senior Program Committee with the general idea of supporting the current trend of hybridization between CP and machine learning, and the increasing importance of 'algorithms' in everybody's life. This volume includes one-page abstracts of their talks. The conference also included four tutorials and three satellite workshops, whose topics are listed in this volume. The doctoral program gave Ph.D. students an opportunity to present their work to more senior researchers, to meet with an assigned mentor for advice on their research and early career, to attend special invited talks, and to interact with each other. Doctoral program papers went through an internal reviewing process that allowed young scientists to familiarize themselves with reviewing, discussion, and with the usage of EasyChair, the usual CP conference submissions management tool.

The program chairs are grateful to the many people that made this conference such a success. First of all, we are grateful to the authors who provided the material from which the conference is made. Then to the senior Program Committee members who helped us in several of the crucial phases of the conference organization, be it for tutorial and invited speaker selection or for reviews, rebuttal, and discussion management, meta-review writing, and participation in live remote meetings for final acceptance decisions (sometimes at extreme hours). The chairs are also extremely liable to the members of the program committees. By filtering the most novel and original contributions and maintaining high standards of quality in rigor and writing quality, their work is essential for both authors and the community at large. The chairs also address a very special thanks to the authors of various additional reviews that were needed to give all papers enough reviews from qualified persons.

Of course, there is a whole team standing around us, who directly managed specific aspects of the conference: Pierre Schaus (Application track chair), Michele Lombardi and Tias Guns (CP, Data Science and Machine Learning track), Arnaud Gotlieb and Nadjib Lazaar (Testing and Verification track), Ferdinando Fioretto and William Yeoh (Multi-agent and Parallel CP track), Michela Milano and Barry O'Sullivan (Computational Sustainability track), François Fages and Sylvain Soliman (CP and Life Sciences track), Javier Larrosa (workshop and tutorial chair), and Charlotte Truchet (publicity chair).

We would also like to thank the Association for Constraint Programming (ACP). The ACP has been managing the conference for fifteen years now, the conference benefits from very helpful organization support. The program chairs are grateful for the help of the ACP president (Maria Garcia de la Banda) and the ACP conference coordinator (Claude-Guy Quimper) for their support and availability when we needed them.

The conference would not have been possible without the great job done by all the people involved in the local organization. The program chairs heavily relied on the local chair (David Bergman) to provide support for all the special contributors to the CP 2019 program such as invited talks and tutorial speakers and to speedily announce program updates on the conference website. David was supported in this endeavor by Ugochukwu Etudo, Tamilla Triantoro, Niam Yaraghi, Mohsen Emadikhiav, Teng Huang, Saharnaz Mehrani, and Arvind Raghunathan. We would therefore also like to thank the institutions that supported them during the organization: the University of Connecticut first, but also the Quinnipiac University and the Mitsubishi Electric Research Lab.

We acknowledge the generous support of all our sponsors. They include, at the time of this writing:

- *The Artificial Intelligence Journal* (Elsevier)
- Cosling (a French CP startup)
- Huawei
- IBM Research
- AIMMS
- Mitsubishi Electric Research Laboratories Inc.
- The Optimization Firm
- Springer

Thanks to these donations, the local chairs have been able to make CP even better, especially by supporting some of the doctoral program expenses.

July 2019

Thomas Schiex
Simon de Givry

Tutorials and Workshops

Tutorials

Complete Characterisation of Tractable Constraint Languages

Andrei Bulatov   School of Computing Science, Simon Fraser University, Canada

Planning/Scheduling with CP Optimizer

Philippe Laborie   IBM CPLEX Optimization Studio, IBM France Lab, France

Graphical Models and Constraint Satisfaction Problems

Tomas Werner   Center for Machine Perception, Czech Technical University, Czech Republic

Building a Fast CSP Solver based on SAT

Neng-Fa Zhou   CUNY Brooklyn College and Graduate Center, USA

Workshops

Constraint Modeling and Reformulation (ModRef 2019)

Kevin Leo   Monash University, Australia
Alan Frisch   University of York, UK

Progress Towards the Holy Grail (PTHG 2019)

Eugene Freuder   University College Cork, Ireland

Constraint Solving and Special Purpose Hardware Architectures

J. Christopher Beck   University of Toronto, Canada
Merve Bodur   University of Toronto, Canada
Carleton Coffrin   Los Alamos National Laboratory, USA

Organization

Program Chairs

Simon de Givry	INRA, France
Thomas Schiex	INRA, France

Conference Chair

David Bergman   University of Connecticut, USA

Application Track Chair

Pierre Schaus   Université Catholique de Louvain, Belgium

Computational Sustainability Track Chairs

Michela Milano	University of Bologna, Italy
Barry O'Sullivan	Insight Center for Data Analytics, University College, Ireland

CP and Life Sciences Track Chairs

François Fages	Inria, France
Sylvain Soliman	Inria, France

CP, Data Science and Machine Learning Track Chairs

Michele Lombardi	University of Bologna, Italy
Tias Guns	VUB Brussels, Belgium

Multi-agent and Parallel CP Track Chairs

Ferdinando Fioretto	Georgia Institute of Technology, USA
William Yeoh	Washington University in St Louis, USA

Testing and Verification Track Chairs

Arnaud Gotlieb	SIMULA Research Laboratory, Norway
Nadjib Lazaar	LIRMM Montpellier, France

Publicity Chair

Charlotte Truchet   Université de Nantes, France

Doctoral Program Chairs

Emir Demirović   University of Melbourne, Australia
Ciaran McCreesh   University of Glasgow, Scotland

Local Chairs

Ugochukwu Etudo   University of Connecticut, USA
Tamilla Triantoro   Quinnipiac University, USA
Niam Yaraghi   University of Connecticut, USA

Website Chairs

Mohsen Emadikhiav   University of Connecticut, USA
Teng Huang   University of Connecticut, USA

Sponsorship Chairs

Saharnaz Mehrani   University of Connecticut, USA
Arvind Raghunathan   Mitsubishi Electric Research Lab, USA

Senior Program Committee

Christian Bessiere   LIRMM, CNRS, University of Montpellier, France
Martin Cooper   IRIT - Universite Paul Sabatier, France
Maria Garcia de la Banda Monash University, Australia
Joao Marques-Silva   Universidade de Lisboa, Portugal
Nina Narodytska   VMware Research, USA
Claude-Guy Quimper   Laval University, Canada
Christine Solnon   LIRIS CNRS UMR 5205, INSA Lyon, France
Willem-Jan Van Hoeve   Carnegie Mellon University, USA

Program Committee

Deepak Ajwani   University College Dublin, Ireland
Özgür Akgün   University of St Andrews, UK
Roberto Amadini   The University of Melbourne, Australia
Carlos Ansótegui   Universitat de Lleida, Spain
Gilles Audemard   CRIL, Lens, France
Behrouz Babaki   Polytechnique Montreal, Canada
Fahiem Bacchus   University of Toronto, Canada
Sebastien Bardin   CEA LIST, France

Roman Barták   Charles University, Czech Republic
Chris Beck   University of Toronto, Canada
Nicolas Beldiceanu   IMT Atlantique (LS2N), France
David Bergman   University of Connecticut, USA
Filippo Bistaffa   IIIA-CSIC, Spain
Andrea Borghesi   University of Bologna, Italy
Ken Brown   University College Cork, Ireland
Hadrien Cambazard   G-SCOP, University Joseph Fourier, France
Quentin Cappart   Université Catholique de Louvain, Belgium
Clément Carbonnel   CNRS, France
Mats Carlsson   SICS, Sweden
Roberto Castañeda Lozano   SICS, Sweden
Berthe Y. Choueiry   University of Nebraska-Lincoln, USA
Andre Augusto Cire   University of Toronto, Canada
David Cohen   Royal Holloway, University of London, UK
Bruno Cremilleux   Universite de Caen Normandie, France
Victor Dalmau   University Pompeu Fabra, Spain
Rina Dechter   University of California – Irvine, USA
Sophie Demassey   CMA, MINES ParisTech, France
Agostino Dovier   University di UDINE, Italy
Catherine Dubois   ENSIIE-Samovar, France
Sebastijan Dumancic   Katholieke Universiteit Leuven, Belgium
Guillaume Escamocher   Insight Centre for Data Analytics, Ireland
Pierre Flener   Uppsala University, Sweden
Andrea Formisano   Università di Perugia, Italy
Graeme Gange   Monash University, Australia
Carmen Gervet   Université de Montpellier, France
Tal Grinshpoun   Ariel University, Israel
Stefano Gualandi   Università degli studi di Pavia, Italy
Djamal Habet   LIS, France
Jin-Kao Hao   University of Angers, France
Emmanuel Hébrard   LAAS, CNRS, France
Matthias Heizmann   University of Freiburg, Germany
John Hooker   Carnegie Mellon University, USA
Hiroshi Hosobe   Hosei University, Japan
Alexey Ignatiev   Universidade de Lisboa, Portugal
Mikolas Janota   INESC-ID, IST, University of Lisbon, Portugal
Peter Jeavons   University of Oxford, UK
George Katsirelos   MIAT, INRA, France
Philip Kilby   Data61, Australian National University, Australia
Joris Kinable   Eindhoven University of Technology, The Netherlands
Lars Kotthoff   University of Wyoming, USA
T. K. Satish Kumar   University of Southern California, USA
Philippe Laborie   IBM, France
Jean Marie Lagniez   CRIL, University of Artois, France
Christophe Lecoutre   CRIL, University of Artois, France

Jimmy Lee	The Chinese University of Hong Kong, SAR China
Kevin Leo	Monash University, Australia
Xavier Lorca	IMT Mines Albi, France
Samir Loudni	Université de Caen Basse-Normandie, France
Ines Lynce	INESC-ID, IST, Universidade de Lisboa, Portugal
Feifei Ma	Chinese Academy of Sciences, China
Michael Maher	Reasoning Research Institute, Australia
Radu Marinescu	IBM, Ireland
Ciaran McCreesh	University of Glasgow, UK
Christopher Mears	Redbubble, Australia
Amnon Meisels	Ben Gurion University of the Negev, Israel
Laurent Michel	University of Connecticut, USA
Ian Miguel	University of St Andrews, UK
Thierry Moisan	Element AI, Canada
Samba Ndojh Ndiaye	Liris, France
Peter Nightingale	University of York, UK
Anastasia Paparrizou	CRIL-CNRS, University of Artois, France
Andrea Passerini	University of Trento, Italy
Justin Pearson	Uppsala University, Sweden
Marie Pelleau	Université Côte d'Azur, CNRS, I3S, France
Guillaume Perez	University of Nice-Sophia Antipolis, I3S, France
Laurent Perron	Google France, France
Gilles Pesant	Polytechnique Montréal, Canada
Thierry Petit	IMT Atlantique, France and WPI, USA
Gauthier Picard	Ecole des Mines de Saint-Etienne, France
Cédric Piette	CRIL, University of Artois, France
Andreas Podelski	University of Freiburg, Germany
Enrico Pontelli	New Mexico State University, USA
Cédric Pralet	ONERA Toulouse, France
Steve Prestwich	Insight Centre for Data Analytics, Ireland
Patrick Prosser	Glasgow University, UK
Jean-Charles Regin	University Nice-Sophia Antipolis, I3S, CNRS, France
Andrea Rendl	Satalia, UK
Emma Rollon	Universitat Politècnica de Catalunya, Spain
Louis-Martin Rousseau	Polytechnique, Canada
Olivier Roussel	CRIL, CNRS UMR 8188, France
Hana Rudová	Masaryk University, Czech Republic
Domenico Salvagnin	University of Padova, Italy
Scott Sanner	University of Toronto, Canada
Pierre Schaus	Université Catholique de Louvain, Belgium
Paul Shaw	IBM, France
Mohamed Siala	Insight Centre for Data Analytics, Ireland
Anne Siegel	Irisa, France
Laurent Simon	Labri, Bordeaux Institute of Technology, France

Helmut Simonis	Insight Centre for Data Analytics, Ireland
Sylvain Soliman	Inria, France
Kostas Stergiou	University of Western Macedonia, Greece
Peter Stuckey	Monash University, Australia
Peter J. Stuckey	Monash University, Australia
Stefan Szeider	Vienna University of Technology, Austria
Guido Tack	Monash University, Australia
Cyril Terrioux	LIS, UMR CNRS 7020, France
Kevin Tierney	University of Bielefeld, Germany
Gilles Trombettoni	LIRMM, University of Montpellier, France
Peter van Beek	University of Waterloo, Canada
Pascal Van Hentenryck	Georgia Institute of Technology, USA
Elise Vareilles	Mines Albi, France
Philippe Vismara	LIRMM, SupAgro, France
Christel Vrain	LIFO, University of Orléans, France
Mark Wallace	Monash University, Australia
Toby Walsh	The University of New South Wales, Australia
Nic Wilson	Insight UCC, Cork, Ireland
Lebbah Yahia	University of Oran 1, Algeria
Roland Yap	National University of Singapore, Singapore
Harel Yedidsion	Ben-Gurion University, Israel
Yingqian Zhang	Eindhoven University of Technology, The Netherlands
Roie Zivan	Ben Gurion University of the Negev, Israel
Luis Quesada	Insight Centre for Data Analytics, University College Cork, Ireland
Christian Schulte	KTH Royal Institute of Technology, Sweden

Additional Reviewers

Aribi, Noureddine	Lhomme, Olivier
Belaid, Mohammed Said	Li, Chu-Min
Björdal, Gustav	Montmirail, Valentin
Bonacina, Ilario	Sais, Lakhdar
Choi, Arthur	Slivovsky, Friedrich
Hebrard, Emmanuel	Tardivo, Fabio
Hoenicke, Jochen	Tentrup, Leander
Ihler, Alexander	Tiwari, Ashish
Kolter, Zico	Trimble, James
Lammich, Peter	Tsouros, Dimosthenis C.
Langenfeld, Vincent	Vella, Flavio

Invited Talks

Using Constraints in Machine Learning

Ian Davidson

Department of Computer Science, University of California, Davis, USA
davidson@cs.ucdavis.edu

Abstract. We will give a broad overview of our decade long effort to add constraints to machine learning. We will begin by exploring motivating examples of the need for constraints from applications in social network analysis, medical imaging, and intelligent tutoring systems. We then discuss how constraints can be used such as for encoding domain knowledge, for transfer learning, and for adding humans to the machine learning loop.

We overview our results on encoding constraints in terms of their computational difficulty in general and for encoding in different types of solvers including SAT and CP solvers. With the benefit of hindsight, we will then discuss successful and unsuccessful formulations we have worked on in the past from clustering, regression, block modeling, and outlier detection.

We will conclude by overviewing future new directions and challenges for using CP and other discrete solvers in machine learning such as encoding rules a priori for deep learning, post-processing results for explanation, and encoding notions of fairness.

Discrete Optimization and Machine Learning for Sustainability

Bistra Dilkina

Viterbi School of Engineering, University of Southern California, USA
dilkina@usc.edu

Abstract. My research focuses on advancing the state of the art in combinatorial optimization techniques for solving real-world large-scale problems, particularly ones that arise in sustainability areas such as biodiversity conservation planning and urban planning. The work I will present is at the intersection of discrete optimization and machine learning. One key area of research is designing machine-learning-driven combinatorial optimization algorithms, by leveraging the plethora of data generated by solving distributions of real world optimization problems.

Verification and Explanation of Deep Neural Networks

Nina Narodytska

VMware Research, Palo Alto, CA, USA
n.narodytska@gmail.com

Abstract. Deep neural networks are among the most successful artificial intelligence technologies making an impact in a variety of practical applications. However, many concerns were raised about the 'magical' power of these networks. It is disturbing that we are clearly lacking an understanding of the decision making process behind this technology. Therefore, a natural question is whether we can trust decisions that neural networks make.

There are two ways to address this problem that are closely related. The first approach is to define properties that we expect a neural network to satisfy. Verifying whether a neural network fulfills these properties sheds light on the properties of the function that it represents. Verification guarantees can reassure the user that the network has an expected behavior. The second approach is to better understand the decision making process of neural networks. Namely, the user can require that a neural network decision must be accompanied by an explanation for this decision. Such explanations help the user to understand the decision making process of the network function.

In this talk, we consider both research directions. We take a logic-based approach to analysis of neural networks, where the network is represented in a logical formalism, like Boolean Satisfiability (SAT) or Satisfiability Modulo Theories (SMT). From this standpoint, we overview the progress in verification and explainability. In particular, we will discuss recent progress in verification of neural networks, focusing on a special class of neural networks – Binarized Neural Networks – that can be represented and analyzed using Boolean Satisfiability. We discuss how we can take advantage of the training procedure and the structure of the network to speed up verification. In particular, we demonstrate that the choice of the training procedure can have significant impact on scalability of the network verification procedure. For the explainability, we present our work on producing logical explanations for machine learning model decisions. We also explain how logic-based tools can be used to verify the quality of explanations produced by well-known explainer tools.

AI and Robust Optimization for Social Good

Phebe Vayanos

Center for Artificial Intelligence in Society, University of Southern California,
USA
phebe.vayanos@usc.edu

Abstract. In the last decades, significant advances have been made in AI and optimization. Recently, systems relying on these technologies are being transitioned to the field with the potential of having tremendous positive influences on people and society. With increase in the scale and diversity of deployment of AI- and optimization-driven algorithms in the open world come several challenges including the need for tractability and resilience, issues of data scarcity and bias, information endogeneity, ethical considerations, and issues of shared responsibility between humans and algorithms. In this talk, we focus on the problems of homelessness, wildlife conservation, and public health in vulnerable communities, and present research advances in AI and robust optimization to address one key cross-cutting question: how to effectively allocate scarce intervention resources in these domains while accounting for the challenges of open world deployment? We will show concrete improvements over the state of the art in these domains based on both real world data and deployments in the LA area. We are convinced that, by pushing this line of research, AI and robust optimization can play a crucial role to help fight injustice and solve complex problems facing our society.

Contents

Application Track

Multi-agent and Parallel CP Track

Testing and Verification Track

CP and Data Science Track

Computational Sustainability Track

CP and Life Sciences Track

Technical Track

Instance Generation via Generator Instances

Özgür Akgün, Nguyen Dang[⊠], Ian Miguel, András Z. Salamon,
and Christopher Stone

School of Computer Science, University of St Andrews, St Andrews, UK
{ozgur.akgun,nttd,ijm,Andras.Salamon,cls29}@st-andrews.ac.uk

Abstract. Access to good benchmark instances is always desirable when
developing new algorithms, new constraint models, or when comparing
existing ones. Hand-written instances are of limited utility and are time-
consuming to produce. A common method for generating instances is
constructing special purpose programs for each class of problems. This
can be better than manually producing instances, but developing such
instance generators also has drawbacks. In this paper, we present a
method for generating *graded* instances completely automatically starting from a class-level problem specification. A graded instance in our
present setting is one which is neither too easy nor too difficult for a
given solver. We start from an abstract problem specification written
in the ESSENCE language and provide a system to transform the problem specification, via automated type-specific rewriting rules, into a new
abstract specification which we call a generator specification. The generator specification is itself parameterised by a number of integer parameters; these are used to characterise a certain region of the parameter
space. The solutions of each such generator instance form valid problem
instances. We use the parameter tuner irace to explore the space of possible generator parameters, aiming to find parameter values that yield
graded instances. We perform an empirical evaluation of our system for
five problem classes from CSPlib, demonstrating promising results.

Keywords: Automated modelling · Instance generation ·
Parameter tuning

1 Introduction

In constraint programming, each problem class is defined by a problem specification; many different specifications are possible for the same problem class.
A problem specification identifies a class of combinatorial structures, and lists
constraints that these structures must satisfy. A solution is a structure satisfying all constraints. Problem specifications usually also have formal parameters,
which are variables for which the specification does not assign values but are not
intended to be part of the search for solutions. Values for such formal parameters

© Springer Nature Switzerland AG 2019
T. Schiex and S. de Givry (Eds.): CP 2019, LNCS 11802, pp. 3–19, 2019.
https://doi.org/10.1007/978-3-030-30048-7_1

are provided separately, and the specification together with a particular choice of values for these formal parameters defines a problem instance.

Instance generation is the task of choosing particular values for the formal parameters of a problem instance, and is often a key component of published work when existing benchmarks are inadequate or missing. Our goal is to automate instance generation. We aim to automatically create parameter files containing definitions of the formal parameters of a problem specification, from the high level problem specification itself, and without human intervention.

We automate instance generation by rewriting a high level constraint specification in the ESSENCE language [7] into a sequence of generator instances for the problem class. Values for the parameters of the generator specification are chosen based on the high level types in the problem specification. A solution to a generator instance is a valid parameter file defining a problem instance. We use irace [15], a popular tool for the automatic configuration of algorithms, to search the space of generator parameters for regions where "graded instances" exist. Graded instances have specific properties; in this work they are satisfiable, and neither too trivial nor too difficult to be solved. However, our methodology does not depend on a specific definition of grading, and can be applied more generally. We first prove the soundness of our rewriting scheme. The system is then empirically evaluated over 5 different problem classes that contain different combinations of integers, functions, matrices, relations and sets of sets. We show the viability of our system and the efficacy of the parameter tuning against randomised search over all problem classes.

2   Related Work

In combinatorial optimisation a wide variety of custom instance generators have been described. These are used to construct synthetic instances for problem classes where too few benchmarks are available. In just the constraint programming literature generators have been proposed for many problem classes, including quasigroup completion [4], curriculum planning [17], graph isomorphism [26], realtime scheduling [11], and bike sharing [6]. Different evolutionary methods have also been proposed to find instances for binary CSPs [18], Quadratic Knapsack [13], and TSP [23]. In particular, Ullrich et al. specified problem classes with a formal language, and used this system to evolve instances for TSP, MaxSAT, and Load Allocation [24]. Efforts have also been made to extend existing repositories of classification problems via automated instance generation [19].

Instance generators are typically built to support other parts of the research, such as verifying robustness of models. However, a generator often requires significant effort to develop, and it cannot be applied to new problem classes without major modifications [24]. A generator is typically controlled by means of parameters, and a further challenge of instance generation is to find regions of parameter values where an instance generator can reliably create interesting instances.

Gent et al. developed parametric generators of instances for several problem classes [8]. They developed a semi-automated prototype to produce instances for discriminating among potential models for a given high-level specification. Their system requires manual rewriting of the domains when there are dependencies between parameters, and does not support all of ESSENCE. In contrast, our system works in a completely automated fashion, for all ESSENCE types, and supports dependencies between formal parameters.

We use the irace system to sample intelligently from the space of instances. irace is a general-purpose tool for automatic configuration of an algorithm's parameters, and its effectiveness has been shown in a wide range of applications [5,12,14,15]. Our system uses irace to find values of the generator parameters covering graded instances.

Our generator instance method could be applied to many constraint modelling languages such as MiniZinc [20], Zinc [16], Essence Prime [22], or OPL [25]. In this paper we focus on the ESSENCE language [7] because of its support for high level types, and since the open-source CONJURE system [1–3] provides a convenient basis on which to build an automated instance generation system. We exploit the high level types of ESSENCE to guide the rewriting process.

3   Background

We now introduce notation used in the remainder of the paper.

A *problem class* is the set of problem instances of interest. A *problem specification* is a description of a problem class in a constraint specification language. A problem specification defines the types but not the values of several *formal parameter* variables. An assignment of specific values to the formal parameters is called an *input*, and a *parameter file* contains an input. A variable that occurs in the problem specification, but which neither occurs within the scope of a quantifier over that variable, nor is a formal parameter, is called a *decision variable*. We refer to a specification together with an input as an *instance*. A *solution* to an instance is an assignment of values to the decision variables in the instance. An instance is *satisfiable* if it has a solution. If all input values are of the correct type for the corresponding formal parameters, then the input is *valid*. An *valid instance* consists of a specification and a valid input for that specification. Valid instances may have many, one, or no solutions. For optimisation problems, we further wish to search among satisfying solutions to find those of high quality, where quality is determined by an expression to be optimised.

We use the abstract constraint specification language ESSENCE [7]. This comprises formal parameters (`given`), which may themselves be constrained (`where`); the combinatorial objects to be found (`find`); constraints the objects must satisfy (`such that`); identifiers declared (`letting`); and an optional objective function (`min/maximising`). ESSENCE supports *abstract* decision variables, such as multiset, relation and function, as well as *nested* types, such as multiset of sets.

We seek *graded* instances. With this we mean instances that satisfy predefined criteria. The criteria should be tailored to the use to which the instances will be put. In this work, we require graded instances to be neither too easy nor too difficult. To ensure an instance is not too easy, we require that the back-end solver (in our case, Minion [9]) takes at least 10 s to decide the instance. To ensure an instance is not too difficult, we exclude instances for which the solver has not returned a solution in 5 min. Our choices of grading criteria were guided by our computational budget and available resources, and so in this work we have chosen to accept only satisfiable instances as graded. We do not advocate a specific definition of grading, and other criteria for grading would be reasonable, such as "the instance is decided or solved to optimality by at least one solver from a portfolio of solvers in a reasonable amount of time".

A key step of our method is a process of automatic rewriting, discussed in more detail in Sect. 4.1. Briefly, the rewriting steps are:

1. remove all constraints (such that statements) and decision variables (find statements),
2. replace all input parameters (given statements) with decision variables (find statements) and type specific constraints, and
3. promote parameter constraints (where statements) to constraints.

We call the result of this process a *generator specification*.

Definition 1. *A generator instance consists of a generator specification together with a particular choice of generator parameters, which restrict the domains of decision variables appearing in the generator instance.*

Rewriting is one step in an iterative process. The choice of generator parameters is performed automatically using the parameter tuning tool irace. Solutions to the generator instance are then filtered according to our grading criteria, retaining graded instances. We want the rewriting procedure to have the following two properties: *soundness* (the solutions of the generator instance should always be valid inputs for the instance), and *completeness* (every valid instance should be obtainable as a possible solution of the generator instance). We now discuss the semantics of generator instances and our approach.

Variables may represent tuples, and for clarity of presentation we take some liberties with the corresponding ESSENCE syntax. When referring to a specification s with variables v, we omit the variables that occur within the scope of a quantifier, and partition the remaining variables so that $s(x \mid y)$ denotes a specification s with formal parameters x and decision variables y, both of which are generally tuples. With $s(x := a \mid y)$ we denote the specification $s'(\mid y)$ (with no formal parameters and only decision variables) which is obtained from s by substituting the tuple of formal parameters x by a fixed tuple of values a.

Start with an Essence specification of the form

$$s(x \mid y) := \texttt{given } x : D \texttt{ where } h(x) \texttt{ find } y : E \texttt{ such that } f(x,y)$$

where the specification has formal parameters x and decision variables y. We are interested in valid inputs a, such that $s(x := a \mid y)$ is a valid instance. Here domain D may be a product of component domains, of arbitrarily nested types as allowed in the ESSENCE language. Now let

$$s'(\mid x) := \texttt{find } x : D \texttt{ such that } h(x)$$

be a specification obtained from $s(x)$ by our rewriting process, which drops the original constraints $f(x, y)$, replaces the given by a find, and modifies where statements into such that statements, leaving a specification with no formal parameters but only decision variables (Note that many possible but equivalent specifications are possible for s'). In principle we could search for a solution $x := a$ to this specification $s'(|\ x)$, as this would be a valid input for $s(x \mid y)$, yielding the instance $s(x := a \mid y)$. Such search seldom finds graded instances in a reasonable amount of time, unless more guidance is provided.

Thus, we want to introduce a new parameter p with domain P to structure our search for instances. We then rewrite the specification $s(x \mid y)$ differently, as

$$s''(p \mid x) := \text{find } x : D(p) \text{ such that } c(p, x)$$

so that as the values assigned to the formal parameters p vary, the solutions to the instance $s''(p := q \mid x)$ form valid inputs to $s(x \mid y)$. The specification $s''(p \mid x)$ will be our generator specification, instead of $s'(|\ x)$. We can then treat P as a space of parameters, and explore this space with a parameter tuning tool.

The types or domain expressions of the formal parameters x with domain D in the specification $s'(|\ x)$ may have a lot of structure and be quite complex. Exploring such parameter spaces successfully is a challenging problem. We therefore aim to simplify our task of instance generation by replacing these structured domains by the usually smaller domains $D(p)$ in the new specification $s''(p \mid x)$, and automatically incorporate this structural information into the constraints $c(p, x)$ instead; the constraints $c(p, x)$ include both the constraints $h(x)$ and also the additional constraints to capture structural information. Like D, the parameter domain P is usually a product of domains, but for P these are usually just intervals of reals, ranges of integers, or Booleans.

4 Methodology

In Fig. 1 we show how our system turns an abstract problem specification into concrete problem instances with the use of rewriting rules and an iterated sequence of tuned generator instances. The steps of the automated process are:

1. Start with a specification of a problem in the ESSENCE language.
2. Rewrite the problem specification into a generator specification (Sect. 4.1).
3. Create a configuration file for the parameter tuner irace.
4. irace searches for promising values of the generator parameters (Sect. 4.3).
5. At each iteration the current generator instance is used to create multiple problem instances which are solved by Savile Row [21].
6. The time to solve an instance and its satisfiability are used as feedback to irace about the quality of the current parameters.
7. At the end of the process several problem instances are generated.

The rest of this section describes the details of this process, correctness of the rewriting procedure, how we use tuning based on instance difficulty, the problem classes we studied, and our experimental setup.

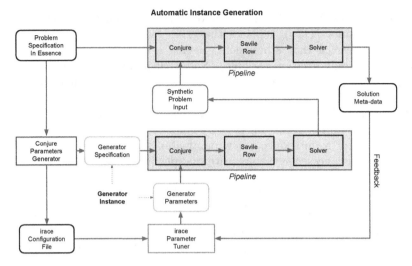

Fig. 1. ESSENCE specifications (*top-left*) are fed to the CONJURE parameter generator (*left-red*). Here they are rewritten into a generator specification and a configuration file for irace (*bottom-green*), which selects parameter values to generate a synthetic instance (*centre-purple*). Solution meta-data are used to inform the tuner. (Color figure online)

4.1 Rewriting Rules

For each ESSENCE type we deploy a set of rules that transform a `given` state-ment into a different ESSENCE statement or set of statements that captures the problem of finding valid input parameters for the initial `given`. Whenever a given type is nested inside other types, such as an input parameter, the rewriting rules are applied recursively until an explicit numerical value is obtained.

4.1.1 Rewriting `int`

For every integer domain, we generate two configurator parameters, `middle` and `delta`. The domains of these configurator parameters are identical to that of the original integer domain. If the original domain is not finite, we use `MININT` and `MAXINT` values as bounds, which are to be provided to CONJURE. Default values for `MININT` and `MAXINT` are 0 and 50, respectively. For an integer decision variable x we generate the following constraints to relate it to the corresponding `middle` and `delta` parameters: `x >= middle - delta` and `x <= middle + delta`.

```
given x : int(1..50)
```

is rewritten as:

```
given x_middle : int(1..50)
given x_delta : int(0..24)
find x : int(1..50)
such that x >= x_middle - x_delta, x <= x_middle + x_delta
```

4.1.2   Rewriting `function`

For every parameter with a `function` domain, we produce a decision variable
that has a finite function domain and additional constraints to ensure it can
only be assigned to the allowed values. Total function domains are rewritten as
`function` (without the `total` attribute) and add an extra constraint to ensure
the function is defined (NB `defined(f)` returns the set of elements of the range
of function `f` that have an image) up to the value required.

```
given d : int(1..10)
given f : function (total) int(1..d) --> int(1..50)
```

is rewritten as:

```
given d_middle : int(1..10)
given d_delta : int(0..4)
find d : int(1..10)
such that d >= d_middle - d_delta, d <= d_middle + d_delta
given f_range_middle : int(1..50)
given f_range_delta : int(0..24)
find f : function int(1..10) --> int(1..50)
such that
    forAll i : int(1..10) .
        i >= 1 /\ i <= d <-> i in defined(f),
    forAll i in defined(f) .
        f(i) >= f_range_middle - f_range_delta /\
        f(i) <= f_range_middle + f_range_delta
```

4.1.3   Rewriting `matrix`

Each `matrix` is rewritten into a `function` and the rewriting rules for functions
are utilised.

4.1.4   Rewriting `relation`

For relations we generate two configurator parameters that bound the cardinality
of the relations, two that bound the left-hand side values of the relation (`R_1`),
and another two for the right-hand side values (`R_2`).

```
letting DOM1 be domain int: (1..10)
letting DOM2 be domain int: (1..50)
given R: relation of (DOM1*DOM2)
```

is rewritten as:

```
given R_cardMiddle : int(1..50)
given R_cardDelta : int(0..24)
given R_1_middle : int(1..10)
given R_1_delta : int(0..4)
given R_2_middle : int(1..50)
```

```
given R_2_delta : int(0..24)
find R: relation (maxSize 50) of (int(1..10) * int(1..50))
such that
    |R| >= R_cardMiddle - R_cardDelta /\
    |R| <= R_cardMiddle + R_cardDelta,
    forAll i in defined(R) .
        i[1] >= R_1_middle - R_1_delta /\
        i[1] <= R_1_middle + R_1_delta /\ i[1] <= 10 /\
        i[2] >= R_2_middle - R_2_delta /\
        i[2] <= R_2_middle + R_2_delta /\ i[2] <= 50
```

4.1.5 Rewriting set

We discuss the case of a set of set. Here we generate a pair of configurator parameters for the cardinality of the outer set with the usual bounds, then for the cardinality of the inner set we use a much smaller delta and use the size of the set as middle. Finally another pair of parameters bounds the size of the innermost set. The outer cardinality and the innermost bounds parameters are omitted as they are equivalent to the ones for relation and function, respectively.

```
letting DOM be domain int: (1..50)
given S : set of set (size 2) of DOM
```

is rewritten as:

```
<<middle/delta cardinality parameters as in relation>>
<<middle/delta parameters as in function>>
given S_inner_cardMiddle: int(2) ·
given S_inner_cardDelta: int(0..3)
find S: set of set (minSize 2, maxSize 2) of int(1..50)
such that
    <<middle/delta cardinality bound as in relation>>
    <<middle/delta bounds as in functions>>
    forAll s1 in S .
        |s1| >= S_inner_cardMiddle - S_inner_cardDelta /\
        |s1| <= S_inner_cardMiddle + S_inner_cardDelta /\
        |s1| >= 2 /\ |s1| <= 2 /\ forAll s2 in s1 . s2 <= 50
```

4.2 Correctness of Instance Generation via Generator Instances

We need to prove that the rewriting that CONJURE does to turn an ESSENCE specification into a generator instance is sound, in that rewriting should always produce an instance that is a valid input to the specification. We show soundness by means of a decomposition based on types; we illustrate the proof for the case of total functions and leave the remaining cases to the full version of the paper. We also wish rewriting to be complete, in that every possible instance for the specification should be an output of the instance generator specification as long as it is given the right parameter file as input, but this is only possible for instances that satisfy some additional assumptions.

We illustrate the rewriting process with the following example, which demonstrates the rewriting of the function type for a restricted instance. Given the specification $s(\mathtt{d}, \mathtt{f} \mid \mathtt{x})$ as in the example Sect. 4.1.2 our system rewrites this into the generator specification

$s''(\mathtt{d\_middle}, \mathtt{d\_delta}, \mathtt{f\_range\_middle}, \mathtt{f\_range\_delta} \mid \mathtt{d}, \mathtt{f})$.

We have built our system to ensure soundness by design.

Proposition 1. *The semantics of our rewriting rule for* function *types is sound.*

Proof. Consider a solution of the generator instance

$$s''(\mathtt{d\_middle} := u, \mathtt{d\_delta} := v, \mathtt{f\_range\_middle} := r, \mathtt{f\_range\_delta} := s \mid \mathtt{d}, \mathtt{f})$$

where the values u, v, r, s are provided in a parameter file, created by our system. This solution consists of an integer $\mathtt{d}$ in the range $\mathtt{int(1..10)}$ and a function $\mathtt{f}$ with domain $\mathtt{int(1..10)}$ and codomain $\mathtt{int(1..50)}$. The constraints force $\mathtt{f}$ to be defined over the entire range $\mathtt{int(1..d)}$, and for its values to be in the range

$$\mathtt{int((f\_range\_middle - f\_range\_delta)..(f\_range\_middle + f\_range\_delta))}.$$

Moreover $\mathtt{f\_range\_middle} - \mathtt{f\_range\_delta} \geq 1$ must hold as a consequence of the choices made by the system for $\mathtt{f\_range\_middle}$ and $\mathtt{f\_range\_delta}$. Similarly the system ensures that $\mathtt{f\_range\_middle} + \mathtt{f\_range\_delta} \leq 50$. Hence $\mathtt{f}$ is a total function with domain $\mathtt{int(1..d)}$ and codomain $\mathtt{int(1..50)}$. Therefore $s(\mathtt{d}, \mathtt{f} \mid \mathtt{x})$ together with a solution of the generator instance is a valid instance. $\square$

The other types can be dealt with similarly; in particular, proofs for nested types follow a standard compositional style.

In contrast, it seems challenging to ensure completeness. One issue is infinite domains: when a formal parameter of a specification has a type that allows an infinite domain, then any restriction of this domain to a finite set means that the rewriting process cannot be complete. However, our current system is built on CONJURE and requires finite domains for all decision variables. Our generator specifications therefore restrict all domains to be finite, and in such cases completeness is necessarily lost. For specifications where the domains of the formal parameters are all finite, it seems possible to guarantee completeness. Parameters that are dependent can be another obstacle to achieving completeness. To avoid this issue the generator parameters must all be sampled independently and the rewriting process must ensure that no dependencies between parameters are introduced. We leave issues of completeness to further work.

4.3   Tuning Instance Difficulty

Posing the problem of finding valid instances as ESSENCE statements is a fundamental step but not sufficient for the reliable creation of problem instances.

Efficient and effective searching in the instance space for graded instances is not a trivial task. We solve this problem by utilising the tuning tool irace. In this section, we first describe the tuning procedure of irace. We then explain how we have applied irace, including details of the input to irace and the feedback provided by each generator's evaluation to guide the search of irace.

4.3.1   The Tuning Procedure of irace

We give a brief summary of the specific tuning procedure implemented by irace and explain why such an automatic algorithm configurator is a good choice for our system in the next section. For a detailed description of irace and its applications, readers are referred to [15].

The algorithm configuration problem irace tackles is as follows: given a parameterised algorithm A and a problem instance set I, we want to find algorithm configurations of A that optimise a performance metric defined on I, such as minimising the average solving time of A across all instances in I. The main idea of irace is using *racing*, a machine learning technique, in an iterated fashion to efficiently use the tuning budget. Each iteration of irace is a *race*. At the first iteration, a set of configurations is randomly generated and these are evaluated on a number of instances. A statistical test is then applied to eliminate the statistically significantly worse configurations. The remaining configurations continue to be tested on more instances before the statistical test is applied again. At the end of the race, the surviving configurations are used to update a sampling model. This model is then used to generate a set of new configurations for the next iteration (race). This process repeats until the tuning budget is exhausted. The search mechanism of irace allows it to focus more on the regions of promising configurations: the more promising a configuration is, the more instances it is evaluated on and the more accurate the estimate of its performance over the whole instance set I will be. This is particularly useful when I is a large set and/or A is a stochastic algorithm.

4.3.2   Using irace to Find Graded Instances

In our instance generation context, the parameterised algorithm A is our generator instance. Each input for the generator instance, which we call a generator *configuration*, will cover a part of the instance space. The instance set I in our context is a set of random seeds. The search procedure of irace enables efficient usage of the tuning budget, as the more promising an instance region covered by a configuration proves, the more instances will be generated from it.

Paired with each ESSENCE generator specification there is a configuration file that is utilised by irace to tune the parameters of the generator specification. The configuration file is automatically created by CONJURE and defines a generator configuration.

Given a random seed, an evaluation of a generator configuration involves two steps. First, a problem instance is generated by solving the generator instance using CONJURE, Savile Row, and Minion. The generator configuration normally covers several instances, and the random seed is passed to Minion for deciding which instance is to be returned.

Second, the generated instance is solved by Minion, and its satisfaction property and the solving time are recorded. We use these values to assign a score to the generator configuration. The highest score is given if the instance satisfies our grading criteria, so that irace is guided to move towards the generator's configuration spaces where graded instances lie. The assignment of scores depends on the specific definition of instance grading. In our case, we define graded instances as satisfiable (SAT) and solvable by Minion within $[10, 300]$ seconds. We also place a time limit of 5 min on Savile Row for the translation from the ESSENCE instance parameter to Minion input format. A score of 0 is given if the generated instance is either UNSAT, or too difficult (Minion times out), or too large (Savile Row times out). If the instance is SAT but too easy (solvable by Minion in less than 10 s), the Minion solving time is returned as the score. If the instance satisfies our grading criteria, a score of 10 is returned. The scale of the scores is not important, as the default choice for the statistical test used in irace is the Friedman test, a non-parametric test where scores are converted to ranks before being compared. Following tuning, we collect the set of graded instances generated. irace also returns a number of promising generator configurations. These configurations can be kept for when we want to sample more graded instances that are similar to the ones produced by the tuning procedure.

4.4   Problem Classes

CSPlib is a diverse collection of combinatorial search problems, covering ancient puzzles, operational research, and group theory [10]. Most of these problems have ESSENCE specifications. To test our system we have selected representative problems that span most of the ESSENCE types used for formal parameters in CSPlib. We now briefly describe each of these problems (with CSPlib problem numbers).

Template Design (2): The objective is to minimise the wastage in a printing process where the number of templates, the number of design variations and the number of slots are given, while satisfying the demand. The formal parameters are 3 integers and 1 total function.

The Rehearsal Problem (39): The objective is to produce a schedule for a set of musicians that have to practice pieces with specified durations in groups. The goal is to minimise the total amount of time the musicians are waiting to play. The formal parameters are 2 integers, 1 total bijective function and 1 relation.

A Distribution Problem with Wagner-Whitin Costs (40): The objective is to find an ordering policy minimising overall cost, given the number of products, their cost, maximum stock available, their demand, holding costs, and the distribution hierarchy. The formal parameters are 4 integers and 4 matrices.

Synchronous Optical Networking (SONET) Problem (56): Consider a set of nodes and a demand value for each pair of nodes. A ring connects nodes and a

Table 1. Number of graded instances produced for each problem class and parameter search method within a budget of 1000 evaluations.

CSPlib	Problem name	Types	Problem kind	irace Linear	random Linear	irace Log	random Log
2	Template Design	3 integer, 1 function	Optimisation	**788**	491	464	49
39	Rehearsal	2 integers, 1 function, 1 relation	Optimisation	10	0	**25**	0
40	Wagner-Whitin	4 integers, 4 matrices	Optimisation	48	4	**60**	2
56	SONET	3 integers, 1 set of sets	Optimisation	37	7	**78**	40
135	Van der Waerden Numbers	3 integers	Satisfaction	**121**	64	33	18

node can be installed into the ring using an add-drop multiplexer (ADM). Network traffic can be routed between two nodes only if they are on the same ring. The objective is to minimise the number of ADMs to install while satisfying all demands. The formal parameters are 3 integers and 1 set of sets of integers.

Van der Waerden Numbers **(135)**: The goal is to decide if a given number n is smaller than the Van der Waerden number predefined by a number of colors and an arithmetic length. The problem has 3 formal integer parameters.

4.5 Experimental Setup

We demonstrate our methodology on the five problem classes described in Sect. 4.4. A budget of 1000 generator configuration evaluations is given to the tuning. To illustrate the tuning's efficiency, we also run the same experiment with uniformly randomly sampling using the same budget.

The system parameter MAXINT defines the maximum value for any unbounded integer parameters. Here we set it to 50. We leave for future work the questions about the impact of this parameter and the tuning budget on the effectiveness of the system, and how to set them properly given a specific problem class.

We also consider two options for sampling each generator parameter's values, both of which are supported by irace. The first is *linear-scale sampling*, where all values in the domain are treated equally at the start of tuning (or during the whole random search). The second is *logarithmic-scale sampling*, where the logarithms of the lower and upper bounds of the parameter domains are calculated first, and a value is sampled from this new domain before being converted back into the original range. The logarithmic scale makes smaller ranges finer-grained and vice versa, and can potentially help search in scenarios where larger parameter values tend to make instances become either too large or too difficult. This will be demonstrated in our experimental results in the next section.

Each generated instance is solved using Minion for 5 random seeds. During an evaluation of a generator configuration, as soon as the generated instance violates the criteria on one of the seeds, the evaluation is stopped, and the violated run is used as a result for scoring the generator configuration. We use this early-stopping mechanism to maximise the information gained per CPU-hour, as little information is gained from multiple runs on an uninteresting instance.

Experiments were run on two servers, one with 40-core Intel Xeon E5-2640 2.4 GHz, and one with 64-core AMD Opteron 6376 2.3 GHz. All experiments for the same problem class were performed on the same server. Each experiment used between 7 and 95 CPU core hours, depending on problem class and search variants. The experiments we report here used 700 CPU core hours in total.

5   Results and Analysis

In Table 1 we report the number of graded instances found by the four search variants: irace or random search in combination with linear or logarithmic scale-sampling. Across the five problem classes, the winner is always an irace tuning variant. irace with linear-scale sampling works best on Template Design and Van der Waerden Numbers, while irace with logarithmic-scale sampling is able to find more graded instances for Rehearsal, Wagner-Whitin Distribution, and SONET.

In Table 2 we juxtapose plots of the progress over time with the total numbers of instances produced during the process for each problem class, divided into categories. It can be seen that irace vastly outperforms randomised search. In all cases, during the first half of the tuning budget, the difference in performance between irace and random search is not always clearly visible as irace is still in its exploration mode (the few first iterations/races where the sampling model of irace was still initialised). However, by the second half of the budget the tuning has gained some knowledge about the promising regions of the generator configuration space, and irace starts showing a significant boost in the number of graded instances found compared with random search.

In the case of the Rehearsal Problem, where we generate relatively fewer instances compared with other classes, the plot shows that by the end of the tuning budget the system is just picking up pace and it is fair to expect that with more iterations it would produce significantly more instances.

Looking at the category results in Table 2, we can infer some knowledge about the instance space of a specific problem class based on those statistics and the difference in performance between the two scales for sampling (linear vs logarithmic). For example, in the Template Design case, most generated instances are SAT, and are either too easy or graded. The larger number of too easy instances found by the logarithmic scale sampling suggests a strong correlation between the domains of the generator parameters and easiness of the instances generated. Smaller generator configuration values mostly cover SAT and easy-to-solve instances, while larger configurations cover more difficult instances. Since we prefer sufficiently difficult instances to too easy instances, this explains why linear sampling works better than log-scale for this particular problem. A similar

Table 2. Progress of the four search variants for each problem class, with a budget of 1000 evaluations. Each plot shows how the number of graded instances found grows as the number of evaluations increases. The table displays the number of instances for each problem class, instance category, and search variant. (SR refers to Savile Row.)

Progress	Instance categories	irace Linear	irace Log	random Linear	random Log
002-TemplateDesign	Minion timeout	14	17	44	44
	SAT too easy	217	557	521	910
	SR timeout	0	0	0	11
	UNSAT	0	0	0	0
	graded	**788**	464	491	49
	no instance	0	0	0	0
039-Rehearsal	Minion timeout	109	57	14	26
	SAT too easy	74	154	3	18
	SR timeout	0	0	0	0
	UNSAT	629	306	662	556
	graded	10	**25**	0	2
	no instance	152	438	321	399
040-DistributionWagnerWhitin	Minion timeout	97	98	16	47
	SAT too easy	139	256	9	21
	SR timeout	0	0	0	0
	UNSAT	215	242	134	317
	graded	48	**60**	4	2
	no instance	517	352	838	613
056-SONET	Minion timeout	216	198	284	237
	SAT too easy	169	508	19	230
	SR timeout	0	0	0	0
	UNSAT	2	19	4	38
	graded	37	**79**	7	40
	no instance	607	244	692	489
135-VanDerWaerden	Minion timeout	283	179	555	77
	SAT too easy	571	546	282	439
	SR timeout	0	0	0	12
	UNSAT	136	258	162	458
	graded	**126**	94	64	18
	no instance	0	0	0	14

explanation can be applied for the Van der Waerden case. However, the large number of too-difficult instances suggests that `MAXINT=50` is probably too large for this problem, and reducing this parameter value could potentially boost performance of the search within our limited budget. Another example is Rehearsal where the statistics indicate a strong correlation between the numbers of UNSAT and too-difficult instances, and between the number of too easy instances and infeasible generator configurations. These suggest that a smaller `MAXINT` value combined with linear sampling could potentially improve search performance.

6   Conclusions and Future Work

We have developed a system that automates the production of graded instances for combinatorial optimisation and decision problems. Our system creates a generator specification from an abstract problem specification. Generator parameters are explored using the irace parameter tuning system. We demonstrated the soundness of our approach and performed an empirical evaluation over several problem classes. The experiments showed that automated tuning of generator parameters outperforms random sampling for all problem classes under study, and is able to discover significant numbers of graded instances automatically. The system and all data produced by this work is publicly available as a github repository https://github.com/stacs-cp/CP2019-InstanceGen.

Much future work remains. We first would like to extend our approach to generate instances for every problem class in CSPlib, or at least the ones for which exhibiting a valid instance does not involve first solving long-standing open problems. Many of the classes in CSPlib only have trivially easy instances, or have none, and we would like to remedy this situation. We further seek to automate creation of balanced and heterogeneous sets of instances, by refining our system's notion of a graded instance, and by further investigating the diversity of the generated instances. We believe much work also remains in investigating grading of instances more generally. As we saw in Sect. 5, some problem classes are especially amenable to automatic discovery of their features; in particular, we plan to automate the choice of sampling regime based on performance of tuning in its early stages. A comparison with existing hand-crafted instances/instance-generators will also be considered. Furthermore the system can be adapted to find instances that are easy for one solver but challenging for other solvers; we believe automating the generation of such instances would greatly assist those researchers who build solvers to improve performance of their solvers. Another application is to find instances with certain structures that reflect real-world instances. Finally, we intend to work toward automatic instance generation for specifications involving infinite domains.

Acknowledgements. This work is supported by EPSRC grant EP/P015638/1 and used the Cirrus UK National Tier-2 HPC Service at EPCC (http://www.cirrus.ac.uk) funded by the University of Edinburgh and EPSRC (EP/P020267/1).

References

1. Akgün, Ö.: Extensible automated constraint modelling via refinement of abstract problem specifications. Ph.D. thesis, University of St Andrews (2014)
2. Akgun, O., et al.: Automated symmetry breaking and model selection in CONJURE. In: Schulte, C. (ed.) CP 2013. LNCS, vol. 8124, pp. 107–116. Springer, Heidelberg (2013). https://doi.org/10.1007/978-3-642-40627-0_11
3. Akgün, Ö., Miguel, I., Jefferson, C., Frisch, A.M., Hnich, B.: Extensible automated constraint modelling. In: AAAI 2011: Proceedings of the Twenty-Fifth AAAI Conference on Artificial Intelligence, pp. 4–11. AAAI Press (2011). https://www.aaai.org/ocs/index.php/AAAI/AAAI11/paper/viewPaper/3687
4. Barták, R.: On generators of random quasigroup problems. In: Hnich, B., Carlsson, M., Fages, F., Rossi, F. (eds.) CSCLP 2005. LNCS (LNAI), vol. 3978, pp. 164–178. Springer, Heidelberg (2006). https://doi.org/10.1007/11754602_12
5. Bezerra, L.C.T., López-Ibáñez, M., Stützle, T.: Automatic component-wise design of multiobjective evolutionary algorithms. IEEE Trans. Evol. Comput. **20**(3), 403–417 (2016). https://doi.org/10.1109/TEVC.2015.2474158
6. Di Gaspero, L., Rendl, A., Urli, T.: Balancing bike sharing systems with constraint programming. Constraints **21**(2), 318–348 (2016). https://doi.org/10.1007/s10601-015-9182-1
7. Frisch, A.M., Harvey, W., Jefferson, C., Martínez-Hernández, B., Miguel, I.: Essence: a constraint language for specifying combinatorial problems. Constraints **13**(3), 268–306 (2008). https://doi.org/10.1007/s10601-008-9047-y
8. Gent, I.P., et al.: Discriminating instance generation for automated constraint model selection. In: O'Sullivan, B. (ed.) CP 2014. LNCS, vol. 8656, pp. 356–365. Springer, Cham (2014). https://doi.org/10.1007/978-3-319-10428-7_27
9. Gent, I.P., Jefferson, C., Miguel, I.: Minion: a fast scalable constraint solver. In: Proceedings of ECAI 2006, pp. 98–102. IOS Press (2006). http://ebooks.iospress.nl/volumearticle/2658
10. Gent, I.P., Walsh, T.: CSPlib: a benchmark library for constraints. In: Jaffar, J. (ed.) CP 1999. LNCS, vol. 1713, pp. 480–481. Springer, Heidelberg (1999). https://doi.org/10.1007/978-3-540-48085-3_36
11. Gorcitz, R., Kofman, E., Carle, T., Potop-Butucaru, D., de Simone, R.: On the scalability of constraint solving for static/off-line real-time scheduling. In: Sankaranarayanan, S., Vicario, E. (eds.) FORMATS 2015. LNCS, vol. 9268, pp. 108–123. Springer, Cham (2015). https://doi.org/10.1007/978-3-319-22975-1_8
12. Hoos, H.H.: Automated algorithm configuration and parameter tuning. In: Hamadi, Y., Monfroy, E., Saubion, F. (eds.) Autonomous Search, pp. 37–71. Springer, Heidelberg (2011). https://doi.org/10.1007/978-3-642-21434-9_3
13. Julstrom, B.A.: Evolving heuristically difficult instances of combinatorial problems. In: GECCO 2009: Proceedings of the 11th Annual conference on Genetic and evolutionary computation, pp. 279–286. ACM (2009). https://doi.org/10.1145/1569901.1569941
14. Lang, M., Kotthaus, H., Marwedel, P., Weihs, C., Rahnenführer, J., Bischl, B.: Automatic model selection for high-dimensional survival analysis. J. Stat. Comput. Simul. **85**(1), 62–76 (2015). https://doi.org/10.1080/00949655.2014.929131
15. López-Ibáñez, M., Dubois-Lacoste, J., Cáceres, L.P., Birattari, M., Stützle, T.: The irace package: iterated racing for automatic algorithm configuration. Oper. Res. Persp. **3**, 43–58 (2016). https://doi.org/10.1016/j.orp.2016.09.002, http://iridia.ulb.ac.be/irace/

16. Marriott, K., Nethercote, N., Rafeh, R., Stuckey, P.J., Garcia Banda, M., Wallace, M.: The design of the Zinc modelling language. Constraints **13**(3), 229–267 (2008). https://doi.org/10.1007/s10601-008-9041-4

17. Monette, J.N., Schaus, P., Zampelli, S., Deville, Y., Dupont, P.: A CP approach to the balanced academic curriculum problem. In: Seventh International Workshop on Symmetry and Constraint Satisfaction Problems, vol. 7 (2007). https://info.ucl.ac.be/~pschaus/assets/publi/symcon2007_bacp.pdf

18. Moreno-Scott, J.H., Ortiz-Bayliss, J.C., Terashima-Marín, H., Conant-Pablos, S.E.: Challenging heuristics: evolving binary constraint satisfaction problems. In: GECCO 2012: Proceedings of the 14th Annual Conference on Genetic and Evolutionary Computation, ACM (2012). https://doi.org/10.1145/2330163.2330222

19. Muñoz, M.A., Villanova, L., Baatar, D., Smith-Miles, K.: Instance spaces for machine learning classification. Mach. Learn. **107**(1), 109–147 (2018). https://doi.org/10.1007/s10994-017-5629-5

20. Nethercote, N., Stuckey, P.J., Becket, R., Brand, S., Duck, G.J., Tack, G.: MiniZinc: towards a standard CP modelling language. In: Bessière, C. (ed.) CP 2007. LNCS, vol. 4741, pp. 529–543. Springer, Heidelberg (2007). https://doi.org/10.1007/978-3-540-74970-7_38

21. Nightingale, P., Akgün, Ö., Gent, I.P., Jefferson, C., Miguel, I., Spracklen, P.: Automatically improving constraint models in Savile Row. Artif. Intell. **251**, 35–61 (2017). https://doi.org/10.1016/j.artint.2017.07.001

22. Nightingale, P., Rendl, A.: ESSENCE' description 1.6.4 (2016). https://arxiv.org/abs/1601.02865

23. Smith-Miles, K., van Hemert, J.: Discovering the suitability of optimisation algorithms by learning from evolved instances. Ann. Math. Artif. Intell. **61**(2), 87–104 (2011). https://doi.org/10.1007/s10472-011-9230-5

24. Ullrich, M., Weise, T., Awasthi, A., Lässig, J.: A generic problem instance generator for discrete optimization problems. In: GECCO 2018: Proceedings of the Genetic and Evolutionary Computation Conference Companion, pp. 1761–1768. ACM (2018). https://doi.org/10.1145/3205651.3208284

25. Van Hentenryck, P., Michel, L., Perron, L., Régin, J.-C.: Constraint programming in OPL. In: Nadathur, G. (ed.) PPDP 1999. LNCS, vol. 1702, pp. 98–116. Springer, Heidelberg (1999). https://doi.org/10.1007/10704567_6

26. Zampelli, S., Deville, Y., Solnon, C.: Solving subgraph isomorphism problems with constraint programming. Constraints **15**(3), 327–353 (2010). https://doi.org/10.1007/s10601-009-9074-3

Automatic Detection of At-Most-One and Exactly-One Relations for Improved SAT Encodings of Pseudo-Boolean Constraints

Carlos Ansótegui[1], Miquel Bofill[2], Jordi Coll[2(✉)], Nguyen Dang[3],
Juan Luis Esteban[4], Ian Miguel[3], Peter Nightingale[5], András Z. Salamon[3],
Josep Suy[2], and Mateu Villaret[2]

[1] University of Lleida, Lleida, Spain
`carlos@diei.udl.cat`
[2] University of Girona, Girona, Spain
`{miquel.bofill,jordi.coll,josep.suy,mateu.villaret}@imae.udg.edu`
[3] University of St Andrews, St Andrews, UK
`{nttd,ijm,Andras.Salamon}@st-andrews.ac.uk`
[4] Technical University of Catalonia, Barcelona, Spain
`esteban@cs.upc.edu`
[5] University of York, York, UK
`peter.nightingale@york.ac.uk`

Abstract. Pseudo-Boolean (PB) constraints often have a critical role in constraint satisfaction and optimisation problems. Encoding PB constraints to SAT has proven to be an efficient approach in many applications, however care must be taken to encode them compactly and with good propagation properties. It has been shown that at-most-one (AMO) and exactly-one (EO) relations over subsets of the variables can be exploited in various encodings of PB constraints, improving their compactness and solving performance. In this paper we detect AMO and EO relations completely automatically and exploit them to improve SAT encodings that are based on Multi-Valued Decision Diagrams (MDDs). Our experiments show substantial reductions in encoding size and dramatic improvements in solving time thanks to automatic AMO and EO detection.

Keywords: Automatic CSP reformulation · SAT · Pseudo-Boolean · At-most-one constraint

1 Introduction

Solving constraint satisfaction and optimisation problems often requires dealing with Pseudo-Boolean (PB) constraints, either explicitly stated in the original

Work supported by grants TIN2015-66293-R, TIN2016-76573-C2-1/2-P (MINECO/ FEDER, UE), Ayudas para Contratos Predoctorales 2016 (grant number BES2016-076867, funded by MINECO and co-funded by FSE), RTI2018-095609-B-I00 (MICINN/ FEDER, UE), and EPSRC EP/P015638/1.

© Springer Nature Switzerland AG 2019
T. Schiex and S. de Givry (Eds.): CP 2019, LNCS 11802, pp. 20–36, 2019.
https://doi.org/10.1007/978-3-030-30048-7_2

model or as a product of some reformulation process. A successful approach to solving constraint problems is by translation to SAT and the use of SAT solvers. Example tools that support this method include MiniZinc [18,24], Picat [28], and Savile Row [25]. Ideally, such encodings would be compact (in terms of the number of clauses and additional variables) and would have good propagation properties.

In this paper we focus on efficiently translating PB constraints to SAT within Savile Row, which produces a reformulated SAT model from an input constraint model in the Essence Prime language [26]. There exist several approaches for compactly encoding PB constraints to SAT based on different representations, such as Decision Diagrams [2,13], Sequential Weight Counters [17], Generalised Totalisers [19], and Polynomial Watchdog schemes [5].

There are also attempts to exploit collateral constraints to shrink these encodings further [1,8]. In particular, in [8], it is shown how to use existing At-Most-One (AMO) and Exactly-One (EO) relations on subsets of the variables of a PB constraint to obtain very compact decision diagram-based representations. In that work, the authors provide empirical evidence of the utility of using this technique in several scheduling problems. Specifically, they provide specialised SAT Modulo Theories (SMT) encodings exploiting AMO and EO relations. However, these relations are found by hand and are not always obvious.

In this work we propose a technique for exploiting such collateral constraints when encoding PB constraints to SAT in a fully automatic manner. By collateral constraints we mean constraints that are derived from the entire model in some way. They may appear directly in the model, or they may be implied by constraints in the model. One can then use a declarative constraint modelling language and forget about collateral constraints when posting PB constraints. The proposed system is able to automatically identify AMO and EO relations and to take them into account when encoding PB constraints. In particular, we use the approach described in [3] to detect sets of Boolean variables in a SAT formula that model finite-domain variables, which essentially corresponds to detecting the AMO (i.e., cardinality constraints with $\leq$ operator and $k = 1$) and At-Least-One (ALO) relations among a set of Boolean variables. Later, in [7], a method to detect arbitrary cardinality constraints ($k \geq 1$) was introduced. To the best of our knowledge, [7] is the first attempt to apply in practice reformulation techniques through the automatic detection of cardinality constraints. They reformulate the input SAT formula by erasing the clauses entailed by the cardinality constraints detected so far. In our work, we tackle a different goal since our aim is to use the automatically detected cardinality constraints to improve the encoding of more general constraints, specifically PB constraints.

The proposed techniques are embedded in Savile Row. In preparing the SAT encoding Savile Row employs the propagation facilities of the constraint solver Minion [15] in order to identify AMOs, plus a syntactic technique for identifying At-Least-One (ALO) relations (which together with AMOs comprise EO relations). The use of propagation techniques to obtain semantic information has already been used in other scenarios. For example, in [11] unit propagation was used to deduce sub-clauses from implication graphs, and also unit propagation was used in [14] to detect redundant clauses in SAT formulas.

We apply the technique to several problem classes and highlight the characteristics of each regarding the automatically found AMO and EO relations. Our experiments show dramatic improvements of encoding size and solving time.

2   Preliminaries

Essence Prime is typical of solver-independent constraint modelling languages in providing integer and Boolean variable types, as well as multidimensional matrices of these types. It supports arbitrarily nested arithmetic and logical constraint expressions, as well as a suite of global constraints. Savile Row is able to translate any Essence Prime model into SAT, which we define here.

A *Boolean variable* is a variable than can take truth values 0 (false) and 1 (true). A *literal* is a Boolean variable x or its negation $\neg x$. A *clause* is a disjunction of literals. A *propositional formula in conjunctive normal form* (CNF) is a conjunction of clauses. Any propositional formula can be transformed into CNF.

A CNF formula represents a Boolean function, i.e. a function of the form $f : \{0,1\}^n \rightarrow \{0,1\}$. An *assignment* is a mapping of Boolean variables to truth values, which can also be seen as a set of literals (e.g., $\{x = 1, y = 0, z = 0\}$ is usually denoted $\{x, \neg y, \neg z\}$). A *satisfying assignment* of a Boolean function f is an assignment that makes the function evaluate to 1. In particular, an assignment A satisfies a CNF formula F if at least one literal l of each clause in F belongs to A. Such an assignment is called a *model* of the formula.

SAT is the problem of determining if there exists a satisfying assignment for a given propositional formula. Given two formulas F and G, we say that G is a logical consequence of F, written $F \models G$, iff every model of F is also a model of G. We say that two Boolean functions F and G are logically equivalent, denoted $F \equiv G$, if $F \models G$ and $G \models F$.

Unit propagation (UP) is the core deduction mechanism in modern SAT solvers: whenever each literal of a clause but one is false, the remaining literal must be set to true in order to satisfy the clause. We say that G is a logical consequence of F by UP, written $F \models_{UP} G$, iff $F \wedge \neg G$ can be determined to be unsatisfiable by UP.

Savile Row encodes integer variables to provide SAT literals for $(x = a)$ and $(x \leq a)$ for each integer variable x and value a. Each constraint type is then encoded using these SAT literals, as described in [25]. For this work we have added the MDD encoding of PB constraints as defined below.

Definition 1. *A* pseudo-Boolean *(PB) constraint is a Boolean function of the form $\sum_{i=1}^{n} q_i l_i \diamond K$ where K and the q_i are integer constants, l_i are literals, and $\diamond \in \{<, \leq, =, \geq, >\}$.*

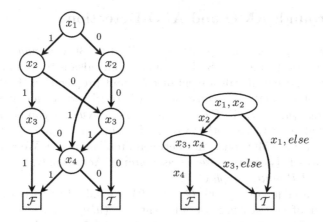

Fig. 1. Left: BDD for $P = 2x_1 + 3x_2 + 4x_3 + 5x_4 \leq 7$; Right: MDD for P, assuming $AMO(x_1, x_2)$ and $AMO(x_3, x_4)$, where each x_i branch means choosing $x_i = 1$, and the *else* branches mean choosing $x_i = 0$ for all x_i in the corresponding source node.

Definition 2. *An* at-most-one *(AMO) constraint is a Boolean function of the form* $\sum_{i=1}^{n} l_i \leq 1$, *where all* l_i *are literals.*

Definition 3. *An* at-least-one *(ALO) constraint is a Boolean function of the form* $\sum_{i=1}^{n} l_i \geq 1$, *where all* l_i *are literals.*

Definition 4. *An* exactly-one *(EO) constraint is a Boolean function of the form* $\sum_{i=1}^{n} l_i = 1$, *where all* l_i *are literals.*

One of the best methods to encode PB constraints to SAT is to use Binary Decision Diagrams (BDDs) [13]. In [2] an even more efficient encoding is given for PB constraints where all coefficients, literals and K are positive and the relational operator is $\leq$. Such a constraint has the important property of being *monotonic decreasing*, i.e. any model remains a model after flipping inputs from 1 to 0. In [8] it is shown how the encoding can be dramatically reduced in size in the presence of AMO constraints over subsets of the variables. The improved encoding is based on Multi-Valued Decision Diagrams (MDDs) and is intended also for monotonic decreasing PB constraints. Figure 1 shows an example of this situation. The number of nodes and edges in the second diagram is substantially reduced, and the number of clauses and variables needed to encode the diagram is reduced accordingly. The input of this encoding is a PB constraint, and a partition of its literals, where each part must satisfy an AMO constraint. We will refer to each of the parts as an AMO group.

An interesting particular case occurs when there are not only AMO constraints, but EO constraints over subsets of the variables in the PB constraint. In this case, the number of variables can be reduced [8]: by subtracting the same integer from all the coefficients of a set of variables in an EO relation, as well as from K, we can make at least one coefficient become zero, and then remove the zero-coefficient terms. The result of reducing the set of variables with an EO relation is also an AMO group.

3   Background: AMO and ALO Detection

In this section we present the approach described in [3] to *semantically* detect AMO and ALO constraints in a SAT formula F. The idea is to compute for each literal in F which other literals are entailed by unit propagation (UP). Then an undirected graph $G = (V, E)$ is constructed, where all vertices $u \in V$ are literals of F and an edge $(u, v) \in E$ iff $F \wedge u \models_{UP} \neg v$, i.e. $F \wedge u \wedge \neg v$ can be determined to be unsatisfiable by UP. In other words, if $(u, v) \in E$ then $F \models (\neg u \vee \neg v)$, therefore there is an AMO constraint between literals u and v. We refer to these AMO constraints between two literals as *mutexes*. Accordingly, we refer to the graph G as the *UP-mutex graph* of F.

Recall that a clique of a graph $G = (V, E)$ is a subset of vertices of G such that every pair of vertices u, v are adjacent, i.e. $(u, v) \in E$. Therefore, every clique $C = (V', E')$ in the UP-mutex graph of a SAT formula F corresponds to an AMO $A = \sum_{v \in V'} v \leq 1$ such that $F \models A$. By construction, we know that there is a mutex between all pairs of literals $u, v \in V'$, hence $F \models u + v \leq 1$ and so $F \models \sum_{v \in V'} v \leq 1$. Thus we can identify all the AMO constraints in a SAT formula F that can be detected by UP by finding the cliques in the UP-mutex graph of F.

In [7] the authors propose an approach to detect cardinality constraints (Boolean functions of the form $\sum_{i=1}^{n} l_i \leq k$ where all l_i are literals and $k \geq 1$ is an integer) which generalize AMO constraints. As pointed out by the authors, this methodology is particularly useful for $k > 2$, compared to other approaches for detecting cardinality constraints.

Given a set of literals L of a formula F we can also automatically detect whether $F \models_{UP} \vee_{l \in L} l$, i.e. F entails by UP an ALO constraint on L, by testing whether $F \wedge \bigwedge_{l \in L} \neg l$ is unsatisfiable by UP.

There are two key details in the procedure we have described to semantically detect the AMO constraints in a SAT formula F. First of all, how do we detect the mutexes, i.e. the level of local consistency (power of propagation) we use to find them. Notice that by enforcing stronger consistency than UP we may identify more mutexes and consequently more AMO constraints. Second, how do we detect the cliques in the UP-mutex graph. Depending on the goal of the particular application, the challenge is to properly address these two key details. In the following section, we adapt this procedure to our context by replacing the SAT formula F with a CSP instance, replacing unit propagation with the propagation of the constraint solver Minion [15].

4   AMO and EO Relations in Savile Row

In this section we describe our approach and how it is integrated into Savile Row. As part of this process we must deal with sum constraints that contain integer terms, negative coefficients, and any comparator $\diamond \in \{<, \leq, =, \neq, \geq, >\}$. The end result is a monotonic decreasing PB constraint and a partition of its literals into AMO groups. This is achieved by a sequence of reformulations, where the AMO groups will arise either from the decomposition of an integer variable, or from the detection of a clique of mutexes in the mutex graph. As described in [25] Savile Row performs two tailoring processes, the first of which uses the constraint solver Minion [15] to filter variable domains, and the second produces output for the desired solver (SAT in this case). Our approach adds mutex detection to Minion, and finds AMO and EO groups during the second tailoring process.

4.1   Mutex Inference

The mutex inference step is performed on Minion's CSP representation of the problem at hand. This representation contains integer constraints that will be transformed into PB constraints later. These integer constraints are of this form $\sum_{i=1}^{n} q_i e_i \diamond K$. An expression e_i may be an integer variable, a Boolean literal, or $(x_i \diamond k_i)$ where x_i is an integer variable or a Boolean literal. Next, any Boolean expressions of the form $(x_i \diamond k_i)$ are replaced with a new Boolean variable b_i and the constraint $b_i \leftrightarrow (x_i \diamond k_i)$ is added to the model. By adding the b_i variables, the mutex detection algorithm is able to see the mutex between $x < 5$ and $x \geq 5$ for example.

Minion is called to perform domain filtering [25] and to find mutexes between literals of Boolean variables. For each Boolean variable b in the CSP, each value of b is assigned in turn and the propagation loop of Minion is called. Consequences of the assignment are propagated through the entire constraint model, including integer variables and global constraints. All assignments of other Boolean variables (to either 0 or 1) by propagation are recorded in the mutex graph G.

Mutex inference is very similar to [3] (described in Sect. 3) with the SAT formula replaced by the CSP, and unit propagation replaced by Minion's propagation algorithms. Comparing propagation power is not straightforward because it depends on the SAT encoding on the one hand, and fine details of propagators on the other. However, there is one key advantage to using the CSP representation: we avoid generating the (potentially very large) encoding of the problem instance without considering AMO and EO relations. See, for example, the Nurse Scheduling Problem (Sect. 5.3) where the encoding that uses AMO and EO relations is ten times smaller than the one without.

4.2   Normalisation

To use the MDD encoding referred to in Sect. 2 we must have *monotonic decreasing* PB constraints in $\leq$ form. Reformulations are required both before and after the AMO and EO groups are constructed. In the first step, all PB and sum constraints are rearranged into the form $\sum_{i=1}^{n} q_i e_i \leq K$ with arithmetic transformations [13].

Terms $q_i e_i$ where e_i is integer are dealt with as follows. Let $q = q_i$ and $e = e_i$. First, if $q < 0$, then $q \leftarrow -q$ and $e \leftarrow -e$. Second, if the smallest possible value c of e is less than 0, then $e \leftarrow e + c$ and K is adjusted by adding qc. Finally, the term qe with n possible values becomes an AMO group of $n - 1$ terms containing $e = k_i$ by enumerating all values k_i except the smallest value, and K is adjusted accordingly.

At this point, all expressions e_i in the constraint are Boolean. All terms $q_i e_i$ where $q_i < 0$ are made positive by replacing with $q_i(1 - \neg e_i)$, then multiplying out and subtracting the constant from both sides. The constraint is now a monotonic decreasing $\leq$ PB constraint, suitable for encoding to SAT via an MDD as described in Sect. 2. However, the next steps may require inverting the polarity of some Boolean expressions e_i in order to match the detected AMOs, losing the normal form. In this case, the normal form will be restored after making the polarities match.

4.3   AMO and EO Detection

For each PB constraint, we take the subgraph $G' = (V', E')$ of the mutex graph G where V' is a set containing both literals of all Boolean variables in the constraint. The algorithm has a list of vertices L, initially containing all vertices in V'. L is sorted by descending degree in G'. A clique cover is constructed by iterating a greedy clique finding algorithm. To construct one clique, the algorithm takes the first vertex from L then adds as many as possible other vertices in the order of L, breaking ties (where the degree is equal) by choosing the vertex whose coefficient is most common within the clique (as a heuristic to reduce the number of outgoing edges of the corresponding nodes in the MDD). Whenever a vertex v is added to a clique, both v and $\neg v$ are removed from L. The end result is a clique cover containing one literal of each Boolean variable in the constraint.

For each clique in the cover, a new AMO or EO group is built as follows. If the negations of literals in the clique correspond with negations in the PB constraint (or the clique has one literal) then we do (1), otherwise (2).

1. The AMO group is constructed directly from the clique. If all literals in the group form an EO corresponding to an integer variable (i.e. literals correspond to $(x = a)$ or $\neg(x \neq a)$ for all values a of some integer variable x), then we can exploit the EO relation to reduce the size of the group. We delete the term(s) with the smallest coefficient c, and subtract c from K and from the other coefficients within the AMO group.
2. If the negation of the term $q_i e_i$ does not match the literal in the clique, the term is rewritten as $q_i(1 - \neg e_i)$ (and rearranged as above), creating a term with a negative coefficient. Once all terms of the group have the appropriate sign, an EO is created by making a new Boolean variable b (constrained to be true iff all expressions e_i in the group are false) and adding a term $0b$ to the group. All coefficients within the group and K are adjusted by subtracting the smallest coefficient. Terms with coefficient zero are removed to create an AMO group.

The result in all cases is an AMO group whose size is at most the size of the clique. In case (1), if an EO is detected then at least one term can be removed relative to the clique. In case (2), if multiple terms have the smallest coefficient then the AMO group is smaller than the clique. Each AMO group detected in this way will be added to the model as an AMO constraint.

We find EO groups by a syntactic check in case (1) above. EO groups can also be detected semantically using propagation (Sect. 3), and the semantic approach may find more EO groups. In our case this would involve calling Minion a second time, with more overhead than the syntactic check.

4.4 Reformulation Example

In this section we give an example of the normalisation and reformulation process that illustrates the described steps and cases. Suppose we have a CSP instance $\mathcal{C}$ with the following variables:

- x which is an integer variable with domain $\{1, 2, 3\}$;
- y which is an integer variable with domain $\{-2, -1, 0, 1\}$; and
- z and t that are Boolean variables.

Suppose $\mathcal{C}$ has the following two constraints to be translated to SAT:

$$C_1 : 2(x = 1) + 4(x = 2) + 3(x = 3) - 3y + 4z + 5t \leq 13$$
$$C_2 : \neg z \vee \neg t$$

Before performing the mutex inference, we replace each of the expressions of the form $(x \diamond k)$ with a Boolean auxiliary variable b, and add the constraint $b \leftrightarrow (x \diamond k)$. C_1 is replaced with the following four constraints:

$$b_1 \leftrightarrow (x = 1)$$
$$b_2 \leftrightarrow (x = 2)$$
$$b_3 \leftrightarrow (x = 3)$$
$$C_3 : 2b_1 + 4b_2 + 3b_3 - 3y + 4z + 5t \leq 13$$

The inference mechanism described in Sect. 4.1 detects the following mutexes, where the first three come from the decomposition of integer variable x, and the last one is due to constraint C_2:

$$\neg b_1 \vee \neg b_2$$
$$\neg b_1 \vee \neg b_3$$
$$\neg b_2 \vee \neg b_3$$
$$\neg z \vee \neg t$$

The following two AMO relations are inferred from the above mutexes:

$$b_1 + b_2 + b_3 \leq 1$$
$$z + t \leq 1$$

These two AMO relations are added to the model as AMO constraints.

An EO relation is detected among b_1, b_2, and b_3, as described in Sect. 4.3. The EO relation is converted into an AMO by removing the term with the smallest coefficient in C_3 (b_1 in this case), and adjusting the coefficients of the other terms (as described in Sect. 4.3). The two Boolean variables z and t form an

AMO group. Finally, the integer variable y with four values will form an AMO group of three terms, as described in Sect. 4.2.

C_3 is reformulated into C_4 as follows:

$$C_4 : 2b_2 + 1b_3 + 9[y = -2] + 6[y = -1] + 3[y = 0] + 4z + 5t \leq 14$$

Note that the right hand side constant has been adjusted to 14, and the coefficients of the terms corresponding to x and y have been adjusted as well. The variables of C_4 are partitioned into the following three AMO groups:

$$\{b_2, b_3\}$$
$$\{[y = -2], [y = -1], [y = 0]\}$$
$$\{z, t\}$$

If the AMO and EO detection process is enabled, the SAT encoding has 18 variables and 33 clauses. Without the detection, it has 33 variables and 53 clauses. The SAT encoding of the MDD derived from C_4 has only 7 clauses, whereas the MDD derived from the constraint without AMO and EO detection (which has only one non-singleton AMO group derived from y) is encoded with 37 clauses.

5   Experimental Evaluation

In this section we evaluate our approach on four diverse case studies: Combinatorial Auctions (CA), the Multi-Mode Resource-Constrained Project Scheduling Problem (MRCPSP), the Nurse Scheduling Problem (NSP), and the Multiple-Choice Multidimensional Knapsack Problem (MMKP). Each of these problem classes have AMO and EO relations that could be identified by expert modellers, and we show that our system is able to identify them without any human effort. The effects on the size and solving time of the resulting SAT formula are dramatic.

All problems except MRCPSP use a PB objective function. To abstract solving performance from any particular optimisation process of the PB objective function, we converted CA, NSP and MMKP problem classes into decision problems. Specifically, we bound the objective function with the best known value of the objective function, so we are searching for a solution that is as good as the best known solution.

For the decision problems CA, NSP, and MMKP, we use the Glucose 4.1 SAT solver [4]. For MRCPSP, where we minimise an integer variable, we use the MaxSAT solver Open-WBO version 2.0 [23], which uses Glucose 4.1 as its core SAT solver. All the experiments were run on an 8GB Intel® Xeon® E3-1220v2 machine at 3.10 GHz. In a preliminary experiment we ran the SAT solver Lingeling (version bcj) [6] on the CA problem and obtained similar results to those reported below with Glucose.

In our experiments we use three configurations. The first (PB) has no AMO or EO detection, however normalisation is always applied when encoding a constraint via an MDD (Sect. 4.2). The second configuration $(PB(AMO))$ performs

AMO detection but not EO detection (i.e. the EO check in step (1) of Sect. 4.3 is switched off). The third configuration (*PB(EO)*) has both AMO and EO detection.

Reported solving times include both reformulation preprocessing and time spent by the SAT solver.

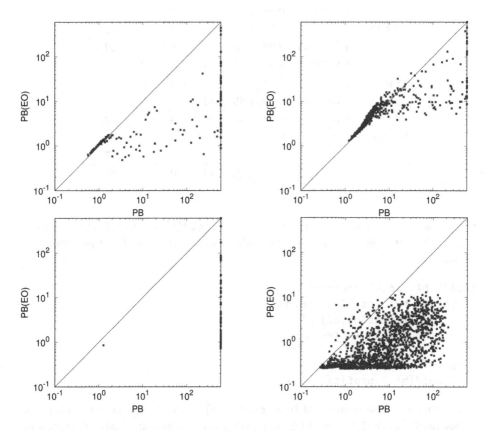

Fig. 2. Scatter plots comparing the median of the solving time among all 10 executions for each instance in the dataset. From left to right and top to bottom: CA, MRCPSP, NSP, MMKP.

5.1   Combinatorial Auctions

The Combinatorial Auctions (CA) problem can be stated as the problem of assigning items to bidders in such a way that the maximum profit is obtained [22]. Every bidder makes an offer for a set of items (a package), and it has to be decided whether to sell the whole package to the bidder. It is not allowed to sell only a proper subset of the demanded items. A natural viewpoint to model

Table 1. Summary statistics of configurations PB, PB(AMO) and PB(EO) for the four case studies. — indicates time out.

problem	setting	Q1	med	Q3	t.o.	vars	clauses
CA	PB	1.11	3.74	—	42	506	1006
	PB(AMO)	0.98	1.40	3.33	0	47	236
	PB(EO)	0.98	1.40	3.33	0	47	236
MRCPSP	PB	2.55	3.68	9.33	29	54	112
	PB(AMO)	2.89	4.70	8.57	8	12	59
	PB(EO)	2.89	4.68	8.41	8	12	57
NSP	PB	—	—	—	199	116	231
	PB(AMO)	1.30	1.85	6.81	4	26	120
	PB(EO)	0.76	0.86	1.26	3	8	22
MMKP	PB	0.82	4.98	21.09	0	31	62
	PB(AMO)	0.33	0.47	1.53	0	3	17
	PB(EO)	0.28	0.39	1.43	0	2	10

the problem is to introduce a Boolean variable `sold[b]` for each package b, that states whether it is sold or not. Then, the decision version of the problem can be stated as:

```
forAll b1: int(1..nBids-1) .
   forAll b2: int(b1+1..nBids) .
      incompBids[b1,b2] ->
      (!sold[b1] \/ !sold[b2]),

(sum b : int(1..nBids) .
     sold[b] * profit[b] ) >= lb
```

where `nBids` is the number of bids, `profit[b]` is the bid value for package b, `incompBids[b1,b2]` is true when two bids have a non-empty intersection, and `lb` is the minimum total profit that is required.

The first constraint ensures that no item is sold in two different packages, or equivalently that every item is sold in at most one package. This will allow Savile Row to detect mutexes between variables `sold[b]` where packages share some item. Typically the sets of packages that contain each particular item will not be disjoint, so the clique cover finding algorithm plays an especially important role when reformulating this problem.

In this work we consider the dataset reported in [9] which was generated using the Combinatorial Auctions Test Suite [22], and have an appropriate complexity to illustrate the effects of our techniques. It consists of 170 instances with the number of bids between 70 and 200. For this problem the syntactic check does not identify any EO relation, so PB(AMO) and PB(EO) are identical.

5.2   MRCPSP

The Multi-mode Resource-Constrained Project Scheduling Problem (MRCPSP) is an iconic problem in the scheduling field [10]. The problem requires deciding a start time (schedule) and an execution mode (schedule of modes) for each job of a project. The jobs are non-preemptive, i.e. they cannot be paused once they have started. Also, the jobs have demands over a set of resources, that can be either renewable, i.e. the amount of resource assigned to a job is recovered once the job finishes, or non-renewable, i.e. availability is not restored when jobs finish. For each job, its duration and its demands depend on the chosen execution mode. The schedule must ensure that a given set of precedence relations between jobs are all satisfied, that the given availability of renewable resources is never surpassed during the execution of the project, and that the given availability of non-renewable resources is enough to supply the demands. Moreover, the project completion time (makespan) must be minimised.

We model the resource constraints as follows. We introduce an auxiliary integer variable mode[j] for each job j, which represents the selected execution mode for job j. To deal with renewable resources constraints we also introduce a Boolean variable jobActive[j,m,t] for each job j, execution mode m and time instant t within a scheduling horizon, which is constrained to be true iff job i is running in mode m at time t. The renewable resource constraints are:

```
forAll t: int(0..horizon) .
forAll res: int(1..resRenew) . (
  sum j: int(1..jobs) .
    sum m: int(1..nModes[j]) .
      jobActive[j,m,t]*resUsage[j,m,res]
  ) <= resLimits[res]
```

We model non-renewable resource constraints as:

```
forAll res : int(resRenew+1..nRes) . (
  sum j: int(1..jobs) .
    sum m: int(1..nModes[j]) .
      (mode[j]=m) * resUsage[j,m,res]
  ) <= resLimits[res]
```

where horizon is a scheduling horizon which accepts a valid schedule (if the instance is satisfiable), 1..resRenew and resRenew+1..nRes are the sets of renewable and non-renewable resources respectively, 1..jobs is the set of all jobs, 1..nModes[j] is the set of available execution modes for job j, resUsage[j,m,res] is the consumption of job j on resource res when it runs in mode m, and resLimits[res] is the availability of resource res.

MRCPSP contains many notions of activity and mode incompatibilities, which allow the reformulation process to find AMO constraints on the variables of resource PB constraints. For instance, every activity must run in exactly one execution mode, and if an activity precedes another they will never run in parallel. Further, two modes of a pair of activities are incompatible if the combined demands for the two modes surpass the availability of some resource.

For this problem we have used the 552 satisfiable instances of the j30 dataset, which is the hardest from PSPLib [21]. These instances contain projects of 30 activities, 3 possible execution modes for each activity, 2 renewable resources and 2 non-renewable resources.

5.3 NSP

The Nurse Scheduling Problem (NSP) is the problem of finding an optimal assignment of nurses to shifts per day considering some coverage and shift preference constraints. There are plenty of variants of this problem depending on the constraints considered [12,27]. In this work we consider the basic version of the problem where solutions must satisfy all shift coverage constraints, i.e. each shift and day must have a certain number of nurses assigned, and must satisfy the constraint that each nurse only works a certain number of days per week, and must minimise the total penalisation according to the preferences of the nurses.

PB constraints appear in the Essence Prime model when bounding the total amount of penalisation allowed. We use integer variable nS[n,d] to state the shift assignment of each nurse n and day d, and the penalisation constraint is as follows:

```
(sum n: int(1..nNurses) .
  sum d: int(1..nDays) .
    sum st: int(1..nShiftTypes) .
    (nS[n,d]=s) * p[n,d,st] ) <= ub
```

where nNurses is the number of nurses, nDays the number of days, nShiftTypes the number of shift types and p[n,d,st] is the penalty of assigning shift st to nurse n on day d. Finally, since we are computing the decision version of NSP, ub is the maximum cost allowed. Notice that EO relations occur among the penalties for each nurse and day, since nS ranges over integer values from 1 to nShiftTypes.

In this work we consider a set of instances from NSPLib, a repository of thousands of NSP instances grouped into classes by several complexity indicators. Details can be found in [27]. We focus on a sample of 200 instances taken uniformly and independently at random from the N25 Set: 25 nurses, 7 days and 4 shift types (including the free shift). Each instance has a minimum number of nurses required per shift and day, and includes the nurses preferences to work on each shift and day (a penalty is between 1 and 4, where 1 is the rank of the most preferred shift).

5.4 MMKP

The Multiple-choice Multidimensional Knapsack Problem (MMKP) is a maximisation problem. Given a set of classes of items and a knapsack with several capacity-bounded dimensions, it is required to pack exactly one item of each class without surpassing the knapsack capacities. Each item of each class has a given profit, and a weight in each dimension. It is also required to maximise the profit of the chosen items [20]. The decision version of the problem requires that the profit is greater than or equal to a lower bound lb.

The PB constraints appear in our Essence Prime model when bounding capacities and profit. We use integer variables item[c] to state which item of class c has been chosen. The constraints are as follows:

```
forAll d: int(1..nDimensions) . (
 sum c: int(1..nClasses) .
  sum i: int(1..classSize) .
   (item[c]=i) * weight[c,i,d]
) <= cap[d],

(sum c: int(1..nClasses) .
  sum i:  int(1..classSize) .
   (item[c]=i) * profit[c,i] ) >= lb
```

where nDimension is the number of dimensions, nClasses is the number of classes, classSize is the number of items in each class (n.b. in this dataset all classes have the same number of items), weight[c,i,d] is the weight of item i of class c for dimension d, cap[d] is the capacity of dimension d, profit[c,i] is the profit of item i of class c and lb is the minimum profit to be achieved.

Notice that EO relations occur because item ranges over integer values from 1 to classSize.

For conducting the experimental evaluation we have chosen the 1983 satisfiable instances from the 2000 instances of dataset (10-5-5-G-R-W) from [16], that contain 10 classes of 5 items each, and the knapsack has 5 dimensions. This dataset turns out to be reasonably hard in comparison to others from the same work that appear to be easy for SAT solvers.

5.5   Experimental Results

Our results in Table 1 show a very significant reduction in the sizes of the SAT formulas for all four studied problems, both in the number of variables and number of clauses, thanks to the AMO and EO detection and reformulation process. The greatest reduction with approach PB(AMO) occurs in CA, where the number of variables is divided by 10 and the number of clauses by 4. In all four problem classes, the reduction in size directly translates to improved solving time. The most extreme case is NSP, in which only one instance is solved within the given timeout if AMO detection is not used, whereas almost all instances are solved with PB(AMO). Only 4 instances reach the time limit with PB(AMO). PB(EO) gives a further size reduction on all problems except CA, and it has a particular impact on NSP, where the additional size reduction reduces the number of clauses by ten times overall.

Figure 2 compares total time (including reformulation and solving) of PB and PB(EO) for every instance of each problem class. The solving time improvements are remarkable for all four problem classes. There are improvements between one and two orders of magnitude in many cases between PB and PB(EO), although there is a small overhead on some of the easiest instances.

6   Conclusion and Future Work

We have presented a fully automatic approach to find and exploit at-most-one (AMO) and exactly-one (EO) relations in SAT encodings of PB constraints. The approach is integrated into Savile Row, a constraint modelling tool that can automatically produce a SAT encoding of any constraint model written in the language Essence Prime. Until now, AMO and EO relations have been exploited for this purpose only in problem-specific encodings constructed by experts. Results show dramatic improvements in SAT formula size and solving time on four problem classes.

In future work we will explore stronger inference mechanisms for the detection of mutexes, which could lead to larger and more effective AMO relations. We also plan to study whether we can reformulate PB constraints more efficiently through detection of cardinality constraints with $k \geq 2$ applying the approach in [7].

References

1. Abío, I., Mayer-Eichberger, V., Stuckey, P.J.: Encoding linear constraints with implication chains to CNF. In: Pesant, G. (ed.) CP 2015. LNCS, vol. 9255, pp. 3–11. Springer, Cham (2015). https://doi.org/10.1007/978-3-319-23219-5_1

2. Abío, I., Nieuwenhuis, R., Oliveras, A., Rodríguez-Carbonell, E., Mayer-Eichberger, V.: A new look at BDDs for pseudo-Boolean constraints. J. Artif. Intell. Res. 443–480 (2012). https://doi.org/10.1613/jair.3653

3. Ansótegui, C.: Complete SAT solvers for Many-Valued CNF Formulas. Ph.D. thesis, University of Lleida (2004)

4. Audemard, G., Simon, L.: On the glucose SAT solver. Int. J. Artif. Intell. Tools **27**(1), 1–25 (2018). https://doi.org/10.1142/S0218213018400018

5. Bailleux, O., Boufkhad, Y., Roussel, O.: New encodings of pseudo-boolean constraints into CNF. In: Kullmann, O. (ed.) SAT 2009. LNCS, vol. 5584, pp. 181–194. Springer, Heidelberg (2009). https://doi.org/10.1007/978-3-642-02777-2_19

6. Biere, A.: Lingeling. SAT Race (2010)

7. Biere, A., Le Berre, D., Lonca, E., Manthey, N.: Detecting cardinality constraints in CNF. In: Sinz, C., Egly, U. (eds.) SAT 2014. LNCS, vol. 8561, pp. 285–301. Springer, Cham (2014). https://doi.org/10.1007/978-3-319-09284-3_22

8. Bofill, M., Coll, J., Suy, J., Villaret, M.: Compact MDDs for pseudo-boolean constraints with at-most-one relations in resource-constrained scheduling problems. In: IJCAI, pp. 555–562 (2017). https://doi.org/10.24963/ijcai.2017/78

9. Bofill, M., Palahí, M., Suy, J., Villaret, M.: Solving intensional weighted CSPs by incremental optimization with BDDs. In: O'Sullivan, B. (ed.) CP 2014. LNCS, vol. 8656, pp. 207–223. Springer, Cham (2014). https://doi.org/10.1007/978-3-319-10428-7_17

10. Brucker, P., Drexl, A., Möhring, R., Neumann, K., Pesch, E.: Resource-constrained project scheduling: notation, classification, models, and methods. Eur. J. Oper. Res. **112**(1), 3–41 (1999). https://doi.org/10.1016/S0377-2217(98)00204-5
11. Darras, S., Dequen, G., Devendeville, L., Mazure, B., Ostrowski, R., Saïs, L.: Using Boolean Constraint Propagation for sub-clauses deduction. In: van Beek, P. (ed.) CP 2005. LNCS, vol. 3709, pp. 757–761. Springer, Heidelberg (2005). https://doi.org/10.1007/11564751_59
12. De Causmaecker, P., Vanden Berghe, G.: A categorisation of nurse rostering problems. J. Sched. **14**(1), 3–16 (2011). https://doi.org/10.1007/s10951-010-0211-z
13. Eén, N., Sorensson, N.: Translating pseudo-boolean constraints into SAT. J. Satisfiability Boolean Model. Comput. **2**, 1–26 (2006). http://satassociation.org/jsat/index.php/jsat/article/view/18
14. Fourdrinoy, O., Grégoire, É., Mazure, B., Saïs, L.: Eliminating redundant clauses in SAT instances. In: Van Hentenryck, P., Wolsey, L. (eds.) CPAIOR 2007. LNCS, vol. 4510, pp. 71–83. Springer, Heidelberg (2007). https://doi.org/10.1007/978-3-540-72397-4_6
15. Gent, I.P., Jefferson, C., Miguel, I.: Minion: a fast scalable constraint solver. In: ECAI: European Conference on Artificial Intelligence. Frontiers in Artificial Intelligence and Applications, vol. 141, pp. 98–102. IOS Press (2006). http://www.booksonline.iospress.nl/Content/View.aspx?piid=1654
16. Han, B., Leblet, J., Simon, G.: Hard multidimensional multiple choice knapsack problems, an empirical study. Comput. Oper. Res. **37**(1), 172–181 (2010). https://doi.org/10.1016/j.cor.2009.04.006
17. Hölldobler, S., Manthey, N., Steinke, P.: A compact encoding of pseudo-boolean constraints into SAT. In: Glimm, B., Krüger, A. (eds.) KI 2012. LNCS (LNAI), vol. 7526, pp. 107–118. Springer, Heidelberg (2012). https://doi.org/10.1007/978-3-642-33347-7_10
18. Huang, J.: Universal booleanization of constraint models. In: Stuckey, P.J. (ed.) CP 2008. LNCS, vol. 5202, pp. 144–158. Springer, Heidelberg (2008). https://doi.org/10.1007/978-3-540-85958-1_10
19. Joshi, S., Martins, R., Manquinho, V.: Generalized totalizer encoding for pseudo-boolean constraints. In: Pesant, G. (ed.) CP 2015. LNCS, vol. 9255, pp. 200–209. Springer, Cham (2015). https://doi.org/10.1007/978-3-319-23219-5_15
20. Kellerer, H., Pferschy, U., Pisinger, D.: Multidimensional knapsack problems. In: Knapsack Problems, pp. 235–283. Springer, Heidelberg (2004). https://doi.org/10.1007/978-3-540-24777-7_9
21. Kolisch, R., Sprecher, A.: PSPLIB - a project scheduling problem library. Eur. J. Oper. Res. **96**(1), 205–216 (1997). https://doi.org/10.1016/S0377-2217(96)00170-1
22. Leyton-Brown, K., Shoham, Y.: A test suite for combinatorial auctions. In: Combinatorial Auctions, chap. 18, pp. 451–478. The MIT Press (2006)
23. Martins, R., Manquinho, V., Lynce, I.: Open-WBO: a modular MaxSAT solver'. In: Sinz, C., Egly, U. (eds.) SAT 2014. LNCS, vol. 8561, pp. 438–445. Springer, Cham (2014). https://doi.org/10.1007/978-3-319-09284-3_33
24. Nethercote, N., Stuckey, P.J., Becket, R., Brand, S., Duck, G.J., Tack, G.: MiniZinc: towards a standard CP modelling language. In: Bessière, C. (ed.) CP 2007. LNCS, vol. 4741, pp. 529–543. Springer, Heidelberg (2007). https://doi.org/10.1007/978-3-540-74970-7_38
25. Nightingale, P., Akgün, Ö., Gent, I.P., Jefferson, C., Miguel, I., Spracklen, P.: Automatically improving constraint models in Savile Row. Artif. Intell. **251**, 35–61 (2017). https://doi.org/10.1016/j.artint.2017.07.001

26. Nightingale, P., Rendl, A.: Essence' description. arXiv:1601.02865 (2016)
27. Vanhoucke, M., Maenhout, B.: NSPLib: a nurse scheduling problem library: a tool to evaluate (meta-)heuristic procedures. In: Brailsford, S., Harper, P. (eds.) Operational research for health policy: making better decisions, pp. 151–165. Peter Lang (2007)
28. Zhou, N.-F., Kjellerstrand, H.: The Picat-SAT compiler. In: Gavanelli, M., Reppy, J. (eds.) PADL 2016. LNCS, vol. 9585, pp. 48–62. Springer, Cham (2016). https://doi.org/10.1007/978-3-319-28228-2_4

Exploring Declarative Local-Search Neighbourhoods with Constraint Programming

Gustav Björdal[1]([⊠]) [iD], Pierre Flener[1] [iD], Justin Pearson[1] [iD],
and Peter J. Stuckey[2] [iD]

[1] Department of Information Technology, Uppsala University, Uppsala, Sweden
{Gustav.Bjordal,Pierre.Flener,Justin.Pearson}@it.uu.se
[2] Faculty of Information Technology, Monash University, Melbourne, Australia
Peter.Stuckey@monash.edu

Abstract. Using constraint programming (CP) to explore a local-search neighbourhood was first tried in the mid 1990s. The advantage is that constraint propagation can quickly rule out uninteresting neighbours, sometimes greatly reducing the number actually probed. However, a CP model of the neighbourhood has to be handcrafted from the model of the problem: this can be difficult and tedious. That research direction appears abandoned since large-neighbourhood search (LNS) and constraint-based local search (CBLS) arose as alternatives that seem easier to use. Recently, the notion of declarative neighbourhood was added to the technology-independent modelling language MiniZinc, for use by any backend to MiniZinc, but currently only used by a CBLS backend. We demonstrate that declarative neighbourhoods are indeed technology-independent by using the old idea of CP-based neighbourhood exploration: we explain how to encode automatically a declarative neighbourhood into a CP model of the neighbourhood. This enables us to lift any CP solver into a local-search backend to MiniZinc. Our prototype is competitive with CP, CBLS, and LNS backends to MiniZinc.

1 Introduction

Technology-independent modelling is an paradigm where we model a problem and choose among solvers of several technologies in order to solve it for given data. This helps avoid early commitment to a technology and solver, and enables the easy comparison of technologies and solvers on the same model. MiniZinc [17] is a technology-independent modelling language, supported by solvers of many technologies, such as constraint programming (CP), lazy clause generation (LCG), integer programming (IP), Boolean satisfiability (SAT), satisfiability modulo theories (SMT), constraint-based local search (CBLS [27]), and hybrids.

Many solvers can work in a black-box way, where we only need to provide a model and the instance data. Some technologies, notably CP and LCG, also allow

T. Schiex and S. de Givry (Eds.): CP 2019, LNCS 11802, pp. 37–53, 2019.
https://doi.org/10.1007/978-3-030-30048-7_3

a search strategy to be attached to a model; in practice, a good search strategy is often key to an efficient solving process. MiniZinc has had from its inception a notation for declaratively indicating a search strategy for CP and LCG solvers. Recently, MiniZinc was extended with a notation [3] for declaratively specifying a local-search *neighbourhood*, that is a set of candidate moves that re-assign some variables within the current valuation of a local-search method. At present, these *declarative neighbourhoods* are only supported by the CBLS backend fzn-oscar-cbls [4]: experiments showed that, as expected, CBLS via MiniZinc can be accelerated via a neighbourhood specification.

In this paper, we revisit the idea of encoding a local-search neighbourhood for a CP solver so that, given the current valuation of the variables in a local-search method, the CP solver finds by systematic search a best neighbour of that current valuation, under some heuristic. The local-search method then moves to that neighbour, using some meta-heuristic for escaping local optima, such as simulated annealing or tabu search. The idea of encoding a neighbourhood was first proposed in [20] and then refined in [25]: exploring a neighbourhood by using a CP solver on such a *neighbourhood model* can lead to the efficient pruning of both infeasible and sub-optimal neighbours, sometimes greatly reducing the number of actually probed neighbours. However, this idea was presented as a methodology, where a neighbourhood model has to be handcrafted from the problem model, and there is limited reusability of encodings between neighbourhoods.

Large-neighbourhood search (LNS) [24] is another popular method for performing local search by using a CP solver. An LNS neighbourhood is constructed by *freezing* some variables, that is fixing them to their values in the current valuation of a local-search method, and it is explored by performing systematic search on the problem model in order to find an improving valuation for the remaining variables. However, this is fundamentally different from the neighbourhoods classically explored by CBLS solvers and ad hoc local-search methods. For example, a relocation neighbourhood (e.g., [22, Chapter 23]) cannot be explored by LNS without also exploring a very large number of neighbours that are not obtainable by relocation moves: LNS does not have the classical notion of move. Another crucial difference is that LNS uses *one* copy of the variables of the problem model and freezes *some* variables in each move, whereas a neighbourhood model uses *two* copies of the variables and freezes *most* variables in each move.

Putting all these ingredients together, the organisation and contributions of this paper are as follows:

- an encoding of a declarative neighbourhood [3] for any CP solver;
- the new global constraint WRITES for encoding local-search moves;
- a good definition of WRITES using constraints available in most CP solvers;
- a recipe for building a local-search backend to MiniZinc from any CP solver;
- evidence that declarative neighbourhoods are technology-independent.

We wrap the paper up with an experimental evaluation of our prototype against CP, CBLS, and LNS backends to MiniZinc, as well as directions for future work.

2   Background

After discussing in Sect. 2.1, with the example of a vehicle routing problem, the technology-independent modelling language MiniZinc and its extension with the notion of declarative neighbourhood [3], conceived for local-search backends, we briefly summarise in Sect. 2.2 the principles of local search (e.g., [14]).

For brevity, we discuss constrained minimisation problems: the maximisation of an objective function amounts to minimising its opposite, and the satisfaction of constraints amounts to satisfying them while minimising a constant.

2.1   MiniZinc and Declarative Neighbourhoods

Conceptually, a MiniZinc model for a constrained minimisation problem is in this paper a tuple $\langle \mathcal{V}, \mathcal{C}, o, \mathcal{S}, \mathcal{N} \rangle$, where $\mathcal{V}$ is the set of variables; $\mathcal{C}$ is the set of constraints on these variables, including their domain membership constraints; the variable $o \in \mathcal{V}$ is the *objective variable*, whose value is to be minimised; $\mathcal{S}$ is the optional annotation for suggesting a systematic-search branching strategy on the variables; and $\mathcal{N}$ is the optional annotation for suggesting a declarative neighbourhood [3].

As a running example, we use the model for the travelling salesperson problem with time windows (TSPTW) used in [2], but extended with a relocation neighbourhood. Given are n locations; an array TravTime, where TravTime[i,j] is the travel time from location i to location j plus the service time at i; and an array ArrWin of arrival-time windows, where ArrWin[i,1] is the earliest arrival time and ArrWin[i,2] the latest arrival time at location i. The objective is to find a shortest Hamiltonian circuit that visits each location exactly once and within its arrival-time window.

Listing 1 has a MiniZinc model for TSPTW with a relocation neighbourhood, with the data above declared in lines 1 to 3. The route is modelled in line 4 by an array Pred, where variable Pred[i] denotes the location visited before location i. The circuit constraint in line 5 requires Pred to represent a Hamiltonian circuit. Location 1 is assumed in line 6 to be the depot, that is the start of the route. The arrival times are modelled in line 7 using the array ArrTime, where variable ArrTime[i] denotes the arrival time at location i. Each arrival time is constrained, in lines 8 to 11, to be at least either the arrival time at the preceding location plus the travel time, or the start of its arrival-time window, whichever is greater, and at most the end of its arrival-time window. The objective is to minimise the travel time of the entire circuit, which is stated in lines 12 and 15.

The relocation neighbourhood is used in the annotation (prefixed by ::) of line 13 and declared in lines 16 to 22. It considers in line 18 all combinations of two locations i and j such that they are distinct and the predecessor of i is not j: this is prescribed by the where pre-condition in line 18 on the elements of the moves set comprehension of candidate moves. Each *candidate move* consists of the composition of three parallel re-assignments that relocates the predecessor of i so that it goes between j and the predecessor of j, as prescribed by lines 19 to 21. The initialisation post-condition of the neighbourhood is that Pred forms

```
1   int: n;  set of int: Loc = 1..n; % number and set of locations
2   array[Loc,Loc] of int: TravTime; % travel times
3   array[Loc,1..2] of int: ArrWin; % arrival-time windows: earliest, latest
4   array[Loc] of var Loc: Pred; % predecessor locations
5   constraint circuit(Pred);
6   int: depot = 1; % location 1 is the depot
7   array[Loc] of var int: ArrTime; % arrival times
8   constraint ArrTime[depot] = ArrWin[depot,1];
9   constraint forall(i in Loc where i != depot)(
10    ArrTime[i] = max(ArrTime[Pred[i]]+TravTime[Pred[i],i], ArrWin[i,1]));
11  constraint forall(i in Loc)(ArrTime[i] <= ArrWin[i,2]);
12  var int: time = sum(i in Loc)(TravTime[Pred[i],i]); % objective variable
13  solve :: use_neighborhood(relocate())
14       :: int_search(Pred,first_fail,indomain_min,complete)
15    minimize time;
16  function ann: relocate() :: neighborhood_definition =
17    initially(circuit(Pred)) /\
18    moves(i, j in Loc where i != j /\ Pred[i] != j)(
19      Pred[i] := Pred[Pred[i]] /\
20      Pred[j] := Pred[i] /\
21      Pred[Pred[i]] := Pred[j] % /\ ensuring(circuit(Pred)) % implied
22    );
```

Listing 1. A MiniZinc model for TSPTW and a relocation neighbourhood.

a Hamiltonian circuit, as prescribed by the `initially` condition in line 17: every (re-)start must be from a valuation satisfying this condition. Together, this initialisation post-condition, the pre-condition on candidate moves, and the nature of the candidate moves imply that each candidate move reaches a valuation of `Pred` that forms a Hamiltonian circuit, so we do not need to include the commented-out `ensuring` post-condition on candidate moves in line 21.

The `initially`, `where`, and `ensuring` conditions of a declarative neighbourhood are constraint satisfaction problems, expressed on the data and variables of the problem model, using the existing and full MiniZinc syntax.

2.2   Local Search

Given a model $\langle \mathcal{V}, \mathcal{C}, o, \mathcal{S}, \mathcal{N} \rangle$ for a constrained minimisation problem, a *local-search method* iteratively maintains a *current valuation* θ that maps each variable in $\mathcal{V}$ to a value in its domain prescribed in $\mathcal{C}$, and that is initialised under some amount of randomisation so as to satisfy the initialisation condition of $\mathcal{N}$, usually a subset of the constraints in $\mathcal{C}$. At each iteration, the local-search method considers the set of candidate moves defined by the neighbourhood $\mathcal{N}(\theta)$: it *selects* under some amount of randomisation a candidate move, as specified by some *heuristic* such as best-improving or first-improving, and *makes* the selected candidate move by updating the current valuation θ accordingly. The idea is that each move made should reduce the value of some *cost function* $\mathrm{cost}(\theta)$, which does not necessarily return the current objective value $\theta(o)$, as seen below.

In order to escape local optima of cost(θ), a *meta-heuristic*, such as simulated annealing or tabu search [12], is used. Together, the neighbourhood $\mathcal{N}$, the heuristic, and the meta-heuristic form the *local-search strategy* of the local-search method. A local-search method typically also involves *restarts*, where periodically the search may be begun again from scratch, in order to avoid being trapped in local minima. It also may use *intensification*, where more search effort is applied around a current valuation that seems promising.

All constraints in $\mathcal{C}$ are to be satisfied. Existing local-search backends to Mini-Zinc automatically choose for each constraint among three ways of handling it:

- A constraint $c \in \mathcal{C}$ can be satisfied when initialising the current valuation θ and its satisfaction can be preserved by all candidate moves. Hence c is *hard*: it cannot be violated during search.
- A constraint c that functionally defines some variable $v \in \mathcal{V}$ in terms of other variables $W \subset \mathcal{V}$ can be made hard by extending every made move on at least one variable in W into also re-assigning v accordingly.
- A constraint c can be made *soft*, meaning it can be violated during search but should be satisfied in the final valuation, by using a *violation function* giving 0 if c is satisfied under θ, and otherwise a positive value that indicates how violated c is under θ. For example, for linear expressions x and y, the linear constraints $x = y$ and $x \leq y$ are softened [27] into $v_1 = |x - y|$ and $v_2 = $ **if** $x \leq y$ **then** 0 **else** $x - y$ **endif**, respectively, defining an introduced *violation variable* v_i.

For example, for the model in Listing 1, a typical way of handling its constraints is: the `circuit` constraint in line 5 is satisfied by every valuation explored during the search; the constraint that functionally defines `time` in line 12 is made hard and moves do not consider changing this variable; the time-window end constraint in line 11 is made soft; finally, although the constraints in lines 8 to 10 define the `ArrTime[i]` variables functionally, it is hard to detect that the definition is not circular: hence they are made soft and moves must consider changing these variables.

Let soft(C, g) denote the constraint set where some constraints in C are softened under some scheme, including a new variable g, denoting the *global violation*, constrained to be the sum of all the introduced violation variables. We assume that the individual violation variables and the global violation variable g are implicitly added to the variable set of the model containing C. Replacing C by soft(C, g) requires changing the model containing C to minimising both the objective variable o and the global violation variable g. For example, in fzn-oscar-cbls [4], a weighted sum $\alpha \cdot o + \beta \cdot g$ is used as the cost function cost(θ), where the values of α and β are dynamically tuned during search.

If too many constraints are made hard, then this may *disconnect* the search space, since local search only moves from one valuation to another via a move of the neighbourhood: it may be that no sequence of moves in the declarative neighbourhood are able to move between two given valuations. If this is the case, then we can seriously weaken the local-search capability to find good valuations.

Given a neighbourhood $\mathcal{N}$, we can partition the variables $\mathcal{V}$ of a model into three sub-sets: $\mathcal{V}_{\mathrm{targ}}$ is the set of variables that are targeted by the moves of $\mathcal{N}$ (such as the array `Pred` in line 4 of Listing 1); $\mathcal{V}_{\mathrm{func}}$ is the set of non-targeted variables that are each functionally defined by some constraint (such as `time` being functionally defined by `Pred` in line 12); and $\mathcal{V}_{\mathrm{aux}}$ is the set of the remaining variables, which we call *auxiliary variables* (such as the array `ArrTime` in line 7). Search must be over $\mathcal{V}_{\mathrm{targ}} \cup \mathcal{V}_{\mathrm{aux}}$, but local search over $\mathcal{V}_{\mathrm{targ}} \cup \mathcal{V}_{\mathrm{aux}}$ was shown in [2] to degrade greatly the performance of CBLS solvers, unless every move on $\mathcal{V}_{\mathrm{targ}}$ is somehow automatically extended by a corresponding re-assignment of $\mathcal{V}_{\mathrm{aux}}$.

3   Encoding a Declarative Neighbourhood as a CP Model

We show how to encode automatically a declarative neighbourhood, specified in MiniZinc, as a CP model, which we call the *neighbourhood model*. We show in Sect. 3.1 how to encode two states of a local-search method, namely its current and next valuations, using variables. We explain in Sect. 3.2 how to encode a move as constraints on these variables. The exploration of the neighbourhood then amounts to solving the neighbourhood model, as discussed in Sect. 3.3.

3.1   Encoding the Current and Next Valuations

Given a MiniZinc model $\langle \mathcal{V}, \mathcal{C}, o, \mathcal{S}, \mathcal{N} \rangle$, we extract the following sets:

- $\mathcal{V}_{\mathrm{gen}}$ has variables for the generators of the `moves` set comprehension of $\mathcal{N}$;
- $\mathcal{M}$ has the move expressions of the `moves` comprehension of $\mathcal{N}$;
- $\mathcal{V}_{\mathrm{targ}} \subseteq \mathcal{V}$ has the *targeted variables* of $\mathcal{V}$, that is those re-assigned in $\mathcal{M}$;
- $\mathcal{C}_{\mathrm{where}}$ has the constraints on $\mathcal{V} \cup \mathcal{V}_{\mathrm{gen}}$ of the `where` pre-condition of $\mathcal{N}$; and
- $\mathcal{C}_{\mathrm{ensure}}$ has the constraints on $\mathcal{V} \cup \mathcal{V}_{\mathrm{gen}}$ of the `ensuring` post-condition of $\mathcal{N}$.

For example, for the model in Listing 1, the set $\mathcal{V}_{\mathrm{gen}}$ has variables, called *generator variables*, for the generators `i` and `j` in line 18; the set $\mathcal{M}$ has the three re-assignments in lines 19 to 21; the set $\mathcal{V}_{\mathrm{targ}}$ has the entire array `Pred` since any variable thereof can be referred to in the left-hand sides of the re-assignments in $\mathcal{M}$; the set $\mathcal{C}_{\mathrm{where}}$ has the two constraints of the `where` pre-condition in line 18; and the set $\mathcal{C}_{\mathrm{ensure}}$ is empty since there is no `ensuring` post-condition.

Since our encoding must reason on the current and next valuations of the variables in a local-search method, we must use in the neighbourhood model two copies of some variables of the given problem model: for a set X of variables, we denote by X^{c} the set of variables corresponding to X in the current valuation, and by X^{n} the set of variables corresponding to X in the next valuation. We use the same notation for individual variables.

The variable set of the *neighbourhood model* is $\mathcal{V}^{\mathrm{c}} \cup \mathcal{V}^{\mathrm{n}} \cup \mathcal{V}_{\mathrm{gen}}^{\mathrm{c}}$. We give its constraint set in Sect. 3.3, after focussing on the constraints encoding a move.

3.2 Encoding a Move

A move is a transition from $\mathcal{V}_{\text{targ}}^{\text{c}}$ to $\mathcal{V}_{\text{targ}}^{\text{n}}$, so we must constrain each variable in $\mathcal{V}_{\text{targ}}^{\text{n}}$ to take the same value as its corresponding variable in $\mathcal{V}_{\text{targ}}^{\text{c}}$, except those re-assigned by the move, which are constrained to take new values accordingly.

Towards encoding this, we introduce the constraint WRITES(O, I, P, V) on two arrays O and I of the same number n of variables and two arrays P and V of the same number m of variables: it holds if and only if O, called the *output array*, is point-wise equal to I, called the *input array*, except that $O[P[j]]$ is constrained to be equal to $V[j]$ for each j in $\{1, \dots, m\}$. We assume that all indexing in this paper starts from 1.

We encode using WRITES the set $\mathcal{M}$ of move expressions. The *basic moves* are $x := y$, $X[i] := y$, $x :=: y$, and $X[i] :=: Y[j]$, specified and encoded as follows:

- $x := y$ means re-assign to x the current value of y, which is encoded as either $x^{\text{n}} = y^{\text{c}}$ or WRITES$([x^{\text{n}}], [x^{\text{c}}], [1], [y^{\text{c}}])$;
- $X[i] := y$ means re-assign to $X[i]$ the current value of y, which is encoded as WRITES$(X^{\text{n}}, X^{\text{c}}, [i^{\text{c}}], [y^{\text{c}}])$;
- $x :=: y$ means swap the current values of x and y, which is encoded as either $x^{\text{n}} = y^{\text{c}} \land y^{\text{n}} = x^{\text{c}}$ or WRITES$([x^{\text{n}}, y^{\text{n}}], [x^{\text{c}}, y^{\text{c}}], [1, 2], [y^{\text{c}}, x^{\text{c}}])$;
- $X[i] :=: Y[j]$ means swap the current values of $X[i]$ and $Y[j]$, which is encoded the way the compound move $X[i] := Y[j] \ \text{/\textbackslash} \ Y[j] := X[i]$ is; see below.

The first and third WRITES-based encodings are only useful when we merge them with others in order to preserve the semantics of moves, as discussed next.

A *compound move* is the parallel composition of basic moves, which is written by overloading the $\text{/\textbackslash}$ logical-and connective. The composition of basic moves that always re-assign different variables, such as $X[i] := u \ \text{/\textbackslash} \ Y[j] := v$ when the arrays X and Y share no variables, is the conjunction of the encodings of the basic moves. However, the composition of basic moves that can re-assign the same variable, such as $X[i] := u \ \text{/\textbackslash} \ X[j] := v$, *must* be encoded by merging the encodings of the basic moves, since WRITES$(X^{\text{n}}, X^{\text{c}}, [i^{\text{c}}], [u^{\text{c}}])$ requires $\forall k \neq i^{\text{c}} : X^{\text{n}}[k] = X^{\text{c}}[k]$, which prevents any value other than $X^{\text{c}}[j^{\text{c}}]$ from being written at index j^{c} by WRITES$(X^{\text{n}}, X^{\text{c}}, [j^{\text{c}}], [v^{\text{c}}])$, unless $j^{\text{c}} = i^{\text{c}}$.

Rules 1, 2 and 3 below show how to merge WRITES constraints; we only give rules for the cases that can appear in our prototype backend to MiniZinc:

Rule 1. The constraints WRITES(O, I, P, V) and WRITES$([x], [y], [1], [v])$, where for some constant index p we have that $O[p]$ is x and $I[p]$ is y, are merged into WRITES$(O, I, P \mathbin{+\!\!+} [p], V \mathbin{+\!\!+} [v])$, where $\mathbin{+\!\!+}$ denotes array concatenation.

Rule 2. The constraints WRITES(O, I, P_1, V_1) and WRITES(O, I, P_2, V_2) on the same output and input arrays are merged into WRITES$(O, I, P_1 \mathbin{+\!\!+} P_2, V_1 \mathbin{+\!\!+} V_2)$.

Rule 3. Consider WRITES(O_1, I_1, P_1, V_1) and WRITES(O_2, I_2, P_2, V_2), where a non-empty set J has the indices j for which there exists an index i such that $O_2[j]$ is $O_1[i]$ and $I_2[j]$ is $I_1[i]$. Let O_2 and I_2 have length n. Let $O' = O_1 \mathbin{+\!\!+} [O_2[k] \mid$

$k \in \{1, \ldots, n\} \setminus J]$ and $I' = I_1 + [I_2[k] \mid k \in \{1, \ldots, n\} \setminus J]$ be the non-redundant mergers of the two O_ℓ arrays and the two I_ℓ arrays, respectively. Let M be the array that maps indices of I_2 to indices of I' defined so that $\forall i \in \{1, \ldots, n\} : O_2[i] = O'[M[i]] \wedge I_2[i] = I'[M[i]]$. Let P_2 and V_2 have length m. Let $P' = P_1 + [M[P_2[i]] \mid i \in \{1, \ldots, m\}]$. The two WRITES constraints above are merged into WRITES$(O', I', P', V_1 + V_2)$.

For example, for the model in Listing 1, the compound move is encoded, after maximal merging, as the single constraint WRITES(Pred^n, Pred^c, [i^c, j^c, Pred[i^c]^c], [Pred[Pred[i^c]^c]^c, Pred[i^c]^c, Pred[j^c]^c]).

Let the maximally merged encodings of the move expressions in $\mathcal{M}$, together with the constraint $\exists v \in \mathcal{V}_{\text{targ}} : v^n \neq v^c$ requiring at least one variable in $\mathcal{V}_{\text{targ}}^n$ to be different from the corresponding one in $\mathcal{V}_{\text{targ}}^c$, form the constraint set $\mathcal{C}_{\text{move}}$.

3.3   The Neighbourhood Model and Neighbourhood Exploration

The *neighbourhood model* has the variable set $\mathcal{V}^c \cup \mathcal{V}^n \cup \mathcal{V}_{\text{gen}}^c$ mentioned in Sect. 3.1 and the following constraint set for channelling between $\mathcal{V}^c$ and $\mathcal{V}^n$:

- the set $\mathcal{C}_{\text{where}}\{\mathcal{V}/\mathcal{V}^c, \mathcal{V}_{\text{gen}}/\mathcal{V}_{\text{gen}}^c\}$, for meeting the **where** pre-condition;
- the set $\mathcal{C}_{\text{move}}$ defined at the end of Sect. 3.2, for encoding a move;
- the set $\text{soft}(\mathcal{C}, g)\{\mathcal{V}/\mathcal{V}^n\}$, for evaluating and pruning neighbours; and
- the set $\mathcal{C}_{\text{ensure}}\{\mathcal{V}/\mathcal{V}^n, \mathcal{V}_{\text{gen}}/\mathcal{V}_{\text{gen}}^c\}$, for meeting the **ensuring** post-condition.

where $R\{X/Y\}$ denotes the copy of the constraint set R where the variables of the set X are point-wise substituted by those of the same-sized set Y of variables.

A declarative neighbourhood can have the union of several **moves** set comprehensions with possibly different pre- and post-conditions, effectively giving the union of sub-neighbourhoods. In order to encode such a neighbourhood, we propose that each sub-neighbourhood be separately encoded in its own neighbourhood model, each being explored under its own instantiation of a CP solver. We believe that the disjunctive encoding of the sub-neighbourhoods would be at most as efficient as encoding and exploring the sub-neighbourhoods separately.

Thus, given the current valuation θ of a local-search method, exploring its neighbourhood amounts to solving the neighbourhood model, but with the additional constraints $\{v^c = \theta(v) \mid v \in \mathcal{V}\}$ for enforcing θ, using a CP solver: either we apply systematic search in order to find one or all neighbours, or we add to the neighbourhood model the objective function that corresponds to the cost function of the local-search method and apply systematic branch-and-bound search in order to find a best neighbour and prune sub-optimal ones on-the-fly.

4   Implementing a Local-Search Solver Using a CP Solver

Since we can explore a declarative neighbourhood using a CP solver, we now show how to lift any CP solver into a local-search backend for MiniZinc. We use OscaR.cp [18] in order to implement our prototype backend, called LS(cp).

```
1  predicate writes(array[int] of var int: O, array[int] of var int: I,
2                   array[int] of var int: P, array[int] of var int: V) =
3    forall(j in index_set(P))(O[P[j]] = V[j]) /\
4    forall(i in index_set(I) where forall(j in index_set(P))(P[j]!=i))(
5      O[i]=I[i]);
```

Listing 2. A straightforward definition of the WRITES constraint in MiniZinc syntax.

```
1  predicate writes(array[int] of var int: O, array[int] of var int: I,
2                   array[int] of var int: P, array[int] of var int: V) =
3  let { int: k = min(index_set(P));
4       array[index_set(I)] of var 0 .. length(P): S;
5  } in forall(i in index_set(I))(S[i] = 0 -> O[i] = I[i] /\
6          forall(j in 1..length(P))(S[i] = j -> P[j+k-1] = i)) /\
7    alldifferent_except_0(S) /\forall(j in index_set(P))(O[P[j]] = V[j]);
```

Listing 3. An improved definition of the WRITES constraint in MiniZinc syntax.

We use only existing components of OscaR.cp (including the FlatZinc parser of the CBLS solver OscaR.cbls [4,7] of the same OscaR framework), provide a good implementation of WRITES (Sect. 4.1), motivate a particular constraint softening scheme (Sect. 4.2), and discuss the control flow (Sect. 4.3).

4.1 Implementation of the **WRITES** Global Constraint

A straightforward definition (or: decomposition) of the WRITES(O, I, P, V) constraint of Sect. 3.2 is given in Listing 2 using MiniZinc syntax. We also propose the improved definition in Listing 3, which reasons on a matching between the variables of P and O: an array S denotes for each index i of I if its element is unchanged in O (when $S_i = 0$) or denotes the value at index j of P that determines its change (when $S_i = j$).The improved definition propagates WRITES($[1..3, 1..3, 1..3]$, $[4, 4, 4]$, $[1..3, 2..3, 2..3], [1, 1, 1]$), where $\ell..u$ denotes a variable of that domain, to WRITES($[1, 1..3, 1..3]$, $[4, 4, 4]$, $[1, 2..3, 2..3], [1, 1, 1]$), whereas the first definition propagates nothing. However, neither achieves domain consistency, namely WRITES($[1, 1, 1], [4, 4, 4], [1, 2..3, 2..3], [1, 1, 1]$). Given that the max-clique problem reduces to achieving domain consistency on a WRITES constraint, domain-consistent propagation is NP-hard and we do not investigate this further in this paper. In practice, we provide special cases in the definition when O and I have length $n \in \{1, 2\}$, capturing the WRITES-free encodings of the $x := y$ and $x :=: y$ moves shown in Sect. 3.2.

4.2 Constraint Softening Scheme

As hard constraints decrease the neighbourhood size and exploration time [20], it can be beneficial to soften only a few constraints, if any. We argue that one should

Algorithm 1. Control flow of a CP-based local-search backend to MiniZinc.

while no time-out **do**	
$\quad \theta := \text{initialise}(\tau_{\text{init}})$	{create a new current valuation}
$\quad$ **while** no time-out and no restart **do** {the meta-heuristic decides when to restart}	
$\quad\quad \theta' := \text{explore}(\theta, \tau_{\text{explo}})$	{select a neighbour}
$\quad\quad \theta := \text{intensify}(\theta', \tau_{\text{intens}})$	{improve the selected neighbour}

soften neither constraints that functionally define variables, nor constraints on auxiliary variables. The former rule is what the MiniZinc CBLS backend fzn-oscar-cbls [4] does, and the latter rule allows us to leverage propagation for determining values of the auxiliary variables, which was shown to be beneficial in [2]. In our LS(cp), the $\text{soft}(C, g)$ operator of Sect. 2.2 softens each linear (in)equality constraint that neither functionally defines some variable nor constrains auxiliary variables, and we change the objective function into the cost function $\alpha \cdot o + \beta \cdot g$, as in [4], but currently statically with $\alpha = 1 = \beta$.

4.3 Control Flow

Given a MiniZinc model $\langle \mathcal{V}, \mathcal{C}, o, \mathcal{S}, \mathcal{N} \rangle$ with a declarative neighbourhood $\mathcal{N}$, our local-search backend consists of three major components—initialisation, exploration, and intensification—which are used under the control flow in Algorithm 1. Each component has its own instantiation of a CP solver for its own model, but the exploration has several in case of sub-neighbourhoods.

Initialisation. Let $\mathcal{C}_{\text{init}}$ denote the constraint set in the initialisation post-condition of $\mathcal{N}$: the *initialisation model* is $\langle \mathcal{V}, \text{soft}(\mathcal{C}, g) \cup \mathcal{C}_{\text{init}}, o + g, \mathcal{S}_{\text{init}}, \_ \rangle$, where the systematic-search strategy $\mathcal{S}_{\text{init}}$ is a randomisation of $\mathcal{S}$, if present, and otherwise a randomising default strategy. The CP solver is limited to return the best solution found within τ_{init} seconds, or, if no solution was found yet, then to return the first solution found thereafter. The CP solution returned by $\text{initialise}(\tau_{\text{init}})$ is used as the initial or re-start valuation θ by the local search.

Exploration. Let $\mathcal{C}_{\text{explo}}$ denote the four constraint sets at the start of Sect. 3.3: the *neighbourhood model* is $\langle \mathcal{V}^{\text{c}} \cup \mathcal{V}^{\text{n}} \cup \mathcal{V}^{\text{c}}_{\text{gen}}, \mathcal{C}_{\text{explo}}, o^{\text{n}} + g^{\text{n}}, \mathcal{S}_{\text{explo}}, \_ \rangle$. We describe everything for a trail-based CP solver. Before applying the constraints $\{v^{\text{c}} = \theta(v) \mid v \in \mathcal{V}\}$ for enforcing the given current valuation θ of the local search, a choice point (recording the current state of the CP solver) is pushed onto the trail of the CP solver. This allows us to backtrack, when the local search has a new current valuation θ, to the choice point before the variables were fixed, by popping the trail, thus reusing rather than re-building the neighbourhood model in the next iteration of local search. Note that in an LCG solver this also allows us to keep all generated nogoods, since any dependence on the current valuation θ is included in the nogood.

The search strategy $\mathcal{S}_{\text{explo}}$ is similar to the one of [25], which argues that branching by domain bisection on the generator variables $\mathcal{V}_{\text{gen}}^{\text{c}}$ propagates better. However, in general this only guarantees fixing the targeted variables $\mathcal{V}_{\text{targ}}^{\text{n}} \subseteq \mathcal{V}^{\text{n}}$ by propagation. We therefore then also branch on the remaining variables.

Many local-search heuristics for selecting a neighbour are easy to implement. For example, for the first-improving heuristic used by our LS(cp), we limit the CP branch-and-bound search to stop after finding a solution within τ_{explo} seconds where $\langle g^{\text{n}}, o^{\text{n}} \rangle$ is lexicographically strictly less than $\langle g^{\text{c}}, o^{\text{c}} \rangle$; if no such solution was found yet, then the best solution found so far is returned, if any. Further, the best-improving heuristic can be implemented by searching exhaustively.

We implement a greedy local-search phase by enforcing g^{c} and o^{c} as upper bounds on g^{n} and o^{n}: this prunes all non-improving neighbours to θ. After some iterations, once θ is a local minimum, these bounds will empty the neighbourhood and we end the greedy local-search phase by no longer using these bounds. This allows the best-found CP solution to become a non-improving neighbour.

Let σ be the solution returned by the CP solver, if any: $\text{explore}(\theta, \tau_{\text{explo}})$ returns the valuation $\theta' = \{v \mapsto \sigma(v^{\text{n}}) \mid v \in \mathcal{V}\}$ if σ is defined, otherwise $\theta' = \theta$.

Intensification. Let θ' be the local-search valuation of $\mathcal{V} = \mathcal{V}_{\text{targ}} \cup \mathcal{V}_{\text{func}} \cup \mathcal{V}_{\text{aux}}$ returned by the exploration; recall the end of Sect. 2.2 for the semantics of these variable sets. The projection of θ' onto only $\mathcal{V}_{\text{targ}}$ may have several extensions for $\mathcal{V}_{\text{func}}$ and $\mathcal{V}_{\text{aux}}$ that have a better objective value than under θ'. This may happen for example upon a first-improving heuristic. In order to try and improve θ', the function $\text{intensify}(\theta', \tau_{\text{intens}})$ calls a CP solver on the *intensification model* $\langle \mathcal{V}, \text{soft}(\mathcal{C}, g) \cup \{v = \theta'(v) \mid v \in \mathcal{V}_{\text{targ}}\}, o + g, \mathcal{S}, \_ \rangle$, where $\mathcal{S}$ is from the original model. The CP solver is limited to return a best solution found within τ_{intens} seconds, if any. This intensification is essentially a single LNS iteration where $\mathcal{V}_{\text{targ}}$ is frozen and values for $\mathcal{V}_{\text{func}} \cup \mathcal{V}_{\text{aux}}$ are sought.

Let σ be the solution returned by the CP solver, if any: $\text{intensify}(\theta', \tau_{\text{intens}})$ returns the valuation $\theta = \{v \mapsto \sigma(v) \mid v \in \mathcal{V}\}$ if σ is defined, otherwise $\theta = \theta'$.

Meta-Heuristic: Tabu Search, Aspiration, and Restarts. In order to help local search escape local minima, we improve on Algorithm 1 by implementing a tabu search meta-heuristic by extending $\text{explore}(\theta, \tau_{\text{explo}})$ to return also which variables in $\mathcal{V}_{\text{targ}}$ were re-assigned in the selected neighbour: after the latter is intensified into the new current valuation, each re-assigned variable in $\mathcal{V}_{\text{targ}}$ is called *tabu* for $\delta + u$ local-search iterations, where u is taken uniformly at random between 0 and the *tabu tenure* δ. Before starting exploration, each tabu variable in $\mathcal{V}_{\text{targ}}^{\text{n}}$ is required to be equal to its corresponding variable in $\mathcal{V}_{\text{targ}}^{\text{c}}$. We need not do this algorithmically: the constraint $\forall v \in \text{tabu}(\mathcal{V}_{\text{targ}}) : v^{\text{n}} = v^{\text{c}}$, where $\text{tabu}(\mathcal{V}_{\text{targ}})$ denotes the set of tabu variables, is added to the neighbourhood model before starting CP search on it.

We improve this tabu search by adding the *aspiration criterion* that allows re-assigning tabu variables if this yields a new overall best valuation, under lexicographic order on $\langle g, o \rangle$. This is achieved by instead posting the constraint

$g^{n} < \text{best}(g) \vee (g^{n} = \text{best}(g) \wedge o^{n} < \text{best}(o)) \vee \forall v \in \text{tabu}(\mathcal{V}_{\text{targ}}) : v^{n} = v^{c}$, where $\text{best}(v)$ denotes the value of variable v in the overall best valuation.

In order to further help the local search escape local minima, we also implement a *restart mechanism*: if the exploration step does not return a new valuation for γ iterations of local search or does not improve the overall best valuation since the last restart for λ iterations, then a restart is made. Our LS(cp) performs restarts from the initialisation step. Recall that the initialisation step uses randomisation in its branching strategy.

5   Experimental Evaluation

Our aim is to show that declarative neighbourhoods are technology independent and enable lifting any CP solver into also being a MiniZinc local-search backend.

We evaluate our prototype LS(cp) against fzn-oscar-cbls [4], a CBLS backend that uses declarative neighbourhoods (but can be run black-box); Yuck,[1] a CBLS backend that only runs black-box; Gecode [11], a CP backend that uses CP search annotations such as line 14 of Listing 1; and Gecode-lns, an LNS backend that uses Gecode upon adding the `relax_and_reconstruct` annotation, with an 80% probability of freezing for each variable, to the MiniZinc model and the `-restart luby` flag when running the backend. These settings for Gecode-lns were decided upon after observing robust performance in initial experiments. We use two configurations of LS(cp) in order to see the impact of better neighbour pruning at the cost of possibly disconnecting the search space: LS(cp)-soft uses the constraint softening scheme of Sect. 4.2, and LS(cp)-hard sets $\text{soft}(C, g) = C$. For both configurations, we use the parameters $\tau_{\text{init}} = 10$ s, $\tau_{\text{explo}} = 30$ s, $\tau_{\text{intens}} = 10$ s, $\delta = 0.1 \cdot |\mathcal{V}_{\text{targ}}|$, $\gamma = 3 \cdot \delta$, and $\lambda = 1000$ and the first-improving heuristic during exploration, as initial experiments showed these settings were robust. For all backends that use randomisation, we report the best-found and median objective values of 10 independent runs, as well as, in prefixed superscript if non-zero, the number of runs where feasibility was not established. For Gecode, we report the best-found objective of a single run. For each run we use a 15-min timeout, under Linux Ubuntu 18.04 (64 bit) on an Intel Xeon E5520 of 2.27 GHz, with 4 processors of 4 cores each, with 24 GB RAM.

Since we must handcraft declarative neighbourhoods for each model, which requires a good understanding of both the model and the underlying problem, we evaluated LS(cp) on the models, instances, and declarative neighbourhoods of [3], namely steel-mill slab design [23], generalised balanced academic curriculum design (GBAC) [6], car sequencing [8], and community detection [10]. For space reasons, we omit the last two, which give results similar to GBAC. We also evaluated on the TSPTW model in Listing 1, and added declarative neighbourhoods to models used in the MiniZinc Challenge [26] for a capacitated vehicle routing problem (CVRP), a time-dependent travelling salesperson problem (TDTSP), and the seat moving problem. Table 1 has the results, where boldface indicates the best objective value by all backends for the instance of

---

[1] https://github.com/informarte/yuck.

the corresponding row and a "–" indicates that no feasible valuation was found in any run by the backend in the corresponding column.

CVRP. We used the model and instances of the MiniZinc Challenge 2015, except the easiest instance, 'simple2'. We extended the model with a declarative neighbourhood similar to the one in Listing 1, as CVRP is here also modelled using a `circuit` constraint. Gecode-lns was never worse and usually much better than both versions of LS(cp), but they significantly outperformed fzn-oscar-cbls, Yuck, and Gecode on all instances. Yuck and fzn-oscar-cbls here make moves on auxiliary variables (recall the end of Sect. 2.2): this explains why LS(cp) was better. LS(cp)-hard gave better valuations than LS(cp)-soft, which suggests that softening constraints is not important for this model. Hence it is unsurprising that LNS was best.

GBAC. Yuck and fzn-oscar-cbls overall outperformed the other backends, and fzn-oscar-cbls was best, possibly due to the declarative neighbourhoods. For the UD3, UD7, and UD9 instances, LS(cp) did not find any feasible valuation in any run, while Gecode-lns and Gecode found very bad valuations. This indicates that finding a feasible valuation for these instances is difficult using CP-style search, which LS(cp) and Gecode-lns use for initialisation. Clearly, this is an example where softening many constraints is important to find reasonable valuations.

Seat Moving. We used the model and instances of the MiniZinc Challenge 2018, except the hardest instance, 15-12-00, for which no solutions were found by any backend. We extended the model with a declarative neighbourhood defining moves that either swap two variables in a row of a 2D array or re-assigns one. LS(cp) and Gecode-lns performed best, with LS(cp)-soft being the sole best on one instance. Yuck and fzn-oscar-cbls failed to find a feasible valuation for the instances 10-12-00 and 20-20-00, and found poor valuations for the other ones. It would therefore appear that CP-style search is here more suitable for finding initial valuations, and that keeping more (but not all) constraints hard improves local search.

Steel Mill. We used the `hard_steelmill` neighbourhood of [3]. Arguably, fzn-oscar-cbls was best, followed by LS(cp)-hard, LS(cp)-soft, and then Gecode-lns, but each of these approaches wins on some instances. There is no clear pattern here, illustrating the importance of trying multiple technologies on the same model for each instance.

TDTSP. We used the model and instances of the MiniZinc Challenge 2017. We extended the model with a declarative neighbourhood defining moves preserves the satisfaction of an `inverse` constraint. Gecode-lns significantly outperformed all the other backends, and both versions of LS(cp) outperformed fzn-oscar-cbls, Yuck, and Gecode. As with the CVRP model, fzn-oscar-cbls and Yuck here make moves on auxiliary variables, which explains why LS(cp) was better.

TSPTW. We used the model in Listing 1 and medium-sized GendreauDumas-Extended *.001 instances.[2] LS(cp) and Gecode-lns outperformed all the other

---

[2] http://lopez-ibanez.eu/tsptw-instances.

Table 1. Experimental results on various minimisation problems.

| | Declarative neighbourhood | | | | | | Black-box | | CP search annotation | | |
| | LS(cp)-soft | | LS(cp)-hard | | fzn-oscar-cbls | | Yuck | | Gecode-lns | | Gecode |
CVRP	best	med.	best	med.	best	med.	best	med.	best	med.	best
A-n37-k5	773	920	722	773	2530	2736	–	–	**693**	**693**	1673
A-n64-k9	3187	3273	3113	3202	–	–	–	–	**1617**	**1617**	3544
B-n45-k5	1833	2019	1633	1848	3633	4004	–	–	**769**	**769**	2408
P-n16-k8	**450**	455	**450**	**450**	–	–	559	559	**450**	**450**	530
GBAC											
UD1	8577	9776	7635	9722	**438**	624	944	944	31263	31264	45420
UD2	**174**	213	189	217	189	206	289	289	354	376	12305
UD3	–	–	–	–	**191**	267	413	413	37576	37654	57681
UD4	470	1190	974	1190	**401**	472	485	485	904	904	11925
UD5	386	427	368	489	**272**	327	626	626	2039	2244	23028
UD6	124	135	125	144	122	153	154	154	**55**	**55**	9846
UD7	–	–	–	–	**519**	639	745	761	27330	27330	44044
UD8	65	87	74	107	63	86	105	105	**48**	**48**	9472
UD9	–	–	–	–	**463**	572	692	692	29213	29213	44010
UD10	126	176	107	162	81	91	138	138	**53**	**53**	12101
Seat Moving											
10-12-00	**463**	465	464	467	–	–	–	–	735	735	555
10-20-05	**90**	**90**	**90**	**90**	130	132	132	132	**90**	**90**	139
15-20-00	**199**	**199**	**199**	**199**	209	^{7}210	–	–	**199**	**199**	207
20-20-00	**262**	**262**	**262**	**262**	–	–	–	–	**262**	**262**	286
Steel Mill											
bench_3_0	11	13	11	14	12	15	629	629	**8**	**8**	64
bench_3_1	31	46	29	42	**22**	31	–	–	77	77	167
bench_3_2	31	43	**28**	43	42	60	–	–	33	33	83
bench_3_3	36	48	**38**	54	**38**	76	896	896	70	70	326
bench_3_4	20	41	23	31	71	111	–	–	**14**	**14**	38
bench_3_5	**40**	50	46	54	54	56	925	925	98	98	270
bench_3_6	**27**	48	33	58	54	106	–	–	55	55	292
bench_3_7	81	103	82	103	**79**	101	1039	1039	109	109	203
bench_3_8	128	173	132	161	**116**	205	1400	1400	183	183	341
bench_3_9	212	230	**183**	260	205	252	1678	1678	240	240	331
TDTSP											
10_35_20	9114	9764	**9055**	9279	14424	17217	10847	10847	**9055**	**9055**	**9055**
10_42_00	**8421**	**8421**	**8421**	**8421**	15866	18751	9248	9248	**8421**	**8421**	**8421**
10_58_20	11043	11435	10800	10986	15581	19021	12319	12323	**10306**	**10306**	13799
20_26_00	16124	17677	15105	17427	22752	^{5}22961	17906	19956	**12741**	**12741**	18197
20_36_10	15272	16045	15028	15772	22020	^{1}22602	17727	17727	**12308**	**12308**	15051
TSPTW											
n40w120	**434**	**434**	**434**	**434**	–	–	–	–	**434**	**434**	490
n40w140	**328**	**328**	**328**	**328**	–	–	–	–	**328**	**328**	380
n40w160	**348**	**348**	**348**	**348**	–	–	–	–	**348**	**348**	425
n60w120	**384**	**384**	**384**	**384**	–	–	–	–	387	387	513
n100w80	720	^{8}856	**679**	^{8}694	–	–	–	–	–	–	772

backends on all instances. In fact, the best-found objective values on all but the n100w80 instance are optimal. For the latter, Gecode-lns did not find any feasible valuation, but both versions of LS(cp) found feasible valuations in $10 - 8 = 2$ runs. Like with the CVRP and TDTSP models, both fzn-oscar-cbls and Yuck here make moves on auxiliary variables, namely the `ArrTime[i]` ones.

6   Conclusion, Related Work, and Future Work

Conclusion. We have demonstrated that declarative neighbourhoods, which were originally conceived for CBLS backends to MiniZinc, can also be used in order to generate automatically an LS(cp) method for local search, where the neighbourhood is initialised and explored using CP models and a CP solver. In fact, since we propose a decomposition for our new WRITES constraint, nothing in our recipe for lifting any CP solver into a local-search backend to MiniZinc is specific to CP: our recipe equally applies to any IP, SAT, or SMT solver.

While we see a wide variety of behaviour across the benchmarks against CP, CBLS, and LNS backends to MiniZinc, our prototype LS(cp) backend finds the best solutions for the seat moving and TSPTW benchmarks and is competitive on all others, except GBAC, where it sometimes even struggles to find initial valuations. Hence there seems to be a sweet spot for the CP-based exploration of local-search neighbourhoods, where auxiliary variables make CBLS backends slow and non-LNS neighbourhoods are important for local search.

Related Work. We have already discussed the differences between LS(cp) and LNS (large-neighbourhood search, [24]) near the end of Sect. 1

Structured neighbourhood search (SNS, [1]) is a local-search framework for models written in Essence [9]. In SNS, a set of neighbourhoods is automatically inferred from the variables of a model and their types. This is done via a set of predefined rules for each basic variable type and predefined rules for variables of nested types, such as lists of sets of integers. Although the connection is not explicitly made in [1], the SNS framework is defined using the ideas in [20]. Specifically, in SNS, the variables of the given model of a problem are referred to as *active variables*. For each active variable, a corresponding *primary variable* is introduced to represent the next valuation. Each neighbourhood is then expressed as a neighbourhood model connecting the active and primary variables, which is essentially the same approach as in [20] and in this paper. Because multiple neighbourhoods are inferred, a multi-armed-bandit algorithm is used for selecting which neighbourhood to use.

The STOREELEMENT(p, v, I, O) constraint used in [13] for constraint-based testing, with a propagator in [5], is the particular case WRITES($O, I, [p], [v]$), which is equivalent to $O = \text{write}(I, p, v)$ in the theory of arrays [16]. The WRITES(O, I, P, V) constraint encodes a *parallel* series of writes on an array I giving an array $O = \text{write}(\cdots \text{write}(\text{write}(I, p_1, v_1), p_2, v_2) \cdots p_m, v_m)$ with $\forall i, j : p_i = p_j \Rightarrow v_i = v_j$. However, encoding WRITES as a sequence of STOREELEMENT generates m copies of the input array in the encoding, but we only need the initial input array and the final output array.

Future Work. More advanced versions of LNS, such as propagation-guided LNS [19], cost-impact-guided LNS [15], and explanation-guided LNS [21], should be compared with LS(cp) to better understand the problems suitable for LS(cp) and LNS respectively.

Our LS(cp) is limited by initialisation being able to find an initial valuation, just like LNS. While the softening of some constraints makes initialisation more likely to succeed, the finding of an initial valuation is not guaranteed. Furthermore, the initialisation, exploration, and intensification steps all rely on appropriate CP search strategies being used, and our initial experiments indicate that these branching strategies have a significant impact on the performance. Clearly there is scope here for further investigation.

We also need to determine how best to soften constraints and which ones to soften, trading search over larger neighbourhoods for better robustness. Ideally, some form of dynamic adjustment, automatically softening when it is difficult to find improving moves, seems attractive to pursue.

Acknowledgements. We would like to thank the anonymous reviewers for their constructive feedback that helped improve this paper. This work is supported by the Swedish Research Council (VR) through Project Grant 2015-04910.

References

1. Akgün, O., et al.: A framework for constraint based local search using Essence. In: Lang, J. (ed.) IJCAI 2018, pp. 1242–1248. IJCAI Organization (2018)
2. Björdal, G., Flener, P., Pearson, J.: Generating compound moves in local search by hybridisation with complete search. In: Rousseau, L.-M., Stergiou, K. (eds.) CPAIOR 2019. LNCS, vol. 11494, pp. 95–111. Springer, Cham (2019). https://doi.org/10.1007/978-3-030-19212-9_7
3. Björdal, G., Flener, P., Pearson, J., Stuckey, P.J., Tack, G.: Declarative local-search neighbourhoods in MiniZinc. In: Alamaniotis, M., Lagniez, J.M., Lallouet, A. (eds.) ICTAI 2018. pp. 98–105. IEEE Computer Society (2018)
4. Björdal, G., Monette, J.N., Flener, P., Pearson, J.: A constraint-based local search backend for MiniZinc. Constraints 20(3), 325–345 (2015). The fzn-oscar-cbls backend is available at http://optimisation.research.it.uu.se/software
5. Charreteur, F., Botella, B., Gotlieb, A.: Modelling dynamic memory management in constraint-based testing. J. Syst. Softw. 82(11), 1755–1766 (2009)
6. Chiarandini, M., Di Gaspero, L., Gualandi, S., Schaerf, A.: The balanced academic curriculum problem revisited. J. Heuristics 18(1), 119–148 (2012)
7. De Landtsheer, R., Ponsard, C.: OscaR.cbls: an open source framework for constraint-based local search. In: ORBEL-27, the 27th Annual Conference of the Belgian Operational Research Society (2013). https://www.orbel.be/orbel27/pdf/abstract293.pdf, the OscaR.cbls solver is available at https://bitbucket.org/oscarlib/oscar/branch/CBLS
8. Dincbas, M., Simonis, H., Van Hentenryck, P.: Solving the car-sequencing problem in constraint logic programming. In: Kodratoff, Y. (ed.) ECAI 1988, pp. 290–295. Pitman (1988)
9. Frisch, A.M., Grum, M., Jefferson, C., Martinez Hernandez, B., Miguel, I.: The design of Essence: a constraint language for specifying combinatorial problems. In: IJCAI 2007, pp. 80–87. Morgan Kaufmann (2007)

10. Ganji, M., Bailey, J., Stuckey, P.J.: A declarative approach to constrained community detection. In: Beck, J.C. (ed.) CP 2017. LNCS, vol. 10416, pp. 477–494. Springer, Cham (2017). https://doi.org/10.1007/978-3-319-66158-2_31

11. Gecode Team: Gecode: a generic constraint development environment (2018). The Gecode solver and its MiniZinc backend are available at https://www.gecode.org

12. Glover, F., Laguna, M.: Tabu search. In: Modern Heuristic Techniques for Combinatorial Problems, pp. 70–150. Wiley (1993)

13. Gotlieb, A., Botella, B., Watel, M.: INKA: ten years after the first ideas. In: Pollet, Y. (ed.) ICSSEA 2006 (2006)

14. Hoos, H.H., Stützle, T.: Stochastic Local Search: Foundations & Applications. Elsevier/Morgan Kaufmann, San Francisco (2004)

15. Lombardi, M., Schaus, P.: Cost impact guided LNS. In: Simonis, H. (ed.) CPAIOR 2014. LNCS, vol. 8451, pp. 293–300. Springer, Cham (2014). https://doi.org/10.1007/978-3-319-07046-9_21

16. McCarthy, J.: Towards a mathematical science of computation. In: Proceedings of IFIP Congress (1962)

17. Nethercote, N., Stuckey, P.J., Becket, R., Brand, S., Duck, G.J., Tack, G.: MiniZinc: towards a standard CP modelling language. In: Bessière, C. (ed.) CP 2007. LNCS, vol. 4741, pp. 529–543. Springer, Heidelberg (2007), the MiniZinc toolchain is available at https://www.minizinc.org. https://doi.org/10.1007/978-3-540-74970-7_38

18. OscaR Team: OscaR: Scala in OR (2012). https://oscarlib.bitbucket.io

19. Perron, L., Shaw, P., Furnon, V.: Propagation guided large neighborhood search. In: Wallace, M. (ed.) CP 2004. LNCS, vol. 3258, pp. 468–481. Springer, Heidelberg (2004). https://doi.org/10.1007/978-3-540-30201-8_35

20. Pesant, G., Gendreau, M.: A constraint programming framework for local search methods. J. Heuristics **5**(3), 255–279 (1999). Extends a preliminary version at CP 1996, LNCS, vol. 1118, pp. 353–366, Springer (1996)

21. Prud'homme, C., Lorca, X., Jussien, N.: Explanation-based large neighborhood search. Constraints **19**(4), 339–379 (2014)

22. Rossi, F., van Beek, P., Walsh, T. (eds.): Handbook of Constraint Programming. Elsevier, Amsterdam (2006)

23. Schaus, P., Van Hentenryck, P., Monette, J.N., Coffrin, C., Michel, L., Deville, Y.: Solving steel mill slab problems with constraint-based techniques: CP, LNS, and CBLS. Constraints **16**(2), 125–147 (2011)

24. Shaw, P.: Using constraint programming and local search methods to solve vehicle routing problems. In: Maher, M., Puget, J.-F. (eds.) CP 1998. LNCS, vol. 1520, pp. 417–431. Springer, Heidelberg (1998). https://doi.org/10.1007/3-540-49481-2_30

25. Shaw, P., De Backer, B., Furnon, V.: Improved local search for CP toolkits. Ann. Oper. Res. **115**(1–4), 31–50 (2002)

26. Stuckey, P.J., Feydy, T., Schutt, A., Tack, G., Fischer, J.: The MiniZinc challenge 2008–2013. AI Mag. **35**(2), 55–60 (2014). https://www.minizinc.org/challenge.html

27. Van Hentenryck, P., Michel, L.: Constraint-Based Local Search. The MIT Press, Cambridge (2005)

Vehicle Routing by Learning
from Historical Solutions

Rocsildes Canoy$^{(\boxtimes)}$ and Tias Guns

Vrije Universiteit Brussel, Brussels, Belgium
{Rocsildes.Canoy,Tias.Guns}@vub.be

Abstract. The goal of this paper is to investigate a decision support system for vehicle routing, where the routing engine learns from the *subjective* decisions that human planners have made in the past, rather than optimizing a distance-based *objective* criterion. This is an alternative to the practice of formulating a custom VRP for every company with its own routing requirements. Instead, we assume the presence of past vehicle routing solutions over similar sets of customers, and learn to make similar choices. The approach is based on the concept of learning a first-order Markov model, which corresponds to a probabilistic transition matrix, rather than a deterministic distance matrix. This nevertheless allows us to use existing arc routing VRP software in creating the actual route plans. For the learning, we explore different schemes to construct the probabilistic transition matrix. Our results on a use-case with a small transportation company show that our method is able to generate results that are close to the manually created solutions, without needing to characterize all constraints and sub-objectives explicitly. Even in the case of changes in the client sets, our method is able to find solutions that are closer to the actual route plans than when using distances, and hence, solutions that would require fewer manual changes to transform into the actual route plan.

1 Introduction

Route planning at SME companies is constrained by the limited number of vehicles, the capacity of each delivery vehicle, and the scheduling horizon within which all deliveries have to be made. The objective, often implicitly, can include a wide range of company goals including reducing operational costs, minimizing fuel consumption and carbon emissions, as well as optimizing driver familiarity with the routes and maximizing fairness by assigning tours of similar duration to the drivers. Daily plans are often created in a route optimization software that is capable of producing plans that are optimal in terms of route length and travel time. We have observed, however, that in practice, route planners heavily modify the result given by the software, or simply pull out, modify, and reuse an old plan that has been used and known to work in the past. The planners, by performing these modifications, are essentially optimizing with their own set of objectives and personal preferences.

© Springer Nature Switzerland AG 2019
T. Schiex and S. de Givry (Eds.): CP 2019, LNCS 11802, pp. 54–70, 2019.
https://doi.org/10.1007/978-3-030-30048-7_4

The goal of this research is to learn the preferences of the planners when choosing one option over another and to more effectively reuse all of the knowledge and effort that have been put into creating previous plans. Our focus is on intelligent tools that learn from historical data, and can hence manage and recommend similar routes as used in the past.

In collaboration with a small transportation company, one of our initial steps was to analyze their historical data. Close data inspection has confirmed that the route planners often rely on historical data in constructing the daily plans, which is consistent with the observations gathered during company visits.

To learn from historical data, we take inspiration from various machine learning research on route prediction for a single vehicle. Markov models developed from historical data have been applied to driver turn prediction, prediction of the remainder of the route by looking at the previous road segments taken by the driver, and predicting individual road choices given the origin and destination. These studies have produced positive and encouraging results for those tasks. Hence, in this work, we investigate the use of Markov models for predicting the route choices for an entire fleet, and how to use these choices to solve the VRP.

With a first-order Markov model, route optimization can be done by maximizing the product of the probabilities of the arcs taken by the vehicles, which corresponds to maximizing the sum of log likelihoods. Hence, a key property of our approach is that it can use any existing VRP solution method. This is a promising, novel approach to the vehicle routing problem.

This paper's contributions are presented in the succeeding sections as follows. After a brief literature review, we present in Sect. 3 our transition probability matrix reformulation of the VRP. In Sect. 4, we introduce the algorithm for learning the transition matrix from historical data and its different variants. The comparison of the different construction schemes and the experimental results on actual company data are shown in Sect. 5.

2   Related Work

The first mathematical formulation and algorithmic approach to solving the Vehicle Routing Problem (VRP) appeared in 1959 in the paper by Dantzig & Ramser [6] which aimed to find an optimal routing for a fleet of gasoline delivery trucks. Since its introduction, the VRP has become one of the most studied combinatorial optimization problems. Faced on a daily basis by distributors and logistics companies worldwide, the problem has attracted a lot of attention due to its significant economic importance.

A large part of the research effort concerning the VRP has focused on its classical and basic version—the *Capacitated* Vehicle Routing Problem (CVRP). The presumption is that the algorithms developed for CVRP can be extended and applied to more complicated real-world cases [13]. Due to the recent development of new and more efficient optimisation methods, research interest has shifted towards realistic VRP variants known as Rich VRP [4,9]. These problems deal with realistic, and sometimes multi-objective, optimisation functions, uncertainty, and a wide variety of real-life constraints related to time and distance factors, inventory and scheduling, environmental and energy concerns, personal preferences of route planners and drivers, etc. [4].

The VRP becomes increasingly complex as additional sub-objectives and constraints are introduced. The inclusion of preferences, for example, necessitates the difficult, if not impossible, task of formalizing the route planners' knowledge and choice preferences explicitly in terms of constraints and weights. In most cases, it is much easier to get examples and historical solutions rather than to extract explicit decision rules from the planners, as observed by Potvin et al. in the case of vehicle dispatching [17]. One approach is to use learning techniques, particularly learning by examples, to reproduce the planners' decision behavior. To this end, we develop a new method that learns from previous solutions by using a Markov model, and which also simplifies the problem by eliminating the need to characterize preference constraints and sub-objectives explicitly.

Learning from historical solutions has been investigated before within the context of constraint programming, e.g., in the paper of Beldiceanu and Simonis on constraint seeker [1] and model seeker [2], and Picard-Cantin et al. on learning constraint parameters from data, where a Markov chain is used, but for individual constraints [16]. In this respect, our goal is not to learn constraint instantiations, but to learn choice preferences, e.g., as part of the objective. Related to the latter is the work on Constructive Preference Elicitation [8], although that actively queries the user, as does constraint acquisition [3].

Our motivation for Markov models is that they have been previously used in route prediction of individual vehicles. Krumm [12] has developed an algorithm for driver turn prediction using a Markov model. Trained from the driver's historical data, the model makes a probabilistic prediction based on a short sequence of just-driven road segments. Experimental results showed that by looking at the most recent 10 segments into the past, the model can effectively predict the next segment with about 90% accuracy. Ye et al. [19] introduced a route prediction method that can accurately predict an entire route early in the trip. The method is based on Hidden Markov Models (HMM) and also trained from the driver's past history. Another route prediction algorithm that predicts a driving route for a given pair of origin and destination was presented by Wang et al. [18]. Also based on the first-order Markov model, the algorithm uses a probability transition matrix that was constructed to represent the knowledge of the driver's preferred links and routes. Personalized route prediction has been used in transportation systems that provide drivers with real-time traffic information and in intelligent vehicle systems for optimizing energy efficiency in hybrid vehicles [7].

3   Formalisation

3.1   Standard CVRP

In its classical form, the CVRP can be defined as follows. Let $G = (V, A)$ be a graph where $V = \{0, 1, \ldots, N\}$ is the vertex set and $A = \{(i, j) : i, j \in V, i \neq j\}$ is the arc set. The vertex 0 denotes the depot, whereas the other vertices represent the customers to be served. A non-negative cost matrix $\mathbf{C} = [c_{ij}]$ is associated with every arc (i,j), $i \neq j$, where the cost c_{ij} can be instantiated based on true distance, travel time, travel costs or a combination thereof. A homogeneous fleet of m vehicles, each with capacity Q, is available at the depot. Each customer $i \in V$ for $i > 0$ is associated with a known non-negative demand q_i.

The Capacitated Vehicle Routing Problem is to determine a set of least-cost vehicle routes such that

(i) each customer vertex $i \in V$, is visited exactly once by exactly one vehicle;
(ii) each vehicle must start and finish the route at the depot, $i = 0$;
(iii) the sum of demands of each route must not exceed the vehicle capacity Q.

A common way to formulate the CVRP is by using Boolean decision variables which indicate whether a vehicle travels between a pair of vertices in G. Let x_{ij} be such a Boolean decision variable, which takes the value 0 or 1, with $x_{ij} = 1$ when arc (i,j) is traveled and $x_{ij} = 0$ otherwise. The CVRP can then be expressed as the following integer program whose objective is to minimize the total routing cost [14, 15, 20]:

$$(\text{CVRP}) \quad \min \sum_{(i,j) \in A} c_{ij} x_{ij} \tag{3.1}$$

$$\text{subject to} \quad \sum_{j \in V, \, j \neq i} x_{ij} = 1 \qquad i = 1, \ldots, N \tag{3.2}$$

$$\sum_{i \in V, \, i \neq j} x_{ij} = 1 \qquad j = 1, \ldots, N \tag{3.3}$$

$$\sum_{j=1}^{n} x_{0j} \leq m, \tag{3.4}$$

$$\text{if } x_{ij} = 1 \Rightarrow u_i + q_j = u_j \qquad (i,j) \in A : j \neq 0, \, i \neq 0 \tag{3.5}$$

$$q_i \leq u_i \leq Q \qquad i = 1, \ldots, N \tag{3.6}$$

$$x_{ij} \in \{0,1\} \qquad (i,j) \in A \tag{3.7}$$

Constraints (3.2) and (3.3) impose that every customer node must be visited by exactly one vehicle and that exactly one vehicle must leave from each node. Constraint (3.4) limits the number of routes to the size of the fleet, m. In constraint (3.5), u_j denotes the cumulative vehicle load at node j. The constraint plays a dual role—it prevents the formation of subtours, i.e., cycling routes that do not pass through the depot, and together with constraint (3.6), it ensures that the vehicle capacity is not exceeded. While the model does not make explicit which stop belongs to which route, this information can be reconstructed from the active arcs in the solution.

We will consider the case where the exact number of vehicles to use is given, i.e., constraint (3.4) becomes $\sum_{j=1}^{n} x_{0j} = m$. This is the operational setting in which the company works, where work is divided among the vehicles and drivers available on the given day.

3.2   CVRP with Arc Probabilities

In the subsequent section, we will study how to learn, from historical solutions, a Markov model that represents the following probability distribution: $\mathbf{Pr}$(next stop $= j \mid$ current stop $= i$). That is, it represents the probability of moving from a current stop to a next stop.

The goal then, is to find the routing X that is most likely, i.e., the set of routes that maximizes the joint probability over the arcs taken:

$$\max \prod_{(i,j)\in X} \mathbf{Pr}(\text{next stop} = j \mid \text{current stop} = i),$$

The question is how to efficiently search for the most likely routing among all the *valid* routings.

For this, we observe that the first-order Markov model can be represented as a transition probability matrix $\mathbf{T} = [t_{ij}]$, with $t_{ij} = \mathbf{Pr}$(next stop $= j \mid$ current stop $= i$). Furthermore, maximizing $\prod_{(i,j)\in X} t_{ij}$ is equivalent to maximizing the sum of log probabilities: $\max \sum_{(i,j)\in X} log(t_{ij})$.

Formulated with respect to Boolean decision variables x_{ij} as in the CVRP formulation, the goal is to maximize the joint probability:

$$\max \sum_{(i,j)\in A} log(t_{ij})x_{ij}. \tag{3.8}$$

Hence, to find the most likely routing, we can solve the CVRP with, as cost matrix $\mathbf{C} = [c_{ij}]$, the transformed transition probability matrix: $c_{ij} = -log(t_{ij})$.

As a result, any existing CVRP solver can be used to find the most likely solution once the transition probability matrix is learned.

4   Learning Transition Probabilities from Data

We now explain how to learn the transition probability matrix from historical solutions (Sect. 4.1), followed by different ways of using data (Sect. 4.2) and of weighing the instances (Sect. 4.3). Finally, in Sect. 4.4 we discuss how to combine a learned probability matrix with a distance-based probability matrix.

4.1   Constructing the Transition Probability Matrix

To compute the probabilities, we assume given a sequence $\langle h_k \rangle$ of historical instances as input, e.g., ordered by date. Each h_k is a VRP solution over a set of customers S_k. Note that the S_k can change from instance to instance. Let $S_k = \{s_1, s_2, \ldots, s_p\}$ be a given set of customers of solution h_k. A solution, or routing plan, h_k over S_k is a set of routes $\{r_1, \ldots, r_m\}$ servicing each customer in S_k exactly once. Each route r_l starts at the depot, serves some number of customers at most once, and returns to the depot. Using s_0 to denote the depot, r_l can then be represented by a sequence $\langle s_0, s_{l1}, \ldots, s_{lq}, s_0 \rangle$, where $s_{li} \in S_k$ and all s_{li} are distinct.

Algorithm 1. Building a transition matrix from historical instances

---

Input: A sequence of n historical data instances $H = \langle h_1, \ldots, h_n \rangle$ sorted such that h_1 is the oldest and h_n is the most recent instance, a weight w_k per data instance, where the default value is $w_k = 1$ for $k = 1, \ldots, n$, and the Laplace smoothing parameter $\alpha \geq 0$.

1. Extract and gather all the stops visited in H into a set $\Sigma = \{s_0, s_1, \ldots, s_t\}$, where stop s_0 denotes the depot.
2. For each h_k, $k = 1, \ldots, n$, do:
    Construct an adjacency matrix $\mathbf{A}^k_{t+1 \times t+1} = [a^k_{ij}]$, where $a^k_{ij} = 1$ if $(s_i, s_j) \in h_k$, and 0 otherwise.
3. Build the arc transition frequency matrix $\mathbf{F}_{t+1 \times t+1}$ with the weights w_k and the adjacency matrices constructed in Step 2:

$$\mathbf{F} = \sum_{k=1}^{n} w_k \mathbf{A}^k. \tag{4.1}$$

4. Apply the Laplace smoothing technique to get the transition matrix $\mathbf{T}_{t+1 \times t+1}$: For every element t_{ij} of $\mathbf{T}$,

$$t_{ij} = \frac{f_{ij} + \alpha}{N_i + \alpha d},$$

where $d = t + 1$ is the row length (=total number of stops), and $N_i = \sum_{j=1}^{t+1} f_{ij}$ is the row sum.

Output: Transition matrix $\mathbf{T}_{t+1 \times t+1} = [t_{ij}]$, where

$$t_{ij} = \mathbf{Pr}(\text{next stop} = s_j \mid \text{current stop} = s_i)$$
$$= \frac{f_{ij} + \alpha}{\sum_{j=1}^{t+1} f_{ij} + \alpha(t+1)}.$$

Probability Computation. The conditional probability of a vehicle moving to the next stop s_j given its current location s_i can be computed as follows:

$$\mathbf{Pr}(\text{next stop} = s_j \mid \text{current stop} = s_i) = \frac{\mathbf{Pr}(\text{next stop} = s_j, \text{ current stop} = s_i)}{\mathbf{Pr}(\text{current stop} = s_i)},$$

with $\mathbf{Pr}(\text{current stop} = s_i) = \sum_k \mathbf{Pr}(\text{next stop} = s_k, \text{ current stop} = s_i)$. Empirically, the algorithm counts as f_{ij} the number of times (current stop $= s_i$) and (next stop $= s_j$) have occurred together in the historical solutions. We then have

$$\mathbf{Pr}(\text{next stop} = s_j \mid \text{current stop} = s_i) = \frac{f_{ij}}{\sum_k f_{ik}}. \tag{4.2}$$

Laplace Smoothing. To account for the fact that the number of samples may be small, and some f_{ij} may be zero, we can smooth the probabilities using the Laplace smoothing technique [5,11,19]. Laplace smoothing reduces the impact of data sparseness arising in the process of building the transition matrix. Our proposed construction method adopts the technique to deal with arcs with zero probability. As a result of smoothing, these arcs are given a small, non-negative probability, thereby eliminating the zeros in the resulting transition matrix. Conceptually, with α as the smoothing parameter ($\alpha = 0$ corresponds to no smoothing), we add α observations to each event. The probability computation now becomes:

$$\mathbf{Pr}(\text{next stop} = s_j \mid \text{current stop} = s_i) = \frac{f_{ij} + \alpha}{\sum_k f_{ik} + \alpha d}, \qquad (4.3)$$

with d denoting the number of stops $|S|$.

Construction Algorithm. Algorithm 1 shows the algorithm for constructing the probability transition matrix. The dimensions of the matrix, that is, the total set of unique stops, are determined in Step 1. In Step 2, an adjacency matrix is constructed for each historical instance. A frequency matrix is constructed in Step 3 by computing the (weighted) sum of all the adjacency matrices (4.1); by default, $w_k = 1$ for all instances. Finally, during normalisation in Step 4, Laplace smoothing is applied if $\alpha > 0$.

4.2   Evaluation Schemes

In a traditional machine learning setup, the dataset is split into a training set and a test set. The training set is used for training, and the test set for evaluation. This is a **batch evaluation** as all test instances are evaluated in one batch. The best resulting model is then deployed (and should be periodically updated).

Our data has a temporal aspect to it, namely the routing is performed every day. Hence, each day one additional training instance becomes available, allowing us to incrementally grow the data. In this case, we should perform an **incremental evaluation**. The incremental evaluation procedure is depicted in Algorithm 2.

4.3   Weighing Schemes

Training instances are ordered over time, and the set of stops visited can vary from instance to instance. In order to account for this, Algorithm 1 can weigh each of the instances differently during construction of the transition probability matrix (Step 3).

Algorithm 2. Training and testing with an incrementally increasing training set

Input: $H = \langle h_1, \ldots, h_n \rangle$, an ordered sequence of n historical instances.
1. Start from an initial m training instances, e.g., $m = \lfloor 0.75n \rfloor$ for a $75\% - 25\%$ split.
2. For $j = m, \ldots, n - 1$ do:
    2.1. Build the probability transition matrix $\mathbf{T}_j$ on $\langle h_1, \ldots, h_j \rangle$ using Algorithm 1.
    2.2. Solve CVRP using $\mathbf{T}_j$, by using the log transform of equation (3.8).
    2.3. Evaluate the CVRP solution against h_{j+1}.

We propose three weighing schemes, namely, to uniformly distribute weights, to distribute weights according to time, and to distribute weights according to the similarity of the stop sets. Table 1 shows the three schemes and the variants that we will consider.

As before, we assume that the training instances used during matrix construction are ordered chronologically from old to new as $H = \langle h_1, \ldots, h_n \rangle$ and we need to define a weight w_k for each instance h_k.

Table 1. An overview of the proposed weighing schemes

Name	Weights	Squared weights
Uniform (UNIF)	$w_k = 1$	—
Time-based (TIME)	$w_k = k/n$	$w_k = (k/n)^2$
Similarity-based (SIMI)	$w_k = J(h_k, h_{n+1})$	$w_k = J(h_k, h_{n+1})^2$

Uniform Weighing. The first weighing scheme is the default and simply assumes a uniform weight across all instances:

$$w_k = 1, \qquad k = 1, \ldots, n. \tag{4.4}$$

Time-Based Weighing. It is well known that streaming data can have *concept drift* [10], that is, the underlying distribution can change over time. To account for this, we can use a time-based weighing scheme where older instances are given smaller weights, and newer instances larger ones. Using index k as time indicator, we can weigh the instances as:

$$w_k = \frac{k}{n}, \qquad k = 1, \ldots, n. \tag{4.5}$$

This assumes a linearly increasing importance of instances. We can also consider a squared importance $w_k = (k/n)^2$, or an exponential importance, etc.

Similarity-Based Weighing. The stops in each instance typically vary, and the presence or absence of different stops can lead to different decision behaviors. To account for this, we consider a weighing scheme that uses the similarity between the set of stops of the current instance, which is part of the input of the CVRP, and the set of stops of each historical instance. The goal is to assign larger weights to training instances that are more similar to the test instance, and smaller weights if they are less similar.

The similarity of two stop sets can be measured using the Jaccard similarity coefficient. The Jaccard similarity of two sets is defined as the size of the intersection divided by the size of the union of the two sets:

$$J(A, B) = \frac{|A \cap B|}{|A \cup B|} \tag{4.6}$$

for two non-empty sets A and B. The Jaccard similarity coefficient is always a value between 0 (no overlapping elements) and 1 (exactly the same elements).

Given test instance h_{n+1}, we consider the following similarity-based weighing scheme:

$$w_k = J(h_k, h_{n+1}), \qquad k = 1, \ldots, n. \tag{4.7}$$

To further amplify the importance of similarity, we can also use the squared Jaccard similarity $w_k = J(h_k, h_{n+1})^2$, etc.

4.4   Adding Distance-Based Probabilities

The probability matrix captures well what stops often follow each other. However, if a new stop location is added, Laplace smoothing will give an equal probability to all arcs leaving from this new stop. Also in case of rarely visited stops, the probabilities can be uninformative, and in general there can be equal conditional probabilities among the candidate next stops given a current stop.

We know that human planners take the number of kilometers into account when lacking further information. Indeed, this is the basic assumption of the CVRP. Hence, we wish to be able to bias our system to also take distances into account. To do this, we will mix the transition probability matrix built from historical instances with a transition probability matrix based on distances (or any other cost used in a traditional CVRP formulation).

The goal is to give two stops that have a *low* cost between them, e.g., are close to each other, a *high* probability, and to give stops that have a high cost a low probability. Hence, we construct a probability matrix where the likelihood of moving from one stop to the next is inversely proportional to the cost to that next stop, relative to all candidate next stops:

$$d_{ij} = \mathbf{Pr}_{dist}(\text{next stop} = s_j \mid \text{current stop} = s_i) \tag{4.8}$$

$$= \frac{c'_{ij}}{\sum_k c'_{ik}} \tag{4.9}$$

with

$$c'_{ij} = \frac{\sum_k c_{ik}}{c_{ij}}, \tag{4.10}$$

where c_{ij} is the standard cost between stop i and stop j, and c'_{ij} is the inverse of the relative cost, computed in equation (4.10). This is then normalized in Equation (4.9) to obtain valid transition probabilities.

Combining Transition Probability Matrices. Given transition probability matrices $\mathbf{T} = [t_{ij}]$ and $\mathbf{D} = [d_{ij}]$, we can take the convex combination as follows:

$$t'_{ij} = \beta t_{ij} + (1 - \beta)d_{ij}. \tag{4.11}$$

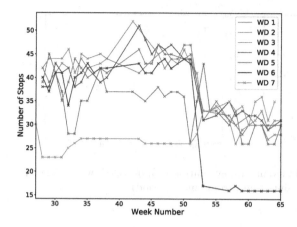

Fig. 1. No. of stops by weekday (WD)

	Before drift		Entire data	
WD	Train	Test	Train	Test
1	14	5	23	7
2	12	5	21	7
3	11	5	19	7
4	13	5	22	7
5	14	5	22	7
6	14	5	22	7
7	15	5	23	7
Total	93	35	152	49

Fig. 2. Training and test set sizes after 75% − 25% split

Taking $\beta = 1$ corresponds to using only the history-based transition probabilities, while $\beta = 0$ will only use distance-based probabilities, with values in between resulting to a combination of the two probabilities.

Note that this approach places no conditions on how the history-based transition matrix $\mathbf{T} = [t_{ij}]$ is computed, and hence is compatible with Laplace smoothing and weighing during the construction of $\mathbf{T}$.

5   Experiments

Description of the Data. The historical data used in the experiments consist of daily route plans collected within a span of nine months. The plans were generated by the route planners and used by the company in their actual operations. Each data is a numbered instance and the entire data is ranked by time. An instance contains the set of stops visited by the fleet, with the stop set divided into sequences corresponding to individual routes.

Data instances are grouped by day-of-week including Saturday and Sunday. This mimics the operational characteristic of the company. The entire data set is composed of 201 instances, equivalent to an average of 29 instances per weekday. The breakdown of the entire data set after the train-test split is shown in Fig. 2. An average of 8.7 vehicles servicing 35.1 stops are used per instance in the data before drift, and 6.4 vehicles (25.4 stops) for the 73 instances after drift (see next paragraph).

Data Visualization. Figure 1 shows the number of customers served per weekday during the entire experimental time period. A *concept drift* is clearly discernible starting Week 53, where a change in stop set size occurs. This observation has prompted us to conduct two separate experiments—one with data from the entire period, and the other using only data from the period before the drift.

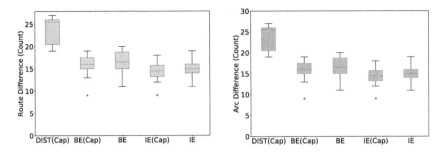

Fig. 3. Batch evaluation (BE) and incremental evaluation (IE) on UNIF with capacity (Cap) and without capacity constraints (data from entire period)

Evaluation Methodology. We made a comparison of the prediction accuracy of the proposed schemes. Performance was evaluated using two evaluation measures, based on two properties of a VRP solution, namely stops and active arcs. *Route Difference* (RD) counts the number of stops that were incorrectly assigned to a different route. Intuitively, RD may be interpreted as an estimate of how many moves between routes are necessary when modifying the predicted solution to match the grouping of stops into routes. To compute route difference, a pairwise comparison of the routes contained in the predicted and test solution is made. The pair with the smallest difference in stops is greedily selected without replacement. RD is the total number of stops that were placed differently. *Arc Difference* (AD) measures the number of arcs traveled in the actual solution but not in the predicted solution. AD is calculated by taking the set difference of the arc sets of the test and predicted solution. Correspondingly, AD gives an estimate of the total number of modifications needed to correct the solution.

Capacity demand estimates for each stop were provided by the company. Note, however, that in our approach, route construction is based primarily on the arc probabilities. This allows for solving the VRP even without capacity constraints. When evaluating, in order to keep the subtour elimination constraint (3.5), each q_i will be assigned a value of 1 while using the number of stops as fictive bound on the vehicle capacities, e.g., $Q = n$.

5.1 Numerical Results

The numerical experiments were performed using Python 3.6.5 and the CPLEX 12.8 solver with the default setting, on a Lenovo ThinkPad X1 Carbon with an Intel Core i7 processor running at 1.8 GHz with 16 GB RAM. The time limit for solving the CVRP is set to 600 s. Unless otherwise stated, $\alpha = 1$ and $\beta = 1$ are used as parameters. Recall from Table 1 that UNIF stands for uniform weighing, TIME for time-based, and SIMI for similarity-based.

Batch Evaluation and Incremental Evaluation on UNIF with and without Capacity Constraints. The first experiment (Fig. 3) was done with UNIF

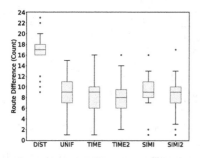
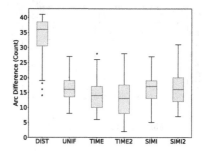

Fig. 4. Route and arc difference (period before drift)

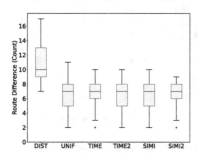
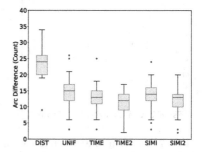

Fig. 5. Route and arc difference (entire period)

to compare the prediction accuracy of batch evaluation and incremental evaluation with and without the capacity (Cap) demand estimates. The motivation is to investigate how UNIF will perform even without the capacity constraints. As a baseline, we included the solution (DIST) obtained by solving the standard distance-based CVRP. Computation was done on a subset of the weekdays with data from the entire period. The subset contains 55 historical instances, split into 41 and 14 for training and testing, respectively.

Results show that DIST is consistently outperformed by the other methods. Moreover, in all cases batch evaluation (BE) performed worse than incremental evaluation (IE). This is likely because IE can incrementally use more data.

As for the computation time, DIST often reached the time limit of 600 s and returned a non-optimal solution, with an average optimality gap of 3.65%. With all the other schemes, it took only an average of 0.096 s to obtain the optimal solution. We observed that the learned matrices are much more sparse (containing more 0 or near-0 values) than the distance matrices.

Remarkably, when using the transition probability matrices, we can even solve the VRP *without* capacity constraints and still get meaningful results. This shows the ability of the method to learn the structure underlying the problem just from the solutions. In all cases, adding capacity constraints, however, does slightly improve the results and especially reduces the variance.

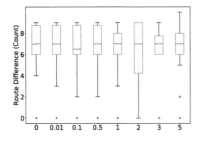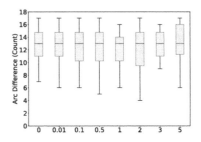

Fig. 6. Route and arc difference for values of Laplace parameter α (entire period)

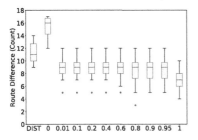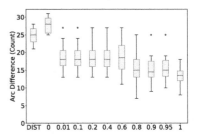

Fig. 7. Route and arc difference for varying values of β (entire period)

Evaluation of Schemes on Historical Data Set. In the next two experiments (Figs. 4 and 5), we tested the proposed schemes: UNIF, TIME, TIME2, SIMI, and SIMI2, with TIME2 and SIMI2 indicating squared weights (see Table 1). As a consequence of the previous experiment, here we used incremental evaluation and also included the capacity constraints.

Figure 4 is on data before drift (week 53 in Fig. 1). It shows that all the proposed schemes gave better estimates than DIST. In all cases, the schemes with the squared weights (TIME2, SIMI2) performed better than their counterparts (TIME, SIMI). While using similarity-based weights (SIMI, SIMI2) did not seem to improve the solutions given by UNIF, time-based weighing (TIME, TIME2) did. Among all schemes, TIME2 gave the most accurate predictions. Hence, more recent routings are more relevant for making choices here.

Results on data from the entire period (Fig. 5) exhibit a slightly different behavior. As before, all the schemes outperformed DIST. In terms of route difference, there is no significant difference in the results. The route difference values seem lower than before the drift, but it should be noted that these instances also involve fewer stops. In terms of arc difference, both TIME and SIMI outperformed UNIF. As before, TIME2 and SIMI2 are better than TIME and SIMI, with TIME2 also giving the most accurate predictions among all schemes.

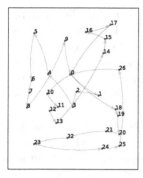 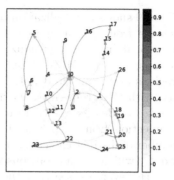 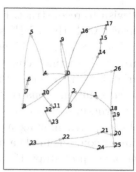

Fig. 8. Actual solution   **Fig. 9.** Learned probabilities, only relevant stops shown   **Fig. 10.** Predicted sol.

5.2   Parameter Sensitivity

Effects of Varying Laplace Parameter Values. To understand the effect of varying the Laplace parameter α (Fig. 6), we perform an experiment on a subset of the weekdays. For simplicity, capacity demands were not taken into account and TIME2 was selected based on the previous experiments. It is interesting to note that TIME2 worked well even with no smoothing ($\alpha = 0$). It can also be observed that the scheme produced stable results with α within the range $[0, 2]$, with a slight improvement discernible at $\alpha = 2$. The accuracy, notably in arc prediction, appeared to diminish for alpha values greater than 2. In general, we see that on our data, Laplace smoothing has very little effect.

Effects of Adding Distance-Based Probabilities. We investigate the potential benefit of combining the learned transition probability matrix with a distance-based probability matrix. Figure 7 shows the result for varying β on a subset of the weekdays. TIME2 was again used and no capacity demands were taken into account. Compared to using the absolute distances (DIST), using only the distance-based probability matrix leads to worse results. This is not entirely surprising as the model loses the ability to compare distance trade-offs between arbitrary arcs, as the probabilities are conditional on a 'current node'.

When combined with the learned probability matrix, we see that even small values of β, i.e., more importance given on the distance-based probabilities, already lead to better results than using pure distances. For values in between 0 and 1 there seems to be little effect, with some improvement in arc difference for higher values. However, the best result is obtained when only the history-based probability matrix is used.

5.3   Detailed Example

A visual example of the way the routes are predicted by the transition matrix can be observed from Figs. 8, 9 and 10. Figure 8 shows the actual solution that we wish to reconstruct. Figure 9 shows a visualization of the probability matrix learned with the UNIF scheme from the previous data of the same weekday. Darker arcs indicate higher probabilities. The visualization shows a clear structure, with distinct connections, e.g., to the furthest stops, but also a higher variability in the denser regions and near the depot. Figure 10 shows our predicted solution, constructed with the probability matrix of Fig. 9. It captures key structural parts and makes trade-offs elsewhere to come up with a global solution, e.g., making a connection from stop 3 to 9. The actual solution, in comparison, has made a number of distinct choices such as a reversed green tour and a swap of stops 16 and 17, which are not obvious to predict by looking at the probability matrix map. However, we see that the routes generally match and that it would require only a small amount of modifications to the predicted solution to obtain the actual solution.

6   Concluding Remarks

One of the crucial first steps in solving vehicle routing problems is explicitly formulating the problem objectives and constraints, as the quality of the solution depends, to a great extent, on this characterization. Oftentimes in practice, the optimization of the route plans takes into account not only time and distance-related factors, but also a multitude of other concerns. Specifying each sub-objective and constraint may be tedious. Moreover, as we have observed in practice, computed solutions seldomly guarantee the satisfaction of the route planners and all involved stakeholders.

We presented an approach to solving the VRP which does not require explicit problem characterization. Inspired by existing research on the application of Markov models to individual route prediction, we developed an approach that learns a probability transition matrix from previous solutions, to predict the routes for an entire fleet. This learned model can be transformed so that any CVRP solver can be used to find the most likely routing. We have shown how the structure of the solution can be learned, resulting in more accurate solutions than using distances alone. The algorithm performs well even without capacity demands, confirming its ability to learn the solution structure. An added advantage is that solving is fast, due to the sparsity of the transition matrix.

This paper shows the potential of learning preferences in VRP from historical solutions. Results on real data have been encouraging, although validation on other real-life data should also be considered, as other data may have more (or less) structure. Our approach could be plugged into existing VRP software today, but as with all predictive techniques there should be human oversight to avoid unwanted bias.

Future work on the routing side will involve applications to richer VRP, e.g., problems involving time windows, multiple deliveries, etc. On the learning side, the use of higher-order Markov models or other probability estimation techniques will be investigated. Also, using separate learned models per vehicle or per driver is worth investigating. Finally, extending the technique so that the user can be actively queried, and learned from, during construction is an interesting direction, e.g., to further reduce the amount of user modifications needed on the predicted solutions.

References

1. Beldiceanu, N., Simonis, H.: A constraint seeker: finding and ranking global constraints from examples. In: Lee, J. (ed.) CP 2011. LNCS, vol. 6876, pp. 12–26. Springer, Heidelberg (2011). https://doi.org/10.1007/978-3-642-23786-7_4

2. Beldiceanu, N., Simonis, H.: A model seeker: extracting global constraint models from positive examples. In: Milano, M. (ed.) CP 2012. LNCS, pp. 141–157. Springer, Heidelberg (2012). https://doi.org/10.1007/978-3-642-33558-7_13

3. Bessiere, C., Koriche, F., Lazaar, N., O'Sullivan, B.: Constraint acquisition. Artif. Intell. **244**, 315–342 (2017)

4. Caceres-Cruz, J., Arias, P., Guimarans, D., Riera, D., Juan, A.A.: Rich vehicle routing problem: survey. ACM Comput. Surv. (CSUR) **47**(2), 32 (2015)

5. Chen, S.F., Goodman, J.: An empirical study of smoothing techniques for language modeling. Comput. Speech Lang. **13**(4), 359–394 (1999)

6. Dantzig, G.B., Ramser, J.H.: The truck dispatching problem. Manage. Sci. **6**(1), 80–91 (1959)

7. Deguchi, Y., Kuroda, K., Shouji, M., Kawabe, T.: HEV charge/discharge control system based on navigation information. Technical report, SAE Technical Paper (2004)

8. Dragone, P., Teso, S., Passerini, A.: Constructive preference elicitation. Front. Robot. AI **4**, 71 (2018)

9. Drexl, M.: Rich vehicle routing in theory and practice. Logistics Res. **5**(1–2), 47–63 (2012)

10. Gama, J., Žliobaitė, I., Bifet, A., Pechenizkiy, M., Bouchachia, A.: A survey on concept drift adaptation. ACM Comput. Surv. (CSUR) **46**(4), 44 (2014)

11. Johnson, W.E.: Probability: the deductive and inductive problems. Mind **41**(164), 409–423 (1932)

12. Krumm, J.: A Markov model for driver turn prediction. In: Withrow, l.L. (eds.) SAE 2008 World Congress, Distinguished Speaker Award, April 2008

13. Laporte, G.: What you should know about the vehicle routing problem. Naval Res. Logistics (NRL) **54**(8), 811–819 (2007)

14. Lau, H.C., Liang, Z.: Pickup and delivery with time windows: algorithms and test case generation. Int. J. Artif. Intell. Tools **11**(03), 455–472 (2002)

15. Munari, P., Dollevoet, T., Spliet, R.: A generalized formulation for vehicle routing problems. arXiv preprint arXiv:1606.01935 (2016)

16. Picard-Cantin, É., Bouchard, M., Quimper, C.-G., Sweeney, J.: Learning parameters for the sequence constraint from solutions. In: Rueher, M. (ed.) CP 2016. LNCS, vol. 9892, pp. 405–420. Springer, Cham (2016). https://doi.org/10.1007/978-3-319-44953-1_26

17. Potvin, J.Y., Dufour, G., Rousseau, J.M.: Learning vehicle dispatching with linear programming models. Comput. Oper. Res. **20**(4), 371–380 (1993)

18. Wang, X., et al.: Building efficient probability transition matrix using machine learning from big data for personalized route prediction. Procedia Comput. Sci. **53**, 284–291 (2015)

19. Ye, N., Wang, Z., Malekian, R., Lin, Q., Wang, R.: A method for driving route predictions based on hidden markov model. Math. Problems Eng. **2015**, 12 (2015)
20. Yu, M., Nagarajan, V., Shen, S.: Minimum makespan vehicle routing problem with compatibility constraints. In: Salvagnin, D., Lombardi, M. (eds.) CPAIOR 2017. LNCS, vol. 10335, pp. 244–253. Springer, Cham (2017). https://doi.org/10.1007/978-3-319-59776-8_20

On Symbolic Approaches for Computing the Matrix Permanent

Supratik Chakraborty[1]([⊠]), Aditya A. Shrotri[2]([⊠]), and Moshe Y. Vardi[2]([⊠])

[1] Indian Institute of Technology Bombay, Mumbai, India
supratik@cse.iitb.ac.in
[2] Rice University, Houston, USA
{Aditya.Aniruddh.Shrotri,vardi}@rice.edu

Abstract. Counting the number of perfect matchings in bipartite graphs, or equivalently computing the permanent of 0-1 matrices, is an important combinatorial problem that has been extensively studied by theoreticians and practitioners alike. The permanent is #P-Complete; hence it is unlikely that a polynomial-time algorithm exists for the problem. Researchers have therefore focused on finding tractable subclasses of matrices for permanent computation. One such subclass that has received much attention is that of sparse matrices i.e. matrices with few entries set to 1, the rest being 0. For this subclass, improved theoretical upper bounds and practically efficient algorithms have been developed. In this paper, we ask whether it is possible to go beyond sparse matrices in our quest for developing scalable techniques for the permanent, and answer this question affirmatively. Our key insight is to represent permanent computation symbolically using Algebraic Decision Diagrams (ADDs). ADD-based techniques naturally use dynamic programming, and hence avoid redundant computation through memoization. This permits exploiting the hidden structure in a large class of matrices that have so far remained beyond the reach of permanent computation techniques. The availability of sophisticated libraries implementing ADDs also makes the task of engineering practical solutions relatively straightforward. While a complete characterization of matrices admitting a compact ADD representation remains open, we provide strong experimental evidence of the effectiveness of our approach for computing the permanent, not just for sparse matrices, but also for dense matrices and for matrices with "similar" rows.

1 Introduction

Constrained counting lies at the heart of several important problems in diverse areas such as performing Bayesian inference [45], measuring resilience of electrical networks [20], counting Kekule structures in chemistry [23], computing

---

Author names are ordered alphabetically by last name and does not indicate contribution.

Work supported in part by NSF grant IIS-1527668, the Data Analysis and Visualization Cyberinfrastructure funded by NSF under grant OCI-0959097 and Rice University, and MHRD IMPRINT-1 Project No. 6537 sponsored by Govt of India.

© Springer Nature Switzerland AG 2019
T. Schiex and S. de Givry (Eds.): CP 2019, LNCS 11802, pp. 71–90, 2019.
https://doi.org/10.1007/978-3-030-30048-7_5

the partition function of monomer-dimer systems [26], and the like. Many of these problems reduce to counting problems on graphs. For instance, learning probabilistic models from data reduces to counting the number of topological sorts of directed acyclic graphs [56], while computing the partition function of a monomer-dimer system reduces to computing the number of perfect matchings of an appropriately defined bipartite graph [26]. In this paper, we focus on the last class of problems – that of counting perfect matchings in bipartite graphs. It is well known that this problem is equivalent to computing the *permanent* of the 0-1 bi-adjacency matrix of the bipartite graph. We refer to these two problems interchangeably in the remainder of the paper.

Given an $n \times n$ matrix A with real-valued entries, the permanent of A is given by $perm(A) = \sum_{\sigma \in S_n} \prod_{i=1}^{n} a_{i,\sigma(i)}$, where S_n denotes the symmetric group of all permutations of $1, \ldots n$. This expression is almost identical to that for the determinant of A; the only difference is that the determinant includes the sign of the permutation in the inner product. Despite the striking resemblance of the two expressions, the complexities of computing the permanent and determinant are vastly different. While the determinant can be computed in time $\mathcal{O}(n^{2.4})$, Valiant [54] showed that computing the permanent of a 0-1 matrix is #P-Complete, making a polynomial-time algorithm unlikely [53]. Further evidence of the hardness of computing the permanent was provided by Cai, Pavan and Sivakumar [11], who showed that the permanent is also hard to compute on average. Dell et al. [19] showed that there can be no algorithm with subexponential time complexity, assuming a weak version of the Exponential Time Hypothesis [3] holds.

The determinant has a nice geometric interpretation: it is the oriented volume of the parallelepiped spanned by the rows of the matrix. The permanent, however, has no simple geometric interpretation. Yet, it finds applications in a wide range of areas. In chemistry, the permanent and the permanental polynomial of the adjacency matrices of fullerenes [32] have attracted much attention over the years [12,13,34]. In constraint programming, solutions to All-Different constraints can be expressed as perfect matchings in a bipartite graph [43]. An estimate of the number of such solutions can be used as a branching heuristic to guide search [42,60]. In physics, permanents can be used to measure quantum entanglement [58] and to compute the partition functions of monomer-dimer systems [26].

Since computing the permanent is hard in general, researchers have attempted to find efficient solutions for either approximate versions of the problem, or for restricted classes of inputs. In this paper, we restrict our attention to exact algorithms for computing the permanent. The asymptotically fastest known exact algorithm for general $n \times n$ matrices is Nijenhuis and Wilf's version of Ryser's algorithm [38,46], which runs in time $\Theta(n \cdot 2^n)$ for all matrices of size n. For matrices with bounded treewidth or clique-width [15,44], Courcelle, Makowsky and Rotics [16] showed that the permanent can be computed in time linear in the size of the matrix, i.e., computing the permanent is Fixed Parameter Tractable (FPT). A large body of work is devoted to developing fast algorithms for sparse matrices, i.e. matrices with only a few entries set to non-zero values [28, 34, 48, 59] in each row. Note that the problem remains #P-Complete even when the input is restricted to matrices with exactly three 1's per row and column [9].

An interesting question to ask is whether we can go beyond sparse matrices in our quest for practically efficient algorithms for the permanent. For example, can we hope for practically efficient algorithms for computing the permanent of *dense* matrices, i.e., matrices with almost all entries non-zero? Can we expect efficiency when the rows of the matrix are "similar", i.e. each row has only a few elements different from any other row (sparse and dense matrices being special cases)? Existing results do not seem to throw much light on these questions. For instance, while certain non-sparse matrices indeed have bounded clique-width, the aforementioned result of Courcelle et al. [14, 16] does not yield practically efficient algorithms as the constants involved are enormous [24]. The hardness of non-sparse instances is underscored by the fact that SAT-based model counters do not scale well on these, despite the fact that years of research and careful engineering have enabled these tools to scale extremely well on a diverse array of problems. We experimented with a variety of CNF-encodings of the permanent on state-of-the-art counters like D4 [33]. Strikingly, no combination of tool and encoding was able to scale to matrices even half the size of those solved by Ryser's approach in the same time, despite the fact that Ryser's approach has exponential complexity even in the best case.

In this paper, we show that practically efficient algorithms for the permanent can indeed be designed for large non-sparse matrices if the matrix is represented compactly and manipulated efficiently using a special class of data structures. Specifically, we propose using *Algebraic Decision Diagrams* [4] (ADDs) to represent matrices, and design a version of Ryser's algorithm to work on this symbolic representation of matrices. This effectively gives us a symbolic version of Ryser's algorithm, as opposed to existing implementations that use an explicit representation of the matrix. ADDs have been studied extensively in the context of formal verification, and sophisticated libraries are available for compact representation of ADDs and efficient implementation of ADD operations [50, 55]. The literature also contains compelling evidence that reasoning based on ADDs and variants scales to large instances of a diverse range of problems in practice, cf. [4, 21]. Our use of ADDs in Ryser's algorithm leverages this progress for computing the permanent. Significantly, there are several sub-classes of matrices that admit compact representations using ADDs, and our algorithm works well for all these classes. Our empirical study provides evidence for the first time that the frontier of practically efficient permanent computation can be pushed well beyond the class of sparse matrices, to the classes of dense matrices and, more generally, to matrices with "similar" rows. Coupled with a technique known as early abstraction, ADDs are able to handle sparse instances as well. In summary, the symbolic approach to permanent computation shows promise for both sparse and dense classes of matrices, which are special cases of a notion of row-similarity.

The rest of the paper is organized as follows: in Sect. 2 we introduce ADDs and other concepts that we will use in this paper. We discuss related work in Sect. 3 and present our algorithm and analyze it in Sect. 4. Our empirical study is presented in Sects. 5 and 6 and we conclude in Sect. 7.

2 Preliminaries

We denote by $A = (a_{ij})$ an $n \times n$ 0-1 matrix, which can also be interpreted as the bi-adjacency matrix of a bipartite graph $G_A = (U \cup V, E)$ with an edge between vertex $i \in U$ and $j \in V$ iff $a_{ij} = 1$. We will denote the ith row of A by r_i. A perfect matching in G_A is a subset $\mathcal{M} \subseteq E$, such that for all $v \in (U \cup V)$, exactly one edge $e \in \mathcal{M}$ is incident on v. We denote by $perm(A)$ the permanent of A, and by $\#PM(G_A)$, the number of perfect matchings in G. A well known fact is that $perm(A) = \#PM(G_A)$, and we will use these concepts interchangeably when clear from context.

2.1 Algebraic Decision Diagrams

Let X be a set of Boolean-valued variables. An Algebraic Decision Diagram (ADD) is a data structure used to compactly represent a function of the form $f : 2^X \to \mathbb{R}$ as a Directed Acyclic Graph (DAG). ADDs were originally proposed as a generalization of Binary Decision Diagrams (BDDs), which can only represent functions of the form $g : 2^X \to \{0, 1\}$. Formally, an ADD is a 4-tuple (X, T, π, G) where X is a set of Boolean variables, the finite set $T \subset \mathbb{R}$ is called the carrier set, $\pi : X \to \mathbb{N}$ is the diagram variable order, and G is a rooted directed acyclic graph satisfying the following three properties:

1. Every terminal node of G is labeled with an element of T.
2. Every non-terminal node of G is labeled with an element of X and has two outgoing edges labeled 0 and 1.
3. For every path in G, the labels of visited non-terminal nodes must occur in increasing order under π.

We use lower case letters $f, g, \dots$ to denote both functions from Booleans to reals as well as the ADDs representing them. Many operations on such functions can be performed in time polynomial in the size of their ADDs. We list some such operations that will be used in our discussion.

- *Product*: The product of two ADDs representing functions $f : 2^X \to \mathbb{R}$ and $g : 2^Y \to \mathbb{R}$ is an ADD representing the function $f \cdot g : 2^{X \cup Y} \to \mathbb{R}$, where $f \cdot g(\tau)$ is defined as $f(\tau \cap X) \cdot g(\tau \cap Y)$ for every $\tau \in 2^{X \cup Y}$,
- *Sum*: Defined in a way similar to the product.
- *If-Then-Else (ITE)*: This is a ternary operation that takes as inputs a BDD f and two ADDs g and h. $ITE(f, g, h)$ represents the function $f \cdot g + \neg f \cdot h$, and the corresponding ADD is obtained by substituting g for the leaf'1' of f and h for the leaf '0', and simplifying the resulting structure.

- *Additive Quantification*: The existential quantification operation for Boolean-valued functions can be extended to real-valued functions by replacing disjunction with addition as follows. The additive quantification of $f : 2^X \to \mathbb{R}$ is denoted as $\exists x.f : 2^{X\setminus\{x\}} \to \mathbb{R}$ and for $\tau \in 2^{X\setminus\{x\}}$, we have $\exists x.f(\tau) = f(\tau) + f(\tau \cup \{x\})$.

ADDs share many properties with BDDs. For example, there is a unique minimal ADD for a given variable order π, called the *canonical ADD*, and minimization can be performed in polynomial time. Similar to BDDs, the variable order can significantly affect the size of the ADD. Hence heuristics for finding good variable orders for BDDs carry over to ADDs as well. ADDs typically have lower *recombination efficiency*, i.e. number of shared nodes, vis-a-vis BDDs. Nevertheless, sharing or recombination of isomorphic sub-graphs in an ADD is known to provide significant practical advantages in representing matrices, vis-a-vis other competing data structures. The reader is referred to [4] for a nice introduction to ADDs and their applications.

2.2 Ryser's Formula

The permanent of A can be calculated by the principle of inclusion-exclusion using Ryser's formula: $perm(A) = (-1)^n \sum_{S \subseteq [n]} (-1)^{|S|} \prod_{i=1}^n \sum_{j \in S} a_{ij}$. Algorithms implementing Ryser's formula on an explicit representation of an arbitrary matrix A (not necessarily sparse) must consider all 2^n subsets of $[n]$. As a consequence, such algorithms have at least exponential complexity. Our experiments show that even the best known existing algorithm implementing Ryser's formula for arbitrary matrices [38], which iterates over the subsets of $[n]$ in Gray-code sequence, consistently times out after 1800 s on a state-of-the-art computing platform when computing the permanent of $n \times n$ matrices, with $n \geq 35$.

3   Related Work

Valiant showed that computing the permanent is #P-complete [54]. Subsequently, researchers have considered restricted sub-classes of inputs in the quest for efficient algorithms for computing the permanent, both from theoretical and practical points of view. We highlight some of the important milestones achieved in this direction.

A seminal result is the Fisher-Temperly-Kastelyn algorithm [29,52], which computes the number of perfect matchings in planar graphs in PTIME. This result was subsequently extended to many other graph classes (c.f. [40]). Following the work of Courcelle et al. a number of different width parameters have been proposed, culminating in the definition of ps-width [47], which is considered to be the most general notion of width [8]. Nevertheless, as with clique-width, it is not clear whether it lends itself to practically efficient algorithms. Bax and Franklin [5] gave a Las Vegas algorithm with better expected time complexity than Ryser's approach, but requiring $\mathcal{O}(2^{n/2})$ space.

For matrices with at most $C \cdot n$ zeros, Servedio and Wan [48] presented a $(2 - \varepsilon)^n$-time and $\mathcal{O}(n)$ space algorithm where ε depends on C. Izumi and Wadayama [28] gave an algorithm that runs in time $\mathcal{O}^*(2^{(1-1/(\Delta \log \Delta))n})$, where Δ is the average degree of a vertex. On the practical side, in a series of papers, Liang, Bai and their co-authors [34,35,59] developed algorithms optimized for computing the permanent of the adjacency matrices of fullerenes, which are 3-regular graphs.

In recent years, practical techniques for propositional model counting (#SAT) have come of age. State-of-the-art exact model counters like DSharp [37] and D4 [33] also incorporate techniques from knowledge compilation. A straightforward reduction of the permanent to #SAT uses a Boolean variable x_{ij} for each 1 in row i and column j of the input matrix $\boldsymbol{A}$, and imposes Exact-One constraints on the variables in each row and column. This gives the formula $F_{perm(A)} = \bigwedge_{i \in [n]} ExactOne(\{x_{ij} : a_{ij} = 1\}) \wedge \bigwedge_{j \in [n]} ExactOne(\{x_{ij} : a_{ij} = 1\})$. Each solution to $F_{perm(A)}$ is a perfect matching in the underlying graph, and so the number of solutions is exactly the permanent of the matrix. A number of different encodings can be used for translating Exact-One constraints to Conjunctive Normal Form (see Sect. 5.1). We perform extensive comparisons of our tool with D4 and DSharp with six such encodings.

4   Representing Ryser's Formula Symbolically

As noted in Sect. 2, an explicit implementation of Ryser's formula iterates over all 2^n subsets of columns and its complexity is in $\Theta(n \cdot 2^n)$. Therefore, any such implementation takes exponential time even in the best case. A natural question to ask is whether we can do better through a careful selection of subsets over which to iterate. This principle was used for the case of sparse matrices by Servedio and Wan [48]. Their idea was to avoid those subsets for which the row-sum represented by the innermost summation in Ryser's formula, is zero for at least one row, since those terms do not contribute to the outer sum in Ryser's formula. Unfortunately, this approach does not help for non-sparse matrices, as very few subsets of columns (if any) will yield a zero row-sum.

It is interesting to ask if we can exploit similarity of rows (instead of sparsity) to our advantage. Consider the ideal case of an $n \times n$ matrix with *identical rows*, where each row has k ($\leq n$) 1s. For any given subset of columns, the row-sum is clearly the same for all rows, and hence the product of all row-sums is simply the nth power of the row-sum of one row. Furthermore, there are only $k + 1$ distinct values (0 through k) of the row-sum, depending on which subset of columns is selected. The number of r-sized column subsets that yield row-sum j is clearly $\binom{k}{j} \cdot \binom{n-k}{r-j}$, for $0 \leq j \leq k$ and $j \leq r \leq n - k + j$. Thus, we can directly compute the permanent of the matrix via Ryser's formula as $perm(\boldsymbol{A}) = (-1)^n \sum_{j=0}^{k} \sum_{r=j}^{n-k+j} (-1)^r \binom{k}{j} \cdot \binom{n-k}{r-j} \cdot j^n$. This equation has a more compact representation than the explicit implementation of Ryser's formula, since the outer summation is over $(k+1).(n-k+1)$ terms instead of 2^n terms.

Drawing motivation from the above example, we propose using memoization to simplify the permanent computation of matrices with similar rows. Specifically, if we compute and store the row-sums for a subset $S_1 \subset [n]$ of columns, then we can potentially reuse this information when computing the row-sums for subsets $S_2 \supset S_1$. We expect storage requirements to be low when the rows are similar, as the partial sums over identical parts of the rows will have a compact representation, as shown above.

While we can attempt to hand-craft a concrete algorithm using this idea, it turns out that ADDs fit the bill perfectly. We introduce Boolean variables x_j for each column $1 \leq j \leq n$ in the matrix. We can represent the summand $(-1)^{|S|} \prod_{i=1}^{n} \sum_{j \in S} a_{ij}$ in Ryser's formula as a function $f_{Ryser} : 2^X \to \mathbb{R}$ where for a subset of columns $\tau \in 2^X$, we have $f_{Ryser}(\tau) = (-1)^{|\tau|} \prod_{i=1}^{n} \sum_{j \in \tau} a_{ij}$. The outer sum in Ryser's formula is then simply the Additive Quantification of f_{Ryser} over all variables in X. The permanent can thus be denoted by the following equation:

$$perm(\boldsymbol{A}) = (-1)^n \cdot \exists x_1, x_2, \ldots x_n . (f_{Ryser}) \tag{1}$$

We can construct an ADD for f_{Ryser} incrementally as follows:

- **Step 1:** For each row r_i in the matrix, construct the Row-Sum ADD $f_{RS}^{r_i}$ such that $f_{RS}^{r_i}(\tau) = \sum_{j:a_{ij}=1} \mathbb{1}_\tau(x_j)$, where $\mathbb{1}_\tau(x_j)$ is the indicator function taking the value 1 if $x_j \in \tau$, and zero otherwise. This ADD can be constructed by using the sum operation on the variables x_j corresponding to the 1 entries in row r_i.
- **Step 2:** Construct the Row-Sum-Product ADD $f_{RSP} = \prod_{i=1}^{n} f_{RS}^{r_i}$ by applying the product operation on all the Row-Sum ADDs.

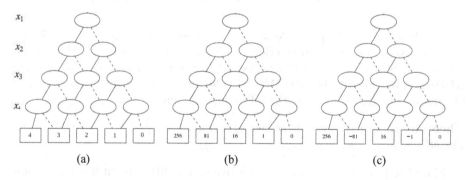

Fig. 1. (a) f_{RS}, (b) f_{RSP} and (c) f_{Ryser} for a 4×4 matrix of all 1s

- **Step 3:** Construct the Parity ADD $f_{PAR} = ITE(\bigoplus_{j=1}^{n} x_j, -1, +1)$, where $\oplus$ represents exclusive-or. This ADD represents the $(-1)^{|S|}$ term in Ryser's formula.
- **Step 4:** Construct $f_{Ryser} = f_{RSP} \cdot f_{PAR}$ using the product operation.

Finally, we can additively quantify out all variables in f_{Ryser} and multiply the result by $(-1)^n$ to get the permanent, as given by Eq. 1.

The size of the ADD f_{RSP} will be the smallest when the ADDs $f_{RS}^{r_i}$ are exactly the same for all rows r_i, i.e. when all rows of the matrix are identical. In this case, the ADDs $f_{RS}^{r_i}$ and f_{RSP} will be isomorphic; the values at the leaves of f_{RSP} will simply be the nth power of the values at the corresponding leaves of $f_{RS}^{r_i}$. An example illustrating this for a 4×4 matrix of all 1s is shown in Fig. 1. Each level of the ADDs in this figure corresponds to a variable (shown on the left) for a column of the matrix. A solid edge represents the 'true' branch while a dotted edge represents the 'false' branch. Observe that sharing of isomorphic subgraphs allows each of these ADDs to have 10 internal nodes and 5 leaves, as opposed to 15 internal nodes and 16 leaves that would be needed for a complete binary tree based representation.

The ADD representation is thus expected to be compact when the rows are "similar". Dense matrices can be thought of as a special case: starting with a matrix of all 1s (which clearly has all rows identical), we change a few 1s to 0s. The same idea can be applied to sparse matrices as well: starting with a matrix of all 0s (once again, identical rows), we change a few 0s to 1s. The case of very sparse matrices is not interesting, however, as the permanent (or equivalently, count of perfect matchings in the corresponding bipartite graph) is small and can be computed by naive enumeration. Interestingly, our experiments show that as we reduce the sparsity of the input matrix, constructing f_{RSP} and f_{Ryser} in a monolithic fashion as discussed above fails to scale, since the sizes of ADDs increase very sharply. Therefore we need additional machinery.

First, we rewrite Eq. 1 in terms of the intermediate ADDs as:

$$perm(\boldsymbol{A}) = (-1)^n \cdot \exists x_1, x_2, \ldots x_n \cdot \left(f_{PAR} \cdot \prod_{i=1}^{n} f_{RS}^{r_i} \right) \qquad (2)$$

We then employ the principle of early abstraction to compute f_{Ryser} incrementally. Note that early abstraction has been used successfully in the past in the context of SAT solving [41], and recently for weighted model counting using ADDs in a technique called ADDMC [1]. The formal statement of the principle of early abstraction is given in the following theorem.

Theorem 1. [1] *Let X and Y be sets of variables and $f : 2^X \to \mathbb{R}$, $g : 2^Y \to \mathbb{R}$. For all $x \in X \setminus Y$, we have $\exists_x(f \cdot g) = (\exists_x(f)) \cdot g$*

Since the product operator is associative and additive quantification is commutative, we can rearrange the terms of Eq. 2 in order to apply early abstraction. This idea is implemented in Algorithm RysersADD, which is motivated by the weighted model counting algorithm in [1].

Algorithm RysersADD takes as input a 0–1 matrix $\boldsymbol{A}$, a diagram variable order π and a cluster rank-order η. η is an ordering of variables which is used to heuristically partition rows of $\boldsymbol{A}$ into clusters using a function clusterRank, where all rows in a cluster get the same rank. Intuitively, rows that are almost

---

Algorithm 1. RysersADD($\boldsymbol{A}, \pi, \eta$)

1: $m \leftarrow max_{x \in X}\ \eta(x)$;
2: **for** $i = m, m-1, \ldots, 1$ **do**
3:     $\kappa_i \leftarrow \{f^r_{RS} : r$ is a row in $\boldsymbol{A}$ and clusterRank$(r, \eta) = i\}$;
4: $f_{Ryser} \leftarrow f_{PAR}$; $\triangleright$ f_{PAR} and each f^r_{RS} are constructed using the diagram variable order π
5: **for** $i = 1, 2, \ldots, m$ **do**
6:     **if** $\kappa_i \neq \emptyset$ **then**
7:         **for** $g \in \kappa_i$ **do**
8:             $f_{Ryser} \leftarrow f_{Ryser} \cdot g$;
9:             **for** $x \in Vars(f_{Ryser})$ **do**
10:                 **if** $x \notin (Vars(\kappa_{i+1}) \cup \ldots \cup Vars(\kappa_m))$ **then**
11:                     $f_{Ryser} \leftarrow \exists_x(f_{Ryser})$
12: **return** $(-1)^n \times f_{Ryser}(\emptyset)$

---

identical are placed in the same cluster, while those that differ significantly are placed in different clusters. Furthermore, the clusters are ordered such that there are non-zero columns in cluster i that are absent in the set of non-zero columns in clusters with rank $> i$. As we will soon see, this facilitates keeping the sizes of ADDs under control by applying early abstraction.

Algorithm RysersADD proceeds by first partitioning the Row-Sum ADDs of the rows $\boldsymbol{A}$ into clusters according to their cluster rank in line 3. Each Row-Sum ADD is constructed according to the diagram variable order π. The ADD f_{Ryser} is constructed incrementally, starting with the Parity ADD in line 4, and multiplying the Row-Sum ADDs in each cluster κ_i in the loop at line 7. However, unlike the monolithic approach, early abstraction is carried out within the loop at line 9. Finally, when the execution reaches line 12, all variables representing columns of the input matrix have been abstracted out. Therefore, f_{Ryser} is an ADD with a single leaf node that contains the (possibly negative) value of the permanent. Following Eq. 2, the algorithm returns the product of $(-1)^n$ and $f_{Rsyer}(\emptyset)$.

The choice of the function clusterRank and the cluster rank-order η significantly affect the performance of the algorithm. A number of heuristics for determining clusterRank and η have been proposed in literature, such as Bucket Elimination [18], and Bouquet's Method [7] for cluster ranking, and MCS [51], LexP [31] and LexM [31] for variable ordering. Further details and a rigorous comparison of these heuristics are presented in [1]. Note that if we assign the same cluster rank to all rows of the input matrix, Algorithm RysersADD reduces to one that constructs all ADDs monolithically, and does not benefit from early abstraction.

4.1 Implementation Details

We implemented Algorithm 1 using the library Sylvan [55] since unlike CUDD [50], Sylvan supports arbitrary precision arithmetic – an essential feature to avoid overflows when the permanent has a large value. Sylvan supports parallelization of ADD operations in a multi-core environment. In order to leverage this capability, we created a parallel version of RysersADD that differs from the sequential version only in that it uses the parallel implementation of ADD operations natively provided by Sylvan. Note that this doesn't require any change to Algorithm RysersADD, except in the call to Sylvan functions. While other non-ADD-based approaches to computing the permanent can be parallelized as well, we emphasize that it is a non-trivial task in general, unlike using Sylvan. We refer to our sequential and parallel implementations for permanent computation as RysersADD and RysersADD-P respectively, in the remainder of the discussion. We implemented our algorithm in C++, compiled under GCC v6.4 with the O3 flag. We measured the wall-times for both algorithms. Sylvan also supports arbitrary precision floating point computation, which makes it easy to extend RysersADD for computing permanent of real-valued matrices. However, we leave a detailed investigation of this for future work.

5 Experimental Methodology

The objective of our empirical study was to evaluate RysersADD and RysersADD-P on randomly generated instances (as done in [35]) and publicly available structured instances (as done in [34,59]) of 0-1 matrices.

5.1 Algorithm Suite

As noted in Sect. 3, a number of different algorithms have been reported in the literature for computing the permanent of sparse matrices. Given resource constraints, it is infeasible to include all of these in our experimental comparisons. This is further complicated by the fact that many of these algorithms appear not to have been implemented (eg: [28,48]), or the code has not been made publicly accessible (eg: [34,59]). A fair comparison would require careful consideration of several parameters like usage of libraries, language of implementation, suitability of hardware etc. We had to arrive at an informed choice of algorithms, which we list below along with our rationale:

- RysersADD and RysersADD-P: For the dense and similar rows cases, we use the monolithic approach as it is sufficient to demonstrate the scalability of our ADD-based approach. For sparse instances, we employ Bouquet's Method (List) [7] clustering heuristic along with MCS cluster rank-order [51] and we keep the diagram variable order the same as the indices of columns in the input matrix (see [1] for details about the heuristics). We arrived at these choices through preliminary experiments. We leave a detailed comparison of all combinations for future work.
- *Explicit Ryser's Algorithm*: We implemented Nijenhuis and Wilf's version [38] of Ryser's formula using Algorithm H from [30] for generating the Gray code sequence. Our implementation, running on a state-of-the-art computing platform (see Sect. 5.2), is able to compute the permanent of all matrices with

$n \leq 25$ in under 5 s. For $n = 30$, the time shoots up to approximately 460 s and for $n \geq 34$, the time taken exceeds 1800 s (time out for our experiments). Since the performance of explicit Ryser's algorithm depends only on the size of the matrix, and is unaffected by its structure, sparsity or row-similarity, this represents a complete characterization of the performance of the explicit Ryser's algorithm. Hence, we do not include it in our plots.

- *Propositional Model Counters*: Model counters that employ techniques from SAT-solving as well as knowledge compilation, have been shown to scale extremely well on large CNF formulas from diverse domains. Years of careful engineering have resulted in counters that can often outperform domain-specific approaches. We used two state-of-the-art exact model counters, viz. D4 [33] and DSharp [37], for our experiments. We experimented with 6 different encodings for At-Most-One constraints: (1) Pairwise [6], (2) Bitwise [6], (3) Sequential Counter [49], (4) Ladder [2,22], (5) Modulo Totalizer [39] and (6) Iterative Totalizer [36]. We also experimented with ADDMC, an ADD-based model counter [1]. However, it failed to scale beyond matrices of size 25; ergo we do not include it in our study.

We were unable to include the parallel #SAT counter countAtom [10] in our experiments, owing to difficulties in setting it up on our compute set-up. However, we could run countAtom on a slightly different set-up with 8 cores instead of 12, and 16 GB memory instead of 48 on a few sampled dense and similar-row matrix instances. Our experiments showed that countAtom timed out on all these cases. We leave a more thorough and scientific comparison with countAtom for future work.

5.2 Experimental Setup

Each experiment (sequential or parallel) had exclusive access to a Westemere node with 12 processor cores running at 2.83 GHz with 48 GB of RAM. We capped memory usage at 42 GB for all tools. We implemented explicit Ryser's algorithm in C++, compiled with GCC v6.4 with O3 flag. The RysersADD and RysersADD-P algorithms were implemented as in Sect. 4.1. RysersADD-P had access to all 12 cores for parallel computation. We used the python library PySAT [27] for encoding matrices into CNF. We set the timeout to 1800 s for all our experiments. For purposes of reporting, we treat a memory out as equivalent to a time out.

5.3 Benchmarks

The parameters used for generating random instances are summarized in Table 1. We do not include matrices with $n < 30$ since the explicit Ryser's algorithm suffices (and often performs the best) for such matrices. The upper bound for n was chosen such that the algorithms in our suite either timed out or came close to timing out. For each combination of parameters, random matrix instances were sampled as follows:

Table 1. Parameters used for generating random matrices

Experiment	Matrix size n	C_f, where $C_f \cdot n$ matrix entries flipped	Starting matrix row density ρ	#Instances	Total benchmarks
Dense	30, 40, 50, 60, 70	1, 1.1, 1.2, 1.3, 1.4	1	20	500
Sparse	30, 40, 50, 60, 70	3.9, 4.3, 4.7, 5.1, 5.5	0	20	500
Similar	40, 50, 60, 70, 80	1, 1.1, 1.2, 1.3, 1.4	0.7, 0.8, 0.9	15	1125

1. We started with an $n \times n$ matrix, where the first row had $\rho \cdot n$ 1s at randomly chosen column positions, and all other rows were copies of the first row.
2. $C_f \cdot n$ randomly chosen entries in the starting matrix are flipped i.e. 0 flipped to 1 and vice versa.

For the dense case, we start with a matrix of all 1s while for the sparse case, we start with a matrix of all 0s, and used intermediate row density values for the similar-rows case. We chose higher values for C_f in the sparse case because for low values, the bipartite graph corresponding to the generated matrix had very few perfect matchings (if any), and these could be simply counted by enumeration. We generated a total of 2125 benchmarks covering a broad range of parameters. For all generated instances, we ensured that there was at least one perfect matching, since the case with zero perfect matchings can be easily solved in polynomial time by algorithms like Hopcroft-Karp [25]. In order to avoid spending inordinately large time on failed experiments, if an algorithm timed out on all generated random instances of a particular size, we also report a time out for that algorithm on all larger instances of that class of matrices. We also double-check this by conducting experiments with the same algorithm on a few randomly chosen larger instances.

The SuiteSparse Matrix Collection [17] is a well known repository of structured sparse matrices that arise from practical applications. We found 26 graphs in this suite with vertex count between 30 and 100, of which 18 had at least one perfect matching. Note that these graphs are not necessarily bipartite; however, their adjacency matrices can be used as benchmarks for computing the permanent. A similar approach was employed in [57] as well.

Fullerenes are carbon molecules whose adjacency matrices have been used extensively by Liang et al. [34,57,59] for comparing tools for the permanent. We were able to find the adjacency matrices of C_{60} and C_{100}, and have used these in our experiments.

6   Results

We first study the variation of running time of RysersADD with the size of
ADDs involved. Then we compare the running times of various algorithms on
sparse, dense and similar-row matrices, as well as on instances from SuiteSparse
Matrix Collection and on adjacency matrices of fullerenes C_{60} and C_{100}. The
total computational effort of our experiments exceeds 2500 h of wall clock time
on dedicated compute nodes.

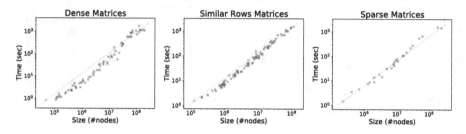

Fig. 2. Comparison of ADD Size vs. Time taken for a subset of random benchmarks

6.1   ADD Size Vs Time Taken by **RysersADD**

In order to validate the hypothesis that the size of the ADD representation
is a crucial determining factor of the performance of RysersADD, we present 3
scatter-plots (Fig. 2) for a subset of 100 instances, of each of the dense, sparse
and similar-rows cases. In each case, the 100 instances cover the entire range of
C_f and n used in Table 1, and we plot times only for instances that didn't time
out. The plots show that there is very strong correlation between the number of
nodes in the ADDs and the time taken for computing the permanent, supporting
our hypothesis.

6.2   Performance on Dense Matrices

We plot the median running time of RysersADD and RysersADD-P against the
matrix size n for dense matrices with $C_f \in \{1, 1.1, 1.2, 1.3\}$ in Fig. 3. We only
show the running times of RysersADD and RysersADD-P, since D4 and DSharp
were unable to solve any instance of size 30 for all 6 encodings. We observe that
the running time of both the ADD-based algorithms increases with C_f. This
trend continues for $C_f = 1.4$, which we omit for lack of space. RysersADD-P is
noticeably faster than RysersADD, indicating that the native parallelism provided
by Sylvan is indeed effective.

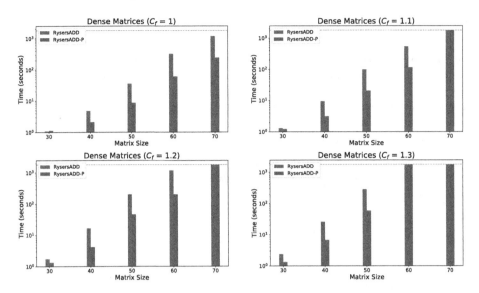

Fig. 3. Performance on Dense Matrices. D4, DSharp (not shown) timeout on all instances

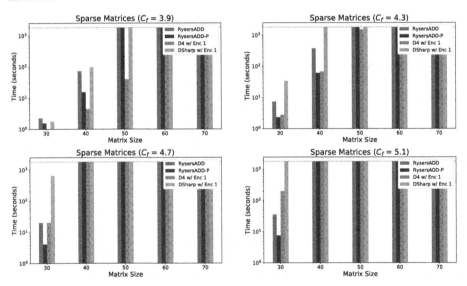

Fig. 4. Performance on Sparse Matrices

6.3 Performance on Sparse Matrices

Fig. 4 depicts the median running times of the algorithms for sparse matrices with $C_f \in \{3.9, 4.3, 4.7, 5.1\}$. We plot the running time of the ADD-based approaches with early abstraction (see Sect. 5.1). Monolithic variants (not shown) time out on all instances with $n \geq 40$. For D4 and DSharp, we plot the running times only for Pairwise encoding of At-Most-One constraints, since our preliminary experiments showed that it substantially outperformed other encodings. We see that D4 is the fastest when sparsity is high i.e. for $C_f \leq 4.3$, but for $C_f \geq 4.7$ the ADD-based methods are the best performers. DSharp is outperformed by the remaining 3 algorithms in general.

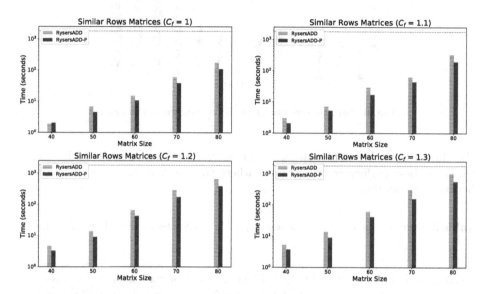

Fig. 5. Performance on similar-rows matrices. D4, DSharp (not shown) timeout on all instances.

6.4 Performance on Similar-Row Matrices

Figure 5 shows plots of the median running time on similar-row matrices with $C_f = \{1, 1.1, 1.2, 1.3\}$. We only present the case when $\rho = 0.8$, since the plots are similar when $\rho \in \{0.7, 0.9\}$. As in the case of dense matrices, D4 and DSharp were unable to solve any instance of size 40, and hence we only show plots for RysersADD and RysersADD-P. The performance of both tools is markedly better than in the case of dense matrices, and they scale to matrices of size 80 within the 1800 s timeout.

6.5 Performance on SuiteSparse Matrix Collection

We report the performance of algorithms RysersADD, RysersADD-P, D4 and DSharp on 13 representative graphs from the SuiteSparse Matrix Collection in Fig. 6. Except for the first 4 instances, which can be solved in under 5 s by all algorithms, we find that D4 is the fastest in general, while the ADD-based algorithms outperform DSharp. Notably, on the instance "can_61", both D4 and DSharp time out while RysersADD and RysersADD-P solve it comfortably within

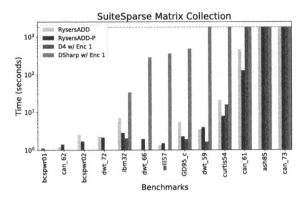

Fig. 6. Performance comparison on structured matrices

the alloted time. We note that the instance "can_61" has roughly $9n$ 1s, while D4 is the best performer on instances where the count of 1s in the matrix lies between $4n$ and $6n$.

Table 2. Running Times on the fullerene C_{60}. EA: Early Abstraction Mono: Monolithic

Tool	D4						DSharp						RysersADD		RysersADD-P	
Encoding/Mode	1	2	3	4	5	6	1	2	3	4	5	6	EA	Mono	EA	Mono
Time (sec)	94.8	150.5	150.6	136	158	156	TimeOut						96.4	TimeOut	57.1	TimeOut

6.6   Performance on Fullerene Adjacency Matrices

We compared the performance of the algorithms on the adjacency matrices of the fullerenes C_{60} and C_{100}. All the algorithms timed out on C_{100}. The results for C_{60} are shown in Table 2. The columns under D4 and DSharp correspond to 6 different encodings of At-Most-One constraints (see Sect. 5.1). It can be seen that RysersADD-P performs the best on this class of matrices, followed by D4. The utility of early abstraction is clearly evident, as the monolithic approach times out in both cases.

Discussion: Our experiments show the effectiveness of the symbolic approach on dense and similar-rows matrices, where neither D4 nor DSharp are able to solve even a single instance. Even for sparse matrices, we see that decreasing sparsity has lesser effect on the performance of ADD-based approaches as compared to D4. This trend is confirmed by "can_61" in the SuiteSparse Matrix Collection as well, where despite the density of 1s being $9n$, RysersADD and RysersADD-P finish well within timeout, unlike D4. In the case of fullerenes, we note that the algorithm in [34] solved C_{60} in 355 s while the one in [59] took 5 s, which are in the vicinity of the times reported in Table 2. While this is not an apples-to-apples comparison owing to differences in the computing platform, it indicates that the performance of general-purpose algorithms like RysersADD and D4 can be comparable to that of application-specific algorithms.

7 Conclusion

In this work we introduced a symbolic algorithm called RysersADD for permanent computation based on augmenting Ryser's formula with Algebraic Decision Diagrams. We demonstrated, through rigorous experimental evaluation, the scalability of RysersADD on both dense and similar-rows matrices, where existing approaches fail. Coupled with the technique of early abstraction [1], RysersADD performs reasonably well even on sparse matrices as compared to dedicated approaches. In fact, it may be possible to optimize the algorithm even further, by evaluating other heuristics used in [1]. We leave this for future work. Our work also re-emphasizes the versatility of ADDs and opens the door for their application to other combinatorial problems.

It is an interesting open problem to obtain a complete characterization of the class of matrices for which ADD representation of Ryser's formula is succinct. Our experimental results for dense matrices hint at the possibility of improved theoretical bounds similar to those obtained in earlier work on sparse matrices. Developing an algorithm for general matrices that is exponentially faster than Ryser's approach remains a long-standing open problem [28], and obtaining better bounds for non-sparse matrices would be an important first step in this direction.

References

1. Dudek, J.M., Phan, V.H.N., Vardi, M.Y.: ADDMC: Exact weighted model counting with algebraic decision diagrams. https://arxiv.org/abs/1907.05000
2. Ansótegui, C., Manyà, F.: Mapping problems with finite-domain variables to problems with boolean variables. In: Hoos, H.H., Mitchell, D.G. (eds.) SAT 2004. LNCS, vol. 3542, pp. 1–15. Springer, Heidelberg (2005). https://doi.org/10.1007/11527695_1
3. Arora, S., Barak, B.: Computational Complexity: A Modern Approach. Cambridge University Press, Cambridge (2009)
4. Bahar, R., et al.: Algebraic decision diagrams and their applications. J. Formal Methods Syst. Des. $10(2/3)$, 171–206 (1997)
5. Bax, E., Franklin, J.: A permanent algorithm with exp $[(n1/3/2\ln{(n)})]$ expected speedup for 0–1 matrices. Algorithmica $32(1)$, 157–162 (2002)
6. Biere, A., Heule, M., van Maaren, H., Walsh, T.: Handbook of Satisfiability: Volume 185 Frontiers in Artificial Intelligence and Applications. IOS Press, Amsterdam (2009)
7. Bouquet, F.: Gestion de la dynamicité et énumération d'impliquants premiers: une approche fondée sur les Diagrammes de Décision Binaire. PhD thesis, Aix-Marseille 1 (1999)
8. Brault-Baron, J., Capelli, F., Mengel, S.: Understanding model counting for β-acyclic CNF-formulas. arXiv preprint arXiv:1405.6043 (2014)

9. Broder, A.Z.: How hard is it to marry at random? (on the approximation of the permanent). In: Proceedings of the Eighteenth Annual ACM Symposium on Theory of Computing, pp. 50–58. ACM (1986)

10. Burchard, J., Schubert, T., Becker, B.: Laissez-faire caching for parallel #SAT solving. In: Heule, M., Weaver, S. (eds.) SAT 2015. LNCS, vol. 9340, pp. 46–61. Springer, Cham (2015). https://doi.org/10.1007/978-3-319-24318-4_5

11. Cai, J.-Y., Pavan, A., Sivakumar, D.: On the hardness of permanent. In: Meinel, C., Tison, S. (eds.) STACS 1999. LNCS, vol. 1563, pp. 90–99. Springer, Heidelberg (1999). https://doi.org/10.1007/3-540-49116-3_8

12. Cash, G.G.: A fast computer algorithm for finding the permanent of adjacency matrices. J. Math. Chem. $18(2)$, 115–119 (1995)

13. Chou, Q., Liang, H., Bai, F.: Computing the permanental polynomial of the high level fullerene C70 with high precision. MATCH Commun. Math. Comput. Chem. 73, 327–336 (2015)

14. Courcelle, B.: Graph rewriting: an algebraic and logic approach. In: Formal Models and Semantics, pp. 193–242. Elsevier (1990)

15. Courcelle, B., Engelfriet, J., Rozenberg, G.: Handle-rewriting hypergraph grammars. J. Comput. Syst. Sci. $46(2)$, 218–270 (1993)

16. Courcelle, B., Makowsky, J.A., Rotics, U.: On the fixed parameter complexity of graph enumeration problems definable in monadic second-order logic. Discrete Appl. Math. $108(1–2)$, 23–52 (2001)

17. Davis, T.A., Hu, Y.: The university of florida sparse matrix collection. ACM Trans. Math. Softw. (TOMS) $38(1)$, 1 (2011)

18. Dechter, R.: Bucket elimination: a unifying framework for reasoning. Artif. Intell. $113(1–2)$, 41–85 (1999)

19. Dell, H., Husfeldt, T., Marx, D., Taslaman, N., Wahlén, M.: Exponential time complexity of the permanent and the tutte polynomial. ACM Trans. Algorithms (TALG) $10(4)$, 21 (2014)

20. Duenas-Osorio, L., Meel, K.S., Paredes, R., Vardi, M.Y.: Counting-based reliability estimation for power-transmission grids. In: AAAI, pp. 4488–4494 (2017)

21. Fujita, M., McGeer, P., Yang, J.-Y.: Multi-terminal binary decision diagrams: an efficient datastructure for matrix representation. Form. Methods Syst. Des. $10(2–3)$, 149–169 (1997)

22. Gent, I.P., Nightingale, P.: A new encoding of all different into SAT. In: International Workshop on Modelling and Reformulating Constraint Satisfaction, pp. 95–110 (2004)

23. Gordon, M., Davison, W.: Theory of resonance topology of fully aromatic hydrocarbons. I. J. Chem. Phys. $20(3)$, 428–435 (1952)

24. Grohe, M.: Descriptive and parameterized complexity. In: Flum, J., Rodriguez-Artalejo, M. (eds.) CSL 1999. LNCS, vol. 1683, pp. 14–31. Springer, Heidelberg (1999). https://doi.org/10.1007/3-540-48168-0_3

25. Hopcroft, J.E., Karp, R.M.: An n^5/2 algorithm for maximum matchings in bipartite graphs. SIAM J. Comput. $2(4)$, 225–231 (1973)

26. Huo, Y., Liang, H., Liu, S.-Q., Bai, F.: Computing monomer-dimer systems through matrix permanent. Phys. Rev. E $77(1)$, 016706 (2008)

27. Ignatiev, A., Morgado, A., Marques-Silva, J.: PySAT: a Python toolkit for prototyping with SAT oracles. In: SAT, pp. 428–437 (2018)

28. Izumi, T., Wadayama, T.: A new direction for counting perfect matchings. In: 2012 IEEE 53rd Annual Symposium on Foundations of Computer Science, pp. 591–598. IEEE (2012)

29. Kasteleyn, P.W.: The statistics of dimers on a lattice: I. The number of dimer arrangements on a quadratic lattice. Physica **27**(12), 1209–1225 (1961)
30. Knuth, D.E.: Generating all n-tuples. The Art of Computer Programming, 4 (2004)
31. Koster, A.M., Bodlaender, H.L., Van Hoesel, S.P.: Treewidth: computational experiments. Electron. Notes Discrete Math. **8**, 54–57 (2001)
32. Kroto, H.W., Heath, J.R., O'Brien, S.C., Curl, R.F., Smalley, R.E.: C60: Buckminsterfullerene. Nature **318**(6042), 162 (1985)
33. Lagniez, J.-M., Marquis, P.: An improved decision-DNNF compiler. In: IJCAI, pp. 667–673 (2017)
34. Liang, H., Bai, F.: A partially structure-preserving algorithm for the permanents of adjacency matrices of fullerenes. Comput. Phys. Commun. **163**(2), 79–84 (2004)
35. Liang, H., Huang, S., Bai, F.: A hybrid algorithm for computing permanents of sparse matrices. Appl. Math. Comput. **172**(2), 708–716 (2006)
36. Martins, R., Joshi, S., Manquinho, V., Lynce, I.: Incremental cardinality constraints for MaxSAT. In: O'Sullivan, B. (ed.) International Conference on Principles and Practice of Constraint Programming, pp. 531–548. Springer, Cham (2014). https://doi.org/10.1007/978-3-319-10428-7_39
37. Muise, C., McIlraith, S.A., Beck, J.C., Hsu, E.: DSHARP: fast d-DNNF compilation with sharpSAT. In: Canadian Conference on Artificial Intelligence (2012)
38. Nijenhuis, A., Wilf, H.S.: Combinatorial Algorithms: for Computers and Calculators. Elsevier (2014)
39. Ogawa, T., Liu, Y., Hasegawa, R., Koshimura, M., Fujita, H.: Modulo based CNF encoding of cardinality constraints and its application to MaxSAT solvers. In: 2013 IEEE 25th International Conference on Tools with Artificial Intelligence, pp. 9–17. IEEE (2013)
40. Okamoto, Y., Uehara, R., Uno, T.: Counting the number of matchings in chordal and chordal bipartite graph classes. In: Paul, C., Habib, M. (eds.) WG 2009. LNCS, vol. 5911, pp. 296–307. Springer, Heidelberg (2010). https://doi.org/10.1007/978-3-642-11409-0_26
41. Pan, G., Vardi, M.Y.: Search vs. symbolic techniques in satisfiability solving. In: Hoos, H.H., Mitchell, D.G. (eds.) SAT 2004. LNCS, vol. 3542, pp. 235–250. Springer, Heidelberg (2005). https://doi.org/10.1007/11527695_19
42. Pesant, G., Quimper, C.-G., Zanarini, A.: Counting-based search: Branching heuristics for constraint satisfaction problems. J. Artif. Intell. Res. **43**, 173–210 (2012)
43. Régin, J.-C.: A filtering algorithm for constraints of difference in CSPs. AAAI **94**, 362–367 (1994)
44. Robertson, N., Seymour, P.D.: Graph minors. III. Planar tree-width. J. Comb. Theory Ser. B **36**(1), 49–64 (1984)
45. Roth, D.: On the hardness of approximate reasoning. Artif. Intell. **82**(1), 273–302 (1996)
46. Ryser, H.: Combinatorial mathematics, the carus mathematical monographs. Mathematical Association of America, no. 4 (1963)
47. Sæther, S.H., Telle, J.A., Vatshelle, M.: Solving #SAT and MaxSAT by dynamic programming. J. Artif. Intell. Res. **54**, 59–82 (2015)
48. Servedio, R.A., Wan, A.: Computing sparse permanents faster. Inf. Process. Lett. **96**(3), 89–92 (2005)
49. Sinz, C.: Towards an optimal CNF encoding of boolean cardinality constraints. In: van Beek, P. (eds) International Conference on Principles and Practice of Constraint Programming, pp. 827–831. Springer, Heidelberg (2005). https://doi.org/10.1007/11564751_73

50. Somenzi, F.: CUDD package, release 2.4.1. http://vlsi.colorado.edu/~fabio/CUDD/
51. Tarjan, R.E., Yannakakis, M.: Simple linear-time algorithms to test chordality of graphs, test acyclicity of hypergraphs, and selectively reduce acyclic hypergraphs. SIAM J. Comput. **13**(3), 566–579 (1984)
52. Temperley, H.N., Fisher, M.E.: Dimer problem in statistical mechanics-an exact result. Philos. Mag. **6**(68), 1061–1063 (1961)
53. Toda, S.: On the computational power of PP and (+)P. In: Proceedings of FOCS, pp. 514–519. IEEE (1989)
54. Valiant, L.: The complexity of enumeration and reliability problems. SIAM J. Comput. **8**(3), 410–421 (1979)
55. van Dijk, T., van de Pol, J.: Sylvan: multi-core framework for decision diagrams. Int. J. Softw. Tools Technol. Transf. **19**(6), 675–696 (2017)
56. Wallace, C., Korb, K.B., Dai, H.: Causal discovery via MML. In: ICML 1996, pp. 516–524 (1996)
57. Wang, L., Liang, H., Bai, F., Huo, Y.: A load balancing strategy for parallel computation of sparse permanents. Numer. Linear Algebra Appl. **19**(6), 1017–1030 (2012)
58. Wei, T.-C., Severini, S.: Matrix permanent and quantum entanglement of permutation invariant states. J. Math. Phys. **51**(9), 092203 (2010)
59. Yue, B., Liang, H., Bai, F.: Improved algorithms for permanent and permanental polynomial of sparse graph. MATCH Commun. Math. Comput. Chem. **69**, 831–842 (2013)
60. Zanarini, A., Pesant, G.: Solution counting algorithms for constraint-centered search heuristics. Constraints **14**(3), 392–413 (2009)

Towards the Characterization of Max-Resolution Transformations of UCSs by UP-Resilience

Mohamed Sami Cherif[(⊠)] and Djamal Habet

Aix Marseille Univ, Université de Toulon, CNRS, LIS, Marseille, France
{mohamed-sami.cherif,djamal.habet}@univ-amu.fr

Abstract. The Max-SAT problem consists in finding an assignment maximizing the number of satisfied clauses. Complete methods for this problem include Branch and Bound (BnB) algorithms which use max-resolution, the inference rule for Max-SAT, to ensure that every computed Inconsistent Subset (IS) is counted only once in the lower bound estimation. However, learning max-resolution transformations can be detrimental to their performance so they are usually selectively learned if they respect certain patterns. In this paper, we focus on recently introduced patterns called Unit Clause Subsets (UCSs). We characterize the transformations of certain UCS patterns using the UP-resilience property. Finally, we explain how our result can help extend the current patterns.

Keywords: Max-resolution · UP-resilience · Unit Clause Subset

1 Introduction

Max-SAT is an optimization extension of the satisfiability (SAT) problem. For a given formula in Conjunctive Normal Form (CNF), it consists in finding an assignment of the variables which maximizes the number of satisfied clauses. Complete methods for this problem include SAT based approaches (e.g. MAXHS [7], OPEN-WBO [13], EVA [14], WPM1 [5]) and Branch and Bound (BnB) algorithms (e.g. AHMAXSAT [3], AKMAXSAT [9], MAXSATZ [10,12]) among others. The former which iteratively call SAT solvers are particularly efficient on industrial instances while the latter are competitive on random and crafted instances.

BnB based approaches construct a search tree and compute, at each node, the Lower Bound (LB) by counting the disjoint Inconsistent Subsets (ISs) of the formula using Simulated Unit Propagation (SUP) [11]. When an IS is found, it is either temporarily deleted or transformed by max-resolution, the inference

---

This work is funded by the French National Research Agency (ANR), reference ANR-16-C40-0028.

T. Schiex and S. de Givry (Eds.): CP 2019, LNCS 11802, pp. 91–107, 2019.
https://doi.org/10.1007/978-3-030-30048-7_6

rule for Max-SAT [6,8], to ensure that it will be counted only once. However, learning max-resolution transformations, i.e., memorizing them in the current subtree (including the current node), may affect negatively the quality of the lower bound estimation [2,4,12]. Therefore, state of the art solvers learn transformations selectively mainly in the form of patterns [12].

Recently, new patterns called Unit Clause Subsets (UCSs) were introduced and empirically studied in [2]. The most significant feature of these patterns is producing unit clauses after the transformation by max-resolution. The empirical study of these patterns lead to the first observations on the relation between max-resolution transformations and the efficiency of the SUP mechanism which is indispensable for the lower bound estimation. These observations were formally stated by the introduction of a new property called UP-resilience [4].

In this paper, we conduct a theoretical study of particular UCS patterns and, more specifically, their relation with UP-resilience: we prove that binary UCSs are UP-resilient and we generalize this result on UCSs where only one clause of any size is involved in the conflict. We also explain how our results can help extend the current patterns by showing that the current mechanisms in BnB solvers can't ensure UP-resilience for these patterns.

This paper is organized as follows. In Sect. 2, we give basic definitions and notations. In Sect. 3, we show how the UP-resilience property highlights the impact of max-resolution transformations on the SUP mechanism. We characterize UCS transformations and we show the limit of the current mechanisms in Sect. 4 and we conclude in Sect. 5.

2   Definitions and Notations

Let X be a set of propositional variables. A literal l is a variable $x \in X$ or its negation $\overline{x}$ and a clause is disjunction of literals, represented as a set of literals. A formula in Conjunctive Normal Form (CNF) is a conjunction of clauses and can be represented as a set of clauses. An assignment $I: X \longrightarrow \{true, false\}$ maps each variable to a Boolean value and is represented as a set of literals. For a given literal l, $var(l)$ denotes the variable appearing in l. A clause c is satisfied by an assignment I if at least one of its literals is satisfied, i.e., $\exists l \in c$ such that $l \in I$. The empty clause $\square$ is always falsified. An Inconsistent Subset (IS) of a formula Φ is an unsatisfiable set of clauses $\psi \subseteq \Phi$. Solving the Max-SAT problem consists in finding an assignment which maximizes the number of satisfied clauses for a given CNF formula.

Definition 1 (Max-resolution [6,8]). *The inference rule for Max-SAT, max-resolution, is defined as follows:*

$$\frac{c = \{x, y_1, ..., y_s\}, c' = \{\overline{x}, z_1, ..., z_t\}}{cr = \{y_1, ..., y_s, z_1, ..., z_t\}, cc_1, ..., cc_t, cc_{t+1}, ..., cc_{t+s}}$$

where the compensation clauses are defined as follows:

$$cc_1 = \{x, y_1, ..., y_s, \overline{z_1}\}$$
$$cc_2 = \{x, y_1, ..., y_s, z_1, \overline{z_2}\}$$

$$...$$

$$cc_t = \{x, y_1, ..., y_s, z_1, ..., z_{t-1}, \overline{z_t}\}$$
$$cc_{t+1} = \{\overline{x}, z_1, ..., z_t, \overline{y_1}\}$$
$$cc_{t+2} = \{\overline{x}, z_1, ..., z_t, y_1, \overline{y_2}\}$$

$$...$$

$$cc_{t+s} = \{\overline{x}, z_1, ..., z_t, y_1, ..., y_2, \overline{y_s}\}$$

Unlike the SAT inference rule, max-resolution replaces the premises in the rule by its conclusions. Furthermore, it produces an equivalent formula, i.e., it preserves the number of unsatisfied clauses for any assignment. The results established in this paper can be easily extended to weighted Max-SAT formulas (hard clauses can be included with infinite weights in the case of partial formulas) using the weighted version of max-resolution introduced in [6].

Notation. Let ψ be an IS of a CNF formula Φ and $S = \langle x_1, ..., x_k \rangle$ be a sequence of variables appearing in ψ. We denote $\Theta(\psi, S)$ the set of clauses obtained from ψ after the application of max-resolution steps in accordance to the sequence S, i.e., $\Theta(\psi, S) = \theta(\theta...(\theta(\psi, x_1), x_2)..., x_k)$ where $\theta(\psi, x)$ denotes the application of the max-resolution step defined above on two clauses c and c' such that $x \in c$ and $\overline{x} \in c'$.

Next, we recall the notion of UP-resilience [4]. The empirical study conducted in [2] shows a correlation between the decrease of the number of propagations, the decrease of the number of detected ISs and the increase of the number of decisions, i.e., if the number of propagations is reduced, then less ISs will be detected and the quality of the LB estimation will be reduced. This observation was stated more clearly in [4] as the fragmentation phenomenon which was the main motivation behind the introduction of the UP-resilience property. This phenomenon, showcased in Example 1, occurs when clauses are fragmented into two (or more) clauses after transformation by max-resolution which may obstruct their exploitation by the SUP mechanism.

Example 1. we consider the IS $\psi = \{\{x_1\}, \{x_2\}, \{x_3\}, \{\overline{x_3}, \overline{x_4}\}, \{\overline{x_1}, \overline{x_2}, x_4\}\}$ detected by the sequence of unit propagations represented in the form of an implication graph [15] in Fig. 1. The max-resolution transformation of this IS with respect to the variable sequence $S = \langle x_4, x_3, x_2, x_1 \rangle$ (in the reverse order of propagation) is given on the right in Fig. 1. If the unique neighbor of x_1 in the implication graph is set to true in the transformed IS, we obtain $\Theta(\psi, S)|_{\{x_4\}} = \{\{x_1, \overline{x_3}\}, \{x_1, x_3\}, \{x_1, x_2\}, \{\overline{x_1}, x_2, \overline{x_3}\}, \{\overline{x_1}, x_2, x_3\}\}$. Clearly, the literal x_1 can't be propagated in $\Theta(\psi, S)|_{\{x_4\}}$. We can produce the resolvent x_1 if we perform a max-resolution step between the clauses $\{x_1, x_3\}$ and $\{x_1, \overline{x_3}\}$ but the SUP mechanism alone cannot ensure the propagation of this literal in the transformed IS even with respect to its neighborhood in the implication graph. We say that the information leading to the propagation of x_1 was fragmented into several compensation clauses.

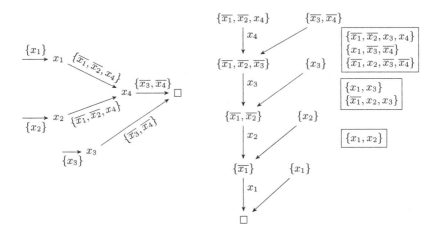

Fig. 1. Implication graph corresponding to a propagation sequence of ψ in Example 1 and its transformation by max-resolution, where compensation clauses for each step are represented in boxes.

As an IS can be detected by different propagation sequences each corresponding to an implication graph [15]. So, before recalling the definition of the UP-resilience property, we give the formal definition of an implication graph of an IS in the context of BnB solvers for MaxSAT and of possible neighborhoods of a literal appearing in an IS.

Definition 2 (Implication graph of an IS). *Let Ψ be an IS of a CNF formula Φ and I an assignment. We suppose that exactly one clause is falsified by I (SUP stopped when the first empty clause is generated). An implication graph of Ψ is a directed acyclic graph $G = (V, A)$ defined as follows:*

- $V = \{l \in I\} \cup \{\diamond_c | c \in \Psi \text{ and } |c| = 1\} \cup \{\square\}$
- $A = \{(l, l', c) \mid l, l' \in I \text{ and } c \in \Psi \text{ is reduced by } l \text{ and propagates } l'\} \bigcup$
  $\quad \{(\diamond_c, l, c) \mid l \in I \text{ and } c = \{l\} \in \Psi\} \bigcup$
  $\quad \{(l, \square, c) \mid l \in I \text{ and } c \in \Psi \text{ is falsified by } I \text{ and } \bar{l} \in \Psi\}$

The directed edges are labeled by clauses and the nodes $\diamond$ are omitted in G.

Definition 3 (Possible neighborhoods [4]). *Let ϕ be a CNF formula and ψ an IS. For a literal l appearing in ψ, we define its possible neighborhoods as $pneigh(l) = \{neigh_G(l)|G = (V, A) \text{ implication graph of } \psi \text{ s.t. } l \in V\}$ where $neigh_G(l)$ denotes the neighbors of l in G. We extend this definition on any set of literals L appearing in ψ as $pneigh(L) = \{\bigcup_{l \in L} neigh_G(l)|G = (V, A) \text{ implication graph of } \psi \text{ s.t. } L \subseteq V\}$.*

Definition 4 (UP-resilience [4]). *Let ϕ be a CNF formula, ψ an IS and S a sequence of variables appearing in ψ. The transformation $\Theta(\psi, S)$ is UP-resilient for a literal l appearing in ψ iff $\forall N \in pneigh_\psi(l)$: $\square \in N$ or l can be propagated* in $\Theta(\psi, S)|_N$ where $\Theta(\psi, S)|_N$ denotes the set of clauses in $\Theta(\psi, S)$ with the literals appearing in N set to true. We say that $\Theta(\psi, S)$ is UP-resilient if it is UP-resilient for all the literals appearing in ψ.

We finish this section by a brief overview of UCS patterns which were introduced and empirically studied in order to extend the learning mechanisms in BnB Max-SAT solvers [2].

Definition 5 (Unit Clause Subset [2]). *Let ϕ be a CNF formula and $k \geq 2$. A k-Unit Clause Subset, denoted k-UCS, is a set of clauses $\{c_1, ..., c_k\} \subseteq \phi$ such that there exists a sequence of max-resolution steps on $c_1, ..., c_k$ that produces a unit clause resolvent. In particular, if $\forall i \in \{1, ..., k\}$ we have $|c_i| = 2$, it is a binary k-UCS, denoted k^b-UCS.*

Example 2. The following patterns:

$$\frac{\{l_1, l_2\}, \{l_1, \overline{l_2}\}}{\{l_1\}} \; (P_1) \qquad \frac{\{l_1, l_2\}, \{l_1, l_3\}, \{\overline{l_2}, \overline{l_3}\}}{\{l_1\}, \{l_1, l_2, l_3\}, \{\overline{l_1}, \overline{l_2}, \overline{l_3}\}} \; (P_2)$$

which are learned in state of the art BnB solvers, correspond respectively to a 2^b-UCS and a 3^b-UCS.

Definition 6 (First Unique Implication Point [15]). *Let G be an implication graph. A Unique Implication Point (UIP) is any node in G such that any path from the literals propagated by unit clauses to the conflict node must pass through it. The First UIP (FUIP) is the UIP closest to the conflict node.*

It is important to note that UCS patterns have a high apparition frequency (in more than 57% of the detected ISs [2]). Furthermore, certain k-UCS patterns are easily detectable by analyzing the implication graph of the obtained IS [2]. Indeed, the clauses which are between the conflict and the FUIP produce a unit resolvent clause if they are transformed by max-resolution in the reverse propagation order. From here on, we will focus on such k-UCS patterns.

3   Preliminaries and Motivation

In this section, we explain how the notion of UP-resilience quantifies the impact of max-resolution on the SUP mechanism and thus on the detection of Inconsistent Subsets. This is highlighted in Property 1 which proves that UP-resilient transformations maintain the propagations which are not necessary anymore to an inconsistent subset. We provide a different proof for this property that is shorter and simpler than the one in [4]. We also show, in Propositions 1 and 2, that the transformations corresponding to patterns (P_1) and (P_2) are UP-resilient which contributes to explain from a theoretical point of view the empirical efficiency of these patterns. We give detailed proofs of these propositions to emphasize the fact that they are valid for any possible order of application of max-resolution, a fact that will be of importance in the discussion of our results in Sect. 4.

Property 1. *Let ϕ be a CNF formula, ψ an IS of ϕ and S a sequence of variables appearing in ψ. For any set of literals L appearing in ψ, if the transformation $\Theta(\psi, S)$ is UP-resilient for L then $\forall N \in pneigh(L) : \square \in N$ or every literal $l \in L$ can be propagated in $\Theta(\psi, S)|_{N \setminus \{l\}}$.*

Proof. We prove this property by induction on $|L| = n$:

- If $n = 1$ then $L = \{l\}$ and the property is verified.
- Suppose the property is true for every set of size n. Let L be of size $n + 1$ and l a literal in L. We set $L' = L \setminus \{l\}$ and let $N \in pneigh(L)$. Clearly, $N = N_1 \cup N_2$ where $N_1 \in pneigh(L')$ and $N_2 \in pneigh(l)$. Moreover, since $|L'| = n$, we know by induction that $\forall N \in pneigh(L') : \square \in N$ or every literal l' in L' can be propagated in $\Theta(\psi, S)|_{N_1 \setminus \{l'\}}$. In particular, $\square \in N_1$ or every literal l' in L' can be propagated in $\Theta(\psi, S)|_{N_1 \setminus \{l'\}}$. Also, The transformation $\Theta(\psi, S)$ is UP-resilient for L and particularly for l and thus, we have $\forall N \in pneigh(l) : \square \in N$ or l can be propagated in $\Theta(\psi, S)|_N$. In particular, $\square \in N_2$ or l can be propagated in $\Theta(\psi, S)|_{N_2}$. Thus, We have the following cases:
  - If $\square \in N_1$ or $\square \in N_2$ then $\square \in N$
  - Else every literal l' in L' and l can be propagated respectively in $\Theta(\psi, S)|_{N_1 \setminus \{l'\}}$ and $\Theta(\psi, S)|_{N_2}$. Therefore, the clauses that ensure the propagation of every literal l' in L' in $\Theta(\psi, S)|_{N_1 \setminus \{l'\}}$ also ensure their propagation in $\Theta(\psi, S)|_{(N_1 \cup N_2) \setminus \{l'\}}$ and, similarly, the clauses that ensure the propagation of l in $\Theta(\psi, S)|_{N_2}$ also ensure its propagation in $\Theta(\psi, S)|_{(N_1 \cup N_2) \setminus \{l\}}$.

  We deduce that $\forall N \in pneigh(L) : \square \in N$ or every literal l in L can be propagated in $\Theta(\psi, S)|_{N \setminus \{l\}}$. ∎

Proposition 1. *Let Φ be a CNF formula, Ψ an IS and $\Psi' \subset \Psi$ such that Ψ' matches the premises of pattern (P_1). Then, the max-resolution transformation described in (P_1) is UP-resilient.*

Proof. $\psi' = \{\{l_1, l_2\}, \{l_1, \overline{l_2}\}\}$. Therefore, there are two possible propagation sequences whose implication graphs are represented in Fig. 2. Since all possible neighborhoods of literals $\overline{l_1}$, l_2 and $\overline{l_2}$ contain the empty clause, the transformation of ψ' as in (P_1), with respect to the only possible variable sequence $S = <var(l_2)>$, is UP-resilient. ∎

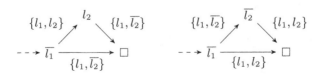

Fig. 2. Implication graphs corresponding to the possible propagation sequences for an IS containing the premises of pattern (P_1).

Proposition 2. *Let Φ be a CNF formula, Ψ an IS and $\Psi' \subset \Psi$ such that Ψ' matches the premises of pattern (P_2). Then, the max-resolution transformation described in (P_2) is UP-resilient.*

Proof. $\psi' = \{\{l_1, l_2\}, \{l_1, l_3\}, \{\overline{l_2}, \overline{l_3}\}\}$. Therefore, there are two possible propagation sequences whose implication graphs are represented in Fig. 3. There are two max-resolution application orders $S_1 = <var(l_2), var(l_3)>$ and $S_2 = <var(l_3), var(l_2)>$ that produce the same transformation described by pattern (P_2). Since all possible neighborhoods of l_2 and $\overline{l_2}$ contain the empty clause, the transformation of ψ by max-resolution is UP-resilient for l_2 and $\overline{l_2}$. We have $pneigh(\overline{l_1}) = \{\{l_3, \square\} \cup pred(\overline{l_1}), \{l_2, l_3\} \cup pred(\overline{l_1})\}$, where $pred(\overline{l_1})$ denotes the predecessors of $\overline{l_1}$, and clearly the clause $c = \{\overline{l_1}, \overline{l_2}, \overline{l_3}\}$ propagates $\overline{l_1}$ when the literals l_2, l_3 in its second neighborhood are set to true. Also, $pneigh(l_3) = \{\{\overline{l_1}, \square\}, \{\overline{l_1}, \overline{l_2}\}\}$ and similarly the clause $c' = \{l_1, l_2, l_3\}$ propagates l_3 when the literals in its neighborhood $\{\overline{l_1}, \overline{l_2}\}$ are set to true.   ∎

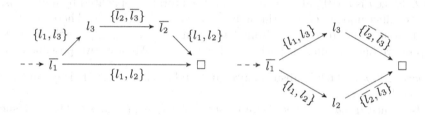

Fig. 3. Implication graphs corresponding to the possible propagation sequences for an IS containing the premises of pattern (P_2).

Corollary 1. *For $k \in \{2, 3\}$, k^b-UCSs are UP-resilient.*

Proof. 2^b-UCSs and 3^b-UCSs are all of the respective forms $\Psi_{2^b} = \{\{l_1, l_2\}, \{l_1, \overline{l_2}\}\}$ and $\Psi_{3^b} = \{\{l_1, l_2\}, \{l_1, l_3\}, \{\overline{l_2}, \overline{l_3}\}\}$ which correspond to the premises of patterns (P_1) and (P_2). Thus, we obtain the wanted result using Propositions 1 and 2.   ∎

The previous results establish that UP-resilient transformations can't negatively impact the SUP mechanism and that the transformations learned in state of the art BnB solvers for Max-SAT in the form of patterns (P_1) and (P_2) are UP-resilient. One major challenge is to use this property to help decide the relevance of application of max-resolution transformations either by devising an efficient algorithm to verify the property on potential transformations or by using it to characterize the transformation of certain patterns. Since checking the property on potential transformations seems computationally costly, we tackle in the next section the second problem by generalizing the result of Corollary 1.

4   Contributions

In this section, we prove that binary k-UCSs are UP-resilient by providing two different orders that ensure the UP-resilience of their transformation by max-resolution. We also show that unlike the given orders, the current used mechanisms can't ensure UP-resilience for these patterns which provides an explanation to the empirical results in [2] and shows that our results can help extend the current used patterns in state of the art solvers. Furthermore, we generalize our result on the resilience of k^b-UCSs to k-UCSs where all clauses are binary except one of any size that is involved in the conflict. We start by proving the following lemma in order to characterize the detected implication graphs of such k-UCSs.

Lemma 1. *Let $k \geq 2$ and ψ be a k-UCS whose clauses are binary except for the conflict clause of size $s \geq 2$, recognized by the FUIP l in an implication graph G of an IS such that $|succ(l)| = s$. Then, there exists exactly s disjoint paths from l to $\Box$ in G.*

Proof. Since l is a UIP, all the paths from the literals propagated by unit clauses to the conflict node in G pass through it. We have $|succ(l)| = s$. Therefore, there are at least s different paths from l to $\Box$ in G. Let $p_1,...,p_s$ be those paths. Suppose we have a different path p_{s+1} from l to $\Box$. We have two possible cases:

- $|pred(\Box)| \neq s$. This is absurd since the conflict clause c is of size s and thus $|pred(\Box)| = s$.
- Else, since $|pred(\Box)| = s$, there exists $l' \neq l \in p_{s+1}$ and $i \in \{1,...,s\}$ such that $l' \in p_i$ and $|pred(l')| > 1$. This is absurd since all clauses of the k-UCS except c are binary.

We deduce that there are exactly s different paths from l to $\Box$ in G. The same argument of the second case ensures that these paths are disjoint.   ∎

   As explained in Sect. 2, when a UCS is detected, we know that the reverse propagation order ensures the production of a unit clause after the transformation but, in general, this is not necessarily true for all the orders. Since this is the main feature of UCS patterns, we must ensure that the introduced orders produce unit clauses. It is important to note that the condition on the successors of the FUIP in Lemma 1 ensures this property for all possible orders. We start by proving the UP-resilience of k^b-UCSs. To this end, we show in the next proposition that the condition on the FUIP successors in Lemma 1 is always verified for k^b-UCSs. Later, when we generalize our result, we only consider the graphs described by Lemma 1, i.e., which verify the condition on the successors of the FUIP.

Proposition 3. *Let $k \geq 2$ and ψ be a k^b-UCS recognized by the FUIP l in an implication graph G of an IS. Then, $|succ(l)| = 2$.*

Proof. Suppose that $|succ(l)| \neq 2$. We have two possible cases:

- if $|succ(l)| > 2$ then, since $|succ(\square)| = 2$, there exists a literal with two predecessors. This is absurd since all the clauses are binary.
- if $|succ(l)| = 1$ then l is not the FUIP which is absurd.   ∎

Definition 7 (Path Resolvent Order). *Let* $p_1 = \langle l, l_1^{p_1}, ..., l_{n_1}^{p_1}, \square \rangle (n_1 \geq 0)$ *and* $p_2 = \langle l, l_1^{p_2}, ..., l_{n_2}^{p_2}, \square \rangle (n_2 \geq 0)$ *denote two disjoint paths from* l *to* $\square$. *The Path Resolvent Order (PRO) of* p_1 *and* p_2 *is defined as* $PRO(p_1, p_2) = \langle var(l_1^{p_1}), ..., var(l_{n_1}^{p_1}), var(l_1^{p_2}), ..., var(l_{n_2}^{p_2}) \rangle$.

Theorem 1. *For any* $k \geq 2$, *the transformation of* k^b-*UCSs with respect to PRO is UP-resilient.*

Proof. Let $k \geq 2$ and ψ be a k^b-UCS recognized by the FUIP l in the implication graph G of an IS. By Lemma 1 and Proposition 3, we know that there are 2 disjoint paths from l to $\square$ in G. Let $p_1 = \langle l, l_1^{p_1}, ..., l_{n_1}^{p_1}, \square \rangle (n_1 \geq 0)$ and $p_2 = \langle l, l_1^{p_2}, ..., l_{n_2}^{p_2}, \square \rangle (n_2 \geq 0)$ denote these paths in G where $n_1 + n_2 = k - 1$. And, suppose w.l.o.g that $l_{n_1}^{p_1} = l'$ is the conflict literal, i.e., the last propagated literal. We have two possible propagation sequences whose implication graphs are G and G' represented in Fig. 4.

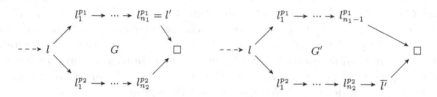

Fig. 4. Implication graphs corresponding to the possible propagation sequences for k^b-UCSs.

We prove that the max-resolution transformation relatively to the order $O = PRO(p_1, p_2)$ is UP-resilient:

- The clause propagating l is not deleted after the transformation by max-resolution relatively to the order O so it clearly propagates l if its predecessors are set to true and thus the transformation by max-resolution relatively to the order O is UP-resilient for l. This argument also applies for the literals that were involved in the propagation of l.
- All possible neighborhoods of literals $l_{n_1}^{p_1} = l'$ and $\overline{l'}$ contain the empty clause. Therefore, the transformation by max-resolution relatively to the order O is UP-resilient for l' and $\overline{l'}$.
- For $i \in \{1, 2\}$, we set $l_0^{p_i} = l$. Every literal $l_j^{p_i}$ such that $1 \leq j < n_i$ admits exactly one neighborhood $neigh(l_j^{p_i}) = \{l_{j-1}^{p_i}, l_{j+1}^{p_i}\}$ that doesn't contain the empty clause. Similarly, for $l_{n_2}^{p_2}$ we have $neigh(l_{n_2}^{p_2}) = \{l_{n_2-1}^{p_i}, \overline{l'}\}$. The max-resolution step on $var(l_j^{p_1})(1 \leq j < n_i)$ is of the form:

$$\frac{\{\bar{l}, l_j^{p_i}\}, \{\overline{l_j^{p_i}}, l_{j+1}^{p_i}\}}{\{\bar{l}, l_{j+1}^{p_i}\}, \{\bar{l}, l_j^{p_i}, \overline{l_{j+1}^{p_i}}\}, \{l, \overline{l_j^{p_i}}, l_{j+1}^{p_i}\}}$$

The clause $c = \{l, \overline{l_j^{p_i}}, l_{j+1}^{p_i}\}$ clearly ensures the propagation of literal $l_{j+1}^{p_i}$ if $l_j^{p_i} \in neigh(l_{j+1}^{p_i})$ is set to true since $\bar{l}$ is propagated by the unit resolvent clause $\{\bar{l}\}$. Also, for $j = 1$, the clause $c' = \{\bar{l}, l_1^{p_1}, \overline{l_2^{p_1}}\}$ ensures the propagation of $l_1^{p_1}$ if $l, l_2^{p_1} \in neigh(l_1^{p_1})$ are set to true. Thus, We deduce that the transformation is UP-resilient for $l_j^{p_i}$ where $1 \leq j \leq n_i$ $(j \neq n_1)$.

We conclude that the transformation of ψ by max-resolution relatively to the order O is UP-resilient. ∎

Definition 8 (Path Resolvent Circular Order). Let $p_1 = \langle l, l_1^{p_1}, ..., l_{n_1}^{p_1}, \square \rangle$ $(n_1 \geq 0)$ and $p_2 = \langle l, l_1^{p_2}, ..., l_{n_2}^{p_2}, \square \rangle (n_2 \geq 0)$ denote two disjoint paths from l to $\square$. The Path Resolvent Circular Order (PRCO) of p_1 and p_2 is defined as $PRCO(p_1, p_2) = \langle var(l_1^{p_1}), ..., var(l_{n_1}^{p_1}), var(l_{n_2}^{p_2}), ..., var(l_1^{p_2}) \rangle$.

Theorem 2. For any $k \geq 2$, the transformation of k^b-UCSs with respect to PRCO is UP-resilient.

Proof. Let $k \geq 2$ and ψ be a k^b-UCS recognized by the FUIP l in the implication graph G of an IS. By Lemma 1 and Proposition 3, let $p_1 = \langle l, l_1^{p_1}, ..., l_{n_1}^{p_1}, \square \rangle (n_1 \geq 0)$ and $p_2 = \langle l, l_1^{p_2}, ..., l_{n_2}^{p_2}, \square \rangle (n_2 \geq 0)$ denote the two disjoint paths from l to $\square$ in G where $n_1 + n_2 = k - 1$. And, suppose w.l.o.g that $l_{n_1}^{p_1} = l'$ is the conflict literal. We have two possible propagation sequences whose implication graphs are G and G' represented in Fig. 4. We prove that the max-resolution transformation relatively to the order $O = PRCO(p_1, p_2)$ is UP-resilient:

- The same arguments in the proof of Theorem 1 ensure the UP-resilience of the transformation respectively to O for $l_j^{p_1} (1 \leq j \leq n_1)$ and $\bar{l'}$ as well as l and all the literals involved in its propagation.
- Every literal $l_j^{p_2}$ such that $1 \leq j \leq n_2$ admits exactly one neighborhood $neigh(l_j^{p_2}) = \{\overline{l_{j-1}^{p_2}}, l_{j+1}^{p_2}\}$ that doesn't contain the empty clause (we set $l_0^{p_2} = l$ and $l_{n_2+1}^{p_2} = \overline{l'}$). The max-resolution step on $var(l_j^{p_2})$ $(j \neq 1)$ is of the form:

$$\frac{\{\bar{l}, \overline{l_j^{p_2}}\}, \{l_j^{p_2}, \overline{l_{j-1}^{p_2}}\}}{\{\bar{l}, \overline{l_{j-1}^{p_2}}\}, \{\bar{l}, \overline{l_j^{p_2}}, l_{j-1}^{p_2}\}, \{l, l_j^{p_2}, \overline{l_{j-1}^{p_2}}\}}$$

The clause $c = \{l, l_j^{p_2}, \overline{l_{j-1}^{p_2}}\}$ clearly ensures the propagation of literal $l_j^{p_2}$ when $l_{j-1}^{p_2} \in neigh(l_j^{p_2})$ is set to true since $\bar{l}$ is propagated by the unit resolvent clause $\{\bar{l}\}$. Also, the clause $c' = \{\bar{l}, \overline{l_2^{p_2}}, l_1^{p_2}\}$, generated by the max-resolution step on $var(l_2^{p_2})$, clearly ensures the propagation of $l_1^{p_2}$ when its neighbors $l, l_2^{p_2} \in neigh(l_1^{p_2})$ are set to true. Thus, the transformation is UP-resilient for $l_j^{p_2}$ where $1 \leq j \leq n_2$.

We conclude that the transformation by max-resolution relatively to the order O is UP-resilient. ∎

There is a major difference between the orders we introduced. Indeed, PRCO ensures a linear input resolution transformation, i.e., at each intermediary max-resolution step we use the resolvent obtained in the previous step and a clause from the detected k^b-UCS. This is not always the case for PRO. The following result is an immediate consequence of either Theorems 1 or 2.

Corollary 2. *For any $k \geq 2$, there exists a UP-resilient transformation of k^b-UCSs.*

Empirical results show that 2^b-UCSs and 3^b-UCSs, which correspond respectively to the patterns $(P1)$ and $(P2)$ have a positive impact on the performance of BnB solvers for Max-SAT [2,10]. The result in Corollary 1 obtained through properties 1 and 2 prove that 2^b-UCSs and 3^b-UCSs are UP-resilient for any given order of application of max-resolution which explains why learning them has a positive impact regardless of the chosen order. This is not the case for k^b-UCSs when $k > 3$. Empirical studies on the AHMAXSAT solver in [2] show that learning 4^b-UCSs and 5^b-UCSs had a major negative impact on its performance. This can be explained by the inadequacy of the max-resolution application orders used in state of the art BnB solvers for k^b-UCSs when $k > 3$. Indeed, it was shown in [4] that the order impacts the UP-resilience of the transformations by comparing the following heuristics:

- Reverse Propagation Order (RPO) which applies max-resolution steps in the reverse order of propagation.
- Smallest Intermediary Resolvent (SIR) which applies the max-resolution steps based on the size of the resolvents between clauses, favoring the smallest ones [1].

In particular, the results show that the average percentage of UP-resilience of the transformations is comparatively higher with SIR. In the case of k^b-UCSs, these orders don't always ensure the UP-resilience property on the transformations. More specifically, the SIR heuristic becomes unusable since all the intermediary resolvents have the same size (binary) as shown in the proofs of Theorems 1 and 2, whereas the Reverse Propagation Order doesn't always ensure the UP-resilience of the transformation as shown in the following example on a 4^b-UCS which can be easily extended to any k^b-UCS for $k > 4$.

Example 3. We consider the IS $\psi = \{\{l\}, \{\bar{l}, l_1\}, \{\bar{l}, l_2\}, \{\bar{l_1}, l_3\}, \{\bar{l_2}, \bar{l_3}\}\}$ detected by one of the possible implication graphs represented on the left in Fig. 5 after the respective propagation of literals l_1, l_2 and l_3 (or $\bar{l_3}$). Clearly, the subset $\psi' = \{\{\bar{l}, l_1\}, \{\bar{l}, l_2\}, \{\bar{l_1}, l_3\}, \{\bar{l_2}, \bar{l_3}\}\} \subset \psi$ is a 4^b-UCS recognized by the FUIP l. The max-resolution transformation of ψ' with respect to RPO which corresponds to the variable sequence $S = \langle var(l_3), var(l_2), var(l_1) \rangle$ is represented on the right in Fig. 5. The literal l_1 has one neighborhood $neigh(l_1) = \{l, l_3\}$ that doesn't contain the empty clause. Clearly, the literal l_1 can't be propagated in $\Theta(\psi, S)|_{neigh(l_1)} = \{\{l_1, \bar{l_2}\}, \{l_1, l_2\}\}$. Similarly, the fragmentation phenomenon also occurs for l_2 and we conclude that the transformation of ψ' relatively to RPO is not UP-resilient.

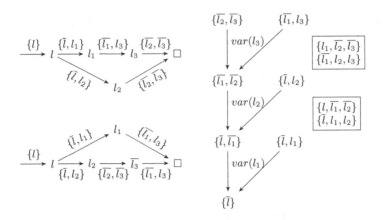

Fig. 5. Implication graphs corresponding to the possible propagation sequences of ψ in Example 3 and the application of max-resolution steps relatively to RPO

Now, we want to generalize our result to k-UCSs where all clauses are binary except one of any size that is involved in the conflict when the implication graph corresponds to the description in Lemma 1. A clause involved in the conflict is either the falsified clause or contains the conflict literal, i.e., the last propagated literal. Unfortunately, although PRCO has the advantage of ensuring a linear input transformation, we couldn't generalize it to obtain the wanted result. Nevertheless, we managed to prove our result using a generalization of PRO to a multitude of paths.

Definition 9 (Multiple Path Resolvent Order). *Let $s \geq 2$ and $p_1 = \langle l, l_1^{p_1}, ..., l_{n_1}^{p_1}, \square \rangle, ..., p_s = \langle l, l_1^{p_s}, ..., l_{n_s}^{p_s}, \square \rangle$ denote s disjoint paths from l to $\square$. The Multiple Path Resolvent Order (MPRO) of $p_1, ..., p_s$ is defined inductively on s as follows:*

- *If $s = 2$, $MPRO(p_1, p_2) = PRO(p_1, p_2)$*
- *Else $MPRO(p_1, ..., p_s) = PRO(\langle l, MPRO(p_1, ..., p_{s-1}), \square \rangle, p_s)$.*

Theorem 3. *Let $k \geq 2$ and ψ be a k-UCS whose clauses are binary except for the conflict clause c of size $|c| = s \geq 3$, recognized by the FUIP l in the implication graph G of an IS such that $|succ(l)| = s$. The transformation of ψ with respect to MPRO is UP-resilient.*

Proof. We suppose w.l.o.g that $c = \{\overline{l_1}, ..., \overline{l_s}\}$. By Lemma 1, there are exactly s disjoint paths $p_1 = \langle l, l_1^{p_1}, ..., l_{n_1}^{p_1}, \square \rangle, ..., p_s = \langle l, l_1^{p_s}, ..., l_{n_s}^{p_s}, \square \rangle$ from l to $\square$ in the implication graph G, represented in Fig. 6, such that $\sum_{i=1}^{s} n_i = k - 1$ and $l_{n_i}^{p_i} = l_i$ for $i \in \{1, ..., s\}$. Other than G, there are exactly $\binom{s-1}{s} = s$ possible implication graphs all similar to the graph G' represented in Fig. 6. We prove that the max-resolution transformation relatively to the order $O = MPRO(p_1, ..., p_s)$ is UP-resilient:

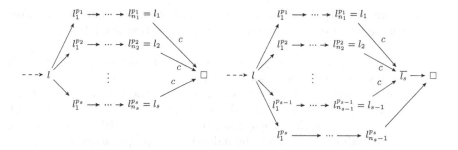

Fig. 6. Implication graphs corresponding to the possible propagation sequences for k-UCSs with binary clauses except for the conflict clause

- The same arguments in the proof of Theorem 1 ensure the UP-resilience of the transformation respectively to O for $l_j^{p_i}$ where $1 \leq i \leq s$ and $1 \leq j < n_i$ as well as l and all the literals involved in its propagation. Furthermore, all the neighborhoods of literals $\overline{l_1}, ..., \overline{l_s}$ contain the empty clause.
- For $i \in \{1, ..., s\}$, $\forall N \in pneigh(l_{n_i}^{p_i})$ $(n_i > 1)$ s.t $\square \notin N$, we have $l_{n_i-1}^{p_i} \in N$ (exists since $n_i > 1$). Clearly, the clause $c = \{l, \overline{l_{n_i-1}^{p_i}}, l_{n_i}^{p_i}\}$ obtained by the application of max-resolution on $var(l_{n_i-1}^{p_i})$ ensures the propagation of $l_{n_i}^{p_i}$ in any of these neighborhoods when $l_{n_i-1}^{p_i}$ is set to true since $\overline{l}$ is propagated by the unit resolvent clause $\{\overline{l}\}$. We deduce that the transformation relatively to the order O is UP-resilient for $l_{n_i}^{p_i}$ where $1 \leq i \leq s$ and $n_i > 1$.
- We still need to prove the UP-resilience of the transformation for literals $l_{n_i}^{p_i} = l_i$ when $n_i = 1$, with respect to their possible neighborhoods $\{l, \overline{l_j}\}$ for $j \in \{1, .., s\} \setminus \{i\}$ not containing the empty clause. For this end, we prove by induction on $|c| \geq 3$ that the compensation clauses produced by the max-resolution steps on $var(l_1), ..., var(l_s)$ ensure the propagation of each literal l_i if we consider the neighborhoods as mentioned above. For simplification, in the first max-resolution step, we replace c by the clause $c' = \{\overline{l}, \overline{l_1}, ..., \overline{l_s}\}$. This doesn't affect our result since we only omit a single clause containing the literal l:
  - If $|c| = 3$, $c = \{\overline{l_1}, \overline{l_2}, \overline{l_3}\}$. The max-resolution steps are represented on the left in Fig. 7 and we can easily check that the compensation clauses ensure the propagation of the literals l_i, for $1 \leq i \leq 3$, if we consider the neighborhoods mentioned above.
  - Suppose the property is true for any clause of size $s \geq 3$. Let $c = \{l_1, ..., l_{s+1}\}$ of size $s + 1$. The first max-resolution step is represented on the right in Fig. 7. The resolvent clause is $\{\overline{l}, l_2, ..., l_{s+1}\}$ and if we consider $c' = \{l_2, ..., l_{s+1}\}$ of size s we ensure by induction the propagation of any literal l_i where $2 \leq i \leq s + 1$ with respect to the neighborhoods $\{l, \overline{l_j}\}$ for $j \in \{2, .., s + 1\} \setminus \{i\}$. Thus, each compensation clause $cc_k = \{\overline{l}, l_1, \overline{l_2}, ..., \overline{l_k}, l_{k+1}\}$ for $k \in \{1, ..., s\}$ ensures the propagation of literal l_1 with respect to the neighborhood $\{l, \overline{l_{k+1}}\}$ since by induction the propagation of literals $l_2, ..., l_k$ is ensured in the same neighborhood.

Now, we prove by induction on $k \in \{1, ..., s\}$ that the clause cc_k ensures the propagation of l_{k+1} with respect to the neighborhood $\{l, \overline{l_1}\}$:

* If $k = 1$, $cc_1 = \{\overline{l}, l_1, l_2\}$ clearly ensures the propagation of l_2 with respect to the neighborhood $\{l, \overline{l_1}\}$.
* Suppose for $1 \leq k' < k \leq s$, $cc_{k'}$ ensures the propagation of $l_{k'+1}$ with respect to the neighborhood $\{l, \overline{l_1}\}$. $cc_k = \{\overline{l}, l_1, \overline{l_2}, ..., \overline{l_k}, l_{k+1}\}$ clearly ensures the propagation of literal l_{k+1} with respect to the neighborhood $\{l, \overline{l_1}\}$ since by induction the propagation of $l_2, ..., l_k$ is ensured in the same neighborhood by the clauses $cc_1, ..., cc_{k-1}$.

We conclude that the transformation by max-resolution relatively to the order O is UP-resilient. ∎

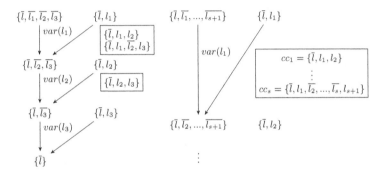

Fig. 7. Application of max-resolution steps on the variables of the non binary clause c by induction on its size

Corollary 3. *Let $k \geq 2$ and ψ be a k-UCS whose clauses are binary except for a single clause c of size $|c| = s \geq 3$ involved in the conflict, recognized by the FUIP l in the implication graph G of an IS such that $succ(l) = s$. There exists a UP-resilient transformation of ψ.*

Proof. If c is the conflict clause then we obtain the result by Theorem 3. Else, c contains the conflict literal and the detected implication graph G has the same form as the second graph represented in Fig. 6. Clearly, there is a propagation sequence where c is falsified, i.e., corresponding to an implication graph G' similar to the first graph represented in Fig. 6. Thus, we deduce the UP-resilience of the transformation with respect to MPRO through the same arguments in the proof of Theorem 3. ∎

The SIR order is defined relatively to the size of the intermediary resolvents. Thus, it may theoretically simulate any order when the sizes of the resolvents are the same or many different orders when many resolvents share the same size which is the case of the studied UCSs. That's why this heuristic remains practically unusable even in the generalized case. Furthermore, RPO doesn't necessarily ensure the UP-resilience of k-UCSs described in the previous corollary. We finish this section by an example that highlights this fact. This example where the non binary clause is tertiary can be easily extended to any size $s > 3$.

Example 4. We consider the IS $\psi = \{\{l\}, \{\bar{l}, l_1\}, \{\bar{l}, l_2\}, \{\bar{l}, l_3\}, \{\overline{l_1}, l_4\}, \{\overline{l_2}, \overline{l_3}, \overline{l_4}\}\}$ (we name the tertiary clause c) detected by the first implication graph represented on the left in Fig. 8 after the respective propagation of literals l_1, l_2, l_3 and l_4. In the second graph on the left in the same figure, we represent another possible propagation sequence which outlines the possible neighborhood of l_4, $neigh(l_4) = \{l_1, \overline{l_3}\}$ not containing the empty clause. Clearly, the subset $\psi' = \psi \setminus \{\{l\}\}$ is a 5-UCS recognized by the FUIP l such that c participates in the conflict and $|succ(l)| = |c| = 3$. The max-resolution transformation of ψ' with respect to RPO which corresponds to the variable sequence $S = \langle var(l_4), var(l_3), var(l_2), var(l_1) \rangle$ is represented on the right in Fig. 8. Clearly, the literal l_4 can't be propagated in $\Theta(\psi, S)|_{neigh(l_4)} = \{\{\bar{l}\}, \{\bar{l}, \overline{l_2}\}, \{l_2, l_4\}, \{\overline{l_2}, l_4\}\}$. We conclude that the transformation of ψ' relatively to RPO is not UP-resilient.

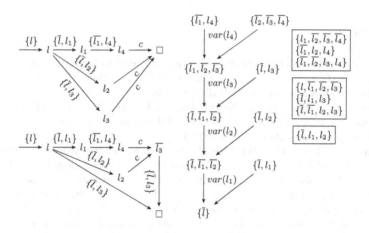

Fig. 8. Implication graphs corresponding to the possible propagation sequences of ψ in Example 4 and the application of max-resolution steps relatively to RPO

5   Conclusion

In this paper, we proved that k^b-UCSs are UP-resilient with respect to two different orders PRO and PRCO. Then, we generalized this result to k-UCSs where all clauses are binary except one of any size involved in the conflict. We showed that unlike our orders, the current mechanisms don't necessarily ensure UP-resilience for these patterns. Thus, our orders can help extend the current patterns used in state of the art BnB solvers.

Until now, UP-resilience was mainly used to explain the impact of max-resolution transformations on the SUP mechanism. To our best knowledge, this is the first work in which this property is used to characterize the transformations by max-resolution in order to decide the relevance of their application. Indeed, this can be a starting point of a new approach to extend max-resolution patterns. In our case, we chose UCS patterns because they present several advantages: the introduction of unit clauses as well as the high frequency of their apparition. We also showed the limits of the current orders of application of max-resolution. In fact, this is the first work in which the proposed orders are introduced relatively to the structure of the implication graphs representing the possible propagation sequences of an IS.

The prospects of our research include the extension of our studies to k-UCSs in general. It also opens a new perspective for finding orders of application of max-resolution that ensure UP-resilience or maximizes its percentage by thoroughly studying the implication graphs corresponding to the propagation sequences of certain ISs. Finally, increasing knowledge about max-resolution can be useful for SAT-based solvers, which are mainly efficient on industrial instances, as some solvers, such as EVA [14], already exploit max-resolution to transform cores returned by SAT solvers.

References

1. Abramé, A., Habet, D.: Efficient application of Max-SAT resolution on inconsistent subsets. In: O'Sullivan, B. (ed.) CP 2014. LNCS, vol. 8656, pp. 92–107. Springer, Cham (2014). https://doi.org/10.1007/978-3-319-10428-7_10
2. Abramé, A., Habet, D.: On the extension of learning for Max-SAT. In: Endriss, U., Leite, J. (eds.) Frontiers in Artificial Intelligence and Applications, vol. 241, pp. 1–10. IOS Press, Amsterdam (2014)
3. Abramé, A., Habet, D.: AHMAXSAT: description and evaluation of a branch and bound Max-SAT solver. J. Satisf. Boolean Model. Computation 9, 89–128 (2015)
4. Abramé, A., Habet, D.: On the resiliency of unit propagation to max-resolution. In: Yang, Q., Wooldridge, M. (eds.) Proceedings of the 24th International Joint Conference on Artificial Intelligence (IJCAI 2015), pp. 268–274. AAAI Press (2015)
5. Ansótegui, C., Bonet, M.L., Levy, J.: Solving (weighted) partial MaxSAT through satisfiability testing. In: Kullmann, O. (ed.) SAT 2009. LNCS, vol. 5584, pp. 427–440. Springer, Heidelberg (2009). https://doi.org/10.1007/978-3-642-02777-2_39
6. Bonet, M.L., Levy, J., Manyà, F.: Resolution for Max-SAT. Artif. Intell. 171(8), 606–618 (2007)
7. Davies, J., Bacchus, F.: Solving MAXSAT by solving a sequence of simpler SAT instances. In: Lee, J. (ed.) CP 2011. LNCS, vol. 6876, pp. 225–239. Springer, Heidelberg (2011). https://doi.org/10.1007/978-3-642-23786-7_19
8. Heras, F., Larrosa, J.: New inference rules for efficient Max-SAT solving. In: Cohn, A. (ed.) Proceedings of the 21st National Conference on Artificial Intelligence (AAAI 2006), pp. 68–73. AAAI Press (2006)
9. Küegel, A.: Improved exact solver for the weighted Max-SAT problem. In: Berre, D.L. (ed.) POS-10. Pragmatics of SAT. EPiC Series in Computing, vol. 8, pp. 15–27. EasyChair (2012)

10. Li, C.M., Manyà, F., Mohamedou, N.O., Planes, J.: Resolution-based lower bounds in MaxSAT. Constraints **15**, 456–484 (2010)
11. Li, C.M., Manyà, F., Planes, J.: Detecting disjoint inconsistent subformulas for computing lower bounds for Max-SAT. In: Proceedings of the 21st National Conference on Artificial Intelligence (AAAI-06), Boston, Massachusetts, vol. 1, pp. 86–91. AAAI Press (2006)
12. Li, C.M., Manyà, F., Planes, J.: New inference rules for Max-SAT. J. Artif. Intell. Res. **30**, 321–359 (2007)
13. Martins, R., Manquinho, V., Lynce, I.: Open-WBO: a modular MaxSAT solver'. In: Sinz, C., Egly, U. (eds.) SAT 2014. LNCS, vol. 8561, pp. 438–445. Springer, Cham (2014). https://doi.org/10.1007/978-3-319-09284-3_33
14. Narodytska, N., Bacchus, F.: Maximum satisfiability using core-guided MaxSAT resolution. In: Proceedings of the Twenty-Eighth Conference on Artificial Intelligence (AAAI-14), pp. 2717–2723. AAAI Press (2014)
15. Marques-Silva, J.P., Sakallah, K.A.: GRASP: a search algorithm for propositional satisfiability. IEEE Trans. Comput. **48**, 506–521 (1999)

Logic-Based Benders Decomposition for Super Solutions: An Application to the Kidney Exchange Problem

Danuta Sorina Chisca[1](✉), Michele Lombardi[2](✉), Michela Milano[2](✉),
and Barry O'Sullivan[1](✉)

[1] Insight Centre for Data Analytics, University College Cork, Cork, Ireland
{sorina.chisca,barry.osullivan}@insight-centre.org
[2] DISI, University of Bologna, Bologna, Italy
{michele.lombardi2,michela.milano}@unibo.it

Abstract. When optimising under uncertainty, it is desirable that solutions are robust to unexpected disruptions and changes. A possible formalisation of robustness is given by super solutions. An assignment to a set of decision variables is an (a, b, c) super solution if any change involving at most a variables can be repaired by changing at most b other variables; the repair solution should have a cost of at most c units worse than a non-robust optimum. We propose a method exploiting Logic Based Benders Decomposition to find super solutions to an optimisation problem for generic disruptions. The master deals with the original problem, while subproblems try to find repair solutions for each possible disruption. As a case study, we consider the Kidney Exchange Problem, for which our method scales dramatically better on realistic instances than a reformulation-based approach from the literature.

1 Introduction

Dealing with uncertainty in optimization is increasingly recognized as a key necessity for tackling real-world problems [28]. One such example is given by kidney exchange programs which provide broader access to transplants, by allowing incompatible patient-donor pairs to exchange donors. Centralised kidney exchange programs exist in many countries, including the US, the Netherlands and the UK [24], and require that a complex optimisation problem is solves at fixed intervals (the Kidney Exchange Problem - KEP) to find the best matching. About one per thousand European citizens suffers from end-stage renal disease.[1]

Since most algorithms use a virtual crossmatch to identify compatibility, different types of failures can occur that will prevent a donor from donating his/her kidney to the target recipient. Surprisingly, most planned matches fail to go to transplant, 93% of matches fail [10]. Planned exchanges may fail for a number of reasons, making robust solutions for the KEP highly desirable.

---

[1] http://www.enckep-cost.eu/page/introduction.

© Springer Nature Switzerland AG 2019
T. Schiex and S. de Givry (Eds.): CP 2019, LNCS 11802, pp. 108–125, 2019.
https://doi.org/10.1007/978-3-030-30048-7_7

Cycles cannot be executed in parts because if someone backs out of a cycle, then someone has lost a kidney unnecessarily. If a failure occurs in a cycle, then the entire cycle does not take place. Ideally, for any exchange lost, one wishes to have some ready alternative, making super solutions particularly appealing in this context.

Due to improved computational power and advances in optimisation techniques, approaches to stochastic problem-solving are becoming more viable and effective. Most of such methods rely on some form of probabilistic or statistical model, which requires reliable knowledge from the domain experts or substantial amounts of data. Sometimes, however, such probabilistic information is not available or cannot be used (due to fairness concerns), or it is desirable to be resilient against any of a given set of disruptions. In such cases, super solutions provide a fully combinatorial approach to robust decision making. An assignment x of decision variables is a super solution if any disruption causing a change in a bounded number of those variables can be countered by changing a bounded number of other variables.

Uncertainty and robustness have been studied before. For example, in [12,29] the authors generalise the notion of robust solutions and fault tolerance using a SAT encoding. More recently, robustness has been studied in combinatorial problems, such as jobshop scheduling [8,14,15] and combinatorial auctions [16,17].

There are two main approaches to find super solutions: reformulation and search-based algorithms. The reformulation approach is to extend the initial instance in such way that the extended version finds both a robust solution as well as its repairs [12]. Unfortunately, this approach has poor scalability. The search-based approach requires low-level modifications to constraint solvers and is, therefore, difficult to implement and maintain. These factors have historically limited the practical applicability of the super solutions concept in practice.

In this paper, we introduce a new method to build super solutions for the Kidney exchange problem, inspired by Logic-Based Benders Decomposition (LBBD). The main idea is to search for a solution that can be repaired with minimum perturbation when a disruption occurs. Therefore, the approach broadens the scope of super solutions to include disruptions in a wide range of forms, i.e. pairs, edges. Our method is more scalable than a reformulation-based one, and easier to manage than specialised algorithms. The method also provides somewhat of an anytime approach, since robust solutions for *some* disruptions are available in case of an early stop. The approach can also be parallelised. We use the KEP as a case study, and compare our methods against a reformulation-based approach.

The paper is structured as follows. In Sect. 2 we cover the necessary background and briefly survey related work. In Sect. 3 we describe our method. In Sect. 4 we ground it on the KEP and discuss an alternative, reformulation-based, model. We provide experimental results in Sect. 5 and concluding remarks in Sect. 6.

2   Background and Related Work

In this section we recall the main definitions and approaches related to super solutions, the KEP, and LBBD.

2.1   Super Solutions

The concept of super models and super solution was first introduced by [12]. Formally, an assignment of the decision variables x is an (a, b) super solution if, in case of a loss of values for any set of at most a variables, it is possible to construct an alternative solution by reassigning those a variables and changing at most b others. More flexibility can be added by specifying which variables are subject to breaks, the so called *break-set*, and which values and variables can be used for the repair [15]. Notice that a robust solution is not necessarily an optimal solution, however in many real world situation it is worth sacrificing some optimality for a solution that is resilient to change; this is often referred to as the *the price of robustness* [11].

The super solutions framework was generalized in [18] to (a, b, c) super solutions. The c parameter constrains the cost of the robust and repair solutions to be at most c units (or equivalently a factor c) from the cost c^* of an optimal, non-robust solution. After c^* is known, finding an (a, b, c) super solution is equivalent to finding an (a, b) super solution for the original problem with the cost bounding constraint.

Super solutions are obtained in [14, 15, 18] via specialized constraint programming algorithms. Those employ local consistency as a sufficient condition to check robustness, and to prune a branch of the search tree when such check fails. When all variables are assigned, the check turns also into a necessary condition, thus ensuring soundness. All such algorithms require low-level modifications to a constraint solver, making the approach difficult to deploy and to maintain. The special case of $(1, 0)$ super solutions (which remain solutions once any variable changes value) had been considered earlier as *supermodels* [12], and as *fault tolerant solutions* for Constraint Satisfaction Problems. In both cases, a robust solution was found by means of a reformulated model, corresponding to the problem of finding a robust solution *and all its repair solutions*. This approach can be implemented on top of any off-the-shelf solver, but unfortunately has poor scalability.

In [29] the authors study the complexity of finding σ-models given a SAT formulation. There are other papers on fault tolerant solutions using SAT encodings such as [5] which improves the work of [12] using a weighted Partial MaxSAT formulation. Their approach strengthens the complexity to $\mathcal{O}(n^a)$ for each handled variable change, where n is the number of variables, and a is the number of breaks, instead of $\mathcal{O}(n^{a+b})$ of [12], where b is the number of repairs. However, our method is more general and can deal with variables with non-binary domains, or disruptions that may affect multiple variables at the same time. We use a baseline formulation that has the same asymptotic space complexity, i.e.

$\mathcal{O}(n^a)$, for handling each variable change as [5]. Robust solutions for combinatorial auctions are studied in [17,19]. The work of [7] introduces recoverable team formation (RTF) in the coalition context.

2.2   The Kidney Exchange Problem

The KEP can be formulated as a non-bipartite matching problem on a directed graph (V, A). Nodes in the set V correspond to patient/donor pairs, or to "altruistic" donors willing to donate a kidney with no return. An arc between nodes i and j is present in the A set if the donor at node i is compatible with patient at node j, and a cycle (or chain starting from an altruistic donor) corresponds a viable set of exchanges. The goal is to maximize a utility function, which is often the number of transplants.

Cycles may involve any number of nodes, in which case they are referred to as k-way exchanges. However, dealing with more than three pairs is hard in practice, since all the operations must take place simultaneously to avoid donor withdrawal after their pair received the organ. The KEP is known to be NP-complete [1] for $k \geq 3$. Solution approaches for the KEP have been proposed in [3,4,6,23,24].

A classical Mathematical Programming model for the KEP is the *cycle formulation*. Let $\mathcal{C} = [1, \ldots, n_c]$ denote the set of indices of all cycles in a graph (chains starting from altruistic donors are equivalent to cycles). Let x_i be a binary variable such that $x_i = 1$ iff the exchanges corresponding to cycle i are selected. Then, the formulation is given by:

$$\max z = \sum_{i \in \mathcal{C}} w_i x_i \qquad \textbf{(CF)} \qquad (1)$$

$$\text{s.t.} \sum_{i \in \mathcal{C}:k \in C_i} x_i \leq 1 \qquad \forall k \in V \qquad (2)$$

$$x_i \in \{0,1\} \qquad \forall i \in \mathcal{C} \qquad (3)$$

where Eq. (2) prevents the selection of two cycles involving the same node. The w_i parameters typically represent the number of exchanges associated to each cycle: with this convention, the goal of the model is to maximize the number of transplants. Due to the requirement to enumerate all cycles, the formulation does not scale, but this can be tackled via Column Generation [10,13,21,27], or by switching to alternative models, i.e. compact formulations [2,9].

Not all the planned exchanges translate into actual transplants: some may fail due to last-minute clinical tests, patient withdrawal, or worsening health conditions. Failure-aware approaches in the KEP have been studied in [10,22,26], by adjusting the weights of the cycles or including some failure probabilities. However, those approaches tend to favor more reliable exchanges, rather than to define repair actions in case of failures: for this reason they are complementary, rather than a replacement, for super solutions.

2.3 Logic-Based Benders Decomposition

Our method relies on Logic-Based Benders Decomposition (LBBD) [20], an iterative method that breaks down a decision making task into a *master problem*
and a *subproblem*, over different subsets of variables. At each iteration, the solution produced by the master is fed to the subproblem, which tries to extend
it into a full assignment. If this is not feasible, a procedure generates one or
more constraints (*cuts*) that are added in the master formulation. However, the
LBBD method does not have a standard template for the production of valid
cuts. Instead, they must be tailored to the problem at hand, typically based
on knowledge of its structure. These cuts invalidate a set of master solutions,
always including the current one. If the problem cost does not depend on the
subproblem variables, then once all subproblems are feasible, the current master
solution is also optimal. In the opposite case, cuts should be generated even for
feasible subproblems, until the master becomes infeasible.

3 Super Solutions via Benders Decomposition

We define a method to obtain super solutions for problems in the generic form:

$$\min z = f(x) \qquad \textbf{(MP)} \qquad (4)$$
$$\text{subject to } x \in X \qquad (5)$$

where x is a vector of n decision variables, $f(x)$ is a cost function, and the X set
corresponds to the feasible decision space, which can be expressed by any means
(constraints, equations, logical clauses, etc.).

3.1 Generic Disruptions

Our method supports, and it is best described in terms of, generic disruptions.
These are characterized via a triplet $\langle D, E, P \rangle$, where $D(\hat{x})$ denotes the set of
all possible disruptions for a given solution $\hat{x}$, by associating each of them to an
index d. For example, in classical $(1, b)$ super solutions:

$$D(\hat{x}) = \{1, \ldots, n\}. \qquad (6)$$

In other words, in this case there is one potential disruption per variable. In
general, *a disruption is always uniquely identified by a* $(d, \hat{x})$ *pair*. P is a set of
predicates $P_{d,\hat{x}}(x)$, each corresponding to the preconditions of disruption $(d, \hat{x})$.
For example, in $(1, b)$ super solutions:

$$P_{d,\hat{x}}(x) = [\![x_d = \hat{x}_d]\!] \qquad (7)$$

where $[\![\cdot]\!]$ denotes the truth value of a predicate. Informally, the precondition
represents a condition that should hold for disruption $(d, \hat{x})$ to happen. E is a

set of predicates $E_{d,\hat{x}}(x)$, each corresponding to the effects of disruption $(d, \hat{x})$. For example, in $(1, b)$ super solutions:

$$E_{d,\hat{x}}(x) = [\![x_d \neq \hat{x}_d]\!]. \tag{8}$$

As another example, often only a particular set of variables in the solution may be subject to change and these are said to be members of the break-set. For each variable in the break-set, a repair-set is required that comprises the set of variables whose values may change to provide another solution. For this case we can write the Eqs. (6), (7), (8) as follows:

$$D(\hat{x}) = \mathcal{B} \tag{9}$$

$$P_{d,\hat{x}}(x) = [\![x_d = \hat{x}_d]\!] \tag{10}$$

$$E_{d,\hat{x}}(x) = [\![x_d \neq \hat{x}_d]\!] \wedge \bigwedge_{i \notin \mathcal{R}_d} [\![x_i = \hat{x}_i]\!] \tag{11}$$

where $\mathcal{B}$ is the break-set and $\mathcal{R}_d$ is the repair set associated to the disruption. The precondition predicates are as in the previous case. Concerning the effect predicates, a disruption of the variables of the break-set prevents the variable x_d from taking the value it had in the $\hat{x}$ solution. Moreover, since the disruption can be handled by changing only the variables in the repair set, all other variables are forced to maintain their value.

The framework can handle more general cases, and with particular disruptions with domain-specific preconditions and domain-specific effects.

Remark 1. If $P_{d,\hat{x}}(x)$ holds for solution x, then $E_{d,\hat{x}}(x)$ holds in any repair solution for disruption $(d, \hat{x})$, where d denotes a single possible disruption.

The result holds by construction, since as long as the disruption identified by the pair $(d, \hat{x})$ applies, all of its repair solutions should take its effects into account. As in all super solution approaches, we assume that repairs can change at most b variables. The c parameter from (a, b, c) super solutions can be supported as described in Sect. 2.

3.2 Decomposition Scheme

We exploit the basic properties of super solutions to obtain an LBBD scheme, and more importantly cut generation procedures, that can be applied to any target problem. The method will inherit some classical benefits of LBBD, such as the ability to use different solvers for the master and the subproblem, and provides some unique benefits. We task our master problem to find a solution, and the subproblem to identify how to repair every disruption. This choice provides two advantages: (a) since disruptions in super solutions are unrelated, we can handle each of them *in a separate subproblem*; and (b) having multiple subproblems allows us to *determine in a procedural fashion which disruptions should be handled*: this is precisely what makes our method capable of dealing with generic disruptions.

Formally, *the master is the same as the target problem*, i.e. **MP**. Given a master solution $\hat{x}$ and a disruption index d, the *basic form of one subproblem* is:

$$E_{d,\hat{x}}(y) \qquad\qquad \textbf{(SP}_0\textbf{)} \qquad\qquad (12)$$

$$y \in X \qquad\qquad\qquad (13)$$

where the y variables correspond one-to-one to the master variables and are used to define the repair solution. Hence, the subproblem is the same as the original problem, without the cost function and with the addition of the disruption effects. Since the subproblem has no impact on the cost of the master solution, if at any point all subproblems are feasible, the current $\hat{x}$ represents an optimal robust solution.

Pseudo-code for a basic version of this process is provided in Algorithm 1, whose name stands for "Benders Decomposition For Super Solutions". Note that, while stopping every iteration at the first infeasible subproblem is enough for convergence, processing all potential disruptions may lead to more cuts and be beneficial in the long run. We will show the difference of these two options in Sect. 5.

3.3   Basic Cuts

A *trivial cut* can be obtained by observing that there are two options for dealing with a non-repairable disruption: (a) making sure that it cannot occur; and (b) changing some assignments in the current solution. Hence, a valid cut for a disruption $(d, \hat{x})$ which lead to an unfeasible subproblem is:

$$\neg P_{d,\hat{x}}(x) \vee [\![H(x, \hat{x}) \geq 1]\!] \qquad\qquad (14)$$

where H is the Hamming distance, i.e.

Algorithm 1. BD4SS

Require: Triplet $\langle D, E, P \rangle$
  repeat
    solve master problem to find $\hat{x}$
    if master infeasible **then break**
    $robust = \top$
    for $d \in D(\hat{x})$ **do**
      if the subproblem for $(\hat{x}, d)$ is infeasible **then**
        $robust = \bot$
        generate cuts for $(\hat{x}, d)$
        break
  until *robust*

$$H(x', x'') = \sum_{i=1}^{n} [\![x_i' \neq x_i'']\!] \, . \tag{15}$$

Equation (14) states that either the precondition for d should be violated, or at least one change should be made with respect to $\hat{x}$.

An improved cut can be obtained by modifying the subproblem so that its objective is to minimize the distance between the repair solution and the current master solution, i.e.

$$\min \beta = H(y, \hat{x}) \qquad \textbf{(SP)} \tag{16}$$
$$\text{s.t. } E_{d,\hat{x}}(y) \tag{17}$$
$$y \in X. \tag{18}$$

If the optimal subproblem cost β^* is not greater than b, then a repair solution exists. This counts as a "feasible" result for the purpose of Algorithm 1. However, if $\beta^* > b$, then we know that any repair solution will need to change at least β^* variables. Based on this observation, we can derive a tightened version of Eq. (14), which serves as our *basic cut*

$$\neg P_{d,\hat{x}}(x) \vee [\![H(x, \hat{x}) \geq \beta^* - b]\!] \, . \tag{19}$$

Lemma 1. *Any new master solution should either violate the precondition $P_{d,\hat{x}}(x)$, or $\beta^* - b$ is the minimum number of variables that needs to be changed.*

Proof. In the case of a violation of the precondition for $(d, \hat{x})$, then the cut is correct. Otherwise, $P_{d,\hat{x}}(x)$ holds and finding a repair solution is necessary. Due to Remark 1, we have that:

$$P_{d,\hat{x}}(x) \Rightarrow E_{d,\hat{x}}(y). \tag{20}$$

Hence, the same $E_{d,\hat{x}}(y)$ will be part of any possible subproblem for the disruption. This means that *all such subproblems will share the same feasible space.* Let such feasible space be $X_{d,\hat{x}}$. We know that

$$\min\{H(y, \hat{x}) : y \in X_{d,\hat{x}}\} \geq \beta^*. \tag{21}$$

Then, due to triangular inequality in the Hamming distance

$$\min\{H(y, x) + H(x, \hat{x}) : y \in X_{d,\hat{x}}\} \geq \beta^*. \tag{22}$$

We wish for x to be repairable, hence the maximum value for $H(y, x)$ is b. From this, we obtain

$$\min\{b + H(x, \hat{x}) : y \in X_{d,\hat{x}}\} \geq \beta^*. \tag{23}$$

And, therefore,

$$\min\{H(x, \hat{x}) : y \in X_{d,\hat{x}}\} \geq \beta^* - b \tag{24}$$

which proves correctness. $\square$

3.4   Strengthened Cuts

Our basic cuts can be strengthened by proving that a *subset* of assignments in the master solution is enough to prevent repairability. Generic cut strengthening procedures for Bender Decomposition exist, but they are computationally very expensive. Here, we leverage the structure of super solutions to obtain insights into which variables should be included in the cut. In particular, given a set of indices I, let H_I be the Hamming distance restricted to I, i.e.

$$H_I(x', x'') = \sum_{i \in I} [\![x'_i \neq x''_i]\!]. \tag{25}$$

Given an infeasible subproblem, we will try to identify a subset of indices I such that the minimum H_I distance from the master solution $\hat{x}$ is still greater than b. We do this heuristically and iteratively, by starting from the variables that are different from $\hat{x}$ in the solution of **SP**, and then by solving the modified subproblem:

$$\min H_I(y, \hat{x}) + \frac{1}{n} H_{\overline{I}}(y, \hat{x}) \qquad \text{(SP}_\mathbf{I}) \tag{26}$$

$$\text{subject to } E_{d,\hat{x}}(y) \tag{27}$$

$$y \in X. \tag{28}$$

where $\overline{I}$ is the set of variable indices that are not in I. The cost function is a lexicographic composition, whose main goal is to minimize the distance $H_I(y, \hat{x})$. Let y^* be the optimal solution of this subproblem. If $H_I(y^*, \hat{x}) > b$, we can generate the *strengthened cut*

$$\neg P_{d,\hat{x}}(x) \vee [\![H_I(x, \hat{x}) \geq b - \beta_I^*]\!] \tag{29}$$

where $\beta_I^* = H_I(y^*, \hat{x})$. We can then expand the I set by including all variables that take different values in y^* and $\hat{x}$. The term $1/n H_{\overline{I}}(y, \hat{x})$ exists to reduce the number of variables that are added to I in this fashion at each step.

Algorithm 2. STRENGTHEN

Require: A disruption $(d, \hat{x})$
   solve **SP** to find y^* and β^*
   $I = \{i \in [1..n] : y_i^* \neq \hat{x}_i\}$
   $\beta_I^* = 0$
   repeat
     Solve **SP_I** to find y^*
     if $H_I(y^*, \hat{x}) > \max(b, \beta_I^*)$ **then**
       $\beta_I^* = H_I(y^*, \hat{x})$
       Generate the cut from Equation (29)
   until $\beta_I^* = \beta^*$

Pseudo-code for the procedure is given in Algorithm 2, and adds two optimizations. First, it generates cuts only when the value of $H_I(y^*, \hat{x})$ increases:

since the I set grows monotonically, subsequent iterations with the same H_I distance result in redundant cuts. Second, the process is stopped as soon as β_I^* becomes equal to β^*, which is guaranteed to happen when all variable indices are included in the I set.

3.5 Combined Cuts

A third cut generation method can be defined via a linear combination of pairs of existing cuts. In particular, let us suppose we have:

$$\neg P_{d_k,\hat{x}^k}(x) \vee \left[\!\left[H_{I^k}(x,\hat{x}^k) \geq b - \beta_{I^k}^* \right]\!\right] \tag{30}$$

$$\neg P_{d_h,\hat{x}^h}(x) \vee \left[\!\left[H_{I^k}(x,\hat{x}^h) \geq b - \beta_{I^h}^* \right]\!\right]. \tag{31}$$

The cuts are built for disruptions $(d_k, \hat{x}^k)$ and $(d_h, \hat{x}^h)$, respectively, and are expressed in the form of Eq. (29), which subsumes that of Eq. (19). First, we observe that any inequality concerning a H_I distance holds true if we expand the set I. Hence, we can write:

$$\neg P_{d_k,\hat{x}^k}(x) \vee \left[\!\left[H_{I^{kh}}(x,\hat{x}^k) \geq b - \beta_{I^k}^* \right]\!\right] \tag{32}$$

$$\neg P_{d_h,\hat{x}^h}(x) \vee \left[\!\left[H_{I^{kh}}(x,\hat{x}^h) \geq b - \beta_{I^h}^* \right]\!\right] \tag{33}$$

where both Hamming distances are defined on the same set $I^{kh} = I^k \cup I^h$. Due to triangular inequality, we have that:

$$H_{I^{kh}}(x,\hat{x}^k) + H_{I^{kh}}(x,\hat{x}^h) \geq H_{I^{kh}}(\hat{x}^k,\hat{x}^h). \tag{34}$$

Therefore, the two cuts can always be merged to yield a valid *combined cut*:

$$\neg P_{d_k,\hat{x}^k}(x) \vee \neg P_{d_h,\hat{x}^h}(x) \vee$$
$$\left[\!\left[H_{I^{kh}}(x,\hat{x}^k) + H_{I^{kh}}(x,\hat{x}^h) \geq H_{I^{kh}}(\hat{x}^k,\hat{x}^h) \right]\!\right] \tag{35}$$

which is non-redundant if $H_{I^{kh}}(\hat{x}^k,\hat{x}^h) > 2b - \beta_{I^k}^* - \beta_{I^h}^*$. A simple procedure for generating all non-redundant combined cuts is provided in Algorithm 3, where n_c is assumed to be the number of basic or strengthened cuts generated so far.

Algorithm 3. COMBINE

Require: $C = \{(d_k, \hat{x}^k, I^k, \beta_{I^k}^*) : \forall k \in [1..n_c]\}$
   for $k \in [1..n_c - 1]$ **do**
      for $h \in [k + 1, n_c]$ **do**
         if $H_{I^{kh}}(\hat{x}^k, \hat{x}^h) > 2b - \beta_{I^k}^* - \beta_{I^h}^*$ **then**
            Generate the cut from Equation (35)

Some Remarks. The master and the subproblem for our decomposition correspond very closely to the original target problem in terms of both structure and size. This makes them easy to implement (in contrast to specialized search methods), and typically not harder to solve than the target problem (in contrast to reformulated models).

Finding super solutions, however, remains very challenging, as it should be expected of any method providing strong robustness guarantees. Each iteration of the method requires to solve multiple (typically NP-hard) problems. Moreover, as the number of iterations grows, the master tends to become more difficult due to the accumulated cuts.

The use of procedural code for many sections of our methods provides a substantial degree of flexibility. In particular, it allows our method to deal with generic disruptions. The available flexibility could be further exploited: for example, one could conceive a hybrid chance-constraint/super solution method to provide robustness against disruptions with a given total probability of occurrence.

Another advantage of the approach is that, in case of early stops, one can still access a master solution that is robust against *some* disruptions. This can be achieved via reformulated models, but not via specialized search algorithms, at least not trivially.

4   A Case Study on the Kidney Exchange Problem

We now proceed to demonstrate our method on the Kidney Exchange Problem (KEP), using the cycle formulation from Sect. 2 as a basis. For sake of simplicity and of maximal compatibility with earlier super-solution formulations, we are interested in finding $(1, b, c)$ super solutions that are repairable with respect to the loss of *cycles*. Hence we have:

$$D(\hat{x}) = \{d \in [1..n_c] : \hat{x}_d = 1\} \tag{36}$$

$$E_{d,\hat{x}}(x) = [\![x_d = 0]\!] \tag{37}$$

$$P_{d,\hat{x}}(x) = [\![x_d = 1]\!] \tag{38}$$

where n_c is the number of cycles, and a disruption index d corresponds to a cycle. Each cut is in the form:

$$\neg P_{d,\hat{x}}(x) \vee [\![H_I(x, \hat{x}) \geq b - \beta_I^*]\!] \tag{39}$$

and can be translated into the mathematical constraint:

$$\sum_{\substack{i \in [1..n] \\ \hat{x}_i = 0}} x_i + \sum_{\substack{i \in [1..n] \\ \hat{x}_i = 1}} (1 - x_i) \geq (b - \beta_I^*)(1 - x_d). \tag{40}$$

The right hand side is non-trivial, i.e. larger than 0, iff x_d is equal to 0, if the precondition for the disruption is violated. This information is sufficient to apply all the techniques described in Sect. 3 to the KEP.

Since we are dealing with $(1, b, c)$ super solutions, we need to include an additional cost-bounding constraint both in the master and in the subproblems:

$$\sum_{i \in C} w_i x_i \geq c^* - c \tag{41}$$

where c^* is the cost of an optimal non-robust solution. Intuitively, in a robust solution we do not want to lose too many lives, which makes sense from a medical perspective.

We choose to use the $(1, b, c)$ configuration to align with prior work on super solutions in CP, see [14–16]. However, it is not particularly difficult to obtain $\langle D, E, P \rangle$ triplets and cut expressions for other disruptions, including the loss of multiple nodes, edges, or cycles. Below we show the example of robust solution when a node disruption may occur:

$$D(\hat{x}) = \{d \in [1..V] : \sum_{c \in C : d \in c} \hat{x}_d \geq 1\} \tag{42}$$

$$P_{d,\hat{x}}(x) = \left[\!\!\left[\sum_{c \in C, d \in c} x_d \geq 1 \right]\!\!\right] \tag{43}$$

$$E_{d,\hat{x}}(x) = \left[\!\!\left[\sum_{c \in C, d \in c} x_d = 0 \right]\!\!\right] \tag{44}$$

where V is the set of pairs. The disruption in this case is associated with each node d that is part of at least one selected cycle, and its effect is to forbid the selection of other cycles that it is part of. In real life, it is more realistic to consider a node disruption, since it happens quite often that a pair pulls out from the program due to worsening health conditions.

4.1 Model Reformulation

As a basis for comparison we obtained a model reformulation by generalising the approach from [31]. In particular, we generate a model with a "main" vector of n decision variables x, and a vector of n repair variables y_i for each x_i. Overall, the model has $n \times (n + 1)$ variables, and corresponds to the problem of finding an $(1, b)$ super solution and all its repair solutions:

$$\max z = \sum_{i \in \mathcal{C}} w_i x_i \qquad \textbf{(FTF}_{\textbf{KEP}}\textbf{)} \qquad (45)$$

$$\text{s.t.} \sum_{i \in \mathcal{C}: k \in C_i} x_i \leq 1 \qquad \forall k \in V \qquad (46)$$

$$\sum_{j \in \mathcal{C}: k \in C_j} y_{ij} \leq 1 \qquad \forall i \in \mathcal{C}, \forall k \in V \qquad (47)$$

$$y_{ii} = 1 - x_i \qquad \forall i \in \mathcal{C} \qquad (48)$$

$$x_i \in \{0, 1\} \qquad \forall i \in \mathcal{C} \qquad (49)$$

$$y_{ij} \in \{0, 1\} \qquad \forall i, j \in \mathcal{C} \qquad (50)$$

where Eqs. (46) and (47) ensure that neither the super solution nor any repair solution can use a node twice. Equation (48) corresponds to the effects of the disruption. We then need to add constraints to limit the Hamming distance between the super solution and the repair solutions. In particular, for each $i \in \mathcal{C}$ we have:

$$\sum_{j \in \mathcal{C}} y_{ij}(1 - x_j) + \sum_{j \in \mathcal{C}} (1 - y_{ij}) x_j \leq b + |\mathcal{C}|(1 - x_i)$$

which is a quadratic constraint, with a right-hand side term that is non-trivial iff $x_i = 1$, i.e. if cycle i is used. Finally, we need one last set of constraints to handle the cost bound:

$$\sum_{j \in \mathcal{C}} w_j y_{ij} \geq c^* - c \qquad \forall i \in \mathcal{C} \qquad (51)$$

Overall, the reformulated model is not only considerably larger, but also non-linear, although a big-M based linearization is, of course, possible.

5   Experiments

In this section we present our experimental evaluation, which was designed with two main goals: first, to compare the effectiveness of our method against one with a similar applicability and flexibility, i.e. the reformulation from Sect. 4; and, second, testing the effectiveness of different cut generation procedures.

We use the benchmarks from [25], which were obtained via the state-of-the-art donor pool generation method described in [30]. We consider all transplants equally worthy, hence the weight of each cycle corresponds to its number of exchanges. The maximum length of a cycle is 3. We use the docplex library[2] to model the problems which uses IBM® Decision Optimization CPLEX solver. All the tests run on Intel Xeon E5430 processors with Linux.

---

[2] https://pypi.org/project/docplex/.

5.1 Choosing the (a, b, c) Parameters

As mentioned in Sect. 4, we focus on finding $(1, b, c)$ super solutions, where disruptions correspond to (single) selected cycles becoming non-viable. Therefore, we have $a = 1$, and we are left with two main parameters to tune, i.e. b and c.

The b parameter refers to how many variables can be changed in a repair solution excluding, as in classical super solution approaches, the variables that have lost their values. In our formulation, this requires to exclude such variables in the computation of the Hamming distance. For most problems, finding a repair solution by changing just one variable (i.e. $b = 1$) is a difficult task, and the KEP is no exception. Interestingly, however, we found that setting $b = 2$ was enough for the solution of the cycle formulation to be robust for most of the instances in our benchmarks. Therefore, we chose to limit our experiments to $b = 1$.

The use of the cost-bounding parameter c is in this case mandatory. In fact, any KEP solution remains feasible if a cycle is removed: hence, without some cost bound, the loss of a cycle could be countered by doing nothing at all. In general, if k is the largest size of an allowed exchange, then any solution is trivially (a, b) robust, unless we require not losing more than $ak - 1$ units from the non-robust optimum (i.e. $c \leq ak - 1$). In our case, there are therefore only two reasonable values for c, i.e. 1 and 2. We chose to keep $c = 2$ in our experiments.

5.2 Results

We performed experiments over groups of 10 instances with number of nodes $n_{pairs} \in \{16, 32, 64, 128\}$. We solved them via the reformulation-based model, and via our Benders Decomposition approach, using different cut generation procedures, and always stopping the process after $5 \times n_{pairs}$ master iterations. We investigated approaches that stop each iteration of Algorithm 1 after the first infeasible subproblem (with the _1st suffix), and approaches that process all disruptions (with the _all suffix). The _all-suffix approaches will generate more cuts: this may reduce the number of iterations, but require more time and makes the master more difficult to solve. We considered the following combination of the cuts from Sect. 3: only basic cuts, identifiable from the cut1 prefix; basic cuts + combined cuts, identifiable by cut2; strengthened cuts, identifiable by cut3; and strengthened cuts + combined cuts, identifiable by cut4. Strengthened cuts are expensive to produce, but may reduce the number of iterations. Combined cuts are cheap to obtain, but their large number may make the master more difficult to solve.

The results of the experiments are reported in Tables 1, 2, Figs. 1 and 2. The tables report, for each approach and each size: (a) the fraction of instances for which a robust solution was found; (b) the fraction of instances for which robustness was proved infeasible; and (c) the fraction of open instances, for which the iteration limit was reached. In particular Table 1 contains the results for the reformulation approach and the first infeasible subproblem approach, and Table 2 contains only the approaches that process all disruptions. Overall, the reformulation scales very poorly, leading to consistent memory problems for sizes

Table 1. Solutions of the Reformulation and Benders decomposition

#Pairs	Reform	Benders decomposition											
		cut1_1st			cut2_1st			cut3_1st			cut4_1st		
	%sol	%sol	nosol	open	%sol	nosol	open	%sol	nosol	open	%sol	nosol	open
16	50	100	0	0	100	0	0	100	0	0	100	0	0
32	10	80	0	20	80	0	20	80	20	0	80	20	0
64	-	60	0	40	60	0	40	70	30	0	70	30	0
128	-	90	0	10	90	0	10	90	0	10	90	0	10

Table 2. Solutions of Benders decomposition

#Pairs	Benders decomposition											
	cut1_all			cut2_all			cut3_all			cut4_all		
	%sol	nosol	open	%sol	nosol	open	%sol	nosol	open	%sol	nosol	open
16	100	0	0	100	0	0	100	0	0	100	0	0
32	80	0	20	70	0	30	80	20	0	80	20	0
64	60	0	40	60	30	10	70	30	0	70	30	0
128	90	0	10	90	10	0	100	0	0	90	10	0

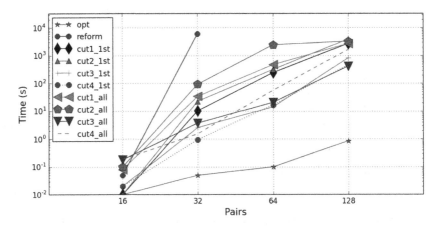

Fig. 1. Resolution time with $b = 1$ log scale

larger than 32. Using combined cuts does not have a large impact on the number
of open instances but it does have a huge impact on increasing the size of the
instances, whereas strengthened cuts are very effective. Processing all disruption
at each iteration seems to be slightly detrimental.

In Fig. 1 we show the average solution time of all approaches on a logarithmic
scale, including the non-robust cycle formulation, referred to as optimal. The
time for the reformulation grows very quickly with the number of pairs, and the
series has only two data points. The approaches using strengthened cuts tend

to be the most effective, while the time for methods using combined cuts are consistently higher than those who do not generate them.

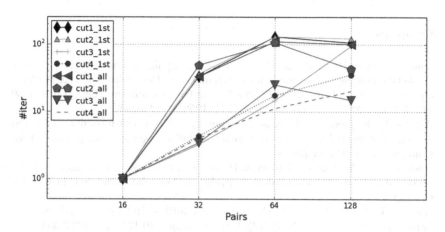

Fig. 2. Number of master's iterations with b = 1 log scale

In Fig. 2 we plot the average number of master iterations for the Benders Decomposition, again on a logarithmic scale. Strengthened cuts are the clear winner. The _all-suffix approaches do sometimes iterate less than the corresponding _1st-suffix ones, although this is not always the case. The reason is that adding more cuts may force the master to produce different sequences of solutions, and therefore have a non-monotonic effect on the number of iterations. Combined cuts seem to have a more consistent effect, although not a dramatic one: this suggests that a more difficult master is the main reason for their increased solution time, and that controlling the number of generated cuts per iteration may make them more useful.

6   Conclusion and Future Work

We have proposed a Benders Decomposition approach to obtain super solutions for the kidney exchange problem that dramatically outperforms reformulation-based approaches in terms of scalability, and is more flexible and much easier to apply compared to specialised algorithms from the literature. As part of future work, we believe our contributions can be applied to other real-world satisfaction and optimisation problems, which we are investigating at the moment. The method is not without drawbacks and, in our current research, many improvement directions are still open, such as defined using multiple subproblems, or principled approaches to limit cut generation, improvements to the anytime features of the approach, and hybridisation with probabilistic models.

Acknowledgements. The Insight Centre for Data Analytics is supported by Science Foundation Ireland under Grant Number SFI/12/RC/2289, which is co-funded under the European Regional Development Fund.

References

1. Abraham, D.J., Blum, A., Sandholm, T.: Clearing algorithms for barter exchange markets: enabling nationwide kidney exchanges. In: Proceedings 8th ACM Conference on Electronic Commerce (EC-2007), San Diego, California, USA, pp. 295–304, 11–15 June 2007. https://doi.org/10.1145/1250910.1250954
2. Alvelos, F., Klimentova, X., Rais, A., Viana, A.: A compact formulation for maximizing the expected number of transplants in kidney exchange programs. J. Phys.: Conf. Ser. **616**, 012011 (2015)
3. Anderson, R., Ashlagi, I., Gamarnik, D., Roth, A.E.: Finding long chains in kidney exchange using the traveling salesman problem. Proc. Nat. Acad. Sci. **112**(3), 663–668 (2015)
4. Ashlagi, I., Roth, A.E.: New challenges in multihospital kidney exchange. Am. Econ. Rev. **102**(3), 354–359 (2012). https://doi.org/10.1257/aer.102.3.354
5. Bofill, M., Busquets, D., Muñoz, V., Villaret, M.: Reformulation based maxsat robustness. Constraints **18**(2), 202–235 (2013). https://doi.org/10.1007/s10601-012-9130-2
6. Constantino, M., Klimentova, X., Viana, A., Rais, A.: New insights on integer-programming models for the kidney exchange problem. Eur. J. Oper. Res. **231**(1), 57–68 (2013). https://doi.org/10.1016/j.ejor.2013.05.025
7. Demirović, E., Schwind, N., Okimoto, T., Inoue, K.: Recoverable team formation: building teams resilient to change. In: Proceedings of the 17th International Conference on Autonomous Agents and MultiAgent Systems, pp. 1362–1370. International Foundation for Autonomous Agents and Multiagent Systems (2018)
8. Derrien, A., Petit, T., Zampelli, S.: A declarative paradigm for robust cumulative scheduling. In: O'Sullivan, B. (ed.) CP 2014. LNCS, vol. 8656, pp. 298–306. Springer, Cham (2014). https://doi.org/10.1007/978-3-319-10428-7_23
9. Dickerson, J.P., Manlove, D.F., Plaut, B., Sandholm, T., Trimble, J.: Position-indexed formulations for kidney exchange. In: Proceedings of EC, pp. 25–42. ACM (2016)
10. Dickerson, J.P., Procaccia, A.D., Sandholm, T.: Failure-aware kidney exchange. In: Proceedings of the Fourteenth ACM Conference on Electronic Commerce, pp. 323–340. ACM (2013)
11. Dickerson, J.P., Procaccia, A.D., Sandholm, T.: Price of fairness in kidney exchange. In: Proceedings of the 2014 International Conference on Autonomous Agents and Multi-Agent Systems, pp. 1013–1020. International Foundation for Autonomous Agents and Multiagent Systems (2014)
12. Ginsberg, M.L., Parkes, A.J., Roy, A.: Supermodels and robustness. In: AAAI/IAAI, pp. 334–339 (1998)
13. Glorie, K.M., van de Klundert, J.J., Wagelmans, A.P.: Kidney exchange with long chains: an efficient pricing algorithm for clearing barter exchanges with branch-and-price. Manufact. Serv. Oper. Manage. **16**(4), 498–512 (2014)
14. Hebrard, E., Hnich, B., Walsh, T.: Super CSPs. In: Proceedings of the CP-03 Workshop on "Handling Change and Uncertainty", Cork, Ireland (2003)

15. Hebrard, E., Hnich, B., Walsh, T.: Super solutions in constraint programming. In: Régin, J.-C., Rueher, M. (eds.) CPAIOR 2004. LNCS, vol. 3011, pp. 157–172. Springer, Heidelberg (2004). https://doi.org/10.1007/978-3-540-24664-0_11
16. Holland, A., O'Sullivan, B.: Super solutions for combinatorial auctions. In: Faltings, B.V., Petcu, A., Fages, F., Rossi, F. (eds.) CSCLP 2004. LNCS (LNAI), vol. 3419, pp. 187–200. Springer, Heidelberg (2005). https://doi.org/10.1007/11402763_14
17. Holland, A., O'Sullivan, B.: Robust solutions for combinatorial auctions. In: Proceedings 6th ACM Conference on Electronic Commerce (EC-2005), Vancouver, BC, Canada, pp. 183–192, 5–8 June 2005. https://doi.org/10.1145/1064009.1064029
18. Holland, A., O'Sullivan, B.: Weighted super solutions for constraint programs. In: Proceedings, The Twentieth National Conference on Artificial Intelligence and the Seventeenth Innovative Applications of Artificial Intelligence Conference, Pittsburgh, Pennsylvania, USA, pp. 378–383, 9–13 July 2005
19. Holland, A., O'Sullivan, B.: Truthful risk-managed combinatorial auctions. In: Proceedings of the 20th International Joint Conference on Artificial Intelligence IJCAI 2007, pp. 1315–1320. Morgan Kaufmann Publishers Inc., San Francisco (2007). http://dl.acm.org/citation.cfm?id=1625275.1625488
20. Hooker, J.N., Ottosson, G.: Logic-based benders decomposition. Math. Programm. **96**(1), 33–60 (2003)
21. Klimentova, X., Alvelos, F., Viana, A.: A new branch-and-price approach for the kidney exchange problem. In: Murgante, B., Misra, S., Rocha, A.M.A.C., Torre, C., Rocha, J.G., Falcão, M.I., Taniar, D., Apduhan, B.O., Gervasi, O. (eds.) ICCSA 2014. LNCS, vol. 8580, pp. 237–252. Springer, Cham (2014). https://doi.org/10.1007/978-3-319-09129-7_18
22. Klimentova, X., Pedroso, J.P., Viana, A.: Maximising expectation of the number of transplants in kidney exchange programmes. Comput. Oper. Res. **73**, 1–11 (2016)
23. Mak-Hau, V.: On the kidney exchange problem: cardinality constrained cycle and chain problems on directed graphs: a survey of integer programming approaches. J. Comb. Optim. **33**, 1–25 (2015).https://doi.org/10.1007/s10878-015-9932-4
24. Manlove, D.F., O'Malley, G.: Paired and altruistic kidney donation in the UK: algorithms and experimentation. ACM J. Exp. Algorithmics **19**(1) (2014). https://doi.org/10.1145/2670129
25. Mattei, N., Walsh, T.: PREFLIB: a library for preferences HTTP://WWW.PREFLIB.ORG. In: Perny, P., Pirlot, M., Tsoukiàs, A. (eds.) ADT 2013. LNCS (LNAI), vol. 8176, pp. 259–270. Springer, Heidelberg (2013). https://doi.org/10.1007/978-3-642-41575-3_20
26. Pedroso, J.P.: Maximizing expectation on vertex-disjoint cycle packing. In: Murgante, B., et al. (eds.) ICCSA 2014. LNCS, vol. 8580, pp. 32–46. Springer, Cham (2014). https://doi.org/10.1007/978-3-319-09129-7_3
27. Plaut, B., Dickerson, J.P., Sandholm, T.: Fast optimal clearing of capped-chain barter exchanges. In: Proceedings of AAAI, pp. 601–607 (2016)
28. Powell, W.B.: A unified framework for optimization under uncertainty. In: Optimization Challenges in Complex, Networked and Risky Systems, pp. 45–83. INFORMS (2016)
29. Roy, A.: Fault tolerant boolean satisfiability. J. Artif. Intell. Res. **25**, 503–527 (2006)
30. Saidman, S.L., Roth, A.E., Sönmez, T., Ünver, M.U., Delmonico, F.L.: Increasing the opportunity of live kidney donation by matching for two-and three-way exchanges. Transplantation **81**(5), 773–782 (2006)
31. Weigel, R., Bliek, C., Faltings, B.V.: On reformulation of constraint satisfaction problems (extended version). Artificial Intelligence, Citeseer (1998)

Exploiting Glue Clauses to Design Effective CDCL Branching Heuristics

Md Solimul Chowdhury[(⊠)], Martin Müller, and Jia-Huai You

Department of Computing Science, University of Alberta,
Edmonton, Alberta, Canada
{mdsolimu,mmueller,jyou}@ualberta.ca

Abstract. In conflict-directed clause learning (CDCL) SAT solving, a state-of-the-art criterion to measure the importance of a learned clause is called *literal block distance* (LBD), which is the number of distinct decision levels in the clause. The lower the LBD score of a learned clause, the better is its quality. The learned clauses with LBD score of 2, called *glue clauses*, are known to possess high pruning power. In this work, we relate glue clauses to decision variables. First, we show experimentally that branching decisions with variables appearing in glue clauses, called *glue variables*, are more conflict efficient than with nonglue variables. This observation motivated the development of a structure-aware CDCL variable bumping scheme, which increases the heuristic score of a glue variable based on its appearance count in the glue clauses that are learned so far by the search. Empirical evaluation shows the effectiveness of the new method on the main track instances from SAT Competitions 2017 and 2018 with four state-of-the-art CDCL SAT solvers. Finally, we show that the frequency of learned clauses that are glue clauses can be used as a reliable indicator of solving efficiency for some instances, for which the standard performance metrics fail to provide a consistent explanation.

Keywords: CDCL SAT · Branching heuristics · Glue clauses

1 Introduction

Given a formula $\mathcal{F}$ of boolean variables, the task of SAT solving is to determine a variable assignment that satisfies $\mathcal{F}$ or to report the unsatisfiability of $\mathcal{F}$ in case no such assignment exists. SAT is known to be NP-complete [5]. Despite the hardness, modern CDCL SAT solvers can solve large real-world problems from important domains, such as hardware design verification [8], software debugging [4], planning [21], and encryption [18,23], sometimes with surprising efficiency. This is the result of a careful combination of its key components, such as preprocessing [6,10] and inprocessing [11,17], robust branching heuristics [13,14,19], efficient restart policies [2,20], intelligent conflict analysis [22], and effective clause learning [19].

Clause learning prunes search space. As conflict discovery is the only way to learn clauses, the rate of discovery is critical for CDCL SAT solvers. As a large

© Springer Nature Switzerland AG 2019
T. Schiex and S. de Givry (Eds.): CP 2019, LNCS 11802, pp. 126–143, 2019.
https://doi.org/10.1007/978-3-030-30048-7_8

amount of learned clauses reduces the overall performance, the management of the learned clause database also becomes a key component of a modern CDCL SAT solver [19,22].

In earlier CDCL SAT solvers, the size and recent activities of learned clauses were the dominant criteria for determining the relevance of learned clauses [7]. The CDCL SAT solver Glucose [1] was the first to apply a new measure called *literal block distance* (LBD), which indicates the number of distinct decision levels in a learned clause. The learned clauses with LBD score of 2, called *glue clauses*, are of particular interest [1,20] because a glue clause connects a block of closely related variables, and thus a relatively small number of decisions are needed to make it a *unit clause* (i.e., a clause that has all but one literals assigned under the current partial assignment). A glue clause therefore may cause a faster generation of conflicts within fewer numbers of decisions, which leads to pruning of the search space. Simply put, glue clauses have higher potential to reduce search space more quickly than other learned clauses. For this reason, all modern CDCL SAT solvers permanently store glue clauses.

Inspired by the intuitive characteristics of glue clauses, we ask the following question: Can glue clauses be used to help re-rank decision variables to improve search efficiency? We call the decision variables that have appeared in at least one glue clause up to the current search state *glue variables*, and others *nonglue variables*.

The main contributions of this paper are:

- We conduct an experiment using the 750 instances from the main track of SAT Competition 2017 and 2018 (abbreviated as SAT-2017 and SAT-2018, respectively) with four state-of-the-art CDCL SAT solvers: glucose 4.1[1] (just called Glucose), MAPLECOMSPS_PURE_LRB[2] (abbreviated as MapleLRB), Maple_LCM_Dist[3] (abbreviated as MLD, winner of SAT-2017) and MapleLCMDistChronoBT[4] (abbreviated as MLD_CBT, winner of SAT-2018). Our experiment shows that decisions with glue variables are more conflict efficient than those with nonglue variables. Furthermore, glue variables are picked up by CDCL branching heuristics disproportionately more often.
- We design a structure-aware variable score bumping method called *Glue Bumping* (GB), which dynamically bumps activity score of a glue variable based on its current activity score and (normalized) *glue level*, which is a measure of the count of glue clauses in which the variable appears. The method is simple to implement.
- We implemented the GB method on top of the same four SAT solvers mentioned above. For the 750 instances from SAT-2017 and SAT-2018, all GB extensions solve more instances than the baselines and achieve lower

---

[1] https://www.labri.fr/perso/lsimon/glucose/.

[2] https://sites.google.com/a/gsd.uwaterloo.ca/maplesat/.

[3] https://baldur.iti.kit.edu/sat-competition-2017/solvers/.

[4] http://sat2018.forsyte.tuwien.ac.at/solvers/main_and_glucose_hack/.

PAR-2 scores[5]. One of our extended solver solves 9 additional instances over the instances from SAT-2017. According to [2], this level of performance gain closely resembles to the introduction of a critical feature, which is remarkable, given the simplicity of the new method.

– We provide evidence that the frequency of glue clauses in learned clauses may serve as a reliable indicator of solving efficiency. In [16], the authors reported correlations between solving efficiency of branching heuristics and standard metrics based on the global learning rate (GLR) and average LBD (aLBD) scores - higher solving efficiency is indicated by higher average GLR and lower average aLBD. We show that these two measures do not provide a consistent explanation of solving efficiency for some subsets of SAT-2017 and SAT-2018, for which the correlations are highly expected to hold. However, using a new measure based on the frequency of learned clauses that are glue, we are able to provide a consistent explanation.

The next section provides preliminaries. Section 3 reports an experiment on the role of glue variables in CDCL SAT solving, which motivates the design of a bumping scheme in Sect. 4. Section 5 reports an experimental analysis. In Sect. 7 we explain why our standard bumping scheme does not work very well for Glucose and how to fix the issue. Section 8 reports some additional experimental results with the GB method. Section 9 is about related work and future directions can be found in Sect. 10.

2   Preliminaries

2.1   Inner Working of a CDCL Solver

A CDCL SAT solver works by extending an initially empty *partial assignment* using two operations in an interleaving fashion: a *branching decision* and *unit propagation* (UP). A branching decision selects an unassigned variable by using a branching heuristic and assigns a boolean value to it. Following a branching decision, UP simplifies $\mathcal{F}$ by deducing a new set of implied variable assignments. UP may lead to a conflict due to a falsified or conflicting clause. *Conflict analysis* determines the root cause of a conflict and generates a *learned clause* that is added to $\mathcal{F}$ to prevent the conflict from reappearing in the future, thereby pruning the search. Search continues from a *backjumping level* computed from the learned clause. We refer the reader to [3] for more details on CDCL SAT solving.

2.2   Terminologies

We review some terminologies used in this paper.

---

[5] A metric used in SAT competitions. Defined as the sum of all runtimes for solved instances $+ 2 * timeout$ for unsolved instances; lowest score wins.

- **Activity Based Branching Heuristics:** The standard CDCL branching heuristic, such as VSIDS [19], LRB [14] and CHB [13], maintains an activity score for each variable of a given formula. During the search, a variable's involvement in conflicts contribute to the increments of its activity score. At any given state of the search, the activity score of a variable measures its involvement in the recent conflicts.
- **Global Learning Rate** (GLR): This is defined as $\frac{n_c}{n_d}$, where n_c is the number of conflicts generated in n_d decisions [16], i.e., GLR measures the average number of conflicts that a solver generates per decision.
- **Literal Block Distance (LBD):** The LBD of a learned clause θ indicates the number of distinct decision levels in θ [1]. If LBD$(\theta) = k$, then θ contains k propagation blocks, where each block has been propagated within the same branching decision. Intuitively, variables in a block are closely related. Learned clauses with lower LBD score tend to have higher quality.
- **Glue Clauses:** These are the learned clauses with LBD score of 2 [1], which have the potential for fast propagations of truth values under a partial assignment.

Let $\mathcal{F}$ be a SAT formula. Suppose a CDCL solver Ψ is solving $\mathcal{F}$ and s is its current search state. At s, Ψ has taken $d > 0$ decisions and has learned a set of glue clauses. A *glue variable* is a variable that has appeared in at least one glue clause up to the search state s. Other variables that have not appeared in a glue clause are called *nonglue variables*. A *glue decision* is the branching decision that selects a glue variable and a *nonglue decision* is the branching decision that selects a nonglue variable. Suppose that until s, Ψ has taken gd glue decisions (resp. ngd nonglue decisions) which generated gc conflicts (resp. ngc conflicts).

- *Learning Rate* (LR): In contrast with GLR (global learning rate) where the rate of conflict generation is over all decisions, we are also interested in such rates over glue decisions only or over nonglue decisions only, up to a search state. *LR with glue decisions* is defined as $\frac{gc}{gd}$, while *LR with nonglue decisions* is defined as $\frac{ngc}{ngd}$.
- *Average LBD* (aLBD): This is the average LBD score per conflict generated solely by glue decisions or solely by nonglue decisions. Let $sumLBD_{gc}$ (resp. $sumLBD_{ngc}$) be the sum of LBD scores of the learned clauses derived from those gc (resp. ngc) conflicts. The *aLBD with glue decisions* (resp. *nonglue decisions)* is defined as $\frac{sumLBD_{gc}}{gc}$ (resp. $\frac{sumLBD_{ngc}}{ngc}$).

3   Conflict Efficiency of Glue Variables

In this section, we report an experiment that studies the role played by glue variables in CDCL SAT solving, which shows that glue decisions are more conflict efficient (i.e., achieve higher average LR and lower average aLBD, in general) than nonglue decisions and the branching heuristics of modern CDCL SAT solvers exhibit bias towards selection of glue variables over nonglue variables.

The solvers in this experiment are Glucose, MapleLRB, MLD, and MLD_CBT. The branching heuristics used in the first two solvers are, respectively, VSIDS [19] and LRB [14]. For the next two, the branching heuristics are based on a combination of three heuristics, VSIDS, LRB, and Dist [24].

We run all 750 instances used in the main track of SAT-2017 (350 instances) and 2018 (400 instances) with 5000 s timeout limit per instance. We instrumented the four solvers to collect the following statistics for each instance: (i) the numbers of glue and nonglue decisions, (ii) LR and aLBD for both glue and nonglue decisions, and (iii) the numbers of glue and nonglue variables. For each instance, all the measurements are taken at the final search state (i.e., either after satisfiability/unsatisfiability is determined or after timeout). All experiments are run on a Linux workstation with 64 GB RAM and processor clock speed of 2.40 GHZ.

3.1 Conflict Generation Power of Glue Variables

Table 1 shows a comparison of average LR and average aLBD for glue and nonglue decisions, grouped by satisfiable, unsatisfiable and unsolved instances. Comparing column D1 and D2, on average, all solvers achieve significantly higher LR with glue decisions. For all three categories of instances, MLD and MLD_CBT achieve significantly lower average LBD (compare columns E1 and E2) for glue decisions. For Glucose and MapleLRB, the numbers under E1 and E2 are largely comparable, without showing significant gaps.

Table 1. Comparison of average LR (higher is better) and average aLBD (lower is better) for glue and nonglue decisions.

(A) Systems	(B) Type	(C) #Inst	(D) Average of Learning Rate (LR)		(E) Average of aLBD	
			(D1) Glue decisions	(D2) Nonglue decisions	(E1) Glue decisions	(E2) Nonglue decisions
Glucose	SAT	180	**0.55**	0.41	18.44	**18.18**
	UNSAT	191	**0.56**	0.44	**11.2**	11.4
	Unsolved	379	**0.57**	0.48	**24.76**	25.48
MapleLRB	SAT	194	**0.47**	0.38	20.18	**19.25**
	UNSAT	190	**0.58**	0.46	**11.92**	12.39
	Unsolved	366	**0.48**	0.44	34.86	**33.39**
MLD	SAT	235	**0.47**	0.19	**31.76**	40.55
	UNSAT	207	**0.59**	0.27	**12.8**	30.1
	Unsolved	308	**0.52**	0.37	**24.23**	34.09
MLD_CBT	SAT	238	**0.51**	0.21	**32.1**	41.9
	UNSAT	215	**0.61**	0.27	**13.17**	24.74
	Unsolved	297	**0.53**	0.37	**25.25**	36.7

To confirm that the average values for these 2 measures reported in Table 1 reflect the actual distribution of these measures, we plot the LR and aLBD values for the 750 instances for the four solvers.

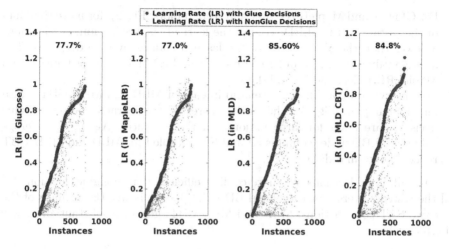

Fig. 1. Comparison of LR values for glue and nonglue decisions. Instances are sorted by the LR values of glue decisions. The number at the top of each plot represents the percentage of instances, for which LR of glue decisions are higher than LR of nonglue decisions.

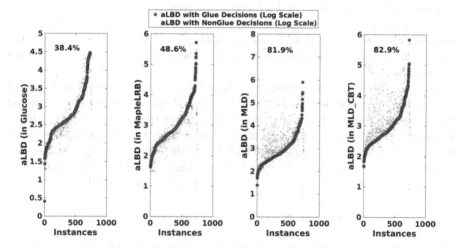

Fig. 2. Comparison aLBD scores (in Log Scale). Instances are sorted by the aLBD of glue decisions. The number at the top of each plot represents the percentage of instances, for which aLBD of glue decisions are lower than aLBD of nonglue decisions.

Figure 1 shows per instance LR values for both glue and nonglue decisions for the four solvers in four subplots. For all solvers and for large majority of the instances, glue decisions achieve higher LR than nonglue decisions.

Figure 2 shows per instance aLBD scores (in Log scale) for the 750 instances for glue and nonglue decisions.

- For Glucose and MapleLRB (first and second plots, Fig. 2), for more than half of the instances, the aLBD score of the learned clauses by nonglue decisions is lower than the aLBD score of the learned clauses by glue decisions. The average values of aLBD under columns E1 and E2 in Table 1 for Glucose and MapleLRB reflect the ground data.
- We observe quite a different scenario in case of MLD and MLD_CBT (third and fourth plot, Fig. 2). The aLBD scores of the learned clauses by glue decisions are lower for large majority of the instances. Again, the average values of aLBD under columns E1 and E2 in Table 1 for MLD and MLD_CBT reflect the ground data.

Overall, glue decisions are more conflict efficient than nonglue decisions for all the tested solvers. For average aLBD with glue decisions, the winners of the last two SAT competitions, MLD and MLD_CBT, generate substantially lower (better) values.

3.2   Selection Bias of Glue Variables

We are interested in the question: Do conflict guided CDCL branching heuristics exhibit any bias towards glue variables over nonglue variables?

Given a SAT formula $\mathcal{F}$ and a solver Ψ, we define *glue fraction* (GF) (resp. *nonglue fraction* (NF)) as the fraction of variables in $\mathcal{F}$ that are glue (resp. nonglue) variables, after Ψ ends its run with $\mathcal{F}$. GF (resp. NF) measures the pool size of glue (resp. nonglue) variables in $\mathcal{F}$ with respect to the total number of variables in $\mathcal{F}$.

Over the 750 instances, column B of Table 2 shows the average GF and average percentage of glue decisions and column C shows the average nonglue fraction and the average percentage of nonglue decisions. It shows that for all the four solvers, on average, the pool size of glue variables is significantly smaller than the pool size of nonglue variables (columns B1 and C1). For all the four solvers, on average, glue decisions relative to glue variables pool size are higher (column B2) than nonglue decisions (column C2) relative to the nonglue variables pool size.

Table 2. Biased selection of glue variables

(A) Systems	(B) Average for glue variable		(C) Average for nonglue variables	
	GF (B1)	Glue decisions % (B2)	NF (C1)	Nonglue decisions % (C2)
Glucose	0.25	65.43%	0.75	34.57%
MapleLRB	0.21	63.14%	0.69	36.86%
MLD	0.22	47.60%	0.78	52.60%
MLD_CBT	0.22	48.76%	0.78	51.24%

In summary, the four state-of-the-art CDCL SAT solvers make a much larger percentage of glue decisions against relatively smaller pools of glue variables. This shows the bias of these solvers towards selecting glue variables in branching decisions.

4 Activity Score Bumping for Glue Variables

From the above analysis, it is clear that decisions with glue variables are more conflict efficient than with nonglue variables. An interesting question is how we can exploit this empirical characteristic for more efficient SAT solving. Here, we present a score bumping method, called *Glue Bump* (GB), which bumps the activity score of glue variables. The amount of bumping for a glue variable depends on the appearance count of that variable in glue clauses and its current activity score.

Glue Level. Let G be the set of learned glue clauses until search state s. The *glue level* of a glue variable v, denoted $gl(v)$, is defined to be the number of glue clauses in G in which v appears.[6] A higher glue level indicates higher potential to create conflicts.

4.1 The GB Method

By using the current activity scores and (normalized) glue levels of glue variables (we will comment on normalization shortly), the GB method bumps the activity scores of glue variables. This gives higher preference to recently active glue variables with high glue levels. The GB method is simple to implement and conveniently integrates with activity based standard CDCL heuristics.

The GB method modifies a CDCL SAT solver Ψ by adding the following two procedures, which are called at different states of the search. We denote by Ψ^{gb} the GB extension of the baseline solver Ψ.

Alg. 1: Increase Glue Level	**Alg. 2: Bump Glue Variable**
Input: *A newly learned glue clause* θ	**Input:** *A glue variable* v
1 **For** $i \leftarrow 1$ *to* $|\theta|$	
2 $v \leftarrow varAt(\theta, i)$	1 $bf_v \leftarrow activity(v) * \left(\frac{gl(v)}{|G|}\right)$
3 $gl(v) \leftarrow gl(v) + 1$	2 $activity(v) \leftarrow activity(v) + bf_v$
4 **End**	

Increase Glue Level: Whenever Ψ^{gb} learns a new glue clause θ, it invokes Algorithm 1. For each variable v in θ, the glue level of v is increased by 1 (line 3).

Bump Glue Variable: Algorithm 2 bumps a glue variable v. It computes the *bumping factor* for v, denote bf_v, by combining both of the current activity score and normalized glue level of v (line 1). The bumping is performed by adding the bumping factor of v to the activity score of v, which becomes the new activity score for v (line 2).

Glue Level Normalization: The glue level of a glue variable can grow unboundedly with the discovery of more and more glue clauses. The activity score of a glue variable also grows, but at a different rate. Thus scaling the glue level is necessary.

---

[6] We omit the parameter s since the glue level of a variable is always computed w.r.t. a underlying search state by default, without confusion.

We normalize $gl(v)$ to $(0,1]^7$ by

$$\frac{gl(v)}{|G|}$$

where G is the set of glue clauses discovered by the search so far. The normalization scales the glue levels of glue variables by the *total number* of glue clauses discovered by the search so far.

Delayed Bumping of Glue Variables: Ψ^{gb} does not perform the bumping of v right after its hosting clause θ is discovered. It delays the bumping (i.e., the invocation of the *Bump Glue Variable* procedure) of v until it is unassigned by backtracking. This is a subtle point which we explain below.

- The glue clause θ is the latest learned clause and all the variables in θ including v are assigned at the current search state. At this stage, any score bumping that v receive would not be used until it gets unassigned.
- Let $T = d^e - d^s > 0$ be the decision window starting from the decision d^s that generates θ and ending at the decision d^e in which v gets unassigned. Within T, the search may generate more glue clauses in some of which v may appear. Furthermore, v may get involved in several conflicts during T and may have its activity score increased. It is clear that the bumping factor of v computed at d^e reflects a more recent measure than the one computed at d^s. By delaying the bumping of v until d^e when v has just got unassigned and become a candidate variable for branching, the GB method boosts the activity score of v by a more recent bumping factor.

5   Implementation and Experiments

5.1   Implementation

We implemented the GB method on top of the CDCL SAT solvers Glucose, MapleLRB, MLD, and MLD_CBT and call the extended solvers Glucose^{gb}, MapleLRB^{gb}, MLD^{gb}, and MLD_CBT^{gb}, respectively. The baseline solvers do not distinguish between glue and nonglue variables, except Glucose, which bumps activity scores of variables that are propagated from a glue clause.

In Glucose^{gb} and MapleLRB^{gb}, on the unassignment of a glue variable, the GB method updates the activity score of that glue variable by VSIDS and LRB, respectively, which are the heuristics used in their baselines. As remarked earlier, the baseline solvers MLD and MLD_CBT employ three heuristics, namely DIST, VSIDS and LRB, which are activated at different phases of the search. At any given phase, on the unassignment of a glue variable, MLD^{gb} and MLD_CBT^{gb} update the activity score of that glue variable for the currently active heuristic at that phase.

---

[7] At a given state of the search, a given glue variable v appears in at least one glue clause. So, the glue level of v (which is the count of number of glue clauses in which v appears), $gl(v) > 0$. After dividing $gl(v)$ with $|G|$, the normalized glue level remains larger than 0. Hence, the normalization normalizes the glue level within the range $(0,1]$.

5.2 Experiments

We conduct our experiments with four extended solvers with the same set of 750 instances on the same machine with 5000 s timeout per instance. Here, we present comparisons between the extended solvers and their counterpart baselines in terms of solved instances, solved time and PAR-2 score.

Table 3. Comparison of the four baseline solvers with their GB extensions for the instances from SAT-2017 and SAT-2018. The PAR-2 scores are scaled down by the factor of $\frac{1}{10,000}$.

Systems	SAT Comp-17				SAT Comp-18				Combined			
	SAT	UNSAT	Total	PAR-2	SAT	UNSAT	Total	PAR-2	SAT	UNSAT	Total	PAR-2
Glucose	83	96	179	1893	97	95	192	2274	180	191	371	4167
Glucose^{gb}	86 (+3)	96 (+0)	182 (+3)	1868	96 (-1)	97 (+2)	193 (+0)	2273	182 (+2)	193 (+2)	375 (+4)	4141
MapleLRB	80	95	175	1897	114	95	209	2069	194	190	384	3966
MapleLRB^{gb}	87 (+7)	97 (+2)	184 (+9)	1824	117 (+3)	96 (+1)	213 (+4)	2027	204 (+10)	193 (+3)	397 (+13)	3851
MLD	99	106	205	1635	136	101	237	1807	235	207	442	3442
MLD^{gb}	103 (+4)	107 (+1)	210 (+5)	1593	143 (+7)	102 (+1)	245 (+8)	1725	246 (+11)	209 (+2)	455 (+13)	3318
MLD_CBT	103	113	216	1565	135	102	237	1800	238	215	453	3365
MLD_CBT^{gb}	102 (-1)	114 (+1)	216 (+0)	1539	138 (+3)	101 (-1)	239 (+2)	1756	240 (+2)	215 (+0)	455 (+2)	3295

Solved Instances Comparison. Table 3 compares the four extended solvers with their baselines. Both MapleLRB^{gb} and MLD^{gb} solves 13 more instances (9 SAT, 4 UNSAT for the former and 11 SAT, 2 UNSAT for the latter). Glucose^{gb} solves 4 more instances (2 SAT, 2 UNSAT), and MLD_CBT^{gb} solves 2 additional instances (both SAT).

According to Audemard and Simon [2], solving 10 or more instances on a fixed set of instances from a competition by using a new technique, generally shows a critical feature. MapleLRB^{gb} solves 9 more instances over the instances from SAT-2017 and MLD^{gb} solves 8 additional instances over the instances from SAT-2018. The gains with MapleLRB^{gb} and MLD^{gb} are significant and closely resemble to the introduction of a critical feature.

Solve Time Comparison. Figure 3 compares the performance of Glucose^{gb} (blue line), MapleLRB^{gb} (red line), MLD^{gb} (yellow line) and MLD_CBT^{gb} (purple line) against their baselines. This figure plots the difference in the number of instances solved as a function of time. At most points in time, each of MapleLRB^{gb}, MLD^{gb}, and MLD_CBT^{gb} solves more problems. This is particularly pronounced for MLD^{gb} (yellow line) at earlier time points, for MLD_CBT^{gb} (purple line) on mid range time points. The improvement for MapleLRB^{gb} (red line) remains steady, with a brief downward slope in the middle. Glucose^{gb} performs slightly worse than Glucose at most of the times.

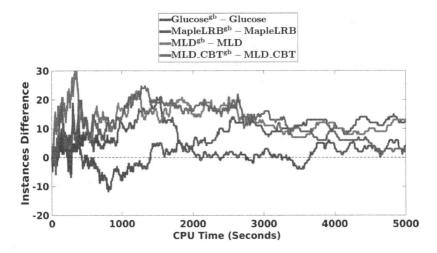

Fig. 3. Solve time comparisons. For any point above 0 in the vertical axis, our extensions solve more instances than their baselines at the time point in the horizontal axis. (Color figure online)

PAR-2 Score Comparison. In SAT competitions, solvers are ranked based on their PAR-2 scores. A PAR-2 score is computed as the sum of all runtimes for solved instances $+ 2*timeout$ for unsolved instances; solvers of lower PAR-2 scores are better.

Table 3 shows that all our extended versions achieve a lower PAR-2 score than the baselines for all the problem sets. Overall, the percentage of PAR-2 score reductions (computed from the last column of Table 3) with MLD^{gb}, $MapleLRB^{gb}$ and MLD_CBT^{gb} are 3.73%, 2.98% and 2.12%, respectively, which are considered significant with respect to SAT competition. For example, in SAT-2018 the winning solver has a PAR-2 score which is a reduction of only 0.81% over the runner-up.[8]

$Glucose^{gb}$ also lowers the PAR-2 score but only by 0.60%. The improvement is less impressive than with other three GB extensions. In Sect. 7, we will discuss the reason and show that this performance gap is not an indication of ineffectiveness of the GB method.

Finally in this section, as many benchmarks in SAT-2017/SAT-2018 are of industrial strength, we provide the information about the benchmark families, for which our GB method is particularly efficient. Table 4 lists those benchmark families for which our GB extended solvers solve at least 2 more instances than their baselines.

---

[8] http://sat2018.forsyte.tuwien.ac.at/index.php?cat=rules.

Table 4. Benchmark families for which the GB extended solvers solve at least two more instances than their baselines.

GB extensions	Benchmarks/SAT Comp	Solved by baseline	Solved by GB extensions	% Improvements
Glucose^{gb}	Integer Prefix/2017	28	32 (+4)	14.32%
	Soos/2018	8	11 (+3)	37.5%
	Ofer/2018	9	11 (+2)	18.18%
MapleLRB^{gb}	T/2017	28	31 (+3)	9.67%
	Integer Prefix/2017	27	30 (+3)	10.00%
	Klieber/2017	17	19 (+2)	10.52%
	Chen/2018	2	4 (+2)	50.00%
	Ofer/2018	5	7 (+2)	28.57%
	Scheel/2018	18	20 (+2)	10.00%
MLD^{gb}	ak128/2017	11	13 (+2)	15.38%
	Heule/2018	16	20 (+4)	20.00%
MLD_CBT^{gb}	Xiao/2018	7	9 (+2)	22.22%
	Collatz/2018	7	10 (+3)	30.30%

6 A New Measure of Solving Efficiency

In [16], the authors show that on average, better branching heuristics have higher GLR values and lower average LBD (aLBD) scores of the learned clauses. In Table 5, we compare our extended solvers and their baselines in terms of the average GLR values and average aLBD scores. All the solvers with GB extension generate conflicts at about the same rate as their corresponding baselines and achieve slightly smaller average aLBD scores. These results are largely consistent with [16].

Of course, one can pick up some subset of the benchmarks and show that the standard metrics that are based on average GLR and average aLBD may not be always applicable. On the other hand, for some subsets of benchmarks it may be highly expected that these metrics should be re-enforced. In this section, we select two subsets of this kind, but surprisingly the standard metrics do not provide a consistent explanation; they even lead to opposite conclusions. However, we show that a simple new measure, based on the fraction of learned clauses that are glue clauses, provides a consistent explanation of solving efficiency.

Table 5. Comparison of average GLR and aLBD score for GB extension solvers and baselines over the 750 test instances.

Systems	Glucose	Glucose^{gb}	MapleLRB	MapleLRB^{gb}	MLD	MLD^{gb}	MLD_CBT	MLD_CBT^{gb}
Avg. GLR	0.49	0.49	0.48	0.48	0.40	0.40	0.40	**0.41**
Avg. aLBD	20.09	**19.93**	24.88	**24.79**	27.73	**27.36**	27.59	**27.26**

6.1 Metrics for Solving Efficiency

We define a new performance metric called *Glue to Learned* (G2L). Then we present an analysis with three metrics, two standard ones and G2L on two different types of instances, where the baseline heuristics and their GB extensions show opposite strengths.

Glue to Learned (G2L). G2L represents the fraction of learned clauses that are glue clauses. More precisely, it is defined by $\frac{\#glue\_clauses}{\#learned\_clauses}$, where our solver Ψ has learned $\#learned\_clauses$ clauses for a given run on a given formula, among which $\#glue\_clauses$ are glue clauses.

Relating G2L to Solving Efficiency. The performance of branching heuristics correlates well with average GLR and the average aLBD scores at large scale. However, these two metrics fail to explain the performance of the baseline heuristics and their GB extensions for two specially designed subsets of instances from SAT-2017 and SAT-2018:

- $\mathbf{GB_{exclusive}}$: These instances are solved by Ψ^{gb}, but not by Ψ.
- $\mathbf{Baseline_{exclusive}}$: These instances are solved by Ψ, but not by Ψ^{gb}.

Table 6 compares the four baseline solvers and their GB extensions in terms of average GLR, average aLBD, and average G2L for $\mathbf{GB_{exclusive}}$ and $\mathbf{Baseline_{exclusive}}$ instances. For these two types of instances, it is expected that the solving efficiency will positively (resp. negatively) correlate with average GLR (resp. average aLBD).

We observe:

- Average GLR: For instances from $\mathbf{GB_{exclusive}}$ (Column C) and $\mathbf{Baseline_{exclusive}}$ (Column D), the better branching heuristics have lower average GLR values. This is surprising since the performance of branching heuristics is negatively correlated with average GLR values. This is highly inconsistent with the results reported in [16].
- Average aLBD: In both $\mathbf{GB_{exclusive}}$ and $\mathbf{Baseline_{exclusive}}$, the better heuristics have lower average aLBD in Glucose and MapleLRB based systems. This is consistent with the results from [16]. However, in MLD and MLD_CBT

Table 6. Comparison between baselines and their GB extensions for average GLR, average aLBD and average G2L for instance sets $\mathbf{GB_{exclusive}}$ and $\mathbf{Baseline_{exclusive}}$; Column B shows the heuristics employed for the systems in column A, where $\{x\}^{gb}$ in column B is the GB extension of baseline heuristic x. Column C (resp. Column D) shows three metrics: avg. GLR, avg. aLBD and avg. G2L for instance category $\mathbf{GB_{exclusive}}$ (resp. $\mathbf{Baseline_{exclusive}}$), where the sub-column #inst shows the number of $\mathbf{GB_{exclusive}}$ (resp. $\mathbf{Baseline_{exclusive}}$) instances for which we are comparing the heuristics in Column B.

(A) Systems	(B) Employed heuristics	(C) $\mathbf{GB_{exclusive}}$				(D) $\mathbf{Baseline_{exclusive}}$			
		#inst	Avg. GLR	Avg. aLBD	Avg. G2L	#inst	Avg. GLR	Avg. aLBD	Avg. G2L
Glucose	{VSIDS}	33	**0.56**	28.60	0.0005	29	0.59	**18.52**	**0.0015**
Glucose^{gb}	{VSIDS}^{gb}		0.53	**24.69**	**0.0016**		0.62	20.14	0.00078
MapleLRB	{LRB}	27	**0.50**	26.06	0.00073	14	0.47	**30.75**	**0.00046**
MapleLRB^{gb}	{LRB}^{gb}		0.46	**20.38**	**0.00126**		0.48	32.02	0.00037
MLD	{Dist/VSIDS/LRB}	28	**0.55**	**23.60**	0.00029	15	0.53	26.70	**0.0011**
MLD^{gb}	{Dist/VSIDS/LRB}^{gb}		0.51	26.04	**0.00032**		0.58	**23.21**	0.0009
MLD_CBT	{Dist,VSIDS,LRB}	26	**0.49**	**26.08**	0.0006	24	0.51	29.64	**0.00065**
MLD_CBT^{gb}	{Dist/VSIDS/LRB}^{gb}		0.43	36.24	**0.0011**		0.55	**25.42**	0.00037

based systems, the better branching heuristics have higher average aLBD scores, which is inconsistent with the results of [16].

- Average G2L: For both **GB$_{\text{exclusive}}$** and **Baseline$_{\text{exclusive}}$**, the better heuristics always achieve higher average G2L values. The biggest difference in G2L is 220% (0.0016–0.0005) for VSIDS and VSIDS^{gb} in Glucose and Glucose^{gb} for the **GB$_{\text{exclusive}}$**. We observe a significantly larger average G2L values for all the other cases as well (compare the bold values in avg. G2L subcolumn with the values not in bold, for both columns C and D in Table 6).

To summarize, for instances for which one heuristic is better than the other, the correlation between the performance of branching heuristics and average GLR and average aLBD is not always consistent with the results of [16]. The average value of the new metric G2L positively correlates with the performance of the branching heuristics in each case.

7   Effect of Glue Level Normalization

Earlier, we noticed that Glucose^{gb} showed less improvement than the other GB extensions. Compared to its baseline, Glucose^{gb} solves 4 additional instances, lowers the PAR-2 score only by 0.60% (Table 3), and solves instances at a slower rate than its baseline at most time points (Fig. 3).

Unlike the other 3 baseline solvers used in our experiments, the baseline solver Glucose already bumps variables that are propagated from glue clauses by using VSIDS [1]. These variables are a subset of what we call glue variables. Thus in Glucose^{gb}, these variables get bumped from two sources: from GB bumping and from VSIDS. We hypothesize that the relatively weak performance of Glucose^{gb} comes from this imbalance.

We tested this hypothesis by changing the glue level normalization method in GB to decrease the bumping factor in Algorithm 2. For a given glue variable v, instead of dividing $gl(v)$ by $|G|$, we divide by a bigger factor: $\frac{gl(v)}{\sum_{\theta \in G} len(\theta)}$, where $len(\theta)$ is the number of variables in the glue clause θ. The sum is the total number of the glue variables discovered so far in the search. If the average length of the glue clauses in G is n, then in this version, $gl(v)$ is scaled-down n times more than before.

We repeated our experiment with this version. Over the 750 instances from SAT-2017 and 2018, Glucose^{gb} now solves 11 more instances than Glucose and and lowers the PAR-2 score by 2.86%. For the other three GB extensions, this reduction does not work well.

8   Additional Experimental Results

8.1   Results with Benchmarks from SAT-2016

We performed an additional experiment with all of our GB extended solvers for the bumping factor $\frac{gl(v)}{|G|}$ for the benchmark instances from SAT-2016[9]. In the below, we summarize the results:

- Both Glucose and its GB extension solve equal number of problems (SAT 64, UNSAT 123, total 187).
- MapleLRB^{gb} solves (SAT 69, UNSAT 102, total 171) equal number of instances as its baseline *MapleLRB* (SAT 67, UNSAT 104, total 171).
- MLD^{gb} solves 2 more instances (SAT 73, UNSAT 135, total 208) than MLD (SAT 69, UNSAT 137, total 206).
- MLD_CBT^{gb} solves 2 less instances (SAT 66, UNSAT 137, total 203) than its baseline MLD_CBT (SAT 65, UNSAT 140, total 205).

For this benchmark set, the GB method does not work as well as it works for the benchmarks from SAT-2017 and 2018. Further tuning of the GB method is expected to improve the performance of the GB extended solvers on this benchmark set.

8.2   Experiment with Non-Delayed Bumping

We performed a smaller scale experiment with MLD over the 350 instances from SAT competition-2017, where we bump the score of the glue variables as soon as their hosting glue clause is learned (i.e., without delaying the bumping). MLD, with this version of glue variable bumping, solves 2 more UNSAT instances, but 2 less SAT instances than the baseline. As this non-delayed bumping did not appear to be promising with MLD, we did not perform any further experiment.

9   Related Work

As remarked earlier, Glucose [1] explicitly increases the activity scores of variables of the learned clause that were propagated by a glue clause. In their work, the bumping was based on VSIDS score bumping scheme. In contrast, we increase the activity scores of all variables that appear in glue clauses based on their normalized glue level.

In [12], the authors studied the behavior of Glucose with respect to *eigencentrality*, a precomputed static measure of ranking of the variables in industrial SAT instances. They show that the branched and propagated variables in Glucose have high eigencentrality and compared to the variables that appear in

---

[9] A total of 483 instances (283 applications, 200 crafted) after removing 17 duplicate instances between SAT-2016 and SAT-2017.

conflict clauses, the variables that appear in learned clauses are more eigencentral. In contrast, we dynamically characterize glue and nonglue variables within the course of a search and show that decisions with glue variables are more conflict efficient than decisions with nonglue variables.

The authors of [15] show that the VSIDS heuristic branches disproportionately more often on variables that are bridges between communities. Here, we have shown that CDCL heuristics branch disproportionately more often on glue variables with respect to their relatively smaller pool size.

In [9], the authors exploit the *betweeness centrality* measure of variables in industrial SAT formulas to design new heuristics. This measure is precomputed for a given instance. In contrast, we compute the normalized glue level of the variables dynamically during the search.

10 Summary and Future Work

In this work, we showed experimentally that decisions with variables appearing in glue clauses are more conflict efficient than decisions with other variables, and state-of-the-art CDCL SAT solvers tend to make glue decisions more often. Motivated by these observations, we developed a structure-aware CDCL variable bumping scheme, which increases the heuristic score of a glue variable based on the frequency of its appearance in glue clauses. Our empirical evaluation showed the effectiveness of the new method on the main track instances from SAT-2017 and SAT-2018 with four state-of-the-art CDCL SAT solvers. Lastly, we found that for some subsets of SAT-2017 and SAT-2018 benchmarks, our experimental data are surprisingly inconsistent with the standard performance metrics based on GLR and average LBD. We showed that for these subsets of benchmarks, the measure based on the fraction of learned clauses that are glue clauses provides a consistent explanation of our experimental data.

A number of questions deserve further considerations. The first is on the relationships between normalized glue level and other centrality measures, such as eigencentrality or betweenness centrality. The notion of glue level is central in our glue bumping scheme. Can we design clause deletion heuristics based on the notion of glue level? A similar question can be asked for the G2L metric: can we design more efficient branching heuristics based on this measure of solving efficiency?

Acknowledgements. We thank the anonymous reviewers for their valuable advice. This research is supported by Natural Sciences and Engineering Research Council of Canada (NSERC) PGS Doctoral award, President's Doctoral Prize of Distinction (PDPD), Alberta Innovates Graduate Student Scholarship (AIGSS), and NSERC discovery grant.

References

1. Audemard, G., Simon, L.: Predicting learnt clauses quality in modern SAT solvers. In: Proceedings of IJCAI 2009, pp. 399–404 (2009)

2. Audemard, G., Simon, L.: Refining restarts strategies for SAT and UNSAT. In: Milano, M. (ed.) CP 2012. LNCS, pp. 118–126. Springer, Heidelberg (2012). https://doi.org/10.1007/978-3-642-33558-7_11

3. Biere, A., Heule, M., Maaren, H.V., Walsh, T.: Handbook of Satisfiability: Volume 185 Frontiers in Artificial Intelligence and Applications. IOS Press, Amsterdam (2009)

4. Cadar, C., Ganesh, V., Pawlowski, P.M., Dill, D.L., Engler, D.R.: EXE: automatically generating inputs of death. In: Proceedings of CCS 2006, pp. 322–335 (2006)

5. Cook, S.A.: The complexity of theorem-proving procedures. In: Proceedings of the 3rd Annual ACM Symposium on Theory of Computing 1971, pp. 151–158 (1971)

6. Eén, N., Biere, A.: Effective preprocessing in SAT through variable and clause elimination. In: Bacchus, F., Walsh, T. (eds.) SAT 2005. LNCS, vol. 3569, pp. 61–75. Springer, Heidelberg (2005). https://doi.org/10.1007/11499107_5

7. Eén, N., Sörensson, N.: An extensible SAT-solver. In: Giunchiglia, E., Tacchella, A. (eds.) SAT 2003. LNCS, vol. 2919, pp. 502–518. Springer, Heidelberg (2004). https://doi.org/10.1007/978-3-540-24605-3_37

8. Gupta, A., Ganai, M.K., Wang, C.: SAT-based verification methods and applications in hardware verification. In: Bernardo, M., Cimatti, A. (eds.) SFM 2006. LNCS, vol. 3965, pp. 108–143. Springer, Heidelberg (2006). https://doi.org/10.1007/11757283_5

9. Jamali, S., Mitchell, D.: Centrality-based improvements to CDCL heuristics. In: Beyersdorff, O., Wintersteiger, C.M. (eds.) SAT 2018. LNCS, vol. 10929, pp. 122–131. Springer, Cham (2018). https://doi.org/10.1007/978-3-319-94144-8_8

10. Järvisalo, M., Biere, A., Heule, M.: Blocked clause elimination. In: Esparza, J., Majumdar, R. (eds.) TACAS 2010. LNCS, vol. 6015, pp. 129–144. Springer, Heidelberg (2010). https://doi.org/10.1007/978-3-642-12002-2_10

11. Järvisalo, M., Heule, M.J.H., Biere, A.: Inprocessing rules. In: Gramlich, B., Miller, D., Sattler, U. (eds.) IJCAR 2012. LNCS (LNAI), vol. 7364, pp. 355–370. Springer, Heidelberg (2012). https://doi.org/10.1007/978-3-642-31365-3_28

12. Katsirelos, G., Simon, L.: Eigenvector centrality in industrial SAT instances. In: Milano, M. (ed.) CP 2012. LNCS, pp. 348–356. Springer, Heidelberg (2012). https://doi.org/10.1007/978-3-642-33558-7_27

13. Liang, J.H., Ganesh, V., Poupart, P., Czarnecki, K.: Exponential recency weighted average branching heuristic for SAT solvers. In: Proceedings of AAAI 2016, pp. 3434–3440 (2016)

14. Liang, J.H., Ganesh, V., Poupart, P., Czarnecki, K.: Learning rate based branching heuristic for SAT solvers. In: Creignou, N., Le Berre, D. (eds.) SAT 2016. LNCS, vol. 9710, pp. 123–140. Springer, Cham (2016). https://doi.org/10.1007/978-3-319-40970-2_9

15. Liang, J.H., Ganesh, V., Zulkoski, E., Zaman, A., Czarnecki, K.: Understanding VSIDS branching heuristics in conflict-driven clause-learning SAT solvers. In: Proceedings of Haifa Verification Conference, HVC 2015, pp. 225–241 (2015)

16. Liang, J.H., Hari Govind, V.K., Poupart, P., Czarnecki, K., Ganesh, V.: An empirical study of branching heuristics through the lens of global learning rate. In: Gaspers, S., Walsh, T. (eds.) SAT 2017. LNCS, vol. 10491, pp. 119–135. Springer, Cham (2017). https://doi.org/10.1007/978-3-319-66263-3_8

17. Luo, M., Li, C.-M., Xiao, F., Manyà, F., Lü, Z.: An effective learnt clause minimization approach for CDCL SAT solvers. In: Proceedings of IJCAI 2017, pp. 703–711 (2017)

18. Massacci, F., Marraro, L.: Logical cryptanalysis as a SAT problem. J. Autom. Reasoning **24**(1/2), 165–203 (2000)

19. Moskewicz, M.W., Madigan, C.F., Zhao, Y., Zhang, L., Malik, S.: Chaff: engineering an efficient SAT solver. In: Proceedings of Design Automation Conference, DAC 2001, pp. 530–535 (2001)
20. Oh, C.: Between SAT and UNSAT: the fundamental difference in CDCL SAT. In: Heule, M., Weaver, S. (eds.) SAT 2015. LNCS, vol. 9340, pp. 307–323. Springer, Cham (2015). https://doi.org/10.1007/978-3-319-24318-4_23
21. Rintanen, J.: Engineering efficient planners with SAT. In: Proceedings of ECAI 2012, pp. 684–689 (2012)
22. Marques Silva, J.P., Sakallah, K.A.: GRASP: a search algorithm for propositional satisfiability. IEEE Trans. Comput. **48**(5), 506–521 (1999)
23. Soos, M., Nohl, K., Castelluccia, C.: Extending SAT solvers to cryptographic problems. In: Kullmann, O. (ed.) SAT 2009. LNCS, vol. 5584, pp. 244–257. Springer, Heidelberg (2009). https://doi.org/10.1007/978-3-642-02777-2_24
24. Xiao, F., Luo, M., Li, C.-M., Manya, F., Lu, Z.: MapleLRB_LCM, Maple_LCM, Maple_LCM_Dist, MapleLRB_LCMoccRestart and Glucose3.0+width in sat competition 2017. In: Proceedings of SAT Competition 2017, pp. 22–23 (2017)

Industrial Size Job Shop Scheduling Tackled by Present Day CP Solvers

Giacomo Da Col and Erich C. Teppan[(✉)]

Alpen-Adria Universität Klagenfurt, 9020 Klagenfurt, Austria
{giacomo.da,erich.teppan}@aau.at

Abstract. The job shop scheduling problem (JSSP) is an abstraction of industrial scheduling and has been studied since the dawn of the computer era. Its combinatorial nature makes it easily expressible as a constraint satisfaction problem. Nevertheless, in the last decade, there has been a hiatus in the research on this topic from the constraint community; even when this problem is addressed, the target instances are from benchmarks that are more than 20 years old. And yet, constraint solvers have continued to evolve and the standards of today's industry have drastically changed. Our aim is to close this research gap by testing the capabilities of the best available CP solvers on the JSSP. We target not only the classic benchmarks from the literature but also a new benchmark of large-scale instances reflecting nowadays industrial scenarios. Furthermore, we analyze different encodings of the JSSP to measure the impact of high-level structures (such as interval variables and no-overlap constraints) on the problem solution. The solvers considered are OR-Tools, Google's open-source solver and winner of the last MiniZinc Challenge, and IBM's CP Optimizer, a proprietary solver targeted towards industrial scheduling problems.

Keywords: Constraint programming · Job shop scheduling · JSSP · OR-Tools · CP Optimizer · Large-scale benchmark

1 Introduction

The job shop scheduling problem (JSSP) is among the first combinatorial problems ever studied [11]. Due to its relevant application to, both, computer and manufacturing systems, it is one of the most studied and analyzed.

The problem is presented as a set of jobs that must be processed by a set of machines. In the classical formulation, every job has to go through each machine exactly once. The processing of a job by one machine is called operation and the processing time is called duration. Every job has a specific ordering of operations that must be respected. An admissible solution for an instance of this problem is a sequence of operations on every machine where there is no time overlap between two operations in the same machine and the ordering of the operations is respected. The most typical optimization criterium is the minimization of the

© Springer Nature Switzerland AG 2019
T. Schiex and S. de Givry (Eds.): CP 2019, LNCS 11802, pp. 144–160, 2019.
https://doi.org/10.1007/978-3-030-30048-7_9

makespan, *i.e.* the time interval between the start of the first operation and the end of the last.

The structure of the problem makes it easily representable as a constraint satisfaction problem. In fact, there have been many successful applications of constraint-based approaches to this problem over the past years, *e.g.* [2,5,14].

Among the various CP techniques, a particularly successful one is Large Neighborhood Search (LNS) [7]. This method consists of an iterative process of relaxation and re-optimization of the problem, progressively selecting the most promising partial schedules to improve the final solution. This idea was also applied to MIP approaches (in the form of Relaxation Induced Neighborhood Search [4]). Hybrid CP-MIP methods have been considered as well, in order to take the best of the two worlds [13], and in some cases it has been shown that hybrid approaches perform better than MIP alone [6].

Despite these advancements in constraint solving, the last decade has experienced a decrease of research interest of CP applied to the classic JSSP. The major contributor in this period was IBM, which stole the scene with their proprietary CP Solver, CP Optimizer. This solver has been capable of finding better solutions for many JSSP instances from the classic benchmarks [19]. They also use a hybrid CP-MIP approach in case of non-regular objective functions, as in scheduling problems with earliness costs [8].

This evolution of solving capabilities, however, does not correspond to an evolution of benchmarks instances, which are almost the same as twenty years ago. In the meantime, industrial standards have changed to the point that scheduling problems from modern manufacturing systems can easily require up to 2000 jobs to be scheduled on 100 machines [3,17]. In comparison, the biggest instance of the Taillard benchmark [16], which reflected real dimensions of industrial problems in 1993 and it is still among the largest available, has 50 jobs on 20 machines. To close this gap, we created a new benchmark of JSSP instances, based on Taillard's specification, but in line with today's industrial scenarios.

The aim of this paper is twofold. First, we want to close the gap on the JSSP research, testing the capabilities of the best available CP solvers on, both, classic benchmarks and our large-scale one. Second, we want to investigate how different design choices affect the search process. In fact, while CP allows a compact representation, it still offers multiple ways to encode a problem. In particular, we focus on the application of high-level structures like interval variables and no-overlap global constraints.

As anticipated, one of the most successful CP solvers on scheduling problems is CP Optimizer (abbreviated CPO). To find a worthy opponent, we took the winner of the last years MiniZinc challenge[1]. The MiniZinc challenge is a recurring competition where all the constraint solvers that support the MiniZinc modeling language [12] compete on various combinatorial problems, including scheduling. OR-Tools[2] (ORT), an open-source solver developed by Google, won the gold medal in all categories in 2018. While preliminary studies done on

---

[1] https://www.minizinc.org/challenge2018/challenge.html.
[2] https://developers.google.com/optimization/.

global constraints suggest the advantage of CP Optimizer over OR-Tools [9], a direct comparison of these two solvers on the JSSP has never been conducted, especially on industrial-size instances.

2   Experimental Setup

We implement three encodings for the JSSP to measure the effectiveness of high-level constraint structures:

- A Naive encoding, which uses primitive CP constraints and integer variables;
- A SemiNaive encoding, which takes advantage of the interval variables, particularly well suited to represent operations in JSSP;
- An Advanced encoding, which combines the interval variables with global constraints designed for scheduling problems, such as the no-overlap constraint.

The comparison is carried out measuring the quality of the final solution (makespan) and the time needed to reach that solution. Each solver is tested with all three encodings on all benchmark instances. The search procedure runs on a single core.

On the classic instances, the time allowed for each instance is 20 min, instantiation time included; this time limit complies with the MiniZinc challenge rules.

On the large-scale instances, the time allowed is 6 h, instantiation time included. The time limit is extended because the size of large-scale instances demands more time to have a thorough exploration of the search space. Nevertheless, given that these instances were generated with industrial scenarios in mind, we selected a time limit which would allow the calculation to be completed overnight, to minimize gaps in the production flow. Concerning the solvers' version, we use version 12.8.0 for CP Optimizer and version 6.10.6025 for OR-Tools. Since CPO does not support MiniZinc as modeling language, but both solvers offer Java APIs, we decided to use Java to interface with the solvers, to avoid the bias that different modeling languages might introduce. We conducted a short preliminary experiment to test the various solver configurations. Concerning CPO, we found the default configuration to be the most performant. In ORT, we tested both the classic CP Solver and the new CP-SAT Solver, finding the latter to be better than the former in most cases.

The experiment is conducted on a system equipped with a 2 GHz AMD EPYC 7551P 32 Cores CPU and 128 GB of RAM. Each run was performed on a single core with a maximum cap of 10 GB of RAM.

2.1   Problem Instances

Our tests on the various models are conducted on the classic benchmark and the large-scale benchmark. All the instances of both benchmarks are rectangular JSSP instances. This means that every job has to go through all the machines, therefore every job will have a number of operations equal to the total number of machines and every machine will have assigned a number of operations equal to the total number of jobs. The classic benchmark consists of a selection of problem instances and comprises the most used JSSP benchmarks in the literature. In particular, we used the same problem instances selected for the MiniZinc benchmark[3]:

- **FT:** This is one of the oldest benchmarks for JSSP [11]. It includes 3 problem instances of sizes 6×6, 10×10 and 20×5. The square instance 10×10 is famous for remaining unsolved for more than 20 years.
- **LA:** This benchmark contains 40 problem instances from 10×5 to 30×10 [10].
- **ABZ:** 5 problem instances from the work about shifting bottleneck by [1].
- **ORB:** 10 problem instances proposed by [2].
- **YN:** 1 randomly generated problem instance of size 20×20 [20].
- **SWV:** A set of 14 problem instances from [15].
- **VW:** 1 instance from [18].

The large-scale benchmark consists of 90 instances that we generated following the Taillard benchmark specification [16]. We generated our own benchmark instead of using Taillard's because our aim is to verify the performance of the two solvers on problem instances with a size comparable with modern industrial problems[4]. The benchmark is structured with instances of increasing sizes, as follows:

- 10 instances 10×10 (*i.e.* 10 jobs to be scheduled on 10 machines, for a total of 100 operations)
- 10 instances 10×100 (*i.e.* 10 jobs to be scheduled on 100 machines, for a total of 1000 operations)
- 10 instances 10×1000 (10000 operations)
- 10 instances 100×10 (1000 operations)
- 10 instances 100×100 (10000 operations)
- 10 instances 100×1000 (100000 operations)
- 10 instances 1000×10 (10000 operations)
- 10 instances 1000×100 (100000 operations)
- 10 instances 1000×1000 (1000000 operations)

2.2 Encodings

One of the main advantages of adopting a constraint programming approach is that it allows a compact and formal definition of the problem, which is easily maintainable and adaptable to sudden changes in the problem configuration. Nevertheless, various decisions can be made during the modeling process to affect the search phase. In particular, we are interested in the impact of high-level structures and constraints on the solving capabilities of solvers, *i.e.* :

---

[3] https://github.com/MiniZinc/minizinc-benchmarks/tree/master/jobshop.
[4] complete encodings and benchmarks are available at https://goo.gl/qarP3m.

– Interval variable: a special type of variable well suited to represent job operations in scheduling. This variable incorporates a start time, an end time and a duration, and automatically enforces a duration constraint, such that $start + duration = end$;
– No overlap constraint: given a sequence of interval variables $v_1 \ldots v_n$, if v_i starts before v_j, then v_j cannot start before the end of v_i (for every variable index $i, j \in \{1 \ldots n\} \mid i \neq j$). This constraint is designed to work with interval variables, and cannot be used without them.

INPUT
opDurations : IntegerArray[1..numJobs][1..numMachines]
opSuccessors : IntegerArray[1..numJobs][1..numMachines]

VARIABLES
opStarts : IntegerVariableArray[1..numJobs][1..numMachines]
opEnds   : IntegerVariableArray[1..numJobs][1..numMachines]

CONSTRAINTS
opStarts[j][m] + opDurations[j][m] = opEnds[j][m],
$\forall j \in \{1, \ldots, numJobs\}, \forall m \in \{1, \ldots, numMachines\}$

opEnds[j][m] $\leq$ opStarts[j][opSuccessors[j][m]],
$\forall j \in \{1, \ldots, numJobs\}, \forall m \in \{1, \ldots, numMachines\}$
with $opSuccessors[j][m] \neq NULL$

opEnds[j][m] $\leq$ opStarts[k][m] $\vee$ opEnds[k][m] $\leq$ opStarts[j][m],
$\forall j \in \{1, \ldots, numJobs\}, \forall k \in \{1, \ldots, numJobs\}, \forall m \in \{1, \ldots, numMachines\}$
with $j \neq k$

OBJECTIVE
minimize max($\{end \mid end \in opEnds\}$)

Encoding 1. Naive encoding for the job shop scheduling problem

Based on these constructs, we created three encodings for the JSSP:

The **Naive** encoding does not take advantage of any of the specialized structures, and relies on integer variables and primitive constraints to model the JSSP. Encoding 1 shows the model of the naive encoding. Keeping in mind that all the adopted instances are rectangular, we use a matrix structure with a number of rows equal to the number of jobs and a number of columns equal to the number of machines of an instance. In the input data there are two such matrices, one to store the durations and one for the succession of the operations of each job on the machines. In the model, *opStarts* and *opEnds* are matrices of integer variables that store respectively the start and end variables of each operation. The first constraint imposes that for each operation of each job, *start + duration = end*. The second constraint enforces the precedence relation between operations within a job (as long as a certain operation has a successor). The third constraint ensures that on each machine, for every couple of operations o_j and o_k (with j different from k), either o_j comes before than o_k or viceversa, without any overlap. The objective function aims to minimize the largest of the *opEnds* (*i.e.* the makespan).

INPUT
opDurations : IntegerArray[1..numJobs][1..numMachines]
opSuccessors : IntegerArray[1..numJobs][1..numMachines]

VARIABLES
ops : IntervalVariableArray[1..numJobs][1..numMachines]
with $ops[j][m].duration = opDurations[j][m]$
$\forall j \in \{1, \ldots, numJobs\}, \forall m \in \{1, \ldots, numMachines\}$

CONSTRAINTS
ops[j][m].end $\leq$ ops[j][opSuccessors[j][m]].start
$\forall j \in \{1, \ldots, numJobs\}, \forall m \in \{1, \ldots, numMachines\}$
with $opSuccessors[j][m] \neq NULL$

ops[j][m].end $\leq$ ops[k][m].start $\lor$ ops[k][m].end $\leq$ ops[j][m].start,
$\forall j \in \{1, \ldots, numJobs\}, \forall k \in \{1, \ldots, numJobs\}, \forall m \in \{1, \ldots, numMachines\}$
with $j \neq k$

OBJECTIVE
minimize max($\{op.end | op \in ops\}$)

Encoding 2. SemiNaive encoding for the job shop scheduling problem

The **SemiNaive** encoding makes use of interval variables to encode job operations. The model described in Encoding 2 is more compact compared to the Naive encoding. In fact, the interval variables already contain the information about start and end of each operation, as well as the duration constraints. Therefore, the only constraints that have to be explicitly expressed are the precedence constraints and the no-overlap constraints, done by manually instantiating a disjunctive constraint for every couple of operations in the machine, as in the Naive encoding.

The **Advanced** encoding (Encoding 3) also exploits interval variables, like the SemiNaive encoding, but instead of a quadratic number of primitive constraints per machine, one no overlap global constraint is used. Given that this procedure condenses two iterations over the number of jobs, a wise implementation of such procedure can lead to a great speed-up of both the instantiation and the solving process.

INPUT
opDurations : IntegerArray[1..numJobs][1..numMachines]
opSuccessors : IntegerArray[1..numJobs][1..numMachines]

VARIABLES
ops : IntervalVariableArray[1..numJobs][1..numMachines]
with $ops[j][m].duration = opDurations[j][m]$
$\forall j \in \{1,\ldots,numJobs\}, \forall m \in \{1,\ldots,numMachines\}$

CONSTRAINTS
ops[j][m].end $\leq$ ops[j][opSuccessors[j][m]].start
$\forall j \in \{1,\ldots,numJobs\}, \forall m \in \{1,\ldots,numMachines\}$
with $opSuccessors[j][m] \neq NULL$

noOverlap($\{op|op \in ops[1..numJobs][m]\}$),
$\forall m \in \{1,\ldots,numMachines\}$

OBJECTIVE
minimize max($\{op.end|op \in ops\}$)

Encoding 3. Advanced encoding for the job shop scheduling problem

3   Results

This section illustrates the results of the experiment carried out on the classic benchmark and on the large-scale benchmark. The tests are conducted on pairs of solver-encoding configurations, *e.g.* OR-Tools solver using Naive encoding, CP Optimizer solver using Advanced encoding, and so on. For the sake of synthesis, from now on we will refer to any solver-encoding configuration simply as a system.

3.1   Results of Classic Benchmark

The results of the classic benchmark are summarized in Table 1. In the makespan columns, the listed values correspond to the best solution achieved after 20 min of computation (or earlier, if the optimal solution is detected). The best solutions found are highlighted in bold. The last column indicates whether the best solution found is optimal or not. In the solving time columns, the time is measured in seconds. The best values are highlighted in bold.

Table 1. Results of the experiment on the classic benchmarks.

Problem Instances	Makespan						Solving Time						optimal
	CP Optimizer			OR-Tools			CP Optimizer			OR-Tools			
	naive	seminaive	advanced	naive	seminaive	advanced	naive	seminaive	advanced	naive	seminaive	advanced	optimal
abz5	1238	1234	1234	1234	1234	1234	1200	6.32	2.19	2.03	2.01	1.75	Yes
abz6	943	943	943	943	943	943	1200	1.25	0.68	0.43	1.19	0.7	Yes
abz7	685	700	656	689	671	660	1200	1200	1170.24	1200	1200	1200	Yes
abz8	700	705	682	706	685	679	1200	1200	1200	1200	1200	1200	No
abz9	728	739	685	703	687	695	1200	1200	1200	1200	1200	1200	No
ft06	55	55	55	55	55	55	0.04	0.04	0	0.03	0.02	0.01	Yes
ft10	930	930	930	930	930	930	1200	22.81	3.76	5.42	4.4	4.85	Yes
ft20	1196	1197	1165	1206	1174	1165	1200	1200	1.36	1200	1200	4.88	Yes
la01	666	666	666	666	666	666	1200	0.66	0	0.25	0.25	0.08	Yes
la02	655	655	655	655	655	655	1200	2.25	0.3	0.51	0.46	0.04	Yes
la03	597	597	597	597	597	597	816.38	2.14	0.07	0.53	0.52	0.06	Yes
la04	590	590	590	590	590	590	1200	1.27	0.34	0.19	0.35	0.18	Yes
la05	593	593	593	593	593	593	1200	1.5	0	0.76	0.64	0.02	Yes
la06	926	926	926	926	926	926	1200	1200	0	1200	1200	1.04	Yes
la07	890	890	890	890	890	890	1200	1200	0.02	1200	1200	0.11	Yes
la08	863	863	863	863	863	863	1200	1200	0.02	1200	1200	0.26	Yes
la09	951	951	951	951	951	951	1200	1200	0	1200	1200	0.48	Yes
la10	958	958	958	958	958	958	1200	1200	0	1200	1200	0.87	Yes
la11	1222	1222	1222	1222	1222	1222	1200	1200	0.01	1200	1200	0.7	Yes
la12	1039	1039	1039	1039	1039	1039	1200	1200	0.14	1200	1200	0.64	Yes
la13	1150	1150	1150	1150	1150	1150	1200	1200	0.02	1200	1200	3.02	Yes
la14	1292	1292	1292	1292	1292	1292	1200	1200	0.01	1200	1200	1.92	Yes
la15	1207	1207	1207	1207	1207	1207	1200	1200	0.14	1200	1200	5.61	Yes
la16	945	945	945	945	945	945	1200	2.38	1.45	0.73	0.95	0.6	Yes
la17	784	784	784	784	784	784	1200	1.71	1.12	0.54	1.7	0.32	Yes
la18	848	848	848	848	848	848	93.2	1.67	0.9	0.55	1.85	0.88	Yes
la19	842	842	842	842	842	842	1200	3.83	2.93	0.93	1.44	1.65	Yes
la20	902	902	902	902	902	902	1200	2.03	1.6	0.55	2.49	0.7	Yes
la21	1059	1051	1046	1048	1046	1046	1200	1200	22.56	1200	1200	83.03	Yes
la22	932	927	927	927	927	927	1200	173.47	5.28	103.93	80.29	6.62	Yes
la23	1032	1032	1032	1032	1032	1032	1200	303.5	0.12	118.82	257.44	2.8	Yes
la24	969	935	935	935	935	935	1200	314.42	15.42	113.99	47.07	24.5	Yes
la25	982	977	977	977	977	977	1200	218.82	14.63	60.28	47.45	18.82	Yes
la26	1218	1246	1218	1237	1218	1218	1200	1200	7.35	1200	1200	78.47	Yes
la27	1293	1275	1235	1289	1266	1235	1200	1200	129.38	1200	1200	509.21	Yes
la28	1297	1248	1216	1250	1224	1216	1200	1200	17.53	1200	1200	13.97	Yes
la29	1198	1226	1152	1236	1191	1153	1200	1200	1200	1200	1200	1200	No
la30	1426	1355	1355	1355	1355	1355	1200	1200	0.28	1200	1200	20.79	Yes
la31	1791	1784	1784	1841	1784	1784	1200	1200	0.44	1200	1200	23.74	Yes
la32	1853	1850	1850	1882	1850	1850	1200	1200	0.04	1200	1200	29.09	Yes
la33	1747	1719	1719	1746	1719	1719	1200	1200	0.25	1200	1200	14.16	Yes
la34	1793	1721	1721	1781	1746	1721	1200	1200	1.57	1200	1200	69.39	Yes
la35	1888	1898	1888	1922	1888	1888	1200	1200	0.24	1200	1200	25.14	Yes
la36	1281	1268	1268	1268	1268	1268	1200	108.6	10.37	32.34	40.16	10.79	Yes
la37	1399	1397	1397	1397	1397	1397	1200	330.43	4.03	179.52	158.46	8.5	Yes
la38	1202	1196	1196	1196	1196	1196	1200	856.21	84	462.06	135.24	260.72	Yes
la39	1248	1233	1233	1233	1233	1233	1200	110	5.93	51.89	40.1	13.81	Yes
la40	1240	1222	1222	1222	1222	1222	1200	707.27	9.87	396.79	187.83	52.46	Yes
orb01	1079	1059	1059	1059	1059	1059	1200	167.49	7.09	110.33	38.87	22.69	Yes
orb02	888	888	888	888	888	888	1200	3.6	2.23	0.92	1.56	1.9	Yes
orb03	1005	1005	1005	1005	1005	1005	1200	48.16	6.54	56.56	23.45	20.6	Yes
orb04	1011	1005	1005	1005	1005	1005	1200	5.03	2.73	2.99	1.96	2.86	Yes
orb05	887	887	887	887	887	887	1200	7.98	3.62	1.81	2.71	2.4	Yes
orb06	1023	1010	1010	1010	1010	1010	1200	41.35	4.64	15.61	10.04	8.59	Yes
orb07	397	397	397	397	397	397	1200	4.77	1.57	1.53	1.25	1.33	Yes
orb08	899	899	899	899	899	899	1200	9.78	1.09	2.45	2.29	1.36	Yes
orb09	934	934	934	934	934	934	1200	8.48	1.19	2.36	3.92	1.38	Yes
orb10	944	944	944	944	944	944	1200	4.68	0.72	1.57	2.6	1.8	Yes
swv01	1517	1544	1445	1541	1483	1412	1200	1200	1200	1200	1200	1200	No
swv02	1620	1590	1491	1571	1502	1475	1200	1200	1200	1200	1200	901.91	Yes
swv03	1491	1582	1420	1547	1493	1410	1200	1200	1200	1200	1200	1200	No
swv04	1568	1667	1520	1597	1553	1482	1200	1200	1200	1200	1200	1200	No
swv05	1537	1618	1424	1559	1518	1436	1200	1200	1134.75	1200	1200	1200	Yes
swv06	1947	1886	1728	1841	1784	1746	1200	1200	1200	1200	1200	1200	No
swv07	1699	1816	1672	1759	1711	1677	1200	1200	1200	1200	1200	1200	No
swv08	2072	2063	1785	1920	1881	1855	1200	1200	1200	1200	1200	1200	No
swv09	1942	1841	1713	1825	1779	1715	1200	1200	1200	1200	1200	1200	No
swv10	1953	1974	1823	1920	1872	1807	1200	1200	1200	1200	1200	1200	No
swv11	3472	3468	3041	3815	3696	3317	1200	1200	1200	1200	1200	1200	No
swv12	3447	3282	3114	3753	3794	3358	1200	1200	1200	1200	1200	1200	No
swv13	3616	3477	3205	3934	3938	3421	1200	1200	1200	1200	1200	1200	No
swv14	3474	3293	3032	3832	3807	3162	1200	1200	1200	1200	1200	1200	No
vw3x3	256	256	256	256	256	256	0	0	0	0	0	0	Yes
yn4	1048	1054	980	1030	1005	994	1200	1200	1200	1200	1200	1200	No

While perfect for a complete view on the classic benchmark experiment, Table 1 falls short when it comes to direct comparisons of systems. Figure 1 offers a more eye-friendly summary of the experiment. Every cell of the map represents the *difference* between number of instances where the row system beats the column system *and* number of instances where the row system is beaten by the column one. In this case, **to beat** means to achieve a makespan **strictly** lower or, in case of a draw, to achieve the same makespan faster. in case of a draw in both makespan and solving time (*e.g.* instance **vw3 × 3**), the corresponding instance is not counted. The scale goes from −74 (the corresponding row system is beaten in all the instances by the column one) to 74 (the system beats the corresponding column one in every instance). Hence, the map has to be read row-wise, and the "greener" the row, the better the corresponding system.

It is noticeable that the greenest of the rows corresponds to CPO Advanced, which is better than the other CPO systems in almost all the instances, and scores a positive 35 against ORT Advanced, winning on 53 instances and losing on 18. ORT Advanced also achieves good performance against the Naive and SemiNaive encodings, going "in the red" only against CPO Advanced. The worst performer is unarguably CPO Naive, which is defeat on almost all the instances by the Advanced encodings and on more than 50% of the instances by the other systems. In general, CPO performs better than ORT with the Advanced encoding, but poorly with the other encodings. If we consider the percentage of instances solved optimally, the difference between the two solvers is even narrower on the Advanced encodings, with CPO solving optimally 77.03% of the instances and ORT following at 75.68%. In both solvers, there is an improvement from Naive to SemiNaive encoding (better results on about 60% of the instances), but the improvement is even larger from SemiNaive to Advanced (better on 89% of instances for ORT and 99% for CPO).

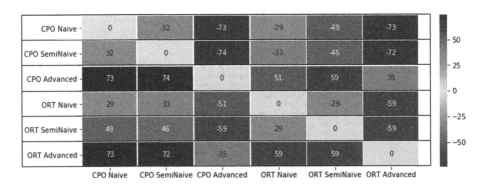

Fig. 1. Map on the pairwise confrontation between systems on the classic benchmarks. To be read row-wise, the greener the better. (Color figure online)

3.2   Results of the Large-Scale Benchmark

The results of the large-scale benchmark are summarized in Table 2. The results are grouped by instance size. Given that there are 10 instances per group, every value represents the mean of the final makespans of the corresponding group. In the first two groups, every system is able to solve all the instances optimally. From the 10×1000 group on, we do not refer anymore to optimal solutions, because it is not possible for any of the systems to solve all the instances optimally. On 10×1000, all systems converge to the same solution besides ORT SemiNaive, which reaches almost double the makespan compared to the others.

OR-Tools provides worst solutions on the 100×10 and 100×100 instances; in fact, in CP Optimizer the Advanced and the SemiNaive encodings achieve the best results among the other systems, with even the Naive encoding achieving better results than the OR-Tools ones. In the remaining instances, the role of the global constraint becomes crucial. In fact, it is possible to solve those instances only with the Advanced encodings, while with Naive and SemiNaive encodings it is not manageable to even instantiate the problem. The high number of operations, and in particular the number of jobs, makes the number of disjunctive constraints explode. Concerning the biggest instances 1000×1000, only CP Optimizer is able to find a solution in each of the 10 instances. OR-Tools is able to instantiate the problem instances but does not manage to find any solutions within the timeout of 6 h.

Table 2. Average makespan over the 9 instance groups of the large-scale benchmark.

Problem instances	CP Optimizer			OR-Tools		
	Naive	SemiNaive	Advanced	Naive	SemiNaive	Advanced
10×10	8169.9	8169.9	8169.9	8169.9	8169.9	8169.9
10×100	55224.5	55224.5	55224.5	55224.5	55224.5	55224.5
10×1000	514393.3	514393.3	514393.3	514393.3	1139284.3	514393.3
100×10	55084.9	54858.6	54858.6	59623.6	82619.6	55493.5
100×100	93827.1	89311.4	80570.5	141524.5	3327955.9	117332.8
100×1000	Timeout	Timeout	545687.7	Timeout	Timeout	604119.2
1000×10	Timeout	Timeout	515429.7	Timeout	Timeout	550534.4
1000×100	Timeout	Timeout	536403.6	Timeout	Timeout	686811.1
1000×1000	Timeout	Timeout	1017974.1	Timeout	Timeout	Timeout

We analyzed also the evolution of intermediate solutions during the search, in order to have a better understanding of the solving process. Figures 2, 3 and 4 offer a detailed view of the search process: every plot shows the results of one instance class, showing time (in seconds) on the x-axis and makespan on the y-axis. In each plot, every line is linked with a system and, conforming to Table 2, represents the mean makespan of the intermediate solutions at each time x. Basically, Table 2 shows the situation when $x = timeout$, while the plots show the evolution of the search as x varies from 0 to $timeout$.

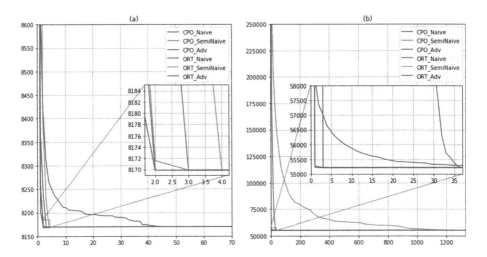

Fig. 2. (a) Plot of instance group 10×10 (b) Plot of instance group 10×100.

Figure 2 shows the results regarding 10×10 (a) and 10×100 (b), the only instance groups where it was possible to achieve optimal makespan in all the instances. In (a) the time interval is limited between 0 and 70 s, since all the systems converge hastily to the optimal solution in all cases. The slowest is CPO Naive, achieving the optimal makespan on all instances after 70 s. Concerning the other systems, we can see that both CPO Advanced and ORT Naive converge to the optimum after 2 s, ORT Advanced after 3 and ORT SemiNaive after 4. In (b) the slowest system is ORT SemiNaive, reaching optimal results after 1330 s. The other ORT encodings follow at 77 s for the Advanced and 37 for the Naive. All CPO encodings reached the optimum within 3 s.

The fact that in OR-Tools the Naive encoding is faster than both SemiNaive and Advanced is peculiar. Concerning the Advanced encoding, this is due to one particular instance ($10 \times 100\_4$) which needs 77 s to be solved by ORT Advanced and 32 by ORT Naive. However, the average solving time of the 10×100 group (without instance 4) is 15 s for ORT Advanced and 25 for ORT Naive. Thus, we can see this case as an outlier. Concerning ORT SemiNaive, we see a general performance decline in all the instances of this benchmark. The reason for this behavior is more complex and will be treated in detail in Sect. 4.

Figure 3 illustrates the last instance group with 10 jobs (10×1000) and all groups with 100 jobs. In Fig. 3(a) we omitted the legend for space reasons, but we maintained the line colors consistent across plots, therefore the legend from any other plot can be used as a reference. The worst performer is ORT SemiNaive, which is visibly far from all other systems. In fact, the improvement of the solutions is very slow compared to the others and, albeit continuing until the timeout

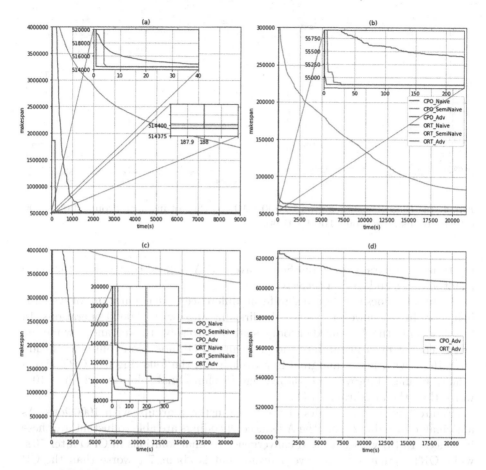

Fig. 3. (a) Plot of instance group **10 × 1000** (b) Plot of instance group **100 × 10** (c) Plot of instance group **100 × 100** (d) Plot of instance group **100 × 1000**. (Color figure online)

occurs, it stops being effective after around 8000 s (about 2 h and 20 min). The other systems are much faster and they all converge to the same result: ORT Naive in 1484 s, CPO Naive in 188 s, ORT Advanced in 202 (although already at 40 s is just 0.08% off CPO's result, and at 188 is 0.002% off) and finally CPO SemiNaive in 10 s and CPO Advanced in 4.

In Fig. 3(b) the trend continues, with ORT SemiNaive being the worst performer, followed by ORT Naive, which comes as close as 8% off CPO Advanced after 195 s, and then does not improve this result. ORT Advanced arrives at 10% off CPO Advanced already after 100 s, and then continues to slowly improve the solutions until 1.16% off. CPO Naive behaves in a similar way, being able to find solutions just 1% off in the first 30 s, improving them until 0.41% off when

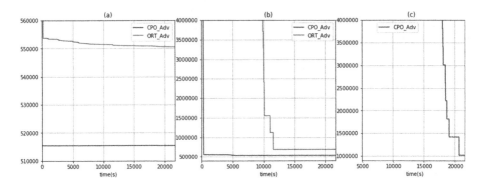

Fig. 4. (a) Plot of instance group **1000 × 10** (b) Plot of instance group **1000 × 100** (c) Plot of instance group **1000 × 1000**.

the timeout occurs. Both CPO SemiNaive and Advanced converge to the same result after 27 and 4 s respectively.

Figure 3(c) does not show a significant difference from (b) in terms of systems' behavior. In (d) however, we see the first instance with 100000 operations. In this case, only the Advanced encodings are able to solve the instances, while the others cannot even instantiate the problem. CPO finds good solutions already after 160 s, with just a marginal improvement in the following 6 h. ORT shows a more regular improvement, on average reaching final solutions which is 10% worse than the ones from CPO.

Figure 4 shows the results of the three instance groups with 1000 jobs. As anticipated in Table 2, only the Advanced encodings are able to instantiate these instances. In (a) we see a rapid convergence to final solutions in CPO after 70 s, while ORT manages to achieve a result that is about 7% worse than the CP Optimizer one. In (b), CP Optimizer achieves the average makespan of 536454.1 in 11 s, with an improvement in the following 6 h of just 400.5 makespan on average (about 0.07% improvement). OR-Tools, after the 3-hour mark, manages to greatly improve its solutions, achieving a result 28% above CPO's.

Finally, (c) shows the test on the largest instances of the benchmark. As we can see, CP Optimizer finds first solutions after 5 h of computation, and then gradually improves its results until it reaches an average makespan of 1017974.1. Although it is hard to judge these results without any comparison from other solvers, it is still remarkable that CP Optimizer tackles all 10 instances with 1 million operations within 6 h. Moreover, based on the plots of all the other instance groups, the fast convergence to a value often corresponds to a (near) optimal solution. This belief is strengthened by a further test that we launched on the largest instance group, doubling the solving time. In 7 out of 10 instances, there was no improvement in the solutions. In the remaining 3, the improvement was less than 0.01% of the makespan.

In the large-scale benchmark, CP Optimizer dominates the scene, being both faster and better than the OR-Tools counterparts, and managing to solve instances up to one million operations. OR Tools, albeit with higher makespans, is able to solve all the instances of the large benchmark up to 100 thousand operations, and is even able to instantiate (but not solve) the biggest instances **1000 × 1000**. In general, we can confirm the conclusions on the classic benchmark on the benefits of interval variables and no-overlap constraint[5]. Moreover, we can add that the global constraints are not only helpful during the solving process, but also crucial to instantiate the larger instances.

4   Discussion

In our experiment we thoroughly tested two state of the art CP solvers with different encodings on a vast selection of instances. The best overall combination of solver-encoding is, without a doubt, CP Optimizer with the Advanced encoding. OR-Tools with Advanced encoding performed not much worse than CPO on the classic benchmark but was not able to keep the same pace on the large-scale instances. To understand why, we looked at the different implementation choices of the two solvers.

In CPO the interval variables are represented compactly using primitive types to express the bounds ($start_{min}$, $start_{max}$, end_{min}, end_{max}) and few other parameters to deal with optional intervals and no-fixed lengths (which are not used in our JSSP scenario). On the other hand, ORT instantiates two integer variables for start and end bounds in addition to the interval variable (for each operation). Therefore, ORT has effectively three times the number of variables of CPO, slowing down the propagation of constraints.

The two solvers use a similar search strategy based on large neighborhood search (LNS), which iteratively relax and re-optimize the problem instance. To decide what to fix and what to vary on the partial schedules, CPO uses portfolio strategies in combination with machine learning to converge to the best neighborhoods [8]; ORT uses a less sophisticated method with random variables, random constraints and local neighborhood in the var-constraint graph. Furthermore, CPO uses a "plan B" strategy called failure directed search (FDS), which is triggered when LNS is not able to improve the current solution [19].

While negligible on the small instances, these differences appear to be particularly impactful on the large problem instances, as it appears from our experiment. In fact, considering the Naive and SemiNaive encoding, OR-Tools performs even better than CPO on the classic benchmarks. Even looking at the **10 × 10** instances of the large-scale benchmark (which are the most similar in size to the classic benchmarks), we see that ORT Naive is at pair with CPO Advanced (Fig. 2(a)) and much better than CPO Naive to converge to the optimum. However, CPO takes the lead already with the **10 × 100** instances, which, albeit being among the smallest of our benchmark, are still bigger than any of the instances of the classic benchmark.

A particularly unexpected performance is registered by the SemiNaive encoding on OR-Tools, which is among the best performer on the classic benchmark,

---

[5] With the exception of ORT SemiNaive.

but is by far the worst in the large-scale one, even compared with the Naive encoding on ORT. This unforeseen behavior is explained by the fact that in the SemiNaive encoding the interval variables ensure that start and end values always respect the duration, without an explicit constraint. However, in ORT, the relaxation process will pick up the equality constraint on the Naive encoding, but not the interval equivalent. Thus, even if the Naive encoding has to instantiate more constraints, it can take full advantage of the relaxation process which, as the results show, becomes more impactful the bigger the instances. To be useful in OR-Tools, the interval variables have to be coupled with the no overlap constraint, which is also picked up in the relaxation process.

In CP Optimizer, on the other hand, there is no such issue because the interval variables are also treated during the relaxation process. In particular, each interval is treated as four numerical variables, while the duration of the interval is imposed with a "precedence" constraint with a delay between the start and the end of the interval. As a result, the use of interval variables is always beneficial in CP Optimizer, both in the classic and the large-scale benchmark.

5   Conclusions

In this paper, we tested two of the best current CP Solvers, OR-Tools and CP Optimizer. The experiment was conducted on, both, a set of famous benchmark instances from literature as well as a large-scale benchmark generated by us, testing three encodings of increasing sophistication for each solver. Based on the results, we can draw the following conclusions:

- CPO with the Advanced encoding was the best overall solver-encoding system, followed by OR-Tools Advanced.
- The difference between ORT and CPO is marginal in small instances (e.g. classic benchmark) but becomes more significant the larger the problem instances.
- The use of interval variables in the model brings a tangible benefit compared to a naive approach, both in terms of solving time and quality of solution, but does not help during the instantiation of the problem.
- The use of global constraints further improves solving time and solution quality, and makes possible to instantiate large problem instances.

A more efficient implementation of the interval variables, an additional strategy to avoid the local optima and more a sophisticated algorithm to select the best partial schedules played a major role in the supremacy of CP Optimizer, with was able to solve instances up to 1 million operations. However, one has to consider that CP Optimizer is a proprietary solution mainly targeted at solving scheduling problems. Despite being an open-source solver, OR-Tools performance was able to solve instances up to 100000 operations, which means that it could be applied even for real-world industrial problems.

References

1. Adams, J., Balas, E., Zawack, D.: The shifting bottleneck procedure for job shop scheduling. Manage. Sci. **34**(3), 391–401 (1988). http://www.jstor.org/stable/2632051
2. Applegate, D., Cook, W.: A computational study of the job-shop scheduling problem. ORSA J. Comput. **3**(2), 149–156 (1991). https://doi.org/10.1287/ijoc.3.2.149
3. Da Col, G., Teppan, E.C.: Declarative decomposition and dispatching for large-scale job-shop scheduling. In: Friedrich, G., Helmert, M., Wotawa, F. (eds.) KI 2016. LNCS (LNAI), vol. 9904, pp. 134–140. Springer, Cham (2016). https://doi.org/10.1007/978-3-319-46073-4_11
4. Danna, E., Rothberg, E., Le Pape, C.: Integrating mixed integer programming and local search: a case study on job-shop scheduling problems. In: Fifth International Workshop on Integration of AI and OR techniques in Constraint Programming for Combinatorial Optimisation Problems (CP-AI-OR'2003), pp. 65–79 (2003)
5. Fox, M.S., Allen, B.P., Strohm, G.: Job-shop scheduling: an investigation in constraint-directed reasoning. In: AAAI, pp. 155–158 (1982)
6. Ku, W.Y., Beck, J.C.: Mixed integer programming models for job shop scheduling: a computational analysis. Comput. Oper. Res. **73**, 165–173 (2016)
7. Laborie, P., Godard, D.: Self-adapting large neighborhood search: application to single-mode scheduling problems. In: Proceedings MISTA-07, Paris, vol. 8 (2007)
8. Laborie, P., Rogerie, J.: Temporal linear relaxation in IBM ILOG CP optimizer. J. Sched. **19**(4), 391–400 (2016)
9. Laborie, P., Rogerie, J., Shaw, P., Vilím, P.: IBM ILOG CP optimizer for scheduling. Constraints **23**(2), 210–250 (2018)
10. Lawrence, S.: Resource constrained project scheduling: an experimental investigation of heuristic scheduling techniques (Supplement). Carnegie-Mellon University, Graduate School of Industrial Administration (1984)
11. Muth, J., Thompson, G.: Industrial Scheduling. International Series in Management. Prentice-Hall, New Jersey (1963)
12. Nethercote, N., Stuckey, P.J., Becket, R., Brand, S., Duck, G.J., Tack, G.: MiniZinc: towards a standard CP modelling language. In: Bessière, C. (ed.) CP 2007. LNCS, vol. 4741, pp. 529–543. Springer, Heidelberg (2007). https://doi.org/10.1007/978-3-540-74970-7_38
13. Refalo, P.: Linear formulation of constraint programming models and hybrid solvers. In: Dechter, R. (ed.) CP 2000. LNCS, vol. 1894, pp. 369–383. Springer, Heidelberg (2000). https://doi.org/10.1007/3-540-45349-0_27
14. Sadeh, N.M., Fox, M.S.: Variable and value ordering heuristics for the job shop scheduling constraint satisfaction problem. Artif. Intell. **86**, 1–41 (1996)
15. Storer, R.H., Wu, S.D., Vaccari, R.: New search spaces for sequencing problems with application to job shop scheduling. Manage. Sci. **38**(10), 1495–1509 (1992)
16. Taillard, E.: Benchmarks for basic scheduling problems. Eur. J. Oper. Res. **64**(2), 278–285 (1993)
17. Teppan, E.C., Da Col, G.: Automatic generation of dispatching rules for large job shops by means of genetic algorithms. In: 8th International Workshop on Combinations of Intelligent Methods and Applications (CIMA 2018), pp. 43–57 (2018)
18. Vazquez, M., Whitley, L.D.: A comparison of genetic algorithms for the dynamic job shop scheduling problem. In: 2nd Annual Conference on Genetic and Evolutionary Computation, pp. 1011–1018. Morgan Kaufmann Publishers Inc. (2000)

19. Vilím, P., Laborie, P., Shaw, P.: Failure-directed search for constraint-based scheduling. In: Michel, L. (ed.) CPAIOR 2015. LNCS, vol. 9075, pp. 437–453. Springer, Cham (2015). https://doi.org/10.1007/978-3-319-18008-3_30
20. Yamada, T., Nakano, R.: A genetic algorithm applicable to large-scale job-shop problems. In: PPSN, pp. 283–292 (1992)

Dual Hashing-Based Algorithms
for Discrete Integration

Alexis de Colnet[1]([⊠]) and Kuldeep S. Meel[2]

[1] CNRS, CRIL UMR 8188, Lens, France
decolnet@cril.fr
[2] School of Computing, National University of Singapore, Singapore, Singapore

Abstract. Given a boolean formula F and a weight function ρ, the problem of discrete integration seeks to compute the weight of F, defined as the sum of the weights of satisfying assignments. Discrete integration, also known as weighted model counting, is a fundamental problem in computer science with wide variety of applications ranging from machine learning and statistics to physics and infrastructure reliability. Given the intractability of the exact variant, the problem of approximate weighted model counting has been subject to intense theoretical and practical investigations over the years.

The primary contribution of this paper is to investigate development of algorithmic approaches for discrete integration. Our framework allows us to derive two different algorithms: WISH, which was already discovered by Ermon et al. [8], and a new algorithm: SWITCH. We argue that these algorithms can be seen as dual to each other, in the sense that their complexities differ only by a permutation of certain parameters. Indeed we show that, for F defined over n variables, a weight function ρ that can be represented using p bits, and a confidence parameter δ, there is a function f and an NP oracle such that WISH makes $\mathcal{O}\left(f(n, p, \delta)\right)$ calls to NP oracle while SWITCH makes $\mathcal{O}\left(f(p, n, \delta)\right)$ calls. We find $f(x, y, \delta)$ polynomial in x, y and $1/\delta$, more specifically $f(x, y, \delta) = x \log(y) \log(x/\delta)$. We first focus on striking similarities of both the design process and structure of the two algorithms but then show that despite this quasi-symmetry, the analysis yields time complexities dual to each other. Another contribution of this paper is the use of 3-wise property independence of XOR based hash functions in the analysis of WISH and SWITCH. To the best of our knowledge, this is the first usage of 3-wise independence in deriving stronger concentration bounds and we hope our usage can be generalized to other applications.

The original version of this chapter was revised: The acknowledgement was modified. The correction to this chapter is available at https://doi.org/10.1007/978-3-030-30048-7_45

The author list has been sorted alphabetically by last name; this order should not be used to determine the extent of authors' contributions.

The work was performed during first author's stay at NUS.

The full version along with Appendix is available at https://github.com/meelgroup/dualhashing.

T. Schiex and S. de Givry (Eds.): CP 2019, LNCS 11802, pp. 161–176, 2019.
https://doi.org/10.1007/978-3-030-30048-7_10

1   Introduction

Given a set of constraints F and a weight function ρ that assigns a non-negative weight to every assignment of values to variables, the problem of discrete integration seeks to compute the weight of F, defined as the sum of weights of its satisfying assignments. If every assignment has weight 1, the corresponding problem is often simply called *model counting*. For clarity of presentation, we use *unweighted model counting* to denote this variant. Discrete integration is a fundamental problem in computer science. A wide variety of problems such as probabilistic inference [14], partition function of graphical models, permanent of a matrix [18], un-reliability of a network [13] can be reduced to discrete integration.

In his seminal work, Valiant [18] established the complexity of discrete integration as #P-complete for all polynomially computable weight functions, where #P is the complexity class comprised of counting problems whose decision variant lies in NP. Given the computational intractability of discrete integration, approximate variants have been subject of intense theoretical and practical investigations over the past few decades.

Approaches to discrete integration can be classified into three categories: variational techniques, sampling techniques, and hashing-based techniques. Inspired from statistical physics, variational methods often scale to large instances but do not provide guarantees on the computed estimates [17,19]. Sampling-based techniques focus on approximation of the discrete integral via sampling from the probability distribution induced by the boolean formula and the weight function [11]. The estimation of rigorous bounds, however, requires exponential mixing times for the underlying chains and therefore, practical implementations such as those based on Markov Chain Monte Carlo methods [2] or randomized branching choices [9] fail to provide rigorous estimates [7,12]. Recently, hashing-based techniques have emerged as a promising alternative to variational and sampling techniques to provide rigorous approximation guarantees [4,5,8]. The hashing-based algorithm WISH seeks to utilize progress made in combinatorial solving over the past two decades and to this end, the problem of discrete integration is reduced to linear number of optimization queries subject to randomly generated parity constraints [8].

The primary contribution of this paper is to investigate the development of algorithmic approaches for discrete integration. Our framework allows us to derive two different algorithms, which can be seen as dual to each other: WISH, which was already discovered by Ermon et al. [8], and a new algorithm: SWITCH. In particular, WISH reduces the problem of discrete integration to optimization queries while SWITCH proceeds via reduction to unweighted model counting. Both WISH and SWITCH compute constant factor approximations with arbitrarily high probability $1 - \delta$ via usage of universal hash functions, a concept invented by Carter and Wegman in their seminal work [3]. We first focus on the design process of WISH and SWITCH. We study discrete integration through the framework of general integration and reduce the task to optimization and counting subproblems. Then we present WISH and SWITCH as hashing-based algorithms solving the aforementionned subproblems to approximate a discrete integral. Finally we analyse these algorithms, proving that both compute constant factor approximations of the integral with high probability. However we show that they have dual time complexities in the sense that, for F defined over n variables and a weight function ρ that can be represented using p bits, there is a function f and an NP oracle such that WISH makes $\mathcal{O}\left(f(n, p, \delta)\right)$ calls to NP oracle while SWITCH makes $\mathcal{O}\left(f(p, n, \delta)\right)$. We find $f(x, y, \delta)$ polynomial in x, y and $1/\delta$, more specifically $f : x, y, \delta \mapsto x \log(y) \log(x/\delta)$.

Another contribution of this paper is the use of 3-wise property independence of XOR based hash functions in the analysis of WISH and SWITCH. To the best of our knowledge, this is the first usage of 3-wise independence in deriving stronger concentration bounds. The hardness of usage of 3-independence for concentration bounds is well documented by absence of such analyses (c.f.: wonderful blogpost by Mihai Pătraşcu:[1]).

The duality obtained may not seem surprising in retrospect but such has not been the case for the past few years. The prior work has often, without complete evidence, asserted that the corresponding dual approach would be inferior both theoretically and empirically [4, 8]. Our work, in turn, contradicts such assertions and shows that the two approaches indeed have dual time complexity from theoretical perspective and empirical analysis will be key in determining their usefulness. Since the work on development of MaxSAT solvers that support XORs and SAT solvers that support XORs and Pseudo-Boolean (PB) constraints is in its infancy; our work provides a strong argument for the need and potential of both of these solvers as queries generated by WISH require MaxSAT solvers with the ability to handle XORs while the queries by SWITCH requires SAT solvers that support XORs and PB constraints.

The rest of the paper is organized as follows. We introduce notations and preliminaries in Sect. 2. We then provide general framework for discrete integration in Sect. 3, which is employed to derive the aforementioned algorithms, WISH and SWITCH, in Sect. 4. We finally conclude in Sect. 5.

2   Preliminaries and Notations

Let F be a boolean formula over n variables. Let X be the set of variables appearing in F. A literal is a variable x or its negation $\neg x$. An assignment σ of all n variables is a *satisfying assignment* or *witness* of F if it makes F evaluate to *true*, which we note $\sigma \models F$. We note $\#F$ the number of witnesses of F.

Weight Function. Let $\rho : \{0, 1\}^n \to \mathbb{Q}_+$ be the *weight function* mapping each truth assignment to a positive value such that

- $\forall \sigma \in \{0, 1\}^n$, weight $\rho(\sigma)$ is computable in polynomial time
- $\forall \sigma \in \{0, 1\}^n$, weight $\rho(\sigma)$ is written in binary representation with less than p bits.

---

[1] http://infoweekly.blogspot.com/2010/01/moments.html.

We extend the weight function to sets of truth assignments and boolean formulas. Let Y be a subset of $\{0,1\}^n$, the weight of Y is defined as the cumulative weight of the truth assignments in Y: $\rho(Y) = \sum_{\sigma \in Y} \rho(\sigma)$. By definition the weight of the empty set is 0. The weight of a formula F is defined as the cumulative weight of its witnesses $\rho(F) = \sum_{\sigma \models F} \rho(\sigma)$. For notational clarity, we overload ρ to indicate weight of an assignment, set of assignments, and formula depending on the context.

Given a formula F and weight function ρ, we define the *effective weight function* w as the restriction of ρ to the witnesses of F

$$w(\sigma) = \begin{cases} \rho(\sigma) & \text{if } \sigma \models F \\ 0 & \text{otherwise} \end{cases}$$

We will note $w_{\min} = \min_{\sigma \models F} w(\sigma)$ and $w_{\max} = \max_{\sigma \models F} w(\sigma)$ the minimum and maximum weights of a witness of F. Due to the hypothesis on ρ we have $w_{\max} \leq 2^p$ and $w_{\min} \geq 2^{-p}$ if F is satisfiable. Note that the expression for the weight of a formula can be rewritten $\rho(F) = \sum_{\sigma \in \{0,1\}^n} w(\sigma)$.

Tail Function. Dual to the effective weight function is the *tail function* τ. It is defined from the space of weights to $\mathbb{N}$. The tail function on some weight u counts the number of truth assignments heavier than u (i.e. of weight greater than u).

$$\tau(u) = \left| \{\sigma \in \{0,1\}^n : w(\sigma) \geq u\} \right|$$

For notational clarity we extend the tail function to truth assignments using the notation $\tau(\sigma)$ for $\tau(w(\sigma))$. Note that

1. The tail function is non-increasing.
2. The maximum tail is $\tau(0) = 2^n$.
3. For any $0 < u \leq w_{\min}$ there is $\tau(u) = \#F$.
4. If $u > w_{\max}$ then $\tau(u)$ evaluates to 0, but the minimal non-zero tail $\tau(w_{\max})$ is not necessarily 1 since more than one truth assignment can weight $w_{\max}$.

MPE-MAP Queries. Following standard definitions, MPE (*most probable explanation*) corresponds to solving $\max(\rho(\sigma) : \sigma \models F)$, which is to find $w_{\max}$. It is worth noting that MPE is related to another query: MAP (*maximum a posteriori*), and different communities use different definitions for MAP and MPE, to the extent that what one community calls MAP is called MPE by another [4,8].

(ε, δ)-Approximation Algorithms. Given computational intractability of computing $\rho(F)$, we are interested in approximation schemes. For a tolerance $\varepsilon > 0$ and a confidence $\delta > 0$, an algorithm $\mathcal{A}$ generates a (ε, δ)-approximation of W if it returns a quantity in $\left[W(1+\varepsilon)^{-1}, (1+\varepsilon)W \right]$ with probability at least $1 - \delta$.

$$\Pr\left[(1+\varepsilon)^{-1}W \leq \mathcal{A}(F, \rho, \varepsilon, \delta) \leq (1+\varepsilon)W \right] \geq 1 - \delta$$

3-Universal Hash Functions. We focus on hashing-based methods to approximate $\rho(F)$. We use particular classes of hash functions based on parity constraints. A constraint specifies a set of indices S from $[n]$ and a bucket value β in $\{0,1\}$. The assignment σ is said to satisfy the constraint if the xored value of its coordinates on S matches β, or more formally if $\bigoplus_{i \in S} \sigma[i] = \beta$, where $\oplus$ denotes the "xor" operation. Using the binary vector representation of subsets S in $\{0,1\}^n$, one can rewrite the left hand side of the constraint as a scalar product in the field $\mathbb{F}_2^n$ which addition and product operations are, respectively, the "xor" and the "and" operations. Therefore we will use matrix representations when applying several constraints. For m given constraints represented with the matrix $A \in \{0,1\}^{m \times n}$ and the vector of bucket values $b \in \{0,1\}^m$, σ satisfies all m constraints if $A\sigma = b$, or equivalently $A\sigma \oplus b = 0$. A hash function h from $\{0,1\}^n$ to $\{0,1\}^m$ is defined by a collection of m constraints embedded in A and b. An assignment σ is hashed through h to $h(\sigma) = A\sigma \oplus b$. So the i-th component of $h(\sigma)$ is

$$h(\sigma)[i] = b_i \oplus \bigoplus_{j=1}^{n} A[i,j]\sigma[j]$$

Let $H_{\mathrm{xor}}(n,m)$ be the class of all such hash functions from $\{0,1\}^n$ to $\{0,1\}^m$.

$$H_{\mathrm{xor}}(n,m) = \{\sigma \mapsto A\sigma \oplus b : A \in \{0,1\}^{m \times n}, b \in \{0,1\}^m\}$$

We note $h \xleftarrow{R} H_{\mathrm{xor}}(n,m)$ the action of choosing a hash function uniformly at random from $H_{\mathrm{xor}}(n,m)$, which is equivalent to sampling A from $\mathcal{B}_{1/2}(m,n)$ and b from $\mathcal{B}_{1/2}(m)$. Hash functions in $H_{\mathrm{xor}}(n,m)$ have uniformity property, meaning that for all y in $\{0,1\}^m$ and σ in $\{0,1\}^n$, there is

$$\Pr\left[h \xleftarrow{R} H_{\mathrm{xor}}(n,m) : h(\sigma) = y\right] = \frac{1}{2^m}$$

It was also shown in [10] that they display 3-wise independence property, meaning that for all three images y_1, y_2, y_3 in $\{0,1\}^m$ and for all three distinct assignments σ_1, σ_2, σ_3 in $\{0,1\}^n$, there is

$$\Pr\left[h \xleftarrow{R} H_{\mathrm{xor}}(n,m) : h(\sigma_1) = y_1 \text{ and } h(\sigma_2) = y_2 \text{ and } h(\sigma_3) = y_3\right] = \frac{1}{2^{3m}}$$

They do not display independence at higher order. For instance for 4-wise independence, consider three assignments $\sigma_1, \sigma_2, \sigma_3$ and four images y_1, y_2, y_3, y_4. Define $\sigma_4 = \sigma_1 \oplus \sigma_2 \oplus \sigma_3$ and see that if $h(\sigma_i) = y_i$ for $i \in \{1,2,3\}$, then $h(\sigma_4) = y_1 \oplus y_2 \oplus y_3$. So the probability for all four assignments to be projected on their respective images is null when $y_4 \neq y_1 \oplus y_2 \oplus y_3$.

3   A Framework for Discrete Integration

This section presents a framework for discrete integration. Methods from this framework follow a two-steps strategy:

1. translate the task of discrete integration into an integration problem for a real non-increasing function
2. apply a method to approximate the integral of a real function

In the first step, we specifically ask for a non-increasing function so that we can ensure constant factor approximations when estimating its integral. Examples of approximation methods for real function integrals are the upper and lower rectangles approximations or Monte Carlo integrators.

3.1   From Discrete Integration to Real Function Integration

Given F a boolean formula and a weight function ρ, let $u_1, \cdots, u_K$ be all possible weights taken by the satisfying assignments of F. To obtain the discrete integral $\rho(F)$, i.e. the sum of weights of satisfying assignments of F, one can gather assignments in packets of same effective weight and sum over these packets. For the weight u_i, the pre-image $w^{-1}(u_i)$ is the set of all witnesses of F mapped to u_i by ρ. So the discrete integral can be written

$$\rho(F) = \sum_{i=1}^{K} u_i |w^{-1}(u_i)| \tag{1}$$

We observe the following *tail transformation*:

- For $i < K$, there is $|w^{-1}(u_i)| = \tau(u_i) - \tau(u_{i+1})$
- In the case $i = K$, there is $|w^{-1}(u_K)| = \tau(u_K)$

Applying this transformation to Eq. (1) gives:

$$\rho(F) = u_K \tau(u_K) + \sum_{i=1}^{K-1} u_i \left(\tau(u_i) - \tau(u_{i+1}) \right) \tag{2}$$

and after rearranging the terms:

$$\rho(F) = u_1 \tau(u_1) + \sum_{i=1}^{K-1} \tau(u_{i+1}) \left(u_{i+1} - u_i \right) \tag{3}$$

These two representations of the discrete integral have a graphical interpretation: draw the curve of τ as a function of the weight, and observe that both $\tau(u_{i+1})(u_{i+1} - u_i)$ and $u_i(\tau(u_i) - \tau(u_{i+1}))$ are areas of rectangles under the curve as illustrated in Fig. 1. Equation (2) decomposes the integral into rectangles built along the τ axis while Eq. (3) is a decomposition into rectangles built along the w axis.

The discrete integral $\rho(F)$ is the area under the curve of τ.

$$\rho(F) = \int \tau(u)du \tag{4}$$

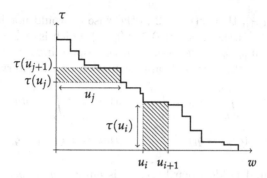

Fig. 1. Decomposition into rectangle areas

The effective weight function w can be expressed as a function of the tails which extension to $\mathbb{R}_+$ is $w : t \mapsto \max_\sigma(w(\sigma) : \tau(\sigma) \geq t)$. Graphically, one can just rotate the graph of τ to obtain that of w and see that:

$$\rho(F) = \int w(t)dt \tag{5}$$

Both (4) and (5) are integrals of non-increasing functions defined over $\mathbb{R}_+$ and of finite support.

3.2   From Discrete Integration to Optimization

Direct computation of any form previously obtained is intractable. We resort to approximations of $\rho(F)$ when it is written as (4) or (5). Given that τ and w are staircase functions, rectangles approximation seems to be the only method fitted to approximate their integrals. First we apply the method on Eq. (4). The first step is the partition of the weight axis into linearly many intervals. We split the axis at the quantile weights, defined as followed:

Definition 1. *The 2^i-th quantile weight of the weight distribution is the maximal weight q_i such that $\tau(q_i) \geq 2^i$.*

The quantile weights $q_0, \cdots, q_n$ are all well-defined, and form a non-increasing sequence. Consecutive quantile weights can be equal. For instance if F has $< 2^m$ witnesses for some $m < n$, then $q_m = q_{m+1} = \cdots = q_n = 0$. Note that for each quantile weight q_i, there exists some truth assignment σ such that $q_i = w(\sigma)$.

The partition of integral (4) at the quantile weights gives:

$$\rho(F) = q_n 2^n + \sum_{i=1}^{n} \int_{q_i}^{q_{i-1}} \tau(u)du$$

where $\int_{q_i}^{q_{i-1}}$ represents the integral on $]q_i, q_{i-1}]$. Since the weight q_n does not lie in any interval we add the term $q_n \tau(q_n) = q_n 2^n$ manually.

If u is in $]q_i, q_{i-1}]$, then $\tau(u) < 2^i$, otherwise q_i would not be the maximal weight of tail $\geq 2^i$. Furthermore $\tau(u) \geq \tau(q_{i-1})$ which is $\geq 2^{i-1}$ by definition. So for each weight in $]q_i, q_{i-1}]$ we bound the corresponding tail within a factor of 2. Figure 2 illustrates this rectangle approximation on the interval $]q_{n-1}, q_{n-2}]$.

$$2^{i-1} \int_{q_i}^{q_{i-1}} du \leq \int_{q_i}^{q_{i-1}} \tau(u) du \leq 2^i \int_{q_i}^{q_{i-1}} du$$

$$2^{i-1} (q_{i-1} - q_i) \leq \int_{q_i}^{q_{i-1}} \tau(u) du \leq 2^i (q_{i-1} - q_i)$$

Note that the bound holds when $]q_i, q_{i-1}]$ is empty ($q_i = q_{i-1}$). Summing all bounds together and rearranging the terms to get rid of differences of quantiles, we obtain:

$$q_0 + \sum_{i=0}^{n-1} q_{i+1} 2^i \leq \rho(F) \leq q_0 + \sum_{i=0}^{n-1} q_i 2^i$$

The two bounds are within a ratio of 2 of each other because the integral on each interval was bounded within a ratio of 2. Let us choose the lower bound to be our first estimate of $\rho(F)$ and name it $W_1 = q_0 + \sum_{i=0}^{n-1} q_{i+1} 2^i$. We have

$$W_1 \leq \rho(F) \leq 2W_1 \tag{6}$$

Given $q_0, \cdots, q_n$, the estimate W_1 can be computed in polynomial time. For all i, the weight q_i is, by definition, the solution of the following optimization problem:

$$q_i = \max \left\{ w(\sigma) : \tau(w(\sigma)) \geq 2^i \right\}$$

So the approximation of the discrete integral $\rho(F)$ has been reduced to $n + 1$ optimization sub-problems.

3.3   From Discrete Integration to Counting

To find W_1 we have done rectangles approximation on Eq. (4). In this section we investigate the estimate resulting from a similar approximation on Eq. (5). The first step is the partition of the tail axis. We will assume, for notational clarity, that $w_{\max} \leq 1$. This bound is legitimate in the context of probabilistic inferences [14], and the results of this paper can be extended to any arbitrary but fixed bound. Recall that the weights are written with p bits in binary representation, so the bounds $w_{\max} \leq 2^p$ and $w_{\min} \geq 2^{-p}$ are always valid. For our partition, we define the splitting tails as followed:

Definition 2. *The i-th splitting tail τ_i is the tail at weight $1/2^i$: $\tau_i = \tau(1/2^i)$.*

Given the assumption on the range value of w, the interesting tails are $\tau_0, \cdots, \tau_p$. They form a non-decreasing sequence. The partition of integral (5) at the splitting tails gives:

$$\rho(F) = \tau_0 + \sum_{i=0}^{p-1} \int_{\tau_i}^{\tau_{i+1}} w(t) dt$$

where $\int_{\tau_i}^{\tau_{i+1}}$ represents the integral on $]\tau_i, \tau_{i+1}]$. Since the tail τ_0 does not lie in any interval we add the term $\tau_0 w(\tau_0)$ manually. τ_0 is the number of assignments heavier than weight 1. Either there are no such assignment and $\tau_0 = 0$, or there are some, in which case $w(\tau_0) = w_{\max} = 1$. In both cases we find that $\tau_0 w(\tau_0) = \tau_0$. If t is in $]\tau_i, \tau_{i+1}]$, then $2^{-i-1} \leq w(t) \leq 2^{-i}$. So for each tail in $]\tau_i, \tau_{i+1}]$, we bound the corresponding weight within a factor of 2. Figure 3 illustrates this rectangle approximation on the interval $]\tau_1, \tau_2]$.

$$2^{-i-1} \int_{\tau_i}^{\tau_{i+1}} dt \leq \int_{\tau_i}^{\tau_{i+1}} w(t)dt \leq 2^{-i} \int_{\tau_i}^{\tau_{i+1}} dt$$

$$2^{-i-1} \left(\tau_{i+1} - \tau_i\right) \leq \int_{\tau_i}^{\tau_{i+1}} w(t)dt \leq 2^{-i} \left(\tau_{i+1} - \tau_i\right)$$

Note that the bound holds when $]\tau_i, \tau_{i+1}]$ is empty ($\tau_i = \tau_{i+1}$). Summing all bounds together and rearranging the terms to get rid of differences of tails, we obtain:

$$\tau_p 2^{-p} + \sum_{i=0}^{p-1} \tau_i 2^{-(i+1)} \leq \rho(F) \leq \tau_p 2^{-p} + \sum_{i=0}^{p-1} \tau_{i+1} 2^{-(i+1)}$$

The two bounds are within a ratio of 2 of each other because the integral on each interval was bounded within a ratio of 2. Let us choose the lower bound to be our first estimate of $\rho(F)$ and name it $W_2 = \tau_p 2^{-p} + \sum_{i=0}^{p-1} \tau_i 2^{-(i+1)}$. We have

$$W_2 \leq \rho(F) \leq 2W_2 \tag{7}$$

Given $\tau_0, \cdots, \tau_p$, the estimate W_2 can be computed in polynomial time. For all i, the tail τ_i is, by definition, the solution of the following counting problem:

$$\tau_i = \left|\{\sigma : w(\sigma) \geq 2^{-i}\}\right|$$

So the approximation of the discrete integral $\rho(F)$ has been reduced to $p + 1$ counting sub-problems.

3.4   On the Limitations of the Estimates

The two estimates W_1 and W_2 are not only similar in terms of construction but also in terms of theoretical guarantees and limitations. Both are lower bounds of $\rho(F)$ and approximate $\rho(F)$ within a ratio of 2. Furthermore, both W_1 and W_2 use some unknown quantities, respectively the weights $q_0, \cdots, q_n$ and the tails $\tau_0, \cdots, \tau_p$. These are to be approximated.

Assuming that for positive some ε and δ we have an algorithm $\mathcal{A}$ returning C, a (δ, ε)-approximations of W_1 (resp. W_2). Then with probability at least $1 - \delta$, C is a bounded estimate of $\rho(F)$ such that $\frac{\rho(F)}{2(1+\varepsilon)} \leq C \leq (1 + \varepsilon)\rho(F)$. In any case, the quality of the estimate is capped: the best approximation interval possible is $[\rho(F)/2, \rho(F)]$. However, note that we obtained 2-approximations of $\rho(F)$ using base-2 partitions of the tail and weight axis for the rectangles approximations. If we use base-β partitions instead, with $\beta < 2$, we can improve our estimates. For instance for $\beta = 1 + \varepsilon$, we partition the weight axis at the $(1 + \varepsilon)^i$-th quantile weights and the tail axis at the tails $\tau((1 + \varepsilon)^{-i})$. Rectangles approximations W_1' and W_2' are then both in $[\rho(F), (1 + \epsilon)\rho(F)]$. Now, algorithm $\mathcal{A}$ returns some quantity in $\left[\rho(F)/(1 + \varepsilon)^2, (1 + \varepsilon)\rho(F)\right]$ with probability at least $1 - \delta$. So $\mathcal{A}$ computes a $(3\varepsilon, \delta)$-approximation of $\rho(F)$ (for $\varepsilon < 1$ we have $(1 + \varepsilon)^{-2} \geq (1 + 3\varepsilon)^{-1}$).

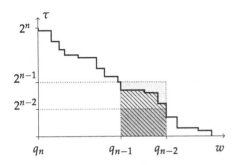

Fig. 2. Rectangles approximation on Eq. (2)

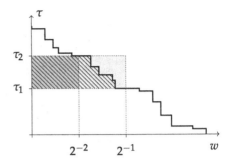

Fig. 3. Rectangles approximation on Eq. (3)

4   Algorithms

In this section we present two algorithms to approximate the discrete integral $\rho(F)$. The first one approximates W_1. It uses a hashing-based approach to approximate solutions for the optimization/MPE sub-problems described in Sect. 3.2. Given the lack of any approach to find 2^i-th quantiles, hashing is used to reduce the task to that of standard optimization. The second algorithm approximates W_2 and also implements a strategy based on hashing functions to approximately solve the counting sub-problems described in Sect. 3.3, this choice is motivated by the success of hashing-based technique for model counting [5]. Since it is known that $(1 + \varepsilon)$-approximations can be obtained from constant factor approximations by standard amplification techniques [8], we will focus on obtaining constant factor approximations. Possibilities of extension of the algorithms to reach arbitrary precision approximations following the strategy of Sect. 3.4 will be discussed in Sect. 4.4.

4.1   An NP Oracle

The procedure for discrete integration via optimization was first discovered by Ermon et al. [8]. They expressed the complexity of their algorithm as the number of calls to an MPE oracle. It is customary to express complexity with respect to oracles corresponding to decision problems. Therefore, we express complexities in terms of invocations of an NP oracle. The oracle is a system capable of solving a decision problem in $\mathcal{O}(1)$ time. In this paper, the oracle is given a boolean formula F, a weight function ρ computable in polynomial time and a real number u, and returns *YES* if and only if there exists a satisfying assignment of F of weight greater than u. More formally, it solves the problem $\mathrm{SAT}(F \wedge \{\rho(\sigma) \geq u\})$ where the constraint $\rho(\sigma) \geq u$ is not necessarily boolean. Given a solution σ, the condition $\sigma \models F$ can be tested in polynomial time, and so is $\rho(\sigma) \geq u$ by hypothesis on ρ. So the decision problem is in NP.

4.2   Discrete Integration by Optimization: WISH

We present a modified version of the algorithm of Ermon et al. [8]: WISH (Weighted Integration and Sum by Hashing). When comparing our version of WISH to the original, we will refer to the latter as WISH_EGSS, from the initials of its authors. WISH takes in a formula F, a weight function ρ, and a confidence parameter δ, and returns an estimate for W_1.

Algorithm 1. WISH(F, ρ, δ)

1: $T \leftarrow \lceil 128 \ln(2n/\delta) \rceil$
2: **for all** $1 \leq t \leq T$ **do**
3:      $A_0 \leftarrow [\,]$, $b_0 \leftarrow [\,]$
4:      **for all** $0 \leq i \leq n$ **do**
5:          Sample constraint C_i in $\{0,1\}^n$ and β_i in $\{0,1\}$
6:          $A_{i+1} \leftarrow \mathrm{concat}(A_i, C_i)$, $b_{i+1} \leftarrow \mathrm{concat}(b_i, \beta_i)$
            Let m_i^t be $\max_\sigma(w(\sigma) : A_{i+1}\sigma \oplus b_{i+1} = 0)$
7:          $\widehat{m}_i^t \leftarrow 2^\kappa$ where $\kappa = \max(k : 2^k \leq m_i^t)$
8:      **end for**
9: **end for**
10: $\forall i$, $\widehat{q}_i \leftarrow \mathrm{Median}(\widehat{m}_i^1, \cdots, \widehat{m}_i^T)$
11: **return** $\sqrt{2}\left(\widehat{q}_0 + \sum_{i=0}^{n-1} \widehat{q}_{i+1} 2^i\right)$

WISH's main task is to estimate the quantiles weights q_i. The general idea is to use hash functions projecting the truth assignments into 2^i buckets and to take the heaviest assignment in a random bucket. Hash functions are built adding xor constraints incrementally: $\mathrm{concat}(A, C)$ adds the line C to the matrix of constraints A. By uniformity property, an arbitrary bucket contains in expectation 2^{n-i} truth assignments after i constraints. Since there are roughly 2^{i-1} assignments heavier than q_{i-1}, the expected amount mapped to the chosen bucket should be close to zero. So hopefully the heaviest weight of the bucket (noted m_i^t for the t-th run) is in $[q_i, q_{i-1}]$, and it is chosen as candidate for the estimate $\widehat{q}_i$. The following lemma gives guarantees on the range of the heaviest weight of a bucket (proof is deferred to appendix).

Lemma 1. *For all t in $[\![1, T]\!]$ and all i in $[\![1, n]\!]$, there is*

$$Pr\left[m_i^t \geq q_i\right] \geq \left(\frac{3}{4}\right)^2 \quad and \quad Pr\left[m_i^t \leq q_{i-1}\right] \geq \left(\frac{3}{4}\right)^2$$

Iterating this process T times and taking the median candidate amplifies the confidence of the estimation.

Lemma 2. *For $i > 0$, let $\widehat{q_i}$ be the median of $m_i^1, \cdots, m_i^T$ resulting from the T independent iterations. And let I_i be the interval $[q_i, q_{i-1}]$. There is:*

$$\Pr\left[\widehat{q_i} \in I_i\right] \geq 1 - 2\exp\left(-\frac{T}{\alpha}\right)$$

where $\alpha = 2^7 = 128$.

In the algorithm, $\widehat{q_i}$ is actually not the median of $m_i^1, \cdots, m_i^T$ but the median of their 2-approximations $\widehat{m}_i^1, \cdots, \widehat{m}_i^T$. With this modification, the statement of the lemma holds for $I_i = [q_i/2, q_{i-1}]$. An estimate of the integral is finally generated using the formula for W_1 and replacing the q_i by their estimates $\widehat{q_i}$.

We make several contributions to WISH_EGSS in WISH. We first reduce the MPE queries employed in WISH_EGSS to find the weights m_i^t to binary searches using the NP oracle queries. Weights are written with p bits so finding m_i^t takes $\mathcal{O}(p)$ oracle queries. However we prefer the approximate variant in which the binary search explores $[\![0, p]\!]$ to find $\kappa = \lfloor \log(m_i^t) \rfloor$ and returns 2^κ. This variant reduces the cost to $\mathcal{O}(\log(p))$ queries while approximating m_i^t within a factor of 2. A second contribution is the usage of dependence among different hash functions: hash functions of $i+1$ constraints are no longer sampled independently but built upon hash functions of i constraints (hence the concat$(\cdot,\cdot)$ function). Our last contribution is the significant improvement of WISH_EGSS's guarantees: the original analysis used only the pairwise independence of hash functions, but we use 3-wise independence to obtain the improved Lemma 1. This lemma ultimately allows us to prove that WISH approximates $\rho(F)$ within a factor of 8, while the initial factor was 256. The reduce factor is still quite large but greatly accelerates the amplification process described in [8].

Theorem 1. *For any $\delta > 0$, WISH (F, ρ, δ) makes $\mathcal{O}\left(n \log(p) \log(n/\delta)\right)$ calls to NP oracle and returns an approximation of $\rho(F)$ within $\left[\rho(F)/(2\sqrt{2}), 2\sqrt{2}\rho(F)\right]$ with probability at least $1 - \delta$.*

4.3   Discrete Integration by Counting: SWITCH

We now describe an algorithm for discrete integration that utilizes the reduction to counting sub-problems. We call the algorithm SWITCH (Sum of Weights and Integral via Threshold Counting and Hashing). SWITCH takes in a formula F, a weight function ρ, and a confidence parameter δ, and returns an estimate for W_2. SWITCH's main task is to estimate the tails $\tau_i = \tau(2^{-i})$. The core idea is to view tails as cardinals of some subsets of witnesses of F and use hashing to estimate these cardinalities. The approximation method is very similar to previous hashing-based techniques [5,16]. For a given subset of size τ_i, we successively apply constraints until its projection on an arbitrary bucket is empty. Each new randomly sampled constraint halves the remaining set in expectation, so the number of constraints necessary to reach the empty set can be viewed as a good approximation of $l_i = \log(\tau_i)$ and its power of 2 approaches τ_i.

Algorithm 2. SWITCH(F, ρ, δ)

1: $T \leftarrow \lceil 128 \ln(4p/\delta) \rceil$
2: **for all** $1 \leq t \leq T$ **do**
3:    Sample A in $\{0,1\}^{n \times n}$ and b in $\{0,1\}^n$
4:    **for all** $0 \leq i \leq p$ **do**
5:        $\widehat{l_i}^t \leftarrow \max \{ k \mid \exists\, \sigma$ such that $\sigma \models F$, $\rho(\sigma) \geq 2^{-i}$ and $A_k \sigma \oplus b_k = 0 \}$
         where $A_k \leftarrow A[1..k]$ and $b_k \leftarrow b[1..k]$
6:    **end for**
7: **end for**
8: $\forall i$, $\widehat{l_i} \leftarrow \mathrm{Median}(\widehat{l_i}^1, \cdots, \widehat{l_i}^T)$, $\widehat{\tau_i} = 2^{\widehat{l_i}}$
   (handle cases $\tau_i = 0$ and $\tau_i = 1$ exactly)
9: **return** $\sqrt{2} \left(\widehat{\tau_p} 2^{-p} + \sum_{i=0}^{p-1} \widehat{\tau_i} 2^{-i-1} \right)$

Lemma 3. *If $l_i > 0$ ($\tau_i > 1$), then there is for all t in $[\![1, T]\!]$:*

$$\Pr\left[\widehat{l_i}^t \leq \lceil l_i \rceil \right] \geq \left(\frac{3}{4} \right)^2 \qquad and \qquad \Pr\left[\widehat{l_i}^t \geq \lfloor l_i \rfloor \right] \geq \left(\frac{3}{4} \right)^2$$

One may note that τ_i is not necessarily a power of 2, so our method of approximating logarithms by integers is imprecise and the estimation error is amplified as a power of 2. Furthermore, there are two cases not handled by the lemma

- The case $\tau_i = 0$ ($l_i = -\infty$): there are no witness of F of weight greater than 2^{-i}. One call to the NP oracle is enough to spot this case.
- The case $\tau_i = 1$ ($l_i = 0$): a set of 1 element stays intact after 1 constraint with probability $1/2$, so we overestimate is size with probability $1/2$. This case is spotted with two NP oracle queries (adding a block clause before the second query).

Assuming we are not in any such cases, we amplify the confidence on the estimates of l_i repeating the process T times and choosing the median candidate.

Lemma 4. *Let $\widehat{l_i}$ be the median of the T independent $\widehat{l_i}^1, \cdots, \widehat{l_i}^T$. Assume $\tau_i > 1$ and let J_i be the interval $[\lfloor \log(\tau_i) \rfloor, \lceil \log(\tau_i) \rceil]$. There is:*

$$\Pr\left[\widehat{l_i} \in J_i \right] \geq 1 - 2 \exp\left(-\frac{T}{\alpha} \right)$$

where $\alpha = 2^7 = 128$. Therefore $\widehat{\tau_i} = 2^{\widehat{l_i}}$ is an estimate of τ_i that lies in $\left[\frac{\tau_i}{2}, 2\tau_i \right]$ with same probability.

The tails estimates are finally used to compute an estimate of W_2.

The NP oracle is called to check if a set is empty after application of constraints. When approximating l_i line 5, the constraints are taken from the same set of n constraints stored in A and b (A_k and b_k representing the first k lines of A and b). Consequently, if there are witnesses satisfying $A_j\sigma \oplus b_j = 0$ for some j, they also satisfy $A_i\sigma \oplus b_i = 0$ for all $i \leq j$. Similarly if no witness satisfy $A_j\sigma \oplus b_j = 0$, none satisfy $A_i\sigma \oplus b_i = 0$ for $i \geq j$. So to find how many constraints are enough to empty the set of witnesses of F heavier than 2^{-i}, one can proceed by binary search in $[\![0, n]\!]$. Following this idea, the procedure line 5 makes $\mathcal{O}(\log(n))$ calls to the NP oracle.

Theorem 2. *For any $\delta > 0$, SWITCH (F, ρ, δ) makes $\mathcal{O}(p\log(n)\log(p/\delta))$ calls to NP oracle and returns a approximation of $\rho(F)$ within $\left[\rho(F)/(2\sqrt{2}), 2\sqrt{2}\rho(F)\right]$ with probability at least $1 - \delta$.*

Theorems 1 and 2 show that WISH and SWITCH approximate the discrete integral within a factor of 8 but have dual complexities, in the sense that there is a function f, such that WISH makes $\mathcal{O}(f(n, p, \delta))$ NP oracle calls against $\mathcal{O}(f(p, n, \delta))$ calls for SWITCH. We have found this function to be $f(n, p, \delta) = n\log(p)\log(n/\delta)$. Furthermore, the analysis shows that the constants hidden by the $\mathcal{O}$ notation are of same order of magnitude. So depending on the value of n and p, one may prefer one algorithm to the other.

4.4   On the Extension to Arbitrary Precision Algorithms

In the discussion on the limitation of the estimates Sect. 3.4, we pointed out that generating approximations of W_1 or W_2, one could not hope for better than approximations of the discrete integral within a factor 2. This capped approximation factor comes from doing base 2 partitions of the integration axis when defining the quantiles weights q_i and the splitting tails τ_i. We explained that using base β (β in $]0, 1[$) partitions, one could easily find estimates of arbitrarily close approximations.

Typically WISH and SWITCH should be adapted so as to ensure arbitrarily close approximations of the quantiles weights and splitting tails defined from base β partitions. Both algorithm rely on hash functions which particularity is to halve cardinality with each new constraint added. For WISH, such hash functions are fitted when computing base 2 quantile weights, because the tails are also halved from one quantile to the next. But for base β quantile weights, we have yet to find how to adapt the algorithm. Another alternative would be to use Stockmeyer's trick for converting an algorithm $\mathcal{A}$ returning constant factor approximation into arbitrary precision algorithm by invoking $\mathcal{A}$ on multiple copies of F [1,8].

For SWITCH, there exists hashing-base algorithms for approximate model counting which return (ε, δ)-approximations [5,6,15]. SWITCH can be adapted taking inspiration from these algorithms so as to generate arbitrarily close approximations of all base β splitting tails.

5 Conclusion

In this paper, we provide a framework for developing algorithms for approximate discrete integration. In this framework, discrete integrals are transformed into integrals of non-increasing real functions which are subsequently approximated using classical methods. We build two algorithms from this framework: we demonstrate how transformations over the discrete integral give rise to two different algorithmic approaches. One approach, WISH, relies on usage of optimization queries while the other, SWITCH, reduces the problem of discrete integration to that of several unweighted counting problems. The analysis that lead to these two reductions were shown to be very alike, in that they follow similar steps in the transformation of the discrete integral. The similarity extends to the algorithms as we have shown that SWITCH makes $\mathcal{O}(p\log(n)\log(p/\delta))$ calls to NP oracle in contrast to $\mathcal{O}(n\log(p)\log(n/\delta))$ calls in the context of WISH, so that the two complexities are dual on n and p. This result provides insight on deciding which approach to use depending on the context, as the approach expected to do fewer oracle queries depends on n and p.

It would be of interest to understand empirical performance comparison of WISH and SWITCH and we hope that the aforementioned algorithmic approaches will motivate practitioners to develop the underlying required solvers: (i) SAT solvers capable of handling XOR and PB constraints, and (ii) MaxSAT solvers capable of handling XOR constraints. The current MaxSAT solvers and the CNF-PB solvers handle these XOR constraints blasting them into CNF after performing top-level Gaussian elimination. The recent success of BIRD framework owing to a tighter integration of CNF and XOR solving for CNF-XOR formulas motivates the tighter integration of (i) XOR and PB constraints, and (ii) MaxSAT solving with XOR constraints [15].

Acknowledgements. This research has been supported in part by the National Research Foundation Singapore under its AI Singapore Programme [Award Number: AISG-RP-2018-005] and the NUS ODPRT Grant [R-252-000-685-133].

References

1. Bellare, M., Petrank, E.: Making zero-knowledge provers efficient. In: Proceedings of the 24th Annual Symposium on the Theory of Computing. ACM Citeseer (1992)
2. Brooks, S., Gelman, A., Jones, G., Meng, X.L.: Handbook of Markov Chain Monte Carlo. Chapman & Hall/CRC, Hoboken (2011)
3. Carter, J.L., Wegman, M.N.: Universal classes of hash functions. J. Comput. Syst. Sci. **18**, 143–154 (1977)
4. Chakraborty, S., Fremont, D.J., Meel, K.S., Seshia, S.A., Vardi, M.Y.: Distribution-aware sampling and weighted model counting for sat. In: Proceedings of AAAI, pp. 1722–1730 (2014)
5. Chakraborty, S., Meel, K.S., Vardi, M.Y.: A scalable approximate model counter. In: Schulte, C. (ed.) CP 2013. LNCS, vol. 8124, pp. 200–216. Springer, Heidelberg (2013). https://doi.org/10.1007/978-3-642-40627-0_18

6. Chakraborty, S., Meel, K.S., Vardi, M.Y.: Algorithmic improvements in approximate counting for probabilistic inference: from linear to logarithmic SAT calls. In: Proceedings of IJCAI (2016)
7. Ermon, S., Gomes, C., Sabharwal, A., Selman, B.: Embed and project: discrete sampling with universal hashing. In: Proceedings of NIPS, pp. 2085–2093 (2013)
8. Ermon, S., Gomes, C., Sabharwal, A., Selman, B.: Taming the curse of dimensionality: discrete integration by hashing and optimization. In: Proceedings of ICML, pp. 334–342 (2013)
9. Gogate, V., Dechter, R.: Approximate counting by sampling the backtrack-free search space. In: Proceedings of the AAAI, vol. 22, p. 198 (2007)
10. Gomes, C., Sabharwal, A., Selman, B.: Near-uniform sampling of combinatorial spaces using XOR constraints. In: Proceedings of NIPS, pp. 481–488 (2006)
11. Jerrum, M.R., Sinclair, A.: The markov chain monte carlo method: an approach to approximate counting and integration. In: Approximation Algorithms for NP-Hard Problems, pp. 482–520 (1996)
12. Kitchen, N., Kuehlmann, A.: Stimulus generation for constrained random simulation. In: Proceedings of ICCAD, pp. 258–265 (2007)
13. Paredes, R., Duenas-Osorio, L., Meel, K.S., Vardi, M.Y.: Network reliability estimation in theory and practice. In: Reliability Engineering and System Safety (2018)
14. Roth, D.: On the hardness of approximate reasoning. Artif. Intell. **82**(1–2), 273–302 (1996)
15. Soos, M., Meel, K.S.: Bird: Engineering an efficient CNF-XOR sat solver and its applications to approximate model counting. In: Proceedings of AAAI Conference on Artificial Intelligence (AAAI 2019) (2019)
16. Stockmeyer, L.: The complexity of approximate counting. In: Proceedings of STOC, pp. 118–126 (1983)
17. Tzikas, D.G., Likas, A.C., Galatsanos, N.P.: The variational approximation for Bayesian inference. IEEE Sig. Process. Mag. **25**(6), 131–146 (2008)
18. Valiant, L.G.: The complexity of computing the permanent. Theoret. Comput. Sci. **8**, 189–201 (1977)
19. Wainwright, M.J., Jordan, M.I.: Graphical models, exponential families, and variational inference. Found. Trends Mach. Learn. **1**(1–2), 1–305 (2008)

Techniques Inspired by Local Search for Incomplete MaxSAT and the Linear Algorithm: Varying Resolution and Solution-Guided Search

Emir Demirović[1]($\boxtimes$) ⓘ and Peter J. Stuckey[2,3] ⓘ

[1] University of Melbourne, Melbourne, Australia
emir.demirovic@unimelb.edu.au
[2] Monash University, Melbourne, Australia
[3] Data61 CSIRO, Melbourne, Australia

Abstract. We present a MaxSAT algorithm designed to find high-quality solutions when faced with a tight time budget, e.g. five minutes. The motivation stems from the fact that, for many practical applications, time resources are limited and thus a 'good solution' suffices. We identify three weaknesses of the linear MaxSAT algorithm that prevent it from effectively computing low-violation solutions early in the search and develop a novel approach inspired by local search to address these issues. Our varying resolution method initially considers a rough view of the soft clauses (*low resolution*) and with time refines and adds the remaining constraints until the original problem is solved (*high resolution*). In addition, we combine the technique with solution-guided search. We experimentally evaluate our approach on test bed benchmarks from the MaxSAT Evaluation 2018 and show that improvements can be achieved over the baseline linear MaxSAT algorithm.

Keywords: MaxSAT · Solution-guided search · Incomplete MaxSAT

1 Introduction

Satisfiability (SAT) is a fundamental and well-known problem in computer science. Given a Boolean formula, it is concerned in determining the existence of a satisfying interpretation. Its optimisation variant, Maximum Boolean satisfiability (MaxSAT), deals with computing the interpretation that maximises satisfiability. Given the tremendous improvements in solving technology, MaxSAT has found a wide range of applications in the field of combinatorial optimisation, such as timetabling [2,13], planning, and scheduling. See [5,24] for more details.

Substantial research efforts in the MaxSAT community have been directed towards *complete* MaxSAT solving, i.e. developing algorithms that exhaustively explore the search space. In theory, these techniques guarantee to compute the optimum solution. While this is a clear strong point and has proven to be effective

© Springer Nature Switzerland AG 2019
T. Schiex and S. de Givry (Eds.): CP 2019, LNCS 11802, pp. 177–194, 2019.
https://doi.org/10.1007/978-3-030-30048-7_11

for a number of problems, for large and difficult problems, such as high school timetabling, computing the optimum solution with current technology cannot be done within a reasonable time frame.

As an alternative, *incomplete* algorithms relax the optimality criteria with the aim of providing a suitable trade-off between computational time and solution quality. It is not uncommon for complete MaxSAT algorithms to provide intermediary solutions, playing the role of both complete and incomplete approaches. However, the main focus is laid on *proving* optimality *later* rather than computing *good* solutions *early* in the search.

There has been growing interest in incomplete algorithms in recent years, with a surge of new methods at the recent MaxSAT Evaluation 2018. It has been observed that better anytime performance can be achieved when algorithmic design decisions are centred around finding high-quality solutions quickly. The algorithm presented in this paper follows this line of work.

The first step towards designing an efficient algorithm is to understand the underlying issues and limitations that are preventing current incomplete algorithms from effectively computing good solutions early in the search. We focus on the linear MaxSAT algorithm, an upper-bounding method which repeatedly calls a Satisfiability (SAT) solver, each time imposing constraints to find a solution better than previously found. We identified three core problems with the linear MaxSAT algorithm: scalability, lack of guidance towards good solutions, and a tendency to focus on poor regions of the search space.

We designed an algorithm that aims to address these issues. There are two key components to our approach: (1) a novel *varying resolution* technique and (2) directed search around the currently best-known solution. The former simplifies the formula to roughly approximate the original instance (*low resolution*) and with time refines its view until the original constraints are rebuilt (*high resolution*). The benefits are two-fold: from a strategic side, it aims to satisfy the high impact constraints early in the search, and from a practical side, it allows the linear MaxSAT algorithm to scale by reducing memory requirements. The second key component directs the solver to provide incremental improvements to the currently best-known solution. While this technique has been used in other works [6,7], we provide a subtle yet impactful variation that provides notable improvements for our purposes of incomplete solving. When the two key techniques are combined, better results are obtained over the baseline linear MaxSAT algorithm on benchmarks from the MaxSAT Evaluation 2018.

To summarise, our contributions are as follows:

- We identify three core issues with the linear MaxSAT algorithm that hinders it in computing high-quality solution early in the search.
- We develop a novel varying resolution approach and combine it with a more effective solution-guided search strategy.
- We experimentally evaluate of our algorithm in the context of *incomplete* MaxSAT solving and study the impact of each individual component. Our results demonstrate that varying resolution and solution-guided search

provide improvements over the baseline. We note that our approach was ranked as the *best* performing solver in the incomplete weighted 300 s track of the MaxSAT Evaluation 2018.

2   Preliminaries

SAT and MaxSAT. The Satisfiability problem (SAT) is concerned with deciding whether or not there exists an assignment of truth values to variables such that a given propositional logic formula is satisfied. A *literal* l is a Boolean variable x or its negation $\neg x$. A *clause* c is a disjunction of literals, $c \equiv l_1 \vee l_2 \vee \cdots \vee l_n$. A propositional formula is, for our purposes, a set of clauses understood as their conjunction, thus in *conjunctive normal form*. An assignment θ is a mapping from a set of Boolean variables $x \in vars(\theta)$ to a value *true* or *false*. We extend θ to map negative literals, by defining $\theta(\neg x) = \neg\theta(x)$. An assignment θ satisfies a clause c, written $\theta \models c$, if for some literal l in the clause c, $\theta(l) = true$. In Partial Weighted MaxSAT, clauses are partitioned into hard H and soft S clauses. Each soft clause c is given a weight $w(c)$. The goal is to find an assignment that satisfies the hard clauses and minimises the weighted sum of the unsatisfied soft clauses. An alternative viewpoint for MaxSAT [9], which we adopt throughout this paper, is to associate an *objective variable* with each soft clause and state the problem as satisfying hard clauses while minimising the weighted sum of objective variables. See [10] for more information on SAT and MaxSAT.

CDCL Solvers for SAT [31]. The state of the art for solving SAT problems is based on conflict driven clause learning. The key components are unit propagation, activity based search, and clause learning. Unit propagation of a set of clauses P and a partial assignment θ, repeatedly finds a clause $c \equiv l_1 \vee l_2 \vee \cdots l_n \in P$ where $\theta(l_i) = false, 1 \leq i < n$ for all literals and extends θ so that $\theta(l_n) = true$. The literal has c recorded as its reason for becoming true. If $\theta(l) = false$ for all literals l_i in c the solver detects unsatisfiability. The SAT solver applies unit propagation to extend an initially empty assignment θ. Afterwards, the solver chooses a literal and extends θ to make the literal *true* (treating it as an assumption) and applies unit propagation again. The choice of literal is usually based on the variables that have been in the most recent failures. On detecting unsatisfiability, the solver performs conflict analysis to create a nogood/learned clause which is added to the set of clauses to be solved. Solving continues until either a satisfying assignment is discovered, or unsatisfiability is proven.

Phase Saving [27,29]. SAT solvers repeatedly make decisions on both branching variables and values. Variables are chosen based on their recent activity in conflicts (VSIDS scheme [25]). A wide-spread approach for truth value assignment is based on phase saving [27], where the solver selects the most recently used value in the search for the variable. Therefore, after backtracking, the solver aims to return to its previous state as closely as possible. Hence, clauses learnt about the previous region of the search space will still be relevant.

The Generalised Totaliser [19]. The Pseudo-Boolean constraint $\sum w_i \cdot x_i <$ k is converted into propositional logic by encoding a binary tree, where the leaf nodes are the input variable. Each parent node contains weighted variables that represent partial sums of its children. The root contains the variables that represent the total sum of the input variables. The desired constraint is obtained by forcing violating output variables to false. The encoding roughly depends on the number of distinct weights, as this is related to the number of possible partial sums. Thus, the encoding is *pseudo-polynomial*, but does not depend on the magnitude of the weights, which can be seen as a unique advantage.

Algorithm 1: The Linear algorithm for MaxSAT

Input: A set of hard H clauses and objective variables X. Each $x_i \in X$ is associated with a weight $w(x_i)$.

Output: An optimal solution θ^* minimising $\sum_{x_i \in X} w(x_i) \cdot x_i$

1 **begin**
2     $\theta^* \longleftarrow \emptyset$
3     $P \longleftarrow H$
4   **while** $\exists\, \theta, \forall c \in P.\theta \models c$ **do**
5        $\theta^* \longleftarrow \theta$
6        $k \longleftarrow cost(\theta, X)$
7        $P \longleftarrow P \cup (\sum_{x_i \in X} w(x_i) \cdot x_i < k)$
8   **return** θ^*

Example 1. Consider the encoding of the pseudo-Boolean constraint $8x_1 + 5x_2 + 3x_3 + x_4 < 9$. We create a node n representing the sum $n = 8x_1 + 5x_2$ defined by Booleans $[\![n \geq 5]\!]$ and $[\![n \geq 8]\!]$ and clauses $x_1 \rightarrow [\![n \geq 8]\!]$, $x_2 \rightarrow [\![n \geq 5]\!]$ and $x_1 \wedge x_2 \rightarrow false$. The last clause encodes the fact that the partial sum is already too big. Similarly we create a node m representing the sum $m = 3x_3 + x_4$ using Booleans $[\![n \geq 1]\!]$, $[\![n \geq 3]\!]$, $[\![n \geq 4]\!]$ and clauses $x_3 \rightarrow [\![n \geq 3]\!]$, $x_4 \rightarrow [\![n \geq 1]\!]$ and $x_3 \wedge x_4 \rightarrow [\![n \geq 4]\!]$. The root node s encoding the entire sum is encoded using Booleans $[\![s \geq 1]\!]$, $[\![s \geq 3]\!]$, $[\![s \geq 4]\!]$, $[\![s \geq 5]\!]$, $[\![s \geq 8]\!]$ and the clauses $[\![m \geq 1]\!] \rightarrow [\![s \geq 1]\!]$, $[\![m \geq 3]\!] \rightarrow [\![s \geq 3]\!]$, $[\![m \geq 4]\!] \rightarrow [\![s \geq 4]\!]$, $[\![n \geq 5]\!] \rightarrow [\![s \geq 5]\!]$, $[\![n \geq 5]\!] \wedge [\![m \geq 1]\!] \rightarrow [\![s \geq 6]\!]$, $[\![n \geq 8]\!] \rightarrow [\![s \geq 8]\!]$, $[\![n \geq 5]\!] \wedge [\![n \geq 3]\!] \rightarrow [\![s \geq 8]\!]$, $[\![n \geq 5]\!] \wedge [\![m \geq 4]\!] \rightarrow false$, $[\![n \geq 8]\!] \wedge [\![m \geq 1]\!] \rightarrow false$, $[\![n \geq 8]\!] \wedge [\![m \geq 3]\!] \rightarrow false$, and $[\![n \geq 8]\!] \wedge [\![m \geq 4]\!] \rightarrow false$ In fact the s literals are not needed, they are included to show the general process of building a node from two children. We only need to keep the clauses encoding incompatible combinations of n and m

Note that we encode the constraint $800x_1 + 500x_2 + 300x_3 + 100x_4 < 900$ identically. □

The Linear MaxSAT Algorithm [14,20,21,23]. The optimal solution to a MaxSAT instance can be obtained by solving a series of SAT problems. This is depicted in Algorithm 1. It makes repeated calls to a SAT solver. After each call, it adds a pseudo-Boolean constraint to the formula that enforces the formula to only admit solutions that have a cost strictly lower than the current best solution. In Sect. 5, we discuss different encodings for the pseudo-Boolean constraint and in the rest of the paper focus our attention on the generalised totaliser encoding (see above). The algorithm iterates until it proves unsatisfiability, in which case the optimal solution was computed in the previous iteration.

3 Algorithm

Our algorithm is designed for short run times, e.g. five minutes. The assumption is that proving optimality within the given time frame is infeasible. Thus, the aim is to find 'good solutions' early during the search. The main challenge is to determine a strategy which can identify where the 'good' solutions reside in the search space and ensure scalability across a wide range of benchmarks.

Our approach is based on the linear MaxSAT algorithm. This method was chosen as it was the best performing solvers in the *incomplete unweighted 60 s* track of the MaxSAT Evaluation 2017. In addition, it has shown competitive resulting for certain applications, e.g. high school timetabling [13]. Two techniques play a key role in our algorithm: (1) a novel *varying resolution* approach and (2) directed search around the currently best-known solution.

3.1 Issues with the Linear MaxSAT Algorithm

To obtain a better understanding of our algorithm, it is important to note the core issues with the linear MaxSAT algorithm. We identified three main issues in the context of incomplete saving: (1) scalability and sensitivity to the values of the weights of the objective variables, (2) lack of a strategy to guide the search towards solutions with low objective value, and (3) proneness to falling in "local optima", i.e. excessively spending efforts proving unsatisfiability in a certain region of the search space rather than exploring a different part of the search space. The varying resolution approach aims to address the first two points, while the directed search tackles the third point and partially the second. These issues are described in greater detail below.

Issue #1: Scalability and Weight-Value Sensitivity. The linear MaxSAT algorithm encodes a single large pseudo-Boolean constraint, which is directly dependent on the values of the weights of the objective variables.

The generalised totaliser pseudo-Boolean encoding [19], used in this work, roughly depends on the number of unique values of the weights. Therefore, when faced with a large number of diverse weights, the number of clauses and auxiliary variables required to encode the pseudo-Boolean can be prohibitively high. As a result, the pseudo-Boolean encoding can dominate the algorithmic performance and become a bottleneck. Other encodings suffer from related issues.

We note that MaxSAT algorithms that do not require explicitly encoding the pseudo-Boolean constraints are largely unaffected by the variety in weights, e.g. core-guided approaches. On a related note, WPM3 [6] uses the splitting rule to allow using an encoding where the weights are equal, i.e. cardinality constraints. However, these algorithms focus on increasing the lower bound rather than computing good solutions early in the search.

Issue #2: Lack of Guidance Towards Good Solutions. In each iteration of the linear algorithm, the SAT solver merely seeks to find a satisfying assignment and not necessarily a solution with low cost. Therefore, there is no guidance towards good solutions, which might lead the algorithm to spend excessive time searching in areas that potentially fine-tune small improvements to the objective even though the crucial soft constraints are left unattended.

Issue #3: Tendency to Focus on Poor Regions of the Search Space. This issue is linked to the underlying value-selection heuristic of the SAT algorithm: phase saving. While phase saving is known to be effective for pure satisfiability problems, it can introduce undesired behaviour when used in the linear MaxSAT algorithm. The problem stems from the fact that upon conflict detection and backtracking, phase saving aims to drive the search back into a similar region of the search space as before. This is systematically done through value-assignments for variables: once a new variable is selected, the value most recently used for that variable will be assigned to it. As a result, once the algorithm reaches a region of the search space where there are no better solutions, it will effectively spend its efforts in proving unsatisfiability. Unfortunately, this can be time-consuming and does not lead to finding good solutions quickly.

3.2 Our Approach

There are two key components in our algorithm: the varying resolution approach and solution-guided search, complementary techniques that aim to address the identified issues of the linear MaxSAT algorithm. The former ensures scalability (Issue #1) and guides the search towards good solutions on a high level (Issue #2), while the latter provides incremental improvements to the current best solution (Issues #2 and #3). These components are built into the linear MaxSAT algorithm and exhibit a high degree of synergy, resulting in a better algorithm than the baseline linear MaxSAT algorithm.

Key Component #1: Varying Resolution Approach. The aim of this part is to address Issue #1 and #2. It starts by viewing the MaxSAT formula in *low resolution* by decreasing the weights for all constraints. The weights reduced to zero are removed. After the resulting problem is solved, the weight values are increased (*increase the resolution*), a portion of the previously ignored constraints are added, and the problem is resolved. This process iterates until the problem is viewed in *high resolution*, i.e. the original formula is restored and solved. Weight adjustment results in a heuristic that approximates the formula and reduces the memory requirements, which in turns offers speed-ups. In theory, the procedure preserves completeness, i.e. does not remove any optimal solution, but in practice, only a few iterations of the algorithm are executed within the allocated time resources. We discuss related approaches, namely stratification for core-guided approaches [4] and weight-clustering [18], in Sect. 5.

Algorithm 2: Compute the initial cutoff value

Input: A set of objective variables X, a mapping $w : X \to \mathbf{N}$, and the threshold coefficient $\beta \in [0, 1]$.

Output: Initial cutoff value d

```
 1 begin
 2 │   S ⟵ ∑_{x∈X} w(x)
 3 │   k ⟵ max_{x∈X}{dec_digits(w(x))}
 4 │   for i = 1..k do
 5 │   │   frac_sum[i] ⟵ ∑_{x∈X∧dec_digits(x)=i} w(x)
 6 │   d ⟵ k
 7 │   for i = 1..k do
 8 │   │   if frac_sum[i] ÷ S ≥ β then
 9 │   │   │   d ⟵ i
10 │   │   │   break
11 │   return 10^(d−1)
```

To explain our algorithm in detail, we first discuss the initial cutoff value computation, present the varying resolution approach, and lastly describe our modification to the linear MaxSAT algorithm.

Initial Cutoff Value Computation. Algorithm 2 describes the procedure. The goal is to determine a cutoff threshold that partitions the objective variables into low- and high-weighted variables. It first computes: S - the sum of the weights, k - the number of decimal digits used to represent the largest weight, and $frac\_sum$ - the array where $frac\_sum[i]$ represents the sum of weights with exactly i decimal digits. Note that the number of digits is computed as $dec\_digits(x) = \lfloor log_{10}(x) \rfloor + 1$. Afterwards, the cutoff value is chosen based on the total contribution of weights with precisely d digits with respect to the overall MaxSAT problem. The smallest value d that meets the specified threshold β is selected, or the default value k if no such value exists. The cutoff value is returned as $10^{(d-1)}$. The intuition is that the cutoff point discriminates weights between those that contribute significantly towards the objective and those that do not. The parameter $\beta \in [0, 1]$ regulates the sensitivity of the division: lower/higher values for β lead to lower/higher cutoff values.

Example 2. Consider the formula with $X = \{x_i : i \in \{0, 1, ..., 8\}\}$ and $w = \{w_0 \mapsto 1200, w_1 \mapsto 800, w_2 \mapsto 700, w_3 \mapsto 500, w_4 \mapsto 50, w_5 \mapsto 15, w_6 \mapsto 9, w_7 \mapsto 8, w_8 \mapsto 2\}$ and parameter $\beta = 0.20$. The sum of weights is 3284 and $frac\_sum = \{1 \mapsto 19, 2 \mapsto 65, 3 \mapsto 2000, 4 \mapsto 1200\}$. The inner *if* condition is not satisfied for $i \in \{1, 2\}$, as neither $\frac{19}{3284} \geq 0.20$ nor $\frac{65}{3284} \geq 0.20$, but will trigger for $i = 3$ since $\frac{2000}{3284} \geq 0.20$. Therefore, $d = 3$ and the returned cutoff is 100.   □

Varying Resolution. Algorithm 3 gives an overview. The algorithm starts by computing the initial *cutoff value* (Algorithm 2). Iteratively, a new MaxSAT formula is built, where the hard constraints are as in the original formula, and the weights of objective variables are divided by the cutoff value (rounded down). The new formulation, along with the best solution found so far θ^* and the original formula, are used to initialise the linear MaxSAT algorithm. The best solution is used by the solution-guided search component, while the original formula is required due to our previously discussed modification of the linear MaxSAT algorithm (see next subsections for both points). After the resulting formula is solved, the cutoff value is decreased and the process is repeated until the original formula is solved. Note that if the sum of the weights is lower than a given parameter α, the varying resolution approach is deemed unnecessary, i.e. the cutoff is set to one and the algorithm proceeds as a linear MaxSAT algorithm with solution-guided search (see component #2). The procedure can be viewed as a search by exponentially decreasing steps, where the approximate objective function is refined at an exponential rate each iteration until the original objective is restored.

Example 3. (continued) Let $\alpha = 1000$. As $\sum_{x_i} w(x_i) \geq \alpha$, the cutoff d is set to 100 (see Example 2). Therefore, $w' = \{x_0 \mapsto 12, x_1 \mapsto 8, x_2 \mapsto 7, x_3 \mapsto 5, x_4 \mapsto 0, x_5 \mapsto 0, x_6 \mapsto 0, x_7 \mapsto 0, x_8 \mapsto 0\}$, and $X' = \{x_0, x_1, x_2, x_3\}$. After the simplified formula is solved, d is decreased to 10 and the process is repeated. □

Learned clauses are kept as usual during the search within each individual iteration, but the SAT solver is rebuilt at the beginning of each iteration (Algorithm 3, line 11). Learned clauses are not shared in between iterations of varying resolution, as learned clauses in one iteration might refer to auxiliary variables in the pseudo-Boolean encoding that are no longer present in the next iteration.

The approximate objective function requires fewer auxiliary variables and clauses than the original pseudo-Boolean constraint. Recall that the size of the generalised totaliser encoding [19] is related to the number of unique weight values, e.g. the smallest encoding is obtained if all weights are the same value. Dividing the weights by the cutoff results in fewer unique weights, leading to a smaller encoding, which in turn reduces the memory requirements. Note that varying resolution is designed for shorter run times and thus are not particularly suitable for longer runtimes, i.e. the last iteration of varying resolution is the standard linear MaxSAT algorithm.

Observation 1. *Given an initial constraint* $I \equiv \sum_{x \in X} w(x) \cdot x \leq ub - 1$, *the* r *rounded version of the constraint is given by* $I_r \equiv \sum_{x \in X} \lfloor \frac{w(x)}{r} \rfloor \cdot \leq \lfloor \frac{ub-1}{r} \rfloor$. *It follows that* $I \models I_r$, *i.e. all solutions of* I *are also solutions of* I_r, *since* I_r *is a Gomory cut [16] derived from* I. *Hence, adding this constraint does not exclude any optimal solution to the original problem.*

Observation 2. *The varying resolution algorithm is complete regardless of the choice for parameters* α *and* β.

Observation 3. *The varying resolution is anytime, i.e. it provides intermediary results during its execution.*

Linear MaxSAT Modification. The standard linear MaxSAT algorithm is modified as follows. It additionally stores the original MaxSAT formula and the best

---

Algorithm 3: The Varying Resolution Approach

Input: A set of hard H clauses and objective variables X, a mapping $w : X \rightarrow \mathbf{N}$, and threshold coefficients $\alpha \in N$ and $\beta \in [0, 1]$

Output: An optimised solution θ^*

1 **begin**
2     $\theta^* \longleftarrow \emptyset$
3   **if** $\sum_{x \in X} w(x) \geq \alpha$ **then**
4        $d \longleftarrow compute\_initial\_cutoff(X, \beta)$
5   **else**
6        $d \longleftarrow 1$
7     $cutoff \longleftarrow 10^{d-1}$
8   **while** $cutoff \geq 1$ **do**
9        $w'(x) = \lfloor \frac{w(c)}{cutoff} \rfloor$
10        $X' = \{x : x \in X \wedge w'(x) > 0\}$
11        $solver \longleftarrow initialiseMaxSAT(H, X, w, X', w')$
12        $solver.setInitialSolution(\theta^*)$
13        $\theta^* \longleftarrow solver.solve()$
14        $cutoff \longleftarrow \lfloor \frac{cutoff}{10} \rfloor$
15   **return** θ^*

---

solutions with respect to the current and the original MaxSAT formula. Note that an assignment with a lower objective value for the simplified problem in the varying resolution approach does not necessarily lead to a better solution for the original problem. Therefore, once a new assignment is computed, its cost is computed with respect to the original formula, and it is kept as the globally best solution if its cost is lower than the previous best solution. Regardless of the outcome, the algorithm proceeds as usual, i.e. adds the upper bound with respect to the newly found locally best solution. Thus, it optimises its current problem, but only updates the global solution if it is better with respect to the original MaxSAT formula.

Example 4. Considering the formula within the varying resolution approach with $d = 10$: $w' = \{x_0 \mapsto 120, x_1 \mapsto 80, x_2 \mapsto 70, x_3 \mapsto 50, x_4 \mapsto 5, x_5 \mapsto 1, x_6 \mapsto 0, x_7 \mapsto 0, x_8 \mapsto 0\}$. The linear MaxSAT algorithm is called and assume it finds the solution θ that only violates x_5. The objective value of θ is 1 locally and 15 globally. Both values are kept as these are the best values found in their respective categories. The pseudo-Boolean constraint $\sum w'(x) < 1$ is added to the MaxSAT formula and the SAT solver is called again. Now assume the solver finds a new solution that violates x_6 and x_7. The solution is kept as the best local solution ($\sum w'(x) = 0$), but the best global is not updated ($\sum w(x) = 17 \geq 15$). As locally no further improvements can be made, the linear MaxSAT algorithm stops, leading to a new iteration of the varying resolution approach with $d = 1$ where θ is passed as the initial solution. $\square$

Key Component #2: Solution-Guided Search. In local search, *intensification* aims to provide improvements to the solution by searching through a neighbourhood of solutions *close* to the current solution. Thus, a better solution is found by iteratively performing small incremental changes to the currently considered solution, driving the solution into a (local) minima. The essence of this idea can be captured in a complete search algorithm by using the following value-selection heuristic: once a branching variable has been chosen, assign the value to the variable that it assumes in the best-known solution. Hence, the search progresses *close* to the best-known solution, resembling local search. In our algorithm, we use this value-selection heuristic, as it partially addresses issues #2 and #3 of the linear MaxSAT algorithm.

Similar techniques were used under various names, e.g. solution-based phase saving [1,6,12], solution-guided search [7], and large neighbourhood search [30]. The phase saving strategy used in WPM3 [6] is the closest to our work. The difference is subtle yet impactful: we apply solution-guided search to *all* variables in the MaxSAT formula, including auxiliary variables introduced by the pseudo-Boolean encoding, as opposed to only considering variables that appear in the original MaxSAT formula as in WPM3 [6]. For our experimental setting, our strategy proved to be more effective, but we note that WPM3 considered a different setting for their phase saving, i.e. it was considered for solving subproblems generated during the search with the aim of increasing the lower bound.

4   Experimental Results

We performed a detailed computational study to empirically evaluate the effect of varying resolution and solution-guided search.

4.1   Setting

Our setting is the same as in incomplete track of the MaxSAT Evaluation 2018. Thus, we consider unweighted and weighted benchmarks with 60 and 300 s timeouts, for a total of four separate settings. The evaluation uses industrial and application benchmarks. The comparisons are performed on a total of 153 and 172 unweighted and weighted benchmarks, respectively. The experiments were performed on the StarExec cluster, allocating 32 GB of RAM per benchmark.

Scoring. The scoring of a solver for the incomplete track is the sum of scores s_i for each instance. For instance i, a solver finding a solution with objective o_i is awarded score $s_i = bo_i/o_i$ where bo_i is the best objective found by any solver on that instance during the 60 and 300 s runs. If a solver finds no solution for an instance i, the corresponding score is $s_i = 0$. The best solution is taken from the 300 s track.

Our Solver: LinSBPS. We implemented varying resolution and solution-guided search in Open-WBO [23], an open-source MaxSAT solver.

Other Solvers. The remaining solvers used in the evaluation are discussed in more detail in the Sect. 5.

4.2 Results and Discussions

We provide experiments that support our previous claims. The same timeouts and benchmarks are used across experiments. Note that the *score* metric is relative to the solvers considered, i.e. the score for a particular benchmark depends on the best solution computed by the considered solvers. Hence, the score values may differ in different experiments.

Effect of Varying Resolution and Solution-Guided Search. In Table 1a we compare the performance of our techniques compared to the baseline linear algorithm. We consider four variants, depending on whether varying resolution and solution-guided search is used. Note that varying resolution is only used for weighted benchmarks which are deemed as *large enough*, as detailed in Algorithm 3. For the considered benchmark set, varying resolution was used on 89 out of 172 benchmarks (51%).

Each component, varying resolution and solution-guided search, improves the baseline. The best approach is obtained by combining both techniques.

Number of Unique Weights Produced. Varying resolution reduces the number of unique weights in the benchmarks, thus leading to more compact encodings with the GTE. On average, the number of distinct weights drops from 1342 to 29 in the first iteration of varying resolution.

Number of Objective Variables Considered. The underlying MaxSAT formula is simplified with varying resolution. Nevertheless, most of the objective variables are still taken into account, even in the first iteration. On average, the algorithm considers 87% percent of the total number of objective variables in the first iteration of varying resolution.

Number of Iterations Performed. For the benchmarks that use varying resolution, on average, 1.19 iterations were executed with 6.04 iterations needed to restore the original formula.

GTE vs. Adder Pseudo-Boolean Encoding. One of the benefits of varying resolution comes from its ability to produce a smaller pseudo-Boolean encoding. As an alternative, in Table 1b we consider the adder encoding, which represents numbers in binary form and encodes binary adders. This allows for a significant reduction in the encoding size, at the expense of arc consistency.

Our comparison was done only on the benchmarks that use varying resolution. However, while effective for complete solving [20], the adder encoding shows weaker performance for incomplete solving. We believe the loss of arc consistency for the adder encodings forces the solver to spend more time in search, which is detrimental given the tight time budget.

Solution-Guided Search. In Table 1c we compare our variant of solution-guided search with the phase saving strategy used in WPM3 [6]. The difference in the techniques is subtle yet impactful. Nevertheless, regardless of the variant chosen, incrementally improving an existing solution proved to be beneficial for incomplete MaxSAT solving. Compared to WPM3 [6], our variant considers all variables and not only the original variables. This proved to be advantageous for our setting. As discussed previously in Sect. 3.2, our experimental setting differs from the one considered in WPM3, and hence it must be emphasised that our claims only hold for our particular case of incomplete solving with the linear algorithm. The auxiliary variables in the formula are implied by the original variables in the pseudo-Boolean constraint. Thus, following the idea of remaining *close* to the best solution, all variables must be considered. Setting a different value to an auxiliary variable reflects on the original variables, which was undesirable in our setting.

Parameter Choice. The parameter α defines the size of the benchmark required to activate varying resolution, while β is used to discriminate between more and less important weights. The final values chosen in the solver are $\alpha = 5 \cdot 10^5$ and $\beta = 0.05$. Note that no parameter tuning was performed. The parameter choice discussion that follows is presented as a post-analysis.

Varying resolution is activated when the sum of weights in a benchmark exceeds the threshold α. To study other possible choices for α, we sort the considered benchmarks by the sum of their weights. The 83rd smallest value is $489 \cdot 10^3$. However, the 73rd and 93rd are $71 \cdot 10^3$ and $1603 \cdot 10^3$, exhibiting substantial differences in values. Thus, α can be varied significantly with little effect. Therefore, we selected $\alpha = 5 \cdot 10^5$ in an *ad hoc* manner and decided not to fine-tune the parameter on the previous competition benchmark set, as doing so would likely lead to overfitting.

Parameter β was kept low since our intention was to discard low-valued weights that increase the encoding size but do not provide a significant difference in the objective. Thus, we selected $\beta = 0.05$, i.e. we stop discarding weights with d digits if their contribution is at least 5%.

Comparison with the MaxSAT Evaluation 2018 Solvers. In Table 2 we show the results from the MaxSAT Evaluation 2018 as a comparison with other state-of-the-art incomplete MaxSAT solvers. Our solver, LinSBPS, uses the techniques described in this paper. Our approach can be further improved using *core-boosting* [8] as a preprocessing step, but the main aim of these experiments is to demonstrate the effectiveness of the techniques presented in this paper.

Weighted Track. Our algorithm achieved the *best* rank in the 300 s category. A detailed view of these results is given in Fig. 1 (top), which shows the distribution of scores per instance. For each solver, the scores for every instance is computed, the resulting array is sorted, and then plotted as a curve. We can see that our approach provides highly competitive results for the majority of the benchmarks, with only a handful of cases where the score is below 0.8. This illustrates the robustness of our technique when handling a diverse set of benchmarks. For the 60 s track, our approach is ranked second.

Our approach takes into account most of the objective variables. Open-WBO-Inc-BMO, as a solver with comparable performance, in contrast, initially aggressively optimises the most important constraints. This seems to provide better results for 60 s runs. However, as more time is allocated, our approach is able to exploit a broader view of the problem, while the other approach keeps optimising a rough approximation.

Unweighted Track. Our approach solely relies on solution-guided search to provide improvements over the baseline for the unweighted track. Nevertheless, our method ranked second and third in the 60 and 300 s track, respectively. From Fig. 1 (bottom), we can see that there is a higher deviation in solver performance depending on the benchmark. While for the weighted benchmarks our approach achieved consistently good results when compared with others, for the unweighted track there are no robust solvers: the score distributions are scattered across the interval [0.1, 1] for each solver. We believe this is because it is harder to identify the key constraints for unweighted compared to the weighted instances, and thus there is a higher fluctuation between the results. The best performing solver in the unweighted track, SATLike, is a local search solver specialised in exploring different areas of the search space quickly rather than using sophisticated reasoning technique such as CDCL, which could explain its effectiveness for these benchmarks.

Table 1. Comparison of different variants of our approach. 300 s. (a) The effect of each individual component; (b) Comparison with the adder encoding; (c) Comparison with solution-guided search used in WPM3: SGS(OV).

(a)		(b)		(c)	
Solver	Score	Solver	Score	Solver	Score
VR+SGS	162.00	**VR+SGS**	161.55	**VR+SGS**	161.03
SGS	144.46	VR	140.05	VR+SGS(OV)	148.73
VR	140.47	Adder+SGS	129.87	SGS	143.81
Baseline	128.8	Baseline+Adder	125.26	SGS(OV)	132.43

Table 2. Results from the MaxSAT Evaluation 2018. The *score* listed for solvers

(a) Weighted 60 s		(b) Weighted 300 s		(c) Unweighted 60 s		(c) Unweighted 300 s	
Solver	Score	Solver	Score	Solver	Score	Solver	Score
Open-WBO-Inc-BMO	0.810	LinSBPS	0.900	SATLike-c	0.735	SATLike-c	0.854
LinSBPS	0.799	Open-WBO-Inc-BMO	0.842	LinSBPS	0.705	maxroster	0.829
maxroster	0.773	maxroster	0.804	SATLike	0.675	LinSBPS	0.782
Open-WBO-Inc-Cluster	0.743	Open-WBO-Inc-Cluster	0.762	Open-WBO-Inc-OBV	0.654	SATLike	0.702
SATLike-c	0.696	SATLike-c	0.747	Open-WBO-Inc-MCS	0.631	Open-WBO-Inc-OBV	0.842
Open-WBO-Gluc	0.669	SATLike	0.702	Open-WBO-Gluc	0.612	Open-WBO-Inc-MCS	0.762
SATLike	0.661	Open-WBO-Gluc	0.68	Open-WBO-Riss	0.564	Open-WBO-Gluc	0.68
Open-WBO-Riss	0.638	Open-WBO-Riss	0.663	maxroster	0.541	Open-WBO-Riss	0.663

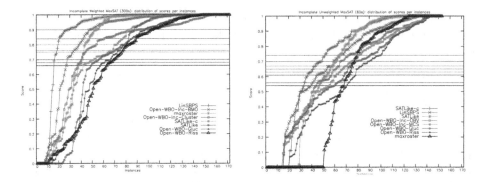

Fig. 1. Detailed results for the weighted 300 s (left) and unweighted 60 s (right) track. Image courtesy of the MaxSAT Evaluation 2018.

5   Related Work

Local Search. These methods share a common pattern: start by generating a random assignment and iteratively select a variable from an unsatisfied clause and flip its assignment. Complex reasoning mechanisms are typically not employed as in complete solvers, e.g. CDCL. Rather, the success of the methods comes from their ability to rapidly explore a large number of solutions through the use of specialised data structures, careful implementation, and heuristics.

In the recent MaxSAT Evaluation, *SATLike* [22] won the unweighted incomplete track and demonstrated good performance in the weighted track of the recent MaxSAT Evaluation 2018. Its key component is a novel weighting scheme that dynamically changes the weights of clauses during the search.

Complete Techniques for Incomplete Solving. In some cases, complete algorithms report intermediate solutions and thus can take the role of incomplete methods. The linear MaxSAT [14] solvers fall into this category and approaches differ in the way the upper bounding constraint is handled: *SAT4J* [21] uses linear propagators, a technique from constraint programming to avoid explicitly encoding the constraint into Boolean formula, while *QMaxSAT* [20] and Open-WBO-Gluc use the adder and GTE pseudo-Boolean encoding, respectively. The winner of the weighted incomplete track in 2017, *maxroster* [32], uses a stochastic solver to produce an initial solution before applying complete-based techniques.

Core-guided approaches [3,15,17,26] consider an initial SAT formula where the soft clauses are treated as hard clauses. Iteratively, a SAT solver is used to compute either a satisfying assignment, which would represent the optimal solution, or an *unsatisfiable core*, i.e. a subset of clauses that cannot simultaneously be satisfied. The core is used to rewrite the formula, e.g. relax the formula by allowing at most one of the clauses from the core to be unsatisfied. *Hitting-set* approaches [11,28] utilise unsatisfiable cores to separate MaxSAT solving into a SAT and an integer programming component. These approaches are inherently lower bounding and in their pure form do not produce any solution other than the optimum, rendering them inapplicable to incomplete scenarios. However, when combined with other techniques, unsatisfiable cores can be used for anytime algorithms. For example, WPM3 [6] is a core-guided solver with a stratified approach [4], where initially only a subset of the soft clauses with the highest weights are considered, and after satisfiability is detected, a portion of the remaining soft clauses are added and the process is repeated. Obtaining cores with high weights contribute towards faster lower bounds, and during this process upper bounds are additionally computed. This was the best incomplete solver in the 2016 evaluation. At a high level, our strategy resembles the stratified method, but the underlying solving process and the reasoning behind the techniques make a clear distinction between the approaches.

Core-boosting [8] has recently been proposed to improve linear MaxSAT algorithms. The main idea is to run the linear MaxSAT algorithm after performing core-based rewriting for a limited time. The resulting formula has fewer soft clauses, which simplifies the pseudo-Boolean constraint required in the linear algorithm. Core-boosting can be seen as a form of preprocessing and can be combined with other linear algorithms, such as the one presented in this paper.

Incomplete Weight-Relaxation. Inc-BMO and Inc-Cluster [18] from the MaxSAT Evaluation bear the most similarity to our approach.

These methods cluster the objective variables. Each variable in a cluster is reassigned a *representative weight* as follows: the array of weights is sorted and the difference between adjacent elements is computed. The top $k - 1$ indices with the highest differences are selected, effectively partitioning the weights into k clusters. Each variable within a cluster is reassigned the arithmetic mean of the weights in the group. As a result, there are at most k different weights values.

The two approaches, Inc-BMO and Inc-Cluster, differ in the next step. In Inc-Cluster, a linear MaxSAT algorithm is applied to the new formula. In Inc-BMO, the resulting formula is solved as a lexicographical optimisation problem: the problem is solved considering only the variables with the highest weight, the sum of their violations is fixed to the computed value, and the process is repeated with the second-highest weighted variables, and so forth.

There are two main reasons for the success of these methods: (1) reducing the number of distinct weights results in more compact pseudo-Boolean encodings, increasing performance, and (2) Inc-BMO aggressively optimises the most important constraints. Note that, once the problem is simplified, the effects are irreversible. Therefore, each clustering choice plays an important role. The methods excel for problems where the clustering can be done effectively.

To further illustrate the difference with our approach, consider the MaxSAT problem from Example 2. For $k = 2$, the clustering algorithms partitions the weights into $P_1 = \{w_0, w_1, w_2, w_3\}$ and $P_2 = X - P_1$. Thus, violations within each cluster are treated equally. This is not an issue at the start of the algorithm. However, as the algorithm progresses, some form of refinement is necessary to provide better results. Varying resolution with a cutoff of 100 initially considers only P_1 as BMO-INC, but is able to differentiate between violations, i.e. $w' = \{12, 8, 7, 5\}$. Note that each iteration thereafter refines the formula.

6 Conclusion

We developed a novel approach to incomplete MaxSAT solving consisting of two key components: our *varying resolution* approach, and solution-guided search. The former initially views the problem in low-resolution and with time refines the constraints until the formula is solved in high resolution, i.e. the original problem. Solution-guided search provides incremental improvements by searching close to the current best solution. Overall, our algorithm has proven to be highly effective for short runtimes, placing first in the incomplete weighted 300 s track of the MaxSAT Evaluation 2018.

Acknowledgements. We would like to thank the anonymous reviewers for their valuable feedback in preparing the final version of this paper.

References

1. Abío, I., Deters, M., Nieuwenhuis, R., Stuckey, P.J.: Reducing chaos in SAT-like search: finding solutions close to a given one. In: Sakallah, K.A., Simon, L. (eds.) SAT 2011. LNCS, vol. 6695, pp. 273–286. Springer, Heidelberg (2011). https://doi.org/10.1007/978-3-642-21581-0_22

2. Achá, R.A., Nieuwenhuis, R.: Curriculum-based course timetabling with SAT and MaxSAT. Ann. Oper. Res. **218**(1), 71–91 (2014)

3. Alviano, M., Dodaro, C., Ricca, F.: A MaxSAT algorithm using cardinality constraints of bounded size. In: Proceedings of IJCAI 2015 (2015)

4. Ansótegui, C., Bonet, M.L., Gabàs, J., Levy, J.: Improving SAT-based weighted MaxSAT solvers. In: Milano, M. (ed.) CP 2012. LNCS, pp. 86–101. Springer, Heidelberg (2012). https://doi.org/10.1007/978-3-642-33558-7_9

5. Ansótegui, C., Bonet, M.L., Levy, J.: Solving (weighted) partial MaxSAT through satisfiability testing. In: Proceedings of SAT 2009, pp. 427–440 (2009)

6. Ansótegui, C., Gabàs, J.: WPM3: an (in)complete algorithm for weighted partial MaxSAT. Artif. Intell. J. **250**, 37–57 (2017)

7. Beck, J.C.: Solution-guided multi-point constructive search for job shop scheduling. J. Artif. Intell. Res. **29**, 49–77 (2007)

8. Berg, J., Demirović, E., Stuckey, P.J.: Core-boosted linear search for incomplete MaxSAT. In: Rousseau, L.-M., Stergiou, K. (eds.) CPAIOR 2019. LNCS, vol. 11494, pp. 39–56. Springer, Cham (2019). https://doi.org/10.1007/978-3-030-19212-9_3

9. Berg, J., Järvisalo, M.: Unifying reasoning and core-guided search for maximum satisfiability. In: Calimeri, F., Leone, N., Manna, M. (eds.) JELIA 2019. LNCS (LNAI), vol. 11468, pp. 287–303. Springer, Cham (2019). https://doi.org/10.1007/978-3-030-19570-0_19

10. Biere, A., Heule, M., van Maaren, H., Walsh, T. (eds.): Handbook of Satisfiability, volume 185 of Frontiers in Artificial Intelligence and Applications. IOS Press, Amsterdam (2009)

11. Davies, J., Bacchus, F.: Solving MAXSAT by solving a sequence of simpler SAT instances. In: Lee, J. (ed.) CP 2011. LNCS, vol. 6876, pp. 225–239. Springer, Heidelberg (2011). https://doi.org/10.1007/978-3-642-23786-7_19

12. Demirović, E., Chu, G., Stuckey, P.J.: Solution-based phase saving for CP: a value-selection heuristic to simulate local search behavior in complete solvers. In: Hooker, J. (ed.) CP 2018. LNCS, vol. 11008, pp. 99–108. Springer, Cham (2018). https://doi.org/10.1007/978-3-319-98334-9_7

13. Demirović, E., Musliu, N.: MaxSAT-based large neighborhood search for high school timetabling. Comput. Oper. Res. **78**, 172–180 (2017)

14. Eén, N., Sorensson, N.: Translating pseudo-boolean constraints into SAT. J. Satisfiability Boolean Model. Comput. **2**, 1–26 (2006)

15. Fu, Z., Malik, S.: On solving the partial MAX-SAT problem. In: Biere, A., Gomes, C.P. (eds.) SAT 2006. LNCS, vol. 4121, pp. 252–265. Springer, Heidelberg (2006). https://doi.org/10.1007/11814948_25

16. Gomory, R.E.: Outline of an algorithm for integer solutions of linear programs. Bull. Am. Math. Soc. **64**, 275–278 (1958)

17. Heras, F., Morgado, A., Marques-Silva, J.: Core-guided binary search algorithms for maximum satisfiability. In: Proceedings of AAAI 2011 (2011)

18. Joshi, S., Kumar, P., Martins, R., Rao, S.: Approximation strategies for incomplete MaxSAT. In: Hooker, J. (ed.) CP 2018. LNCS, vol. 11008, pp. 219–228. Springer, Cham (2018). https://doi.org/10.1007/978-3-319-98334-9_15

19. Joshi, S., Martins, R., Manquinho, V.: Generalized totalizer encoding for pseudo-boolean constraints. In: Pesant, G. (ed.) CP 2015. LNCS, vol. 9255, pp. 200–209. Springer, Cham (2015). https://doi.org/10.1007/978-3-319-23219-5_15

20. Koshimura, M., Zhang, T., Fujita, H., Hasegawa, R.: QMaxSAT: a partial max-sat solver. J. Satisfiability Boolean Model. Comput. **8**, 95–100 (2012)

21. Le Berre, D., Parrain, A.: The SAT4J library, release 2.2 system description. J. Satisfiability, Boolean Model. Comput. **7**, 59–64 (2010)

22. Lei, Z., Cai, S.: Solving (weighted) partial MaxSAT by dynamic local search for SAT. In: Proceedings of IJCAI, pp. 1346–1352 (2018)

23. Martins, R., Manquinho, V., Lynce, I.: Open-WBO: a modular MaxSAT solver'. In: Sinz, C., Egly, U. (eds.) SAT 2014. LNCS, vol. 8561, pp. 438–445. Springer, Cham (2014). https://doi.org/10.1007/978-3-319-09284-3_33

24. Morgado, A., Heras, F., Liffiton, M.H., Planes, J., Marques-Silva, J.: Iterative and core-guided maxsat solving: a survey and assessment. Constraints **18**(4), 478–534 (2013)

25. Moskewicz, M.W., Madigan, C.F., Zhao, Y., Zhang, L., Malik, S.: Chaff: engineering an efficient SAT solver. In: Proceedings of DAC 2001, pp. 530–535 (2001)

26. Narodytska, N., Bacchus, F.: Maximum satisfiability using core-guided MaxSAT resolution. In: Proceedings of AAAI 2014 (2014)

27. Pipatsrisawat, K., Darwiche, A.: A lightweight component caching scheme for satisfiability solvers. In: Marques-Silva, J., Sakallah, K.A. (eds.) SAT 2007. LNCS, vol. 4501, pp. 294–299. Springer, Heidelberg (2007). https://doi.org/10.1007/978-3-540-72788-0_28

28. Saikko, P., Berg, J., Järvisalo, M.: LMHS: a SAT-IP hybrid MaxSAT solver. In: Creignou, N., Le Berre, D. (eds.) SAT 2016. LNCS, vol. 9710, pp. 539–546. Springer, Cham (2016). https://doi.org/10.1007/978-3-319-40970-2_34

29. Sellmann, M.: Disco-Novo-GoGo: integrating local search and complete search with restarts. In: Proceedings of AAAI 2006 (2006)

30. Shaw, P.: Using constraint programming and local search methods to solve vehicle routing problems. In: Maher, M., Puget, J.-F. (eds.) CP 1998. LNCS, vol. 1520, pp. 417–431. Springer, Heidelberg (1998). https://doi.org/10.1007/3-540-49481-2_30
31. Marques Silva, J.P., Lynce, I., Malik, S.: Conflict-driven clause learning SAT solvers. In: Biere, et al. [10], pp. 131–153
32. Sugawara, T.: Maxroster: solver description. In: MaxSAT evaluation 2017, p. 12 (2017)

A Join-Based Hybrid Parameter
for Constraint Satisfaction

Robert Ganian[1], Sebastian Ordyniak[2]([✉]), and Stefan Szeider[1]

[1] Algorithms and Complexity Group, TU Wien, Vienna, Austria
[2] Algorithms Group, University of Sheffield, Sheffield, UK
sordyniak@gmail.com

Abstract. We propose *joinwidth*, a new complexity parameter for the Constraint Satisfaction Problem (CSP). The definition of joinwidth is based on the arrangement of basic operations on relations (joins, projections, and pruning), which inherently reflects the steps required to solve the instance. We use joinwidth to obtain polynomial-time algorithms (if a corresponding decomposition is provided in the input) as well as fixed-parameter algorithms (if no such decomposition is provided) for solving the CSP.

Joinwidth is a *hybrid* parameter, as it takes both the graphical structure as well as the constraint relations that appear in the instance into account. It has, therefore, the potential to capture larger classes of tractable instances than purely *structural* parameters like hypertree width and the more general fractional hypertree width (fhtw). Indeed, we show that any class of instances of bounded fhtw also has bounded joinwidth, and that there exist classes of instances of bounded joinwidth and unbounded fhtw, so bounded joinwidth properly generalizes bounded fhtw. We further show that bounded joinwidth also properly generalizes several other known hybrid restrictions, such as fhtw with degree constraints and functional dependencies. In this sense, bounded joinwidth can be seen as a unifying principle that explains the tractability of several seemingly unrelated classes of CSP instances.

1 Introduction

The Constraint Satisfaction Problem (CSP) is a central and generic computational problem that provides a common framework for many theoretical and practical applications in AI and other areas of Computer Science [31]. An instance of the CSP consists of a collection of variables that must be assigned values subject to constraints, where each constraint is given in terms of a relation whose tuples specify the allowed combinations of values for specified variables.

CSP is NP-complete in general. A central line of research is concerned with the identification of classes of instances for which the CSP can be solved in polynomial time. The two main approaches are to define classes either in terms of the

---

Robert Ganian acknowledges support by the Austrian Science Fund (FWF, Project P31336) and is also affiliated with FI MUNI, Czech Republic.

T. Schiex and S. de Givry (Eds.): CP 2019, LNCS 11802, pp. 195–212, 2019.
https://doi.org/10.1007/978-3-030-30048-7_12

constraint relations that may occur in the instance (*syntactic restrictions*; see, e.g., [4]), or in terms of the constraint hypergraph associated with the instance (*structural restrictions*; see, e.g., [18]). There are also several prominent proposals for utilizing simultaneously syntactic and structural restrictions called *hybrid restrictions* (see, e.g., [6–8,28]).

Grohe and Marx [20] showed that CSP is polynomial-time tractable whenever the constraint hypergraph has bounded *fractional hypertree width*, which strictly generalizes previous tractability results based on hypertree width [15] and acyclic queries [33]. Bounded fractional hypertree width is the most general known structural restriction that gives rise to polynomial-time tractability of CSP.

Our Contribution: Joinwidth. We propose a new hybrid restriction for the CSP, the width parameter *joinwidth*, which is based on the arrangement of basic relational operations along a tree, and not on hypertree decompositions. Interestingly, as we will show, our notion strictly generalizes (i) bounded fractional hypertree width, (ii) recently introduced extensions of fractional hypertree width with degree constraints and functional dependencies [24], (iii) various prominent hybrid restrictions [5], as well as (iv) tractable classes based on *functionality* and *root sets* [5,9,10]. Hence, joinwidth gives rise to a common framework that captures several different tractable classes considered in the past. Moreover, none of the other hybrid parameters that we are aware of [8], such as classes based on the Broken Triangle Property or topological minors [6,7] and directional rank [28], generalize fractional hypertree width and hence all of them are either less general or orthogonal to joinwidth.

Joinwidth is based on the arrangement of the constraints on the leaves of a rooted binary tree which we call a *join decomposition*. The join decomposition indicates the order in which relational joins are formed, where one proceeds in a bottom-up fashion from the leaves to the root, labeling a node by the join of the relations at its children, and projecting away variables that do not occur in relations to be processed later. Join decompositions are related to (structural) *branch decompositions* of hypergraphs, where the hyperedges are arranged on the leaves of the tree [2,19,29]. Related notions have been considered in the context of query optimization [1,22]. However, the basic form of join decompositions using only relational joins and projections is still a weak notion that cannot be used to tackle instances of bounded fractional hypertree width efficiently. We identify a further operation that—in conjunction with relational joins and projections—gives rise to the powerful new concept of joinwidth that captures and extends the various known tractable classes mentioned. This third operation *prunes* away all the tuples from an intermediate relation that are inconsistent with a relation to be processed later.

A join decomposition of a CSP instance specifies the order in which the above three operations are applied, and its *width* is the smallest real number w such that each relation appearing within the join decomposition has at most m^w many tuples (where m is the maximum number of tuples appearing in any constraint relation of the CSP instance under consideration). The joinwidth of a

CSP instance is the smallest width over all its join decompositions. Observe that joinwidth is a hybrid parameter—it depends on both the graphical structure as well as the constraint relations appearing in the instance.

Exploiting Joinwidth. Similarly to other width parameters, also the property that a class of CSP instances has bounded joinwidth can only be exploited for CSP solving if a decomposition (in our case a join decomposition) witnessing the bounded width is provided as part of the input. While such a join decomposition can be computed efficiently from a fractional hypertree decomposition or when the CSP instance belongs to a tractable class based on functionality or root sets mentioned earlier, we show that computing an optimal join decomposition is NP-hard in general, mirroring the corresponding NP-hardness of computing optimal fractional hypertree decompositions [13].

However, this obstacle disappears if we move from the viewpoint of polynomial-time tractability to *fixed-parameter tractability* (FPT). Under the FPT viewpoint, one considers classes of instances **I** that can be solved by a fixed-parameter algorithm—an algorithm running in time $f(k)|\mathbf{I}|^{O(1)}$, where k is the parameter (typically the number of variables or constraints), $|\mathbf{I}|$ is the size of the instance, and f is a computable function [14,16,17]. We note that it is natural to assume that k is much smaller than $|\mathbf{I}|$ in typical cases. The use of fixed-parameter tractability is well motivated in the CSP setting; see, for instance, Marx's discussion on this topic [27].

Here, we obtain two single-exponential fixed-parameter algorithms for instances of bounded joinwidth (i.e., algorithms with a running time of $2^{O(k)}$. $|\mathbf{I}|^{O(1)}$): one where k is the number of variables, and the other when k is the number of constraints. In this setting, we do not require an associated join decomposition to be provided with the input.

Under the FPT viewpoint, Marx [27] previously introduced the structural parameter *submodular width* (bounded submodular width is equivalent to bounded *adaptive width* [26]), which is strictly more general than fractional hypertree width, but when bounded only gives rise to fixed-parameter tractability and not polynomial-time tractability of CSP. In fact, Marx showed that assuming the Exponential Time Hypothesis [21], bounded submodular width is the most general purely structural restriction that yields fixed-parameter tractability for CSP. However, as joinwidth is a hybrid parameter, it can (and we show that it does) remain bounded even on instances of unbounded submodular width—and the same holds also for the recently introduced extensions of submodular width based on functional dependencies and degree bounds [24].

Roadmap. After presenting the required preliminaries on (hyper-)graphs, CSP, and fractional hypertree width in Sect. 2, we introduce and motivate join decompositions and joinwidth in Sect. 3. We establish some fundamental properties of join decompositions, provide our tractability result for CSP for the case when a join decomposition is given as part of the input, and then obtain our NP-hardness result for computing join decompositions of constant width. Section 4 provides an in-depth justification for the various design choices underlying join decompositions; among others, we show that the pruning step is required if the aim is to generalize fractional hypertree width. Our algorithmic applications for joinwidth are presented in Sect. 5: for instance, we show that joinwidth generalizes fractional hypertree width, but also other known (and hybrid) parameters such as functionality, root sets, and Turán sets. Section 6 contains our fixed-parameter tractability results for classes of CSP instances with bounded joinwidth. Finally, in Sect. 7, we compare the algorithmic power of joinwidth to the power of algorithms which rely on the unrestricted use of join and projection operations.

2   Preliminaries

We will use standard graph terminology [11]. An *undirected graph* G is a pair (V, E), where V or $V(G)$ is the vertex set and E or $E(G)$ is the edge set. All our graphs are simple and loopless. For a tree T we use $L(T)$ to denote the set of its leaves. For $i \in \mathbb{N}$, we let $[i] = \{1, \ldots, i\}$.

Hypergraphs. Similarly to graphs, a *hypergraph* H is a pair (V, E) where V or $V(H)$ is its vertex set and E or $E(H) \subseteq 2^V$ is its set of hyperedges. We denote by $H[V']$ the hypergraph *induced* on the vertices in $V' \subseteq V$, i.e., the hypergraph with vertex set V' and edge set $\{e \cap V' : e \in E\}$.

The Constraint Satisfaction Problem. Let D be a set and n and n' be natural numbers. An n-ary relation on D is a subset of D^n. For a tuple $t \in D^n$, we denote by $t[i]$, the i-th entry of t, where $1 \leq i \leq n$. For two tuples $t \in D^n$ and $t' \in D^{n'}$, we denote by $t \circ t'$, the concatenation of t and t'.

An instance of a *constraint satisfaction problem* (CSP) $\mathbf{I}$ is a triple $\langle V, D, C \rangle$, where V is a finite set of variables over a finite set (domain) D, and C is a set of constraints. A *constraint* $c \in C$ consists of a *scope*, denoted by $S(c)$, which is a completely ordered subset of V, and a relation, denoted by $R(c)$, which is a $|S(c)|$-ary relation on D. We let $|c|$ denote the number of tuples in $R(c)$ and $|\mathbf{I}| = |V| + |D| + \sum_{c \in C} |c|$. Without loss of generality, we assume that each variable occurs in the scope of at least one constraint.

A *solution* for $\mathbf{I}$ is an assignment $\theta : V \rightarrow D$ of the variables in V to domain values (from D) such that for every constraint $c \in C$ with scope $S(c) = (v_1, \ldots, v_{|S(c)|})$, the relation R contains the tuple $\theta(S(c)) = (\theta(v_1), \ldots, \theta(v_{|S(c)|}))$. We denote by SOL($\mathbf{I}$) the constraint containing all solutions of $\mathbf{I}$, i.e., the constraint with scope $V = \{v_1, \ldots, v_n\}$, whose relation contains one tuple $(\theta(v_1), \ldots, \theta(v_n))$ for every solution θ of $\mathbf{I}$. The task in CSP is to decide whether the instance $\mathbf{I}$ has at least one solution or in other words whether SOL($\mathbf{I}$) $\neq \emptyset$. Here and in the following we will for convenience (and with a slight abuse of notation) sometimes treat constraints like sets of tuples.

For a variable $v \in S(c)$ and a tuple $t \in R(c)$, we denote by $t[v]$, the i-th entry of t, where i is the position of v in $S(c)$. Let V' be a subset of V and let V'' be all the variables that appear in V' and $S(c)$. With a slight abuse of notation, we denote by $S(c) \cap V'$, the sequence $S(c)$ restricted to the variables in V' and we denote by $t[V']$ the tuple $(t[v_1], \ldots, t[v_{|V''|}])$, where $S(c) \cap V' = (v_1, \ldots, v_{|V''|})$.

Let c and c' be two constraints of $\mathbf{I}$. We denote by $S(c) \cup S(c')$, the ordered set (i.e., tuple) $S(c) \circ (S(c') \setminus S(c))$. The *(natural) join* between c and c', denoted by $c \bowtie c'$, is the constraint with scope $S(c) \cup S(c')$ containing all tuples $t \circ t'[S(c') \setminus S(c)]$ such that $t \in R(c)$, $t' \in R(c')$, and $t[S(c) \cap S(c')] = t'[S(c) \cap S(c')]$. The *projection* of c to V', denoted by $\pi_{V'}(c)$, is the constraint with scope $S(c) \cap V'$, whose relation contains all tuples $t[V']$ with $t \in R(c)$. We note that if c contains at least one tuple, then projecting it onto a set V' with $V' \cap S(c) = \emptyset$ results in the constraint with an empty scope and a relation containing the empty tuple (i.e., a tautological constraint). On the other hand, if $R(c)$ is the relation containing the empty tuple, then every projection of c will also result in a relation containing the empty tuple.

For a CSP instance $\mathbf{I} = \langle V, D, C \rangle$ we sometimes denote by $V(\mathbf{I})$, $D(\mathbf{I})$, $C(\mathbf{I})$, and $\sharp_{tup}(\mathbf{I})$ its set of variables V, its domain D, its set of constraints C, and the maximum number of tuples in any constraint relation of $\mathbf{I}$, respectively. For a subset $V' \subseteq V$, we will also use $\mathbf{I}[V']$ to denote the sub-instance of $\mathbf{I}$ induced by the variables in $V' \subseteq V$, i.e., $\mathbf{I}[V'] = \langle V', D, \{ \pi_{V'}(c) : c \in C \} \rangle$. The *hypergraph* $H(\mathbf{I})$ of a CSP instance $\mathbf{I} = \langle V, D, C \rangle$ is the hypergraph with vertex set V and edge set $\{ S(c) : c \in C \}$.

It is well known that for every instance $\mathbf{I}$ and every instance $\mathbf{I}'$ obtained by either (1) replacing two constraints in $C(\mathbf{I})$ by their natural join or (2) adding a projection of a constraint in $C(\mathbf{I})$, it holds that $\mathrm{SOL}(\mathbf{I}) = \mathrm{SOL}(\mathbf{I}')$. As a consequence, $\mathrm{SOL}(\mathbf{I})$ can be computed by performing, e.g., a sequence of joins over all the constraints in C.

Fractional Hypertree Width. Let H be a hypergraph. A *fractional edge cover* for H is a mapping $\gamma : E(H) \to \mathbb{R}$ such that $\sum_{e \in E(H) \wedge v \in e} \gamma(e) \geq 1$ for every $v \in V(H)$. The *weight* of γ, denoted by $w(\gamma)$, is the number $\sum_{e \in E(H)} \gamma(e)$. The *fractional edge cover number* of H, denoted by $\mathrm{fec}(H)$, is the smallest weight of any fractional edge cover of H.

A *fractional hypertree decomposition* $\mathcal{T}$ of H is a triple $\mathcal{T} = (T, (B_t)_{t \in V(T)}, (\gamma_t)_{t \in V(T)})$, where $(T, (B_t)_{t \in V(T)})$ is a tree decomposition [12,30] of H and $(\gamma_t)_{t \in V(T)}$ is a family of mappings from $E(H)$ to $\mathbb{R}$ such that for every $t \in V(T)$, it holds that γ_t is a fractional edge cover for $H[B_t]$. We call the sets B_t the *bags* and the mappings γ_t the *fractional guards* of the decomposition. The *width* of $\mathcal{T}$ is the maximum $w(\gamma_t)$ over all $t \in V(T)$. The *fractional hypertree width* of H, denoted by $\mathrm{fhtw}(H)$, is the minimum width of any fractional hypertree decomposition of H. Finally, the *fractional hypertree width* of a CSP instance $\mathbf{I}$, denoted by $\mathrm{fhtw}(\mathbf{I})$, is equal to $\mathrm{fhtw}(H(\mathbf{I}))$.

Proposition 1. *Let $\mathbf{I}$ be a CSP instance with hypergraph H and let $\mathcal{T} = (T, (B_t)_{t \in V(T)}, (\gamma_t)_{t \in V(T)})$ be a fractional hypertree decomposition of H of width at most ω. For every node $t \in V(T)$ and every subset $B \subseteq B_t$, it holds that $|\mathrm{SOL}(\mathbf{I}[B])| \leq (\sharp_{\mathrm{tup}}(\mathbf{I}))^{\omega}$.*

3   Join Decompositions and Joinwidth

This section introduces two notions that are central to our contribution: *join decompositions* and *joinwidth*. In the following, let us consider an arbitrary CSP instance $\mathbf{I} = \langle V, D, C \rangle$.

Definition 2. *A* join decomposition *for* $\mathbf{I}$ *is a pair* (J, ϱ), *where* J *is a rooted binary tree and* ϱ *is a bijection between the leaves* $L(J)$ *of* J *and* C.

Let j be a node of J. We denote by J_j the subtree of J rooted at j and we denote by $X(j)$, $V(j)$, $\overline{V}(j)$, and $S(j)$ the (unordered) sets $\{\varrho(\ell) : \ell \in L(J_j)\}$, $\bigcup_{c \in X(j)} S(c)$, $\bigcup_{c \notin X(j)} S(c)$, and $V(j) \cap \overline{V}(j)$, respectively; infuitively, $X(j)$ is the set of constraints that occur in the subtree rooted at j, $V(j)$ is the set of variables that occur in the scope of constraints in $X(j)$, $\overline{V}(j)$ is the set of variables that occur in the scope of constraints not in $X(j)$, and $S(j)$ is the set of variables that occur in $V(j)$ and $\overline{V}(j)$. In some cases, we will also consider *linear join decompositions*, which are join decompositions where every inner node is adjacent to at least one leaf.

Semantics of Join Decompositions. Intuitively, every internal node of a join decomposition represents a join operation that is carried out over the constraints obtained for the two children; in this way, a join decomposition can be seen as a procedure for performing joins, with the aim of determining whether $\mathrm{SOL}(\mathbf{I})$ is non-empty (i.e., solving the CSP instance $\mathbf{I}$). Crucially, the running time of such a procedure depends on the size of the constraints obtained and stored by the algorithm which performs such joins. The aim of this subsection is to formally define and substantiate an algorithmic procedure which uses join decompositions to solve CSP.

A naive way of implementing the above idea would be to simply compute and store the natural join at each node of the join decomposition and proceed up to the root; see for instance the work of [3]. Formally, we can recursively define a constraint $C_{naive}(j)$ for every node $j \in V(J)$ as follows. If j is a leaf, then $C_{naive}(j) = \varrho(j)$. Otherwise $C_{naive}(j)$ is equal to $C_{naive}(j_1) \bowtie C_{naive}(j_2)$, where j_1 and j_2 are the two children of j in J. It is easy to see that this approach can create large constraints even for very simple instances of CSP: for example, at the root r of T it holds that $\mathrm{SOL}(\mathbf{I}) = C_{naive}(r)$, and hence $C_{naive}(r)$ would have superpolynomial size for every instance of CSP with a superpolynomial number of solutions. In particular, an algorithm which computes and stores $C_{naive}(j)$ would never run in polynomial time for CSP instances with a superpolynomial number of solutions.

An efficient way of joining constraints along a join decomposition is to only store projections of constraints onto those variables that are still relevant for constraints which have yet to appear; this idea has been used, e.g., in algorithms which exploit hypertree width [15]. To formalize this, let $C_{proj}(j)$ be recursively defined for every node $j \in V(T)$ as follows. If j is a leaf, then $C_{proj}(j) = \pi_{\overline{V}(j)}(\varrho(j))$. Otherwise $C_{proj}(j)$ is equal to $\left(\pi_{\overline{V}(j)}(C_{proj}(j_1) \bowtie C_{proj}(j_2)) \right)$, where

j_1 and j_2 are the two children of j in J. In this case, $\mathbf{I}$ is a YES-instance if and only if $C_{proj}(r)$ does not contain the empty relation. Clearly, for every node j of J it holds that $C_{proj}(j)$ has at most as many tuples as $C_{naive}(j)$, but can have arbitrarily fewer tuples; in particular, an algorithm which uses join decompositions to compute C_{proj} in a bottom-up fashion can solve CSP instances in polynomial time even if they have a superpolynomial number of solutions (see also Observation 8).

However, the above approach still does not capture the algorithmic power offered by dynamically computing joins along a join decomposition. In particular, similarly as has been done in the evaluation algorithm for fractional edge cover [20, Theorem 3.5], we can further reduce the size of each constraint $C_{proj}(j)$ computed in the above procedure by *pruning* all tuples that would immediately violate a constraint c in $\mathbf{I}$ (and, in particular, in $C \setminus C(j)$). To formalize this operation, we let $\mathsf{prune}(c)$ denote the *pruned constraint* w.r.t. $\mathbf{I}$, i.e., $\mathsf{prune}(c)$ is obtained from c by removing all tuples $t \in R(c)$ such that there is a constraint $c' \in C$ with $t[S(c')] \notin \pi_{S(c)}(c')$. This leads us to our final notion of dynamically computed constraints: for a node j, we let $C(j) = \mathsf{prune}(C_{proj}(j))$. We note that this, perhaps inconspicuous, notion of pruning is in fact critical—without it, one cannot use join decompositions to efficiently solve instances of small fractional hypertree width or even small fractional edge cover. A more in-depth discussion on this topic is provided in Sect. 4.

We can now proceed to formally define the considered width measures.

Definition 3. *Let $\mathcal{J} = (J, \varrho)$ be a join decomposition for $\mathbf{I}$ and let $j \in V(J)$. The* joinwidth *of j, denoted $\mathsf{jw}(j)$, is the smallest real number ω such that $|C(j)| \leq (\sharp_{\mathsf{tup}}(\mathbf{I}))^\omega$, i.e., $\omega = \log_{\sharp_{\mathsf{tup}}(\mathbf{I})} |C(j)|$. The joinwidth of $\mathcal{J}$ (denoted $\mathsf{jw}(\mathcal{J})$) is then the maximum $\mathsf{jw}(j)$ over all $j \in V(J)$. Finally, the joinwidth of $\mathbf{I}$ (denoted $\mathsf{jw}(\mathbf{I})$) is the minimum $\mathsf{jw}(\mathcal{J})$ over all join decompositions $\mathcal{J}$ for $\mathbf{I}$.*

In general terms, an instance $\mathbf{I}$ has joinwidth ω if it admits a join decomposition where the number of tuples of the produced constraints never increases beyond the ω-th power of the size of the largest relation in $\mathbf{I}$. Analogously as above, we denote by $\mathsf{ljw}(\mathbf{I})$ the minimum joinwidth of any linear join decomposition of a CSP instance $\mathbf{I}$.

Example 4. *Let $N \in \mathbb{N}$ and consider the CSP instance $\mathbf{I}$ having three variables a, b, and c and three constraints x, y, and z with scopes (a, b), (b, c), and (a, c), respectively. Assume furthermore that the relations of all three constraints are identical and contain all tuples $(1, i)$ and $(i, 1)$ for every $i \in [N]$. Refer also to Fig. 1 for an illustration of the example. Then $|x| = |y| = |z| = \sharp_{\mathsf{tup}}(\mathbf{I}) = 2N - 1$ and due to the symmetry of $\mathbf{I}$ any join-tree $\mathcal{J}$ of $\mathbf{I}$ has the same joinwidth, which (as we will show) is equal to 1. To see this consider for instance the join-tree $\mathcal{J}$ that has one inner node j joining x and y and a root node r joining $C(j)$ and z. Then $\mathsf{jw}(\ell) = 1$ for any leaf node ℓ of $\mathcal{J}$. Moreover $|C(j)| = |\mathsf{prune}(C_{\mathrm{proj}}(j))| = |z| = \sharp_{\mathsf{tup}}(\mathbf{I})$ since the pruning step removes all tuples from $C_{\mathrm{proj}}(j)$ that are not in z and consequently $C(r) = z$ and $\mathsf{jw}(\mathcal{J}) = 1$. Note that in this example $\mathsf{jw}(\mathbf{I}) = 1 < \mathsf{fhtw}(\mathbf{I}) = 3/2$.*

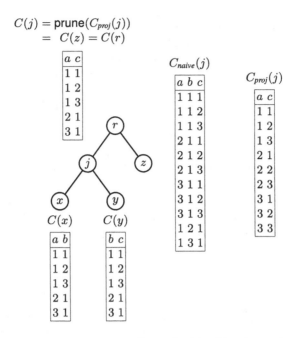

Fig. 1. The join decomposition given in Example 4 for $N = 3$ together with the intermediate constraints obtained for the node j

Finally, we remark that one could in principle also define joinwidth in terms of a (rather tedious and technically involved) variant of hypertree decompositions. However, the inherent algorithmic nature of join-trees makes them much better suited for the definition of joinwidth.

Properties of Join Decompositions. Our first task is to formalize the intuition behind the constraints $C(j)$ computed when proceeding through the join tree.

Lemma 5. *Let (J, ϱ) be a join decomposition for $\mathbf{I} = \langle V, D, C \rangle$ and let $j \in V(J)$. Then $C(j) = \pi_{\overline{V}(j)}(\mathrm{SOL}(\mathbf{I}'))$, where $\mathbf{I}' = \mathbf{I}[V(j)]$.*

Proof. We prove the lemma by leaf-to-root induction along J. If j is a leaf such that $\varrho(j) = c$, then $C(j)$ is the constraint obtained from c by projecting onto $\overline{V}(j)$ and then applying pruning with respect to $\mathbf{I}$. Crucially, pruning c w.r.t. $\mathbf{I}'$ produces the same result as pruning c w.r.t. $\mathbf{I}$. Since pruning cannot remove tuples which occur in $\mathrm{SOL}(\mathbf{I}')$, each tuple in $\mathrm{SOL}(\mathbf{I}')$ must also occur in $C(j)$ (as a projection onto $\overline{V}(j)$). On the other hand, consider a tuple α in $C(j)$ and assume for a contradiction that α is not present in $\pi_{\overline{V}(j)}(\mathrm{SOL}(\mathbf{I}'))$. Since variables outside of $\overline{V}(j)$ do not occur in the scopes of constraints other than c, this means that there would exist a constraint c' in $\mathbf{I}'$ which is not satisfied by an assignment corresponding to α—but in that case α would be removed from $C(j)$ via pruning. Hence $C(j) = \pi_{\overline{V}(j)}(\mathrm{SOL}(\mathbf{I}'))$ holds for every leaf in T.

For the induction step, consider a node j with children j_1 and j_2 (with their corresponding instances being $\mathbf{I}'_1$ and $\mathbf{I}'_2$, respectively), and recall that $C(j)$ is obtained from $C(j_1) \bowtie C(j_2)$ by projecting onto $\overline{V}(j)$ and then pruning (w.r.t. $\mathbf{I}$ or, equivalently, w.r.t. $\mathbf{I}'$). We will also implicitly use the fact that $\overline{V}(j) \subseteq \overline{V}(j_1) \cup \overline{V}(j_2)$ and $V(j) = V(j_1) \cup V(j_2)$. First, consider for a contradiction that there exists a tuple β in $\mathrm{SOL}(\mathbf{I}'[\overline{V}(j)])$ which does not occur in $C(j)$. Clearly, β could not have been removed by pruning, and hence this would mean that there exists no tuple in $C(j_1) \bowtie C(j_2)$ which results in β after projection onto $\overline{V}(j)$; in particular, w.l.o.g. we may assume that every tuple in $C(j_1)$ differs from β in (the assignment of) at least one variable. However, since β occurs in $\mathrm{SOL}(\mathbf{I}'[\overline{V}(j)])$, there must exist at least one tuple, say β', which occurs in $\mathrm{SOL}(\mathbf{I}')$, and consequently there exists a tuple in $\mathrm{SOL}(\mathbf{I}'_1)$ which matches β in (the assignment of) all variables. At this point, we have reached a contradiction with the inductive assumption that $\mathrm{SOL}(\mathbf{I}'_1[\overline{V}(j_1)]) = C(j_1)$.

For the final case, consider a tuple γ in $C(j)$ and assume for a contradiction that γ is not present in $\pi_{\overline{V}(j)}(\mathrm{SOL}(\mathbf{I}'))$. This means that there exists at least one constraint, say c', in $\pi_{\overline{V}(j)}(\mathrm{SOL}(\mathbf{I}'))$ which would be invalidated by (an assignment corresponding to) γ. Let us assume that c' occurs in the subtree rooted in j_2, and let γ_1 be an arbitrary "projection" of γ onto $\overline{V}(j_1)$. Since $\overline{V}(j_1) \supseteq S(c')$, this means that γ_1 would have been removed from $C(j_1)$ by pruning; in particular, we see that there exists no tuple γ_1 in $C(j_1)$ which could produce γ in a join, contradicting our assumptions about γ. By putting everything together, we conclude that indeed $C(j) = \pi_{\overline{V}(j)}(\mathrm{SOL}(\mathbf{I}'))$. $\qquad\square$

Next, we show how join decompositions can be used to solve CSP.

Theorem 6. CSP *can be solved in time* $\mathcal{O}(|\mathbf{I}|^{2\omega+4})$ *provided that a join decomposition of width at most* ω *is given in the input.*

Proof. Let $\mathcal{J} = (J, \varrho)$ be the provided join decomposition of width ω for $\mathbf{I}$. As noted before, the algorithm for solving $\mathbf{I}$ computes $C(j)$ for every $j \in V(J)$ in a bottom-up manner. Since J has exactly $2|C| - 1$ nodes, it remains to analyse the maximum time required to compute $C(j)$ for any node of J. If j is a leaf, then $C(j) = \mathsf{prune}(\pi_{S(j)}(\varrho(j)))$ and since the time required to compute the projection $P = \pi_{S(j)}(\varrho(j))$ from $\varrho(j)$ is at most $\mathcal{O}(\sharp_{tup}(\mathbf{I})|S(j)|)$ and the time required to compute the pruned constraint $\mathsf{prune}(P)$ from P is at most $\mathcal{O}(|C|(\sharp_{tup}(\mathbf{I}))^2|S(j)|)$, we obtain that $C(j)$ can be computed in time $\mathcal{O}(|C|(\sharp_{tup}(\mathbf{I}))^2|S(j)|) \in \mathcal{O}(|\mathbf{I}|^4)$. Moreover, if j is an inner node with children j_1 and j_2, then $C(j) = \mathsf{prune}(\pi_{S(j)}(C(j_1) \bowtie C(j_2)))$ and since we require at most $\mathcal{O}((\sharp_{tup}(\mathbf{I}))^{2\omega}|S(j_1) \cup S(j_2)|)$ time to compute the join $Q = C(j_1) \bowtie C(j_2)$ from $C(j_1)$ and $C(j_2)$, at most $\mathcal{O}((\sharp_{tup}(\mathbf{I}))^{2\omega}|S(j)|)$ time to compute the projection $P = \pi_{S(j)}(Q)$ from Q, and at most $\mathcal{O}(|C|(\sharp_{tup}(\mathbf{I}))^{2\omega+1} \cdot |S(j)|)$ time to compute the pruned constraint $\mathsf{prune}(P)$ from P, we obtain $\mathcal{O}(|C|(\sharp_{tup}(\mathbf{I}))^{2\omega+1} \cdot |S(j_1) \cup S(j_2)|) = \mathcal{O}(|\mathbf{I}|^{2\omega+3})$ as the total time required to compute $C(j)$. Multiplying the time required to compute $C(j)$ for an inner node $j \in V(J)$ with the number of nodes of T yields the running time stated in the lemma. $\qquad\square$

Computing Join Decompositions. Next, let us address the problem of computing join decompositions of bounded joinwidth, formalized as follows.

ω-JOIN DECOMPOSITION
Input: A CSP instance $\mathbf{I}$.
Question: Compute a join decomposition for $\mathbf{I}$ of width at most ω, or correctly determine that $\mathsf{jw}(\mathbf{I}) > \omega$.

We show that ω-JOIN DECOMPOSITION is NP-hard even for width $\omega = 1$. This is similar to fractional hypertree width, where it was only very recently shown that deciding whether $\mathsf{fhtw}(\mathbf{I}) \leq 2$ is NP-hard [13], settling a question which had been open for about a decade. Our proof is, however, entirely different from the corresponding hardness proof for fractional hypertree width and uses a reduction from the NP-complete BRANCHWIDTH problem [32].

Theorem 7. 1-JOIN DECOMPOSITION *is NP-hard, even on Boolean CSP instances.*

4   Justifying Joinwidth

Below, we substantiate the use of both pruning and projections in our definition of join decomposition. In particular, we show that using pruning and projections allows the joinwidth to be significantly lower than if we were to consider joins carried out via C_{naive} or C_{proj}. More importantly, we show that join decompositions without pruning do not cover CSP instances with bounded fractional edge cover (and by extension bounded fractional hypertree width). To formalize this, let $\mathsf{jw}_{naive}(\mathbf{I})$ and $\mathsf{jw}_{proj}(\mathbf{I})$ be defined analogously as $\mathsf{jw}(\mathbf{I})$, with the distinction being that these measure the width in terms of C_{naive} and C_{proj} instead of C.

We also justify the use of trees for join decompositions by showing that there is an arbitrary difference between linear join decompositions (which precisely correspond to simple sequences of joins) and join decompositions.

Observation 8. *For every integer ω there exists a CSP instance $\mathbf{I}_\omega$ such that* $\mathsf{jw}_{\mathrm{proj}}(\mathbf{I}_\omega) \leq 1$, *but* $\mathsf{jw}_{\mathrm{naive}}(\mathbf{I}_\omega) \geq \omega$.

Proof. Consider the CSP instance $\mathbf{I}_\omega$ with variables $x, v_1, \ldots, v_\omega$ and for each $i \in [\omega]$ a constraint c_i with scope $\{x, v_i\}$ containing the tuples $\langle 0, 1 \rangle$ and $\langle 0, 0 \rangle$. Since $\mathsf{SOL}(\mathbf{I}_\omega)$ contains 2^ω tuples, it follows that every join decomposition with root r must have $|C(r)_{naive}| = (\sharp_{tup}(\mathbf{I}))^\omega = 2^\omega$ tuples, hence $\mathsf{jw}_{naive}(\mathbf{I}_\omega) \geq \omega$.

On the other hand, consider a linear join decomposition which introduces the constraints in an arbitrary order. Then for each inner node j, it holds that $S(j) = \{x\}$ and in particular $C_{proj}(j)$ contains a single tuple (0) over scope $\{x\}$. We conclude that $\mathsf{jw}_{proj}(\mathbf{I}_\omega) \leq 1$. $\square$

The next proposition justifies the use of trees instead of just linear join decompositions. Its proof employs an interesting connection between branchwidth and joinwidth.

Proposition 9. *For every integer ω there exists a CSP instance $\mathbf{I}_\omega$ such that* $\mathrm{jw}(\mathbf{I}_\omega) \leq 1$ *but* $\mathrm{ljw}(\mathbf{I}_\omega) \geq \omega$.

The next proposition shows not only that pruning can significantly reduce the size of stored constraints, but also that without pruning (i.e., with projections alone) one cannot hope to generalize structural parameters such as fractional hypertree width.

Proposition 10. *For every integer ω there exists a CSP instance $\mathbf{I}_\omega$ with hypergraph H^ω such that* $\mathrm{jw}(\mathbf{I}_\omega) \leq 2$ *and* $\mathrm{fec}(H^\omega) \leq 2$ *(and hence also* $\mathrm{fhtw}(H^\omega) \leq 2$*), but* $\mathrm{jw}_{\mathrm{proj}}(\mathbf{I}_\omega) \geq \omega$.

We believe that the above results are of general interest, as they provide useful insights into how to best utilize the joining of constraints.

5 Tractable Classes

Here, we show that join decompositions of small width not only allow us to solve a wide range of CSP instances, but also provide a unifying reason for the tractability of previously established structural parameters and tractable classes.

5.1 Fractional Hypertree Width

We begin by showing that joinwidth is a strictly more general parameter than fractional hypertree width. We start with a simple example showing that the joinwidth of a CSP instance can be arbitrarily smaller than its fractional hypertree width. Indeed, this holds for any structural parameter ψ measured purely on the hypergraph representation, i.e., we say that ψ is a *structural parameter* if $\psi(\mathbf{I}) = \psi(H(\mathbf{I}))$ for any CSP instance $\mathbf{I}$. Examples for structural parameters include fractional and generalized hypertree width, but also *submodular width* [27].

Observation 11. *Let ψ be any structural parameter such that for every ω there is a CSP instance with $\psi(\mathbf{I}) = \psi(H(\mathbf{I})) \geq \omega$. Then for every ω there is a CSP instance $\mathbf{I}_\omega$ with $\mathrm{jw}(\mathbf{I}_\omega) \leq 1$ but $\psi(\mathbf{I}_\omega) \geq \omega$.*

The following theorem shows that, for the case of fractional hypertree width, the opposite of the above observation is not true.

Theorem 12. *For every CSP instance $\mathbf{I}$, it holds that $\mathrm{jw}(\mathbf{I}) \leq \mathrm{fhtw}(\mathbf{I})$.*

Proof. Let H be the hypergraph of the given CSP instance $\mathbf{I} = (V, D, C)$ and let $\mathcal{T} = (T, (B_t)_{t \in V(T)}, (\gamma_t)_{t \in V(T)})$ be an optimal fractional hypertree decomposition of H. We prove the theorem by constructing a join decomposition $\mathcal{J} = (J, \varrho)$ for $\mathbf{I}$, whose width is at most $\mathrm{fhtw}(H)$. Let $\alpha : E(H) \to V(T)$ be some function from the edges of H to the nodes of T such that $e \subseteq B_{\alpha(e)}$ for every $e \in E(H)$.

Note that such a function always exists, because $(T, (B_t)_{t \in V(T)})$ is a tree decomposition of H. We denote by $\alpha^{-1}(t)$ the set $\{ e \in E(H) : \alpha(e) = t \}$.

The construction of $\mathcal{J}$ now proceeds in two steps. First we construct a partial join decomposition $\mathcal{J}^t = (J^t, \varrho^t)$ for $\mathbf{I}$ that covers only the constraints in $\alpha^{-1}(t)$, for every $t \in V(T)$. Second, we show how to combine all the partial join decompositions into the join decomposition $\mathcal{J}$ for $\mathbf{I}$ of width at most $\mathsf{fhtw}(H)$.

Let $t \in V(T)$ and let $\mathcal{J}^t = (J^t, \varrho^t)$ be an arbitrary partial join decomposition for $\mathbf{I}$ that covers the constraints in $\alpha^{-1}(t)$. Let us consider an arbitrary node $j \in V(J^t)$. By Lemma 5, we know that $C(j) = \pi_{S(j)}(\mathrm{SOL}(\mathbf{I}[V(j)]))$. Moreover, the fact that $\bigcup_{e \in \alpha^{-1}(t)} e \subseteq B_t$ implies $V(j) \subseteq B_t$. Since $|\pi_{S(j)}(\mathrm{SOL}(\mathbf{I}[V(j)]))| \leq |\mathrm{SOL}(\mathbf{I}[V(j)])|$, by invoking Proposition 1 we obtain that $|C(j)| \leq |\mathrm{SOL}(\mathbf{I}[V(j)])| \leq \sharp_{tup}(\mathbf{I})^{\mathsf{fhtw}(\mathbf{I})}$. Hence we conclude that $\mathsf{jw}(j) \leq \mathsf{fhtw}(H)$.

Next, we show how to combine the partial join decompositions $\mathcal{J}^t$ into the join decomposition $\mathcal{J}$ for $\mathbf{I}$. We will do this via a bottom-up algorithm that computes a (combined) partial join decomposition $\mathcal{F}^t = (F^t, \rho^t)$ (for every node $t \in V(T)$) that covers all constraints in $\alpha^{-1}(T_t) = \bigcup_{t \in V(T)} \alpha^{-1}(t)$. Initially, we set $\mathcal{F}^l = \mathcal{J}^l$ for every leaf $l \in L(T)$. For a non-leaf $t \in V(T)$ with children $t_1, \ldots, t_\ell$ in T, we obtain $\mathcal{F}^t$ from the already computed partial join decompositions $\mathcal{F}^{t_1}, \ldots, \mathcal{F}^{t_\ell}$ as follows. Let P be a path on the new vertices $p_1, \ldots, p_\ell$ and let r_t and $r_{t_1}, \ldots, r_{t_\ell}$ be the root nodes of J^t and $F^{t_1}, \ldots, F^{t_\ell}$, respectively. Then we obtain F^t from the disjoint union of $P, J^t, F^{t_1}, \ldots, F^{t_\ell}$ after adding an edge between r_t and p_1 and an edge between r_{t_i} and p_i for every i with $1 \leq i \leq \ell$ and setting p_ℓ to be the root of F^t. Moreover, ρ^t is obtained as the combination (i.e., union) of the functions $\varrho^t, \rho^{t_1}, \ldots, \rho^{t_\ell}$. Observe that because α assigns every hyperedge to precisely one bag of T, it holds that every constraint assigned to T_t is mapped to precisely one leaf of $\mathcal{F}^t$. At this point, all that remains is to show that $\mathcal{F}^t$ has joinwidth at most $\mathsf{fhtw}(H)$.

Since we have already argued that $|\mathrm{SOL}(\mathbf{I}[V(j)])| \leq \sharp_{tup}(\mathbf{I})^{\mathsf{fhtw}(H)}$ for every node j of $\mathcal{J}^t$ and moreover we can assume that the same holds for every node j of $\mathcal{F}^{t_1}, \ldots, \mathcal{F}^{t_\ell}$ by the induction hypothesis, it only remains to show that the same holds for the nodes $p_1, \ldots, p_\ell$. First, observe that since $(T, (B_t)_{t \in V(T)})$ is a tree decomposition of H, it holds that $S(r_t), S(r_{t_1}), \ldots, S(r_{t_\ell}) \subseteq B_t$. Indeed, consider for a contradiction that, w.l.o.g., there exists a variable $x \in S(r_{t_1}) \setminus B_t$. Then there must exist a hyperedge $e_1 \ni x$ mapped to t_1 or one of its descendants, and another hyperedge $e_i \ni x$ mapped to some node t' that is neither t_1 nor one of its descendants. But then both $B_{t'}$ and B_{t_1} must contain x, and so B_t must contain x as well. Moreover, since $S(p_i) \subseteq (S(r_t) \cup S(r_{t_1}) \cup \cdots \cup S(r_{t_\ell}))$ for every $i \in [\ell]$, it follows that $S(p_i) \subseteq B_t$ as well.

Finally, recall that $C(p_i) = \pi_{S(p_i)}(\mathrm{SOL}(\mathbf{I}[V(p_i)]))$ by Lemma 5, and observe that $|\pi_{S(p_i)}(\mathrm{SOL}(\mathbf{I}[V(p_i)]))| \leq |\mathrm{SOL}(\mathbf{I}[S(p_i)])|$. Then by Proposition 1 combined with the fact that $S(p_i) \subseteq B_t$, we obtain $|C(p_i)| \leq |\mathrm{SOL}(\mathbf{I}[S(p_i)])| \leq \sharp_{tup}(\mathbf{I})^{\mathsf{fhtw}(H)}$, which implies that the width of p_i is indeed at most $\mathsf{fhtw}(H)$. $\quad\square$

5.2   Functionality and Root Sets

Consider a CSP instance $\mathbf{I} = \langle V, D, C \rangle$ with $n = |V|$. We say that a constraint $c \in C$ is *functional* on variable $v \in V$ if c does not contain two tuples that differ *only* at variable v; more formally, for every t and $t' \in R(c)$ it holds that if $t[v] \neq t'[v]$, then there exists a variable $z \in S(c)$ distinct from v such that $t[z] \neq t'[z]$. The instance $\mathbf{I}$ is then called *functional* if there exists a variable ordering $v_1 < \cdots < v_n$ such that, for each $i \in [n]$, there exists a constraint $c \in C$ such that $\pi_{\{v_1,\ldots,v_i\}}(c)$ is functional on v_i. Observe that every CSP instance that is functional can admit at most 1 solution [5]; this restriction can be relaxed through the notion of *root sets*, which can be seen as variable sets that form "exceptions" to functionality. Formally, a variable set Q is a root set if there exists a variable ordering $v_1 < \cdots < v_n$ such that, for each $i \in [n]$ where $v_i \notin Q$, there exists a constraint $c \in C$ such that $\pi_{\{v_1,\ldots,v_i\}}(c)$ is functional on v_i; we say that Q is *witnessed* by the variable order $v_1 < \cdots < v_n$.

Functionality and root sets were studied for Boolean CSP [9,10]. Cohen et al. [5] later extended these notions to the CSP with larger domains. Our aim in this section is twofold: (1) generalize root sets through the introduction of *constraint root sets* and (2) show that bounded-size constraint root sets (and also root sets) form a special case of bounded joinwidth. Before we proceed, it will be useful to show that one can always assume the root set to occur at the beginning of the variable ordering.

Observation 13. *Let Q be a root set in $\mathbf{I}$ witnessed by a variable order α, assume a fixed arbitrary ordering on Q, and let the set $V' = V(\mathbf{I}) \setminus Q$ be ordered based on the placement of its variables in α. Then Q is also witnessed by the variable order $\alpha' = Q \circ V'$.*

For ease of presentation, we will say that $\mathbf{I}$ is *k-rooted* if k is the minimum integer such that $\mathbf{I}$ has a root set of size k. It is easy to see, and also follows from the work of David [9] and Cohen et al. [5], that for every fixed k the class of k-rooted CSP instances is polynomial-time solvable: generally speaking, one can first loop through and test all variable-subsets of size at most k to find a root Q, and then loop through all assignments $Q \to D$ to get a set of functional CSP instances, each of which can be solved separately in linear time.

While even 1-rooted CSP instance can have unbounded fractional hypertree width (see also the discussion of Cohen et al. [5]), the class of k-rooted CSP instances for a fixed value k is, in some sense, not very robust. Indeed, consider the CSP instance $\mathcal{W} = \langle \{v_1, \ldots, v_n\}, \{0, 1\}, \{c\} \rangle$ where c ensures that precisely a single variable is set to 1 (i.e., its relation can be seen as an $n \times n$ identity matrix). In spite of its triviality, it is easy to verify that $\mathcal{W}$ is not k-rooted for any $k < n - 2$.

Let us now consider the following alternative to measuring the size of root sets in a CSP instance $\mathbf{I}$. A constraint set P is a *constraint-root set* if $\bigcup_{c \in P} S(c)$ is a root set, and $\mathbf{I}$ is then called *k-constraint-rooted* if k is the minimum integer such that $\mathbf{I}$ has a constraint-root set of size k. Since we can assume that each variable occurs in at least one constraint, every k-rooted CSP also has a constraint-root set of size at most k; on the other hand, the aforementioned example of $\mathcal{W}$ shows that an instance can be 1-constraint-rooted while not being k-rooted for any small k. The following result, which we prove by using join decompositions and joinwidth, thus gives rise to strictly larger tractable classes than those obtained via root sets:

Proposition 14. *For every fixed $k \in \mathbb{N}$, every k-constraint-rooted CSP instance has joinwidth at most k and can be solved in time $|\mathbf{I}|^{\mathcal{O}(k)}$.*

Proof. Consider a CSP instance $\mathbf{I}$ with a constraint-root set P of size k. We argue that $\mathbf{I}$ has a linear join decomposition of width at most k where the elements of P occur as the leaves farthest from the root. Indeed, consider the linear join decomposition (J, ϱ) constructed in a bottom-up manner, as follows. First, we start by gradually adding the constraints in P as the initial leaves. At each step after that, consider a node j which is the top-most constructed node in the join decomposition. By definition, there must exist a variable v and a constraint c such that $\pi_{\bigcup_{c \in P} S(c)}(c)$ is functional on v. Moreover, this implies that $|\pi_{\bigcup_{c \in P} S(c)}(c) \bowtie C(j)| \leq |C(j)|$, and thus $|c \bowtie C(j)| \leq |C(j)|$. Hence this procedure does not increase the size of constraints at nodes after the initial k constraints, immediately resulting in the desired bound of k on the width of (J, ϱ).

To complete the proof, observe that a join decomposition with the properties outlined above can be found in time at most $|\mathbf{I}|^{\mathcal{O}(k)}$: indeed, it suffices to branch over all k-element subsets of $C(\mathbf{I})$ and test whether the union of their scopes is functional using, e.g., the result of Cohen et al. [5, Corollary 1]. Once we have such a join decomposition, we can solve the instance by invoking Theorem 6. □

As a final remark, we note that the class of k-constraint rooted CSP instances naturally includes all instances which contain k constraints that are in conflict (i.e., which cannot all be satisfied at the same time).

5.3   Other Tractable Classes

Here, we identify some other classes of tractable CSP instances with bounded joinwidth. First of all, we consider CSP instances such that introducing their variables in an arbitrary order always results in a subinstance with polynomially many solutions. In particular, we call a CSP instance $\mathbf{I}$ *hereditarily k-bounded* if for every subset V' of its variables it holds that $|\text{SOL}(\mathbf{I}[V'])| \leq \sharp_{tup}(\mathbf{I})^k$. Examples of hereditarily k-bounded CSP instances include k-Turan CSPs [5, page 12] and CSP instances with fractional edge covers of weight k [20].

Proposition 15. *The class of hereditarily k-bounded CSP instances has join-width at most k and can be solved in time at most $\mathcal{O}(|\mathbf{I}|^k)$.*

Proof. Consider an arbitrary linear join decomposition (J, ϱ). By definition, for each $j \in V(J)$ it holds that $|\text{SOL}(\mathbf{I}[V(j)])| \leq \sharp_{tup}(\mathbf{I})^k$. Then $|\pi_{\overline{V}(j)}(\text{SOL}(\mathbf{I}[V(j)]))| \leq \sharp_{tup}(\mathbf{I})^k$, and by Lemma 5 we obtain $|C(j)| \leq \sharp_{tup}(\mathbf{I})^k$, as required. □

Another example of a tractable class of CSP instances that we can solve using joinwidth are instances where all constraints interact in a way which forces a unique assignment of the variables. In particular, we say that a CSP $\mathbf{I} = \langle V, D, C \rangle$ is *unique at depth k* if for each constraint $c \in C$ there exists a *fixing set* $C' \subseteq C$ such that $c \in C'$, $|C'| \leq k$, and $|(\bowtie_{c' \in C'} c')| \leq 1$.

Proposition 16. *The class of CSP instances which are unique at depth k has joinwidth at most k and can be solved in time at most* $|\mathbf{I}|^{\mathcal{O}(k)}$.

6   Solving Bounded-Width Instances

This section investigates the tractability of CSP instances whose joinwidth is bounded by a fixed constant ω. In particular, one can investigate two notions of tractability. The first one is the classical notion of *polynomial-time tractability*, which asks for an algorithm of the form $|\mathbf{I}|^{\mathcal{O}(1)}$. In this setting, the complexity of CSP instances of bounded joinwidth remains an important open problem. Note that the NP-hardness of the ω-JOIN DECOMPOSITION problem established in Theorem 7 does not exclude polynomial-time tractability for CSP instances of bounded joinwidth. For instance, tractability could still be obtained with a suitable approximation algorithm for computing join decompositions (as it is the case for fractional hypertreewidth [25]) or by using an algorithm that does not require a join decomposition of bounded width as input.

The second notion of tractability we consider is called *fixed-parameter tractability* and asks for an algorithm of the form $f(k) \cdot |\mathbf{I}|^{\mathcal{O}(1)}$, where k is a numerical parameter capturing a certain natural measure of $\mathbf{I}$. Prominently, Marx investigated the fixed-parameter tractability of CSP and showed that CSP instances whose hypergraphs have bounded *submodular width* [27] are fixed-parameter tractable when k is the number of variables. Moreover, Marx showed that submodular width is the most general structural property *among those measured purely on hypergraphs* with this property.

Here, we obtain two single-exponential fixed-parameter algorithms for CSP instances of bounded joinwidth (i.e., algorithms with a running time of $2^{\mathcal{O}(k)} \cdot |\mathbf{I}|^{\mathcal{O}(1)}$): one where k is the number of variables, and the other where k is the number of constraints. Since there exist classes of instances of bounded join-width and unbounded submodular width (see Observation 11), this expands the frontiers of (fixed-parameter) tractability for CSP.

Parameterization by Number of Constraints. To solve the case where k is the number of constraints, our primary aim is to obtain a join decomposition of width at most ω, i.e., solve the ω-JOIN DECOMPOSITION problem defined in Sect. 3. Indeed, once that is done we can solve the instance by Theorem 6.

Theorem 17. ω-JOIN DECOMPOSITION *can be solved in time* $\mathcal{O}(4^{|C|} + 2^{|C|}|\mathbf{I}|^{2\omega+1})$ *and is hence fixed-parameter tractable parameterized by* $|C|$, *for a CSP instance* $\mathbf{I} = \langle V, D, C \rangle$.

From Theorems 17 and 6 we immediately obtain:

Corollary 18. *A CSP instance* **I** *with k constraints and joinwidth at most ω can be solved in time $2^{\mathcal{O}(k)} \cdot |\mathbf{I}|^{\mathcal{O}(\omega)}$.*

Parameterization by Number of Variables. Note that Corollary 18 immediately establishes fixed-parameter tractability for the problem when k is the number of variables (instead of the number of constraints), because one can assume that $|C| \leq 2^{|V|}$ for every CSP instance $\mathbf{I} = (V, D, C)$. However, the resulting algorithm would be double-exponential in $|V|$. The following theorem shows that this can be avoided by designing a dedicated algorithm for CSP parameterized by the number of variables. The main idea behind both algorithms is dynamic programming, however, in contrast to the algorithm for $|C|$, the table entries for the fpt-algorithm for $|V|$ correspond to subsets of V instead of subsets of C. Interestingly, the fpt-algorithm for $|V|$ does not explicitly construct a join decomposition, but only implicitly relies on the existence of one.

Theorem 19. *A CSP instance* **I** *with k variables and joinwidth at most ω can be solved in time $2^{\mathcal{O}(k)} \cdot |\mathbf{I}|^{\mathcal{O}(\omega)}$.*

7   Beyond Join Decompositions

Due to their natural and "mathematically clean" definition, one might be tempted to think that join decompositions capture all the algorithmic power offered by join and projection operations. It turns out that this is not the case, i.e., we show that if one is allowed to use join and projections in an arbitrary manner (instead of the more natural but also more restrictive way in which they are used within join decompositions) one can solve CSP instances that are out-of-reach even for join decompositions. This is interesting as it points towards the possibility of potentially more powerful parameters based on join and projections than joinwidth.

Theorem 20. *For every ω, there exists a CSP instance* $\mathbf{I}_\omega$ *that can be solved in time $\mathcal{O}(|\mathbf{I}|^4)$ using only join and projection operations but* $\mathsf{jw}(\mathbf{I}_\omega) \geq \omega$.

8   Conclusions and Outlook

The main contribution of our paper is the introduction of the notion of a join decomposition and the associated parameter joinwidth (Definitions 2 and 3). These notions are natural as they are entirely based on fundamental operations of relational algebra: joins, projections, and pruning (which can equivalently be stated in terms of semijoins). It is also worth noting that our algorithms seamlessly extend to settings where each variable has its own domain (this can be modeled, e.g., by unary constraints).

Our results give rise to several interesting directions for future work. We believe that Theorem 6 can be generalized to other problems, such as #CSP or the FAQ-Problem [23]. Theorem 7 gives rise to the question of whether there exists a polynomial-time approximation algorithm for computing join decompositions of suboptimal joinwidth, similar to Marx's algorithm for fractional hypertree-width [25].

Observation 11 shows that submodular width is not more general than joinwidth. We conjecture that also the converse direction holds, i.e., that the two parameters are actually incomparable. Motivated by Theorem 20, one could try to define a natural parameter that captures the full generality of join and projection operations, or to at least define a parameter that is more general than join decompositions without sacrificing the simplicity of the definition.

References

1. Ahmed, R., Sen, R., Poess, M., Chakkappen, S.: Of snowstorms and bushy trees. Proc. VLDB Endowment **7**(13), 1452–1461 (2014)
2. Alekhnovich, M., Razborov, A.A.: Satisfiability, branch-width and Tseitin tautologies. In: Proceedings of the 43rd Annual IEEE Symposium on Foundations of Computer Science (FOCS 2002), pp. 593–603 (2002)
3. Atserias, A., Grohe, M., Marx, D.: Size bounds and query plans for relational joins. SIAM J. Comput. **42**(4), 1737–1767 (2013)
4. Bulatov, A.A., Jeavons, P., Krokhin, A.A.: Classifying the complexity of constraints using finite algebras. SIAM J. Comput. **34**(3), 720–742 (2005)
5. Cohen, D.A., Cooper, M.C., Green, M.J., Marx, D.: On guaranteeing polynomially bounded search tree size. In: Lee, J. (ed.) CP 2011. LNCS, vol. 6876, pp. 160–171. Springer, Heidelberg (2011). https://doi.org/10.1007/978-3-642-23786-7_14
6. Cohen, D.A., Cooper, M.C., Jeavons, P.G., Zivny, S.: Binary constraint satisfaction problems defined by excluded topological minors. Inf. Comput. **264**, 12–31 (2019)
7. Cooper, M.C., Duchein, A., El Mouelhi, A., Escamocher, G., Terrioux, C., Zanuttini, B.: Broken triangles: From value merging to a tractable class of general-arity constraint satisfaction problems. Artif. Intell. **234**, 196–218 (2016)
8. Cooper, M.C., Zivny, S.: Hybrid tractable classes of constraint problems. In: The Constraint Satisfaction Problem, volume 7 of Dagstuhl Follow-Ups, pp. 113–135. Schloss Dagstuhl - Leibniz-Zentrum fuer Informatik (2017)
9. David, P.: Using pivot consistency to decompose and solve functional CSPs. J. Artif. Intell. Res. **2**, 447–474 (1995)
10. Deville, Y., Van Hentenryck, P.: An efficient arc consistency algorithm for a class of CSP problems. In: Proceedings of the 12th International Joint Conference on Artificial Intelligence, Sydney, Australia, pp. 325–330, 24–30 August 1991
11. Diestel, R.: Graph Theory: GTM, 4th edn., vol. 173. Springer, Heidelberg (2017). https://doi.org/10.1007/978-3-662-53622-3
12. Downey, R.G., Fellows, M.R.: Fundamentals of Parameterized Complexity. Texts in Computer Science. Springer, London (2013). https://doi.org/10.1007/978-1-4471-5559-1
13. Fischl, W., Gottlob, G., Pichler, R.: General and fractional hypertree decompositions: hard and easy cases. In: Proceedings of the 37th ACM SIGMOD-SIGACT-SIGAI Symposium on Principles of Database Systems, Houston, TX, USA, pp. 17–32. ACM, 10–15 June 2018

14. Flum, J., Grohe, M.: Parameterized Complexity Theory. Texts in Theoretical Computer Science, vol. XIV. An EATCS Series. Springer, Berlin (2006). https://doi.org/10.1007/3-540-29953-X

15. Gottlob, G., Leone, N., Scarcello, F.: Hypertree decompositions and tractable queries. J. Comput. Syst. Sci. **64**(3), 579–627 (2002)

16. Gottlob, G., Szeider, S.: Fixed-parameter algorithms for artificial intelligence, constraint satisfaction, and database problems. Comput. J. **51**(3), 303–325 (2006)

17. Grohe, M.: Parameterized complexity for the database theorist. SIGMOD Rec. **31**(4), 86–96 (2002)

18. Grohe, M.: The complexity of homomorphism and constraint satisfaction problems seen from the other side. J. ACM **54**(1), 1–24 (2007)

19. Grohe, M.: Logic, graphs, and algorithms. In: Logic and Automata: History and Perspectives. Texts in Logic and Games, vol. 2, pp. 357–422. Amsterdam University Press (2007)

20. Grohe, M., Marx, D.: Constraint solving via fractional edge covers. ACM Trans. Algorithms **11**(1), 4:1–4:20 (2014). http://doi.acm.org/10.1145/2636918

21. Impagliazzo, R., Paturi, R., Zane, F.: Which problems have strongly exponential complexity? J. Comput. Syst. Sci. **63**(4), 512–530 (2001)

22. Ioannidis, Y.E., Kang, Y.C.: Left-deep vs. bushy trees: an analysis of strategy spaces and its implications for query optimization. In: Proceedings of the 1991 ACM SIGMOD International Conference on Management of Data, pp. 168–177. ACM Press (1991)

23. Khamis, M.A., Ngo, H.Q., Rudra, A.: FAQ: questions asked frequently. In: Proceedings of the 35th ACM SIGMOD-SIGACT-SIGAI Symposium on Principles of Database Systems, PODS 2016, San Francisco, CA, USA, pp. 3–28. ACM, 26 June–01 July 2016

24. Khamis, M.A., Ngo, H.Q., Suciu, D.: What do Shannon-type inequalities, submodular width, and disjunctive datalog have to do with one another? In: Proceedings of the 36th ACM SIGMOD-SIGACT-SIGAI Symposium on Principles of Database Systems, PODS 2017, Chicago, IL, USA, pp. 429–444. ACM, 14–19 May 2017

25. Marx, D.: Approximating fractional hypertree width. ACM Trans. Algorithms **6**(2), 17 (2010). Art. 29

26. Marx, D.: Tractable structures for constraint satisfaction with truth tables. Theory Comput. Syst. **48**(3), 444–464 (2011)

27. Marx, D.: Tractable hypergraph properties for constraint satisfaction and conjunctive queries. J. ACM **60**(6), 42:1–42:51 (2013)

28. Naanaa, W.: Unifying and extending hybrid tractable classes of csps. J. Exp. Theor. Artif. Intell. **25**(4), 407–424 (2013)

29. Robertson, N., Seymour, P.D.: Graph minors. I. excluding a forest. J. Combin. Theory Ser. B **35**(1), 39–61 (1983)

30. Robertson, N., Seymour, P.D.: Graph minors. II. algorithmic aspects of tree-width. J. Algorithms **7**(3), 309–322 (1986)

31. Rossi, F., van Beek, P., Walsh, T. (eds.): Handbook of Constraint Programming. Elsevier, New York (2006)

32. Seymour, P.D., Thomas, R.: Call routing and the ratcatcher. Combinatorica **14**(2), 217–241 (1994)

33. Yannakakis, M.: Algorithms for acyclic database schemes. In: Proceedings of 7th International Conference Very Large Data Bases, Cannes, France, pp. 81–94. IEEE Computer Society, 9–11 September 1981

An Incremental SAT-Based Approach to the Graph Colouring Problem

Gael Glorian[1(✉)], Jean-Marie Lagniez[1], Valentin Montmirail[2], and Nicolas Szczepanski[1]

[1] CRIL, Artois University and CNRS, 62300 Lens, France
{glorian,lagniez,szczepanski}@cril.fr
[2] I3S, Côte d'Azur University and CNRS, Nice, France
valentin.montmirail@univ-cotedazur.fr

Abstract. We propose and evaluate a new *CNF encoding* based on *Zykov's tree* for computing the *chromatic number* of a graph. Zykov algorithms are branch-and-bound procedures, that branch on pairings of vertices that express whether or not two non-adjacent vertices have the same colour. Thus, vertices with the same colour are contracted whereas edges are added between vertices when they have different colours. Such pairings make possible the use of a well-known recurrence relation, that states that the chromatic number of a graph cannot be lower than the chromatic number of its subgraphs. Our encoding associates with any graph and integer k a CNF formula that is satisfiable if and only if the chromatic number of the graph is at least k. We first show that any colouring satisfying a complete pairing always required a fixed number of colours. Then, we establish a *CNF encoding that counts* the number of colours required by a pairing. However, due to a large number of clauses required to encode transitivity constraints on pairings, a direct encoding does not scale well in practice. To avoid this pitfall, we designed a CEGAR-based (Counter-Example Guided Abstraction Refinement) approach that only encodes a part of the problem and then adds the missing constraints in an *incremental way* until a *valid solution* with k colours is found or the unsatisfiability of the problem is proven, meaning that the chromatic number of the graph is greater than k. We show that our encoding scheme performs in many cases significantly better than the state-of-the-art approaches to the graph colouring problem.

Keywords: Chromatic number · Zykov · CEGAR · SAT encoding

1 Introduction

Graph colouring is the problem of assigning a minimum number of colours to all vertices of a graph such that no adjacent vertices, *i.e.* vertices that are linked by an edge, receive the same colour. The smallest number of colours needed to colour the graph is called the *chromatic number* of the graph. The problem appears in

© Springer Nature Switzerland AG 2019
T. Schiex and S. de Givry (Eds.): CP 2019, LNCS 11802, pp. 213–231, 2019.
https://doi.org/10.1007/978-3-030-30048-7_13

a variety of areas included (but not limited to) scheduling problem [1], sudokus [2], register allocation used in compiler optimization [3], sports scheduling [4] and exam timetabling [5].

Determining the chromatic number of a graph is an NP-hard task. Several computational approaches for the colouring problem, that prove empirically viable for many instances, have been pointed out [6–13]. They can be divided into two categories: complete methods and incomplete methods. Incomplete methods are usually based on greedy or meta-heuristic algorithms and are able to deal with graph containing a large number of vertices. Nevertheless, such methods are only able to find bounds that can be far from the optimal solution. Complete approaches are commonly based on the branch-and-bound paradigm and are able to guarantee that the returned solution is optimal. In this work, we propose a complete approach to compute the chromatic number of a graph.

Recently, the authors of [13] proposed a hybrid CP/SAT approach, called gc-cdcl, using new lower bound and branching heuristic, and that is so far the most efficient approach to the graph colouring problem. In the CP-based approaches cited before, seeing all the colours as a *domain* for each vertex is common. However, because of the interchangeability of colours, such a representation leads to many symmetries which have to be broken. As done previously [7,13], we propose to take advantage of Zykov's tree to break symmetries. More precisely, our CNF encoding does not represent the allowed colours but choose to encode with propositional variables the fact that two vertices are coloured in the same way. Thus, instead of colouring the graph, our encoding tries to pair vertices between them. We demonstrate that, once all vertices are correctly paired, the number of colours required to colour the graph, while satisfying the pairing, is fixed and can be computed efficiently by a Boolean circuit. Then, we present a CNF encoding of this Boolean circuit together with the set of clauses representing the constraints ensuring the pairing correctness, allows us to represent as a whole the k-Colouring decision problem. We empirically tested Partial MaxSAT solvers, as well as linear and binary searches on the number of colours, using this encoding to see the performances against state-of-the-art approaches. Unfortunately, we quickly observed that this encoding cannot scale on large graphs: a cubic number of clauses are needed to ensure the correctness of pairings.

To make our approach scalable in practice, we propose a CEGAR approach using our new CNF encoding. The idea is as follows: instead of designing an equisatisfiable propositional formula, we generate an *under-abstraction* (a formula which is under-constrained, also called *relaxation* in other domains). If this under-abstraction is unsatisfiable, then, by construction the original formula is unsatisfiable; otherwise, the SAT solver outputs a model that can then be checked in polynomial time. It could be the case that the approach is lucky and the model of the under-abstraction is also a model of the original formula, in which case the problem is solved. In general, the under-abstraction is continually refined, *i.e.*, it comes closer to the original formula and, in the worst-case, will eventually become equisatisfiable with the original formula after a finite number of refinements. Notably, CEGAR has been successfully proposed in many prob-

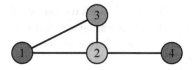

Fig. 1. Illustration of a graph G with 4 vertices such that $\chi(G) = 3$.

lems such as Bounded Model Checking [14], Satisfiability Modulo Theory [15], Planning [16], the Hamiltonian Cycle Problem [17], and more recently RCC8-Reasoning [18] and Minimal S5 Satisfiability Problem [19].

Abstracting decision problems with a CEGAR-under approach is well-known in the SAT community. However, the CP/OR community is more familiar with the Logic-based Benders decomposition (LBBD) [20], which can be viewed as the CEGAR-under approach for optimization. It is used in many domains where one wants to abstract and solve an optimization problem. LBBD approaches are orders of magnitude faster than state-of-the-art MIP for all problems where it has been applied [21–23]. One could also see the CEGAR-under approach as a Lazy-SMT approach [24,25], where the problem-specific knowledge extracted from the abstraction is used to guide the refinement process, instead of a theory solver.

The paper is organized as follows: after a few preliminaries, we present our new CNF based encoding and demonstrate its soundness and completeness. Then, we demonstrate how it can be adapted to be used in a CEGAR-based approach. Then, we show empirically that first: the direct SAT encoding, either with a linear/binary search on the number of colours or via a MaxSAT solver is quite competitive against the state-of-the-art approaches for minimal graph colouring, but, more importantly, that the CEGAR-based approach outperforms all the tested approaches on the benchmarks that have been considered.

2 Preliminaries

2.1 k-Colouring Problem

An undirected graph is a pair $G = (V, E)$, where V it the set of $|V| = n$ vertices (or nodes) and $E \subseteq V \times V$ a set of edges. A sub-graph $G' = (V', E')$ of $G = (V, E)$ is a graph such that $V' \subseteq V$ and $E' \subseteq E$. Let us note $G' = G \setminus V'$ the sub-graph G' obtained by removing from G vertices of V', i.e. $G' = (V \setminus V', \{(u, v) \in E \mid \{u, v\} \cap V' \neq \emptyset\})$. The contraction G/uv of a graph G is the graph obtained by removing any edge containing the vertices u and v, and by merging the vertices. $G + uv$ is the graph G with the edge (u, v) added.

A graph colouring problem aims to assign colours to certain elements of a graph subject to certain constraints. Vertex colouring is the most common graph colouring problem and is defined as follows: given an undirected graph $G = (V, E)$ and an integer k (number of colours), find a mapping $c : V \mapsto \{1, 2, \ldots, k\}$ that associates, each vertex $i \in V$ of G, a colour $c(i)$ so that no adjacent vertex $j \in V$

shares the same colour (*i.e.* $\forall (i,j) \in E$ we have $c(i) \neq c(j)$). A mapping c which verifies that $c(i) \neq c(j), \forall (i,j) \in E$ is called a valid colouring. We assume that V contains only integers from 1 to n. Moreover, for obvious reasons, we suppose that $\nexists i \in V$ such $(i,i) \in E$.

The most common type of vertex colouring seeks to minimise the number of colours for a given graph. The smallest number of colours needed for a graph G is called its **chromatic number** and is denoted by $\chi(G)$. An illustration of a graph G such that $\chi(G) = 3$ is given in Fig. 1. The problem of finding a minimum colouring for a graph is known to be NP-hard [26]. In fact, graph colouring is even NP-hard to approximate in specific scenarios [27]. Because its NP-hardness, k-colouring problem can be naturally translated into CNF.

2.2   Logical Preliminaries and CEGAR Framework

Let $\mathcal{L}$ be a standard Boolean logical language built on a finite set $\mathbb{P}$ of Boolean variables and usual connectives (namely, $\wedge$, $\vee$, $\neg$, $\Rightarrow$ and $\Leftrightarrow$ standing for conjunction, disjunction, negation, material implication and equivalence, respectively). Formulas will be noted using lower-case Greek letters such as $\alpha, \beta, \ldots$ Regarding the semantics aspect of the propositional logic, an interpretation $\mathcal{I}$ assigns valuation from $\{1, 0\}$ to every Boolean variable, thus, following usual compositional rules, to all formulas of $\mathcal{L}$. We denote by $\mathcal{I}(l)$ is 1 if l is satisfied by $\mathcal{I}$, and 0 otherwise. A formula α is *satisfiable* (also called consistent) when there exists at least one interpretation that satisfies α, i.e., that makes α true: such an interpretation is called a model of α and is represented by the set of variables that it satisfies. If a formula is false for any interpretation, this formula is *unsatisfiable*. $\models$ denotes deduction, i.e., $\alpha \models \beta$ denotes that β is a logical consequence of α, namely that β is satisfied in all models of α. Without loss of generality, any formula in $\mathcal{L}$ can be represented (while preserving satisfiability) in conjonctive normal form (CNF) i.e., as a conjunction of clauses [28], where a clause is a finite disjunction of literals and where a literal is a Boolean variable that can be negated.

Example 1 (Basic graph colouring encoding). Let $G = (V, E)$ be a graph, the following CNF formula encodes the problem of deciding if it is possible to colour the graph G with at most k colours (x_{vj} is true when the vertex v takes colour j):

$$\bigwedge_{v \in V} (\bigvee_{i=1}^{k} x_{vi} \wedge \bigwedge_{1 \leq i < j \leq k} (\neg x_{vi} \vee \neg x_{vj})) \wedge \bigwedge_{(u,v) \in E} (\bigwedge_{i=1}^{k} (\neg x_{ui} \vee \neg x_{vi}))$$

Counter-Example-Guided Abstraction Refinement (CEGAR) is an incremental way to decide the satisfiability of problems. It has been originally designed for model checking [14], *i.e.*, to answer questions such as "Does $\alpha \models \beta$ hold?" or, likewise, "Is $\phi = (\alpha \wedge \neg \beta)$ unsatisfiable?", where α describes a system and β a property. For such highly structured problems, it is often the case that only a small part of the formula is needed to answer the question. The keystone of CEGAR is to replace ϕ by an abstraction ϕ', easier to solve in practice. There

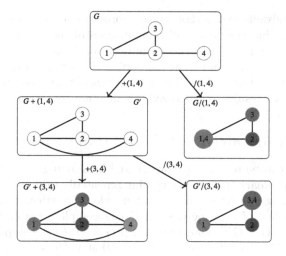

Fig. 2. Zykov's tree for the graph in Fig. 1

are two kinds of abstractions: an over-abstraction (resp. under-abstraction) of ϕ is a formula $\hat{\phi}$ (resp. $\check{\phi}$) such that $\hat{\phi} \models \phi$ (resp. $\phi \models \check{\phi}$) holds. $\hat{\phi}$ has at most as many models as ϕ and $\check{\phi}$ has at least as many models as ϕ.

Roughly, the CEGAR-based methods work on an abstraction of the original model, which is the current problem targeted by the solver. If it is an over-abstraction (resp. under-abstraction) which is proven satisfiable (resp. unsatisfiable), then the initial problem is also proven satisfiable (resp. unsatisfiable). Otherwise, the result returned by the solver can be spurious, and in this case, several situations may arise. If it is possible to check that the outcome is a solution to the initial problem, in the positive case, then the initial problem is also solved. It is also possible to decide the problem with the current result when the abstraction is equisatisfiable to the input problem. In all other situations, the CEGAR method refines the abstraction using information from the outcome in order to carefully select the next abstraction.

3   From Colouring to Zykov, and Vice Versa

A straightforward algorithm for deciding whether or not a graph can be coloured with k colours is to search among all mappings from the set of vertices to the set of colours (brute force). This algorithm, despite being correct, is inefficient for all but the smallest input graphs. Its lack of effectiveness can be partially explained by the fact that colours are interchangeable. When a colouring is incorrect, conflicting colourings exist that can be obtained by permutation. This observation can reveal, partially, why CP solvers are not able to decide the k-colouring problem for *big* graphs when the basic encoding (see Sect. 1) is used. Even if CP solvers are efficient on a wide range of problems, it is also known that they do not perform well on symmetric problems [29]. To break symmetries, [7] propose to take advantage of Zykov's tree in order to add conflict clauses on the fly, in such a way that they cover all permutations of the colours.

We now move on towards the definitions of Zykov trees. Let consider $G = (V, E)$ a graph, and x and y two non-adjacent vertices of G. In any proper colouring of G, either x and y have different colours or they have the same colour. Thus, a well-known result, so-called Zykov's deletion-contraction[1] recurrence is defined as follows:

$$\chi(G) = \min\{\chi(G/(ij)), \chi(G + (ij)))\} \; \forall (i, j) \notin E \qquad (1)$$

Zykov's tree can be recursively constructed by starting with the single node G, the root of our binary tree, and branching repeatedly using *vertex-contraction* on one side and *edge-deletion* on the other side on vertices that are not yet connected. Each leaf of a Zykov tree for G is a complete graph. Of course, we cannot branch on G if G is complete, and if $G = (V, E)$ is complete, it is easy to show that $\chi(G) = |V|$. By Eq. 1, we know that $\chi(G)$ is the minimum value among all leaves of a Zykov tree for G. Let us consider the graph of Figs. 1, 2 shows the related Zykov tree.

The search space of the coloured graphs visited when using Zykov's tree is more succinct than the one visited by an approach that branches on colours. Indeed, it is enough to observe that methods that branch on colours implicitly construct a Zykov tree. Except for the first vertex, each time a new vertex is coloured, it can be contracted with all the vertices previously coloured in the same way; and an edge can be added between the newly coloured one and the vertices we already coloured differently. When the colouring c gives colour at each vertex, each pair of vertices is either contracted or an edge is added between them. The resulting graph is then complete, and it is present in the Zykov's tree as leaf. Because the tree is constructed in a deterministic manner, it is easy to show that with each colouring c it is possible to associate only one leaf of the Zykov tree. Note that the opposite does not hold: each permutation c' of a colouring c leads to the same Zykov's tree leaf. Even if Zykov's trees contain symmetric nodes, there are fewer symmetries in them than in the search space explored by methods that colour vertices.

As explained in [7,13], it is possible to explore the search space represented by Zykov's tree using a CNF encoding. For a graph $G = (V, E)$, this encoding considers a set S of Boolean variables s_{ij} for all $i, j \in V$ that are used to pair the vertices together. Because edges are not oriented, we only consider s_{ij} s.t. $i < j$. In the case when $i > j$, s_{ij} is a renaming for s_{ji}. A variable s_{ij} set to true means that i and j are coloured in the same way (contraction). When set to false, it means there exists an edge between i and j (deletion). The set of clauses $\mathrm{tr}(G)$ consists in unit literals $\neg s_{ij}$ for all $(i, j) \in E$, and a cubic number of clauses to ensure path consistency between Boolean variables. For every triplet i, j and k, we have to encode $s_{i,j} \wedge s_{j,k} \Rightarrow s_{i,k}$ (transitivity) and $s_{i,j} \wedge s_{i,k} \Rightarrow s_{j,k}$ (Euclideanity). In the following, we call *pairing* an assignment that gives a value for each s_{ij}.

---

[1] The historical name is misleading: it either merges vertices or adds non-existing edges.

A *valid pairing* is a pairing that satisfies both path consistency and unit clauses. Let $G = (V, E)$ be a graph, the clauses $\mathtt{tr}(G)$ that encode valid pairings are given by:

$$\mathtt{tr}(G) := \neg s_{ij} \ \forall (i, j) \in E \wedge \mathtt{transitivity}() \wedge \mathtt{euclideanity}()$$

$$\mathtt{transitivity}(i, j, k) := (\neg s_{ij} \vee \neg s_{jk} \vee s_{ik})$$

$$\mathtt{euclideanity}(i, j, k) := (\neg s_{ij} \vee \neg s_{ik} \vee s_{jk})$$

$$\mathtt{transitivity}() := \mathtt{transitivity}(i, j, k) \ \forall i, j, k \in V s.t \ (i < j) \ and \ (j < k)$$

$$\mathtt{euclideanity}() := \mathtt{euclideanity}(i, j, k) \ \forall i, j, k \in V s.t \ (i < j) \ and \ (j < k)$$

Example 2. The following CNF formula encodes the Zykov search space induced by the graph given in Fig. 1:

$$\mathtt{tr}(G) = \neg s_{12} \wedge \neg s_{13} \wedge \neg s_{23} \wedge \neg s_{24}$$
$$\wedge (\neg s_{12} \vee \neg s_{23} \vee s_{13}) \wedge (\neg s_{13} \vee \neg s_{23} \vee s_{12}) \wedge (\neg s_{13} \vee \neg s_{12} \vee s_{23})$$
$$\wedge (\neg s_{12} \vee \neg s_{24} \vee s_{14}) \wedge (\neg s_{12} \vee \neg s_{14} \vee s_{24}) \wedge (\neg s_{14} \vee \neg s_{24} \vee s_{12})$$
$$\wedge (\neg s_{23} \vee \neg s_{34} \vee s_{24}) \wedge (\neg s_{23} \vee \neg s_{24} \vee s_{34}) \wedge (\neg s_{24} \vee \neg s_{34} \vee s_{23})$$
$$\wedge (\neg s_{13} \vee \neg s_{34} \vee s_{14}) \wedge (\neg s_{13} \vee \neg s_{14} \vee s_{34}) \wedge (\neg s_{14} \vee \neg s_{34} \vee s_{13})$$

which is, after unit propagation: $\mathtt{tr}(G) = (\neg s_{14} \vee \neg s_{34})$. As saw earlier on Fig. 2, either we assign s_{14} to true, and s_{34} to false, which gives us the graph $G/(1, 4)$, or we assign s_{14} to false, and s_{34} to true, and thus we obtain the graph $G + (1, 4)/(3, 4)$, or finally, we assign both variables to true, which gives us the final leaf, the graph $G + (1, 4) + (3, 4)$.

Thus, searching among the Zykov's tree leaves, amounts to searching among the set of valid pairings. Unfortunately, the previous encoding does not give the number of colours associated with a valid pairing. If we look back at the Zykov's tree, every leaf is a complete graph; their chromatic number is their number of vertices. In our case, the graph is not explicitly constructed and the information is missing. However, Property 1 shows that it is enough to know which vertices are paired together to compute the number of colours needed while respecting a specific pairing. The general idea is that if we try to colour the graph vertex by vertex, following the information contained in the pairing, then an additional colour is required for the vertex j exactly when all the already coloured vertices i are such that the s_{ij} are false. Thus, it is enough to consider vertices in a given order to compute the number of required colours.

Property 1. Let us consider $G = (V, E)$ a graph s.t. $|V| > 0$ and S the set of pairing variables associated to G. If $I_S \in 2^S$ is a valid pairing, then the number of colours needed to colour G w.r.t. I_S is given by the following formula:

$$\Psi(I_S) = 1 + \sum_{j=2}^{n} min(I_S(s_{ij}) \ s.t. \ s_{ij} \in S \ and \ i < j)$$

Proof. Let us demonstrate this result by structural induction on the number of nodes.

Base case: Show that the statement is true for the sub-graph $G \setminus \{v_2, v_3, \ldots, v_n\}$. It is clear that the number of colours needed to colour a graph with only one node is $\Psi(I_S) = 1$.

Inductive step: Show that if this property holds for $G \setminus \{v_n\}$, then it also holds for G. Let us consider $S' = \{s_{ij} \in S \text{ s.t. } i < j \leq n-1\}$. Using the induction hypothesis, we have $\Psi(I_{S'}) = 1 + \sum_{j=2}^{n-1} min(I_{S'}(s_{ij}) \text{ s.t. } s_{ij} \in S' \text{ and } i < j)$. By construction of S', we also have $\Psi(I_{S'}) = 1 + \sum_{j=2}^{n-1} min(I_S(s_{ij}) \text{ s.t. } s_{ij} \in S \text{ and } i < j)$. Now, let us consider G with the associated pairing I_S. It is easy to show that if there exists $s_{in} \in S$ that is true under I_S and s.t. $i < n$, then no additional colour is required (actually, the vertex v_n can be coloured as the vertices v_i). Otherwise, if $\forall s_{in} \in S$ s.t. $i < n$ the value of $I_S(s_{in})$ is false, then it is impossible to colour the last node with an already used colour. Consequently, the number of additional colours needed when considering the last vertex is $min(I_S(s_{in}) \text{ s.t. } s_{in} \in S \text{ and } i < j)$, and then we have:

$$\Psi(I_S) = \Psi(I_{S'}) + min(I_S(s_{in}) \text{ s.t. } s_{in} \in S \text{ and } i < j)$$
$$= 1 + \sum_{j=2}^{n-1} min(I_S(s_{ij}) \text{ s.t. } s_{ij} \in S \text{ and } i < j) + min(I_S(s_{in}) \text{ s.t. } s_{in} \in S \text{ and } i < j)$$
$$= 1 + \sum_{j=2}^{n} min(I_S(s_{ij}) \text{ s.t. } s_{ij} \in S \text{ and } i < j) \qquad \square$$

It is well known that computing the minimum value of a Boolean vector can be encoded as an AND gate. We can rewrite the previous sum as one on the set of Boolean variables $C = \{c_2, c_3, \ldots, c_n\}$ s.t. $\Psi(I_S) = 1 + \sum_{j=2}^{n} c_j$ where c_j is defined $\forall 1 < j \leq n$ as follows:

$$c_j \Leftrightarrow \bigwedge_{s_{ij} \in S \text{ and } i<j} \neg s_{ij} \qquad (2)$$

By considering $\text{tr}(G)$ and the constraint generated by Eq. 2, it is possible to compute the chromatic number of a given graph G by considering the minimisation problem that consists in satisfying a maximum number of c_i to false. This problem can be encoded as a partial MaxSAT problem where the hard clauses Σ are given by $(\text{tr}(G) \wedge$ Eq. 2) and the soft clauses Δ are given by the units literals $c_i \in C$. Thus, the minimum number of colours required is $1 + \text{MaxSAT}(\Sigma, \Delta)$.

To deal with the decision problem that consists in deciding if the chromatic number of a given graph is at least k, it is enough to consider the CNF formula composed with the clauses of $(\text{tr}(G) \wedge$ Eq. 2) and the set of clauses that encodes $(\sum_{j=2}^{n} c_j \leq k - 1)$ which can be represented using classical encoding, such as a Cardinality Network Encoding [30]. In the following, we consider different Partial MaxSAT solvers to determine whether the approach that consists in minimising the number of c_j assigned to true is a competitive approach to the *minimum graph colouring problem*. However, by looking at the CNF encoding, we can observe that transitivity() and euclideanity() add a cubic number of ternary clauses which slow down the whole approach (see Sect. 6). One way to circumvent them is to use a CEGAR version of the encoding and find a way to refine it by adding as few clauses as possible while minimising the number of SAT calls.

4   A CEGAR Version of the Encoding

The main concern with our encoding is the number of clauses needed to guarantee pairing validity. To overcome this difficulty, we propose to relax transitivity and Euclideanity constraints and incrementally execute a SAT solver on an under-approximation $\check{\phi}$ of the problem. If the solution violates some transitivity or Euclideanity constraints, we prevent them in the new abstraction by adding clauses. To compute the violated constraints, we should consider each triple (likely a large number) and check its consistency with the result. Such an approach is clearly impractical when the number of vertices grows.

To avoid this pitfall, we propose to colour the graph using the information contained in the returned pairing λ. Indeed, we assign each vertex u with the set $c[u]$ of the k possible colours. Then, we consider each vertex u incrementally in the natural order and, when it is possible ($c[u] \neq \emptyset$), select an available colour i for it in $c[u]$. Afterward, the pairings are used in order to colour as u every vertex v such that $s_{uv} \in \lambda$. If v cannot be coloured as u because $i \notin c[v]$, then $c[v]$ becomes empty and we consider a vertex w that removes i in $c[v]$. In such a case, transitivity and Euclideanity constraints on (u, v, w) are added in $\check{\phi}$. We next consider every vertex v s.t. $\neg s_{uv} \in \lambda$ and remove from $c[v]$ the colour i. If $c[v]$ becomes empty, we search for a vertex w that forces v to be coloured with the colour i. In the case when such w exists, the transitivity or/and Euclideanity constraints on the triple (u, v, w) are missing and must be added. In both cases, the solver is run once more on the updated under-approximation and the all process is repeated until spurious triples can be identified. Algorithm 1 gives the pseudo-code of our checking method. The following property shows that if check(λ, G, k) returns $\emptyset$ then it is possible, following λ, to colour G with k colours.

Property 2. Let $G = (V, E)$ a graph, k an integer and λ an interpretation that satisfies an under-abstraction of $tr(G)$ that contains at least: the unit clauses $\neg s_{uv}$ for all $(u, v) \in E$, the clauses encoding Eq. 2 and the clauses that encodes $\sum_{c_i}^{n} c_i \leq k - 1$. If $T = $ check(λ, G, k) returns $\emptyset$ then it is possible, following λ, to colour G with k colours. Otherwise, $\exists (i, j, k) \in T$ then $\lambda \not\models$ transitivity$(i, j, k) \wedge$ euclideanity(i, j, k).

Proof. First, let us demonstrate that if $\exists (i, j, k) \in T$ then $\lambda \not\models$ transitivity $(i, j, k) \wedge$ euclideanity(i, j, k). Let us consider the two cases where a triple t can be added:

- t is added line 10, that means we have two vertices u and v s.t. $u < v$, $c[u] = \{i\}$, $\neg s_{uv} \in \lambda$ and $c[v] \setminus \{i\} = \emptyset$. By the if condition (line 9), we also have $\exists w$ s.t. $w < u \ s_{vw} \in \lambda$ and $c[w] = c[u]$. Since $c[u] = \{i\}$ and $w < u$ then

---

Algorithm 1. check(λ s.t $\lambda \models \breve{\phi}, G = (V, E)$ a graph, k an integer) : T a set

---

1 $T \leftarrow \emptyset$; c a map;
2 **for** $u \leftarrow 1$ **to** $|V|$ **do** $c[u] \leftarrow \{1, 2, \ldots, k\}$;
3 **for** $u \leftarrow 1$ **to** $|V|$ **do**
4   **if** $c[u] \neq \emptyset$ **then**
5      $c[u] = \{i\}$ s.t. $i \in c[u]$;
6   **for** $v \leftarrow u + 1$ **to** $|V|$ **do**
7   **if** $\neg s_{uv} \in \lambda$ **then**
8          $c[v] = c[v] \setminus c[u]$;
9   **if** $c[v] = \emptyset$ **and** $\exists w$ s.t. $w < u$, $s_{wv} \in \lambda$ and $c[w] = c[u]$ **then**
10           $T \leftarrow T \cup \{(u, v, w)\}$
11   **else**
12         $c[v] = c[v] \cap c[u]$;
13   **if** $c[v] = \emptyset$ **then**
14   let w s.t. $w < u$, $s_{vw} \in \lambda$ and $c[w] \neq c[u]$ or $\neg s_{wv} \in \lambda$ and $c[w] = c[u]$;
15           $T \leftarrow T \cup \{(u, v, w)\}$;

16 **return** T;

$s_{wu} \in \lambda$ (otherwise i would have been removed from $c[u]$ by w). Thus, we have $\{s_{wu}, s_{wv}, \neg s_{uv}\} \subseteq \lambda$ which implies that $\lambda \not\models euclideanity(w, u, v)$.

- t is added line 15, that means we have two vertices u and v s.t. $u < v$, $c[u] = \{i\}$, $s_{uv} \in \lambda$ and $i \notin c[v]$. $i \notin c[v]$ means that i has been previously removed (line 8 or line 12) when some vertex $w < u$ has been considered. Two cases have to be considered. Either $c[w] = \{i\}$, in this case $\neg s_{wv} \in \lambda$ and we necessary have $\{\neg s_{wv}, s_{wu}, s_{uv}\} \subseteq \lambda$ ($s_{wu} \in \lambda$ because $i \in c[u]$ and $w < u$). Or $c[w] \neq \{i\}$, which implies that $s_{wv} \in \lambda$ and then we have $\{s_{wv}, \neg s_{wu}, s_{uv}\} \subseteq \lambda$ (similarly, $\neg s_{wu} \in \lambda$ because $i \in c[u]$ and $w < u$). In both cases, we have $\lambda \not\models transitivity(u, v, w) \wedge euclideanity(u, v, w)$.

Therefore, if $\exists (i, j, k) \in T$ then $\lambda \not\models transitivity(i, j, k) \wedge euclideanity(i, j, k)$. Now, let us demonstrate that if check(λ, G) returns $\emptyset$ then λ makes possible the construction of a k-colouring for $G = (V, E)$. First, we show that if T is empty then all the vertices v are coloured in $c[v]$, i.e. $\nexists v \in V$ s.t. $c[v] = \emptyset$. Towards a contradiction, suppose that after the for-loop (line 16) $\exists v \in V$ s.t. $c[v] = \emptyset$ and $T = \emptyset$. By considering the first emptied vertex v, it is easy to show that the only situation, where $c[v]$ can be emptied whereas no triple are added in T, is line 10 where no vertex w can be found when u is considered. Indeed, if $c[v]$ becomes empty at line 12, then a triple is added immediately (lines 13–15). Otherwise, if we can find $w < u$ s.t. $s_{wv} \in \lambda$ and $c[w] = c[\bar{u}]$ then a triple is necessary added. Therefore, $\forall w < v$ we have $\neg s_{wv} \in \lambda$ and then by Eq. 2 we also have $c_u \in \lambda$. Thus, since $\sum_{i=2}^{n} c_i <= k - 1$ must be satisfied by λ then $\sum_{i=2}^{v-1} c_i <= k - 2$ should be satisfied by λ as well. Since $c[v]$ is empty and $\neg s_{wv} \in \lambda$ for all $w < v$, then there exist k vertices w_j s.t. $c[w_j] = \{j\}$ for each $j \in \{1, 2, \ldots, k\}$. Thus, for each colour

j it is possible to determine the vertex w'_j that has been assigned first to the colour j.

Let us show that $\forall w'_j, \neg s_{ww'_j} \in \lambda, \forall w < w'_j$. Towards a contradiction, let us suppose that $\exists w$ s.t. $s_{ww'_j} \in \lambda$. Since the vertices are considered in the natural order, w is coloured before w'_j. But since w'_j is the first vertex assigned to j, we have $c[w] \neq \{j\}$. Thus, since $w < v$, we have $c[w] \neq \emptyset$, and thus the following instructions are executed. Because $s_{ww'_j} \in \lambda$ the else part of the if/else instruction (lines 11–15) should have deleted the colour j from $c[w'_j]$, making impossible to colour w'_j in j. Consequently, $\forall w'_j$ we have $\neg s_{ww'_j} \in \lambda, \forall w < w'_j$.

Therefore, there exist k vertices w'_j s.t. $\neg s_{ww'_j} \in \lambda, \forall w < w'_j$. Since vertex 1 is necessary coloured first, it is easy to show that $w'_1 = 1$. By construction, $\breve{\phi}$ encodes Eq. 2, and then we have $\lambda \models \bigwedge_{j=2}^{n}(c_j \Leftrightarrow \bigwedge_{s_{ij} \in S \text{ and } i<j} \neg s_{ij})$. Consequently, λ satisfies c_v at least $k - 1$ literals c_i s.t. $i < v$ and then $\lambda \not\models \sum_{i=2}^{v-1} c_i <= k - 2$, proving the claim.

We conclude by proving that c is a k-colouring for G. Since $1 \leq i \leq k$, stating that c is a valid colouring necessary implies it is a k-colouring. Thus, it is enough to show that c is a valid colouring. Towards a contradiction, suppose c is not valid. Then, there exist two vertices u and v s.t. $c[u] = c[v] = \{i\}$ and $(u,v) \in E$. Without loss of generality, suppose that $u < v$. Because $(u,v) \in E$, $\neg s_{uv}$ is a unit clause of $\breve{\phi}$ and therefore $\neg s_{uv} \in \lambda$. Thus, when u is coloured (line 4), the set $c[v]$ is updated w.r.t. λ. Since $\neg s_{uv} \in \lambda$, i should be removed from $c[v]$ (line 8) and $c[v]$ is necessary different from $\{i\}$. To conclude, if $T = \emptyset$ then c is a valid colouring of G and a k-colouring of G. □

5 Related Work

Brelaz' Dsatur (*degree of saturation*) [11] greedy algorithm, is one of the oldest but still successful technique for graph colouring. It works as follows: for each vertex v, we compute the *degree of saturation* of v and we use this value and the degree of each vertex to determine an order to colour the vertices. This heuristic is used within a branch-and-bound algorithm with one variable per vertex whose domain is the set of possible colours.

Another way to compute the chromatic number of a graph is to take advantage of its NP-hardness and use the constraint programming paradigm. Since it is trivial to encode colouring problems into propositional logic, several SAT-based approaches have been proposed. To be efficient in practice, such approaches also add constraints to break symmetries or to represent explicitly the information between non-adjacent vertices [7,12]. Let us cite color6 that is one of the most efficient solvers [12].

Schaafsma *et al.* [7] CP approach is very clever but unfortunately, we cannot compare our approach to it. As explained in [13]: "*We could not compare our method to the method of Schaafsma et al. directly. [...] Firstly, the algorithm is restricted to instances with at most 32 colours. Secondly it solves the satisfiability problem $\chi(G) \leq K$ and uses a file converter. Finally, the changes made to Minisat's code do not seem to be robust.*". However, their approach deserves some explanations. The authors also exploit Zykov's contraction. They introduce additional variables e_{ij} in the encoding if the vertices i and j should be merged using Zykov's contraction. However, a fundamental difference with our approach is that they encode the colours as a CP domain. For each colour c and each vertex v, a Boolean variable x_{vc} stating that the vertex v has the colour c is used. Even though they propose symmetry breaking [31] to speed-up their approach, they, unfortunately, suffer lack of efficiency when the number of colours is large.

The second work, closely related to our own one, is the most recent (and most efficient) approach to the minimum graph colouring problem, namely gc-cdcl [13]. Unlike Schaafsma et al. [7], they do not need to encode colours as a CP domain since they have a variable for every non-edge in the input graph. Again, as in [7] and our approach, they rely on Zykov's recurrence.

Hebrard and Katsirelos [13] made the same analysis as us about the performance of Schaafsma et al. They propose a CP hybridisation introducing peculiar propagators to enforce constraints on the bounds. Contrastingly, we have developed a complete SAT-based approach when the constraints are relaxed by performing a CEGAR-based search on the spurious examples that the SAT solver may find.

6   Experimental Evaluation

To assess our approach, we created a tool called Picasso. Picasso is an open-source solver (written in C++)[2]. In the following, we compare different versions of our method:

- **Full *Decision***, full encoding that decides the bound by using sum constraints and oracle calls (ascending (**1toN**), descending (**Nto1**) and binary search (**Dicho**));
- **Full *MaxSAT solver***, full encoding in combination with a MaxSAT solver;
- **CEGAR *Decision***, the relaxed counterpart of the full encoding that decides the bound by using sum constraints and oracle calls (ascending (**1toN**), descending (**Nto1**) and binary search (**Dicho**).

We used, as SAT solver, glucose (4.0) [32,33], in incremental mode (with its caching activated). We also tried several Partial MaxSAT solvers, such as: maxHS-b [34], mscg2015b [35], RC2-B [36] and MSUnCore [37]. We selected MaxSAT solvers which have shown good performances in the 2018 MaxSAT competition [38]. We considered the state-of-the-art approaches for graph colouring according to [13]: gc-cdcl [13], color6 [12] and DSatur [11]. We used instances from a colouring webpage[3], the "Graph coloring" and the "Quasi-random coloring" problems. This leads to a list of 159 instances. The experiments ran on a 4 cores Xeon at 3.3 GHz with CentOS 7.0. The memory limit was set to 32GB and the runtime limit to 900 seconds per solver per benchmark.

---

[2] The source are accessible at: https://github.com/Mystelven/picasso.

[3] https://mat.tepper.cmu.edu/COLOR03/.

6.1 Overall Evaluation of the Different Methods

We start with an overall evaluation of the effectiveness of each approach. The results are presented in Fig. 3 under the form of cactus plots. It makes explicit the number of instances solved in a given amount of time per instance. As expected, the methods using our full encoding are not very effective. Even if the methods using MaxSAT solvers are more efficient than the one that computes the bound, our best version is no more effective than the weaker state-of-the-art approach. Actually, having a closer look at their behaviour, the lack of efficiency does not come from SAT solver, but is due to the encoding itself. 97.3% of the translations have been solved. Hence, there is definitely a bottleneck here due to the translation.

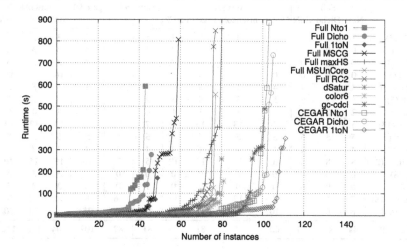

Fig. 3. Cactus-plot of the runtime.

One can observe in Fig. 3, that CEGAR approaches outperform state-of-the-art approaches. They manage to solve more instances than `gc-cdcl`, which solves 102 instances out of 159, which was definitely the best overall approach, as explained in [13]. It is important to note that the results do not depend on the way the optimisation problem is solved: the three types of search perform better than `gc-cdcl`. We can also note that, either in Full or in CEGAR mode, it seems that the 1toN approaches perform better than their Nto1 and Dicho counterparts. This can be explained because the chromatic number is generally far from the number of vertices. Thus, it is better to start from 1 than to start from the number of vertices. Table 1 reports the results regarding the number of problems solved depending on the family of the instance under consideration. As we can see, CEGAR approaches work well on all the families except for the **other** and **random** categories, where `color6` outperforms it. On the random benchmarks, the six unsolved instances are due to the SAT solving phase. These problems seem to have a *random nature* for which CDCL SAT solvers are ill-suited, whereas `color6` seems to deal with them extremely well. In the `other` category, it is more sparse, we do not lose on one big category but few instances here and there, except somewhat for the `school` one, which represents Class Scheduling Problems. All the state-of-the-art approaches solve these instances except us. In our case, it seems that verifying solutions with the checker is time-consuming. It returns only a few triples each time, therefore lacking time to solve the instance.

Table 1. Number of instances of each sub-family solved by the approaches in consideration, in **bold** the best approach for each sub-family.

| | Full | | | MaxSAT | | | | dSatur | color6 | gc-cdcl | CEGAR | | |
	Nto1	Dicho	1toN	MSCG	msuncore	maxHS	RC2				Nto1	Dicho	1toN
DSJ (15)	3	3	3	4	4	6	4	6	2	4	3	3	4
fpsol2 (3)	0	0	0	0	0	0	0	3	2	3	3	3	3
inithx (3)	0	0	0	0	0	0	0	3	1	3	3	3	3
le450 (12)	0	0	0	0	0	0	0	4	4	8	10	10	10
mulsol (5)	4	4	4	4	5	5	5	5	4	5	5	5	5
book (5)	5	5	5	5	5	5	5	2	3	5	5	5	5
miles (5)	2	2	2	1	1	1	1	5	4	5	4	4	5
queen (13)	5	8	8	5	13	13	13	7	11	10	13	13	13
myciel (5)	5	5	5	5	5	5	5	5	5	5	5	5	5
mugg (4)	4	4	4	4	4	4	4	4	4	4	4	4	4
insertion (25)	8	8	8	10	12	12	12	5	7	15	15	15	19
wap (8)	0	0	0	0	0	0	0	0	0	1	3	3	3
qg (4)	0	0	0	0	0	0	0	1	0	3	3	3	3
random (18)	2	2	5	12	12	13	14	14	16	6	10	10	10
other (45)	6	6	6	10	17	17	15	18	25	25	18	20	20
Total (159)	44	47	50	60	78	81	78	81	82	102	104	106	112

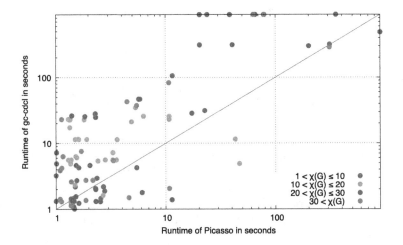

Fig. 4. Picasso vs. `gc-cdcl`.

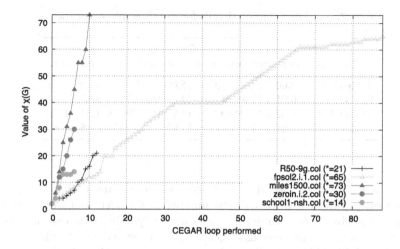

Fig. 5. Progression of Picasso towards the chromatic number of instances.

We also report a comparison between `gc-cdcl` and CEGAR 1toN (now denoted Picasso) in Fig. 4. Each dot corresponds to a colouring instance. The x-axis of the figure represents the computation time needed to compute the chromatic number when Picasso is considered, while the y-axis depicts the time needed to compute the chromatic number when `gc-cdcl` is considered. As we can see, there is a clear trend above the diagonal, especially on the instances solved in less than 10 s by Picasso which require almost a hundred seconds for `gc-cdcl`. Moreover, it seems that it does not depend on the chromatic number that needs to be found. Indeed, one could think that the higher the chromatic number is, more CEGAR loops in Picasso and therefore the worst the overall performances. However, as depicted in Fig. 4 the chromatic-number of the instance does not have much influence. Actually, the models returned by glucose help to refine quickly the bound and therefore provide a tremendous speed-up in comparison to a simple increasing loop. This is a result already known for people working with NP-oracles: an oracle able to output a model can provide to provide a large speed-up compared to one answering only yes or no [39]. Indeed, there are problem solvable with a polynomial number of calls to an oracle that can be solved with a logarithmic number of calls to an oracle outputing a model.

6.2 Analysis on the CEGAR Behaviour

Now that we have analysed how good Picasso in CEGAR mode is, let us see how it is behaving exactly, $i.e.$ let us determine where is the bottleneck in our approach. First, let us take a look at the number of CEGAR loops that Picasso is performing. We got that:

$$min = 2; \ 1^{st}qu. = 3; \ median = 6; \ mean = 10.3; \ 3^{rd}qu. = 11; \ max = 88$$

Table 2. Time distribution details for the three phases in Picasso

Time (s)	Encoding			Checking			Solving		
	min	med	max	min	med	max	min	med	max
$1 < \chi(G) \leq 10$	0.00	3.99	32.34	0.01	15.60	152.45	0.00	3.20	72.35
$10 < \chi(G) \leq 20$	0.01	4.11	45.94	0.01	13.67	161.97	0.00	3.11	83.32
$20 < \chi(G) \leq 30$	0.02	4.24	79.30	0.01	12.90	163.41	0.00	1.27	105.23
$30 < \chi(G)$	0.01	5.60	71.66	0.01	17.92	222.54	0.00	2.45	95.66

This shows that Picasso does not loop so many times. Figure 5 reports for some categories of benchmarks the number of CEGAR loops (x-axis) with respect to the chromatic number of the instance (y-axis). One can observe on this picture that the instances have generally a chromatic number higher than the number of CEGAR loops. The sole exception is fpsol2.1.1.col which has a chromatic number of 65, where we identified that, at each step, the model does not provide any information on the chromatic number. The models returned by the SAT solver help to quickly improve the bound. This implies that in many cases, checker provides a quite precise refinement and not just one spurious triple at a time.

Finally, we report in Table 2 the cumulative time spent by the different phases of CEGAR (encoding, checking and solving) with respect to the chromatic number of the instance considered. One can observe that, on the instances that we managed to solve, the SAT encoding is not really time-consuming, neither is the SAT solving phase. Except for a few cases, we can observe that the checking phase definitely is the bottleneck of the approach. This is not really surprising. Indeed, Glucose [32] is a very efficient SAT solver, and we used the Cardinality Network Encoding from open-wbo [40] which allows us to refine the bound by adding as few clauses as possible. Therefore, the only time-consuming task is for each satisfiable answer from Glucose to check the model returned and determine whether clauses must be added. The median times may look relatively low, however, the reader must keeping in mind Fig. 3. Indeed, most of the instances solved by Picasso are solved in less than 100 s. If we have a look at the instances for which a time-out was reached, it turns out that we spent in median 89.3% of the time to check models. To be convinced that the checker is indeed a bottleneck, we tried to implement a naive one, which consists in testing whether for all i, j and k we have that the constraints transitivity(i, j, k) ∧ euclideanity(i, j, k) are respected. With such a checker, *CEGAR 1toN* solves 100 instances, *CEGAR Dicho* solves 96 and *CEGAR Nto1* solves 95 instances.

7   Conclusion

In this paper, we proposed a new approach for solving the minimal graph colouring problem using an under-abstraction refinement approach within the CEGAR framework. We showed that our encoding is sound and complete and we implemented our approach in the solver Picasso. We compared our solver with the state-of-the-art solvers for the graph colouring problem, on a wide range of benchmarks of different size and difficulty. We conclude that a basic direct-encoding approach is not competitive; many of the available benchmarks are large and require a lot of clauses to be encoded. However, by considering clauses carefully, our CEGAR approach outperforms the other solvers on most of the benchmarks. As future we plan to avoid checking unmodified sub-graphs twice by flagging some nodes, *i.e.*, checking only the part which was modified due to the previous assignment. Moreover, extending the CEGAR approach into a RECAR (Recursively Explore and Check Abstraction Refinement) one [41] could be interesting. The unit propagation of the embedded SAT solver would be stronger, and it could provide us quickly a good upper-bound. Such double-abstraction functions could make the binary search much faster and improve the overall performance. This is an exciting perspective for future work.

Acknowledgements. Part of this work was supported by the French Ministry for Higher Education and Research, the Haut-de-France Regional Council through the "Contrat de Plan État Région (CPER) DATA" and by the IDEX UCA^{JEDI}.

References

1. Marx, D.: Graph colouring problems and their applications in scheduling. Periodica Polytech. Electr. Eng. (Arch.) **48**(1–2), 11–16 (2004)
2. Lewis, R.M.R.: A Guide to Graph Colouring - Algorithms and Applications. Springer, Cham (2016). https://doi.org/10.1007/978-3-319-25730-3
3. Chaitin, G.J.: Register allocation and spilling via graph coloring (with retrospective). In: McKinley, K.S. (ed.) 20 Years of the ACM SIGPLAN Conference on Programming Language Design and Implementation 1979–1999, A Selection, pp. 66–74. ACM (1982)
4. Lewis, R., Thompson, J.M.: On the application of graph colouring techniques in round-robin sports scheduling. Comput. OR **38**(1), 190–204 (2011)
5. Hussin, B., Basari, A.S.H., Shibghatullah, A.S., Asmai, S.A., Othman, N.S.: Exam timetabling using graph colouring approach. In: 2011 IEEE Conference on Open Systems, pp. 133–138 (2011)
6. Gelder, A.V.: Another look at graph coloring via propositional satisfiability. Discrete Appl. Math. **156**(2), 230–243 (2008)
7. Schaafsma, B., Heule, M.J.H., van Maaren, H.: Dynamic symmetry breaking by simulating zykov contraction. In: Kullmann, O. (ed.) SAT 2009. LNCS, vol. 5584, pp. 223–236. Springer, Heidelberg (2009). https://doi.org/10.1007/978-3-642-02777-2_22
8. Caramia, M., Dell'Olmo, P.: Coloring graphs by iterated local search traversing feasible and infeasible solutions. Discrete Appl. Math. **156**(2), 201–217 (2008)
9. Dowsland, K.A., Thompson, J.M.: An improved ant colony optimisation heuristic for graph colouring. Discrete Appl. Math. **156**(3), 313–324 (2008)
10. Galinier, P., Hertz, A., Zufferey, N.: An adaptive memory algorithm for the k-coloring problem. Discrete Appl. Math. **156**(2), 267–279 (2008)

11. Brélaz, D.: New methods to color the vertices of a graph. Commun. ACM **22**(4), 251–256 (1979)
12. Zhou, Z., Li, C.M., Huang, C., Xu, R.: An exact algorithm with learning for the graph coloring problem. Comput. OR **51**, 282–301 (2014)
13. Hebrard, E., Katsirelos, G.: Clause learning and new bounds for graph coloring. In: Hooker, J. (ed.) CP 2018. LNCS, vol. 11008, pp. 179–194. Springer, Cham (2018). https://doi.org/10.1007/978-3-319-98334-9_12
14. Clarke, E.M., Grumberg, O., Jha, S., Lu, Y., Veith, H.: Counter example-guided abstraction refinement for symbolic model checking. J. ACM **50**(5) (2003)
15. Brummayer, R., Biere, A.: Effective bit-width and under-approximation. In: Moreno-Díaz, R., Pichler, F., Quesada-Arencibia, A. (eds.) EUROCAST 2009. LNCS, vol. 5717, pp. 304–311. Springer, Heidelberg (2009). https://doi.org/10.1007/978-3-642-04772-5_40
16. Seipp, J., Helmert, M.: Counterexample-guided cartesian abstraction refinement for classical planning. J. Artif. Intell. Res. **62**, 535–577 (2018)
17. Soh, T., Le Berre, D., Roussel, S., Banbara, M., Tamura, N.: Incremental SAT-based method with native boolean cardinality handling for the hamiltonian cycle problem. In: Fermé, E., Leite, J. (eds.) JELIA 2014. LNCS (LNAI), vol. 8761, pp. 684–693. Springer, Cham (2014). https://doi.org/10.1007/978-3-319-11558-0_52
18. Glorian, G., Lagniez, J.-M., Montmirail, V., Sioutis, M.: An incremental SAT-based approach to reason efficiently on qualitative constraint networks. In: Hooker, J. (ed.) CP 2018. LNCS, vol. 11008, pp. 160–178. Springer, Cham (2018). https://doi.org/10.1007/978-3-319-98334-9_11
19. Lagniez, J.-M., Le Berre, D., de Lima, T., Montmirail, V.: An assumption-based approach for solving the minimal S5-satisfiability problem. In: Galmiche, D., Schulz, S., Sebastiani, R. (eds.) IJCAR 2018. LNCS (LNAI), vol. 10900, pp. 1–18. Springer, Cham (2018). https://doi.org/10.1007/978-3-319-94205-6_1
20. Hooker, J.N.: Logic-based methods for optimization. In: Borning, A. (ed.) PPCP 1994. LNCS, vol. 874, pp. 336–349. Springer, Heidelberg (1994). https://doi.org/10.1007/3-540-58601-6_111
21. Chu, Y., Xia, Q.: A hybrid algorithm for a class of resource constrained scheduling problems. In: Barták, R., Milano, M. (eds.) CPAIOR 2005. LNCS, vol. 3524, pp. 110–124. Springer, Heidelberg (2005). https://doi.org/10.1007/11493853_10
22. Hooker, J.N.: A hybrid method for the planning and scheduling. Constraints **10**(4) (2005)
23. Tran, T.T., Beck, J.C.: Logic-based benders decomposition for alternative resource scheduling with sequence dependent setups. In: Raedt, L.D., (eds.) ECAI 2012–20th European Conference on Artificial Intelligence. Including Prestigious Applications of Artificial Intelligence (PAIS-2012) System Demonstrations Track, Montpellier, France, 27–31 August 2012, Volume 242 of Frontiers in Artificial Intelligence and Applications, pp. 774–779. IOS Press (2012)
24. de Moura, L., Rueß, H., Sorea, M.: Lazy theorem proving for bounded model checking over infinite domains. In: Voronkov, A. (ed.) CADE 2002. LNCS (LNAI), vol. 2392, pp. 438–455. Springer, Heidelberg (2002). https://doi.org/10.1007/3-540-45620-1_35
25. Ji, X., Ma, F.: An efficient lazy SMT solver for nonlinear numerical constraints. In: Reddy, S., Drira, K., (eds.) 21st IEEE International Workshop on Enabling Technologies: Infrastructure for Collaborative Enterprises, WETICE 2012, Toulouse, France, 25–27 June 2012, pp. 324–329. IEEE Computer Society (2012)
26. Gebremedhin, A.H., Manne, F., Pothen, A.: What color is your Jacobian? Graph coloring for computing derivatives. SIAM Rev. **47**(4), 629–705 (2005)

27. Zuckerman, D.: Linear degree extractors and the inapproximability of max clique and chromatic number. Theor. Comput. **3**(1), 103–128 (2007)
28. Tseitin, G.S.: On the complexity of derivation in propositional calculus. In: Siekmann, J.H., Wrightson, G. (eds.) Automation of Reasoning. Symbolic Computation (Artificial Intelligence), pp. 466–483. Springer, Heidelberg (1983). https://doi.org/10.1007/978-3-642-81955-1_28
29. Gent, I.P., Petrie, K.E., Puget, J.: Symmetry in constraint programming. In: Handbook of Constraint Programming, pp. 329–376 (2006)
30. Asín, R., Nieuwenhuis, R., Oliveras, A., Rodríguez-Carbonell, E.: Cardinality networks: a theoretical and empirical study. Constraints **16**(2), 195–221 (2011)
31. Roney-Dougal, C.M., Gent, I.P., Kelsey, T., Linton, S.: Tractable symmetry breaking using restricted search trees. In: de Mántaras, R.L., Saitta, L., (eds.) Proceedings of the 16th Eureopean Conference on Artificial Intelligence, ECAI 2004, including Prestigious Applicants of Intelligent Systems, PAIS 2004, Valencia, Spain, 22–27 August 2004, pp. 211–215. IOS Press (2004)
32. Audemard, G., Lagniez, J.-M., Simon, L.: Improving glucose for incremental SAT solving with assumptions: application to MUS extraction. In: Järvisalo, M., Van Gelder, A. (eds.) SAT 2013. LNCS, vol. 7962, pp. 309–317. Springer, Heidelberg (2013). https://doi.org/10.1007/978-3-642-39071-5_23
33. Audemard, G., Simon, L.: Predicting learnt clauses quality in modern SAT solvers. In: Boutilier, C., (ed.) IJCAI 2009, Proceedings of the 21st International Joint Conference on Artificial Intelligence, Pasadena, California, USA, 11–17 July 2009, pp. 399–404 (2009)
34. Davies, J., Bacchus, F.: Exploiting the power of MIP solvers in MAXSAT. In: Järvisalo, M., Van Gelder, A. (eds.) SAT 2013. LNCS, vol. 7962, pp. 166–181. Springer, Heidelberg (2013). https://doi.org/10.1007/978-3-642-39071-5_13
35. Morgado, A., Ignatiev, A., Marques-Silva, J.: MSCG: robust core-guided MaxSAT solving. JSAT **9**, 129–134 (2014)
36. Ignatiev, A., Morgado, A., Marques-Silva, J.: PySAT: a python toolkit for prototyping with SAT oracles. In: Beyersdorff, O., Wintersteiger, C.M. (eds.) SAT 2018. LNCS, vol. 10929, pp. 428–437. Springer, Cham (2018). https://doi.org/10.1007/978-3-319-94144-8_26
37. Heras, F., Morgado, A., Marques-Silva, J.: Core-guided binary search algorithms for maximum satisfiability. In: Burgard, W., Roth, D. (eds.) Proceedings of the Twenty-Fifth AAAI Conference on Artificial Intelligence, AAAI 2011, San Francisco, California, USA, 7–11 August 2011. AAAI Press (2011)
38. Bacchus, F., Järvisalo, M., Martins, R.: Max-SAT 2018: Thirteen Max-SAT Evaluation (2018). https://maxsat-evaluations.github.io/2018/
39. Janota, M., Marques-Silva, J.: On the query complexity of selecting minimal sets for monotone predicates. Artif. Intell. **233**, 73–83 (2016)
40. Martins, R., Manquinho, V., Lynce, I.: Open-WBO: a modular MaxSAT solver. In: Sinz, C., Egly, U. (eds.) SAT 2014. LNCS, vol. 8561, pp. 438–445. Springer, Cham (2014). https://doi.org/10.1007/978-3-319-09284-3_33
41. Lagniez, J., Le Berre, D., de Lima, T., Montmirail, V.: A recursive shortcut for CEGAR: application to the modal logic K satisfiability problem. In: Sierra, C. (ed.) Proceedings of the Twenty-Sixth International Joint Conference on Artificial Intelligence, IJCAI 2017, Melbourne, Australia, 19–25 August 2017, pp. 674–680 (2017). ijcai.org

Constraint-Based Techniques in Stochastic Local Search MaxSAT Solving

Andreia P. Guerreiro[1], Miguel Terra-Neves[1,2], Inês Lynce[1], José Rui Figueira[3], and Vasco Manquinho[1(✉)]

[1] INESC-ID, Instituto Superior Técnico, Universidade de Lisboa, Lisbon, Portugal
{andreia,ines,vmm}@sat.inesc-id.pt
[2] OutSystems, Lisbon, Portugal
miguel.neves@outsystems.com
[3] CEG-IST, Instituto Superior Técnico, Universidade de Lisboa, Lisbon, Portugal
figueira@tecnico.ulisboa.pt

Abstract. The recent improvements in solving Maximum Satisfiability (MaxSAT) problems has allowed the usage of MaxSAT in several application domains. However, it has been observed that finding an optimal solution in a reasonable amount of time remains a challenge. Moreover, in many applications it is enough to provide a good approximation of the optimum. Recently, new local search algorithms have been shown to be successful in approximating the optimum in MaxSAT problems. Nevertheless, these local search algorithms fail in finding feasible solutions to highly constrained instances. In this paper, we propose two constraint-based techniques for improving local search MaxSAT solvers. Firstly, an unsatisfiability-based algorithm is used to guide the local search solver into the feasible region of the search space. Secondly, given a partial assignment, we perform Minimal Correction Subsets (MCS) enumeration in order to improve upon the best solution found by the local search solver. Experimental results using a large set of instances from the MaxSAT evaluation 2018 show the effectiveness of our approach.

Keywords: Maximum Satisfiability · Local search · Incomplete algorithms

1 Introduction

Over the last decade, a new generation of algorithms for Maximum Satisfiability (MaxSAT) problems has been proposed [1,13,40,42,44]. These new MaxSAT algorithms are usually based on iterative calls to a highly efficient Propositional Satisfiability (SAT) solver. For many industrial benchmarks, the current state-of-the-art MaxSAT algorithms are several orders of magnitude faster than branch-and-bound MaxSAT algorithms [35]. As a result, MaxSAT has been used extensively in many application domains, such as timetabling [4], fault localization in C programs [25], and design debugging [45], among others [20].

© Springer Nature Switzerland AG 2019
T. Schiex and S. de Givry (Eds.): CP 2019, LNCS 11802, pp. 232–250, 2019.
https://doi.org/10.1007/978-3-030-30048-7_14

Despite the success of the new generation of MaxSAT solvers, there is still a wide range of large-scale applications where such solvers fail to prove optimality within a reasonable amount of time. In fact, in some applications, time is so crucial that it suffices to quickly find a good approximation to the optimum [48]. As a result, several incomplete MaxSAT algorithms [3,12,33,36,47] have been proposed with the aim of finding good solutions within a limited time window.

It is well-known that stochastic local search (SLS) solvers are known to be competitive when the problem instances are easy to satisfy. On the other hand, SAT-based algorithms are more effective in problem instances with more constraints, while having more difficulties to deal with the optimization problem.

In the context of SAT solving, there is a large literature on combining SAT-based procedures and SLS techniques (e.g. [5,6,18,34,51]). Moreover, the integration of SAT-based complete algorithms and SLS algorithms has already been proposed for MaxSAT [28]. However, the proposed approaches for MaxSAT are mostly similar to a portfolio of solvers running in parallel or having some predefined criteria where either an SLS or a SAT solver are used. As a result, the integration of SLS and SAT-based techniques is limited.

In this paper, we propose an effective integration of SAT-based techniques in a SLS solver for MaxSAT. In our solver, the control of the solving process changes from SAT-based procedures to stochastic procedures and vice-versa. At each step, each procedure tries to build upon the information received from the other, instead of being independent procedures. The main contributions of the paper are as follows: (1) a new unsatisfiability-based algorithm to correct the SLS current assignment into a feasible solution, (2) a new improvement procedure based on Minimal Correction Subset (MCS) enumeration limited to the context of the SLS solver, and (3) an extensive experimental evaluation that shows the effectiveness of the newly proposed ideas.

The paper is organized as follows. Section 2 introduces the SAT and MaxSAT problems, as well as the notion of unsatisfiable cores and MCSes. Next, Sect. 3 reviews state of the art approximation algorithms for MaxSAT based on complete algorithms and SLS solvers. In Sect. 4, the new unsatisfiability-based procedure for assignment correction is presented, as well as a description of the assignment improvement procedure. Section 5 presents a comparison of our new solver with the top performing solvers on the incomplete track of the 2018 MaxSAT Evaluation. Finally, the paper concludes in Sect. 6.

2 Preliminaries

This section describes the Maximum Satisfiability (MaxSAT) problem, as well as the notions of unsatisfiable core and Minimal Correction Subsets (MCSes). Additional background information and definitions are also provided.

2.1 Maximum Satisfiability

A propositional formula in Conjunctive Normal Form (CNF), defined over a set $X = \{x_1, x_2, \ldots, x_n\}$ of n Boolean variables, is a conjunction of clauses, where a clause is a disjunction of literals. A literal is either a variable x_i or its complement $\bar{x}_i$. A complete assignment is a function $\nu : X \rightarrow \{0,1\}$ that associates each variable in X with a Boolean value. Given an assignment ν, a literal x_i (respectively $\bar{x}_i$) is said to be satisfied if $\nu(x_i) = 1$ (respectively $\nu(x_i) = 0$). A clause is said to be satisfied by ν if any of its literals is satisfied. Otherwise, it is said to be unsatisfied. A formula ϕ is satisfied by ν if all its clauses are satisfied. On the other hand, if any of the clauses in ϕ is unsatisfied by ν, then ϕ is unsatisfied. Given a CNF formula ϕ, the Propositional Satisfiability (SAT) problem consists of finding a truth assignment ν such that ϕ is satisfied, or prove that no assignment exist that satisfies ϕ.

The Maximum Satisfiability (MaxSAT) problem is an optimization version of the SAT problem and several versions of MaxSAT can be used [35]. In the context of this paper, we focus on the partial MaxSAT problem where clauses in a CNF formula $\phi = \phi_h \cup \phi_s$ are labeled as hard (ϕ_h) or soft (ϕ_s). The goal of partial MaxSAT problems is to find an assignment ν that satisfies all hard clauses in ϕ_h, while minimizing the number of unsatisfied soft clauses in ϕ_s. In weighted partial MaxSAT problems, a positive integer weight is associated with each soft clause and the goal is to satisfy all hard clauses, while minimizing the total weight of unsatisfied soft clauses.

If an assignment ν satisfies all hard clauses, then we say that ν is a feasible assignment. Otherwise, we say that ν is infeasible. In this paper, it is assumed that the set of hard clauses ϕ_h can be satisfied, i.e. there is always a feasible assignment for a given MaxSAT problem instance. Otherwise, the MaxSAT formula would be unsatisfiable.

Throughout the paper, the set notation is used for clauses and CNF formulas. In particular, a CNF formula is seen as a set of clauses and a clause as a set of literals. Finally, we extend the notation of satisfiability of a clause and a set of clauses by an assignment ν. If c_i is a clause satisfied by ν, then $\nu(c_i) = 1$, otherwise $\nu(c_i) = 0$. Let ϕ denote a set of clauses. If assignment ν satisfies ϕ, then $\nu(\phi) = 1$, otherwise $\nu(\phi) = 0$.

Example 1. Consider the following weighted partial MaxSAT formula $\phi = \phi_h \cup \phi_s$ where $\phi_h = \{(x_1 \vee x_2 \vee \bar{x}_3), (x_2 \vee x_3), (\bar{x}_1 \vee \bar{x}_2 \vee \bar{x}_3)\}$ and $\phi_s = \{((\bar{x}_1), 1), ((\bar{x}_2), 3), ((\bar{x}_3), 1)\}$. Note that the positive weight associated with each soft clause denotes the cost of not satisfying the clause. In this case, the assignment $\nu = \{x_1 = 1, x_2 = 0, x_3 = 1\}$ is an optimal solution with a cost of 2, since soft clauses $(\bar{x}_1)$ and $(\bar{x}_3)$ are not satisfied by ν.

2.2 Unsatisfiable Cores and Minimal Correction Subsets

Let ϕ be an unsatisfiable formula. A subset $\phi_C \subseteq \phi$ is an unsatisfiable core of ϕ if and only if ϕ_C is also unsatisfiable. Several techniques exist in the literature for computing unsatisfiable cores (e.g. [15,22]) and current state of the art SAT solvers are able to identify an unsatisfiable core of ϕ.

Example 2. Consider the following CNF formula $\phi = \{(x_1 \vee x_2 \vee \bar{x}_3), (x_2 \vee x_3), (\bar{x}_1 \vee \bar{x}_2 \vee \bar{x}_3), (\bar{x}_1), (\bar{x}_2), (\bar{x}_3)\}$. One unsatisfiable core of ϕ would be $\phi_C = \{(x_2 \vee x_3), (\bar{x}_2), (\bar{x}_3)\}$, since this subset of clauses of ϕ is unsatisfiable.

Let ϕ_h and ϕ_s be the sets of hard and soft clauses, respectively, such that ϕ_h is satisfiable and $\phi_h \cup \phi_s$ is unsatisfiable. A subset $C \subseteq \phi_s$ is a Minimal Correction Subset (MCS) if and only if $\phi_h \cup (\phi_s \setminus C)$ is satisfiable and $\phi_h \cup (\phi_s \setminus C) \cup \{c\}$ is unsatisfiable for all $c \in C$.

Observe that MCS algorithms [8,19,39,41] easily provide an approximation to the optimal solution of a MaxSAT instance. An MCS algorithm provides an assignment ν that satisfies $\phi_h \cup (\phi_s \setminus C)$. Let $f(C)$ denote the sum of the weights of the clauses in C. Since ν satisfies all hard clauses, its cost will be $f(C)$, thus providing an approximation to the optimum of the MaxSAT instance. In fact, solving a MaxSAT instance can be reduced to finding an MCS with minimum value of $f(C)$ [9].

Example 3. Consider again the weighted partial MaxSAT formula from Example 1, where $\phi_h = \{(x_1 \vee x_2 \vee \bar{x}_3), (x_2 \vee x_3), (\bar{x}_1 \vee \bar{x}_2 \vee \bar{x}_3)\}$ and $\phi_s = \{((\bar{x}_1), 1), ((\bar{x}_2), 3), ((\bar{x}_3), 1)\}$. This formula has two MCSs: $C_1 = \{(\bar{x}_1), (\bar{x}_3)\}$ and $C_2 = \{(\bar{x}_2)\}$. Observe that the cost of C_1 is 2, while the cost of C_2 is 3. Actually, an assignment that satisfies $\phi_h \cup (\phi_s \setminus C_1)$ is an optimal assignment of ϕ since C_1 is the lowest cost MCS. On the other hand, an assignment that satisfies $\phi_h \cup (\phi_s \setminus C_2)$ is an approximation on the optimum of ϕ.

3   Algorithms to Approximate MaxSAT

This section briefly reviews algorithms that can approximate the optimal solution of MaxSAT instances. First, we refer to complete SAT-based algorithms for MaxSAT that can be adapted to provide an approximate solution. Next, stochastic approaches are presented with focus on stochastic local search algorithms.

3.1   SAT-Based Algorithms

Current state-of-the-art complete algorithms for MaxSAT rely on iterative calls to a SAT solver. One possible approach is to use the linear Sat-Unsat algorithm that performs a linear search on the total weight of unsatisfied soft clauses. These algorithms start by solving the hard clauses using a SAT solver. Next, whenever a solution is found, a new pseudo-Boolean constraint[1] is added, such that solutions with a higher or equal cost are excluded. The algorithm stops when the SAT solver returns unsatisfiable. Hence, the last solution found is an optimal solution to the MaxSAT formula.

In large instances, the performance of these algorithms starts to degrade due to large weights in soft clauses, or when the number of soft clauses is very large. Recently, incomplete algorithms have been proposed where only a subset of soft

---

[1] In the case of partial MaxSAT instances, a cardinality constraint is used.

clauses is considered at each iteration, or the weights are approximated [14,27] to allow a more effective encoding of the Pseudo-Boolean or cardinality constraints.

While linear Sat-Unsat algorithms perform the search refining an upper bound on the optimal solutions, linear Unsat-Sat MaxSAT algorithms iteratively refine a lower bound [2,21,38]. In unsatisfiability-based MaxSAT algorithms, the lower bound is refined by iteratively finding unsatisfiable cores and the first satisfiable SAT call returns an optimal solution to the MaxSAT instance.

Unsatisfiability-based MaxSAT algorithms can also provide upper bounds [1, 3,43] by applying a stratified approach, i.e. only a subset of soft clauses with higher weights are considered. The remaining soft clauses are added iteratively to the solver, after the subproblem considering higher weights has been solved. Observe that any MaxSAT algorithm that maintains an upper bound on the optimum can provide an approximate solution. Nevertheless, in many problem instances, it is hard to quickly find a good quality approximation to the optimum.

3.2   Stochastic Algorithms

Stochastic local search (SLS) algorithms for SAT and MaxSAT have been developed in the past [23,46,50]. These algorithms are inherently incomplete, since they are unable to prove unsatisfiability of SAT problems or prove that an assignment is an optimal solution to a MaxSAT instance. Nevertheless, for randomly generated instances, SLS algorithms have been shown to be very effective at finding very good approximations to the optimal solution. In fact, SLS algorithms have been used to quickly find a tight upper bound to MaxSAT instances so that a subsequent branch and bound algorithm could be more effective in pruning the search space [29].

Given a MaxSAT instance $\phi = \phi_h \cup \phi_s$, SLS algorithms start by defining a random assignment ν to all problem variables. While ν does not satisfy all hard clauses ϕ_h, an unsatisfied hard clause $c_i \in \phi_h$ is selected and ν is updated by flipping the value of a variable in c_i. Hence, c_i becomes satisfied by ν. Next, if ν satisfies all hard clauses, then the algorithm focus on minimizing the weight of unsatisfied soft clauses ϕ_s by flipping assignments in ν. There is a plethora of heuristics to implement this generic SLS approach. Recently, new SLS algorithms and techniques have been proposed such as CCLS [37], CCEHC [36], Ramp [17], and maxroster [47], among others [11,12,33]. In the MaxSAT Evaluation 2018, SATLike [33] was one of the best performing solvers in the incomplete solver track. This was particularly surprising, since no randomly generated instances were selected for the MaxSAT evaluation [7].

Algorithm 1 presents the pseudo-code for SATLike [33]. This algorithm maintains a weight associated to each hard clause in ϕ_h and each soft clause in ϕ_s. Initially, hard clauses have weight 1 and soft clauses are associated with its weight in the MaxSAT instance (line 1). After using a procedure based on unit propagation to compute an initial assignment to ν (line 2), the algorithm performs several iterations until a given cutoff limit is reached. At each iteration, if ν satisfies all hard clauses and improves upon the best previous assignment, then ν is saved (lines 5–6). Let $score(x_i)$ denote the weight increase in satisfied

---

Algorithm 1: SATLike Algorithm

Input: $\phi = \phi_h \cup \phi_s$, *cutoff*
Output: satisfying assignment to ϕ

1   InitializeClauseWeights(ϕ_h, ϕ_s)
2   $\nu \leftarrow$ InitializeAssignment($Vars(\phi)$)
3   $\nu_{best} \leftarrow \emptyset$
4   **while** (*#iterations < cutoff*) **do**
5   **if** $((\nu(\phi_h) = 1) \wedge (\text{Cost}(\phi_s, \nu) < \text{Cost}(\phi_s, \nu_{best})))$ **then**
6       $\lfloor \; \nu_{best} \leftarrow \nu$ // A better solution is found
7     $D \leftarrow \{x_i \in Vars(\phi) | score(x_i) > 0\}$
8   **if** $(D \neq \emptyset))$ **then**
9       $\lfloor \; x_s \leftarrow$ SelectBMS(D)
10   **else**
11   UpdateClauseWeights(ϕ_h, ϕ_s, ν)
12   **if** $(\nu(\phi_h) = 1)$ **then**
13         $\lfloor \; c_s \leftarrow$ RandomSelect($\{c_i : c_i \in \phi_s \wedge \nu(c_i) = 0\}$)
14   **else**
15         $\lfloor \; c_s \leftarrow$ RandomSelect($\{c_i : c_i \in \phi_h \wedge \nu(c_i) = 0\}$)
16       $x_s \leftarrow$ SelectMaxScore($Vars(c_s)$)
17     $\nu \leftarrow$ Flip(ν, x_s)   // Flip value of x_s in ν
18   **return** (ν_{best})   // Returns the best assignment found

---

clauses resulting from flipping x_i. If there are variables that would improve ν with respect to the current clause weight (i.e. variables with positive score), then a variable is selected to be flipped according to a *best from multiple selections (BMS)* strategy (line 9)[2]. Otherwise, the algorithm is at a local minima and the current clause weights are updated according to the strategy defined in [33] (line 11). Next, if ν satisfies all the hard clauses, then an unsatisfied soft clause is selected (line 13). Otherwise, an unsatisfied hard clause is selected instead (line 15). The variable to be flipped is the one with the highest score in the selected clause (line 16). Finally, the best solution found is returned when the cutoff limit is reached (line 18). The cutoff limit depends on a predefined maximum number of iterations without improvement. That is, the algorithm ends when it is unable to find a satisfiable assignment, or when it fails to improve upon the best feasible solution found, within a given number of iterations (in the implementation the limit is of 10^7 iterations). For this purpose, the iteration counter is set to zero whenever ν_{best} is updated.

4 Using SAT Techniques in Local Search

The idea of integrating SAT techniques in SLS algorithms for MaxSAT is not new. For example, solver MiniWalk [28] used SAT solver MiniSat [16] to guide the SLS algorithm WalkSAT [46]. In this case, the SLS algorithm and the SAT solver

---

[2] We refer to the literature for further details [10,33].

were run in parallel using a shared memory array such that the SLS algorithm would not flip a variable x_i if it would result in a complement assignment to the assignment in the SAT solver. The goal is to use the SAT solver to deal with the hard clauses, so that the SLS algorithm can focus on the optimization of the soft clauses.

Nevertheless, despite some exchange of information in MiniWalk, the SLS algorithm and the SAT solver are run in parallel and mostly in an independent fashion, similar to a parallel portfolio solver. In this paper, the goal is to have a SLS algorithm where SAT-based techniques are effectively used to correct and improve the current assignment in the SLS algorithm. Although correction procedures have already been proposed in evolutionary algorithms for multi-objective optimization [24], this paper proposes a novel procedure where unsatisfiable cores are used to identify sets of variable assignments that need to be changed.

Let $\text{SAT}(\phi, \mathcal{A}, budget)$ denote a call to a SAT solver where ϕ is a CNF formula, $\mathcal{A}$ is a set of literals considered as assumptions, and $budget$ is a positive value. A SAT solver call returns a triple (st, ϕ_C, ν) where st denotes the solver return status (SAT, UNSAT or UNRES). If the solver returns SAT, then ν contains a satisfiable assignment to ϕ. On the other hand, if the solver returns UNSAT, then ϕ_C contains an unsatisfiable core. Note that ϕ might be satisfiable, but the solver might still return UNSAT due to the set of assumptions $\mathcal{A}$. This occurs when there are no models of ϕ where all assumption literals in $\mathcal{A}$ are satisfied. Therefore, ϕ_C might contain a subset of clauses from ϕ and literals from $\mathcal{A}$. Finally, the solver returns UNRES if during the SAT call the number of conflicts reaches the defined $budget$. Observe that if $budget$ is set to $+\infty$, then the SAT call does not return UNRES. However, in our context, a conflict limit will be set to avoid the solver to take too much time in a SAT call.

One of the shortcomings of SLS algorithms is that these solvers have difficulties in dealing with highly constrained formulas. Therefore, it might be the case that the SLS algorithm is unable to satisfy ϕ_h or gets stuck in some local minima. In these cases, using SAT-based techniques to find a satisfiable assignment to ϕ_h would be beneficial.

4.1 Assignment Correction

Consider the case when the SLS algorithm is unable to change from an unsatisfiable assignment ν into a better assignment. Algorithm 2 describes our unsatisfiability-based algorithm which performs a correction to ν in order to guide the SLS algorithm to the feasible region of the search space.

First, we start by building a set of assumption literals $\mathcal{A}$ corresponding to the assignment ν (lines 1–3). Next, a SAT call on the set of hard clauses ϕ_h is made (line 4). Clearly, if ν is not feasible, then this call returns UNSAT and ϕ_C contains an unsatisfiable core. Therefore, while a satisfiable assignment is not found, the assumption literals that occur in ϕ_C are deemed responsible for the UNSAT status, removed from $\mathcal{A}$ (line 7) and a new SAT call is made (line 8). Observe that a conflict limit is defined for the correction procedure. Hence, after

---

Algorithm 2: Assignment Correction Algorithm

Input: $\phi = \phi_h \cup \phi_s$, ν, $confLimit$
Output: satisfying assignment to ϕ
1 $\mathcal{A} \leftarrow \emptyset$
2 **foreach** $(x_i \in Vars(\phi))$ **do**
3   $\quad \mathcal{A} \leftarrow \mathcal{A} \cup \{(\nu(x_i) = 1 \ ? \ x_i : \bar{x}_i)\}$
4 $(st, \phi_C, \nu_{new}) \leftarrow \text{SAT}(\phi_h, \mathcal{A}, confLimit)$
5 $confBudget \leftarrow confLimit - satSolverConflicts$
6 **while** $(st \neq SAT \wedge confBudget > 0)$ **do**
7   $\quad \mathcal{A} \leftarrow \{l_j : l_j \in \mathcal{A} \wedge l_j \notin \phi_C\}$ $\qquad$ // remove literals in unsat core
8   $\quad (st, \phi_C, \nu_{new}) \leftarrow \text{SAT}(\phi_h, \mathcal{A}, confBudget)$
9   $\quad confBudget \leftarrow confBudget - satSolverConflicts$
10 **if** $(st \neq SAT)$ **then**
11 $\quad confBudget \leftarrow confLimit/10$
12 $\quad$ **while** $(st \neq SAT \wedge |\mathcal{A}| > 0)$ **do**
13   $\quad\quad \mathcal{A} \leftarrow \text{ChooseRandom}(\mathcal{A}, 0.5)$
14   $\quad\quad (st, \phi_C, \nu_{new}) \leftarrow \text{SAT}(\phi_h, \mathcal{A}, confBudget)$
15 $\nu_{new} \leftarrow \text{Improve}(\phi, \mathcal{A}, \nu_{new}, confLimit)$ $\qquad$ // MCS Enumeration
16 **return** (ν_{new}) $\qquad\qquad\qquad\qquad$ // Returns the best assignment found

---

each SAT call, the conflict budget is reduced by the number of conflicts in the last SAT call.

Note that if ϕ_h is satisfiable, then a satisfiable assignment is eventually found. However, since the number of conflicts is limited at each SAT call, it is possible that the conflict budget is not enough to find a satisfiable assignment. If this is the case, then our algorithm applies a similar procedure with a more aggressive strategy (lines 12–14) where at each iteration 50% of the literals in $\mathcal{A}$ are removed (line 13). Since the correction procedure only depends on the hard clauses, there is no guarantee regarding its quality. As a result, we also apply a SAT-based improvement procedure (line 15) detailed in Algorithm 3.

4.2 Assignment Improvement

Let $\phi = \phi_h \cup \phi_s$ be a MaxSAT formula and $\text{MCS}(\phi_h, \phi_s, budget)$ denote a call to an MCS algorithm where $budget$ is a positive value. An MCS solver call returns a pair (st, ν) where st denotes the return status. If the return status st is SAT, then ν denotes an assignment that satisfies $\phi_h \cup (\phi_s \setminus C)$ where C is an MCS of ϕ. Therefore, ν provides an approximation to the optimal solution of ϕ (see Sect. 2.2). Otherwise, either st is UNSAT if ϕ_h is not satisfiable or st is UNRES if the budget conflict limit is reached.

Algorithm 3 describes our improvement algorithm. Given a MaxSAT instance ϕ, a set of assumptions $\mathcal{A}$, a satisfiable assignment ν, and the conflict budget $ConfBudget$, the goal of this algorithm is to find a better quality solution for ϕ through an MCS enumeration procedure.

Algorithm 3: Assignment Improvement Algorithm using MCS enumeration

---

Input: $\phi = \phi_h \cup \phi_s$, $\mathcal{A}$, ν, *confBudget*
Output: satisfying assignment to ϕ

1 $\nu_{new} \leftarrow \nu$
2 $\phi_w \leftarrow \phi_h \cup \{(l_j) : l_j \in \mathcal{A}\}$
3 **while** (*confBudget* > 0) **do**
4   (st, ν) $\leftarrow$ MCS(ϕ_w, ϕ_s, *confBudget*)
5   **if** ((st = *SAT*) $\wedge$ (Cost(ϕ_s, ν) < Cost(ϕ_s, ν_{new}))) **then**
6         $\lfloor$ $\nu_{new} \leftarrow \nu$
7   **if** (st = *UNSAT*) **then**
8         $\lfloor$ **return** (ν_{new})   // All MCSs found
9      $\phi_w \leftarrow \phi_w \cup$ BlockingClause(ϕ_w, ϕ_s, ν)
10   *confBudget* $\leftarrow$ *confBudget* − *mcsSolverConflicts*
11 **return** (ν_{new})   // Returns the best assignment found

---

Algorithm 4: Assignment Improvement Algorithm using Linear Sat-Unsat

---

Input: $\phi = \phi_h \cup \phi_s$, $\mathcal{A}$, ν, *confBudget*
Output: satisfying assignment to ϕ

1 $\phi_w \leftarrow \phi_h \cup \{(l_j) : l_j \in \mathcal{A}\}$
2 (st, ν_{new}) $\leftarrow$ LinearSat-Unsat(ϕ_w, ϕ_s, *confBudget*)
3 **if** ((st = *SAT*) $\wedge$ (Cost(ϕ_s, ν) < Cost(ϕ_s, ν_{new}))) **then**
4      $\lfloor$ **return** ν_{new} // Linear Sat-Unsat found a better solution
5 **else**
6      $\lfloor$ **return** ν

The algorithm starts by building a working formula ϕ_w from the set of hard clauses ϕ_h and the set of assumptions $\mathcal{A}$ (line 2). Next, the algorithm iterates over all MCSes of ϕ, constrained to the set of assumptions $\mathcal{A}$ and returns the best assignment found (lines 3–10). Each time a new MCS is found, a new clause is added to ϕ_w to prevent the enumeration of the same MCS later on. This new clause (also known as blocking clause) forces at least one of the current variable assignments to have the opposite value (line 9). Finally, observe that, at each iteration, the conflict budget decreases depending on the number of conflicts used in the MCS algorithm.

Notice that the set of literals $\mathcal{A}$ restricts the MCS enumeration procedure. As a result, Algorithm 3 performs a localized MCS enumeration. Moreover, there is no guarantee that the MCSes found by this procedure are MCSes of the original MaxSAT formula ϕ, since the literals in $\mathcal{A}$ must all be satisfied in each MCS call. The main idea is to quickly perform a localized improvement in order to find a better solution than ν.

Many different improvement procedures can be devised, including the usage of complete methods. Algorithm 4 is an alternative to the improvement algorithm where the MCS enumeration is replaced with a call to a Linear Sat-Unsat algorithm (LSU). Observe that the call to the LSU algorithm is limited to a number of conflicts (line 2). Additionally, all literals in $\mathcal{A}$ are forced to be satisfied. Hence, the LSU call is also restricted to a localized region of the search space. If the LSU algorithm finds a feasible assignment, then st equals SAT. In that case, we check whether the assignment found by the LSU algorithm improves upon the previous solution ν and the best solution is returned. Finally, we note that any complete MaxSAT algorithm that is able to produce an approximation to the optimal solution (see Sect. 3.1) could be used instead of LSU.

4.3 Solvers

Two new solvers were developed: sls-mcs and sls-lsu. In sls-mcs, the SATLike solver (Algorithm 1)[3] is extended with the assignment correction algorithm (Algorithm 2) and the assignment improvement algorithm based on MCS enumeration (Algorithm 3). The difference from sls-mcs to sls-lsu is on the assignment improvement algorithm. In sls-lsu, the linear sat-unsat assignment improvement algorithm (Algorithm 4) is used.

Both sls-mcs and sls-lsu use the Glucose SAT solver (version 4.1) on the assignment correction procedure. Moreover, the CLD [39] algorithm is used as the MCS algorithm in sls-mcs. However, for weighted instances, the stratified CLD algorithm [49] is used. In sls-lsu, the linear sat-unsat algorithm is the one available at the open-wbo open source MaxSAT solver. In both sls-mcs and sls-lsu, the assignment correction/improvement algorithm is called just before line 5 in Algorithm 1 if SATlike has reached half of the maximum number of iterations without improvement. In such a case, the correction algorithm is called if the current assignment ν does not satisfy all hard clauses, otherwise the improvement algorithm is directly called with approximately half of the literals in the current assignment ν as assumptions. These assumption literals are randomly chosen from ν using the same procedure as in line 13 in Algorithm 2. Note that the iteration counter in Algorithm 1 is set to zero if ν_{best} is updated after the call to the correction/improvement algorithm.

5 Experimental Results

This section evaluates the effectiveness of the ideas proposed in the paper. The SATLike solver serves as our baseline solver. Nevertheless, we also compare sls-mcs and sls-lsu against the best performing solvers at the incomplete track of the last MaxSAT evaluation. No complete solver is included in this comparison because our preliminary results show that running LSU or enumerating MCSes can be hard for several of the instances used, which led to a poor performance. The solvers used in our experimental evaluation are as follows:

– SATLike: Stochastic local search solver described in Algorithm 1 [33].

---

[3] The source code of SATLike is publicly available at the 2018 MaxSAT evaluation https://maxsat-evaluations.github.io/2018/descriptions.html.

- `SATLike-c`: Version of `SATLike` submitted to the 2018 MaxSAT Evaluation. Initially, the `SATLike` algorithm is applied. If during the first 50 s, `SATLike` does not find a feasible solution, then the Linear Sat-Unsat complete algorithm from the `open-wbo` solver is used [32].
- `LinSBPS`: Linear sat-unsat algorithm with solution phase saving. In weighted instances, the algorithm starts by building a MaxSAT formula where all soft clause weights are divided by a large constant β. After finding an optimal solution for this formula, a new formula is build where the weights are divided by a new constant β' such that $\beta' < \beta$. The process is repeated until the original formula is solved ($\beta = 1$) [14].
- `maxroster`: This solver starts by applying the stochastic local search solver `Ramp` [17] for 6 s. Next, a complete MaxSAT solver is applied. MSU3 is used for partial MaxSAT, while OLL is used for weighted instances [47].
- `Open-WBO-Inc`: Another two-stage solver that starts by applying an incomplete algorithm, followed by the complete linear sat-unsat procedure of the `open-wbo` solver. For unweighted instances, the incomplete solver can be based on MCSes (`Open-WBO-Inc-MCS`) or based on bit vector optimization (`Open-WBO-Inc-OBV`). For weighted instances, the incomplete solver can be based on modifications on the weights of soft clauses and clustering (`Open--WBO-Inc-Cluster`) or partitioning of soft clauses (`Open-WBO-Inc-BMO`) [26].

All experimental results were obtained on a server with processor Intel(R) Xeon(R) CPU E5-2630 v2 @ 2.60GHz with 64GB of memory. The benchmark set corresponds to the one used in the 2018 MaxSAT evaluation for the incomplete track[4]. The benchmark set contains 153 partial MaxSAT problem instances, and 172 weighted partial MaxSAT problem instances. As in the 2018 MaxSAT competition, two time limits were considered: 60 s and 300 s. For each time limit, each solver was executed 7 times with each instance. Whenever `satlike-c`, `satlike`, `sls-mcs`, and `sls-slu` algorithms reach the *cutoff* stopping criteria before the time limit runs out, the algorithm is called again, and the best solution found among all calls is returned. The conflict limits of the correction and the improvement algorithms in `sls-mcs` and `sls-lsu` were set to 10^5.

5.1 Partial MaxSAT

Table 1 shows the number of instances for which the final solution of `sls-mcs` and `sls-lsu` solvers were produced by the local search part, and how many were produced by the correction and the improvement part. Tables 2 and 3 summarize the pairwise comparisons between solvers for the 60 and 300-s time limit scenarios, respectively. Two variants of the MCS-based local search solver are considered, one as described in Sect. 4.3 (`sls-mcs`), and another (`sls-mcs2`) that does not consider the assumptions $\mathcal{A}$ as hard clauses in the the MCS enumeration procedure, i.e., it implements Algorithm 3 with line 2 replaced by $\phi_w \leftarrow \phi_h$. Each one of the new solvers (`sls-mcs`, `sls-mcs2` and `sls-lsu`) is compared against

---

[4] Instances available at https://maxsat-evaluations.github.io/2018/.

Table 1. Number of instances for which the best solution was produced by the local search (sls), and by the correction/improvement algorithm (mcs/lsu).

Time limit	sls-mcs		sls-lsu	
	sls	mcs	sls	lsu
60 s	62	53	65	49
300 s	48	86	50	84

Table 2. Partial MaxSAT. Versus table (row wins, ties, column wins). Time limit 60 s.

	sls-mcs	sls-mcs2	sls-lsu
satlike	(19,71,**63**)	(24,88,**41**)	(23,78,**52**)
satlike-c	(40,70,**43**)	(**52**,75,26)	(**50**,75,28)
LinSBPS	(52,23,**78**)	(57,23,**73**)	(54,24,**75**)
maxroster	(39,36,**78**)	(44,36,**73**)	(40,37,**76**)
Open-WBO-Inc-mcs	(43,20,**90**)	(45,22,**86**)	(47,19,**87**)
Open-WBO-Inc-obv	(50,19,**84**)	(51,20,**82**)	(51,21,**81**)
sls-mcs	–	(**42**,90,21)	(**29**,113,11)
sls-mcs2	(21,90,**42**)	–	(23,97,**33**)
sls-lsu	(11,113,**29**)	(**33**,97,23)	–

every other solver considering the median value obtained for each instance. Each table cell contains a triple, (b, e, w), that represents the number of instances for which the solver in that row found a better (b), equal (e), or worse (w) quality solution than the solver in that column. Note that when both solvers are unable to find a feasible assignment, that fact is counted as a tie (e).

Table 1 shows that the correction and the improvement algorithms contribute with almost the same amount of final solutions as the local search part in the 60-s scenario, and almost twice as much in the 300-s scenario. Compared to one another, the mcs and the slu-based improvement algorithms contribute with nearly the same amount of final solutions to sls-mcs and sls-lsu solvers, respectively. Tables 2 and 3 show that these solvers found equally good solutions for about two thirds of the instances, while for most of the remaining ones, sls-mcs found better solutions than sls-lsu. In comparison to the other solvers, the number of times sls-mcs outperformed the other solvers was always higher than the number of times sls-lsu did. However, compared to most solvers in the 60-s scenario, the difference is very small - it is of only 1 to 3 instances. This means that, for most of the instances for which sls-mcs finds a better solution than sls-lsu, either sls-lsu provides a solution that is also better than the

Table 3. Partial MaxSAT. Versus table (row wins, ties, column wins). Time limit 300 s.

	sls-mcs	sls-mcs2	sls-lsu
satlike	(23,69,**61**)	(29,64,**60**)	(30,69,**54**)
satlike-c	(**48**,68,37)	(**50**,66,37)	(**62**,69,22)
LinSBPS	(53,23,**77**)	(46,34,**73**)	(56,25,**72**)
maxroster	(58,29,**66**)	(**60**,34,59)	(**66**,30,57)
Open-WBO-Inc-mcs	(31,17,**105**)	(27,21,**105**)	(42,18,**93**)
Open-WBO-Inc-obv	(35,17,**101**)	(33,21,**99**)	(44,20,**89**)
sls-mcs	–	(**43**,76,34)	(**40**,106,7)
sls-mcs2	(34,76,**43**)	–	(**51**,79,23)
sls-lsu	(7,106,**40**)	(23,79,**51**)	–

solution found by other solvers, or the solution found by sls-mcs is still not good enough.

The sls-mcs solver clearly improves upon satlike, as it obtained better quality solutions in about 60 instances in both 60 and 300-s scenarios and was worse than satlike in less than 24 instances. Of those 60 instances, satlike was unable to find a feasible solution in about 20 and 30 of them, for the 60 and the 300-s scenario, respectively. This means that not only the correction algorithm was able to help the local search algorithm reach the feasible region, but also the improvement algorithm helped finding better feasible solutions. Compared to satlike-c, sls-mcs is competitive in the 60-s scenario, and is slightly worse in the 300-s scenario. This is not surprising as in the cases where satlike cannot find a feasible solution in the first 48 s, the additional 4 min are fully used by the complete solver. Comparing the two MCS-based local search solvers, sls-mcs had a better performance than sls-mcs2.

Comparing to any other solver in the 60-s scenario, sls-mcs and sls-lsu outperformed all of them. Apart from maxroster, that becomes more competitive, and satlike-c, this remains true for the 300-s scenario.

Overall, the results show that both the correction and the improvement algorithms are useful to the local search algorithm. The former plays an important role for highly constrained problems, for which a feasible solution is difficult to find through local search, and the latter can improve even further upon the solution found.

5.2   Weighted Partial MaxSAT

As the MCS-based local search solver achieved better results for the partial MaxSAT problem than the one based on LSU, only the former was tested for the weighted partial MaxSAT problem. In this scenario, a stratified MCS algorithm is used to enumerate MCSes, where the step of partitioning the set of soft clauses

Table 4. Number of instances of WPMS for which the best solution was produced by the local search (sls), and by the correction/improvement algorithm (mcs/mcs2).

Time limit	sls-mcs		sls-mcs2	
	sls	mcs	sls	mcs2
60 s	72	70	54	88
300 s	59	93	44	110

Table 5. Weighted partial MaxSAT. Versus table (row wins, ties, column wins). Time limit 60 s.

	sls-mcs	sls-mcs2
satlike	(31,91,**50**)	(27,66,**79**)
satlike-c	(**52**,75,45)	(40,52,**80**)
LinSBPS	(**108**,9,55)	(**104**,13,55)
maxroster	(**91**,23,58)	(**77**,23,72)
Open-WBO-Inc-BMO	(**99**,10,63)	(**89**,17,66)
Open-WBO-Inc-cluster	(70,10,**92**)	(41,15,**116**)
sls-mcs	–	(27,79,**66**)
sls-mcs2	(**66**,79,27)	–

is performed only once at the beginning. Tables 4, 5 and 6, are analogous to Tables 1, 2 and 3, respectively.

As in the partial MaxSAT problem, the correction and improvement algorithms contribute directly to solutions reported by the solver(s). They are responsible for at least half, and up to two thirds, of the solutions reported by the solvers (see Table 4). Solvers sls-mcs and sls-mcs2 had similar performance in about half of the instances in the 60-s scenario, and in about one third of the instances in the 300-s scenario (see Tables 5 and 6). Moreover, sls-mcs2 found better solutions than sls-mcs in about two thirds of the remaining instances in both scenario. The correction/improvement algorithms in sls-mcs work in a more restricted search space than in sls-mcs2 because they start with an already defined partial assignment (through the assumptions). This is advantageous when SATlike's current assignment is reasonably good, but when SATlike does not perform so well (in the weighted scenario), it seems more advantageous not to consider its current assignment (as in sls-mcs2).

Compared to satlike, and despite sls-mcs performing better in the 60-s scenario, contrary to what was expected its performance decayed in the 300-s scenario. Conversely, sls-mcs2 performed much better than satlike, and even

Table 6. Weighted partial MaxSAT. Versus table (row wins, ties, column wins). Time limit 300 s.

	sls-mcs	sls-mcs2
satlike	(**70**,40,62)	(51,32,**89**)
satlike-c	(**90**,32,50)	(62,24,**86**)
LinSBPS	(**111**,12,49)	(**98**,19,55)
maxroster	(**97**,19,56)	(**81**,28,63)
Open-WBO-Inc-BMO	(**102**,8,62)	(**85**,16,71)
Open-WBO-Inc-cluster	(68,9,**95**)	(37,16,**119**)
sls-mcs	–	(36,46,**90**)
sls-mcs2	(**90**,46,36)	–

better than `satlike-c`. Compared to the other solvers in the two time-limit scenarios, `sls-mcs` showed a weaker performance, except when compared to `Open-WBO-Inc-cluster`. On the other hand, `sls-mcs2` had, in general, a better performance but still only outperforms `Open-WBO-Inc-cluster`.

Overall, `sls-mcs` has a poor performance, particularly against `satlike`. On the other hand, `sls-mcs2` showed a performance superior to `sls-mcs`, and was more competitive to the remaining solvers. This contrast may be indicative that the local search part is being too biased towards some regions of the search space, and that may be restricting too much the search space of the improvement algorithm after the assignment correction step. The inferior performance of `satlike` and `satlike-c` in the 2018 MaxSAT evaluation for weighted MaxSAT problem instances reinforces the conjecture. Moreover, `sls-mcs` does not make full use of stratification, as it does not take into account that, by forcing the assumption to be satisfied, some of the soft clauses are also satisfied. Thus, considering only the remaining soft clauses in the stratification process should lead to better results.

6   Conclusions and Future Work

In this paper, we propose the integration of SAT-based algorithms into a state of the art SLS solver for MaxSAT, where the solving process changes iteratively between the SLS and SAT-based procedures. A novel algorithm based on the identification of unsatisfiable cores is used for assignment correction of the SLS procedure. As a result, the SLS solver is guided into the feasible area of the search space, thus improving the search process in the SLS solver. Moreover, assignment improvement procedures are also devised and integrated into the SLS solver. Experimental results show the effectiveness of our approach, as the new incomplete MaxSAT solver is able to quickly find better approximations for a larger number of problem instances than other state of the art incomplete MaxSAT solvers.

For future work, we plan to extend the usage of unsatisfiable cores in SLS solvers, since other procedures can be devised where the unsatisfiable cores would guide the SLS algorithm. Furthermore, a more dynamic interaction between the SLS procedure and the SAT-based procedure should be tried. Finally, current results show that SLS algorithms for weighted MaxSAT can be greatly improved. Currently, SLS solvers still spend many iterations trying to satisfy the hard clauses. However, a tighter integration of SAT-based procedures would enable the SLS algorithm to focus on the optimization part of the problem.

Acknowledgments. This work was supported by national funds through FCT with references UID/CEC/50021/2019, PTDC/CCI-COM/31198/2017 and DSAIPA/AI/0044/2018.

References

1. Ansótegui, C., Bonet, M.L., Gabàs, J., Levy, J.: Improving WPM2 for (Weighted) partial MaxSAT. In: Schulte, C. (ed.) CP 2013. LNCS, vol. 8124, pp. 117–132. Springer, Heidelberg (2013). https://doi.org/10.1007/978-3-642-40627-0_12

2. Ansótegui, C., Bonet, M.L., Levy, J.: Solving (Weighted) partial MaxSAT through satisfiability testing. In: Kullmann [30], pp. 427–440

3. Ansótegui, C., Gabàs, J.: WPM3: an (in)complete algorithm for weighted partial maxsat. Artif. Intell. **250**, 37–57 (2017)

4. Asín, R., Nieuwenhuis, R.: Curriculum-based course timetabling with SAT and MaxSAT. Ann. Oper. Res. **218**(1), 71–91 (2014)

5. Audemard, G., Lagniez, J.-M., Mazure, B., Saïs, L.: Boosting local search thanks to CDCL. In: Fermüller, C.G., Voronkov, A. (eds.) LPAR 2010. LNCS, vol. 6397, pp. 474–488. Springer, Heidelberg (2010). https://doi.org/10.1007/978-3-642-16242-8_34

6. Audemard, G., Simon, L.: GUNSAT: a greedy local search algorithm for unsatisfiability. In: Proceedings of the 20th International Joint Conference on Artificial Intelligence, pp. 2256–2261 (2007)

7. Bacchus, F., Järvisalo, M.J., Martins, R., et al.: MaxSAT evaluation 2018 (2018)

8. Bailey, J., Stuckey, P.J.: Discovery of minimal unsatisfiable subsets of constraints using hitting set dualization. In: Hermenegildo, M.V., Cabeza, D. (eds.) PADL 2005. LNCS, vol. 3350, pp. 174–186. Springer, Heidelberg (2005). https://doi.org/10.1007/978-3-540-30557-6_14

9. Birnbaum, E., Lozinskii, E.: Consistent subsets of inconsistent systems: structure and behaviour. J. Exp. Theor. Artif. Intell. **15**(1), 25–46 (2003)

10. Cai, S.: Balance between complexity and quality: local search for minimum vertex cover in massive graphs. In: Twenty-Fourth International Joint Conference on Artificial Intelligence (2015)

11. Cai, S., Luo, C., Thornton, J., Su, K.: Tailoring local search for partial maxsat. In: Brodley, C.E., Stone, P. (eds.) Proceedings of the Twenty-Eighth AAAI Conference on Artificial Intelligence, Québec City, Québec, Canada, 27–31 July 2014, pp. 2623–2629. AAAI Press (2014)

12. Cai, S., Luo, C., Zhang, H.: From decimation to local search and back: a new approach to maxsat. In: Sierra, C. (ed.) Proceedings of the Twenty-Sixth International Joint Conference on Artificial Intelligence, IJCAI 2017, Melbourne, Australia, 19–25 August 2017, pp. 571–577 (2017). ijcai.org

13. Davies, J., Bacchus, F.: Exploiting the power of MIP solvers in MAXSAT. In: Järvisalo, M., Van Gelder, A. (eds.) SAT 2013. LNCS, vol. 7962, pp. 166–181. Springer, Heidelberg (2013). https://doi.org/10.1007/978-3-642-39071-5_13

14. Demirovic, E., Stuckey, P.J.: LinSBPS. MaxSAT Evaluation 2018: Solver and Benchmark Descriptions, volume B-2018-2 of Department of Computer Science Series of Publications B, University of Helsinki, pp. 8–9 (2018)

15. Dershowitz, N., Hanna, Z., Nadel, A.: A scalable algorithm for minimal unsatisfiable core extraction. In: Biere, A., Gomes, C.P. (eds.) SAT 2006. LNCS, vol. 4121, pp. 36–41. Springer, Heidelberg (2006). https://doi.org/10.1007/11814948_5

16. Een, N.: MiniSat: a sat solver with conflict-clause minimization. In: Proceedings SAT-05: 8th International Conference on Theory and Applications of Satisfiability Testing, pp. 502–518 (2005)

17. Fan, Y., Ma, Z., Su, K., Sattar, A., Li, C.: Ramp: a local search solver based on make-positive variables. MaxSAT Evaluation (2016)

18. Fang, L., Hsiao, M.S.: A new hybrid solution to boost SAT solver performance. In: Design, Automation and Test in Europe Conference, pp. 1307–1313 (2007)

19. Felfernig, A., Schubert, M., Zehentner, C.: An efficient diagnosis algorithm for inconsistent constraint sets. Artif. Intell. Eng. Des. Anal. Manuf. **26**(1), 53–62 (2012)

20. Feng, Y., Bastani, O., Martins, R., Dillig, I., Anand, S.: Automated synthesis of semantic malware signatures using maximum satisfiability. In: 24th Annual Network and Distributed System Security Symposium, NDSS 2017, San Diego, California, USA, 26 February–1 March 2017. The Internet Society (2017)

21. Fu, Z., Malik, S.: On solving the partial MAX-SAT problem. In: Biere, A., Gomes, C.P. (eds.) SAT 2006. LNCS, vol. 4121, pp. 252–265. Springer, Heidelberg (2006). https://doi.org/10.1007/11814948_25

22. Goldberg, E.I., Novikov, Y.: Verification of proofs of unsatisfiability for CNF formulas. In: Conference and Exposition on Design, Automation and Test in Europe, pp. 10886–10891 (2003)

23. Gu, J.: Efficient local search for very large-scale satisfiability problems. ACM SIGART Bull. **3**(1), 8–12 (1992)

24. Henard, C., Papadakis, M., Harman, M., Traon, Y.L.: Combining multi-objective search and constraint solving for configuring large software product lines. In: International Conference on Software Engineering, pp. 517–528 (2015)

25. Jose, M., Majumdar, R.: Cause clue clauses: error localization using maximum satisfiability. In: Programming Language Design and Implementation, pp. 437–446. ACM (2011)

26. Joshi, S., Kumar, P., Manquinho, V., Martins, R., Nadel, A., Rao, S.: Open-WBO-Inc in MaxSAT evaluation 2018. MaxSAT Evaluation 2018: Solver and Benchmark Descriptions, volume B-2018-2 of Department of Computer Science Series of Publications B, University of Helsinki, pp. 16–17 (2018)

27. Joshi, S., Kumar, P., Martins, R., Rao, S.: Approximation strategies for incomplete MaxSAT. In: Hooker, J. (ed.) CP 2018. LNCS, vol. 11008, pp. 219–228. Springer, Cham (2018). https://doi.org/10.1007/978-3-319-98334-9_15

28. Kroc, L., Sabharwal, A., Gomes, C.P., Selman, B.: Integrating systematic and local search paradigms: a new strategy for MaxSAT. In: Boutilier, C. (ed.) IJCAI 2009, Proceedings of the 21st International Joint Conference on Artificial Intelligence, Pasadena, California, USA, 11–17 July 2009, pp. 544–551 (2009)

29. Kugel, A.: akmaxsat and akmaxsat_ls solver description. Technical report, MaxSAT Evaluation 2012 Solver Descriptions (2012)

30. Kullmann, O. (ed.): International Conference on Theory and Applications ofSatisfiability Testing, LNCS, vol. 5584. Springer, Heidelberg (2009). https://doi.org/10.1007/978-3-642-02777-2

31. Lang, J. (ed.): Proceedings of the Twenty-Seventh International Joint Conference on Artificial Intelligence, IJCAI 2018, 13–19 July 2018, Stockholm, Sweden (2018). ijcai.org

32. Lei, Z., Cai, S.: SATlike-c. MaxSAT Evaluation 2018: Solver and Benchmark Descriptions, volume B-2018-2 of Department of Computer Science Series of Publications B, University of Helsinki, pp. 24–25 (2018)

33. Lei, Z., Cai, S.: Solving (weighted) partial maxsat by dynamic local search for SAT. In: Lang [33], pp. 1346–1352

34. Letombe, F., Marques-Silva, J.: Hybrid incremental algorithms for booleansatisfiability. Int. J. Artif. Intell. Tools **21**(6) (2012). https://doi.org/10.1142/S021821301250025X

35. Li, C.M., Manyà, F.: MaxSAT, hard and soft constraints. In: Handbook of Satisfiability, pp. 613–631. IOS Press (2009)

36. Luo, C., Cai, S., Su, K., Huang, W.: CCEHC: an efficient local search algorithm for weighted partial maximum satisfiability. Artif. Intell. **243**, 26–44 (2017)

37. Luo, C., Cai, S., Wu, W., Jie, Z., Su, K.: CCLS: an efficient local search algorithm for weighted maximum satisfiability. IEEE Trans. Computers **64**(7), 1830–1843 (2015)

38. Manquinho, V., Marques-Silva, J., Planes, J.: Algorithms for Weighted Boolean Optimization. In: Kullmann [30], pp. 495–508

39. Marques-Silva, J., Heras, F., Janota, M., Previti, A., Belov, A.: On Computing Minimal Correction Subsets. In: International Joint Conference on Artificial Intelligence, pp. 615–622 (2013)

40. Martins, R., Joshi, S., Manquinho, V., Lynce, I.: Incremental cardinality constraints for MaxSAT. In: O'Sullivan, B. (ed.) CP 2014. LNCS, vol. 8656, pp. 531–548. Springer, Cham (2014). https://doi.org/10.1007/978-3-319-10428-7_39

41. Mencía, C., Previti, A., Marques-Silva, J.: Literal-based MCS extraction. In: International Joint Conference on Artificial Intelligence, pp. 1973–1979 (2015)

42. Morgado, A., Dodaro, C., Marques-Silva, J.: Core-guided MaxSAT with soft cardinality constraints. In: O'Sullivan, B. (ed.) CP 2014. LNCS, vol. 8656, pp. 564–573. Springer, Cham (2014). https://doi.org/10.1007/978-3-319-10428-7_41

43. Morgado, A., Ignatiev, A., Marques-Silva, J.: MSCG: robust core-guided maxsat solving. JSAT **9**, 129–134 (2014)

44. Narodytska, N., Bacchus, F.: Maximum satisfiability using core-guided MaxSAT resolution. In: AAAI Conference on Artificial Intelligence, pp. 2717–2723. AAAI Press (2014)

45. Safarpour, S., Mangassarian, H., Veneris, A.G., Liffiton, M.H., Sakallah, K.A.: Improved design debugging using maximum satisfiability. In: Formal Methods in Computer-Aided Design, pp. 13–19. IEEE Computer Society (2007)

46. Selman, B., Kautz, H.A., Cohen, B.: Local search strategies for satisfiability testing. In: Johnson, D.S., Trick, M.A. (eds.) Cliques, Coloring, and Satisfiability, Proceedings of a DIMACS Workshop, New Brunswick, New Jersey, USA, 11–13 October 1993. DIMACS Series in Discrete Mathematics and Theoretical Computer Science, vol. 26, pp. 521–532. DIMACS/AMS (1993)

47. Sugawara, T.: Maxroster: solver description. MaxSAT Eval. **2017**, 12 (2017)

48. Terra-Neves, M., Machado, N., Lynce, I., Manquinho, V.: Concurrency debugging with maxSMT. In: AAAI Conference on Artificial Intelligence. AAAI Press (2019)

49. Terra-Neves, M., Lynce, I., Manquinho, V.M.: Stratification for constraint-based multi-objective combinatorial optimization. In: Lang [31], pp. 1376–1382
50. Tompkins, D.A.D., Hoos, H.H.: UBCSAT: an implementation and experimentation environment for SLS algorithms for SAT & MAX-SAT. In: The Seventh International Conference on Theory and Applications of Satisfiability Testing, SAT 2004, 10–13 May 2004, Vancouver, BC, Canada, Online Proceedings (2004)
51. Zhang, J., Zhang, H.: Combining local search and backtracking techniques for constraint satisfaction. In: Proceedings of the Thirteenth National Conference on Artificial Intelligence and Eighth Innovative Applications of Artificial Intelligence Conference, pp. 369–374 (1996)

Trimming Graphs Using Clausal Proof Optimization

Marijn J. H. Heule[(⊠)]

Department of Computer Science, The University of Texas, Austin, USA
marijn@heule.nl

Abstract. We present a method to gradually compute a smaller and smaller unsatisfiable core of a propositional formula by minimizing proofs of unsatisfiability. The goal is to compute a minimal unsatisfiable core that is relatively small compared to other minimal unsatisfiable cores of the same formula. We try to achieve this goal by postponing deletion of arbitrary clauses from the formula as long as possible—in contrast to existing minimal unsatisfiable core algorithms. We applied this method to reduce the smallest known unit-distance graph with chromatic number 5 from 553 vertices and 2 720 edges to 529 vertices and 2 670 edges.

1 Introduction

Today's satisfiability (SAT) solvers can not only determine whether a propositional formula can be satisfied, but they can also produce a certificate in case no satisfying assignments exists. These certificates, known as proofs of unsatisfiability, can be used for multiple purposes ranging from checking the correctness of the unsatisfiability claim [8,9,14,16,26] to computing interpolants [11]. In this paper, we focus on another application of proofs of unsatisfiability: computing an unsatisfiable core of the formula [1,3,15,27]. We observed that the size of proofs tends to correlate to the size the corresponding unsatisfiable cores: the smaller the proof, the smaller the unsatisfiable core. We present a method to exploit this relation by computing a smaller and smaller proof of unsatisfiability to compute a small unsatisfiable core. This method was developed to improve the upper bound of the smallest unit-distance graph with chromatic number 5, which is currently a Polymath project. Details about the problem and this project are described below. The presented method was developed as existing techniques performed poorly on this application. Yet it could help with other applications that use unsatisfiable cores too—which we plan to study in the near future.

The chromatic number of the plane, a problem first proposed by Nelson in 1950 [25], asks how many colors are needed to color all points of the plane such that no two points at distance 1 from each other have the same color. Early results showed that at least four and at most seven colors are required. By the de Bruijn–Erdős theorem, the chromatic number of the plane is the largest possible chromatic number of a finite unit-distance graph [4]. The Moser Spindle, a unit-distance graph with 7 vertices and 11 edges, shows the lower bound [24], while the upper bound is shown by a 7-coloring of the entire plane by Isbell [25].

© Springer Nature Switzerland AG 2019
T. Schiex and S. de Givry (Eds.): CP 2019, LNCS 11802, pp. 251–267, 2019.
https://doi.org/10.1007/978-3-030-30048-7_15

In a breakthrough for this problem in April 2018, Aubrey de Grey improved the lower bound by providing a unit-distance graph with $1\,581$ vertices with chromatic number 5 [10]. This discovery by de Grey started a Polymath project to find smaller graphs. The current record is a graph with 553 vertices and $2\,720$ edges [13]. We present a new technique to construct a large unit-distance graph with chromatic number 5, which we reduce with the proposed method to a graph with "only" 529 vertices and $2\,670$ edges. This graph is much more symmetric compared to earlier small unit-distance graphs with chromatic number 5. The total costs to compute this graph were roughly $100\,000$ CPU hours.

2 Preliminaries

Propositional Formulas. We will minimize graphs on the propositional level. We consider formulas in *conjunctive normal form* (CNF), which are defined as follows. A *literal* is either a variable x (a *positive literal*) or the negation $\overline{x}$ of a variable x (a *negative literal*). The *complement* $\overline{l}$ of a literal l is defined as $\overline{l} = \overline{x}$ if $l = x$ and $\overline{l} = x$ if $l = \overline{x}$. For a literal l, $var(l)$ denotes the variable of l. A *clause* is a disjunction of literals and a *formula* is a conjunction of clauses.

An *assignment* is a function from a set of variables to the truth values 1 (*true*) and 0 (*false*). A literal l is *satisfied* by an assignment α if l is positive and $\alpha(var(l)) = 1$ or if it is negative and $\alpha(var(l)) = 0$. A literal is *falsified* by an assignment if its complement is satisfied by the assignment. A clause is satisfied by an assignment α if it contains a literal that is satisfied by α. A formula is satisfied by an assignment α if all its clauses are satisfied by α. A formula is *satisfiable* if there exists an assignment that satisfies it and *unsatisfiable* otherwise.

For a formula F and assignment α, we denote by $F|\alpha$ a reduced copy of F without clauses satisfied by α and literals falsified by α. A *unit clause* is a clause with only one literal. The result of applying the *unit clause rule* to a formula F is the formula $F|l$ where (l) is a unit clause in F. The iterated application of this rule to a formula, until no unit clauses are left, is called *unit propagation*. If unit propagation yields the empty clause $\perp$, we say that it derived a *conflict*.

Clausal Proofs. A clause C is *redundant* with respect to a formula F if F and $F \wedge C$ are satisfiability equivalent. For instance, the clause $C = (x \vee y)$ is redundant with respect to the formula $F = (\overline{x} \vee \overline{y})$ since F and $F \wedge C$ are satisfiability equivalent (although they are not logically equivalent). This redundancy notion allows us to add redundant clauses to a formula while preserving satisfiability.

Given a formula $F = \{C_1, \ldots, C_m\}$, a *clausal derivation* of a clause C_n from F is a sequence $C_{m+1}, \ldots, C_n$ of clauses. Such a sequence gives rise to formulas $F_m, F_{m+1}, \ldots, F_n$, where $F_i = \{C_1, \ldots, C_i\}$. We call F_i the *accumulated formula* corresponding to the i-th proof step. A clausal derivation is *correct* if every clause C_i $(i > m)$ is redundant with respect to the formula F_{i-1} and if this redundancy can be checked in polynomial time with respect to the size of the proof. A clausal derivation is a *proof* of a formula F if it derives the unsatisfiable empty clause. Clearly, since every clause-addition step preserves satisfiability, and since the empty clause is always false, a proof of F certifies the unsatisfiability of F.

Checking the correctness of a clause C_i in a derivation consists of computing a justification why C_i is redundant with respect by formula F_{i-1}. The most commonly used method for this purpose is *reverse unit propagation* (RUP): Let α be the assignment that falsifies all literals in C_i. Clause C_i has the RUP property if and only if unit propagation on $F_{i-1}|\alpha$ results in a conflict. In this case the justification of C_i consists of all clauses that were required to derive the conflict. Clausal proofs that can be validated using this method are called RUP proofs. Most SAT-solving techniques can be compactly expressed as RUP.

Example 1. Consider the formula below consisting of 3 variables and 7 clauses:

$$F := (\overline{y} \vee z) \wedge (x \vee z) \wedge (\overline{x} \vee y) \wedge (\overline{x} \vee \overline{y}) \wedge (x \vee \overline{y}) \wedge (y \vee \overline{z}) \wedge (x \vee \overline{z})$$

A clausal proof of F is $\overline{y}, z, \bot$. A justification of this proof is shown in Fig. 1. This justification shows that $\overline{y}$ and z do not depend on each other. As a consequence, swapping them results in another correct proof. Notice that clause $(x \vee \overline{z})$ is not used in this justification and it is thus not part of the core of F.

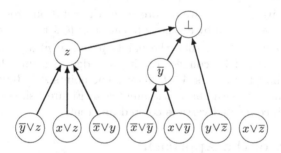

Fig. 1. A justification of the proof of the example formula. Each clause in the proof depends on its incoming arcs. The clauses without incoming arcs represent the formula.

In practice, clausal proofs also contain deletion information. The presence of deletion information significantly reduces the cost to compute a justification. Clausal proofs, which can be validated using the RUP method and include deletion information, are known as DRUP proofs. We mostly ignore the deletion information aspect of clausal proofs to simplify the presentation. All techniques discussed in this paper work with deletion information as well.

Chromatic Number of the Plane. The Chromatic Number of the Plane (CNP) [25] asks how many colors are required in a coloring of the plane to ensure that there exists no monochromatic pair of points with distance 1. A *unit-distance graph* is a graph formed from a set of points in the plane by connecting two points by an edge whenever the distance between the two points is exactly one. A lower bound for CNP of k colors can be obtained by showing that a unit-distance graph has chromatic number k.

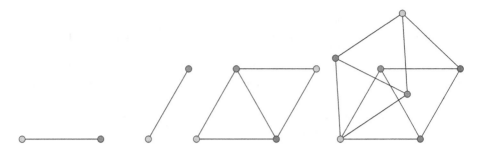

Fig. 2. From left to right: illustrations of unit-distance graphs A, B, $A \oplus B$, and the Moser Spindle. The graphs shown have chromatic number 2, 2, 3, and 4, respectively. The illustrations show valid colorings with the fewest number of colors. (Color figure online)

We will use three operations to construct larger and larger graphs: the Minkowski sum [12], rotation, and merging. Given two sets of points A and B, the Minkowski sum of A and B, denoted by $A \oplus B$, equals $\{a + b \mid a \in A, b \in B\}$. Consider the sets of points $A = \{(0,0), (1,0)\}$ and $B = \{(0,0), (1/2, \sqrt{3}/2)\}$, then $A \oplus B = \{(0,0), (1,0), (1/2, \sqrt{3}/2), (3/2, \sqrt{3}/2)\}$.

Given a positive integer i, we denote by θ_i the rotation around point $(0,0)$ with angle $\arccos(\frac{2i-1}{2i})$ and by θ_i^k the application of θ_i k times. Let p be a point with distance $\sqrt{i}$ from $(0,0)$, then the points p and $\theta_i(p)$ are exactly distance 1 apart and thus would be connected with an edge in a unit-distance graph. Consider again the set of points $A \oplus B$ above. The points $A \oplus B \cup \theta_3(A \oplus B)$ form the Moser Spindle [24] with chromatic number 4. Figure 2 shows visualizations of these sets with connected vertices colored differently.

3   Overview of the Approach

The smallest known unit-distance graph with chromatic number 5 has 553 vertices and 2 720 edges [13]. This graph was found using the following method. Start with a large unit-distance graph G with chromatic number 5. Now reduce the size of that graph by solving the formula that encodes whether the graph can be colored with 4 colors. That formula is unsatisfiable. From the proof of unsatisfiability, an unsatisfiable core can be extracted that represents a subgraph with chromatic number 5. This step is repeated again and again as long as the graph is reduced. In the last step, vertices are randomly eliminated to make the graph vertex-critical: removing any additional vertex introduces a 4-coloring.

In this paper we present two improvements. The most important one is a new method, presented in Sect. 4, to produce short proofs of unsatisfiability. We observed that for the formulas studied in this paper that the shorter the proof, the smaller the unsatisfiable core and thus the smaller the subgraph. The second improvement is the construction of a new large unit-distance graph G that we use as a starting point to find a smaller unit-distance graph with chromatic number 5. The construction of this graph is explained in Sect. 5. These two methods allowed us to find a unit-distance graph with 529 vertices and 2 670 edges. This graph is much more symmetric compared to the graph with 553 vertices.

4 Clausal Proof Optimization

Most SAT solvers can emit a clausal proof of unsatisfiability. There exist several checkers for such proofs, including formally-verified ones [7,19]. We extended the checker DRAT-trim [15] that allows optimizing the clausal proof as well as extracting an unsatisfiable core. One can obtain multiple unsatisfiable cores from a single clausal proof—in contrast to a resolution proof [28]. The existing method works via backward checking [9]: Given a proof of unsatisfiability, the last clause (the empty clause) of the proof is marked. Now the proof is validated in reverse order. For each marked clause it is determined which clauses (occurring earlier in the proof or in the formula) are required for the validation. Those clauses will be marked (if they were not marked already). The order in which unit propagation is applied influences which clauses become marked. Unmarked clauses are not validated. After the proof is verified, the marked clauses in the formula form an unsatisfiable core and the marked clauses in the proof form an optimized proof. We present two new extensions that further reduce the size of the formula.

4.1 Justification Order Shuffling

A clausal proof typically has many different justifications and a justification can typically be converted into many different clausal proofs, i.e., clauses appear in a different order in the sequence. Here we exploit this property by 1) computing a justification for a given clausal proof, 2) removing the clauses that are redundant based on that justification, and 3) shuffle the remaining clauses in the proof based on that justification. These steps are repeated multiple times.

Figure 3 shows the pseudo code of that algorithm. The procedure RemoveRedundancy removes from a given clausal proof P and justification J all the clauses in P that do not occur in any of the justifications of J. Given a justification J, the procedure ShuffleProof produces a random permutation of the clauses in J such that each clause C appears 1) later in the proof than all the clauses in the justification of C and 2) before all clauses that list C in their justification. Additionally, ShuffleProof randomly shuffles the literals of each clause in J.

Clause deletion is not mentioned in the algorithm, but can also be helpful to optimize proofs. A clause C can be deleted in a proof as soon as none of the clauses occurring later in the proof uses C in their justification. On the other hand, one could delete C at a later point in the proof (or not at all) to

OptimizeProof (clausal proof P, formula F)

```
1    do
2        J := ComputeJustification (P, F)
3        J := RemoveRedundancy (J)
4        P := ShuffleProof (J)
5    while (progress)
6    return P
```

Fig. 3. Optimizing a proof by iterative computing a new justification.

allow clauses later in the proof to incorporate C in their justification in the next iteration. We randomly postpone deleting clauses in proofs in a certain window. The window is slightly increased in each next iteration.

4.2   Iterative Trimming the Formula

Given a unsatisfiable formula that encodes the existence of a k-coloring of a graph, an unsatisfiable core of that formula represents a subgraph that cannot be colored with k colors. To find a small subgraph we would like a minimal unsatisfiable core and ideally the smallest minimal unsatisfiable core. Although there has been some research in to the latter [17,20,23], it is already hard to compute a minimal unsatisfiable core. Existing algorithms for computing a minimal unsatisfiable core [21,22] focus more on easy problems. For harder problems it is required to trim the formulas using a preprocessing step [3].

In preliminary experiments we observed that existing algorithms got stuck. It turned out that if a "wrong" vertex is removed from the graph, then proving that the remaining graph still has chromatic number 5 is very expensive. A proof that the initial graph has chromatic number 5 consists of roughly 10,000 clauses. After removing a clause that represents a "wrong" vertex, the proof consists of millions of clauses. We concluded that existing tools are not effective for this application, because they remove clauses arbitrary. This will eventually result in removing a clause representing a "wrong" vertex. Although the checking costs are a serious problem, there is a more problematic issue: as soon as it requires millions of clauses to prove that the graph has chromatic number five, then many vertices are involved in the proof and the minimal unsatisfiable core will be relatively large. As a consequence, this also holds for the graph represented by this core. We address this issue by taking away the elimination of arbitrary clauses. Instead, we only remove clauses via trimming and proof optimization.

Figure 4 shows the pseudo codes of two algorithms to trim a formula: one algorithm, called TrimFormulaPlain, that simply adds proof optimization to the trimming loop and another one, called TrimFormulaInteract, that additionally interacts with the original formula to further optimize the proof. We focus on the latter algorithm, which is one of the main contributions of this paper.

TrimFormulaPlain (formula F)
1   $F_{\text{core}} := F$
2   **do**
3       $P :=$ ComputeProof (F_{core})
4       $P :=$ OptimizeProof (P, F_{core})
5       $F_{\text{core}} :=$ ComputeCore (P, F_{core})
6   **while** (progress)
7   **return** F_{core}

TrimFormulaInteract (formula F)
1   $F_{\text{core}} := F$
2   **do**
3       $P :=$ ComputeProof (F_{core})
4       $P :=$ OptimizeProof (P, F_{core})
5       $P :=$ OptimizeProof (P, F)
6       $F_{\text{core}} :=$ ComputeCore (P, F)
7   **while** (progress)
8   **return** F_{core}

Fig. 4. Pseudo code of two algorithms to trim the size of a formula using proof optimization: TrimFormulaPlain and TrimFormulaInteract. The latter algorithm interacts with the original formula to further optimize the proof.

Algorithm TrimFormulaInteract takes advantage of the following property of (D)RUP proofs: If (D)RUP proof P is a correct proof of formula F, then P is a correct proof of any formula F' such that $F' \supseteq F$. Observe that additional clauses cannot break the RUP check: if unit propagation on F results in a conflict, then unit propagation on F' results in a conflict.

In each step of the main loop of TrimFormulaInteract, we first compute a proof of unsatisfiability of the trimmed formula F_{core}. The size of this proof is crucial for the quality of the trimming. One could therefore solve F_{core} multiple times by shuffling the clauses and select the smallest proof of these runs. Afterwards, this proof is optimized using F_{core} via the algorithm shown in Fig. 3. Next, we use the property discussed above and further optimize the proof using F and the same optimization algorithm. The algorithm has now more options to minimize the proof as $F \supseteq F_{\text{core}}$. Moreover, the algorithm allows for a novel way to compute a smaller core: In an earlier step a clause may have been removed that allows for a small proof of unsatisfiability and/or small unsatisfiable core. Since each step considers again all clauses of F, that clause may be pulled back into F_{core}.

The size of F_{core} does not necessarily decrease with each iteration and may actually increase if a low quality proof is computed in line 3. We repeat the algorithm as long as there is progress. In this case, we measured progress by the reduction of the size of F_{core}.

The result of these trimming algorithms is rarely a minimal unsatisfiable core of the formula. We applied the classical destructive method [5] to reduce F_{core} to a minimal unsatisfiable core. We observed (some details are presented in Sect. 6.2) that the size of the minimal unsatisfiable core can vary significantly based on the selection of the clauses to remove. As a consequence we ran this method multiple (thousands of) times on the cluster to obtain a relatively small minimal unsatisfiable core of F_{core}.

5 Observed Patterns of Points in $\mathbb{Q}[\sqrt{3}, \sqrt{11}] \times \mathbb{Q}[\sqrt{3}, \sqrt{11}]$

The smallest known unit-distance graph with chromatic number 5, called G_{553}, has 553 vertices [13]. Its key component is a set of 420 points embedded in $\mathbb{Q}[\sqrt{3}, \sqrt{11}] \times \mathbb{Q}[\sqrt{3}, \sqrt{11}]$ that have a limited number (19) of the colorings of the points at distance 2 from the origin (central vertex) when coloring the set with 4 colors. Our strategy to compute a small unit-distance graph with chromatic number 5 is finding a small set of vertices with the same property. We explored many large graphs with points in $\mathbb{Q}[\sqrt{3}, \sqrt{11}] \times \mathbb{Q}[\sqrt{3}, \sqrt{11}]$ and computed the size of proofs of unsatisfiability of the formula that determines the existence of a 4-coloring while blocking the limited number of the colorings of the points at distance 2. This section describes how we obtained the large graph with the smallest proof of unsatisfiability that we encountered.

We denote by H_R the graph consisting of i) a regular hexagon with maximal radius R and ii) its center. The points of H_R in the plane are $(0,0)$, $(R,0)$, $(R/2, R\sqrt{3}/2)$, $(-R/2, R\sqrt{3}/2)$, $(-R,0)$, $(-R/2, -R\sqrt{3}/2)$ and $(R/2, -R\sqrt{3}/2)$. Furthermore, we denote by H'_R a copy of H_R rotated by 90 degrees.

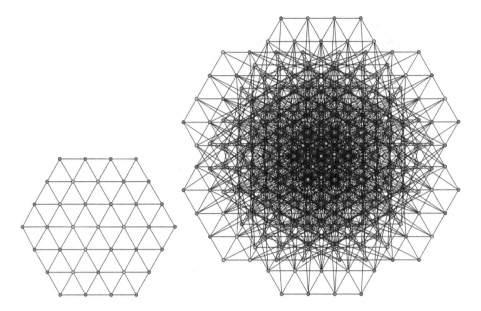

Fig. 5. A 3-coloring of the graph $H_{\frac{1}{3}} \oplus H_{\frac{1}{3}} \oplus H_{\frac{1}{3}}$ (left) and a 4-coloring of the graph $H_{\frac{1}{3}} \oplus H_{\frac{1}{3}} \oplus H_{\frac{1}{3}} \oplus H'_{\frac{\sqrt{3}+\sqrt{11}}{6}}$ (right). (Color figure online)

We observed some interesting patterns when combining the graphs $H_{\frac{1}{3}}$ and $H'_{\frac{\sqrt{3}+\sqrt{11}}{6}}$. Figure 5 (left) shows the graph $H_{\frac{1}{3}} \oplus H_{\frac{1}{3}} \oplus H_{\frac{1}{3}}$, which is a triangular grid with diameter 1. This graph has 37 vertices and 48 edges and can be colored with 3 colors. However, the Minkowski sum of this triangular grid and $H'_{\frac{\sqrt{3}+\sqrt{11}}{6}}$, shown in Fig. 5 (right), is not 3-colorable. Notice that there are many edges between the seven triangular grids. Actually, the graph has 259 vertices and 1 056 edges and most of these edges (720) are between triangular grids. There exist many 4-colorings of this graph and most of them have no observable pattern.

Patterns start to emerge when applying the Minkowski sum again. Figure 6 shows a 4-coloring of the resulting graph $H_{\frac{1}{3}} \oplus H_{\frac{1}{3}} \oplus H_{\frac{1}{3}} \oplus H'_{\frac{\sqrt{3}+\sqrt{11}}{6}} \oplus H'_{\frac{\sqrt{3}+\sqrt{11}}{6}}$. Observe the clustering of vertices with the same color in circles of roughly a diameter of 1 in size. This pattern can be observed in many of the found 4-colorings of this graph, although there also exist some 4-colorings without this pattern. It appears that assigning the same color to nearby vertices is the easiest way to color this graph (using a SAT solver).

Applying the Minkowski sum another time breaks the prior pattern completely, as no more 4-colorings exist with clusters of vertices having the same color. However, new patterns emerge, as can be seen in Fig. 7. For example, notice the reflection in the central vertical axis of the blue and green vertices.

Based on these observations, we experimented with ways to combine $H_{\frac{1}{3}}$ and $H'_{\frac{\sqrt{3}+\sqrt{11}}{6}}$. An effective combination turned out to be unit-distance graph G_{2167}. This graph is constructed as follows. Let C_{13} denote the union of $H_{\frac{1}{3}}$ and

Fig. 6. A 4-coloring of the graph $H_{\frac{1}{3}} \oplus H_{\frac{1}{3}} \oplus H_{\frac{1}{3}} \oplus H'_{\frac{\sqrt{3}+\sqrt{11}}{6}} \oplus H'_{\frac{\sqrt{3}+\sqrt{11}}{6}}$. (Color figure online)

$H'_{\frac{\sqrt{3}+\sqrt{11}}{6}}$. Now G_{2167} equals $C_{13} \oplus C_{13} \oplus C_{13} \oplus C_{13} \oplus C_{13} \oplus C_{13} \oplus C_{13} \oplus C_{13}$ without the points that have a distance larger than 2 from the central vertex. This graph has 2 167 vertices and 16 512 edges and is shown in Fig. 8. Notice that the average vertex degree is larger than 15. This is quite high for a graph with chromatic number 4.

Observe the vertical monochromatic lines in Fig. 8: Points with the same horizontal coordinate have the same color. This pattern appears in many 4-colorings (modulo a rotation of 60 degrees). There are solutions with vertical lines with two colors, but none of the 4-colorings have more colors on a single vertical line (again, modulo a rotation of 60 degrees). The only reason why such solutions can exist is that the construction of G_{2167} does not generate points with distance 1 that have the same horizontal coordinate. There appears no

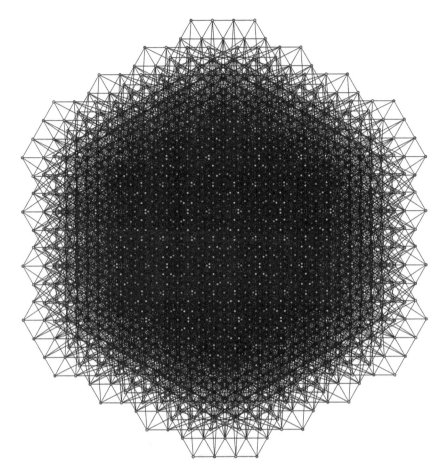

Fig. 7. A 4-coloring of the graph $H_{\frac{1}{3}} \oplus H_{\frac{1}{3}} \oplus H_{\frac{1}{3}} \oplus H'_{\frac{\sqrt{3}+\sqrt{11}}{6}} \oplus H'_{\frac{\sqrt{3}+\sqrt{11}}{6}} \oplus H'_{\frac{\sqrt{3}+\sqrt{11}}{6}}$.
(Color figure online)

obvious way to add such points in a way that the resulting graph has chromatic number 5. Another pattern that can be observed in Fig. 8 is that points with the same vertical coordinate that are 2/3 apart from each other also have the same color. Also any two points that are 1/3 apart have a different color. For example, forcing that any vertex at distance 1/3 from the origin has the same color as the central vertex eliminates all 4-colorings. Hence 1/3 is a so-called virtual-edge in 4-colorings of unit-distance graphs.

6   Small Unit-Distance Graph with Chromatic Number 5

In this section we present our SAT-based approach to improve the smallest known unit-distance graph with chromatic number 5. We first explain how we encode the problem and afterwards apply the new trimming algorithm presented in Sect. 4.2.

Fig. 8. A 4-coloring of graph G_{2167}. (Color figure online)

6.1 Encoding

We can compute the chromatic number of a graph G as follows. Construct two formulas, one asking whether G can be colored with $k-1$ colors, and one whether G can be colored with k colors. Now, G has chromatic number k if and only if the former is unsatisfiable while the latter is satisfiable.

The construction of these two formulas can be achieved using the following encoding [13]: Given a graph $G = (V, E)$ and a parameter k, the encoding uses $k|V|$ boolean variables $x_{v,c}$ with $v \in V$ and $c \in \{1, \ldots, k\}$. These variables have the following meaning: $x_{v,c}$ is true if and only if vertex v has color c. Now we can construct a propositional formula F_k that is satisfiable if and only if G can be colored with k colors:

$$F_k := \bigwedge_{v \in V} (x_{v,1} \vee \cdots \vee x_{v,k}) \wedge \bigwedge_{\{v,w\} \in E} \bigwedge_{c \in \{1,\ldots,k\}} (\overline{x}_{v,c} \vee \overline{x}_{w,c})$$

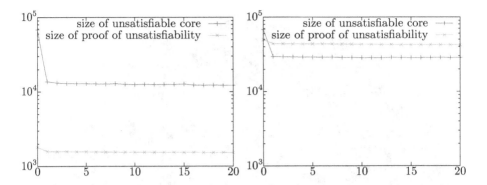

Fig. 9. The size (number of clauses) of the unsatisfiable core and the optimized proof of unsatisfiability (y-axis) of the first twenty steps (x-axis) of the OptimizeProof algorithm, when starting with F_4^+ and the smallest proof (left) or the largest proof (right).

The first type of clauses, called vertex clauses, ensures that each vertex has at least one color, while the second type of clauses, called edge clauses, forces that two connected vertices are colored differently. Additionally, we could include clauses to require that each vertex has at most one color. However, these clauses are redundant and would be eliminated by blocked clause elimination [18], a SAT preprocessing technique. We experimented using formulas with and without blocked clauses. Although the results were quite similar, we had the impression that without blocked clauses is slightly better.

We added symmetry-breaking predicates [6] during all experiments to speed up solving and proof minimization. The color symmetries were broken by fixing the vertex at $(0, 0)$ to the first color, the vertex at $(1, 0)$ to the second color, and the vertex at $(1/2, \sqrt{3}/2)$ to the third color. These three points are at distance 1 from each other and occurred in all our graphs. The speedup is roughly a factor of 24 $(= 4 \cdot 3 \cdot 2)$, when proving the absence of a 4-coloring.

6.2   Reducing the Large Part

The smallest known unit-distance graph with chromatic number 5 has 553 vertices and consists of two parts: a large part with 420 vertices and a small part with 134 vertices. The large part and small part have one vertex in common: the origin. Analysis of these parts [13] showed that they have different purposes: the large part limits the number of valid 4-coloring of 12 vertices at distance 2 from the origin to 19. The small part prevents these 12 vertices to having any of these 19 4-colorings. Some important details are missing from this analysis and they will be discussed later. We focused our effort to search for a small unit-distance graph with chromatic number 5 by looking for a more compact large part.

In the first step, we constructed the formula whether graph G_{2167} has a 4-coloring. Apart from the symmetry-breaking predicates, we added 19 clauses that block the above mentioned 4-colorings that remain in the large part. This formula, called F_4^+, is unsatisfiable and has 8 668 variables and 68 237 clauses. In the next step we produce a proof of unsatisfiability of this formula. We used the SAT solver glucose 3.0 [2] (without preprocessing techniques) for this purpose. This solver allows to randomly initialize the decision heuristics (VSIDS), which is a feature that can easily be added to most SAT solvers. This initialization can have a significant impact on the size of the proof *and* on the size of the core. For example, we solved the formula with 100 different seeds for the initialization. The smallest proof had 1 809 clause addition steps, while the largest proof had 49 838 clause addition steps. The default glucose 3.0, i.e., without decision heuristics initialization, produced a proof with 2 475 clause addition steps.

Figure 9 shows the effect of using the smallest and largest proof as input for the OptimizeProof algorithm, which has been implemented in the DRAT-trim proof checker (available at https://github.com/marijnheule/drat-trim) [15]. In both cases the size of the proof reduction is modest. However, a much smaller unsatisfiable core can be extracted from the optimized smallest proof compared to the optimized largest proof. The smaller core also corresponds to a smaller subgraph (963 versus 1 609 vertices).

We also experimented with the two algorithms presented in Sect. 4.2. Figure 10 shows the size of the subgraph (extracted from the core) for the first 20 iterations with formula F_4^+ as input using TrimFormulaPlain (left) or TrimFormulaInteract (right). Each experiment was run five times. The figure shows that TrimFormulaInteract produces significantly smaller subgraphs. The TrimFormulaPlain algorithm, as shown in Fig. 4, actually performs significantly worse than the performance presented in Fig. 10. This poor performance is caused by the removal of edge clauses and symmetry-breaking predicates from the core. We improved the TrimFormulaPlain algorithm by adding back the removed edge clauses and symmetry-breaking predicates in each iteration.

We studied the resulting graphs and observed that they were close to symmetric: Taking the union of the graph with rotated copies (120 degrees rotation in the origin) added only a few dozen vertices. We decided to check whether

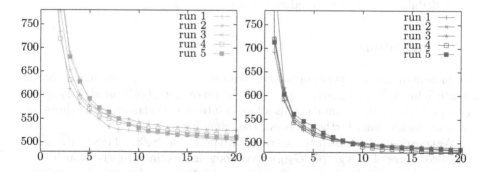

Fig. 10. The size of subgraphs corresponding to the unsatisfiable cores when using the algorithms TrimFormulaPlain (left) and TrimFormulaInteract (right).

this observation could be used to further shrink the large part by taking this union as initial graph (instead of G_{2167}) and rerun the procedure. This turned out to be effective and allowed removing some additional vertices. We ran the entire experiment many times on a cluster Several runs resulted in a graphs with "only" 393 vertices. These graphs turned out to be the same (modulo rotation and reflection). We call this graph L_{393}. One can make L_{393} symmetric, i.e., it maps onto itself when rotating it by 120 degrees along the central vertex, by adding a single vertex.

6.3   Finalizing the Graph

The graph L_{393}, produced in the previous subsection, needs to be extended with a "small part" to establish a unit-distance graph with chromatic number 5. Initially we tried to use the small part of G_{553}. However, the resulting graph is 4-colorable, because L_{393} has fewer connections with that small part compared to the large part of G_{553}. We fixed this as follows: The small part of G_{553} got expanded by merging it with copies that are 60 degrees rotated in the origin. This resulted in a graph with 181 vertices (while the small part of G_{553} has 134 vertices), which we call S_{181}. The union of L_{393} and S_{181} has chromatic number 5. We applied the same techniques as described in the previous subsection to further reduce the size of this graph. This resulted in a graph being the union of L_{393} and a new small part with 137 vertices.

Figure 11 shows the final graph G_{529} consisting of 529 vertices and 2 670 edges. This graph almost maps onto itself when rotating it with 120 degrees in the origin. The figure shows a coloring in which only the origin has the fifth color (white). Such a coloring exists for each vertex as the graph is vertex critical. The shown coloring is a randomly selected one. Observe the clustering of vertices with the same color. This pattern looks similar to the one shown in Fig. 6.

Graph G_{529} is available at https://github.com/marijnheule/CNP-SAT as a list of points in the plane and a list of unit-distance edges. The repository also contains a CNF formula encoding whether G_{529} is 4-colorable and a proof of unsatisfiability that can be validated in a few seconds.

7   Conclusions

We presented a new algorithm to trim a formula by first optimizing a proof of unsatisfiability. The algorithm optimizes the proof using both the shrinking formula and the original formula. This allows reintroducing clauses in the shrinking formula, which could further improve the trimming.

We constructed a unit-distance graph with points in $\mathbb{Q}[\sqrt{3}, \sqrt{11}] \times \mathbb{Q}[\sqrt{3}, \sqrt{11}]$. The 4-colorings of this graph, G_{2167}, have some interesting properties such as 1) many (and in some 4-colorings all) vertices with the same horizontal coordinate have the same color; 2) vertices that are $1/3$ apart having a different color; and 3) vertices with the same vertical coordinate that are $2/3$ apart have the same color. All these properties are for a rotation of G_{2167} by 0, 120, or 240 degrees.

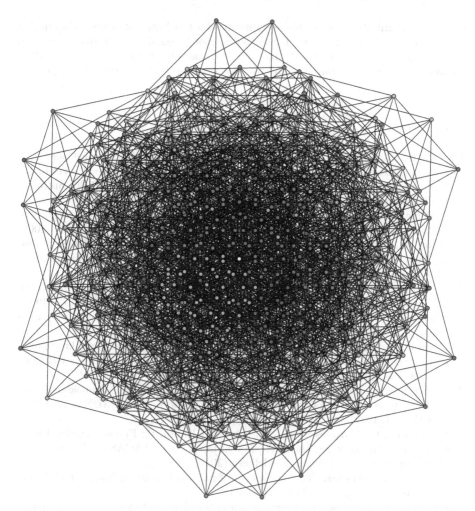

Fig. 11. A 529-vertex unit-distance graph with chromatic number 5. In the shown coloring, only the origin has the fifth color (white). (Color figure online)

By combining the new algorithm and the new graph, we were able to reduce the smallest known unit-distance graph with chromatic number 5 to a graph with 529 vertices and 2670 edges (down from 553 vertices and 2720 edges). This graph is also much more symmetric. It is generally easier to understand why a symmetric object has a certain property compared to an asymmetric object. It may thus provide some insight how to obtain a unit-distance graph with chromatic number 6 (if they exist). Using the techniques in the paper we constructed several graphs with up to 100 000 vertices, but all were 5-colorable.

As future work, we plan to study the effectiveness of the new algorithm on other applications that require minimal unsatisfiable cores.

Acknowledgements. The author is supported by the National Science Foundation (NSF) under grant CCF-1813993. The author acknowledges the Texas Advanced Computing Center (TACC) at The University of Texas at Austin for providing HPC resources that have contributed to the research results reported within this paper.

References

1. Asín, R., Nieuwenhuis, R., Oliveras, A., Rodríguez-Carbonell, E.: Efficient generation of unsatisfiability proofs and cores in SAT. In: Cervesato, I., Veith, H., Voronkov, A. (eds.) LPAR 2008. LNCS (LNAI), vol. 5330, pp. 16–30. Springer, Heidelberg (2008)
2. Audemard, G., Simon, L.: Predicting learnt clauses quality in modern SAT solvers. In: 21st International Joint Conference on Artificial Intelligence, pp. 399–404 (2009)
3. Belov, A., Heule, M.J.H., Marques-Silva, J.: MUS extraction using clausal proofs. In: Sinz, C., Egly, U. (eds.) SAT 2014. LNCS, vol. 8561, pp. 48–57. Springer, Cham (2014)
4. de Bruijn, N.G., Erdős, P.: A colour problem for infinite graphs and a problem in the theory of relations. Nederl. Akad. Wetensch. Proc. Ser. A **54**, 371–373 (1951)
5. Chinneck, J.W., Dravnieks, E.W.: Locating minimal infeasible constraint sets in linear programs. INFORMS J. Comput. **3**, 157–168 (1991)
6. Crawford, J.M., Ginsberg, M.L., Luks, E.M., Roy, A.: Symmetry-breaking predicates for search problems. In: Knowledge Representation and Reasoning - KR 1996, pp. 148–159. Morgan Kaufmann (1996)
7. Cruz-Filipe, L., Heule, M.J.H., Hunt Jr., W.A., Kaufmann, M., Schneider-Kamp, P.: Efficient certified RAT verification. In: de Moura, L. (ed.) CADE 2017. LNCS (LNAI), vol. 10395, pp. 220–236. Springer, Cham (2017)
8. Cruz-Filipe, L., Marques-Silva, J.P., Schneider-Kamp, P.: Efficient certified resolution proof checking. In: Legay, A., Margaria, T. (eds.) TACAS 2017. LNCS, vol. 10205, pp. 118–135. Springer, Heidelberg (2017)
9. Goldberg, E.I., Novikov, Y.: Verification of proofs of unsatisfiability for CNF formulas. In: DATE, pp. 10886–10891 (2003)
10. de Grey, A.D.N.J.: The chromatic number of the plane is at least 5. Geombinatorics **XXVIII**, 18–31 (2018)
11. Gurfinkel, A., Vizel, Y.: DRUPing for interpolants. In: Proceedings of the 14th Conference on Formal Methods in Computer-Aided Design, FMCAD 2014, pp. 19:99–19:106. FMCAD Inc., Austin (2014)
12. Hadwiger, H.: Minkowskische addition und subtraktion beliebiger punktmengen und die theoreme von erhard schmidt. Math. Z. **53**(3), 210–218 (1950)
13. Heule, M.J.H.: Computing small unit-distance graphs with chromatic number 5. Geombinatorics **XXVIII**, 32–50 (2018)
14. Heule, M.J.H., Hunt Jr., W.A., Kaufmann, M., Wetzler, N.D.: Efficient, verified checking of propositional proofs. In: Ayala-Rincón, M., Muñoz, C.A. (eds.) ITP 2017. LNCS, vol. 10499, pp. 269–284. Springer, Cham (2017)
15. Heule, M.J.H., Hunt Jr., W.A., Wetzler, N.D.: Trimming while checking clausal proofs. In: Formal Methods in Computer-Aided Design, pp. 181–188. IEEE (2013)

16. Heule, M.J.H., Hunt Jr., W.A., Wetzler, N.D.: Bridging the gap between easy generation and efficient verification of unsatisfiability proofs. Softw. Test. Verif. Reliab. (STVR) **24**(8), 593–607 (2014)
17. Ignatiev, A., Previti, A., Liffiton, M., Marques-Silva, J.: Smallest MUS extraction with minimal hitting set dualization. In: Pesant, G. (ed.) CP 2015. LNCS, vol. 9255, pp. 173–182. Springer, Cham (2015)
18. Järvisalo, M., Biere, A., Heule, M.J.H.: Simulating circuit-level simplifications on CNF. J. Autom. Reason. **49**(4), 583–619 (2012)
19. Lammich, P.: Efficient verified (UN)SAT certificate checking. In: de Moura, L. (ed.) CADE 2017. LNCS (LNAI), vol. 10395, pp. 237–254. Springer, Cham (2017)
20. Liffiton, M., Mneimneh, M., Lynce, I., Andraus, Z., Marques-Silva, J., Sakallah, K.: A branch and bound algorithm for extracting smallest minimal unsatisfiable subformulas. Constraints **14**(4), 415 (2008)
21. Lynce, I., Marques Silva, J.P.: On computing minimum unsatisfiable cores. In: The Seventh International Conference on Theory and Applications of Satisfiability Testing, SAT 2004, 10–13 May 2004, Vancouver, BC, Canada, Online Proceedings (2004)
22. Marques-Silva, J., Lynce, I.: On improving MUS extraction algorithms. In: Sakallah, K.A., Simon, L. (eds.) SAT 2011. LNCS, vol. 6695, pp. 159–173. Springer, Heidelberg (2011)
23. Mneimneh, M., Lynce, I., Andraus, Z., Marques-Silva, J., Sakallah, K.: A Branch-and-Bound algorithm for extracting smallest minimal unsatisfiable formulas. In: Bacchus, F., Walsh, T. (eds.) SAT 2005. LNCS, vol. 3569, pp. 467–474. Springer, Heidelberg (2005)
24. Moser, L., Moser, W.: Solution to problem 10. Can. Math. Bull. **4**, 187–189 (1961)
25. Soifer, A.: The Mathematical Coloring Book (2008). ISBN-13: 978-0387746401
26. Van Gelder, A.: Verifying RUP proofs of propositional unsatisfiability. In: ISAIM (2008)
27. Wetzler, N.D., Heule, M.J.H., Hunt Jr., W.A.: DRAT-trim: efficient checking and trimming using expressive clausal proofs. In: Sinz, C., Egly, U. (eds.) SAT 2014. LNCS, vol. 8561, pp. 422–429. Springer, Cham (2014)
28. Zhang, L.: Searching for truth: techniques for satisfiability of Boolean formulas. Ph.D. thesis, Princeton University, Princeton, NJ, USA (2003)

Improved Job Sequencing Bounds from Decision Diagrams

John N. Hooker[(⊠)]

Carnegie Mellon University, Pittsburgh, USA
jh38@andrew.cmu.edu

Abstract. We introduce a general method for relaxing decision diagrams that allows one to bound job sequencing problems by solving a
Lagrangian dual problem on a relaxed diagram. We also provide guidelines for identifying problems for which this approach can result in useful
bounds. These same guidelines can be applied to bounding deterministic
dynamic programming problems in general, since decision diagrams rely
on DP formulations. Computational tests show that Lagrangian relaxation on a decision diagram can yield very tight bounds for certain classes
of hard job sequencing problems. For example, it proves for the first time
that the best known solutions for Biskup-Feldman instances are within
a small fraction of 1% of the optimal value, and sometimes optimal.

Keywords: Job sequencing · Decision diagrams ·
Lagrangian relaxation

1 Introduction

In recent years, binary and multivalued decision diagrams (DDs) have emerged
as a useful tool for solving discrete optimization problems [5,6,24]. A key factor in their success has been the development of *relaxed* DDs, which represent
a superset of the feasible solutions of a problem and provide a bound on its
optimal value. While an exact DD representation of a problem tends to grow
exponentially with the size of the problem instance, a relaxed DD can be much
more compact when properly constructed. The tightness of the relaxation can
be controlled by adjusting the maximum allowed width of the DD.

Relaxed DDs are normally used in conjunction with a branching procedure
[5,12], much as is the linear programming (LP) relaxation in an integer programming solver. As branching proceeds, the relaxed diagram provides a progressively
tighter bound. However, combinatorial problems are often solved with heuristic
methods that do not involve branching. This is true, in particular, of job sequencing problems. In such cases it is very useful to have an independently derived
lower bound that can provide an indication of the quality of the solution.

Recent research [20] has found that a relaxed DD can yield good bounds for
hard job sequencing problems without branching. In fact, a surprisingly small
relaxed DD, generally less than 10% the width of an exact DD, can yield a bound

© Springer Nature Switzerland AG 2019
T. Schiex and S. de Givry (Eds.): CP 2019, LNCS 11802, pp. 268–283, 2019.
https://doi.org/10.1007/978-3-030-30048-7_16

equal to the optimal value. On the other hand, since exact DDs grow rapidly with the instance size, relaxed DDs that are 10% of their width likewise grow rapidly. As a result, relaxed DDs of reasonable width tend to provide progressively weaker bounds as the instances scale up.

It is suggested in [20] that Lagrangian relaxation could help strengthen the bounds obtained from smaller relaxed DDs. In this paper, we propose a general technique for relaxing a DD while preserving the ability to obtain Lagrangian bounds from the DD. The relaxed DD is constructed by merging nodes only when they agree on certain state variables that are crucial to forming the Lagrangian relaxation.

We find that for certain types of job sequencing problems, Lagrangian relaxation in relaxed DDs of reasonable width can provide very tight bounds on the optimal value. For example, we prove for the first time that the best known solution values of Biskup-Feldman single-machine scheduling instances are within a small fraction of one percent of the optimum, and sometimes optimal.

Furthermore, we identify general conditions under which Lagrangian relaxation can be implemented in a relaxed DD for purposes of obtaining bounds. The conditions are expressed in terms of structural characteristics of the dynamic programming model that defines the DD. They lead to a new tool for bounding not only job sequencing problems with suitable structure, but general deterministic dynamic programming models that satisfy the conditions.

2   Previous Work

Decision diagrams were introduced as an optimization method by [15, 18]. The idea of a relaxed diagram first appears in [1] as a means of enhancing propagation in constraint programming. Relaxed DDs were first used to obtain optimization bounds in [4, 7]. Connections between DDs and deterministic dynamic programming are discussed in [19].

Bergman, Ciré and van Hoeve first applied Lagrangian relaxation to decision diagrams in [3], where they use it successfully to strengthen bounds for the traveling salesman problem with time windows. They also use Lagrangian relaxation and DDs in [2] to improve constraint propagation.

We advance beyond Bergman et al. [3] in two ways. First, we show how to obtain bounds on tardiness and a variety of other objective functions from a stand-alone relaxed DD. The DD in [3] represents only an all-different constraint and can provide bounds only on total travel time (without taking time windows into account). The DD is embedded in a constraint programming (CP) model that contains the time window constraints. While constraints could be added to the CP model to obtain tardiness and other kinds of bounds from the CP solver, the DD itself cannot provide them. One or more additional state variables are necessary, which results in a more complicated DD than the one used in [3]. Our contribution is to define a new node merger scheme that relaxes such a DD while allowing Lagrangian relaxation to be applied.

Our second contribution is to analyze, in general, when and how Lagrangian relaxation can be combined with DDs. We introduce the concepts of an exact state and an immediate penalty function and use these concepts to formulate sufficient conditions for implementing Lagrangian relaxation in a relaxed DD. This leads to a general method for bounding dynamic programming models that satisfy the conditions. We find that while the method generates impracticably large relaxed DDs for the job sequencing problems in [3,20], it is quite practical for several important types of job sequencing problems.

3 Decision Diagrams

For our purposes, a decision diagram can be defined as a directed, acyclic multi-graph in which the nodes are partitioned into *layers*. Each arc of the graph is directed from a node in layer i to a node in layer $i + 1$ for some $i \in \{1, \dots, n\}$. Layers 1 and $n + 1$ contain a single node, namely the root r and the terminus t, respectively. Each layer i is associated with a finite-domain variable $x_i \in D_i$. The arcs leaving any node in layer i have distinct *labels* in D_i, representing possible values of x_i at that node. A path from r to t defines an assignment to the tuple $x = (x_1, \dots, x_n)$ as indicated by the arc labels on the path. The decision diagram is *weighted* if there is a length (cost) associated with each arc.

Any discrete optimization problem with finite-domain variables can be represented by a weighted decision diagram. The diagram is constructed so that its r–t paths correspond to the feasible solutions of the problem, and the length (cost) of any r–t path is the objective function value of the corresponding solution. If the objective is to minimize, the optimal value is the length of a shortest r–t path. Many different diagrams can represent the same problem, but for a given variable ordering, there is a unique *reduced* diagram that represents it [10,19].

As an example, consider a job sequencing problem with time windows. Each job j begins processing no earlier than the release time r_j and requires processing time p_j. The objective is to minimize total tardiness, where the tardiness of job j is $\max\{0, s_j + p_j - d_j\}$, and d_j is the job's due date. Figure 1 shows a reduced decision diagram for a problem instance with $(r_1, r_2, r_3) = (0, 1, 1)$, $(p_1, p_2, p_3) = (3, 2, 2)$, and $(d_1, d_2, d_3) = (5, 3, 5)$. Variable x_i represents the ith job in the sequence, and arc costs appear in parentheses.

4 Dynamic Programming Models

Decision diagrams most naturally represent problems with a dynamic programming formulation, because in this case a simple top-down compilation procedure yields a DD that represents the problem. A general dynamic programming formulation can be written

$$h_i(\boldsymbol{S}_i) = \min_{x_i \in X_i(\boldsymbol{S}_i)} \left\{ c_i(\boldsymbol{S}_i, x_i) + h_{i+1}\big(\phi_i(\boldsymbol{S}_i, x_i)\big) \right\} \tag{1}$$

Here, $\boldsymbol{S}_i$ is the *state* in stage i of the recursion. Typically the state is a tuple $\boldsymbol{S}_i = (S_{i1}, \dots, S_{ik})$ of *state variables*. Also $X_i(\boldsymbol{S}_i)$ is the set of possible *controls*

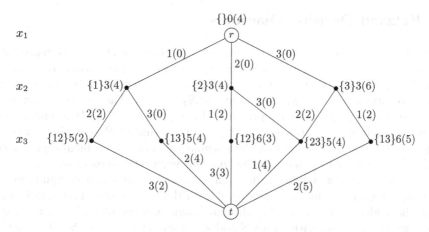

Fig. 1. Decision diagram for a small job sequencing instance, with arc labels and costs shown. States and minimum costs-to-go are indicated at nodes.

(values of x_i) in state S_i, ϕ_i is the *transition function* in stage i, and $c_i(S_i, x_i)$ is the *immediate cost* of control x_i in state S_i. We assume there is single initial state S_1 and a single final state S_{n+1}, so that $h_{n+1}(S_{n+1}) = 0$ and $\phi_n(S_n, x_n) = S_{n+1}$ for all states S_n and controls $x_n \in X_n(S_n)$. The quantity $h_i(S_i)$ is the *cost-to-go* for state S_i in stage i, and an optimal solution has value $h_1(S_1)$.

In the job sequencing problem, the state S_i is the tuple (V_i, t_i), where state variable V_i is the set of jobs scheduled so far, and state variable t_i is the finish time of the last job scheduled. Thus the initial state is $S_1 = (\emptyset, 0)$, and $X_i(S_i)$ is $\{1, \ldots, n\} \setminus V_i$. The transition function $\phi_i(S_i, x_i)$ is given by

$$\phi_i\big((V_i, t_i), x_i\big) = \big(V_i \cup \{x_i\}, \ \max\{r_{x_i}, t_i\} + p_{x_i}\big)$$

The immediate cost is the tardiness that results from scheduling job x_i in state (V_i, t_i). Thus if $\alpha^+ = \max\{0, \alpha\}$, we have

$$c_i\big((V_i, t_i), x_i\big) = \Big(\max\{r_{x_i}, t_i\} + p_{x_i} - d_{x_i} \Big)^+ \tag{2}$$

We recursively construct a decision diagram D for the problem by associating a state with each node of D. The initial state S_1 is associated with the root node t and the final state S_{n+1} with the terminal node t. If state S_i is associated with node u in layer i, then for each $v_i \in X_j(S_i)$ we generate an arc with label v_i leaving u. The arc terminates at a node associated with state $\phi_i(S_i, v_i)$. Nodes on a given layer are identified when they are associated with the same state.

The process is illustrated for the job sequencing example in Fig. 1. Each node is labeled by its state (V_i, t_i), followed (in parentheses) by the minimum cost-to-go at the node. The cost-to-go at the terminus t is zero.

5   Relaxed Decision Diagrams

A weighted decision diagram D' is a *relaxation* of diagram D when D' represents every solution in D with equal or smaller cost, and perhaps other solutions as well. To make this more precise, suppose layers $1, \ldots, n$ of both D and D' correspond to variables $x_1, \ldots, x_n$ with domains $X_1, \ldots, X_n$. Then D' is a relaxation of D if every assignment to x represented by an r–t path P in D is represented by an r–t path in D' with length no greater than that of P. The shortest path length in D' is a lower bound on the optimal value of the problem represented by D. We will refer to a diagram that has not been relaxed as *exact*.

We can construct a relaxed decision diagram in top-down compilation by *merging* some nodes that are associated with different states. The object is to limit the width of the diagram (the maximum number of nodes in a layer). When we merge nodes with states S and T, we associate a state $S \oplus T$ with the resulting node. The operator $\oplus$ is chosen so as to yield a valid relaxation of the given recursion.

The job sequencing problem discussed above uses a relaxation operator

$$(V_i, t_i) \oplus (V_i', t_i') = (V_i \cap V_i', \min\{t_i, t_i'\})$$

V_i is now the set of jobs scheduled along all paths to the current node, and t_i is the earliest finish time of the last scheduled jobs along these paths. The operator is illustrated in Fig. 2, which is the result of merging states $(\{1, 2\}, 6)$ and $(\{2, 3\}, 5)$ in layer 3 of Fig. 1. The relaxed states (V, f) are shown at each node, followed by the minimum cost-to-go in parentheses. The shortest path now has cost 2, which is a lower bound on the optimal cost of 4 in Fig. 1.

Sufficient conditions under which node merger results in a relaxed decision diagram are developed in [20]. A state S_i' *relaxes* a state S_i when (a) all feasible controls in state S_i are feasible in state S_i', and (b) the immediate cost of any

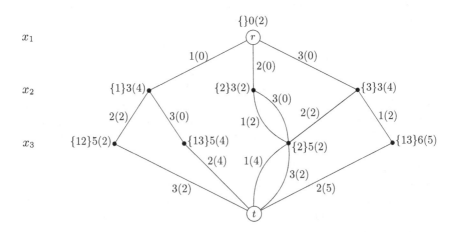

Fig. 2. A relaxation of the decision diagram in Fig. 1.

given feasible control in S_i is no less than its immediate cost in S_i'. That is, $X_i(S_i) \subseteq X_i(S_i')$, and $c_i(S_i, x_i) \geq c_i(S_i', x_i)$ for all $x_i \in X_i(S_i)$.

Theorem 1 ([20]). *If the following conditions are satisfied, the merger of nodes with states S_i and T_i within a decision diagram results in a valid relaxation of the diagram.*

- $S_i \oplus T_i$ *relaxes both* S_i *and* T_i.
- *If state* S_i' *relaxes state* S_i, *then given any control* v *that is feasible in* S_i, $\phi(S_i', v)$ *relaxes* $\phi(S_i, v)$.

All relaxed diagrams we consider are associated with a dynamic programming model that satisfies the conditions of the theorem. The shortest path problem in the relaxed DD requires a modification of the original dynamic programming model that accounts for the merger of states. Also, it is sometimes necessary to use additional state variables to obtain a valid relaxation [5], and so we replace the state vector S_i with a possibly enlarged vector $\bar{S}_i$. The recursive model becomes

$$\bar{h}_i(\bar{S}_i) = \min_{x_i \in X_i(\bar{S}_i)} \left\{ c_i(\bar{S}_i, x_i) + \bar{h}_{i+1}\Big(\rho_{i+1}\big(\phi_i(\bar{S}_i, x_i)\big)\Big) \right\} \tag{3}$$

where $\rho_{i+1}(\bar{S}_{i+1})$ is a relaxation of state $\bar{S}_{i+1}$ that reflects the merger of states in stage $i+1$ of the recursion. In the example, $\bar{S}_i = S_i$, since no additional states are necessary to formulate the relaxation.

It will be convenient to distinguish *exact* from *relaxed* state variables in model (3). A state variable S_{ij} in (3) is exact if any sequence of controls $x_1, \ldots, x_{i-1}$ that leads to a given value of S_{ij} in the original recursion (1) leads to that same value in the relaxed recursion (3). Otherwise S_{ij} is relaxed. In the example, neither V_i nor t_i is exact if any pair of states can be merged. However, if we permit the merger of (V_i, t_i) and (V_i', t_i') only when $t_i = t_i'$, then state variable t_i is exact. In general we have the following, which is easy to show.

Lemma 1. *A state variable S_{ij} is exact if states S_i, S_i' are merged only when $S_{ij} = S_{ij}'$.*

6   Lagrangian Duality and Decision Diagrams

Consider an optimization problem

$$z^* = \min_{x \in \mathcal{X}} \left\{ f(x) \mid g(x) = 0 \right\} \tag{4}$$

where $x = (x_1, \ldots, x_n)$, $g(x) = (g_1(x), \ldots, g_m(x))$ and $0 = (0, \ldots, 0)$. The condition that $x \in \mathcal{X}$ is typically represented by a constraint set. A *Lagrangian relaxation* of (4) has the form

$$\theta(\lambda) = \min_{x \in \mathcal{X}} \left\{ f(x) + \lambda^T g(x) \right\} \tag{5}$$

where $\boldsymbol{\lambda} = (\lambda_1, \ldots, \lambda_m)$. The relaxation *dualizes* the constraints $\boldsymbol{g}(\boldsymbol{x}) = \boldsymbol{0}$. It is easy to show that $\theta(\boldsymbol{\lambda})$ is a lower bound on z^* for any $\boldsymbol{\lambda} \in \mathbb{R}^m$. The *Lagrangian dual* of (4) seeks the tightest bound $\theta(\boldsymbol{\lambda})$:

$$\max_{\boldsymbol{\lambda} \in \mathbb{R}^m} \left\{ \theta(\boldsymbol{\lambda}) \right\} \tag{6}$$

The motivation for using a Lagrangian dual is to obtain tight bonds while dualizing troublesome constraints. If $\boldsymbol{g}(\boldsymbol{x})$ depends on a very small number of state variables, then dualizing the constraint $\boldsymbol{g}(\boldsymbol{x}) = \boldsymbol{0}$ may allow one to solve the problem within time and space constraints.

Lagrangian duality can be illustrated in the minimum tardiness job sequencing problem discussed earlier. The variables $x_1, \ldots, x_n$ should have different values, or equivalently, that each job j should occur in a given solution exactly once. Since a relaxed DD does not enforce this condition, we dualize the constraint $\boldsymbol{g}(\boldsymbol{x}) = \boldsymbol{0}$, where $\boldsymbol{g}(\boldsymbol{x}) = (g_1(\boldsymbol{x}), \ldots, g_n(\boldsymbol{x}))$ and

$$g_j(\boldsymbol{x}) = -1 + \sum_{i=1}^{n} (x_i = j), \quad \text{with } (x_i = j) = \begin{cases} 1 \text{ if } x_i = j \\ 0 \text{ otherwise} \end{cases}$$

As observed in [3], the Lagrangian penalty $\lambda_j g_j(\boldsymbol{x})$ can be represented in a relaxed DD by adding λ_j to the cost of each arc corresponding to control j and subtracting $\sum_j \lambda_j$ from the cost of each arc leaving the root node. Then the length of each r–t path includes the Lagrangian penalty $\boldsymbol{\lambda}^T \boldsymbol{g}(\boldsymbol{x})$ for the corresponding solution $\boldsymbol{x}$, which is zero if $\boldsymbol{x}$ satisfies the all-different constraint.

In general, $\boldsymbol{g}(\boldsymbol{x})$ can be computed recursively in a relaxed DD when there is a vector-valued *immediate penalty function* $\boldsymbol{\gamma}_i(\bar{\boldsymbol{S}}_i^L, x_i)$ for which

$$\boldsymbol{g}(\boldsymbol{x}) = \sum_{i=1}^{n} \boldsymbol{\gamma}_i(\bar{\boldsymbol{S}}_i^L, x_i) \tag{7}$$

where each $\bar{\boldsymbol{S}}_i^L$ is a tuple consisting of *exact* state variables $S_{i\ell}$ for $\ell \in L$. In the example, the constraint function $\boldsymbol{g}(\boldsymbol{x})$ requires no state information, and $\bar{\boldsymbol{S}}_i^L$ is empty. The immediate penalty $\boldsymbol{\gamma}_i((V_i, t_i), x_i)$ can be written simply $\boldsymbol{\gamma}_i(x_i)$, where

$$\gamma_{ij}(x_i) = \begin{cases} (x_i = j) & \text{if } i \in \{2, \ldots, n\} \\ (x_i = j) - 1 & \text{if } i = 1 \end{cases} \tag{8}$$

for $j = 1, \ldots, n$.

Implementing a Lagrangian relaxation (4) in a relaxed DD also requires that the original objective function value $f(\boldsymbol{x})$ be computed as part of the path length. Normally, the path length is only a lower bound on $f(\boldsymbol{x})$. To compute $f(\boldsymbol{x})$ exactly in the relaxed recursion (3), we must have an immediate cost function that depends only on exact state variables. That is, we must have

$$f(\boldsymbol{x}) = \sum_{i=1}^{n} \bar{c}_i(\bar{\boldsymbol{S}}_i^K, x_i) \tag{9}$$

where $\bar{S}_i^K$ is a tuple consisting of exact state variables S_{ik} for $k \in K$. In the example, the immediate cost function (2) depends on the state variable t_i, which must therefore be exact. We therefore have $\bar{S}_i^K = (t_i)$. Due to Lemma 1, we can ensure that t_i is exact by merging nodes only when t_i has the same value in the corresponding states. Thus Lagrangian relaxation can be implemented in the minimum tardiness example if we merge nodes in this fashion.

The above observations can be summed up as follows.

Theorem 2. *Lagrangian relaxation (5) can be implemented in a relaxed DD if there are immediate cost functions $\bar{c}_i(\bar{S}_i^K, x_i)$ for which (9) holds and immediate penalty functions $\gamma_i(\bar{S}_i^L, x_i)$ for which (7) holds, where $\bar{S}_i^K$ and $\bar{S}_i^L$ consist entirely of exact state variables in $\bar{S}_i$. In this case the recursion for computing shortest paths in the relaxed DD becomes*

$$\bar{h}_i(\bar{S}_i, \boldsymbol{\lambda}) = \min_{x_i \in X_i(\bar{S}_i)} \left\{ \bar{c}_i(\bar{S}_i^K, x_i) + \boldsymbol{\lambda}^T \gamma_i(\bar{S}_i^L, x_i) + \bar{h}_{i+1}\Big(\rho_{i+1}\big(\phi_i(\bar{S}_i, x_i)\big), \boldsymbol{\lambda}\Big) \right\}$$

Corollary 1. *Lagrangian relaxation (5) can be implemented in a relaxed DD if nodes are merged only when their states agree on the values of the state variables on which the immediate cost functions and the immediate penalty functions depend. That is, nodes with states $\bar{S}_i$ and $\bar{T}_i$ are merged only when $\bar{S}_i^K = \bar{T}_i^K$ and $\bar{S}_i^L = \bar{T}_i^L$, where K and L are in as Theorem 2.*

7 Problem Classes

We now examine a few classes of job sequencing problems to determine whether they are suitable for Lagrangian relaxation on a relaxed DD. All of these problems have an all-different constraint that is dualized as before using the immediate penalty function (8). Since this function depends on no state variables, the state variables that must be exact are simply those on which the immediate cost function depends. That is, $\bar{S}_i^L$ is empty, and suitability for relaxation depends on which variables are in $\bar{S}_i^K$. We note that even when DD-based Lagrangian relaxation is not suitable for a given problem class, it may be useful when combined with branching, or when the relaxed DD is embedded in a larger model.

7.1 Sequencing with Time Windows

Problems in which jobs with state-independent processing times are sequenced, possibly subject to time windows, are generally conducive to Lagrangian relaxation on DDs. The problem of minimizing total tardiness is discussed above, and computational results are presented in Sect. 8. Minimizing makespan or the number of late jobs is treated similarly.

A popular variation on the problem minimizes the sum of penalized earliness and tardiness with respect to a common due date [8,9,11,16,17,22,25]. Earliness of job j is weighted by α_j and lateness by β_j. The recursive model for an exact DD uses the same state variables (V_i, t_i) as the minimum tardiness problem. However, a valid relaxed DD requires an additional state variable s_i that represents latest start time, while t_i again represents earliest finish time. The transition and immediate cost functions are

$$\bar{\phi}_i\big((V_i, s_i, t_i), x_i\big) = \big(V_i \cup \{x_i\},\ s_i + p_{x_i},\ t_i + p_{x_i}\big)$$
$$\bar{c}_i\big((V_i, s_i, t_i), x_i\big) = \alpha_{x_i}\big(s_i + p_{x_i} - d_{x_i}\big)^+ + \beta_{x_i}\big(t_i + p_{x_i} - d_{x_i}\big)^+$$

The merger operation is

$$(V_i, s_i, t_i) \oplus (V_i', s_i', t_i') = \big(V_i \cap V_i',\ \max\{s_i, s_i'\},\ \min\{t_i, t_i'\}\big)$$

The immediate cost depends on both s_i and t_i, which means that both of these state variables must be exact. This may appear to result in a large relaxed DD, because nodes can be merged only when they agree on both state variables. However, since s_i and t_i are initially equal, and these state variables are exact, they remain equal throughout the relaxed DD construction. The resulting DD is therefore the same that would result if a single state variable were exact. Computational results are presented Sect. 8.

7.2 Time-Dependent Costs and/or Processing Times

Costs and processing times can be time-dependent in two senses: they may depend on the position of each job in the sequence, or on the clock time at which the job is processed. Both senses occur in the literature, and both can be treated with Lagrangian relaxation on DDs.

If the processing time p_{ij} of job j depends on the position i of the job in the sequence, then the immediate cost in the relaxation is

$$\bar{c}_i((V_i, t_i), x_i) = \big(\max\{t_i, r_{x_i}\} + p_{ix_i} - d_{x_i}\big)^+$$

which depends only on the state variable t_i. Since $\bar{c}_i$ is already indexed by the position i, any other element of cost that depends on i is easily incorporated into the function. Thus we need only ensure that states are merged only when they agree on t_i, a condition that is already satisfied in the relaxed model described above for minimum-tardiness sequencing problems.

If the processing time $p_j(s)$ of job j depends on the time s at which job j starts, the immediate cost is

$$\bar{c}_i((V_i, t_i), x_i) = \big(\max\{t_i, r_{x_i}\} + p_{x_i}(\max\{t_i, r_{x_i}\}) - d_{x_i}\big)^+$$

which again depends only on state variable t_i. Any other time-dependent element of cost likewise depends only on t_i, and so states can be merged whenever they agree on t_i.

7.3 Sequence-Dependent Processing Times

We refer to a job j's processing time as sequence dependent when its processing time $p_{j'j}$ depends on the immediately preceding job j' in the sequence. When there are no time windows and the objective is to minimize travel time, the problem is a traveling salesman problem. The state variables are V_i and the immediately preceding job y_i. The transition and immediate cost functions are

$$\bar{\phi}_i\big((V_i, y_i), x_i\big) = \big(V_i \cup \{x_i\}, x_i\big)$$
$$\bar{c}_i\big((V_i, y_i), x_i\big) = p_{y_i x_i} \tag{10}$$

Since the immediate cost depends only on state y_i, nodes can be merged whenever they are reached using the same control. This permits a great deal of reduction in the relaxed DD and suggests that a Lagrangian approach to bounding can be effective. While pure traveling salesman problems are already well solved, a DD-based Lagrangian bounding technique may be useful when there are side constraints.

When there are time windows in the problem, an additional state variable t_i representing the finish time of the previous job is necessary for a stand-alone relaxed DD. The transition function is

$$\bar{\phi}_i\big((V_i, y_i, t_i), x_i\big) = \big(V_i \cup \{x_i\}, x_i, \max\{r_{x_i}, t_i\} + p_{y_i x_i}\big) \tag{11}$$

The immediate cost functions for minimizing travel time and total tardiness, respectively, are

$$\bar{c}_i\big((V_i, y_i, t_i), x_i\big) = (r_{x_i} - t_i)^+ + p_{y_i x_i} \tag{12}$$
$$\bar{c}_i\big((V_i, y_i, t_i), x_i\big) = \big(\max\{r_{x_i}, t_i\} + p_{y_i x_i} - d_{x_i}\big)^+ \tag{13}$$

In either case, the immediate cost depends on two state variables y_i and t_i, and nodes can be merged only when they agree on these variables. This is likely to result in an impracticably large relaxed DD. For example, if there are 50 jobs and a few hundred possible values of t_i, a layer of the relaxed DD could easily expand to tens of thousands of nodes. We confirmed this with preliminary experiments on the Dumas instances [14]. Lagrangian relaxation on a stand-alone relaxed DD therefore does not appear to be a promising approach to bounding TSP problems with time windows. One can, of course, use the simpler DD described by (10) to bound travel time (although not total tardiness), as is done in [3]. However, this relaxation ignores time windows altogether and would yield a weaker bound than (11)–(12).

7.4 State-Dependent Processing Times

We refer to a job's processing times as state dependent when they depend on one or more of the state variables in the recursion, such as the set V_i of jobs already processed. Such a problem is studied in [20], where the processing time is less if a certain job has already been processed. State variables are again V_i and t_i, but to build a relaxed DD we need an additional state variable U_i representing the sets of jobs that have been processed along some path to the current node. The transition function and immediate cost function are

$$\bar{\phi}_i\big((V_i, U_i, t_i), x_i\big) = \big(V_i \cup \{x_i\}, U_i \cup \{x_i\}, \max\{r_{x_i}, t_i\} + p_{x_i}(U_i)\big)$$
$$\bar{c}_i\big((V_i, U_i, t_i), x_i\big) = \big(\max\{r_{x_i}, t_i\} + p_{x_i}(U_i) - d_{x_i}\big)^+$$

where the processing time is $p_{x_i}(U_i)$. The merger operation is

$$(V_i, U_i, t_i) \oplus (V_i', U_i', t_i') = \left(V_i \cap V_i', U_i \cup U_i', \min\{t_i, t_i'\}\right)$$

Since the cost depends on both t_i and U_i, these state variables must be exact, and states can be merged only when they agree on the values of t_i and U_i. This predicts that the relaxed DD will grow rapidly with the number of jobs. DD-based Lagrangian relaxation is therefore not a promising approach to this type of problem.

8   Computational Experiments

8.1   Problem Instances

To assess the quality of bounds obtained from Lagrangian relaxation on relaxed DDs, it is necessary to obtain problem instances with known optimal values, or values that are likely to be close to the optimum. We carry out tests on two well-known sets of instances, corresponding to two sequencing problems identified earlier to be suitable for bounding. One is the set of minimum weighted tardiness instances of Crauwels, Potts and Wassenhove [13], which we refer to as the CPW instances. The other is the Biskup-Feldman collection of minimum weighted earliness-plus-tardiness instances with a common due date [8].

The CPW set consists of 125 instances of each of three sizes: 40 jobs, 50 jobs, and 100 jobs. We compute bounds for first 25 instances in the 40- and 50-job sets. These instances exhibit a wide range of gradually increasing tardiness values, thus providing a diverse selection for testing. Optimal solutions are given in [8] for all of these instances except instance 14 with 40 jobs, and instances 11, 12, 14 and 19 with 50 jobs. Solution values presented in [8] for the unsolved instances are apparently the best known.

The Biskup-Feldman collection includes 10 instances of each size, where the sizes are 10, 20, 50, 100 and 200 jobs. We study the instances with 20, 50 and 100 jobs. The instances specify only the processing times p_j and the earliness/tardiness weights α_j, β_j described earlier. The common due date for all jobs in an instance is not specified. Typical practice is to set the due date equal to $d(h) = \lfloor h \sum_j p_j \rfloor$, where h is a parameter.

We compare our lower bounds against the best known solution values reported in [25]. These authors compute the earliness penalty with respect to $d(h_1)$ and the tardiness penalty with respect to $d(h_2)$ for $h_1 < h_2$, so that the penalty for each job j is $\alpha_j(d(h_1) - t_j)^+ + \beta_j(t_j - d(h_2))^+$, where t_j is the finish time of job j. Heuristics are used in [25] to solve instances with $(h_1, h_2) = (0.1, 0.2), (0.1, 0.3), (0.2, 0.5), (0.3, 0.4)$, and $(0.3, 0.5)$. We study instances with $(h_1, h_2) = (0.1, 0.2)$ and $(0.2, 0.5)$ to provide a look at contrasting cases. To our knowledge, none of these instances have been solved to proven optimality.

8.2 Solving the Lagrangian Dual

We use subgradient optimization to solve the Lagrangian dual problem (6). Each iterate $\boldsymbol{\lambda}^k$ is obtained from the previous by the update formula

$$\boldsymbol{\lambda}^{k+1} = \boldsymbol{\lambda}^k + \sigma_k g(\boldsymbol{x}^k)$$

where $\boldsymbol{x}^k$ is the value of $\boldsymbol{x}$ obtained when computing $\theta(\boldsymbol{\lambda}^k)$ in (5). The art of Lagrangian optimization is choosing the step size σ_k. This choice is avoided in [3] by using the Kelly-Cheney-Goldstein bundle method [21] of deriving $\boldsymbol{\lambda}^k$ from previous iterates. However, Polyak's method [23] seems better suited to our purposes, because it is much easier to implement, and it requires as a parameter only an upper bound θ^* on the optimal value of $\theta(\boldsymbol{\lambda})$. Such a bound is available in practice, because one seeks to estimate how far a known solution value lies from the optimal value, and that solution value is an upper bound θ^*. Polyak's method defines the step size to be

$$\sigma_k = \frac{\theta^* - \theta(\boldsymbol{\lambda}^k)}{\|g(\boldsymbol{x}^k)\|_2^2}$$

8.3 Building the Relaxed Diagram

In previous work [1,4,6], heuristically-selected nodes are merged in each layer after all states obtainable from the previous layer are generated. Since we wish to merge all nodes that agree on t_i, and no others, we merge these nodes as we generate the states, rather than first generating all possible states and then merging nodes. This drastically reduces computation time and results in reasonable widths that gradually increase as layers are created.

8.4 Computational Results

The computational tests were run on a Dell XPS-13 laptop computer with Intel Core i7-6560U (4M cache, 3.2 GHz) and 16 GB memory. Results for the CPW instances appear in Table 1. The table displays optimal (or best known) values and DD-based bounds, as well as the absolute and relative gap between the two. It also shows the maximum width of the relaxed DD (always obtained in layer n), the time required to build the DD, and the time consumed by the subgradient algorithm. Since a subgradient iteration requires only the solution of a shortest path problem in the relaxed DD, we allowed the algorithm to run for 50,000 iterations to obtain as much improvement in the bound as seemed reasonably possible. Due to slow convergence, which is typical for subgradient algorithms, a bound that is nearly as tight can be obtained by executing only, say, 20% as many iterations.

The bounds in Table 1 are reasonably tight. The gap is well below one percent in most cases, and below 0.1% in about a quarter of the cases, although a few of the bounds are rather weak. The optimal value was obtained for one instance.

Table 1. Comparison of bounds with optimal values (target) of CPW instances. Computation times are in seconds.

40 jobs

Instance	Target	Bound	Gap	Percent gap	Max width	Build time	Subgr time
1	913	883	30	3.29%	3163	16	1287
2	1225	1179	46	3.76%	3652	20	1420
3	537	483	54	10.06%	3556	20	1443
4	2094	2047	47	2.24%	3568	20	1427
5	990	980	10	1.01%	3305	18	1312
6	6955	6939	16	0.23%	3588	20	1406
7	6324	6299	25	0.40%	3509	20	1437
8	6865	6743	122	1.78%	3508	20	1393
9	16225	16049	176	1.08%	3699	22	1468
10	9737	9591	146	1.50%	3426	19	1346
11	17465	17417	48	0.27%	3770	23	1493
12	19312	19245	67	0.35%	3644	22	1435
13	29256	29003	253	0.86%	3736	22	1506
14	*14377	14100	277	1.93%	3609	21	1406
15	26914	26755	159	0.59%	3849	23	1554
16	72317	72120	197	0.27%	3418	19	1382
17	78623	78501	122	0.16%	3531	20	1384
18	74310	74131	179	0.24%	3524	20	1431
19	77122	77083	39	0.05%	3407	19	1320
20	63229	63217	12	0.02%	3506	20	1344
21	77774	77754	20	0.03%	3766	22	1433
22	100484	100456	28	0.03%	3489	20	1382
23	135618	135617	1	0.001%	3581	21	1375
24	119947	119914	33	0.03%	3477	19	1295
25	128747	128705	42	0.03%	3597	22	1339

*Best known solution

50 jobs

Instance	Target	Bound	Gap	Percent gap	Max width	Build time	Subgr time
1	2134	2100	34	1.59%	4525	48	2633
2	1996	1864	132	6.61%	4453	53	2856
3	2583	2552	31	1.20%	4703	52	2697
4	2691	2673	18	0.67%	4585	55	2703
5	1518	1342	176	11.59%	4590	52	2658
6	26276	26054	222	0.84%	4490	48	2562
7	11403	11128	275	2.41%	4357	45	2499
8	8499	8490	9	0.11%	4396	46	2501
9	9884	9507	377	3.81%	4696	52	2660
10	10655	10594	61	0.57%	4740	53	2738
11	*43504	43472	32	0.07%	4597	50	2606
12	*36378	36303	75	0.21%	4500	48	2655
13	45383	45310	73	0.16%	4352	47	2521
14	*51785	51702	83	0.16%	4699	52	2656
15	38934	38910	47	0.12%	4650	52	2630
16	87902	87512	390	0.44%	4589	49	2623
17	84260	84066	194	0.23%	4359	45	2526
18	104795	104633	162	0.15%	4448	47	2505
19	*89299	89163	136	0.15%	4609	50	2660
20	72316	72222	94	0.13%	4678	51	2659
21	214546	214476	70	0.03%	4406	47	2580
22	150800	150800	0	0%	4098	39	418
23	224025	223922	103	0.05%	4288	44	2441
24	116015	115990	25	0.02%	4547	49	2620
25	240179	240172	7	0.003%	4639	51	2686

*Best known solution

The gap for instance 14 in the 40-job table suggests that the best known value is probably not optimal, while no such inference can be drawn for the 4 unsolved 50-job instances.

Results for the Biskup-Feldman instances appear in Table 2, where the bounds are compared with the best known solutions. The relaxed DDs are the same for the two sets of due dates $(h_1, h_2) = (0.1, 0.2), (0.2, 0.5)$; only the costs differ. The bounds are very tight, resulting in gaps that are mostly under 0.1%. This indicates that the known solutions are, at worst, very close to optimality. In fact, optimality is proved for 8 instances, which represents the first time that any of these instances have been solved. The bounds may be equal to the optimal value for other instances, since the known values displayed may not be optimal.

Table 2. Comparison of bounds with best known values (target) of Biskup-Feldman instances.

Instance	$(h_1, h_2) = (0.1, 0.2)$					$(h_1, h_2) = (0.2, 0.5)$					Max width	Build time
	Target	Bound	Gap	Percent gap	Subgr time	Target	Bound	Gap	Percent gap	Subgr time		
20 jobs												
1	4089	4089	0	0%	1	1162	1162	0	0%	1	323	0.12
2	8251	8244	7	0.08%	28	2770	2766	4	0.14%	27	287	0.08
3	5881	5877	4	0.07%	27	1675	1669	6	0.36%	28	287	0.08
4	8977	8971	6	0.07%	27	3113	3108	5	0.16%	27	287	0.08
5	4028	4024	4	0.10%	32	1192	1187	5	0.42%	32	341	0.10
6	6306	6288	18	0.29%	26	1557	1557	0	0%	1	271	0.09
7	10204	10204	0	0%	1	3573	3569	4	0.11%	29	305	0.09
8	3742	3739	3	0.08%	25	990	979	11	1.11%	25	267	0.08
9	3317	3310	7	0.21%	21	1056	1055	1	0.09%	22	230	0.07
10	4673	4669	4	0.09%	29	1355	1349	6	0.44%	30	320	0.09
50 jobs												
1	39250	39250	0	0%	16	12754	12752	2	0.02%	501	931	2.8
2	29043	29043	0	0%	191	8468	8463	5	0.06%	524	931	2.9
3	33180	33180	0	0%	300	9935	9935	0	0%	66	836	2.4
4	25856	25847	9	0.03%	549	7373	7335	38	0.52%	521	932	2.8
5	31456	31439	17	0.05%	540	8947	8938	9	0.10%	529	932	3.0
6	33452	33444	8	0.02%	544	10221	10213	8	0.08%	532	932	2.9
7	42234	42228	6	0.01%	491	12002	11981	21	0.17%	465	835	2.4
8	42218	42203	15	0.04%	491	11154	11141	13	0.12%	478	833	2.4
9	33222	33218	4	0.01%	503	10968	10965	3	0.03%	508	884	2.7
10	31492	31481	11	0.03%	529	9652	9650	3	0.03%	522	932	2.9
100 jobs												
1	139573	139556	17	0.01%	4075	39495	39467	28	0.07%	3968	1882	42
2	120484	120465	19	0.02%	4065	35293	35266	27	0.08%	4068	1882	44
3	124325	124289	36	0.03%	3957	38174	38150	24	0.06%	4059	1882	42
4	122901	122876	25	0.02%	3903	35498	35467	31	0.09%	3964	1882	42
5	119115	119101	14	0.01%	3925	34860	34826	34	0.10%	4016	1882	42
6	133545	133536	9	0.007%	3987	35146	35123	23	0.07%	3961	1882	43
7	129849	129830	19	0.01%	4027	39336	39303	33	0.08%	3974	1882	43
8	153965	153958	7	0.005%	3722	44963	44927	36	0.08%	3865	1784	39
9	111474	111466	8	0.007%	3930	31270	31231	39	0.12%	4008	1882	42
10	112799	112792	7	0.006%	3936	34068	34048	20	0.06%	4003	1882	42

9   Conclusion

We have shown how Lagrangian relaxation in a stand-alone relaxed decision diagram can yield tight optimization bounds for certain job sequencing problems. We also characterized problems on which this approach is likely to be effective; namely, problems in which a relaxed DD of reasonable width results from a restricted form of state merger. The restriction is that states may be merged only when they agree on the values of state variables on which the cost function and dualized constraints depend in a recursive formulation of the problem.

Based on this analysis, we observed that job sequencing problems with state-independent processing times and time windows are suitable for this type of bounding, whether one minimizes tardiness, makespan or the number of late jobs. The same is true when processing times are dependent on when the job is processed or its position in the sequence. The traveling salesman problem can also bounded in this fashion. However, the TSP with time windows, as well as problems with state-dependent processing times in general, are normally unsuitable for DD-based Lagrangian relaxation, unless the relaxed diagram is combined with branching or embedded in a larger model.

We ran computational experiments on 110 instances from the well-known Crauwels-Potts-Wassenhove and Biskup-Feldman problem sets, with sizes ranging from 20 to 100 jobs. We found that DD-based Lagrangian relaxation can provide tight bounds for nearly all of these instances. This is especially true of the Biskup-Feldman instances tested, all of which were unsolved prior to this work. We showed that the best known solutions are almost always within a small fraction of one percent of the optimum, and we proved optimality for 8 of the solutions. To our knowledge, these are the first useful bounds that have been obtained for these instances.

More generally, our analysis can be used to identify dynamic programming models that may have a useful relaxation based on relaxed decision diagrams and Lagrangian duality.

References

1. Andersen, H.R., Hadzic, T., Hooker, J.N., Tiedemann, P.: A constraint store based on multivalued decision diagrams. In: Bessière, C. (ed.) CP 2007. LNCS, vol. 4741, pp. 118–132. Springer, Heidelberg (2007). https://doi.org/10.1007/978-3-540-74970-7_11
2. Bergman, D., Cire, A.A., van Hoeve, W.-J.: Improved constraint propagation via lagrangian decomposition. In: Pesant, G. (ed.) CP 2015. LNCS, vol. 9255, pp. 30–38. Springer, Cham (2015). https://doi.org/10.1007/978-3-319-23219-5_3
3. Bergman, D., Ciré, A.A., van Hoeve, W.J.: Lagrangian bounds from decision diagrams. Constraints 20, 346–361 (2015)
4. Bergman, D., Ciré, A.A., van Hoeve, W.J., Hooker, J.N.: Optimization bounds from binary decision diagrams. INFORMS J. Comput. 26, 253–268 (2013)
5. Bergman, D., Ciré, A.A., van Hoeve, W.J., Hooker, J.N.: Discrete optimization with binary decision diagrams. INFORMS J. Comput. 28, 47–66 (2014)

6. Bergman, D., Ciré, A.A., van Hoeve, W.J., Hooker, J.N.: Decision Diagrams for Optimization. Springer, Cham (2016). https://doi.org/10.1007/978-3-319-42849-9
7. Bergman, D., van Hoeve, W.-J., Hooker, J.N.: Manipulating MDD relaxations for combinatorial optimization. In: Achterberg, T., Beck, J.C. (eds.) CPAIOR 2011. LNCS, vol. 6697, pp. 20–35. Springer, Heidelberg (2011). https://doi.org/10.1007/978-3-642-21311-3_5
8. Biskup, D., Feldman, M.: Benchmarks for scheduling on a single machine against restrictive and unrestrictive common due dates. Comput. Oper. Res. **28**, 787–801 (2001)
9. Biskup, D., Feldman, M.: On scheduling around large restrictive common due windows. Eur. J. Oper. Res. **162**, 740–761 (2005)
10. Bryant, R.E.: Graph-based algorithms for boolean function manipulation. IEEE Trans. Comput. **C–35**, 677–691 (1986)
11. Chen, Z.L.: Scheduling and common due date assignment with earliness-tardiness penalties and batch delivery costs. Eur. J. Oper. Res. **93**, 49–60 (1996)
12. Ciré, A.A., van Hoeve, W.J.: Multivalued decision diagrams for sequencing problems. Oper. Res. **61**, 1411–1428 (2013)
13. Crauwels, H., Potts, C., Wassenhove, L.V.: Local search heuristics for the single machine total weighted tardiness scheduling problem. INFORMS J. Comput. **10**, 341–350 (1998)
14. Dumas, Y., Desrosiers, J., Gelinas, E., Solomon, M.M.: An optimal algorithm for the traveling salesman problem with time windows. Oper. Res. **43**, 367–371 (1995)
15. Hadžić, T., Hooker, J.N.: Cost-bounded binary decision diagrams for 0-1 programming. In: Van Hentenryck, P., Wolsey, L. (eds.) CPAIOR 2007. LNCS, vol. 4510, pp. 84–98. Springer, Heidelberg (2007). https://doi.org/10.1007/978-3-540-72397-4_7
16. Hall, N.G., Posner, M.E.: Earliness-tardiness scheduling problems, I: weighted deviation of completion times about a common due date. Oper. Res. **39**, 836–846 (1991)
17. Hall, N.G., Posner, M.E., Sethi, S.P.: Earliness-tardiness scheduling problems, II: weighted deviation of completion times about a restrictive common due date. Oper. Res. **39**, 847–856 (1991)
18. Hooker, J.N.: Discrete global optimization with binary decision diagrams. In: GICOLAG 2006, Vienna, Austria, December 2006
19. Hooker, J.N.: Decision diagrams and dynamic programming. In: Gomes, C., Sellmann, M. (eds.) CPAIOR 2013. LNCS, vol. 7874, pp. 94–110. Springer, Heidelberg (2013). https://doi.org/10.1007/978-3-642-38171-3_7
20. Hooker, J.N.: Job sequencing bounds from decision diagrams. In: Beck, J.C. (ed.) CP 2017. LNCS, vol. 10416, pp. 565–578. Springer, Cham (2017). https://doi.org/10.1007/978-3-319-66158-2_36
21. Lemaréchal, C.: Lagrangian relaxation. In: Jünger, M., Naddef, D. (eds.) Computational Combinatorial Optimization. LNCS, vol. 2241, pp. 112–156. Springer, Heidelberg (2001). https://doi.org/10.1007/3-540-45586-8_4
22. Ow, P.S., Morton, T.E.: The single machine early/tardy problem. Manage. Sci. **35**, 177–191 (1989)
23. Polyak, B.T.: Introduction to Optimization (translated from Russian). Optimization Software, New York (1987)
24. Serra, T., Hooker, J.N.: Compact representation of near-optimal integer programming solutions. Mathe. Program. (to appear)
25. Ying, K.C., Lin, S.W., Lu, C.C.: Effective dynamic dispatching rule and constructive heuristic for solving single-machine scheduling problems with a common due window. Int. J. Prod. Res. **55**, 1707–1719 (2017)

Integration of Structural Constraints into TSP Models

Nicolas Isoart$^{(\boxtimes)}$ and Jean-Charles Régin

Université Côte d'Azur, CNRS, I3S, Sophia Antipolis, France
nicolas.isoart@gmail.com, jcregin@gmail.com

Abstract. Several models based on constraint programming have been proposed to solve the traveling salesman problem (TSP). The most efficient ones, such as the weighted circuit constraint (WCC), mainly rely on the Lagrangian relaxation of the TSP, based on the search for spanning tree or more precisely "1-tree". The weakness of these approaches is that they do not include enough structural constraints and are based almost exclusively on edge costs. The purpose of this paper is to correct this drawback by introducing the Hamiltonian cycle constraint associated with propagators. We propose some properties preventing the existence of a Hamiltonian cycle in a graph or, conversely, properties requiring that certain edges be in the TSP solution set. Notably, we design a propagator based on the research of k-cutsets. The combination of this constraint with the WCC constraint allows us to obtain, for the resolution of the TSP, gains of an order of magnitude for the number of backtracks as well as a strong reduction of the computation time.

Keywords: Global constraint · TSP · Propagator

1 Introduction

The traveling salesman problem (TSP) is an NP-Hard problem. It has many applications and has been motivated by concrete problems, such as school bus routes, logistics, routing, etc. Almost all types of resolution methods (MIP, SAT, CP, evolutionary algorithms, etc.) have been used to solve it. When the graph is Euclidean, the most efficient program is the Concorde software [1]. Unfortunately, it cannot deal with additional constraints that are very present in real-world problems such as Pickup & Delivery, Dial-a-Ride, automatic harvesting, etc.

Solving the TSP is difficult since it involves finding a single cycle passing through all the vertices of a graph such that the sum of the costs of the edges it contains is minimal. It is quite easy to model the fact that each vertex belongs to a cycle. Indeed, it is sufficient that each vertex has at least two distinct neighbors, in other words, each vertex must be the end of at least two edges. Such a result can be obtained by modeling the problem as an assignment problem, which is solved in polynomial time. However, this model is not sufficient to obtain a

© Springer Nature Switzerland AG 2019
T. Schiex and S. de Givry (Eds.): CP 2019, LNCS 11802, pp. 284–299, 2019.
https://doi.org/10.1007/978-3-030-30048-7_17

single cycle in the graph, because the assignment corresponds to a coverage of the vertices by a set of disjoint cycles. From this model, we obtain solutions where each vertex belongs to a cycle, but not to a unique cycle. The covering by a unique cycle can be achieved by imposing that the subgraph generated by the selected edges is connected. The combination of these two aspects is what makes the TSP so difficult.

Unlike the previous approach, a model can be built based on the notion of a connected subgraph. It was exactly the idea of Held and Karp [7,8] who represented this notion by a 1-tree that is formed by a node x, two adjacent edges of x and a spanning tree of the graph without x. A 1-tree such that each node has a degree 2 is a Hamiltonian cycle, and a minimum 1-tree with these constraints is an optimal solution of the TSP. The use of a 1-tree is interesting because a minimum 1-tree is a good lower bound of the TSP. In addition, its computation is strongly related to the computation of a minimum spanning tree.

Held and Karp proposed to relax the degree constraints with a Lagrangian relaxation. More precisely, the cost of the edges are modified in order to integrate the violation of these degree constraints. If a node has a degree strictly greater than 2, then the cost of its adjacent edges are decreased, and if the degree is strictly less than 2 then the cost of its adjacent edges are increased. A convergence towards the optimal solution is obtained by computing a succession of minimum 1-tree based on the Lagrangian relaxation.

The *weighted circuit constraint* (WCC) [2] implements the approach in constraint programming. This constraint can be considered as the state of the art in CP as mentioned by Ducomman et al. [4]: "The best approach regarding the number of instances solved and quality of the bound is the Held and Karp's filtering".

In this paper, we propose to improve the WCC by adding methods for solving Hamiltonian cycles (i.e. TSP without costs). To do this, we consider the work of Cohen and Coudert [3] on the structure of the Hamiltonian cycles carried out for the FHCP Challenge [6]. Figure 1 shows an example in which the structure of the graph is important for the Hamiltonian cycle search. There is no Hamiltonian cycle in this graph, because it is impossible to find a cycle that visits all the vertices that pass only once through node C. Such a graph is said to be 1-connected: there is a vertex in the graph such that its removal disconnects it. We can therefore define a new structural constraint: if a graph is 1-connected, then it does not contain a Hamiltonian cycle.

Fig. 1. Butterfly graph.

This idea can be extended to edges. For instance, consider two edges a_1 and a_2 whose deletion disconnects the graph (i.e. it is 2-edge-connected). If there exists an Hamiltonian cycle then it necessarily contains a_1 and a_2. We propose to study k-edge-connected graphs for $k > 1$, and in particular values $k = 2$ and $k = 3$, which are common in practice. From this study, we defined a general filtering algorithm named k-cutset propagator.

This article is organized as follows: first, we recall some concepts of graph theory. Then, we formally define the structural constraints used in our method of solving the TSP. Next, we define a new data structure called cycled spanning tree, which is used to define a new algorithm to exploit structural constraints. The last part experimentally shows the advantages of our method. Finally, we conclude.

2 Preliminaries

2.1 Definitions

The definitions about graph theory are taken from [12].

A **directed graph** or **digraph** $G = (X, U)$ consists of a **vertex set** X and an **arc set** U, where every arc (u, v) is an ordered pair of distinct vertices. We will denote by $X(G)$ the vertex set of G and by $U(G)$ the arc set of G. The **cost** of an arc is a value associated with the arc. An **undirected graph** is a digraph such that for each arc $(u, v) \in U$, $(u, v) = (v, u)$. If $G_1 = (X_1, U_1)$ and $G_2 = (X_2, U_2)$ are graphs, both undirected or both directed, G_1 is a **subgraph** of G_2 if $V_1 \subseteq V_2$ and $U_1 \subseteq U_2$. A **path** from node v_1 to node v_k in G is a list of nodes $[v_1, ..., v_k]$ such that (v_i, v_{i+1}) is an arc for $i \in [1..k-1]$. The path **contains** node v_i for $i \in [1..k]$ and arc (v_i, v_{i+1}) for $i \in [1..k-1]$. The path is **simple** if all its nodes are distinct. The path is a **cycle** if $k > 1$ and $v_1 = v_k$. A cycle is **Hamiltonian** if $[v_1, ..., v_{k-1}]$ is a simple path and contains every vertex of X. The **length** of a path p, denoted by $length(p)$, is the sum of the costs of the arcs contained in p. For a graph G, a solution to the **traveling salesman problem (TSP)** in G is a Hamiltonian cycle $HC \in G$ minimizing $length(HC)$. An undirected graph G is **connected** if there is a path between each pair of vertices, otherwise it is **disconnected**. The maximum connected subgraphs of G are its **connected components**. A k-edge-connected graph is a graph in which there is no edge set of cardinality strictly less than k disconnecting the graph. A **tree** is a connected graph without a cycle. A tree $T = (X', U')$ is a **spanning tree** of $G = (X, U)$ if $X' = X$ and $U' \subseteq U$. The U' edges are the **tree edges** T and the $U - U'$ edges are the **non-tree edges** T. A **bridge** is an edge such that its removal increases the number of connected components. A partition (S, T) of the vertices of $G = (X, U)$ such that $S \subseteq X$ and $T = X - S$ is a **cut**. The set of edges $(u, v) \in U$ having $u \in S$ and $v \in T$ is the **cutset** of the (S, T) cut. A k-**cutset** is a cutset of cardinality k. A k-cutset is minimum iff there is no subset of the k-cutset that disconnects the graph.

2.2   HCWME : Hamiltonian Cycle with Mandatory Edges

CP-based algorithms solving the TSP tend to:

- Eliminate edges that cannot be part of the optimal solution.
- Define edges belonging to any optimal solution, called mandatory edges.

Since each optimal solution of the TSP is a Hamiltonian cycle, the TSP solution set is a subset of the solutions of the Hamiltonian cycle problem (HCP).

Property 1. Given $G = (X, U)$. If $a \in U$ belongs to all HCP(G) solutions, then a necessarily belongs to all TSP(G) solutions.

Property 2. Given $G = (X, U)$. If HCP(G) has no solution, then TSP(G) has no solution.

As the concept of mandatory arc is introduced, we formulate the Hamiltonian cycle with mandatory edges problem (HCWMEP):
INSTANCE: A graph $G = (X, U)$ and a set of mandatory edges $M \subseteq U$.
QUESTION: Is there a Hamiltonian cycle in G containing all the edges of M?

Since the HCP is an NP-Complete problem, HCWMEP(G, M) is NP-Complete.

3   Structural Constraints

We will use the following notations $G = (X, U)$, $n = |X|$, $m = |U|$, $M \subseteq U$ the set of mandatory edges of G, $\mathcal{P} = \text{HCMWEP}(G, M)$. When not specified we will assume that G is symmetrical, connected and that a k-cutset is minimum.

Proposition 1. *Let K be a k-cutset, then any Hamiltonian cycle C contains an even and strictly positive number of edges from K.*

Proof. Consider a k-cutset of G and C a hamiltonian cycle. The k-cutset partition G into two sets of vertices X_1 and X_2. Let u be our starting vertex in X_1, by definition C visits all the vertices of G and ends up visiting u (its starting vertex). Thus, visiting the vertices of X_2 involves taking one edge of the k-cutset and taking a different one to come back into X_1, at that moment: either all the vertices of X_2 have been visited and we end up joining u without using other edges of the k-cutset, or we have to visit X_2 again and return to X_1, every time we visit X_2 from X_1 we need an edge to go in, and another to go back: this means an even number of edges and the proposition holds.   □

From Proposition 1, we define Properties 3, 4, 5, 6 and 7.

Property 3. If there is $\{a_1, a_2\}$, a 2-cutset in G, then a_1 and a_2 become mandatory: $M \leftarrow M + \{a_1, a_2\}$.

Property 4. If there is a k-cutset with k odd containing $k - 1$ mandatory edges in G, then the non-mandatory edge a is deleted because it cannot be part of a Hamiltonian cycle: $E \leftarrow E - \{a\}$.

Property 5. If there is a k-cutset with k even containing $k - 1$ mandatory edges in G, then the non-mandatory edge a becomes mandatory: $M \leftarrow M + \{a\}$.

Property 6. If there is a 1-cutset in G, then $\mathcal{P}$ has no solution.

Property 7. If there is a k-cutset with k odd containing k mandatory edges in G, then $\mathcal{P}$ has no solution.

Definition 1. *It is said that two problems $\mathcal{P}$ and $\mathcal{P}'$ are equivalent if their solution sets are in bijection. We then note that $\mathcal{P} = \mathcal{P}'$.*

Corollary 1. *Given $\mathcal{P}' = \mathcal{P}$. If one or more of Properties 3, 4, 5, 6 or 7 are applied to $\mathcal{P}$, then $\mathcal{P}$ and $\mathcal{P}'$ remain equivalent.*

Proof. Immediate from Proposition 1. □

We write $a^* \in U$ a mandatory edge.

Example 1:

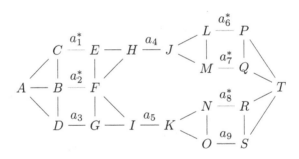

Fig. 2. Graph G_1.

From Properties 3, 4 and 5, we can remove some edges from Fig. 2 and make them mandatory:

- $\{a_4, a_5\}$ is a 2-cutset: if we want to connect the "left" part of H and I to the "right" part of J and K by a cycle we must take (H, J) and (I, K) so a_4 and a_5 become mandatory.
- $\{a_1^*, a_2^*, a_3\}$ is a 3-cutset and with $\{a_1^*, a_2^*\}$ mandatory: it is a cutset with an odd cardinality with an even number of mandatory edges. Then we can delete a_3, because by choosing it the cutset would become a mandatory set of edges with an odd cardinality.

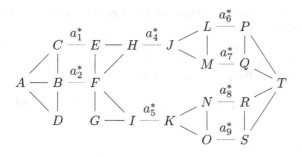

Fig. 3. Application of Properties 3, 4 and 5 on G_1.

- $\{a_6^*, a_7^*, a_8^*, a_9\}$ is a 4-cutset and with $\{a_6^*, a_7^*, a_8^*\}$ mandatory: it is a cutset with an even cardinality with 3 mandatory edges, so a_9 must be mandatory. Figure 3 shows how G_1 is modified when Properties 3, 4 and 5 are applied.

Example 2:

Now, consider Fig. 4. From Property 7, there is no Hamiltonian cycle:

- $\{a_4\}$ is a 1-cutset: there is no Hamiltonian cycle connecting $\{I, J, K\}$ to the other part of the graph.
- $\{a_1^*, a_2^*, a_3^*\}$ forms a 3-cutset with three mandatory edges.

Properties 3, 4, 5, 6 or 7 are based on the cardinality of the cutsets, so it is reasonable to ask how many cutsets a graph can have.

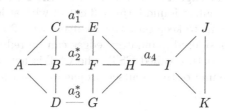

Fig. 4. Graph G_2.

Property 8. The number of cutset of a graph of order n is 2^n.

Proof. Any part $S \subseteq X$ forms an (S, T) cut. The cardinality of the powerset of $S \subseteq X$ is 2^n, so there are 2^n cutsets. □

In the case of the undirected graph, an (S, T) cut has the same cutset as the (T, S) cut. Hence, the number of distinct cutsets is $2^n / 2 = 2^{n-1}$.

In the case of a very dense graph, there is a low probability of satisfying Properties 3, 4, 5, 6 or 7 for a small value of k. Nor does it seem very reasonable to apply these properties with a high value of k for at least two reasons:

- The complexity of the algorithms of k-cutset increases with k because they are enumeration algorithms [14].
- The relationship between the cardinality of the cutset and the number of mandatory edges is strong. The more k increases and the less chance we have of satisfying one of Properties 3, 4, 5, 6 or 7.

Consequently, we propose to study $k = 1$, 2 and 3 with the following behaviors:

- 1-cutsets: raise a fail.
- 2-cutsets: make the edges of the 2-cutset mandatory.
- 3-cutsets: consider only the 3-cutsets with at least 2 mandatory edges. If it contains a non-mandatory edge, then it must be removed, otherwise a fail is raised.

4 k-cutset Propagator

We must be able to find the k-cutsets with $k = 1, 2, 3$ and two mandatory edges for $k = 3$. If we split the problem, finding the 1-cutset, which are actually bridges, can be done with the Tarjan algorithm [11] in $O(m + n)$; finding 2-cutset can be done with the Tsin algorithm [13] in $O(m + n)$. The strength of Tsin's algorithm is that it also allows us to find bridges, so we can manage $k = 1, 2$ at the same time. We now have to manage $k = 3$ with at least two mandatory edges.

By the cut definition, if you remove a k-cutset edge, then it becomes a $(k - 1)$-cutset. We can then propose a first simple algorithm:

For each mandatory edge $a^* \in M$, we look for the 2-cutsets of $G - \{a^*\}$. In this way, each of the 2-cutset found forms a 3-cutset with at least one mandatory edge. It is then sufficient to keep only the 3-cutsets with 2 mandatory edges.

The number of considered mandatory edges can be reduced. To do this, we will build a special structure called CST. The CST is not a required structure for the proper functioning of the k-cutset propagator, just an improvement.

4.1   CST: Cycled Spanning Tree

A CST is a 2-edge-connected subgraph of G such that for each edge a of G there is a cycle in G formed only by edges of the CST and a.

One way to build a CST is to calculate T a spanning tree, then add some edges to T until all the edges, those of T and those outside of T, belong to a CST cycle. Any edge $a \notin T$ belongs to a cycle composed of a and only T edges.

For the edges of T, the CST is built by adding edges to the spanning tree such that each tree edge belongs to a cycle of CST. This can be done in linear time by marking the tree edges each time a cycle is found. More precisely, we consider three graphs at the same time: G the graph, T the spanning tree, and CST the CST, initially equal to T. All tree edges are unmarked. We traverse the non tree edges of G until we find $a_{\overline{T}} = (i, j) \notin T$ such that there is a cycle formed by at least an unmarked edge of T, some edges of T and $a_{\overline{T}}$. We add

$a_{\overline{T}}$ to CST and we mark all the tree edges of C. We repeat this operation until there is no more unmarked edge in T. Clearly, at the end, each tree edge which has been marked belongs to a cycle. In addition, there is at least one tree edge in each cycle, so the number of added edges is bound by n. An example of a construction is shown in Fig. 5. This algorithm can be efficiently implemented, similarly to Kruskal's algorithm, by using a union-find data structure to avoid traversing each edge of each cycle. If we consider first the non mandatory edges for the construction of the spanning tree and for the construction of the CST, then we can expect to reduce the number of mandatory edges in the CST.

W.l.o.g. we assume that G is a connected bridgeless graph. Thus, there exist a CST in G.

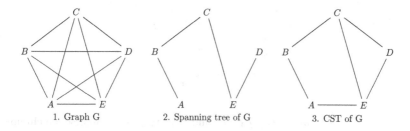

1. Graph G   2. Spanning tree of G   3. CST of G

Fig. 5. Example of building a CST.

Corollary 2. *Given $k > 1$. If there is a k-cutset in G, then at least two edges of the cutset are in the CST.*

Proof. By construction, the CST is connected and covers the graph with cycles. So each cut has a cardinality greater than or equal to two.   □

Corollary 3. *If there is a 3-cutset containing at least two mandatory edges, then at least one mandatory edge belongs to the CST.*

Proof. Immediate from Corollary 2.   □

Definition 2. *The **identification edges** are the mandatory edges for which a 2-cutset algorithm is run.*

From Corollary 3, the simple algorithm can be improved by reducing the number of mandatory edges that are considered. Considering the identification edges as each mandatory edge a^* of CST, the algorithm becomes: for each identification edges, search for the 2-cutsets of $G - \{a^*\}$. For each 3-cutset found, we obtain either a 3-cutset with three mandatory edges, or a 3-cutset with two mandatory edges or a 3-cutset with one mandatory edge. Then, we apply Properties 3, 4, 5, 6 and 7.

Since mandatory edges outside the CST are not considered as identification edges and the edges in the CST are chosen during construction, it is a good idea to minimize the number of mandatory edges in the CST.

4.2   Additional Improvement

The proposed algorithm is highly dependent on the number of identification edges. From Corollary 2, if two edges belong to the same 2-cutset and are mandatory, then they are identification edges. However, when searching for the 3-cutsets with an identification edge, it is not necessary to repeat the search for all the edges forming a 2-cutset with it. More precisely, the problem of searching for 3-cutsets with a^* as an identification edge has the same set of solutions as the problem of searching for 3-cutsets with each of the edges forming a 2-cutset with a^*. Figure 6 illustrates it well since the 2-cutset is a path.

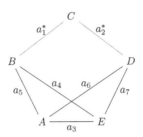

Fig. 6. $\{a_1^*, a_2^*\}$ is a 2-cutset. $\{a_1^*, a_4, a_5\}$ and $\{a_1^*, a_6, a_7\}$ are 3-cutsets including a_1^*. We can deduce that $\{a_2^*, a_4, a_5\}$ and $\{a_2^*, a_6, a_7\}$ are 3-cutsets including a_2^*.

Property 9. Let S_1 be a k-cutset and S_2 be a 2-cutset such that $k > 1$ and $S_2 \not\subseteq S_1$. If $\exists a \in S_1$ such that $a \in S_2$ then $(S_1 \cup S_2) - \{a\}$ forms a k-cutset.

Proof. Given $S_2 = \{a_1, a_2\}$ a 2-cutset and $a_1 \in S_1$. Removing S_1 from the graph disconnects it into two connected components. In the modified graph, $S_2 - \{a_1\} = \{a_2\}$ is a bridge. Removing a_2 further increases the number of connected components: there are now three. If we put back a_1, G is disconnected into two connected components, its cutset is $(S_1 - \{a_1\}) \cup \{a_2\} = (S_1 - \{a_1\}) \cup (S_2 - \{a_1\}) = (S_1 \cup S_2) - \{a_1\}$. Since S_1 is a k-cutset, there is no subset of it that disconnects the graph other than the k-cutset itself. If $(S_1 \cup S_2) - \{a_1\}$ disconnects the graph then it is a k-cutset because we delete and add an edge in a set of initial cardinality k. □

Consider S_1 a 3-cutset, $S_2 = \{a_1, a_2\}$ and S_3 two distinct 2-cutsets. From Property 9 the number of identification edges is reduced:

- If $a_1 \in S_1$, then $(S_1 - \{a_1\}) \cup \{a_2\}$ is a 3-cutset.
- If $a_1 \in S_3$, then $(S_3 - \{a_1\}) \cup \{a_2\}$ is a 2-cutset.

Thus, the set of identification edges is defined by the mandatory edges of the CST that do not belong to any 2-cutset and the subset of edges belonging to all 2-cutsets of G that maximizes its cardinality such that there is no combination of it forming a 2-cutset.

To avoid any inconsistency, all 2-cutsets must be searched before performing the 3-cutset search. Otherwise, there may be a 2-cutset containing at least one non-mandatory edge. This may result in a edge being marked as removable when searching for 3-cutsets while it is necessary for the existence of a Hamiltonian cycle. In addition, deleting an edge in a 3-cutset may create a 2-cutset and so either we perform a 2-cutset search immediately or we wait until the end of the search of all 3-cutsets to make the deletions effective. The first possibility is too time-consuming, a better solution is to postpone the deletions.

With this method we consider a subset of the identification edges. The higher the mandatory number of edges required, the more likely it is that the number of edges considered will be reduced.

Finally, CST has another advantage: it is incremental. Indeed, as long as no CST edges are removed, all edges outside the CST belong to a cycle composed of CST edges, so there is no need to rebuild it.

4.3   Implementation

Algorithm 1 is a possible implementation of the k-cutset filtering. The main function is PROPAGKCUTSET(G,M). Function PROPAG2CUTSET(G, M, set) defines a 2-cutset filtering. Function PROPAG3CUTSET(G, M, a^*) defines a 3-cutset filtering. Both filtering functions use FIND2CUTSET($G, bridge, 2cutsandFound$) which finds all 2-cutsets in G as proposed in [13] with a complexity in $O(n+m)$, this function is used as a black box. Filtering functions also have two subfunctions, bridge() and 2cutsetFound(M, a_1, a_2) describing the behavior that the FIND2CUTSET($G, bridge, 2cutsetFound$) algorithm must have when it finds a bridge or a 2-cutset in G. Function MERGECUTPAIRS(S, set, id) allows the use of the improvement proposed in Sect. 4.2. We will now describe the overall behavior of the algorithm. In Function PROPAGKCUTSET(G, M), we define set as a set of pairs of edges forming 2-cutsets in G. Then, we use the filtering PROPAG2CUTSET(G, M, set) to find and make mandatory all the edges belonging to a 2-cutset in G, the 2-cutsets are stored in set. The id array represents for each edge its 2-cutset identifier. In order to create sets of edges forming 2-cutsets between them Function MERGECUTPAIRS(S, set, id) is called. Each disjoint set will finally have a different identifier and each edge belonging to the same set will have the same identifier. Then, we define an array $visited$ to allow us to consider only one edge per set described above. The $identificationEdges$ set contains the mandatory edges which are in the CST. Then, we consider one edge per set calculated by MERGECUTPAIRS(S, set, id) and all the edges of the CST not being in any set. For each of its edges, the filtering PROPAG3CUTSET(G, M, a^*) is performed, i.e. the Properties 4, 5 and 7 are used. As recommended in Sect. 4.2, deletions are postponed. The final complexity of the Algorithm 1 is $O(k*(n+m))$ where $k <= |M| <= n$. Tsin's algorithm ($O(n + m)$) is called k times.

Algorithm 1. k-CUTSET(G,M)

PROPAGKCUTSET(G, M) :

   set $\leftarrow \emptyset$ /* set of pairs of edges representing the 2-cutset */
   if *not* PROPAG2CUTSET(G,M,set) then return False;
   $\forall e \in U(G) : id[e] \leftarrow nil$ /* contains the 2cutset identifier of each edge */
   MERGECUTPAIRS(identificationEdges,set,id)
   U' $\leftarrow$ U $\forall e \in U(G) : visited[e] \leftarrow False$
   identificationEdges $\leftarrow$ CST(G).GETMANDATORYEDGES()
   for each $a^* \in identificationEdges$ do
      if $id[a^*] = nil$ or $\neg visited[id[a^*]]$ then
         if *not* PROPAG3CUTSET(G, M, U', a^*) then return False;
         if $id[a^*] \neq nil$ then $visited[id[a^*]] \leftarrow$ True;

   $G \leftarrow$ (X,U') /* As deletion are postponed, update G */
   return True

PROPAG2CUTSET(G,M,SET):

   /* Return False if the graph isn't 3-edge-connected */
   define bridge(){**Exit propagation**}
   define 2cutsetFound(M, a_1, a_2){
      if $a_1 \notin M$ then $M \leftarrow M \cup \{a_1\}$;
      if $a_2 \notin M$ then $M \leftarrow M \cup \{a_2\}$;
      set $\leftarrow$ set $\cup$ (a_1, a_2);
   }
   return FIND2CUTSET(G,bridge,2cutsetFound)

PROPAG3CUTSET(G,M,U',a^*):

   /* Return False if the graph contains a 3-cutset with 3 mandatory edges */
   define bridge(){ **continue**; }
   define 2cutsetFound(M, a_1, a_2){
      if $a_1 \in M$ and $a_2 \in M$ then **Exit propagation**;
      else if $a_1 \in M$ then $U' \leftarrow U' - \{a_2\}$;
      else if $a_2 \in M$ then $U' \leftarrow U' - \{a_1\}$;
   }
   $G' \leftarrow (X(G), U(G) - \{a^*\})$
   return FIND2CUTSET$(G'$,bridge,2cutsetFound)

MERGECUTPAIRS(S,set,id):

   cpt $\leftarrow 0$
   for each $(a_1, a_2) \in set$ do
      /* if both a_1 and a_2 do have an identifier */
      if $id[a_1] \neq nil$ and $id[a_2] \neq nil$ then
         /*(a_1, a_2) is a 2-cutset: $id[a_1]$ must be equals to $id[a_2]$: id are merges*/
         for each $s' \in S$ do
            if $id[s'] = id[a_1]$ then
               $id[s'] \leftarrow id[a_2]$

      /* if both a_1 and a_2 do not have an identifier */
      if $id[a_1] = nil$ and $id[a_2] = nil$ then
         $id[a_1] \leftarrow id[a_2] \leftarrow$ cpt
         cpt $\leftarrow$ cpt $+ 1$
      /* if a_2 does not have an identifier and a_1 have one */
      if $id[a_1] \neq nil$ and $id[a_2] = nil$ then
         $id[a_2] \leftarrow id[a_1]$
      /* if a_1 does not have an identifier and a_2 have one */
      if $id[a_1] = nil$ and $id[a_2] \neq nil$ then
         $id[a_1] \leftarrow id[a_2]$

5   Experiments

The algorithms are implemented in Java 11 in a locally developed constraint programming solver. The experiments were performed on a Windows 10 machine using an Intel Core i7-3930K CPU @ 3.20 GHz and 64 GB of RAM. The reference instances are from the TSPLib [9], a library of reference graphs for the TSP. For fairness, we naturally took up the instances given by the state of the art [5]. All instances considered are symmetrical graphs.

We present the results in tables. Each of them reports the solving time in milliseconds. Timeout ($t.o$) is set at 30 min. The number of backtracks is denoted by #bk. Tables include a column expressing the ratio of solving time and number of backtracks.

Table 1. Improvement of k-cutset filtering.

Instance	maxCost (1)		maxCost k-cutsetNotImproved (2)		Ratios (1)/(2)		maxCost k-cutset (3)		Ratios (1)/(3)	
	time	#bk	time	#bk	time	#bk	time	#bk	time	#bk
gr96	13456	14970	3308	1492	4.1	10.0	3064	1492	4.4	10.0
rat99	132	40	321	40	0.4	1.0	196	40	0.7	1.0
kroA100	82296	96252	18594	9442	4.4	10.2	17632	9442	4.7	10.2
kroB100	243514	294148	15736	7286	15.5	40.4	15382	7286	15.8	40.4
kroC100	5937	4238	3677	1540	1.6	2.8	3646	1540	1.6	2.8
kroD100	806	480	944	286	0.9	1.7	819	286	1.0	1.7
kroE100	1213859	1628090	24986	9352	48.6	174.1	22968	9352	52.9	174.1
eil101	309	116	489	112	0.6	1.0	326	112	0.9	1.0
gr120	6610	3872	3089	980	2.1	4.0	2730	980	2.4	4.0
pr124	1876	566	1611	310	1.2	1.8	1530	310	1.2	1.8
bier127	822	402	770	146	1.1	2.8	641	146	1.3	2.8
ch130	18520	11810	7466	2250	2.5	5.2	6465	2250	2.9	5.2
pr136	t.o.	1733604	155283	35150	≥ 11.6	≥ 49.3	137675	35150	≥ 13.1	≥ 49.3
gr137	27828	11968	24223	6788	1.1	1.8	22579	6788	1.2	1.8
pr144	1622	466	1999	434	0.8	1.1	1603	434	1.0	1.1
ch150	11983	5684	6190	1424	1.9	4.0	5314	1424	2.3	4.0
kroA150	1290205	620080	174954	48892	7.4	12.7	171972	48892	7.5	12.7
kroB150	t.o.	791880	1222756	304630	≥ 1.5	≥ 2.6	1124443	304630	≥ 1.6	≥ 2.6
brg180	250527	2957988	t.o.	1000666	≤ 0.1	≤ 3.0	492962	2741812	0.5	1.1
rat195	t.o.	638322	1166767	271352	≥ 1.5	≥ 2.4	980190	271352	≥ 1.8	≥ 2.4
d198	440621	171294	273510	47838	1.6	3.6	179474	47838	2.5	3.6
kroB200	t.o.	647992	1586292	303282	≥ 1.1	≥ 2.1	1432978	303282	≥ 1.3	≥ 2.1
gr202	19681	9812	13385	2282	1.5	4.3	8261	2282	2.4	4.3
pr264	9520	1502	7817	256	1.2	5.9	6852	256	1.4	5.9

Table 1 shows the performance of adding k-cutset filtering to the WCC. This table is composed of three main columns (1, 2 and 3) showing the following results respectively: WCC without k-cutset filtering, WCC with k-cutset filtering without the improvement proposed in Sect. 4.2 and k-cutset improved filtering.

A static strategy such as maxCost, selecting arcs by decreasing costs allows us to compare the performance of the filtering without any disruption due to the strategy. These results show that using structural filtering is very interesting. For example, the search space of pr136 has been reduced by a factor of 49.3 and its solving time by a factor of 11.6 if the improvement proposed in Sect. 4.2 is not considered, by a factor of 13.1 otherwise. Indeed, the number of backtracks is generally reduced by a large factor (mean equal to 14.4, geometric mean equal to 4.3), which allows a good reduction in solving time (mean equal to 5.3, geometric mean equal to 2.4). The improvement allows the results to be refined by further improving the solving times.

We are now considering different strategies, including LCFirst maxCost introduced by Fages et al. [5]. It keeps one of its two extremities for the last branching edge and selects the edges from the neighborhood of the kept node by decreasing costs. It is currently considered the best current strategy for resolving the TSP in CP.

Table 2. Dynamic strategies.

Instance	(1) LCFirst minDeltaDeg		(2) LCFirst minDeltaDeg k-cutset		Ratios (1) / (2)		(3) LCFirst maxCost		(4) LCFirst maxCost k-cutset		Ratios (3) / (4)		(5) LCFirst minRepCost		(6) LCFirst minRepCost k-cutset		Ratios (5) / (6)	
	time	#bk	time	#bk	time	#bk	time	#bk	time	#bk	time	#bk	time	#bk	time	#bk	time	#bk
gr96	2327	1376	744	212	3.1	6.5	1951	1272	3113	1372	0.6	0.9	1534	746	1818	610	0.8	1.2
rat99	291	88	323	80	0.9	1.1	271	56	278	46	1.0	1.2	278	50	256	28	1.1	1.8
kroA100	9092	6278	4315	1846	2.1	3.4	5643	4048	7305	3726	0.8	1.1	3602	1884	3559	1288	1.0	1.5
kroB100	5321	3392	8380	3764	0.6	0.9	6359	4868	23181	10812	0.3	0.5	8232	4022	4419	1514	1.9	2.7
kroC100	2025	1126	2601	1076	0.8	1.0	1434	902	4451	2070	0.3	0.4	693	202	721	160	1.0	1.3
kroD100	868	410	917	290	0.9	1.4	705	286	778	240	0.9	1.2	410	76	453	80	0.9	1.0
kroE100	30414	26932	4304	1776	7.1	15.2	5488	4218	5604	2316	1.0	1.8	7650	3790	3479	1152	2.2	3.3
eil101	302	104	343	86	0.9	1.2	319	74	337	74	0.9	1.0	294	52	279	40	1.1	1.3
gr120	1311	468	685	112	1.9	4.2	1200	548	1791	578	0.7	0.9	1014	312	1062	214	1.0	1.5
pr124	6358	2336	7898	2462	0.8	0.9	1611	448	2387	582	0.7	0.8	1851	424	1415	208	1.3	2.0
bier127	520	128	466	56	1.1	2.3	609	216	728	180	0.8	1.2	533	84	1203	194	0.4	0.4
ch130	6953	3902	5301	1804	1.3	2.2	5287	2726	10243	3682	0.5	0.7	5028	1852	2826	750	1.8	2.5
pr136	19710	9822	28683	7448	0.7	1.3	262470	144980	160126	48370	1.6	6.6	181842	65974	55240	9926	3.3	6.6
gr137	8130	3640	6418	2092	1.3	1.7	5580	2158	13664	4208	0.4	0.5	4953	1548	3053	602	1.6	2.6
pr144	2742	648	3060	668	0.9	1.0	1463	256	1892	316	0.8	0.8	782	88	972	92	0.8	1.0
ch150	7189	2954	4824	1310	1.5	2.3	5100	1988	12350	3514	0.4	0.6	5034	1422	5348	1042	0.9	1.4
kroA150	34168	14996	14197	3874	2.4	3.9	21362	9510	63307	17526	0.3	0.5	14018	3724	8747	1702	1.6	2.2
kroB150	730330	320634	726592	207550	1.0	1.5	799195	373076	1194191	319360	0.7	1.2	1096412	331548	563570	114116	1.9	2.9
brg180	706	86	760	86	0.9	1.0	13423	125018	56323	267004	0.2	0.5	535	62	574	62	0.9	1.0
rat195	60531	17460	110822	25566	0.5	0.7	132012	41758	732018	178312	0.2	0.2	189821	40362	240102	32958	0.8	1.2
d198	26347	7062	27677	5686	1.0	1.2	71567	23740	93713	24048	0.8	1.0	119257	31262	51608	8044	2.3	3.9
kroB200	614139	191058	315601	67666	1.9	2.8	346683	114372	1393679	288336	0.2	0.4	360004	66452	149824	21622	2.4	3.1
gr202	4949	1582	7268	2004	0.7	0.8	8043	3248	7073	1906	1.1	1.7	5285	1066	6007	876	0.9	1.2
pr264	5816	190	6682	290	0.9	0.7	6631	322	7194	278	0.9	1.2	6663	206	6119	122	1.1	1.7
geo mean	6431	2274	5418	1324	1.2	1.7	6788	2911	11559	3490	0.6	0.8	5376	1271	4341	731	1.2	1.7
mean	65856	25695	53703	14075	1.2	1.8	71017	35837	158155	49119	0.4	0.7	83989	23217	46361	8225	1.8	2.8

Surprisingly, Table 2 shows that the k-cutset filtering is not interesting for the LCFirst maxCost strategy. The fact is that for the selected instances, the geometric mean of the solving times increases from 6788 ms to 11559 ms when k-cutset filtering is used. From our experiments, the strategy seems very ad hoc in regards to the propagator of the WCC constraint and in particular to the Lagrangian relaxation. It seems to partially correct the lack of structural constraints of the WCC. However, Fages et al. [5] have proposed other strategies:

Table 3. General results

Instance	(1) LCFirst maxCost		(2) LCFirst minRepCost k-cutset		Ratios (1)/(2)	
	time	#bk	time	#bk	time	#bk
gr96	1951	1272	1818	610	1.1	2.1
rat99	271	56	256	28	1.1	2.0
kroA100	5643	4048	3559	1288	1.6	3.1
kroB100	6359	4868	4419	1514	1.4	3.2
kroC100	1434	902	721	160	2.0	5.6
kroD100	705	286	453	80	1.6	3.6
kroE100	5488	4218	3479	1152	1.6	3.7
eil101	319	74	279	40	1.1	1.9
gr120	1200	548	1062	214	1.1	2.6
pr124	1611	448	1415	208	1.1	2.2
bier127	609	216	1203	194	0.5	1.1
ch130	5287	2726	2826	750	1.9	3.6
pr136	262470	144980	55240	9926	4.8	14.6
gr137	5580	2158	3053	602	1.8	3.6
pr144	1463	256	972	92	1.5	2.8
ch150	5100	1988	5348	1042	1.0	1.9
kroA150	21362	9510	8747	1702	2.4	5.6
kroB150	799195	373076	563570	114116	1.4	3.3
brg180	13423	125018	574	62	23.4	2016.4
rat195	132012	41758	240102	32958	0.5	1.3
d198	71567	23740	51608	8044	1.4	3.0
kroB200	346683	114372	149824	21622	2.3	5.3
gr202	8043	3248	6007	876	1.3	3.7
pr264	6631	322	6119	122	1.1	2.6
geo mean	6788	2911	4341	731	1.6	4.0
mean	71017	35837	46361	8225	2.5	87.4

LCFirst minDeltaDeg and LCFirst minRepCost, with performances comparable to LCFirst maxCost. The strategy LCFirst minRepCost is more suited to our model. It consists in selecting the edges by increasing replacement costs [2] with the LCFirst policy. This strategy has a slightly better sensitivity to the addition of k-cutset filtering and has the advantage of being generally more efficient. Indeed, between LCFirst minDeltaDeg and LCFirst minRepCost, we notice that when the k-cutset filtering is present, the geometric mean of the solving time of LCFirst minDeltaDeg is 5418 while that of LCFirst minRepCost is 4341, which is approximately 25% better. LCFirst minRepCost shows a significant reduction of the search space and a smaller reduction of the reduction time. For example, kroB200 gains a factor of 2.4 on solving time and a factor of 3.1 on the size of the search space.

Table 3 underlines the interest of using a structural filtering such as the k-cutset filtering. In comparison to the state of the art, we reduced the size of the search space for most instances by a very significant factor in order to obtain an improvement in solving time. There is a huge gain (solving time improved by 23.4) for the problem brg180. If we exclude this problem we improve the mean of the solving times by a factor of 1.5 and the mean of the number of backtracks by a factor of 3.6. The number of backtracks is reduced for each instance. The solving time is improved for 92% of the instances.

Note that the interaction of the k-cutset filtering with Lagrangian relaxation is not clear (the WCC is built around Lagrangian relaxation), a more in-depth study will have to be conducted to better understand it. Adding filtering can then disrupt the convergence of the latter and sometimes slow it down [10]. This explains why the gain factor of the number of backtracks is always much higher than that of the solving time.

6   Conclusion

We introduced a new structural constraint in the WCC based on the search for k-cutsets in the graph. The experimental results show the interest of our approach in practice. We observed that the number of backtracks is reduced by an order of magnitude depending on the chosen strategy and resolution times are significantly improved. The interactions between this constraint and the research strategy, as well as between this constraint and the Lagrangian model of the WCC, deserve further study.

Acknowledgements. We would like to thank Pr. Tsin for sending us his 2-cutset search algorithm implementation.

References

1. Applegate, D.L., Bixby, R.E., Chvatal, V., Cook, W.J.: The Traveling Salesman Problem: A Computational Study. Princeton University Press, Princeton (2006)
2. Benchimol, P., Régin, J.-C., Rousseau, L.-M., Rueher, M., van Hoeve, W.-J.: Improving the held and karp approach with constraint programming. In: Lodi, A., Milano, M., Toth, P. (eds.) CPAIOR 2010. LNCS, vol. 6140, pp. 40–44. Springer, Heidelberg (2010). https://doi.org/10.1007/978-3-642-13520-0_6
3. Cohen, N., Coudert, D.: Le défi des 1001 graphes. Interstices, December 2017. https://hal.inria.fr/hal-01662565
4. Ducomman, S., Cambazard, H., Penz, B.: Alternative filtering for the weighted circuit constraint: comparing lower bounds for the TSP and solving TSPTW. In: AAAI (2016)
5. Fages, J.G., Lorca, X., Rousseau, L.M.: The salesman and the tree: the importance of search in CP. Constraints **21**(2), 145–162 (2016)
6. Haythorpe, M.: FHCP challenge set: the first set of structurally difficult instances of the Hamiltonian cycle problem (2019)
7. Held, M., Karp, R.M.: The traveling-salesman problem and minimum spanning trees. Oper. Res. **18**(6), 1138–1162 (1970)

8. Held, M., Karp, R.M.: The traveling-salesman problem and minimum spanning trees: Part ii. Math. Program. **1**(1), 6–25 (1971)
9. Reinelt, G.: TSPLIB-A traveling salesman problem library. ORSA J. Comput. **3**(4), 376–384 (1991)
10. Sellmann, M.: Theoretical foundations of CP-based lagrangian relaxation. In: Wallace, M. (ed.) CP 2004. LNCS, vol. 3258, pp. 634–647. Springer, Heidelberg (2004). https://doi.org/10.1007/978-3-540-30201-8_46
11. Tarjan, R.E.: A note on finding the bridges of a graph. Inf. Process. Lett. **2**, 160–161 (1974)
12. Tarjan, R.E.: Data structures and Network Algorithms. CBMS-NSF Regional Conference Series in Applied Mathematics (1983)
13. Tsin, Y.H.: Yet another optimal algorithm for 3-edge-connectivity. J. Discrete Algorithms **7**(1), 130–146 (2009). Selected papers from the 1st International Workshop on Similarity Search and Applications (SISAP)
14. Yeh, L.P., Wang, B.F., Su, H.H.: Efficient algorithms for the problems of enumerating cuts by non-decreasing weights. Algorithmica **56**(3), 297–312 (2010)

Representing Fitness Landscapes
by Valued Constraints to Understand
the Complexity of Local Search

Artem Kaznatcheev[1,2], David A. Cohen[3], and Peter G. Jeavons[1(✉)]

[1] Department of Computer Science, University of Oxford, Oxford, UK
Peter.Jeavons@cs.ox.ac.uk
[2] Department of Translational Hematology and Oncology Research, Cleveland Clinic,
Cleveland, OH, USA
[3] Department of Computer Science, Royal Holloway, University of London,
Egham, UK

Abstract. Local search is widely used to solve combinatorial optimisation problems and to model biological evolution, but the performance of local search algorithms on different kinds of fitness landscapes is poorly understood. Here we introduce a natural approach to modelling fitness landscapes using valued constraints. This allows us to investigate minimal representations (normal forms) and to consider the effects of the structure of the constraint graph on the tractability of local search. First, we show that for fitness landscapes representable by binary Boolean valued constraints there is a minimal necessary constraint graph that can be easily computed. Second, we consider landscapes as equivalent if they allow the same (improving) local search moves; we show that a minimal normal form still exists, but is NP-hard to compute. Next we consider the complexity of local search on fitness landscapes modelled by valued constraints with restricted forms of constraint graph. In the binary Boolean case, we prove that a tree-structured constraint graph gives a tight quadratic bound on the number of improving moves made by any local search; hence, any landscape that can be represented by such a model will be tractable for local search. We build two families of examples to show that both the conditions in our tractability result are essential. With domain size three, even just a path of binary constraints can model a landscape with an exponentially long sequence of improving moves. With a treewidth two constraint graph, even with a maximum degree of three, binary Boolean constraints can model a landscape with an exponentially long sequence of improving moves.

1 Introduction

Local search techniques are widely used to solve combinatorial optimisation problems, and have been intensively studied since the 1980's [1,13,14,16,20]. They have also played a central role in the theory of biological evolution, ever

© Springer Nature Switzerland AG 2019
T. Schiex and S. de Givry (Eds.): CP 2019, LNCS 11802, pp. 300–316, 2019.
https://doi.org/10.1007/978-3-030-30048-7_18

since Sewall Wright [23] introduced the idea of viewing the evolution of populations of organisms as a local search process over a space of possible genotypes with associated fitness values that became known as a "fitness landscape".

The term *fitness landscape* is now used to designate any structure (A, f, N) consisting of a set of points A, a function f defined on those points, and a neighbourhood function N on those points, that indicates which pairs of points are sufficiently close to be considered neighbours. A point x is said to be *locally optimal* if all neighbours are non-improving (i.e. $\forall y \in N(x)\ f(x) \geq f(y)$) and *globally optimal* if all points are non-improving. The *local search problem* for a fitness landscape is to find such a local optimum. We say the problem is solved by a *local search algorithm* if the only moves allowed in the procedure are from a point x to a point x' with $x' \in N(x)$ and $f(x') > f(x)$.

Many approaches have been developed to try to distinguish fitness landscapes where a local or global optimal point can be found efficiently by local search from those where such optimal points cannot be found efficiently. In the 1980's and 90's these attempts focused on statistical measures such as correlation between function values at various distances and various notions of *ruggedness* [14]. But, by the late 90's there were several studies highlighting the existence of fitness landscapes that were not rugged and yet were hard to optimise. Several new approaches have been developed recently, but the performance of local search algorithms on many kinds of fitness landscapes is still poorly understood [14, 16, 20].

An approach that has not yet been explored in any detail, is to extend the modelling and analysis techniques recently developed for *valued constraint satisfaction problems* [2, 4, 8, 12, 21, 22] to analyse the computational difficulty of local search. In this paper we begin the development of a novel approach to understanding fitness landscapes based on representing those landscapes as valued constraint satisfaction problems (VCSPs), and studying the properties of the associated constraint graphs. In Sect. 3, we show how to efficiently construct a minimal representation (normal form) of fitness landscapes as VCSPs. In Sect. 4, we equate all fitness landscapes that have the same improving local search moves and show that a minimal form still exists for each equivalence class but is, in general, NP-hard to compute. Building on these results, the VCSP representation allows us to classify fitness landscapes in new ways, and hence to distinguish new classes of fitness landscapes with specific properties.

Since the WEIGHTED 2-SAT problem can be cast as a VCSP, finding a locally optimal solution for an arbitrary VCSP is a complete problem for the class of problems known as polynomial local search (PLS) [3, 10, 19]. This means that for a general VCSP it is expected to be computationally intractable even to find a local optimum by any method (not just by a local search algorithm). If we restrict to local search algorithms, then there exist standard constructions that produce families of fitness landscapes where every sequence of improving moves to a local optimum from some starting points is exponentially long [19]. On such landscapes, from such points, any local-search algorithm will require an exponentially long sequence of improving moves to reach a local optimum.

A key goal, therefore, is to identify classes of VCSPs where finding a locally optimal solution by local search is *tractable* (i.e., solvable in polynomial-time). In Sect. 5, we prove that fitness landscapes that can be represented by binary Boolean VCSPs with *tree-structured constraint graphs* can have only quadratically long sequences of improving moves – hence they are tractable for any local search algorithm. This is especially useful for investigating properties of biological evolution, as we discuss in the conclusion.

2   Background, Notation, and General Definitions

We will model the points, A, in our fitness landscapes as assignments to a collection of n variables, indexed by the set $[n] = 1, 2, \ldots, n$, with domains $D_1, \ldots, D_n$. Hence each point corresponds to a vector $x \in D_1 \times \cdots \times D_n$. We will generally focus on uniform domains (i.e., cases where $D = D_1 = \cdots = D_n$), where this simplifies to $x \in D^n$. In particular, we will often be interested in Boolean domains, where $x \in \{0,1\}^n$, so each point can be seen as a bit-vector.

The restriction of a variable assignment to some subset of variables, with indices in a set $S \subseteq [n]$, will be denoted $x[S]$, so $x[S] \in \prod_{j \in S} D_j$. To reference the assignment to the variable at position i, we will usually write x_i unless it is ambiguous, in which case we'll use the more general notation $x[i]$. If we want to modify x by changing a single variable, say the variable at position i, to some element $b \in D_i$, then we'll write $x[i \mapsto b]$.

Given a set of points, A, a **fitness function** on A is defined to be an integer-valued function defined on A, that is, a function $f : A \to \mathbb{Z}$. Because we are modelling fitness, rather than cost, we *maximise* our objective functions in this paper. All results can be carried over directly to the minimisation context.

To complete the definition of a fitness landscape, we will define a **neighbourhood function** on the set of points A to be a function $N : A \to 2^A$. For simplicity, we will assume this function is symmetric in the sense that if $y \in N(x)$, then $x \in N(y)$, and we will call such a pair x and y **adjacent** points. Throughout the paper, we will focus on the case where the set of points A is the set of assignments $D_1 \times \cdots \times D_n$ and N is the **1-flip neighbourhood** defined by $y \in N(x)$ if and only if there is a variable position i such that $x_i \neq y_i$ and this is the only difference (i.e., $\forall j \neq i \ \ x_j = y_j$). In the case of the Boolean domain, the graph of the function N, where the edges are the pairs of adjacent points, is the n-dimensional hypercube.

Definition 1 ([6,7]). *Given any fitness landscape (A, f, N), the corresponding **fitness graph** G has vertex set $V(G) = A$ and directed edge set $E(G) = \{xy \mid y \in N(x) \text{ and } f(y) > f(x)\}$.*

Note that the edges of the fitness graph consist of all pairs of adjacent points which have distinct values of the fitness function, and are oriented from the lower value of the fitness function to the higher value; such directed edges represent the possible moves that can be made by a local search algorithm.

A **(valued) constraint** with scope $S \subseteq [n]$ is a function $C_S : \prod_{j \in S} D_j \to \mathbb{Z}$. The **arity of a constraint** C_S is the size $|S|$ of its scope. For unary and binary constraints we will omit the set notation and just write C_i for $C_{\{i\}}$ or C_{ij} for $C_{\{i,j\}}$. We will represent the values taken by a unary constraint C_i for each domain element by an integer vector of length $|D_i|$, and represent the values taken by a binary constraint C_{ij} for each pair of domain elements by an integer matrix, where x_i selects the row and x_j selects the column. A zero-valued constraint (of any arity) will be denoted by 0.

Definition 2. *An instance of the* valued constraint satisfaction problem (VCSP) *is a set of constraints* $\mathcal{C} = \{C_{S_1}, \ldots, C_{S_m}\}$. *We say that a VCSP-instance* $\mathcal{C}$ **implements** *a fitness function* f *if* $f(x) = \sum_{k=1}^{m} C_{S_k}(x[S_k])$.

The arity of a VCSP-instance is the maximum arity over its constraints; if this maximum arity is 2, then we will call it a *binary* VCSP-instance. The instance-size of a VCSP-instance is the number of bits needed to specify n, m and each constraint.

Given any VCSP-instance $\mathcal{C}$, we can take A as the set of all possible assignments, f as the fitness function implemented by $\mathcal{C}$, and N as the 1-flip neighbourhood, to obtain an associated fitness landscape, (A, f, N), and hence an associated fitness graph, $G_{\mathcal{C}}$, by Definition 1. Note that the vertex set of $G_{\mathcal{C}}$ is the set of possible assignments, A, and hence is exponential in the size of the instance, $\mathcal{C}$, in general. Each *binary* VCSP-instance also has an associated *constraint graph*, defined as follows, whose vertex set is polynomial in the size of the instance:

Definition 3. *Given any binary VCSP-instance* $\mathcal{C}$, *the corresponding* **constraint graph** *has vertices* $V(\mathcal{C}) = [n]$, *edges* $E(\mathcal{C}) = \{ij \mid C_{ij} \in \mathcal{C}, C_{ij} \neq 0\}$, *and constraint-neighbourhood function* $N_{\mathcal{C}}(i) = \{j \mid ij \in E(\mathcal{C})\}$.

3 Magnitude-Equivalence

It is clear from Definition 2 that different VCSP-instances can implement the same fitness function. Consider, the following two small VCSP-instances:

$$\left(x_1\right) - \begin{pmatrix} 1 & 2 \\ 2 & 3 \end{pmatrix} - \left(x_2\right) \quad \text{vs.} \quad C_\emptyset = 1 \qquad \begin{pmatrix} 0 \\ 1 \end{pmatrix} \left(x_1\right) \qquad \qquad \left(x_2\right) \begin{pmatrix} 0 \\ 1 \end{pmatrix}$$

Although these two instances have different constraint graphs, the fitness function they implement is $[f(00), f(01), f(10), f(11)] = [1, 2, 2, 3]$ in both cases. We capture this equivalence with the following definition:

Definition 4. *If two VCSP-instances* $\mathcal{C}_1$ *and* $\mathcal{C}_2$ *implement the same fitness function* f, *then we will say they are* **magnitude-equivalent**.

We will show in this section that for binary Boolean VCSP-instances each equivalence class of magnitude-equivalent VCSP-instances has a *normal form*: a unique, minimal, and easy to compute representative member with special properties.

Definition 5. *A binary Boolean VCSP-instance $\mathcal{C}$ is* **simple** *if every unary constraint has $C_i = \begin{pmatrix} 0 \\ c_i \end{pmatrix}$ and every binary constraint has $C_{ij} = \begin{pmatrix} 0 & 0 \\ 0 & c_{ij} \end{pmatrix}$.*

In drawings of constraint graphs of simple VCSP-instances we will often denote the unary constraint $\begin{pmatrix} 0 \\ c_i \end{pmatrix}$ by c_i, and the binary constraint $\begin{pmatrix} 0 & 0 \\ 0 & c_{ij} \end{pmatrix}$ by c_{ij}.

We now give a direct proof of the following simplification result which is analogous to similar results using constraint propagation in standard VCSP [5].

Theorem 1. *Any binary Boolean VCSP-instance $\mathcal{C}'$ can be transformed into a unique simple VCSP-instance $\mathcal{C}$ that is magnitude-equivalent to $\mathcal{C}'$. Moreover, $\mathcal{C}$ can be constructed from $\mathcal{C}'$ in linear time.*

Proof. First two key observations: (1) Any unary Boolean constraint C_i' can be rewritten as a linear function: $c_i'(x) = (1-x_i)C_i'(0) + x_i C_i'(1)$; and (2) any binary Boolean constraint C_{ij}' can be rewritten as a multilinear polynomial of degree 2: $c_{ij}'(x) = (1 - x_i)(1 - x_j)C_{ij}'(0,0) + (1 - x_i)x_j C_{ij}'(0,1) + x_i(1 - x_j)C_{ij}'(1,0) + x_i x_j C_{ij}'(1,1)$. From this, we can simplify $\mathcal{C}'$ just by simplifying polynomials:

$$f(x) = C_\emptyset' + \sum_{i=1}^n C_i'(x_i) + \sum_{ij \in E(\mathcal{C}')} C_{ij}'(x_i, x_j) = C_\emptyset' + \sum_{i=1}^n c_i'(x) + \sum_{ij \in E(\mathcal{C}')} c_{ij}'(x) \quad (1)$$

$$= C_\emptyset + \sum_{i=1}^n x_i c_i + \sum_{1 \le i < j \le n} x_i x_j c_{ij} \quad (2)$$

where we note that the last part of Eq. 1 is a sum of a constant, some linear functions, and some multilinear polynomials of degree 2, and is thus itself a multilinear polynomial of degree 2 (or less). Equation 2 then follows from Eq. 1 by multiplying out into monomials and then grouping the coefficients of each similar monomial. This can be done in time linear in the number of constraints and the number of bits needed to encode their coefficients. We note that Eq. 2 corresponds to a VCSP-Instance $\mathcal{C}$ comprising a null-term $C_\emptyset$, unary constraints $C_i = \begin{pmatrix} 0 \\ c_i \end{pmatrix}$, and binary constraints $C_{ij} = \begin{pmatrix} 0 & 0 \\ 0 & c_{ij} \end{pmatrix}$.   □

The next result shows that a simple VCSP-instance has the minimal constraint graph of any binary instance that implements the same fitness function:

Theorem 2. *Let $\mathcal{C}$ be a simple binary Boolean VCSP-instance. If the binary Boolean VCSP-instance $\mathcal{C}'$ is magnitude-equivalent to $\mathcal{C}$, then $E(\mathcal{C}) \subseteq E(\mathcal{C}')$.*

Proof. Let $e_i \in \{0,1\}^n$ be a variable assignment that sets the ith variable to one, and all other variables to zero. Similarly, let $e_{ij} \in \{0,1\}^n$ be a variable assignment that sets the ith and jth variables to one, and all other variables to zero. Since $\mathcal{C}$ implements f, we have:

$$f(e_{ij}) - f(e_i) - f(e_j) + f(0^n) = c_{ij} \qquad (3)$$

where we take $c_{ij} = 0$ if $ij \notin E(\mathcal{C})$. Similarly, if $\mathcal{C}'$ also implements f, we have:

$$f(e_{ij}) - f(e_i) - f(e_j) + f(0^n) = C'_{ij}(1,1) - C'_{ij}(1,0) - C'_{ij}(0,1) + C'_{ij}(0,0)$$

If $ij \in E(\mathcal{C})$ then $c_{ij} \neq 0$, so $C'_{ij}(1,1) - C'_{ij}(1,0) - C'_{ij}(0,1) + C'_{ij}(0,0) \neq 0$ and hence $ij \in E(\mathcal{C}')$. □

4 Sign-Equivalence

In the previous section we considered the equivalence class of all VCSP-instances which implement precisely the same fitness function. However, when investigating the performance of local search algorithms, the exact values of the fitness function are not always relevant; it may be sufficient to consider only the fitness graph.

For example, consider a fitness function f, implemented by a VCSP-instance $\mathcal{C}$, where all fitness values are distinct, but there is at least one pair i, j of positions with no constraint C_{ij}. Now consider the new fitness function $f'(x) = 2f(x) + C_{ij}(x_i, x_j)$ where $C_{ij} = [0, 0; 0, 1]$. The fitness graph corresponding to f' is unchanged (since all fitness values given by $2f(x)$ differ by at least 2, every edge is still present in the fitness graph, and no orientations are changed by the new constraint), but we cannot eliminate this new C_{ij} constraint without changing the precise values of the fitness function. To capture this similarity between f and f', we introduce a more abstract equivalence relation:

Definition 6. *If two VCSP-instances $\mathcal{C}_1$ and $\mathcal{C}_2$ give rise to the same fitness graph, then we will say they are **sign-equivalent**.*

As with magnitude-equivalence, we will show that for binary Boolean VCSP-instances it is possible to define a normal form or minimal representative member of each equivalence class of sign-equivalent VCSP-instances with a unique minimal constraint graph. Unfortunately, we will see that, unlike the situation for magnitude-equivalence, this minimum sign-equivalent constraint-graph is NP-hard to compute.

Definition 7. *In a Boolean fitness graph G with vertex set $\{0,1\}^n$, we will say that i **sign-depends** on j if there exists an assignment $x \in \{0,1\}^n$ such that:*

$$xx[i \mapsto \overline{x}_i] \in E(G) \quad but \quad x[j \mapsto \overline{x}_j]x[i \mapsto \overline{x}_i, j \mapsto \overline{x}_j] \notin E(G) \qquad (4)$$

Note that i **sign-depends** on j if and only if, for any fitness function f that corresponds to the fitness graph G, there exists $x \in \{0,1\}^n$ such that:

$$\operatorname{sgn}(f(x[i \mapsto \overline{x}_i]) - f(x)) \neq \operatorname{sgn}(f(x[i \mapsto \overline{x}_i, j \mapsto \overline{x}_j]) - f(x[j \mapsto \overline{x}_j])). \qquad (5)$$

We will say that i and j **sign-interact** if i sign-depends on j or j sign-depends on i (or both). If i and j do not sign-interact then we will say that they are **sign-independent**.

Definition 8. *A simple binary Boolean VCSP-instance C with associated fitness graph G_C is called* **trim** *if for all $ij \in E(C)$, i and j sign-interact in G_C.*

Our sign-equivalent analog of Theorem 1 guarantees a normal form:

Theorem 3. *Any simple binary Boolean VCSP-instance C' can be transformed into a trim VCSP-instance C that is sign-equivalent to C'.*

To prove Theorem 3 we now establish two propositions: Proposition 1 connects the magnitude of constraints with their effect on fitness graphs, and Proposition 2 connects the magnitude of constraints to sign-interaction.

Proposition 1. *Given a simple binary Boolean VCSP-instance C implementing a fitness function f, if removing the constraint C_{ij} changes the corresponding fitness graph, then for at least one $k \in \{i, j\}$ there exists some $x \in \{0,1\}^n$ with $x_i = x_j = 1$ such that:*

$$c_{ij} \geq f(x) - f(x[k \mapsto 0]) > 0 \quad or \quad c_{ij} \leq f(x) - f(x[k \mapsto 0]) < 0 \qquad (6)$$

Proof. Without loss of generality (by swapping i and j in the variable numbering if necessary), we can suppose that $k = i$. Consider two cases:

Case 1 $(c_{ij} > 0)$: If removing C_{ij} changes the fitness graph, then there exists some $x \in \{0,1\}^n$ with $x_i = x_j = 1$ such that:

$$f(x) > f(x[i \mapsto 0]) \quad \text{but} \quad f(x) - c_{ij} \leq f(x[i \mapsto 0]). \qquad (7)$$

We can re-arrange Eq. 7 to get $c_{ij} \geq f(x) - f(x[i \mapsto 0]) > 0$

Case 2 $(c_{ij} < 0)$: This is the same as case 1, except that the direction of the inequalities in Eq. 7 are reversed. □

Proposition 2. *Given a simple binary Boolean VCSP-instance C implementing a fitness function f, if there exists a constraint C_{ij} in C, some assignment $x \in \{0,1\}^n$ with $x_i = x_j = 1$, and some $k \in \{i, j\}$ such that:*

$$c_{ij} \geq f(x) - f(x[k \mapsto 0]) > 0 \quad or \quad c_{ij} \leq f(x) - f(x[k \mapsto 0]) < 0 \qquad (8)$$

then i sign-depends on j in the associated fitness graph G_C.

Proof. As in the proof of Proposition 1, we can suppose that $k = i$ (by swapping i and j in the variable numbering if necessary). Also, as in the proof of Proposition 1, the case for $c_{ij} < 0$ is symmetric (by flipping the direction of inequalities) to $c_{ij} > 0$. Thus, we will just consider the case where $k = i$ and $c_{ij} > 0$:

Given that Eq. 8 tells us that $f(x) > f(x[i \mapsto 0])$ (i.e., that $x[i \mapsto 0]x \in E(G_C)$), to establish that i sign-depends on j per Definition 7, we need to show that $f(x[j \mapsto 0]) \leq f(x[i \mapsto 0, j \mapsto 0])$ (i.e., that $x[i \mapsto 0, j \mapsto 0]x[j \mapsto 0] \notin E(G_C)$). So, let us look at the difference of the latter:

$$f(x[j \mapsto 0]) - f(x[i \mapsto 0, j \mapsto 0]) = f(x) - f(x[i \mapsto 0]) - c_{ij} \leq 0 \qquad (9)$$

where the equality follows from Definition 2 (C implements f) and Definition 5 (C is simple), and the inequality follows from the first part of Eq. 8. □

Proof (of **Theorem** *3).* Note that Eqs. 6 and 8 specify the same conditions, hence the negation of this condition can be used to glue together the contra-positives of Proposition 2 (if i and j are sign-independent then Eq. 6 does not hold) and Proposition 1 (if Eq. 8 does not hold then C'_{ij} can be removed from $\mathcal{C}'$ without changing the corresponding fitness graph). So we can convert $\mathcal{C}'$ to a trim VCSP-instance that is sign-equivalent to $\mathcal{C}'$ by simply removing all $C'_{ij} \in \mathcal{C}'$ where i and j are sign-independent in the associated fitness graph $G_{\mathcal{C}'}$.   □

The next result is the sign-equivalence analog of Theorem 2. It shows that a trim VCSP-instance has the minimal constraint graph of any binary instance with the same associated fitness graph.

Theorem 4. *Let $\mathcal{C}$ be a trim binary Boolean VCSP-instance. If the binary Boolean VCSP-instance $\mathcal{C}'$ is sign-equivalent to $\mathcal{C}$, then $E(\mathcal{C}) \subseteq E(\mathcal{C}')$.*

To prove Theorem 4, we just need to show that constraints between sign-interacting positions cannot be removed while preserving sign-equivalence. That is, we just need the following proposition:

Proposition 3. *Let $\mathcal{C}$ be a binary Boolean VCSP-instance with associated fitness graph $G_{\mathcal{C}}$. If i, j sign-interact in $G_{\mathcal{C}}$, then the constraint C_{ij} in $\mathcal{C}$ is non-zero.*

Proof. Without loss of generality, assume that we have an edge in $G_{\mathcal{C}}$ from $x[i \mapsto \overline{x_i}]$ to x. Thus, the fitness function f implemented by $\mathcal{C}$ must satisfy the following two inequalities:

$$f(x) > f(x[i \mapsto \overline{x_i}]) \quad \text{and} \quad f(x[j \mapsto \overline{x_j}]) \le f(x[i \mapsto \overline{x_i}, j \mapsto \overline{x_j}]) \tag{10}$$

Define $g_i(x_i) = C_i(x_i) + \sum_{k \neq j} C_{ik}(x_i, x_k)$ and similarly for g_j. Also let $K_{ij}(x)$ be the part of f independent of x_i, x_j: i.e., $f(x) = K_{ij}(x) + g_i(x_i) + g_j(x_j) + C_{ij}(x_i, x_j)$. Rewriting (and simplifying) the two parts of Eq. 10, we get:

$$g_i(x_i) + C_{ij}(x_i, x_j) > g_i(\overline{x_i}) + C_{ij}(\overline{x_i}, x_j) \tag{11}$$
$$g_i(x_i) + C_{ij}(x_i, \overline{x_j}) \le g_i(\overline{x_i}) + C_{ij}(\overline{x_i}, \overline{x_j}) \tag{12}$$

These equations can be rotated to sandwich the g_i terms:

$$C_{ij}(x_i, x_j) - C_{ij}(\overline{x_i}, x_j) > g_i(\overline{x_i}) - g_i(x_i) \ge C_{ij}(x_i, \overline{x_j}) - C_{ij}(\overline{x_i}, \overline{x_j}) \tag{13}$$

which simplifies to $C_{ij}(x_i, x_j) - C_{ij}(\overline{x_i}, x_j) > C_{ij}(x_i, \overline{x_j}) - C_{ij}(\overline{x_i}, \overline{x_j})$ and – due to the strict inequality – establishes that C_{ij} is non-zero.   □

However, unlike with magnitude-equivalence, it is NP-hard to determine a minimal sign-equivalent VCSP-instance, as the next result shows:

Theorem 5. *The problem of deciding whether i and j sign-interact in a given simple binary Boolean VCSP-instance is NP-complete.*

Proof. To see that this problem is in NP, note that we can provide a variable assignment x as a certificate and check that under that variable assignment either i sign-depends on j or j sign-depends on i (or both).

We will establish NP-hardness by reduction from the SUBSETSUM problem, which is known to be NP-complete [9]: A set of integers $\{s_1, \ldots, s_n\}$ and a target t is a yes-instance of the SUBSETSUM problem if there exists some subset $S \subseteq [n]$ such that $\sum_{i \in S} s_i = t$.

Now consider a simple binary Boolean VCSP-instance $\mathcal{C}$ on $n + 2$ variables, that implements fitness function f and has associated fitness graph $G_{\mathcal{C}}$, whose constraint graph has the shape of a star, with central variable position $n + 2$:

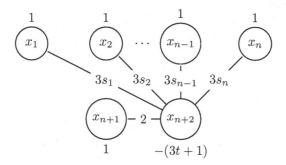

Claim: $\langle \{s_1, \ldots, s_n\}, t \rangle$ is a yes-instance of SUBSETSUM if and only if $n+1$ and $n+2$ sign-interact.

We clearly have that for all $x \in \{0, 1\}^{n+2}$, $f(x[n + 1 \mapsto 1]) > f(x[n + 1 \mapsto 0])$, so $n + 1$ does not sign-depend on $n + 2$. Thus our claim becomes equivalent to verifying the conditions under which $n + 2$ sign-depends on $n + 1$. Let's look at the two directions of the if and only if in the claim:

Case 1 ($\Rightarrow$): If $\langle \{s_1, \ldots, s_n\}, t \rangle \in$ SUBSETSUM, then there is a subset $S \subseteq [n]$ such that $\sum_{i \in S} s_i = t$. Let $e_S \in \{0, 1\}^n$ be the variable assignment such that for any $i \in S$, $e_S[i] = 1$ and for any $j \notin S$, $e_S[j] = 0$. We have that:

$$f(e_S 01) = |S| - 1 \qquad\qquad f(e_S 11) = |S| + 2$$
$$f(e_S 00) = |S| \qquad\qquad\quad f(e_S 10) = |S| + 1$$

By Eq. 5, these imply that $n + 2$ sign-depends on $n + 1$.

Case 2 ($\Leftarrow$): If $\langle \{s_1, \ldots, s_n\}, t \rangle \notin$ SUBSETSUM, then for any $S \subseteq [n]$ we either have $\sum_{i \in S} s_i \leq t - 1$ or $\sum_{i \in S} s_i \geq t + 1$. Thus, given an arbitrary assignment $e_S \in \{0, 1\}$ we have two subcases:

If $\displaystyle\sum_{i \in S} s_i \leq t - 1$ then: $\qquad\qquad$ Or, if $\displaystyle\sum_{i \in S} s_i \geq t - 1$ then:

$$f(e_S 01) - f(e_S 00) \leq -4 \qquad\qquad f(e_S 01) - f(e_S 00) \geq 2$$
$$f(e_S 11) - f(e_S 10) \leq -2 \qquad\qquad f(e_S 11) - f(e_S 10) \geq 4$$

In either subcase, $\mathrm{sgn}(f(e_S 01) - f(e_S 00)) = \mathrm{sgn}(f(e_S 11) - f(e_S 10))$, so by Eq. 5, $n + 2$ does not sign-depend on $n + 1$. $\qquad\qquad\qquad\qquad\qquad\qquad\square$

5   Tree-Structured Boolean VCSP-instances

In this section, we will prove the following:

Theorem 6. *For a binary Boolean VCSP instance C on n variables, if the constraint-graph of C is a tree, then any directed path in the associated fitness graph G_C has length at most $\binom{n}{2} + n$.*

Note that this result bounds the length of *any* directed path in G_C, not just the path taken by a particular local-search algorithm. Thus, on such landscapes even choosing the worst possible sequence of improving moves results in a local optimum being found in polynomial time.

We will show in Sect. 6 that the conditions of being Boolean and tree-structured are essential to obtain a polynomial bound on the length of all paths. To see that the bound in Theorem 6 is the best possible for binary Boolean tree-structured VCSP-instances, consider the example below:

Example 1. (**Path of length $\binom{n}{2} + n$**) Consider the following binary Boolean VCSP-instance C:

$$\left(\!\!\begin{array}{c}x_1\end{array}\!\!\right)\!\cdot\!\begin{pmatrix}1 & 0\\0 & 1\end{pmatrix}\!\!\left(\!\!\begin{array}{c}x_2\end{array}\!\!\right)\!\cdot\!\begin{pmatrix}2 & 0\\0 & 2\end{pmatrix}\!\cdot\!\left(\!\!\begin{array}{c}x_3\end{array}\!\!\right)\!\cdot\!\begin{pmatrix}3 & 0\\0 & 3\end{pmatrix}\!-\cdots-\begin{pmatrix}n-1 & 0\\0 & n-1\end{pmatrix}\!-\!\left(\!\!\begin{array}{c}x_n\end{array}\!\!\right)\!\begin{pmatrix}0\\n\end{pmatrix}$$

To obtain a path of length $\binom{n}{2} + n$ in the corresponding fitness graph G_C, consider an initial variable assignment of $x = (10)^{\frac{n}{2}}$ if n is even and $x = 0(10)^{\frac{n-1}{2}}$ if n is odd, and always select the leftmost variable that is able to flip. This will increase the fitness by 1 at each step, starting from 0 to $\binom{n}{2} + n$.

For example, when $n = 4$, this gives the following sequence of 11 assignments, each of which increases the value of the fitness function by 1:

$$0101 \rightarrow 1101 \rightarrow 1001 \rightarrow 0001 \rightarrow 0011 \rightarrow 0111 \rightarrow 1111 \rightarrow 1110 \rightarrow 1100 \rightarrow 1000 \rightarrow 0000 \quad (14)$$

For the proof of Theorem 6, we introduce some further definitions.

Definition 9. *Given any directed path $p = x^1 \dots x^t \dots x^T$ in a fitness graph G, define the flip function as $m(t) = (i \mapsto b)$ where $x^{t+1} \oplus x^t = e_i$ and $b = x_i^{t+1}$ (i.e., the i-th variable is flipped at time t to value b).*

For illustration, consider the sequence of moves listed in Eq. 14 of Example 1. It corresponds to the following flip function:

t	1	2	3	4	5	6	7	8	9	10
m(t)	$1 \mapsto 1$	$2 \mapsto 0$	$1 \mapsto 0$	$3 \mapsto 1$	$2 \mapsto 1$	$1 \mapsto 1$	$4 \mapsto 0$	$3 \mapsto 0$	$2 \mapsto 0$	$1 \mapsto 0$

To obtain the bound on the length of paths given in Theorem 6, we will identify a structure in the flip function, to bound the maximum possible value for T.

Definition 10. *We say that a flip $m(t') = (j \mapsto c)$* **supports** *a flip $m(t) = (i \mapsto b)$ if $t' < t$ and $C_{ij}(b,c) - C_{ij}(\bar{b},c) > C_{ij}(b,\bar{c}) - C_{ij}(\bar{b},\bar{c})$; if $x_j^t = c$, then the support is said to be* **strong.**

It is useful to note that the inequality on C_{ij} is symmetric in the sense that:

$$C_{ij}(b,c) - C_{ij}(\bar{b},c) > C_{ij}(b,\bar{c}) - C_{ij}(\bar{b},\bar{c})$$
$$\Leftrightarrow C_{ij}(b,c) - C_{ij}(b,\bar{c}) > C_{ij}(\bar{b},c) - C_{ij}(\bar{b},\bar{c}) \tag{15}$$
$$\Leftrightarrow C_{ji}(c,b) - C_{ji}(\bar{c},b) > C_{ji}(c,\bar{b}) - C_{ji}(\bar{c},\bar{b})$$

Definition 11. *Given a binary Boolean VCSP-instance $\mathcal{C}$ implementing fitness function f, the* **fitness contribution** *of the variable at position i in assignment x, restricted to $S \subseteq [n]$ is defined to be:*

$$f_i^S(b|x) = \left\{ \begin{array}{ll} C_i(b) & if\ i \in S \\ 0 & otherwise \end{array} \right\} + \sum_{j \in N_{\mathcal{C}}(i) \cap S} C_{ij}(b, x_j) \tag{16}$$

if $S = [n]$ then we just write f_i rather than $f_i^{[n]}$.

Note that for any path p in G, if $m(t) = (i \mapsto b)$ then $f_i(b|x^t) > f_i(\bar{b}|x^t)$.

We now introduce an *encouragement* relation between a flip and its most recent strong supporting flip, if there is one:

Definition 12. *We say that a flip $m(t) = (i \mapsto b)$ is* **encouraged** *by its most recent strong supporting flip $m(t') = (j \mapsto c)$, and write $(t', j \mapsto c) \leftarrow (t, i \mapsto b)$.*

If there are no strong supporting flips, then we say that a flip $m(t) = (i \mapsto b)$ is **courageous,** *and write $\bot \leftarrow (t, i \mapsto b)$.*

Note that if $(t', j \mapsto c) \leftarrow (t, i \mapsto b)$, then $t' < t$ and $i \in N_{\mathcal{C}}(j)$.

For illustration, consider the sequence of moves listed in Eq. 14 of Example 1. It corresponds to the following encouragement graph:

$$\bot \leftarrow (1,1 \mapsto 1) \qquad\qquad \bot \leftarrow (2,2 \mapsto 0) \quad \leftarrow (3,1 \mapsto 0)$$
$$\bot \leftarrow (4,3 \mapsto 1) \leftarrow (5,2 \mapsto 1) \leftarrow (6,1 \mapsto 1)$$
$$\bot \leftarrow (7,4 \mapsto 0) \leftarrow (8,3 \mapsto 0) \leftarrow (9,2 \mapsto 0) \leftarrow (10,1 \mapsto 0)$$

Proposition 4. *If $(t_1, j \mapsto c) \leftarrow (t_2, i \mapsto b)$ (or if $\bot \leftarrow (t_2, i \mapsto b)$, set $t_1 = 0$) then for all $t_1 < t' \le t_2$ we have $f_i(b|x^{t'}) - f_i(\bar{b}|x^{t'}) \ge f_i(b|x^{t_2}) - f_i(\bar{b}|x^{t_2}) > 0$.*

Proof. Define the set of temporary supports S_w as the set of positions of flips after t_1 that supported $(t_2, i \mapsto b)$ but weren't strong (i.e., they were flipped back by the time we got to t_2): for supportive $(t'', k \mapsto a)$ with $t'' > t_1$ we have $k \in S_w \Rightarrow a \ne x^{t_2}[k]$).

Consider any flip $m(t') = (k \mapsto a)$ for $t' \in [t_1 + 1, t_2 - 1]$. Since it either didn't support $(t_2, i \mapsto b)$ (and so had $C_{ij}(b,a) - C_{ij}(b,\bar{a}) \le C_{ij}(\bar{b},a) - C_{ij}(\bar{b},\bar{a})$ by Eq. 15) or was a temporary support, we have that:

$$f_i^{[n]-S_w}(b|x^{t'+1}) - f_i^{[n]-S_w}(\bar{b}|x^{t'+1}) \le f_i^{[n]-S_w}(b|x^{t'}) - f_i^{[n]-S_w}(\bar{b}|x^{t'}) \tag{17}$$

Thus $\delta_i(t') = f_i^{[n]-S_w}(b|x^{t'}) - f_i^{[n]-S_w}(\overline{b}|x^{t'})$ is monotonically non-increasing in t' over the time interval $[t_1 + 1, t_2]$. So:

$$f_i^{[n]-S_w}(b|x^{t'}) - f_i^{[n]-S_w}(\overline{b}|x^{t'}) \geq f_i^{[n]-S_w}(b|x^{t_2}) - f_i^{[n]-S_w}(\overline{b}|x^{t_2}) \qquad (18)$$

Since every position $k \in S_w$ supported $(t_2, i \mapsto b)$ but is absent in x^{t_2}, we must have $f_i^{S_w}(b|x^{t_2}) - f_i^{S_w}(\overline{b}|x^{t_2}) \leq f_i^{S_w}(b|x^{t'}) - f_i^{S_w}(\overline{b}|x^{t'})$. Noting that $f_i = f_i^{[n]-S_w} + f_i^{S_w}$ then lets us combine this with Eq. 18 (and the fact that $f_i(b|x^{t_2}) > f_i(\overline{b}|x^{t_2})$) to complete the proposition. $\qquad \square$

By Definition 12, each flip can only be encouraged by at most one other flip, so each node in the encouragement graph has out-degree at most one. Directed graphs where each vertex has at most one parent are forests, so the encouragement graph is a forest. This forest has a component for each courageous flip, and we will now show that there are at most n of these:

Proposition 5. *At each variable position i, only the first flip can be courageous.*

Proof. Consider a courageous flip $\perp \leftarrow (t, i \mapsto b)$, by Proposition 4, we know that for all $t' < t$: $f_i(b|x^{t'}) - f_i(\overline{b}|x^{t'}) \geq f_i(b|x^t) - f_i(\overline{b}|x^t) > 0$ Thus, there is no time $t' \leq t$ such that i could have flipped to $\overline{b}$: hence i was always at $\overline{b}$ for $t' \leq t$. So the courageous flip had to be the first flip at that position. $\qquad \square$

We will now prove that an encouragement tree cannot double-back on itself in position (Proposition 6), and that every branch is a branch in position (Proposition 7). When the constraint graph is itself a tree, this will imply that each tree in the encouragement forest is a sub-tree of the constraint graph.

Proposition 6. *If $(t_1, i \mapsto a) \leftarrow (t_2, j \mapsto b) \leftarrow (t_3, k \mapsto c)$ then $i \neq k$.*

Proof. Since $(t_1, i \mapsto a)$ strongly supported $(t_2, j \mapsto b)$, we have $x_i^{t_2} = a$. If, for the sake of contradiction, we assume that $i = k$ then $a = c$ (because if we had $c = \overline{a}$ then the two encouragements would force a contradiction via clashing Eq. 15) and by Proposition 4: $f_i(a|x^{t'}) - f_i(\overline{a}|x^{t'}) \geq f_i(a|x^{t_3}) - f_i(\overline{a}|x^{t_3}) \geq 0$ for all $t_2 < t' \leq t_3$. But this means that i cannot be flipped to $\overline{a}$ and thus $m(t_3) = (i, a)$ is not a legal flip. This is a contradiction and so $i \neq k$. $\qquad \square$

Proposition 7. *For all i, j and $t_1 < t_2 \leq t_3$: if $(t_1, i \mapsto a) \leftarrow (t_2, j \mapsto b)$ and $(t_1, i \mapsto a) \leftarrow (t_3, j \mapsto c)$, then $t_2 = t_3$.*

Proof. From Proposition 4, we can see that for all $t' \in [t_1 + 1, t_3]$, $f_j(c|x^{t'}) - f_j(\overline{c}|x^{t'}) > 0$, so $b = c$ and j couldn't have flipped from c to $\overline{c}$ between t_2 and t_3. Thus, for $(t_2, j \mapsto c)$ to be a legal flip, we must have $t_2 = t_3$. $\qquad \square$

Now, if we look along the arrows then each flip in p is the start of a path of encouraged-by links that ends at one of the n courageous flips.

One final case to exclude is that there might be two encouragement paths that go in the opposite direction over the same positions. This cannot happen:

Proposition 8. *Having both of the following encouragement paths is impossible:*

$$\bot \leftarrow (t_1, i_1 \mapsto b_1) \leftarrow (t_2, i_2 \mapsto b_2) \qquad\qquad \leftarrow \cdots \leftarrow \quad (t_m, i_m \mapsto b_m) \quad (19)$$

$$\bot \leftarrow (s_m, i_m \mapsto c_m) \leftarrow (s_{m-1}, i_{m-1} \mapsto c_{m-1}) \quad \leftarrow \cdots \leftarrow \quad (s_1, i_1 \mapsto c_1) \quad (20)$$

Proof. Without loss of generality (by relabeling), we can assume that $t_1 < s_1$. We can extend this with the following claim:

Claim: If $t_k < s_k$ then $t_{k+1} < s_{k+1}$

Since $(t_k, i_k \mapsto b_k) \leftarrow (t_{k+1}, i_{k+1} \mapsto b_{k+1})$, we have, for all $t \in [t_k + 1, t_{k+1}]$, $x^t[i_k] = b_k$. Thus we can't have i_k flipping in that interval, so $s_k > t_{k+1}$.

But now look at $(s_{k+1}, i_{k+1} \mapsto c_{k+1}) \leftarrow (s_k, i_k \mapsto c_k)$. This shows that we also have, for all $t' \in [s_{k+1} + 1, s_k]$, $x^{t'}[i_{k+1}] = c_{k+1}$. So for both flips at i_{k+1} to happen, we need $s_{k+1} > t_{k+1}$.

Applying the claim repeatedly gets us $t_m < s_m$. But this means that i_m flipped before $m(s_m)$, so by Proposition 5 $(s_m, i_m \mapsto c_m)$ could not have been courageous.   □

*Proof (of **Theorem** 6).* Consider any path p in the fitness graph, and its corresponding flip function m. By the completeness of Definition 12, we know that every flip must have been either courageous or encouraged.

Any encouraged flip is the end-point of a unique (non-zero length) encouragement path in the constraint graph starting from some courageous flip (where Proposition 6 established that they're encouragement paths, not walks; and Proposition 7 established that the encouragement paths are uniquely determined by the variable positions that they pass through.) From Proposition 8, we know that there cannot be two encouragement paths that traverse the same positions but in opposite directions. Thus, there can only be as many non-zero-length encouragement paths as undirected paths in our constraint graph. Since our constraint graph is a tree, an undirected non-zero length path is uniquely determined by its pair of endpoints. Thus, there are at most $\binom{n}{2}$ of these paths.

From Proposition 5, there are at most n courageous flips (encouragement paths of length 0). Thus, our path p must have length at most $n + \binom{n}{2}$.   □

6   Long Paths in Landscapes with Simple Constraint Graphs

In this section we show that the conditions in Theorem 6 are essential. We exhibit binary VCSP-instances with very simple constraint graphs where the associated fitness graphs have exponentially-long directed paths.

Example 2 **(Domain size 3).** Consider a binary VCSP-instance $\mathcal{C}$, with variables $x_n, x_{n-1}, \ldots, x_2, x_1, x_0$, and constraints $\{C_{n,n-1}, \ldots, C_{32}, C_{21}, C_{10}\}$ over the uniform domain $D = \{0, 1, \triangleright\}$, where each constraint C_{ij} is represented by the following matrix:

$$C_{ij} = 3^{i-1} \begin{pmatrix} 1 & 2 & 3 \\ 2 & 3 & 1 \\ 3 & 1 & 2 \end{pmatrix}$$

Even though the constraint graph of $\mathcal{C}$ is just a path of length n, we now show the corresponding fitness graph, $G_{\mathcal{C}}$, contains a directed path of exponential length.

Notice that given two natural numbers $M, M' < 2^n$ written in binary as x^M, $x^{M'} \in \{0,1\}^n$ with the least significant digit as x_0, we have that if $M' > M$ then $f(M') > f(M)$. Thus, counting up in binary from 0^{n+1} to 01^n is monotonically increasing in fitness. However, x^{M+1} is often more than a single flip away from x^M (consider the transition from $x^M = 01^n$ for an extreme example). We handle these multi-flip cases with our third domain value, $\triangleright$, as follows: (1) given $x^M = y01^k$ where $y \in \{0,1\}^{n-k}$, we proceed to replace the 1s in the right-most block of 1s by $\triangleright$, starting from x^M_{k-1} and moving to the right; (2) from $y0\triangleright^k$ we can take a 1-flip to $y1\triangleright^k$ (regardless of $y_0 = 0$ or 1); (3) from $x' = y1\triangleright^k$, we replace the $\triangleright$s by 0s, starting from the rightmost $\triangleright$ (i.e., x'_0) and moving to the left.

This lets our sequence of moves count in binary from 0^{n+1} to 01^n, while using extra steps with $\triangleright$s to make sure all transitions are improving 1-flips; thus, this path in the fitness graph has a length greater than 2^n.

Our final example is a binary Boolean VCSP where the constraint graph has tree-width two and maximum degree three, but the associated fitness graph contains an exponentially long directed path. This example is a simplified and corrected version of a similar example for the MAX-CUT problem, described by Monien and Tscheuschner [15]. Note, however, that by allowing general valued constraints, instead of just MAX-CUT constraints, we are able to reduce the required maximum degree from 4 to 3.

Example 3 (**Tree-width 2**). Consider a binary Boolean VCSP-instance $\mathcal{C}$ with $n = 4K + 1$ variables. The constraint graph contains a sequence of disjoint cycles of length four, linked together by a single additional edge joining each consecutive pair of cycles. The final cycle is replaced by a single variable x_n with unary constraint $\begin{pmatrix} 0 \\ -w_K \end{pmatrix}$. Hence the constraint graph of $\mathcal{C}$ has maximum degree three and treewidth two. The i-th cycle (for $0 \le i \le K - 1$) has the following constraints (where the w_i values are defined recursively with $w_0 = 0$):

To begin the long path all variables are assigned 0, except $x_n = 1$. The path will proceed by always flipping variables in the smallest 4-cycle block possible.

Within each 4-cycle block, let us write the 4 variables by decreasing index as $x_{4i+4}x_{4i+3}x_{4i+2}x_{4i+1}$. We will make the following transitions within each cycle: if

$x_{4(i+1)+1} = 1$ then we'll transition $0000 \rightarrow 1000 \rightarrow 1001 \rightarrow 1101$; if $x_{4(i+1)+1} = 0$ then we'll transition $1101 \rightarrow 0101 \rightarrow 0100 \rightarrow 0110 \rightarrow 0010 \rightarrow 0011 \rightarrow 0001 \rightarrow 0000$. Every time that x_{4i+1} is flipped from 0 to 1 or vice versa, we'll recurse to the $(i-1)$th cycle. Because x_{4i+1} ends up flipping from 1 to 0 twice as often as $x_{4(i+1)+1}$, this means that we double the number of flips in each cycle. Variable x_n will flip once, from 1 to 0, due to the unary constraint, which will cause $x_{4(K-1)+1}$ to flip twice from 1 to 0, which will cause $x_{4(K-2)+1}$ to flip four times from 1 to 0, and so on, until eventually this will cause x_1 to flip 2^K times from 1 to 0. Hence we have an improving path of length greater than 2^K.

7 Conclusion

In this paper, we have considered the broad class of fitness landscapes that can be modelled as arising from the combined effect of simple interactions of a few variables, where each of these interactions is described by an arbitrary valued constraint. Modelling fitness landscapes in this way allows us to classify them in new ways: for example by identifying a minimal constraint graph, and then characterising properties of this constraint graph.

We have shown that when a fitness landscape over Boolean variables has a (minimal) constraint graph that is tree-structured, then finding a local optimum by *any* local search algorithm takes only polynomial time. However, over a slightly larger domain, or allowing even slightly more general constraint graphs, we have shown examples where some local search algorithms can take exponential time to find even a local optimum.

Focusing on the maximum length of improving paths in a fitness graph, rather than the run-time of a particular local search algorithm, lets us use our results in settings where the details of the local search algorithm are unknown or highly contingent.

The most notable example of this is in modeling biological evolution. In the context of a model of biological evolution, each variable assignment represents the values of the alleles at a sequence of genetic loci. The constraint graph can then be interpreted as a gene-interaction network. The notion of sign-interaction that is central to Sect. 4 is based on the biological idea of *sign-epistasis* that is central to the analysis of evolutionary dynamics [6,11,17,18]. In such a model, different local search algorithms correspond to the evolutionary dynamics of populations with different sizes and structures [11]. Since the details of these population structures, and thus the precise evolutionary dynamics, are often unknown (or even potentially unknowable in historic cases), it is very helpful to be able to reason over wide classes of local search algorithms, as we do here.

In settings where locally optimal assignments cannot be efficiently found by any local search algorithm, the computational complexity and the combinatorial structure of the fitness graph can be viewed as an ultimate constraint, that prevents evolution from stabilizing at a local fitness peak [11]; such cases will give rise to *open-ended evolution*. By identifying which families of constraint graphs lead to intractable local search problems, we can therefore classify which forms of gene-interaction network enable open-ended evolution.

Beyond the context of biological evolution, we also believe that the tools for classifying fitness landscapes that we have begun to develop here will allow considerable further progress, and may eventually help to shed more light on the question of why local search algorithms can be extremely effective in practice. Another possible research direction is to use the analysis of constraint graphs and encouragement graphs to design more effective local search algorithms.

Acknowledgments. David A. Cohen was supported by Leverhulme Trust Grant RPG-2018-161.

References

1. Aaronson, S.: Lower bounds for local search by quantum arguments. SIAM J. Comput. **35**(4), 804–824 (2006). https://doi.org/10.1137/S0097539704447237
2. Carbonnel, C., Romero, M., Zivny, S.: The complexity of general-valued CSPs seen from the other side. In: 59th IEEE Annual Symposium on Foundations of Computer Science, FOCS 2018, Paris, France, 7–9 October 2018, pp. 236–246 (2018)
3. Chapdelaine, P., Creignou, N.: The complexity of Boolean constraint satisfaction local search problems. Ann. Math. Artif. Intell. **43**(1–4), 51–63 (2005). https://doi.org/10.1007/s10472-004-9419-y
4. Cohen, D.A., Cooper, M.C., Creed, P., Jeavons, P.G., Zivny, S.: An algebraic theory of complexity for discrete optimization. SIAM J. Comput. **42**(5), 1915–1939 (2013)
5. Cooper, M.C., De Givry, S., Schiex, T.: Optimal soft arc consistency. In: Proceedings of the 20th International Joint Conference on Artifical Intelligence, IJCAI 2007, pp. 68–73 (2007)
6. Crona, K., Greene, D., Barlow, M.: The peaks and geometry of fitness landscapes. J. Theor. Biol. **317**, 1–10 (2013)
7. de Visser, J., Park, S., Krug, J.: Exploring the effect of sex on empirical fitness landscapes. Am. Nat. **174**, S15–S30 (2009)
8. Färnqvist, T.: Constraint optimization problems and bounded tree-width revisited. In: Beldiceanu, N., Jussien, N., Pinson, É. (eds.) CPAIOR 2012. LNCS, vol. 7298, pp. 163–179. Springer, Heidelberg (2012). https://doi.org/10.1007/978-3-642-29828-8_11
9. Garey, M., Johnson, D.: Computers and Intractability: A Guide to the Theory of NP-Completeness. Freeman, San Francisco (1979)
10. Johnson, D., Papadimitriou, C., Yannakakis, M.: How easy is local search? J. Comput. Syst. Sci. **37**, 79–100 (1988)
11. Kaznatcheev, A.: Computational complexity as an ultimate constraint on evolution. Genetics **212**(1), 245–265 (2019)
12. Kolmogorov, V., Zivny, S.: The complexity of conservative valued CSPs. J. ACM **60**(2), 10:1–10:38 (2013). https://doi.org/10.1145/2450142.2450146
13. Llewellyn, D.C., Tovey, C.A., Trick, M.A.: Local optimization on graphs. Discrete Appl. Math. **23**(2), 157–178 (1989)
14. Malan, K.M., Engelbrecht, A.P.: A survey of techniques for characterising fitness landscapes and some possible ways forward. Inf. Sci. **241**, 148–163 (2013). http://www.sciencedirect.com/science/article/pii/S0020025513003125
15. Monien, B., Tscheuschner, T.: On the power of nodes of degree four in the local max-cut problem. In: Calamoneri, T., Diaz, J. (eds.) CIAC 2010. LNCS, vol. 6078, pp. 264–275. Springer, Heidelberg (2010). https://doi.org/10.1007/978-3-642-13073-1_24

16. Ochoa, G., Veerapen, N.: Mapping the global structure of TSP fitness landscapes. J. Heuristics **24**(3), 265–294 (2018)
17. Poelwijk, F., Kiviet, D., Weinreich, D., Tans, S.: Empirical fitness landscapes reveal accessible evolutionary paths. Nature **445**, 383–386 (2007)
18. Poelwijk, F., Sorin, T.N., Kiviet, D., Tans, S.: Reciprocal sign epistasis is a necessary condition for multi-peaked fitness landscapes. J. Theor. Biol. **272**, 141–144 (2011)
19. Schaffer, A., Yannakakis, M.: Simple local search problems that are hard to solve. SIAM J. Comput. **20**(1), 56–87 (1991)
20. Tayarani-Najaran, M., Prügel-Bennett, A.: On the landscape of combinatorial optimization problems. IEEE Trans. Evol. Comput. **18**(3), 420–434 (2014). https://doi.org/10.1109/TEVC.2013.2281502
21. Thapper, J., Zivny, S.: Necessary conditions for tractability of valued CSPs. SIAM J. Discrete Math. **29**(4), 2361–2384 (2015). https://doi.org/10.1137/140990346
22. Thapper, J., Zivny, S.: The complexity of finite-valued CSPs. J. ACM **63**(4), 37:1–37:33 (2016). https://doi.org/10.1145/2974019
23. Wright, S.: The roles of mutation, inbreeding, crossbreeding, and selection in evolution. In: Proceedings of the Sixth International Congress on Genetics, pp. 355–366 (1932)

Estimating the Number of Solutions of Cardinality Constraints Through range and roots Decompositions

Giovanni Lo Bianco[1(✉)], Xavier Lorca[2(✉)], and Charlotte Truchet[3(✉)]

[1] IMT Atlantique, Nantes, France
giovanni.lo-bianco@imt-atlantique.fr
[2] ORKID, Centre de Génie Industriel, IMT Mines Albi, Albi, France
xavier.lorca@mines-albi.fr
[3] Université de Nantes, Nantes, France
truchet.charlotte@univ-nantes.fr

Abstract. This paper introduces a systematic approach for estimating the number of solutions of cardinality constraints. A main difficulty of solutions counting on a specific constraint lies in the fact that it is, in general, at least as hard as developing the constraint and its propagators, as it has been shown on `alldifferent` and `gcc` constraints. This paper introduces a probabilistic model to systematically estimate the number of solutions on a large family of cardinality constraints including `alldifferent`, `nvalue`, `atmost`, etc. Our approach is based on their decomposition into `range` and `roots`, and exhibits a general pattern to derive such estimates based on the edge density of the associated variable-value graph. Our theoretical result is finally implemented within the $maxSD$ search heuristic, that aims at exploring first the area where there are likely more solutions.

Keywords: Cardinality constraints · Counting · Random graphs

1 Introduction

Dealing with a combinatorial problem often leads to the natural question of computing or estimating its number of solutions. Such a question arises, for instance, in several works on probabilistic reasoning and machine learning [8,9], or when exploring the structure of the solution space [17]. Counting solutions has indeed been an active research topic in Constraint Programming, in particular on global constraints [13]. Unfortunately, designing an efficient counting algorithm for a specific constraint is as hard as the constraint development itself. Hence, solution counting methods require customized counting algorithms for bounding, or estimating, the number of solutions for each global constraint. We propose here a systematic method to estimate the number of solutions of most of the cardinality constraints.

© Springer Nature Switzerland AG 2019
T. Schiex and S. de Givry (Eds.): CP 2019, LNCS 11802, pp. 317–332, 2019.
https://doi.org/10.1007/978-3-030-30048-7_19

This article focuses on ten of them: `alldifferent`, `nvalue`, `atmostNValues`, `atleastNValues`, `occurrence`, `atmost`, `atleast`, `among`, `uses`, `disjoint`. They all constrain the number of occurrences of certain values or the number of different values in a solution. They can be mathematically modelled with bipartite graphs. In [13], the problem of counting solutions for `alldifferent` and `gcc` is transformed into counting matchings in these graphs. Solving such problems is very hard: they often belong to the #P-complete complexity class. This is why counting-based search, as presented in [13], are not based on exact counting but on estimations or upper bounds. In this article, we introduce a probabilistic approach to compute such an estimation.

In [2], the authors introduce two new global constraints `range` and `roots`, that can be used to specify many cardinality constraints. In other words, for almost every cardinality constraint, there is an equivalent model using only the more primitive `range` and `roots` constraints (and some arithmetic constraints). This equivalent model is called the decomposition of the initial cardinality constraint. We show how to use the `range` and `roots` decomposition for counting solutions. More precisely, we develop a probabilistic approach to estimate the number of solutions on a `range` and on a `roots` constraint and we derive from it a systematic method to estimate the number of solutions on many cardinality constraints. Compared to [13], we obtain an estimation instead of an upper bound, and we propose a method that can be generalized to a large set of cardinality constraints without redesigning a dedicated model.

Outline: The paper is organized as follows. Section 2 gives an introduction to the `range` and `roots` constraints and some materials to understand the associated bipartite graph model. In Sect. 3, we detail how to count exactly the number of solutions on `range` and `roots` and then we apply a probabilistic model to develop an estimation of the true number of solutions. In Sect. 4, we give the `range` and `roots` decomposition and an estimation of the number of solutions for several cardinality constraints, and we synthesize our estimators under a general formula. In Sect. 5, we experiment our probabilistic estimators within the counting-based strategy `maxSD`.

2   Preliminaries : Introduction to range and roots

In all the article, we will use the following notations. Let $X = \{x_1, \ldots, x_n\}$, the set of variables. For each variable $x_i \in X$, we note D_i its domain, $Y = \bigcup_{i=1}^{n} D_i = \{y_1, \ldots, y_m\}$ the union of the domains and $\mathcal{D} = D_1 \times \ldots \times D_n$, the Cartesian product of the domains. We note $d_i = |D_i|$, the size of the domain of x_i. Given a constraint C on variables X, we write $\mathcal{S}_{C(X)}$ the set of solutions of C for X and we write $\#C(X)$ the number of tuples allowed by C for X.

Cardinality constraints restrict the number of occurrences of particular values taken by set of variables, or the number of values or variables meeting some conditions. Among them, we can list `alldifferent`, `gcc`, `nvalue`, `atleast`, `atmost`. We will come back and define properly these constraints one by one in Sect. 4.

Most of the time, these constraints can be modelled with a bipartite graph, in which we are looking for some mathematical structures, such as matchings for example.

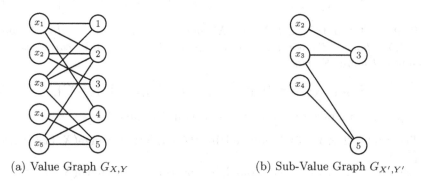

(a) Value Graph $G_{X,Y}$ (b) Sub-Value Graph $G_{X',Y'}$

Fig. 1. Value graph and sub-value graph of Examples 1 and 2.

Definition 1 (Value Graph). *Let* $G_{X,Y} = G(X \cup Y, E)$, *the graph on nodes* $X \cup Y$, *with edges* $E = \{(x_i, y_j) \mid y_j \in D_i\}$. $G_{X,Y}$ *is a bipartite graph representing the domain of each variable. There is an edge between* x_i *and* y_j *iff* $y_j \in D_i$.

Example 1. Let $X = \{x_1, x_2, x_3, x_4, x_5\}$ with $D_1 = \{1, 2, 4\}$, $D_2 = \{2, 3\}$, $D_3 = \{1, 2, 3, 5\}$, $D_4 = \{4, 5\}$ and $D_5 = \{2, 4, 5\}$. We obtain the value graph $G_{X,Y}$ depicted on Fig. 1a.

We also define the sub-value graph induced by two subsets $X' \subseteq X$ and $Y' \subseteq Y$, as the value graph restricted to the considered subset of nodes.

Definition 2 (Sub-Value Graph induced by subsets of X and Y). *Let* $G_{X',Y'} = G(X' \cup Y', E)$, *the value graph of* X' *with* $E = \{(x_i, y_j) \mid y_j \in D'_i = D_i \cap Y'\}$. $G_{X',Y'}$ *is a bipartite graph representing the sub-domain induced by* Y' *of each variable. There is an edge between* x_i *and* y_j *iff* $y_j \in D'_i$.

We will also note $d_i(Y') = |D'_i|$ the size of the domain of x_i restricted to the values of Y'. Example 2 illustrates a sub-value graph of the value graph presented in Example 1.

Example 2. Let $X' = \{x_2, x_3, x_4\} \subseteq X$ and $Y' = \{3, 5\} \subseteq Y$. the sub-value graph induced by X' and Y' is represented in Fig. 1b.

The **range** and **roots** constraints [2] are two auxiliary constraints that can help decomposing a lot of cardinality constraints. In this study, we will use these decomposition to count solutions on cardinality constraints. As the authors wrote in [2], "**range** captures the notion of image of a function and **roots** captures the notion of domain". In this paper, we use alternative definitions for these constraints, equivalent to those of [2] and better suited to our needs.

Definition 3 (range). *Let $X' \subseteq X$ and $Y' \subseteq Y$. The constraint* **range** *(X, X', Y') holds if the values assigned to variables of X' covers **exactly** Y' and not more. Formally:*

$$S_{range(X,X',Y')} = \{(v_1, \ldots, v_n) \in \mathcal{D} \mid \{v_i | x_i \in X'\} = Y'\} \tag{1}$$

Definition 4 (roots). *Let $X' \subseteq X$ and $Y' \subseteq Y$. The constraint* **roots** *(X, X', Y') holds if the variables that are assigned to values of Y' covers **exactly** X' and not more. Formally:*

$$S_{roots(X,X',Y')} = \{(v_1, \ldots, v_n) \in \mathcal{D} \mid \{x_i | v_i \in Y'\} = X'\} \tag{2}$$

Example 3. Let's take the value graph given in Example 1a.

- The tuple $(2, 2, 3, 4, 5)$ is allowed by the constraint **range**$(X, \{x_1, x_2, x_3\}, \{2, 3\})$.
- The tuple $(2, 2, 3, 4, 5)$ is allowed by the constraint **roots**$(X, \{x_1, x_2, x_3\}, \{2, 3\})$.

Note that **range** and **roots** are not exactly reciprocal because every variable must be assigned to a value, but a value is not necessarily assigned to a variable.

3 Counting Solutions on the range and roots Constraints

As developed in [13], counting solutions on cardinality constraints requires dedicated counting algorithm for each constraint. In this section we are interested by computing the number of solutions on the **range** and the **roots** constraints. The idea is then to only use the decomposition of cardinality constraints into these more primitive constraints and to reuse the counting method on **range** and **roots** to count solutions on cardinality constraints.

3.1 Exact Solutions Counting on range and roots

In this subsection, we are interested by exactly computing the number of allowed tuples for a **range** constraint and a **roots** constraint.

Proposition 1. *Let $X' \subseteq X$ and $Y' \subseteq Y$. We note $\overline{X'}$, the complement of X' in X, such that $\overline{X'} \cup X' = X$ and $\overline{X'} \cap X' = \emptyset$. Then, the number of tuples allowed by* **range**(X, X', Y') *is*

$$\#range(X, X', Y') = \#range(X', X', Y') \cdot \prod_{x_i \in \overline{X'}} d_i \tag{3}$$

Proof. On one side, we must consider every possible assignment for the variables of $\overline{X'}$ that are not constrained: $\prod_{x_i \in \overline{X'}} d_i$. And on the other side, we must count every tuples allowed for variables of X', that are constrained, that is simply $\#$**range**(X', X', Y'). The number of tuples is thus the product of these quantities. □

Proposition 1 reduces the problem of counting allowed tuples for every variable in X to only counting tuples for the constrained variables X'. We thus have reduced the problem to counting the number of allowed tuples in the case where every variable and value is constrained.

Proposition 2.

$$\#range(X, X, Y) = \prod_{x_i \in X} d_i - \sum_{Y' \subsetneq Y} \#range(X, X, Y') \tag{4}$$

Proof. Inside $G_{X,Y}$, we must count every possible assignment of variables of X such that every value of Y is covered. To do that, we first count the number of every possible assignment of variables of X in $G_{X,Y}$ (without considering the **range** constraint):

$$\prod_{x_i \in X} d_i$$

And then, we withdraw, one by one, the assignment of X such that Y is not fully covered, that is, for every subset $Y' \subsetneq Y$, the solutions of $\mathbf{range}(X, X, Y')$:

$$\sum_{Y' \subsetneq Y'} \#\mathbf{range}(X, X, Y')$$

Indeed, for two different subsets $Y_1' \neq Y_2' \subsetneq Y$, the sets of allowed tuples $\mathcal{S}_{\mathbf{range}(X,X,Y_1')}$ and $\mathcal{S}_{\mathbf{range}(X,X,Y_2')}$ are necessarily disjoint: there is a value $y_j \in Y$ such that $y_j \in Y_1'$ and $y_j \notin Y_2'$ (or $y_j \in Y_2'$ and $y_j \notin Y_1'$), so the value y_j must be assigned to one of the variable of X to satisfy $\mathbf{range}(X, X, Y_1')$ but none of the variable of X must be assigned to y_j to satisfy $\mathbf{range}(X, X, Y_2')$ (or vice-versa). A solution of $\mathbf{range}(X, X, Y_1')$ cannot be a solution of $\mathbf{range}(X, X, Y_2')$ and vice-versa. No solution are counted twice in $\sum_{Y' \subsetneq Y} \#\mathbf{range}(X, X, Y')$. We have:

$$\#\mathbf{range}(X, X, Y) = \prod_{x_i \in X} d_i - \sum_{Y' \subsetneq Y} \#\mathbf{range}(X, X, Y')$$

$\square$

Remark 1. Proposition 2 can be used in Proposition 1 and we obtain:

$$\#\mathbf{range}(X, X', Y') = \prod_{x_i \in \overline{X'}} d_i \cdot \left(\prod_{x_i \in X'} d_i(Y') - \sum_{Y'' \subsetneq Y'} \#\mathbf{range}(X', X', Y'') \right)$$

This formulae requires to recursively sum and evaluate terms over a exponential-size set and is not tractable in practice (we believe that it is a $\#P-$complete problem). In next subsection, we will give an approximation which is much faster to compute. We now deal with the **roots** constraint.

Proposition 3. *Let $X' \subseteq X$ and $Y' \subseteq Y$. We note $\overline{X'}$, the complement of X' in X and $\overline{Y'}$ the complement of Y' in Y. Then, the number of tuples allowed by* $roots(X, X', Y')$ *is*

$$\#roots(X, X', Y') = \prod_{x_i \in X'} d_i(Y') \cdot \prod_{x_i \in \overline{X'}} d_i(\overline{Y'}) \tag{5}$$

Proof. In order to satisfy $roots(X, X', Y')$, every variable from X' must take a value in Y' and no value from Y' must be assigned to a variable from $\overline{X'}$, that is every variable from $\overline{X'}$ must be assigned to values from $\overline{Y'}$:

- $\prod_{x_i \in X'} d_i(Y')$ represents the number of ways of assigning every variable of X'
- $\prod_{x_i \in \overline{X'}} d_i(\overline{Y'})$ represents the number of ways of assigning every variable of $\overline{X'}$ □

The formula given by Proposition 3 is polynomial to compute. In practice, the formula depends on the subsets X' and Y'. Applying the Erdos-Renyi model on `roots` allows the estimation of $\#roots(X, X', Y')$ using only the sizes of X' and Y', with a linear complexity.

In Sect. 4, we compose these constraints to count solutions on other cardinality constraints.

3.2   Probabilistic Model Applied to range and roots

This subsection presents a probabilistic model for cardinality constraints based on the work of Erdős and Renyi In [5]. The idea is to randomize the domain of the variables. Then, we use this model to get a computable estimation of the number of solutions on `range` and `roots`.

Erdős-Renyi Model Applied to CSP. In [5], Erdős and Renyi studied the existence and the number of perfect matchings on random graphs. Expressed in the vocabulary we introduced above, the idea is to randomize the domain of each variable such that: for all $x_i \in X$ and for all $y_j \in Y$, the event $\{y_j \in D_i\}$ happens with a predefined probability $p \in [0, 1]$ and all such events are **independent**:

$$\mathbb{P}(\{y_j \in D_i\}) = p \in [0, 1] \tag{6}$$

Erdős-Renyi Model Applied to range Constraint. We will study the expectancy of the number of solutions of a `range` constraint within these random graphs. In the case where every variable of X and every value of Y are constrained, the expectancy of $\#range(X, X, Y)$ is a function of n, m and p (as a reminder, $|X| = n$ and $|Y| = m$). More precisely:

Proposition 4. *In the case where every variable of X and every value of Y are constrained, there exists a coefficient $a_{n,m}$ such that:*

$$\mathbb{E}(\#range(X, X, Y)) = a_{n,m} \cdot p^n \tag{7}$$

where $\mathbb{E}\,(\#\boldsymbol{range}(X,X,Y))$ *is the expectancy of* $\#\boldsymbol{range}(X,X,Y)$ *under the hypothesis of the Erdős-Renyi Model.*

Proof. To prove this result, we simply reason with a mathematical induction on $|Y| = m$. Let $|X| = n \in \mathbb{N}$.

Base Case: Let $Y = \{y\}$ be a singleton. In this particular case, an instance $\mathbf{range}(X,X,Y)$ have one allowed tuple, if y is inside every domain D_i, and have zero allowed tuple otherwise. Then,

$$
\begin{aligned}
\mathbb{E}\,(\#\mathbf{range}(X,X,\{y\})) &= 0 * \mathbb{P}\,(\{\mathbf{range}(X,X,\{y\}) \text{ have no solution}\}) \\
&\quad + 1 * \mathbb{P}\,(\{\mathbf{range}(X,X,\{y\}) \text{ have one solution}\}) \\
&= \mathbb{P}\,(\{\mathbf{range}(X,X,\{y\}) \text{ have one solution}\}) \\
&= \mathbb{P}\,(\{\forall x_i \in X, y \in D_i\}) \\
&= \prod_{i=1}^{n} \mathbb{P}\,(\{y \in D_i\})\,, \text{ by hypothesis of independence} \\
&= p^n
\end{aligned}
$$

We thus set $a_{n,1} = 1$, which proves the result.

Inductive Step. We assume that the property is true for all $|Y| = k \in \{1,\dots,m-1\}$: $\forall Y$, such that $1 \le |Y| = k \le m-1, \exists a_{n,k} \in \mathbb{N}$,

$$
\mathbb{E}\,(\#\mathbf{range}(X,X,Y)) = a_{n,k} \cdot p^n.
$$

We want to prove that, under this assumption, for a set Y with $|Y| = m$, there exists $a_{n,m}$ such that $\mathbb{E}\,(\#\mathbf{range}(X,X,Y)) = a_{n,m} \cdot p^n$

According to Proposition 2, we have:

$$
\mathbb{E}\,(\#\mathbf{range}(X,X,Y))
$$

$$
= \mathbb{E}\left(\prod_{x_i \in X} d_i\right) - \sum_{Y' \subset Y} \mathbb{E}\,(\#\mathbf{range}(X,X,Y'))\,, \text{ by linearity of the operator } \mathbb{E}\,(.)
$$

$$
= \mathbb{E}\left(\prod_{x_i \in X} d_i\right) - \sum_{k=1}^{m-1} \binom{m}{k} a_{n,k} \cdot p^n\,, \text{ by hypothesis of induction.}
$$

$$
= \prod_{x_i \in X} \mathbb{E}\,(d_i) - \sum_{k=1}^{m-1} \binom{m}{k} a_{n,k} \cdot p^n\,, \text{ by hypothesis of independence}
$$

$$
= (mp)^n - \sum_{k=1}^{m-1} \binom{m}{k} a_{n,k} \cdot p^n\,, \text{ because } \forall x_i \in X, \mathbb{E}\,(d_i) = mp
$$

$$
= \left(m^n - \sum_{k=1}^{m-1} \binom{m}{k} a_{n,k}\right) \cdot p^n
$$

We have identified the coefficient $a_{n,m}$:

$$a_{n,m} = m^n - \sum_{k=1}^{m-1} \binom{m}{k} a_{n,k} \qquad (8)$$

$\square$

Remarking that $\binom{m}{m} = 1$, we can rewrite 8 as follows:

$$m^n = \sum_{k=1}^{m} \binom{m}{k} a_{n,k} \qquad (9)$$

Also, $\forall n \in \mathbb{N}^+, a_{n,1} = 1$. These coefficients are referenced as the "triangles of numbers" in OEIS.[1] The coefficients $a_{n,m}$ corresponds to the number of possible surjections from a set of cardinal n into a set of cardinal m.[2] There is a non-recursive formula to compute these coefficients. The following results is admitted here. An intuition of the proof is that this results is an application of the inclusion-exclusion principle, see Sect. 1.9. The Twelvefold Way of [18].

Proposition 5. *For $0 < m \leq n$,*

$$a_{n,m} = \sum_{k=0}^{m} (-1)^k \binom{m}{k} (m-k)^n \qquad (10)$$

Proposition 6 is a property of triangle of numbers and will be used to make some simplifications for future mathematical developments.

Proposition 6.

$$a_{n,n} = n! \qquad (11)$$

Proof. $a_{n,n}$ is the number possible surjections from a set of cardinality n into a set of cardinality n, which is actually the number of bijections in that specific case. $\square$

We can now extend Proposition 4 to the case where the **range** constraint only concerns subsets $X' \subseteq X$ and $Y' \subseteq Y$:

Proposition 7. *Let $X' \subseteq X$ and $Y' \subseteq Y$. We note $|X'| = n'$ and $|Y'| = m'$.*

$$\mathbb{E}\left(\#\textbf{range}(X, X', Y')\right) = a_{n',m'} \cdot m^{n-n'} \cdot p^n \qquad (12)$$

---

[1] https://oeis.org/A019538.
[2] $a_{n,m}$ is actually equal to $m! \cdot S_2(n,m)$, where $S_2(n,m)$ is the stirling number of second kind. More information about it can be found in Sect. 1.9 of [18].

Proof. According to Propositions 1 and 4 and by hypothesis of independence:

$$\mathbb{E}\left(\#\mathtt{range}(X, X', Y')\right) = \mathbb{E}\left(\#\mathtt{range}(X', X', Y')\right) \cdot \mathbb{E}\left(\prod_{x_i \in \overline{X'}} d_i\right)$$

$$= a_{n',m'} \cdot p^{n'} \cdot \prod_{x_i \in \overline{X'}} \mathbb{E}\left(d_i\right)$$

$$= a_{n',m'} \cdot p^{n'} \cdot (mp)^{n-n'}$$

$$= a_{n',m'} \cdot m^{n-n'} \cdot p^n$$

$\square$

Erdős-Renyi Model Applied to roots Constraint. We study now the expectancy of the number of solutions of a roots constraint.

Proposition 8. *Let $X' \subseteq X$ and $Y' \subseteq Y$. We note $|X'| = n'$ and $|Y'| = m'$.*

$$\mathbb{E}\left(\#\mathit{roots}(X, X', Y')\right) = m'^{n'} \cdot (m - m')^{n-n'} \cdot p^n \tag{13}$$

Proof. According to Proposition 3 and by hypothesis of independence:

$$\mathbb{E}\left(\#\mathtt{roots}(X, X', Y')\right) = \mathbb{E}\left(\prod_{x_i \in X'} d_i(Y')\right) \cdot \mathbb{E}\left(\prod_{x_i \in \overline{X'}} d_i(\overline{Y'})\right)$$

$$= \prod_{x_i \in X'} \mathbb{E}\left(d_i(Y')\right) \cdot \prod_{x_i \in \overline{X'}} \mathbb{E}\left(d_i(\overline{Y'})\right)$$

$$= (m'p)^{n'} \cdot ((m - m')p)^{n-n'}$$

$$= m'^{n'} \cdot (m - m')^{n-n'} \cdot p^n$$

$\square$

The parameter p corresponds to the density of edges in the value graph. To use the estimators in practice, we need to estimate p: we will later set p to the division of the sum of domains size by the total number of possible edges: $n \cdot m$.

4   Generalization to Cardinality Constraints

This section details, in a systematic way, how to count solutions for many cardinality constraints thanks to their **range** and **roots** decompositions. Due to space limitations, only four constraints are given in detail. For the other six constraints to which our method applies, a synthesis then summarises all the formulae as well as the general computation pattern. Each subsection first recalls the definitions of the considered constraint, then details its decomposition as extracted from [2] and finally provides the formula for the expectancy of its number of solution in our model.

4.1 alldifferent [16]

Definition 5. *A constraint* alldifferent(X) *is satisfied iff each variable* $x_i \in X$ *is instantiated to a value of its domain* D_i *and each value* $y_j \in Y$ *is chosen at most once. We define formally the set of allowed tuples:*

$$S_{alldifferent(X)} = \{(v_1, \ldots, v_n) \in \mathcal{D} \mid \forall i, j \in \{1, \ldots, n\}, i \neq j \Leftrightarrow v_i \neq v_j\} \quad (14)$$

A decomposition of alldifferent with a range constraint is given by the following:

$$\text{alldifferent}(X) \Leftrightarrow \text{range}(X, X, Y') \wedge |Y'| = n$$

From this decomposition, we can deduce a formula for the expectancy of the number solutions on an alldifferent constraint, within the Erdős-Renyi Model.

Proposition 9.

$$\mathbb{E}\left(\#alldifferent(X)\right) = \frac{m!}{(m-n)!} \cdot p^n \quad (15)$$

Proof. According to the decomposition of alldifferent.

$$\#\text{alldifferent}(X) = \sum_{Y' \subseteq Y, |Y'| = n|} \#\text{range}(X, X, Y')$$

Then,

$$\mathbb{E}\left(\#\text{alldifferent}(X)\right) = \sum_{Y' \subseteq Y, |Y'| = n|} \mathbb{E}\left(\#\text{range}(X, X, Y')\right)$$

$$= \binom{m}{n} \cdot a_{n,n} \cdot p^n = \frac{m!}{(m-n)!} \cdot p^n \qquad \square$$

4.2 nvalue [11]

Definition 6. *The constraint* nvalue(X, N) *holds if exactly* N *values from* Y *are assigned to the variables. Formally:*

$$S_{nvalue(X,N)} = \{(v_1, \ldots, v_n) \in \mathcal{D} \mid N = |\{y_j \in Y \mid \exists i \in \{1, \ldots, n\}, v_i = y_j\}|\} \quad (16)$$

A decomposition of nvalue with a range constraint is given by the following:

$$\text{nvalue}(X, N) \Leftrightarrow \text{range}(X, X, Y') \,\&\, |Y'| = N$$

From this decomposition, we can deduce a formula to estimate solutions on a nvalue constraint, within the Erdős-Renyi Model.

Proposition 10. *Let* $N \in \mathbb{N}$,

$$\mathbb{E}\left(\#nvalue(X, N)\right) = \binom{m}{N} \cdot a_{n,N} \cdot p^n \quad (17)$$

Proof. The proof is the same as Proposition 9. $\qquad \square$

We can generalize Proposition 9 to the case where N is a variable. The set of solutions for two different values of N are disjoints, then we can simply sum this estimates on the domain of N to compute an estimate in the general case.

4.3   among [1]

Definition 7 (among). *Let $Y' \subseteq Y$. The constraint $among(X, Y', N)$ holds iff exactly N variables are assigned to value from Y'.*

$$S_{among(X,Y',N)} = \{(v_1, \ldots, v_n) | N = |\{x_i | v_i \in Y'\}|\}$$

The decomposition of among is given by the following equivalence:

$$among(X, Y', N) \Leftrightarrow roots(X, X', Y') \wedge |X'| = N$$

Proposition 11. *Let $m' = |Y'|$ and $N \in \mathbb{N}$,*

$$\mathbb{E}\left(\#among(X, Y', N)\right) = \binom{n}{N} m'^N (m - m')^{n-N} \cdot p^n \qquad (18)$$

Proof. According to the decomposition of among, we can write:

$$\#among(X, Y', N) = \sum_{X' \subseteq X, |X'| = N} \#roots(X, X', Y')$$

Indeed, for two different subsets $X'_1, X'_2 \subseteq X$, the sets of solutions of $roots(X, X'_1, Y')$ and $roots(X, X'_2, Y')$ have an empty intersection, then no solution is counted twice. And:

$$\mathbb{E}\left(\#among(X, Y', N)\right) = \sum_{X' \subseteq X, |X'| = N} \mathbb{E}\left(\#roots(X, X', Y')\right)$$

$$= \sum_{X' \subseteq X, |X'| = N} m'^{m'} (m - m')^{n - |X'|} \cdot p^n, \text{by Proposition 8}$$

$$= \binom{n}{N} m'^{m'} (m - m')^{n-N} \cdot p^n$$

$\square$

In the same way as for nvalue, we can generalize Proposition 11 to the case where N is a variable.

4.4   occurrence [4]

Definition 8 (occurrence). *Let $y \in Y$, the constraint $occurrence(X, y, N)$ holds iff exactly N variables are assigned to value y.*

$$S_{occurrence(X,y,N)} = \{(v_1, \ldots, v_n) | N = |\{x_i | v_i = y\}|\}$$

The decomposition of occurrence is given by the following equivalence:

$$occurrence(X, y, N) \Leftrightarrow roots(X, X', \{y\}) \wedge |X'| = N$$

Table 1. Counting formulae extracted from **range** and **roots** reformulation

Constraint	Formula with $|X| = n$, $|X_1| = n_1$, $|X_2| = n_2$, $|Y| = m$ and $|Y'| = m'$
`alldifferent`(X)	$\frac{m!}{(m-n)!} \cdot p^n$
`among`(X, Y', N)	$\binom{n}{N} m'^N (m - m')^{n-N} \cdot p^n$
`nvalue`(X, N)	$\binom{m}{N} \cdot a_{n,N} \cdot p^n$
`atmostNValues`(X, N)	$\sum_{k=1}^N \binom{m}{k} a_{n,k} \cdot p^n$
`atleastNValues`(X, N)	$\sum_{k=N}^n \binom{m}{k} a_{n,k} \cdot p^n$
`occurrence`(X, y, N)	$\binom{n}{N}(m - 1)^{n-N} \cdot p^n$
`atmost`(X, y, N)	$\sum_{k=1}^N \binom{n}{k}(m - 1)^{n-k} \cdot p^n$
`atleast`(X, y, N)	$\sum_{k=N}^n \binom{n}{k}(m - 1)^{n-k} \cdot p^n$
`uses`(X, X_1, X_2)	$m^{n-n_1-n_2} \cdot \sum_{k=1}^m \binom{m}{k} a_{n_1,k} k^{n_2} \cdot p^n$
`disjoint`(X, X_1, X_2)	$m^{n-n_1-n_2} \cdot \sum_{k=1}^{\min(n_1,m)} \sum_{l=1}^{\min(n_2,m-k)} \binom{m}{k}\binom{m-k}{l} a_{n_1,k} a_{n_2,l} \cdot p^n$

Proposition 12. *Let $N \in \mathbb{N}$,*

$$\mathbb{E}\left(\#\,occurrence(X, y, N)\right) = \binom{n}{N}(m - 1)^{n-N} \cdot p^n \tag{19}$$

Proof. The proof is the same as Proposition 11 in the case where $Y' = \{y\}$ is a singleton. □

Proposition 12 can also be generalized to the case where N is a variable.

4.5   Synthesis

We report the estimators of the number of solutions in Table 1 for several cardinality constraints. We observe a pattern in all these formulae: the estimation of the number of allowed tuples is always p^n multiplied by the number of tuples allowed by the constraint if every domain were equal to the set of values Y (if the value graph were complete). This remark leads to the following Proposition.

Proposition 13. *Let C be a constraint over X with $|X| = n$, Y be the union of the domains and p the edge density in the value graph $G_{X,Y}$, then:*

$$\mathbb{E}\left(\#C\right) = \#C^* \cdot p^n \tag{20}$$

with $\#C^*$ the number of allowed tuples if $G_{X,Y}$ were complete.

Proof. Let $\mathcal{S}_{C^*}$ be the set of allowed tuples if $G_{X,Y}$ were complete. For each $s \in \mathcal{S}_{C^*}$, let Z_s be the random variable such that, $Z_s = 1$ if s is in the set of allowed tuples $\mathcal{S}_C$ of C, and $Z_s = 0$ otherwise. A solution s is an instantiation

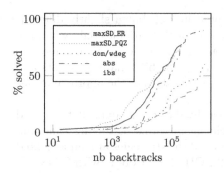 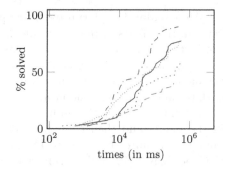

Fig. 2. Performances of `maxSD_ER`, `maxSD_PQZ`, `dom/wdeg`, `ibs` and `abs` on 40 hard Latin Square instances, in number of backtracks (left) and time (right).

of every variable, then, in the Erdős-Renyi Model, $\mathbb{P}(\{Z_s = 1\}) = \mathbb{E}(Z_s) = p^n$. Then,

$$\mathbb{E}(\#C) = \mathbb{E}\left(\sum_{s \in \mathcal{S}_{C^*}} Z_s\right) = \sum_{s \in \mathcal{S}_{C^*}} \mathbb{E}(Z_s) = \#C^* \cdot p^n$$

□

In Sect. 3, we have shown how to count solutions on a **range** and a **roots** constraints and in Sect. 4, how to use the **range** and **roots**/decomposition to estimate the number of solutions on many cardinality constraints. Proposition 13 highlights a general pattern for such estimates. In Sect. 5, we experiment these probabilistic estimators within counting-based heuristics on some problems using cardinality constraints.

5   Experimental Analysis

In this section, we present two problems, on which we have run different heuristics: `maxSD` [13], `dom/wdeg` [3], `abs` (activity-based search) [10] and `ibs` (impact-based search) [15]. This benchmark has been chosen by taking the problems in XSCP, CSPLib, MiniZinc which matched our testing needs: no COP, with cardinality constraints at the core of the problem but no gcc. Also, the lack of knowledge on how to use `maxSD` on problems with several constraints restricts a lot the practical use of the heuristic. These conditions restricted our benchmark to Latin Squares and Sports Tournament Scheduling.

   `maxSD` consists in choosing a pair variable/value based on the estimation of the number of remaining solutions. More precisely, for each constraint, and for each pair variable/value in this constraint, we compute an estimation of the number of remaining allowed tuples and we associated with each pair a solution density. `maxSD` chooses the pair variable/value that maximizes the solution density among every constraint.

We actually do not run `maxSD` as presented in [13], but a slightly different version. It consists in re-computing the ordering of the variables only when the product of the domains size have decreased enough, as suggested in [6]. Here, we set a threshold at 20%. Also, the coefficients $a_{n,m}$, the binomial coefficients and the factorials are computed in advance. The computation of the approximations is thus made in linear time in n.

We first introduce the problem and the cardinality constraints that are used in the model and then compare their efficiency in terms of solving time and number of required backtracks. The instances and the strategies are implemented in Choco solver [14] and we run them on a 2.2 GHz Intel Core i7 with 2.048 GB.

5.1 Latin Square Problem

A Latin Square problem is defined by a $n * n$ grid whose squares each contain an integer from 1 to n such that each integer appears exactly once per row and column [12]. The model uses a matrix of integer variables and an `alldifferent` constraint for each row and each column. We tested on the 40 hard instances used in [13] with n = 30 and 42% of holes (corresponding to the phase transition), generated following [7]. For these instances, we also compare our probabilistic estimator (`maxSD_ER`) with the estimator that is proposed in [13] (`maxSD_PQZ`) for `alldifferent`. We set a time limit to 10 min.

Figures 2 represent the percentage of solved instances in function of the number of required backtracks, and of the solving time. The strategies `maxSD` (for both estimators `maxSD_ER` and `maxSD_PQZ`) and `abs` performed better than `dom/wdeg` and `ibs`. `abs` solved more instances than the two versions of `maxSD`, but required more backtracks. `maxSD` seems to perform better on the easiest instances (in term of number of backtracks). `maxSD_PQZ` has slightly better performances than `maxSD_ER` on the medium instances and have very comparable performances on the hardest ones.

5.2 Sports Tournament Scheduling Problem

This problem is taken from [19] and is presented as follows: the problem is to schedule a tournament of n teams over $n - 1$ weeks, with each week divided into $n/2$ periods, and each period divided into two slots. A tournament must satisfy the following three constraints: every team plays once a week; every team plays at most twice in the same period over the tournament; every team plays every other team. The first and the third constraint are modeled with an `alldifferent` constraint and the second one is modeled with an `atmost` constraints. We run this problem with the different settings: $n \in \{6, 8, 10, 12, 14\}$.

In Table 2, we report the number of backtracks required (and the time required) to solve the problem for different values of n with four different heuristics. Here `maxSD_PQZ` cannot be used as there is no estimator for `atmost` in the previous work of [13]. Consequently, we only focused on our approach `maxSD_ER`. We fixed a time limit to 5 min. We observe that `maxSD_ER` outperforms `abs` and dom/wdeg. For $n \in \{6, 8, 10\}$, maxSD_ER and ibs have similar performances but ibs could not find a solution in less than 5 min for $n = 12$ and $n = 14$.

Table 2. Number of backtracks (time in s) for different settings of n

n	n = 6	n = 8	n = 10	n = 12	n = 14
maxSD_ER	60 (0.239)	**10 (0.707)**	**1056** (3.587)	**74168 (92.396)**	**37883 (128.272)**
ibs	**3 (0.172)**	214 (**0.648**)	1232 (**1.865**)	TO	TO
abs	101 (0.077)	3081 (0.692)	246767 (24.207)	TO	TO
dom/wdeg	89380 (3.829)	TO	TO	TO	TO

We have shown that our probabilistic estimator for alldifferent gives very comparable result than the estimator given in [13] on the Latin Square instances. Also our estimators within maxSD_ER gives better results than ibs, abs and dom/wdeg on the Sport Tournament Scheduling problem.

6   Conclusion

In this paper, we have presented a method to estimate the number of solutions of the range and roots constraints with a probabilistic Erdős-Renyi Model. We can estimate the number of solutions of ten cardinality constraints using their range and roots decompositions. We detailed our method on alldifferent, nvalue, among and occurrence and we report our estimators with atmostNValues, atleastNValues, atmost, atleast, uses and disjoint. We highlighted a general formula to compute such an estimation on cardinality constraints. We have implemented the heuristic maxSD_ER with these new probabilistic estimators and compare their efficiency to dom/wdeg, abs, and ibs.

We think that the main asset of this approach is its systematic nature. We have shown here an application of counting solutions for counting based search. Such an approach could also be used, for example, for uniform random instances generation, probabilistic reasoning or search space structure analysis.

We did not study the gcc constraint in this article, as its decomposition involves several non-disjoint subsets of the variables. Further research includes extending our approach to the case where several range and roots constraints may apply to a common set of variables. This will lead us to estimators of the number of solutions for conjunctions of cardinality constraints, or gcc constraints.

References

1. Beldiceanu, N., Contejean, E.: Introducing global constraints in chip. Math. Comput. Modell. **20**(12), 97–123 (1994). https://doi.org/10.1016/0895-7177(94)90127-9. http://www.sciencedirect.com/science/article/pii/0895717794901279

2. Bessiere, C., Hebrard, E., Hnich, B., Kiziltan, Z., Walsh, T.: Range and roots: two common patterns for specifying and propagating counting and occurrence constraints. Artif. Intell. **173**(11), 1054–1078 (2009). https://doi.org/10.1016/j.artint. 2009.03.001

3. Boussemart, F., Hemery, F., Lecoutre, C., Sais, L.: Boosting systematic search by weighting constraints. In: de Mántaras, R.L., Saitta, L. (eds.) Proceedings of the 16th European Conference on Artificial Intelligence, ECAI 2004, Including Prestigious Applicants of Intelligent Systems, PAIS 2004, Valencia, Spain, 22–27 August 2004, pp. 146–150. IOS Press (2004)

4. Carlsson, M., Fruehwirth, T.: Sicstus PROLOG user's manual 4.3. Books On Demand - Proquest (2014)

5. Erdos, P., Renyi, A.: On random matrices. Publication of the Mathematical Institute of the Hungarian Academy of Science (1963)

6. Gagnon, S., Pesant, G.: Accelerating counting-based search. In: van Hoeve, W.-J. (ed.) CPAIOR 2018. LNCS, vol. 10848, pp. 245–253. Springer, Cham (2018). https://doi.org/10.1007/978-3-319-93031-2_17

7. Gomes, C., Shmoys, D.: Completing quasigroups or latin squares: a structured graph coloring problem, January 2002

8. Gomes, C.P., Hoffmann, J., Sabharwal, A., Selman, B.: From sampling to model counting. In: Proceedings of the 20th International Joint Conference on Artificial Intelligence, IJCAI 2007, Hyderabad, India, 6–12 January 2007, pp. 2293–2299 (2007)

9. Meel, K.S., et al.: Constrained sampling and counting: universal hashing meets SAT solving. CoRR abs/1512.06633 (2015). http://arxiv.org/abs/1512.06633

10. Michel, L., Van Hentenryck, P.: Activity-based search for black-box constraint programming solvers. In: Beldiceanu, N., Jussien, N., Pinson, É. (eds.) CPAIOR 2012. LNCS, vol. 7298, pp. 228–243. Springer, Heidelberg (2012). https://doi.org/10.1007/978-3-642-29828-8_15

11. Pachet, F., Roy, P.: Automatic generation of music programs. In: Jaffar, J. (ed.) CP 1999. LNCS, vol. 1713, pp. 331–345. Springer, Heidelberg (1999). https://doi.org/10.1007/978-3-540-48085-3_24

12. Pesant, G.: CSPLib problem 067: quasigroup completion. http://www.csplib.org/Problems/prob067

13. Pesant, G., Quimper, C., Zanarini, A.: Counting-based search: branching heuristics for constraint satisfaction problems. J. Artif. Intell. Res. **43**, 173–210 (2012). https://doi.org/10.1613/jair.3463

14. Prud'homme, C., Fages, J.G., Lorca, X.: Choco solver documentation. TASC, INRIA Rennes, LINA CNRS UMR 6241, COSLING S.A.S. (2016). http://www.choco-solver.org

15. Refalo, P.: Impact-based search strategies for constraint programming. In: Wallace, M. (ed.) CP 2004. LNCS, vol. 3258, pp. 557–571. Springer, Heidelberg (2004). https://doi.org/10.1007/978-3-540-30201-8_41

16. Régin, J.: A filtering algorithm for constraints of difference in CSPS, pp. 362–367 (1994)

17. Russell, S.J., Norvig, P.: Artificial Intelligence - A Modern Approach, 3rd edn. Pearson Education, London (2010). http://vig.pearsoned.com/store/home/1,1205, store-14563_id-294438,00.html

18. Stanley, R.P.: Enumerative Combinatorics: Volume 1, 2nd edn. Cambridge University Press, Cambridge (2011)

19. Walsh, T.: CSPLib problem 026: sports tournament scheduling. http://www.csplib.org/Problems/prob026

Understanding the Empirical Hardness
of Random Optimisation Problems

Ciaran McCreesh[(✉)] [iD], William Pettersson[iD], and Patrick Prosser[iD]

University of Glasgow, Glasgow, Scotland
{ciaran.mccreesh,william.pettersson,patrick.prosser}@glasgow.ac.uk

Abstract. We look at the empirical complexity of the maximum clique problem, the graph colouring problem, and the maximum satisfiability problem, in randomly generated instances. Although each is NP-hard, we encounter exponential behaviour only with certain choices of instance generation parameters. To explain this, we link the difficulty of optimisation to the difficulty of a small number of decision problems, which are already better-understood through phenomena like phase transitions with associated complexity peaks. However, our results show that individual decision problems can interact in very different ways, leading to different behaviour for each optimisation problem. Finally, we uncover a conflict between anytime and overall behaviour in algorithm design, and discuss the implications for the design of experiments and of search strategies such as variable- and value-ordering heuristics.

1 Introduction

The gap between the best theoretical understanding we have of what makes problems hard and the behaviour witnessed in practice from modern solvers remains vast. For many decision problems in random instances, we have a good general understanding of what happens: as a key parameter is altered, there is often sharp phase transition from satisfiable to unsatisfiable instances, and associated with this is a complexity peak, where instances near the transition are much harder to solve than those far from it on either side [6,16]. (However, this behaviour is not universal—for example, problems involving more than one kind of constraint can exhibit much more complicated behaviour [7,15]).

This paper looks at three optimisation problems: maximum clique, graph colouring, and maximum satisfiability. One view of an optimisation problem is as a sequence of decision problems—but is that all that is needed to understand their behaviour? Previous small-scale experiments [12,17] have only been able to provide an incomplete picture. In this paper we perform experiments on tens of billions of problem instances, which is finally sufficient to comprehensively

This work was supported by the Engineering and Physical Sciences Research Council [grant numbers EP/P026842/1 and EP/P028306/1]. This work used the Cirrus UK National Tier-2 HPC Service at EPCC (http://www.cirrus.ac.uk) funded by the University of Edinburgh and EPSRC (EP/P020267/1).

© Springer Nature Switzerland AG 2019
T. Schiex and S. de Givry (Eds.): CP 2019, LNCS 11802, pp. 333–349, 2019.
https://doi.org/10.1007/978-3-030-30048-7_20

answer the question: yes, there is a link between individual decision and optimisation problems, but these decision problems can interact in many different ways, leading to complex emergent behaviour. Along the way, we uncover interesting implications for the design of search algorithms, and provide lessons for future experimenters. Most interestingly, we identify a trade-off between anytime behaviour and overall behaviour, which could ultimately encourage a rethink of the entire branch and bound paradigm.

1.1   Experimental Setup

Our experiments are performed on the EPCC Cirrus HPC facility, on systems with dual Intel Xeon E5-2695 v4 CPUs and 256 GB RAM, running Centos 7.3.1611, with GCC 7.2.0 as the compiler. These machines are optimised for providing throughput rather than consistent timing measurements, so we avoid measuring runtimes, and instead use whichever natural measure of work each solver provides. Our results therefore do not allow for comparisons between different solvers.

In Sect. 2, we use the Glasgow Subgraph Solver implementation[1] of Prosser's MCSa1 [17]. This is a bit-parallel branch and bound algorithm, which uses a greedy colouring as its bound [20,22]; it can easily be modified to solve the decision problem, rather than the optimisation problem. We measure instance difficulty by counting the number of recursive calls carried out by the algorithm. Later in the section, we also use the MoMC solver [11][2]. MoMC is a more modern branch and bound solver, which incorporates a number of search and inference strategies which are chosen dynamically.

In Sect. 3 we use Trick's implementation[3] of the classic DSATUR branch and bound algorithm [4], and Zhou et al.'s state of the art Color6 solver[4] [24] (which solves only the decision problem). For the Trick solver we measure the number of recursive calls made, whilst for Color6 we measure the number of backtracks. In Sect. 4 we use the Clasp solver[5] version 3.3.4 [8], and we measure the number of decisions made.

2   Maximum Clique

We begin by looking at the maximum clique problem. A clique in a graph is a subset of vertices, each of which is adjacent to every other within the subset, and a maximum clique is one with as many vertices as possible. For random graphs, we use the Erdős-Rényi model: by $G(n, p)$ we mean a graph with n vertices, and

---

[1] https://github.com/ciaranm/glasgow-subgraph-solver.

[2] https://home.mis.u-picardie.fr/~cli/EnglishPage.html. Our experiments uncovered bugs in the published version of this solver—thanks to its authors, our final results use a fixed version of this solver that is not currently publicly available.

[3] https://mat.gsia.cmu.edu/COLOR/color.html.

[4] https://home.mis.u-picardie.fr/~cli/EnglishPage.html.

[5] https://potassco.org/clasp/.

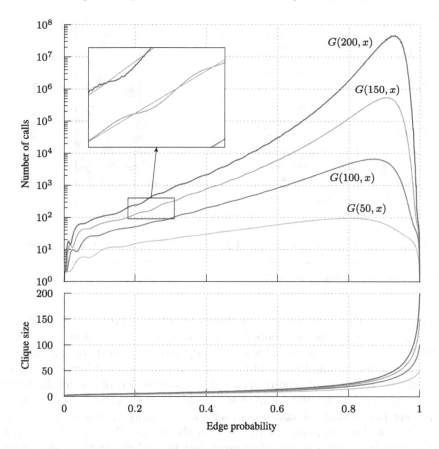

Fig. 1. On top, the difficulty of solving the maximum clique problem in random graphs $G(50, x)$, $G(100, x)$, $G(150, x)$, and $G(200, x)$. Underneath, the mean size of an optimal solution. Density is increased in steps of 0.001 with 100,000 samples per step for the three smaller families, and 1,000 per step for the largest.

an edge between every distinct pair of vertices with probability p. Clique-finding in Erdős-Rényi graphs is known to be exponentially difficult for current clique algorithms [3].

2.1 Maximum Cliques in Random Graphs

In Fig. 1 we show the difficulty of solving the maximum clique problem as we vary the edge probability in Erdős-Rényi graphs with a fixed number of vertices, as well the mean size of an optimal solution. In extremely sparse and extremely dense graphs, the algorithm finds all instances extremely easy, whilst at around densities of 0.8 to 0.96, instances are particularly hard—and unsurprisingly, as the number of vertices increases, all densities get exponentially harder.

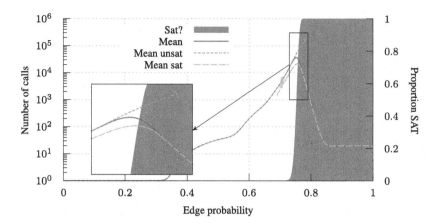

Fig. 2. Does $G(150, x)$ contains a clique of 20 vertices? Mean search effort for satisfiable and unsatisfiable instances are also shown separately. Density is increased in steps of 0.001, with 100,000 samples per step.

These rough trends match up with those presented by Prosser [17]. However, we are using a *much* larger number of instances: we increase density in steps of 0.001, and take 100,000 samples per density step. This scale of experiments reveals a new interesting feature of the plots: the lines are, for lack of a better term, wiggly. This is most readily apparent towards the left of the graph, where several slight peaks and troughs are easily visible by eye, but in fact the wiggles are present throughout the entire plot, with a decreasing "wavelength" as density increases. The remainder of this section shows that these wiggles are not an experimental artifact or sampling error, but instead illustrate an important aspect of the algorithm's behaviour.

2.2   The Clique Decision Problem

To understand what is going on, we first revert to the clique decision problem. In Fig. 2 we ask whether $G(150, x)$ contains a clique of twenty vertices. For very sparse graphs, the answer is obviously no, and the solver can establish this with no search effort. For very dense graphs, the answer is obviously yes, and the solver similarly finds all instances easy. For densities in between 0.691 and 0.782, there is a mix of satisfiable and unsatisfiable instances, but these instances are hard for the solver. For unsatisfiable instances, the higher the density the harder the instance, and the hardest density is 0.780, where all but one of the 100,000 instances sampled are satisfiable. Unexpectedly, for satisfiable instances, we do not get a hard—easy curve, but rather a medium—hard—easy peak, with the hardest density being 0.756 where 62,587 instances were satisfiable. Instances in the "medium" region are extremely rare, however.

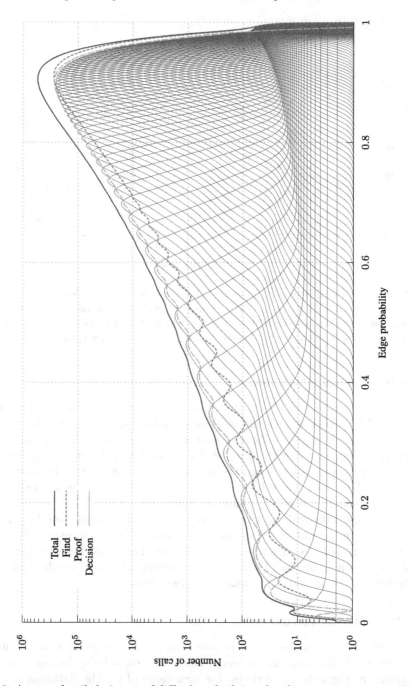

Fig. 3. A more detailed picture of difficulty of solving the clique optimisation problem for $G(150, x)$. Also plotted is the mean search effort to find the optimal solution but not prove its optimality, and the mean search effort needed to prove optimality after the optimal solution is found. Finally, each light line shows the mean search effort for a single decision problem. For each line, density is increased in steps of 0.001, with 100,000 samples per step.

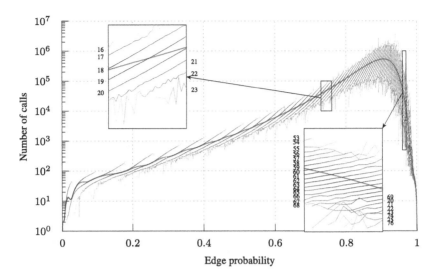

Fig. 4. The mean difficulty of solving the clique optimisation problem for $G(150, x)$, also showing the search effort for each actual optimal size. On each of the individual value lines, darker colours represent exponentially larger sample sizes. Density is increased in steps of 0.001, with 100,000 samples per step. (Color figure online)

2.3 Decision and Optimisation

In Fig. 3 we simultaneously plot the difficulty of every decision problem, and show how this correlates with the total search effort seen in Fig. 1. The "total" line is usually only slightly above whichever decision line is the hardest at a particular density—even at the hardest density of 0.905, the mean gap between the hardest decision problem and the overall cost of solving is only a factor of 2.2. This explains the wiggly lines: they are the result of the gaps between the complexity peaks of different decision problems.

Figure 3 also breaks down the runtimes to show the mean time to find an optimal solution but not prove its optimality, and the time to prove optimality once an optimal solution has already been found. These two lines are perfectly out of phase with each other: densities where finding a solution is relatively easy are the hardest for proving optimality, and vice-versa.

2.4 Difficulty by Actual Solution Size

Another way of grouping results is presented in Fig. 4. Alongside a plot of mean search effort, we also show mean search effort only considering instances where the maximum clique has ω vertices, for each value of ω. The darkness of each line indicates the relative sample size. The plot shows that at any given density, there are several common solution sizes, and the difficulty varies considerably depending upon what the optimal solution size actually is. It also shows that, for any particular maximum clique size ω, there are unusually low densities where occasionally this is the optimum, and these instances are very easy. There are also rare unusually high densities where this is the optimum, and these instances are very hard. Finally, for densities in the middle, instances with solution size ω are common, and are of moderate difficulty. Alternatively, for a given instance, if the maximum clique size is unexpectedly large, the instance will be relatively easy, whilst if it is unexpectedly small, it will be unusually hard.

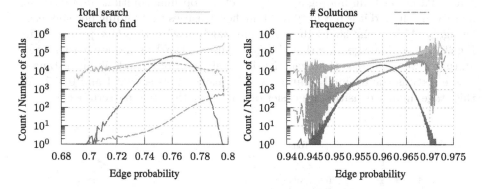

Fig. 5. On the left, looking at only instances where the maximum clique has twenty vertices in $G(150, x)$, and showing the mean search effort, mean time to find but not prove optimality, the frequency of such instances, and the mean number of times a clique of that size occurs in any selected instance. On the right, the same, for a maximum clique of sixty vertices. Density is increased in steps of 0.001 (twenty) or 0.00001 (sixty), with 100,000 samples per step.

2.5   How Common Are Optimal Solutions?

Recall that typically, proving optimality is many times harder than finding an optimal solution. If an instance has an unusually large optimal solution, this should make the proof of optimality much easier. But what about *finding* this unusually large optimum? We might expect that there will only be one optimal solution, if the optimum is unusually large, whilst if the optimum is unusually small, perhaps there are many witnesses to choose from?

In the left-hand plot of Fig. 5 we show that this is the case, looking only at instances where twenty is the optimal solution. We plot the frequency of optimal solutions (how common they are, by density), as well as the effort required to find a first optimal solution but not prove its optimality, and the effort to both find and prove optimality. Finally, we also solve the *maximum* clique enumeration problem, and count how many such optimal solutions exist.

Towards the left of this plot, with densities up to 0.72, instances with a maximum clique size of twenty are rare. Furthermore, the total number of optimal solutions (witnesses) in any given instance is very low, often being one or only a few—and nearly all of our search effort is spent finding the optimal, with optimality proofs being easy. As the density rises, the typical number of optimal solutions per instance also rises, and the time to find but not prove optimality makes up smaller and smaller portions of the overall runtime.

Interestingly, there is not a straightforward inverse relationship between the number of optimal solutions and the amount of time required to find an optimal solution. Rather, finding the unique optimal solution in a lower density graph is somewhat easier than finding any one of several optimal solutions in a medium density graph, and it is not until much higher densities that finding becomes easier again. This is similar to the "medium–hard–easy" complexity peak seen in Fig. 2. One could conjecture that this is because higher densities are harder overall than lower densities. However, the right-hand plot of Fig. 5 looks at instances where sixty is the optimal solution, with densities between 0.94 and 0.975. At this stage, higher densities are easier overall—but the same pattern occurs.

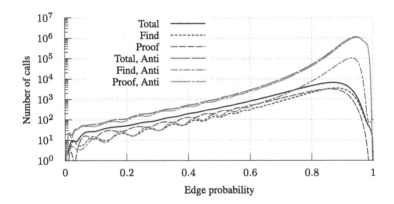

Fig. 6. The mean difficulty of solving the clique optimisation problem for $G(150, x)$, using both the standard search heuristic order for the algorithm, and the opposite search order. For each line, density is increased in steps of 0.001, with 100,000 samples per step.

2.6 Anytime Behaviour

To explain this behaviour, we now demonstrate that the algorithm is in fact not optimised for anytime behaviour, but rather aims to make the proof of optimality as short as possible. McCreesh and Prosser [14] observe that the branching strategy used by this algorithm approximates "smallest domain first" [10], and that (contrary to the claims of the algorithm's designers) it is not good at finding a strong incumbent quickly. So what if we reverse the branching strategy used by the algorithm? Fig. 6 compares the behaviour of the heuristic and the anti-heuristic, showing that the anti-heuristic performs much worse except on the easiest of instances. However, in Fig. 7, we plot the mean size of the first

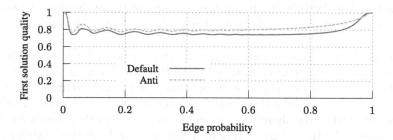

Fig. 7. The mean size of the first solution found expressed as a proportion of the optimal, for $G(100, x)$, using both the standard search heuristic order for the algorithm, and the opposite search order. For each line, density is increased in steps of 0.001, with 100,000 samples per step.

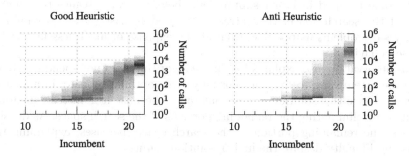

Fig. 8. Comparing the solution quality over time for the two different search orders. We look at instances of $G(100, x)$ where the maximum clique has twenty vertices. Each time a new incumbent is found, we record the search effort so far; each grid point shows the relative frequency of new incumbents of that size during that time window, with darker colours being exponentially more common. We also show termination time, in the final column. Instances drawn from a run with density increased in steps of 0.001, with 100,000 samples per step.

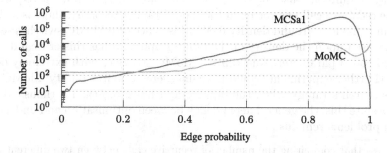

Fig. 9. The difficulty of solving the maximum clique problem in $G(150, x)$ using two different solvers. (This plot does not compare runtimes—the rate of recursive calls per second is much lower in MoMC.)

solution found by both heuristics, as a proportion of the optimal: despite being much worse overall, the anti-heuristic finds a better first solution in nearly all cases.

To understand this seemingly contradictory behaviour, we compare the size of the incumbent as a function of time for the two algorithms. In Fig. 8 we select all the instances of $G(150, x)$ where the optimal solution had twenty vertices, and for both heuristics, record a timepoint for each time the incumbent is improved. We also record when the algorithm terminates, representing this as an incumbent of twenty-one. We then convert this to a heatmap by bucketing, using darker colours to represent exponentially larger buckets. The plot shows us that with the good heuristic, the initial solution size is lower (most commonly fourteen or fifteen) compared to the anti-heuristic (most commonly sixteen to eighteen), but that the anti-heuristic then becomes slower to advance, and slower still to finally prove optimality. This suggests that the anti-heuristic's branching choices cause it to become trapped in larger subproblems before it can advance to a better region of the search space. In contrast, the good heuristic tries to eliminate as many subproblems as possible, even at the expense of much less favourable anytime behaviour.

This observation also explains Fig. 5: the algorithm does not spend nearly all of its time attempting to find an unusually large optimal solution in a sparser instance because this solution is rare, but rather because it is instead spending all of its time eliminating the remaining portions of the search space. As density increases, the remaining portion of the search space increases, explaining the increase in difficulty despite the higher solution counts.

2.7   Solver Independence

What about other solvers? Is what we are seeing merely a quirk of the MCSa1 algorithm, or is it more widespread? In Fig. 9 we repeat parts of Fig. 1, showing the difficulty of solving $G(150, x)$ using the MoMC solver [11].[6] Again, we see wiggles in the curve rather than a smooth straight line, but we also see three other odd features that are not present in the MCSa1 curve. Firstly, MoMC will always require at least 150 recursive calls (and more generally, it requires at least one recursive call per vertex in the input graph). Secondly, there is a sharp change in behaviour around density 0.60—this is because MoMC switches search strategy based upon the density of the input graph, and has this critical density as a hard-coded parameter. And thirdly, MoMC struggles with extremely dense graphs. (Further experiments could have uncovered a fourth oddity: MoMC also switches search strategy when the input graph has more than a thousand vertices). Despite this, the general dependency between optimisation and underlying decision problems remains.

---

[6] We stress that comparing the number of recursive calls between two different algorithms is not a measure of which algorithm is faster—indeed, MoMC performs much more work per recursive call, and on our hardware and on these random instances, is the slower algorithm outwith densities 0.89 to 0.95, despite the lower number of recursive calls.

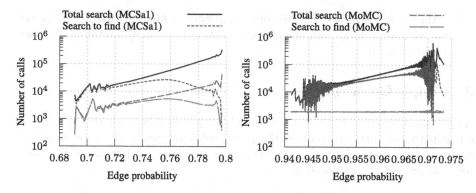

Fig. 10. Repeating Fig. 5 using two solvers. On the left, instances where the optimal solution for $G(150, x)$ has twenty vertices, and on the right, sixty vertices.

In Fig. 10 we repeat parts of Fig. 5, now showing the time to find and the total time for both MCSa1 and MoMC. For solution size twenty instances, the behaviours are remarkably close, despite MoMC using a very different set of search heuristics—in particular, as the number of optimal solutions increases, MoMC initially takes longer to find a witness. For solution size sixty instances, the scaling factor is different, but still MoMC spends nearly all of its time during search having not found an optimal solution, even when witnesses are extremely common.

2.8 Algorithm Design Implications

These results show that clique algorithms have been optimised for proofs of optimality, at the expense of worse anytime behaviour—and also that, if the algorithms were better at finding strong solutions quickly, then their performance would improve considerably on certain instances. It is therefore worth considering whether it is possible to modify these algorithm for both good anytime behaviour, and good overall performance. However, adapting search order heuristics does not appear to help: although doing so can help an algorithm find stronger solutions faster, it then quickly becomes stuck in a subproblem that is hard to eliminate.

Other alternatives may be possible. For example, Maslov et al. [13] apply an iterated local search (ILS) heuristic to generate an initial solution, rather than starting from zero. This technique was also adopted by Tomita et al. [23], who use a different form of local search to prime the incumbent. Both papers describe this as assistance, rather than recognising that their exact algorithms are not optimised for finding strong solutions quickly; both papers also have difficulties selecting a principled amount of time to spend running local search before starting the exact algorithm. Both papers also claim large successes (sometimes being thousands or millions of times faster), particularly on certain families from the standard DIMACS benchmark suite. However, a close inspection of the instances where this happens shows that all come from crafted families that are designed to have unusually large hidden optimal solutions [5,18,19], rather than from application instances.

3   Graph Colouring

Having looked in detail at the maximum clique problem, we now repeat some of our experiments using solvers for the graph colouring problem: we must give a colour to each vertex in a graph, giving adjacent vertices different colours, and using as few colours as possible.

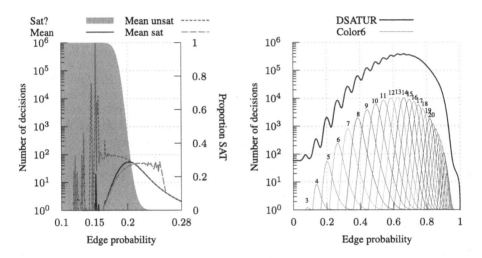

Fig. 11. On the left, the five-colouring phase transition in $G(60, x)$, using the Color6 solver. On the right, the difficulty of each colouring decision problem in $G(60, x)$, using Color6, with outliers removed; the top line is the minimisation problem, using Trick's DSATUR. For each line, density is increased in steps of 0.001, with 100,000 samples per step.

3.1   A Phase Transition, and Outliers

In the left of Fig. 11 we show the difficulty of five-colouring $G(60, x)$, for varying values of x, using the Color6 solver. For densities in between 0.16 and 0.23, we encounter a mix of satisfiable and unsatisfiable instances, and the solver finds the instances more difficult than those outside of this density range. However, at much lower densities, comfortably inside the "satisfiable" region, the mean search effort is extremely variable, and is sometimes far higher than at the complexity peak. Looking more closely at the data shows that for densities between 0.11 and 0.15, between one in ten thousand and one in a hundred thousand instances that we generate are tens of millions of times harder than typical (and this rarity explains why Mann's [12] experiments did not uncover them). Furthermore,

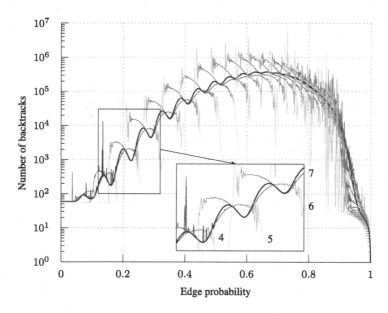

Fig. 12. The difficulty of the minimum colouring problem in $G(60, x)$, using the Trick solver, also showing the means only for instances with each individual optimal solution. Density is increased in steps of 0.001, with 100,000 samples per step.

rather than being entirely satisfiable, these instances are a mix of satisfiable and unsatisfiable. Such instances also occur for other values of the decision problem, although it appears to be even less common as the objective value increases. A similar phenomenon occurs with other random satisfaction problems [1,21], and it could potentially be alleviated by the use of restarts and randomisation [9].

3.2   Branch and Bound

In the right-hand plot of Fig. 11 we show the difficulty of each decision problem together, but exclude these outliers from calculating the means. As for the clique problem, we observe wiggles, with the problem getting easier then harder then easier then harder and so on as we pass successive complexity peaks. The Color6 solver only supports the decision problem. Thus, we also plot the classic DSATUR branch and bound algorithm (whose performance is somewhat worse overall). As with the maximum clique algorithm, the mean complexity line goes from easy to hard to easy over the full range of densities, but this peak has wiggles that line up with the objective values changing.

We also break down the behaviour of the DSATUR solver by optimal solution size: we show this in Fig. 12, in the same style as Fig. 4. Because we are dealing with a minimisation problem, instances that are relatively sparse for their solution size are now found to be harder, rather than easier. And, as with Color6, DSATUR also occasionally finds very sparse instances very hard.

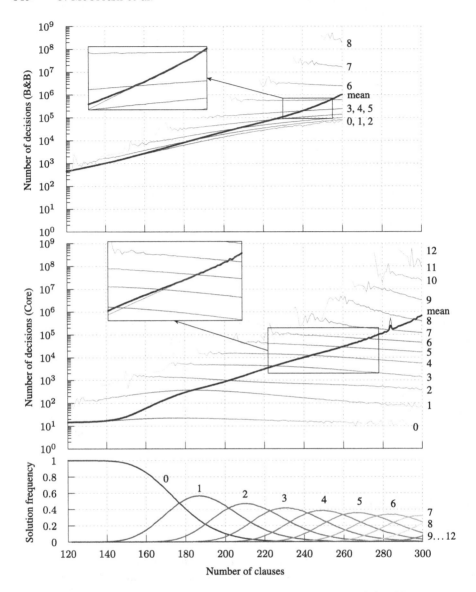

Fig. 13. The difficulty of the maximum 3-satisfiability problem in random instances with 40 variables using Clasp in branch and bound mode (top) and in core-guided mode (middle). The number of clauses is increased in steps of one, and there are 100,000 samples per step; the core-guided plot omits six instances with 290 or more clauses that timed out after one day. Results only for each particular objective value (i.e. the number of unsatisfiable clauses) are also shown as smaller lines in both plots. On the bottom, we show how common each objective value is.

4   Maximum Satisfiability

We finish with a brief look at random maximum satisfiability, or MaxSAT. To generate random MaxSAT instances, we use 40 variables, and a varying number of clauses. Each clause contains three distinct variables chosen uniformly at random, and the polarity of each variable in each clause is also set uniformly at random; all clauses are soft with equal weight. We plot our results in Fig. 13, showing both Clasp's default branch and bound mode, and core-guided optimisation [2], which performs better. Although harder to see, the mean search effort lines for both configuration do exhibit wiggles. Interestingly, the relative difficulty of different instances depends upon the search strategy used—we believe this warrants further experimentation.

5   Discussion and Conclusion

By using very large sample sizes, we have demonstrated that the behaviour of solvers on hard optimisation problems is indeed influenced by the behaviour on individual decision problems—but that these decision problems can interact in many different ways. We also uncovered several interesting phenomena that happened only for one in every ten thousand instances (or even fewer). We therefore encourage future experiments to use similarly large sample sizes if possible, and to consider running many relatively easy experiments instead of a small number of experiments on instances that are as large as possible.

A further advantage of this approach is in uncovering bugs. Indeed, during our experiments, we found that the published version of the MoMC solver produced incorrect results for approximately one in every hundred thousand instances. In fact it was relying upon incorrect reasoning much more frequently than this, but would usually produce the correct answer anyway—the bug only became evident in instances with one or a very small number of witnesses for the optimal clique size, and only if a large combination of events caused the subtree containing these witnesses to be eliminated prematurely and without the witness being found by other means.

Our experiments also uncovered a conflict between designing search order heuristics for anytime behaviour or for overall performance in branch and bound algorithms, which explains why recent exact clique algorithms are using priming with local search algorithms, and which has implications for the design of future solvers. This conflict should also be recognised by experimenters when comparing algorithms in the future—in particular, we would be wary of tables of results that present both "number of instances solved" and "average solution size found" for a single arbitrary choice of timeout.

Acknowledgments. The authors would like to thank Gary Curran, Sonja Kraiczy, David Manlove, Kitty Meeks, and James Trimble for their comments and inspiration.

References

1. Achlioptas, D., Beame, P., Molloy, M.S.O.: A sharp threshold in proof complexity. In: Proceedings on 33rd Annual ACM Symposium on Theory of Computing, 6–8 July 2001, Heraklion, Crete, Greece, pp. 337–346 (2001). https://doi.org/10.1145/380752.380820

2. Andres, B., Kaufmann, B., Matheis, O., Schaub, T.: Unsatisfiability-based optimization in clasp. In: Technical Communications of the 28th International Conference on Logic Programming, ICLP 2012, 4–8 September 2012, Budapest, Hungary, pp. 211–221 (2012). https://doi.org/10.4230/LIPIcs.ICLP.2012.211

3. Atserias, A., Bonacina, I., de Rezende, S.F., Lauria, M., Nordström, J., Razborov, A.A.: Clique is hard on average for regular resolution. In: Proceedings of the 50th Annual ACM SIGACT Symposium on Theory of Computing, STOC 2018, Los Angeles, CA, USA, 25–29 June 2018, pp. 866–877 (2018). https://doi.org/10.1145/3188745.3188856

4. Brélaz, D.: New methods to color vertices of a graph. Commun. ACM **22**(4), 251–256 (1979). https://doi.org/10.1145/359094.359101

5. Brockington, M., Culberson, J.C.: Camouflaging independent sets in quasi-random graphs. In: Cliques, Coloring, and Satisfiability, Proceedings of a DIMACS Workshop, New Brunswick, New Jersey, USA, 11–13 October 1993, pp. 75–88 (1993)

6. Cheeseman, P.C., Kanefsky, B., Taylor, W.M.: Where the really hard problems are. In: Proceedings of the 12th International Joint Conference on Artificial Intelligence, Sydney, Australia, 24–30 August 1991, pp. 331–340 (1991)

7. Dudek, J.M., Meel, K.S., Vardi, M.Y.: The hard problems are almost everywhere for random CNF-XOR formulas. In: Proceedings of the Twenty-Sixth International Joint Conference on Artificial Intelligence, IJCAI 2017, Melbourne, Australia, 19–25 August 2017, pp. 600–606 (2017). https://doi.org/10.24963/ijcai.2017/84

8. Gebser, M., Kaufmann, B., Schaub, T.: Conflict-driven answer set solving: from theory to practice. Artif. Intell. **187**, 52–89 (2012). https://doi.org/10.1016/j.artint.2012.04.001

9. Gomes, C.P., Selman, B., Kautz, H.A.: Boosting combinatorial search through randomization. In: Proceedings of the Fifteenth National Conference on Artificial Intelligence and Tenth Innovative Applications of Artificial Intelligence Conference, AAAI 1998, IAAI 1998, Madison, WI, USA, pp. 431–437 (1998)

10. Haralick, R.M., Elliott, G.L.: Increasing tree search efficiency for constraint satisfaction problems. Artif. Intell. **14**(3), 263–313 (1980). https://doi.org/10.1016/0004-3702(80)90051-X

11. Li, C., Jiang, H., Manyà, F.: On minimization of the number of branches in branch-and-bound algorithms for the maximum clique problem. Comput. OR **84**, 1–15 (2017). https://doi.org/10.1016/j.cor.2017.02.017

12. Mann, Z.Á.: Complexity of coloring random graphs: an experimental study of the hardest region. ACM J. Exp. Algorithmics **23**, 1–3 (2018)

13. Maslov, E., Batsyn, M., Pardalos, P.M.: Speeding up branch and bound algorithms for solving the maximum clique problem. J. Glob. Optim. **59**(1), 1–21 (2014). https://doi.org/10.1007/s10898-013-0075-9

14. McCreesh, C., Prosser, P.: Reducing the branching in a branch and bound algorithm for the maximum clique problem. In: Proceedings of the Principles and Practice of Constraint Programming - 20th International Conference, CP 2014, Lyon, France, 8–12 September 2014, pp. 549–563 (2014). https://doi.org/10.1007/978-3-319-10428-7_40

15. McCreesh, C., Prosser, P., Solnon, C., Trimble, J.: When subgraph isomorphism is really hard, and why this matters for graph databases. J. Artif. Intell. Res. **61**, 723–759 (2018). https://doi.org/10.1613/jair.5768
16. Mitchell, D.G., Selman, B., Levesque, H.J.: Hard and easy distributions of SAT problems. In: Proceedings of the 10th National Conference on Artificial Intelligence, San Jose, CA, USA, 12–16 July 1992, pp. 459–465 (1992)
17. Prosser, P.: Exact algorithms for maximum clique: a computational study. Algorithms **5**(4), 545–587 (2012). https://doi.org/10.3390/a5040545
18. Sanchis, L.A.: Test case construction for the vertex cover problem. In: Proceedings of a DIMACS Workshop on Computational Support for Discrete Mathematics, Piscataway, New Jersey, USA, 12–14 March 1992, pp. 315–326 (1992)
19. Sanchis, L.A.: Generating hard and diverse test sets for NP-hard graph problems. Discrete Appl. Math. **58**(1), 35–66 (1995). https://doi.org/10.1016/0166-218X(93)E0140-T
20. Segundo, P.S., Matía, F., Rodríguez-Losada, D., Hernando, M.: An improved bit parallel exact maximum clique algorithm. Optim. Lett. **7**(3), 467–479 (2013). https://doi.org/10.1007/s11590-011-0431-y
21. Smith, B.M., Grant, S.A.: Modelling exceptionally hard constraint satisfaction problems. In: Smolka, G. (ed.) CP 1997. LNCS, vol. 1330, pp. 182–195. Springer, Heidelberg (1997). https://doi.org/10.1007/BFb0017439
22. Tomita, E., Sutani, Y., Higashi, T., Takahashi, S., Wakatsuki, M.: A simple and faster branch-and-bound algorithm for finding a maximum clique. In: Rahman, M.S., Fujita, S. (eds.) WALCOM 2010. LNCS, vol. 5942, pp. 191–203. Springer, Heidelberg (2010). https://doi.org/10.1007/978-3-642-11440-3_18
23. Tomita, E., Yoshida, K., Hatta, T., Nagao, A., Ito, H., Wakatsuki, M.: A much faster branch-and-bound algorithm for finding a maximum clique. In: Zhu, D., Bereg, S. (eds.) FAW 2016. LNCS, vol. 9711, pp. 215–226. Springer, Cham (2016). https://doi.org/10.1007/978-3-319-39817-4_21
24. Zhou, Z., Li, C.M., Huang, C., Xu, R.: An exact algorithm with learning for the graph coloring problem. Comput. OR **51**, 282–301 (2014). https://doi.org/10.1016/j.cor.2014.05.017

Guarded Constraint Models Define Treewidth Preserving Reductions

David Mitchell[(✉)]

Simon Fraser University, Vancouver, Canada
mitchell@cs.sfu.ca
https://www.cs.sfu.ca/~mitchell

Abstract. Combinatorial problem solving is often carried out by reducing problems to SAT or some other finite domain constraint language. Explicitly defining reductions can be avoided by using so-called "model and solve" systems. In this case the user writes a declarative problem specification in a constraint modelling language, such as MiniZinc. The specification implicitly defines a reduction, which is implemented by the constraint solving system. Unfortunately, reductions can destroy useful instance structure, such has having small treewidth. We show that reductions defined by certain guarded first order formulas preserve bounded treewidth. We also show such reductions can be executed automatically from problem specifications written in a guarded existential second order logic (∃SO) by simple grounding or "flattening" algorithms. Many constraint modelling languages are essentially extensions of ∃SO, and this result applies to natural, useful, fragments of these languages.

Keywords: Constraint modelling language · Reduction · Treewidth

1 Introduction

Application of solvers for finite-domain constraint languages, such as FlatZinc and propositional CNF formulas, requires defining an "encoding", which formally is a reduction from the problem of interest to the target solver language. The exact choice of reduction is important to performance in practice, and considerable time is sometimes spent to find a "good" one. It is often observed that some reductions destroy potentially useful instance structure. The formal study of instance structure in constraint solving goes back at least to Freuder's paper [9] which showed that instances of constraint satisfaction problems (CSPs) having bounded treewidth can be solved in polynomial time. More recently, Samer and Szeider gave a detailed study [19], of conditions under which fixed parameter tractability of CSPs follows from bounded treewidth.

Constraint modelling languages and the solving systems that support them eliminate the need to define a reduction explicitly. Users of these "model and solve" systems write a high level declarative description of their problem, and send that together with a problem instance to the system. Almost all existing

© Springer Nature Switzerland AG 2019
T. Schiex and S. de Givry (Eds.): CP 2019, LNCS 11802, pp. 350–365, 2019.
https://doi.org/10.1007/978-3-030-30048-7_21

systems mapping this pair to a single expression in a "flat" language, which has no quantifiers and limited nesting of operators. For example, the MiniZinc system [16] has several options for this flat language including FlatZinc and propositional CNF formulas. For any problem specification S for a problem P, this map is a reduction (often but not always polynomial time) from problem P to the flat language. This reduction is defined by a combination of the specification and the "flattening" or "grounding" algorithm of the system. The user has some control over this reduction in that they can choose among many possible ways to write S, an activity sometimes called "modelling".

This leads to asking under which conditions the reductions implemented by model-and-solve systems could preserve desirable instance structure. Here we consider the case of treewidth, a widely studied structural measure of "tree-likeness" which has produced many tractability results. In particular, we establish sufficient conditions on S under which an instance I of our problem P is mapped to a CNF formula Γ, such that the treewidth of Γ is bounded by the treewidth of P.

We denote the treewidth of instance I by $\mathrm{TW}(I)$. We say a reduction f between problems is *bounded treewidth preserving* (or just treewidth preserving) if there is a function g, depending only on $\mathrm{TW}(I)$, such that, for every problem instance I, $\mathrm{TW}(f(I)) \leq g(\mathrm{TW}(I))$. We allow the treewidth of the image of I to be larger than that of I, but it must not depend on the size of I. We are interested in when the reduction implemented by a model-and-solve system is treewidth preserving.

To study this question formally, we require a formally defined and sufficiently simple specification. We adopt $\exists\mathrm{SO}$, the existential fragment of classical second order logic, as an abstract constraint modelling language. Many actual constraint modelling languages are essentially extensions of $\exists\mathrm{SO}$ with arithmetic and other features which are convenient for modelling or specifying problems in practice. By Fagin's Theorem [7] $\exists\mathrm{SO}$ can define exactly the problems in the complexity class NP, which seems like a reasonable basis for an initial formal study.

1.1 Contributions

1. We define a family of *guarded reductions*, reductions defined by formulas of first order logic (FO) related to the Packed Fragment of FO, and show that guarded reductions preserve bounded treewidth.
2. We show that, from a "specification" formula Ψ in $\exists\mathrm{SO}$, we can obtain a FO reduction from the NP problem defined by Ψ to SAT.
3. We define a family of guarded $\exists\mathrm{SO}$ formulas, also based on the Packed Fragment of FO. We show that basic grounding or flattening algorithms implement a reduction to SAT that is treewidth preserving when the specification is guarded. More precisely, the indicence treewidth of the CNF formula produced is bounded by a polynomial of the treewidth of the problem instance.
4. We show that from a guarded $\exists\mathrm{SO}$ specification formula Ψ, we can algorithmically obtain an explicit guarded FO reduction from Mod Ψ, the class of models of Ψ, to SAT. Proofs are sketched due to space limitations.

Guarded specifications are very natural, and occur frequently in practice. The essential idea behind the guardedness property is that quantification should be relativized to an input relation. For example, for a problem in which the input is a graph G, a constraint of the form "for every edge e in G ....." is guarded.

The point here not to solve instances of bounded treewidth, but to obtain reductions which apply to all instances and behave well on those with small treewidth. This behaviour is also relevant to instances which are "almost" of small treewidth. A treewidth preserving reduction to SAT does give (in theory) an efficient algorithm for instances of small treewidth. An efficient algorithm for small treewidth instances, in contrast, does not automatically give us a treewidth preserving reduction.

1.2   Organization

Section 2 defines our basic notation regarding structures, associated graphs, and formulas. Sections 3 and 4 are largely expository, giving required background in FO transductions and reductions. Section 5 defines our guarded reductions and gives the proof that guarded reductions preserve bounded treewidth. Section 6 shows how to obtain FO reductions from ∃SO specifications. Section 7 defines guarded specifications and shows that they induce treewidth preserving reductions, and in particular guarded FO reductions. Section 8 concludes with a summary, discussion of related work, etc.

2   Formal Preliminaries

Problems Are Classes of Structures. A decision problem is taken as an isomorphism-closed class of finite relational structures. This view is standard in descriptive complexity theory, and arguably should be used more generally: it is usually more natural to view a graph property as a set of graphs than as a set of strings encoding graphs.

Logic and Notation. We assume the reader is familiar with the syntax and standard model-theoretic semantics of classical logic. In this section we set out our notation and terminology and also give some examples that will aid our exposition later.

A relational vocabulary is a tuple of one or more relation symbols $\bar{R}$. Each symbol R has an arity $ar(R)$. A structure $\mathcal{A}$ for vocabulary τ (or τ-structure), is a tuple $(A, \bar{R}^{\mathcal{A}})$ consisting of a nonempty universe or domain A and a relation $R^{\mathcal{A}} \subseteq A^{ar(R)}$ for each relation symbol $R \in \bar{R}$. The relation $R^{\mathcal{A}}$ is called the interpretation or denotation of R in $\mathcal{A}$. Many authors allow constant symbols in relational vocabularies. Our results would apply also in this case, but we do not include them for simplicity. The size of a structure is the cardinality of its universe. Our structures are all finite, and by default the domain of a size-n structure $\mathcal{A}$ is $A = [n] = \{1, \ldots, n\}$.

Example 1. Let τ_{PL} be the vocabulary $\tau_{PL} = (\mathrm{Form}, \mathrm{SubF}, \mathrm{Atom}, \mathrm{And}, \mathrm{Or}, \mathrm{Not})$ where $\mathrm{Form}, \mathrm{SubF}, \mathrm{Atom}$ are monadic (unary) and $\mathrm{And}, \mathrm{Or}, \mathrm{Not}$ are binary. A τ_{PL}-structure represents a set of formulas of propositional logic: $\mathrm{SubF}(x)$ means x is a subformula; $\mathrm{Form}(x)$ means x is a formula but is not a proper subformula of another (so $\mathrm{Form}^{\mathcal{A}}$ is the set of formulas in $\mathcal{A}$); $\mathrm{Atom}(x)$ means x is an atomic formula; $\mathrm{And}(x, y)$ means x is a conjunction, and y is one of its conjuncts; $\mathrm{Or}(x, y)$ is dual to And; $\mathrm{Not}(x, y)$ means x is the negation of y.

Example 2. The set $\{(q \wedge \neg r), (p \vee \neg t)\}$ of propositional formulas may be represented by τ_{PL}-structure $\mathcal{A}$ where we number subformulas (e.g., by a pre-order traversal of the formula parse trees). So $A = [8]$ and $\mathrm{Form}^{\mathcal{A}} = \{1, 5\}$, $\mathrm{SubF}^{\mathcal{A}} = [8]$; $\mathrm{Atom}^{\mathcal{A}} = \{2, 4, 6, 8\}$; $\mathrm{And}^{\mathcal{A}} = \{(1, 2), (1, 3)\}$; $\mathrm{Or}^{\mathcal{A}} = \{(5, 6), (5, 7)\}$; $\mathrm{Not}^{\mathcal{A}} = \{(3, 4), (7, 8)\}$;

For first order (FO) formula ϕ, we denote by $\mathrm{free}(\phi)$ the set of free FO variables in ϕ, and write $\phi(\bar{x})$ to indicate that the free variables of ϕ are among those in tuple $\bar{x}$. For simplicity we assume all bound variables are distinct. If $\mathcal{A}$ is a τ-structure, $\bar{a} \in A^k$ and $\phi(\bar{x})$ a τ-formula with k free variables $\bar{x}$, we write $\mathcal{A}, \bar{a} \models \phi(\bar{x})$ to say that if the variables $\bar{x}$ in ϕ denote the elements of $\bar{a} \in A^k$ then ϕ is true in $\mathcal{A}$. We write $\phi(\bar{x})^{\mathcal{A}}$, or just $\phi^{\mathcal{A}}$, for the relation defined by ϕ in $\mathcal{A}$. That is, if ϕ has k free variables, $\phi^{\mathcal{A}} = \phi(\bar{x})^{\mathcal{A}} = \{\bar{a} \in A^k \mid \mathcal{A}, \bar{a} \models \phi(\bar{x})\}$. We write $\mathrm{Mod}\ \phi$ for the class of all finite models of a formula ϕ.

Let $\tau = (R_1, \ldots R_m)$ be a vocabulary, $\mathcal{A} = (A, R_1^{\mathcal{A}}, \ldots R_m^{\mathcal{A}})$ a τ-structure, and $(S_1, \ldots S_n)$ a tuple of relation symbols not in τ. If $\mathcal{B} = (A, R_1^{\mathcal{A}}, \ldots R_m^{\mathcal{A}}, S_1^{\mathcal{B}}, \ldots S_n^{\mathcal{B}})$ is a structure for $\tau' = (R_1, \ldots R_m, S_1, \ldots S_n)$, then we call $\mathcal{A}$ the τ-reduct of $\mathcal{B}$ and $\mathcal{B}$ an expansion of $\mathcal{A}$ to τ'.

Graphs and Treewidth of Structures.

Definition 1 (Gaifman graph). *The Gaifman graph of a relational structure $\mathcal{A}$ is the graph $G(\mathcal{A}) = (A, E)$ with vertex set A and $(a, b) \in E$ if and only if there is a tuple in some relation of $\mathcal{A}$ containing both a and b.*

Example 3. The Gaifman graph of $\mathcal{A}$ from Example 2 is $G(\mathcal{A}) = (A, E)$ where $A = \{1, \ldots, 8\}$, and $E = \{(1, 2), (1, 3), (3, 4), (5, 6), (5, 7), (7, 8)\}$. (This is exactly the parse forest for the set of formulas).

The treewidth of a graph or structure is a measure of how "tree-like" it is. Trees have treewidth 1, while the complete graph K_n has treewidth $n - 1$.

Definition 2. (Tree Decomposition; Treewidth).

1. *A tree decomposition of graph $G = (V, E)$ is a labelled tree $T = (U, A, L)$ with (U, A) a tree and L a function from U to 2^V such that*
   (a) For every edge (v, w) of G, there is some $u \in U$ with $v, w \in L(u)$;
   (b) For each vertex v of G, the sub-graph of T induced by the set of tree vertices u with $v \in L(u)$ is connected.
2. *The width of T is 1 less than the maximum cardinality of $L(u)$ for any $u \in U$.*

3. *The treewidth of a graph G, denoted here $\mathrm{TW}(G)$, is the minimum width of any tree decomposition of G.*
4. *The treewidth of a relational structure $\mathcal{A}$ is the treewidth of the Gaifman Graph of $\mathcal{A}$: $\mathrm{TW}(\mathcal{A}) = \mathrm{TW}(G(\mathcal{A}))$.*

Remark 1. Treewidth is often defined as requiring $\cup_{u \in U} L(u) = V$. It is convenient for us to omit this condition, which has no effect on treewidth because if v is independent in G we may have a distinct tree node u with $L(u) = v$.

3   Translation Schemes and Transductions

We use reductions defined by (tuples of) FO formulas. Our terminology approximately follows [13]. In defining translation schemes we allow formulas to contain a finite number of "special" constant symbols not in the vocabulary at hand, which are interpreted as themselves, and are used only for convenience in defining a domain.

Definition 3 (FO Translation Scheme). *Let τ and $\sigma = (R_1, \ldots, R_m)$ be two relational vocabularies and $C = \{c_1, \ldots\}$ a finite set of "special" constant symbols not in τ or σ. Let Φ be a tuple $\Phi = (\phi_0, \phi_1, \ldots, \phi_m)$ of $|\sigma| + 1$ FO formulas over $\tau \cup C$, where the special constants $c_i \in C$ occur only in atoms of the form $x = c_i$. Further, suppose that ϕ_0 has exactly k distinct free variables, and for each relation symbol $R_i \in \sigma$, the number of distinct free variables of the corresponding formula ϕ_i in Φ is exactly $k \cdot \mathrm{ar}(R_i)$. Then Φ is a k-ary τ-σ translation scheme.*

A τ-σ translation scheme Φ defines two functions. One is a (partial) map from τ-structures to σ-structures. This map is often called a *transduction*, generalising the use of the term in formal language theory. The second is a map from σ-formulas to τ-formulas (in model theory called an interpretation of σ in τ) that lets us answer a query about a τ-structure by translating into a query about a σ-structure. A detailed example of a translation scheme will be given in Sect. 4.

Example 4. In the usual interpretation of the complex numbers in the reals we model complex number c with real pair (r_c, i_c), and evaluate a formula over $\mathbb{C}$ by translating it into a formula over $\mathbb{R}$. The related transduction maps $\mathbb{R}$ to $\mathbb{C}$.

Definition 4 (Transduction $\overrightarrow{\Phi}$; Translation $\overleftarrow{\Phi}$). *Let $\sigma = (R_1, \ldots R_m)$ and $\Phi = (\phi_0, \phi_1, \ldots \phi_m)$ be a k-ary τ-σ translation scheme. Then:*

1. *The transduction $\overrightarrow{\Phi}$ is the partial function from τ-structures to σ-structures defined as follows. If $\mathcal{A}$ is a τ-structure and $\phi_0^{\mathcal{A}}$ is not empty, then $\mathcal{B} = \overrightarrow{\Phi}(\mathcal{A})$ is the σ-structure with*
   (a) *universe $B = \{\bar{a} \in (A \cup \{c_i\})^k \mid \mathcal{A}, \bar{a} \models \phi_0\}$;*
   (b) *for each $i \in [1, m]$, $R_i^{\mathcal{B}} = \{\bar{a}_1, \ldots, \bar{a}_i \mid \mathcal{A}, \bar{a}_1, \ldots, \bar{a}_i \models \phi_i\}$*
2. *The translation $\overleftarrow{\Phi}$ is a function from σ-formulas to τ-formulas. We obtain τ-formula $\overleftarrow{\Phi}(\psi)$ from σ-formula ψ by:*

(a) *Replacing each atom $R_i(x_1, \ldots, x_m)$ with $(\wedge_j \phi_0(\bar{x}_j) \wedge \phi_i(\bar{x}_1, \ldots \bar{x}_m))$, where each $\bar{x}_j = (x_{j,1}, \ldots x_{j,k})$ is a k-tuple of new variables;*

(b) *Replacing each existentially quantified subformula $\exists y \psi$ with $\exists \bar{y}(\phi_0(\bar{y}) \wedge \psi)$, where $\bar{y}$ is a k-tuple of new variables. Universally quantified subformulas are relativized in the dual manner.*

If $k > 1$, the universe of $\mathcal{B}$ is a set of tuples of elements of A and $\{c_i\}$, so a τ-formula that defines an r-ary relation in $\mathcal{B}$ has kr free variables. As an aid to reading, we often denote the k-tuples that make up elements of B with $\langle \rangle$.

A fundamental property of translation schemes (standard in expositions of model theory) relates their dual role defining translations and transductions.

Theorem 1 (Fundamental Property of Translation Schemes). *Let Φ be a k-ary τ-σ translation scheme. If $\mathcal{A}$ is a τ-structure for which $\overrightarrow{\Phi}(\mathcal{A})$ is defined, and θ is a σ-formula with r free variables $\bar{x}$, then*

$$\mathcal{A} \models \overleftarrow{\Phi}(\theta)(\bar{y}_1, \ldots \bar{y}_r) \Leftrightarrow \overrightarrow{\Phi}(\mathcal{A}) \models \theta(x_1, \ldots x_r)$$

where $\bar{y}_i$ is the k-tuple of variables corresponding to x_i in the computation of $\overrightarrow{\Phi}$.

4   FO Reductions

FO reductions are poly-time many-one reductions defined by FO transductions. Although they are weak, every problem in NP has a FO reduction to SAT.

Definition 5. *A FO reduction from a class $\mathcal{K}$ of τ-structures to a class $\mathcal{L}$ of σ-structures is a FO τ-σ transduction Φ such that $\mathcal{A} \in \mathcal{K} \Leftrightarrow \overrightarrow{\Phi}(\mathcal{A}) \in \mathcal{L}$.*

Theorem 2 ([12]). *SAT is complete for NP under FO reductions.*

To illustrate we give a translation scheme that defines a FO reduction from Propositional Satisfiability to SAT. This translation scheme may help the reader in understanding the more complex schemes described in Sects. 6 and 7.

The particular reduction is a simplified version of Tseitin's transformation from propositional formulas to CNF [20]. To transform a formula ϕ into a CNF formula of size linear in the size of ϕ, we introduce a new atom for each subformula of ϕ. Then, we write clauses over these new atoms that require the assignments made to these atoms to correspond to the values of their corresponding subformulas when ϕ is evaluated over a satisfying assignment.

Example 5. Applying the Tseitin transformation to the set of formulas $S = \{(q \wedge \neg r), (p \vee \neg t)\}$, yields the set of clauses $\{(x_1), (\neg x_1 \vee x_2), (\neg x_1 \vee x_3), (\neg x_3 \vee \neg x_4), (x_5), (\neg x_5 \vee x_6 \vee x_7), (\neg x_7 \vee \neg x_8)\}$, which is satisfiable if and only if S is. (Here, the numbering of subformulas is the same as used in Example 2).

Let $\tau_{\text{CNF}} = (\text{At}, \text{Cl}, \text{Pos}, \text{Neg})$ be the vocabulary with At, Cl unary and Pos, Neg binary. τ_{CNF}-structures represent propositional CNF formulas, with At the set of atoms, Cl the set of clauses, and $\text{Pos}(a, c)$ (resp. $\text{Neg}(a, c)$) meaning that atom a occurs positively (resp., negatively) in clause c.

Let $\Phi_T = (\phi, \phi_{\text{At}}, \phi_{\text{Cl}}, \phi_{\text{Pos}}, \phi_{\text{Neg}})$, be the translation scheme defined by the following formulas, with special constants {atom,topClause,orClause, andClause, notClause}.

(i) $\phi(\langle s, i \rangle) = [(\text{SubF}(i) \wedge (s = \text{atom})) \vee (\exists j \text{Or}(j, i) \wedge (s = \text{orClause})) \vee (\exists j \text{And}(j, i) \wedge (s = \text{andClause})) \vee (\text{Form}(i) \wedge (s = \text{topClause})) \vee (\text{Form}(i) \wedge (s = \text{notClause}))]$

(ii) $\phi_{\text{At}}(s, i) = [\text{SubF}(i) \wedge (s = \text{atom})]$

(iii) $\phi_{\text{Cl}}(s, i) = [(\exists x \text{Or}(i, x) \wedge (s = \text{orClause})) \vee (\exists x \text{And}(x, i) \wedge (s = \text{andClause})) \vee (\exists x \text{Not}(i, x) \wedge (s = \text{notClause})) \vee (\text{Form}(i) \wedge (s = \text{topClause}))]$

(iv) $\phi_{\text{Neg}}(\langle s, i \rangle, \langle c, j \rangle) = ((s = \text{atom}) \wedge [(\exists x \text{Or}(i, x) \wedge (c = \text{orClause}) \wedge j = i) \vee (\text{And}(i, j) \wedge (c = \text{andClause})) \vee (\text{Not}(j, i) \wedge (c = \text{notClause})) \vee (\exists x \text{Not}(x, i) \wedge j = i \wedge (c = \text{notClause}))])$

(v) $\phi_{\text{Pos}}(\langle s, i \rangle, \langle c, j \rangle) = ((s = \text{atom}) \wedge [(\exists x \text{And}(x, i) \wedge j = i \wedge (c = \text{andClause})) \vee (\text{Or}(i, j) \wedge (c = \text{orClause})) \vee (\text{Form}(i) \wedge i = j \wedge (c = \text{topClause}))])$

Then Φ_T defines a FO transduction that carries out Tseitin's transformation of propositional formulas to CNF formulas. So, $\overrightarrow{\Phi_T}$ is a FO reduction from Propositional Satisfiability to SAT.

Let $\mathcal{A}$ be the τ_{PL}-structure representing a propositional formula ϕ, let $\mathcal{B}$ be the structure $\mathcal{B} = \Phi_T(\mathcal{A})$, and Γ be the CNF formula represented by the τ_{CNF}-structure $\mathcal{B}$. The domain B of $\mathcal{B}$ contains an element $\langle atom, i \rangle$ for each subformula i of ϕ, and these correspond exactly to the atoms in formula Γ. B also contains an element $\langle x, i \rangle$ for each clause in Γ. In each of these elements, the x identifies the role of the clause, and the i identifies the subformula of ϕ that it corresponds to. For example, the domain element $\langle orClause, 3 \rangle$ would be associated with subformula 3 being a disjunction. The roles of clauses are: topClause, notClause, andClause, orClause. In the reduction, we have a single clause for each disjunction, a single clause for each negation, and two clauses for each conjunction in ϕ. The correspondence we use associates the clause for a disjunction or negation with the corresponding subformula, but we associate the two clauses for a conjunction with the two conjuncts. There is also an element for the top clause $\langle topClause, i \rangle$ of each formula i.

Example 6. If $\mathcal{A}$ is the structure of Example 2 then the structure $\mathcal{B} = \overrightarrow{\Phi_T}(\mathcal{A})$ is, in part, as follows. Domain $B = At^{\mathcal{B}} \cup Cl^{\mathcal{B}}$; $At^{\mathcal{B}} = \{\langle atom, 1 \rangle, \langle atom, 2 \rangle, \ldots, \langle atom, 8 \rangle\}$; $Cl^{\mathcal{B}} = \{\langle topClause, 1 \rangle, \langle topClause, 5 \rangle, \langle andClause, 2 \rangle, \langle andClause, 3 \rangle, \langle orClause, 5 \rangle, \langle notClause, 3 \rangle, \langle notClause, 7 \rangle\}$; $Pos^{\mathcal{B}} = \{\langle\langle atom, 1 \rangle, \langle topClause, 1 \rangle\rangle, \langle\langle atom, 5 \rangle, \langle topClause, 5 \rangle\rangle, \ldots\}$; $\ldots$ This represents the CNF formula: $\{(\langle atom, 1 \rangle), (\langle atom, 5 \rangle), (\neg\langle atom, 1 \rangle, \langle atom, 2 \rangle), (\neg\langle atom, 1 \rangle, \langle atom, 3 \rangle), \ldots\}$. Under the map $\langle atom, i \rangle \mapsto x_i$, this is the formula obtained in Example 5.

The well-known meta-theorem tells us that many problems are fixed-parameter tractable for treewidth. MSO is the fragment of second order logic in which second order variables must be monadic.

Theorem 3 (Courcelle [4]). *Every MSO-definable class of structures can be recognized by an algorithm that runs in time $f(w)O(n)$, where n is the size of the encoding of the structure and w the treewidth of the structure, and f is a computable function.*

Example 7. It is straightforward to write an MSO formula, in the vocabulary τ_{CNF}, defining SAT: $\exists S[\forall x(S(x) \rightarrow At(x)) \wedge (\forall y(Cl(y) \rightarrow \exists z((Pos(z,y) \wedge S(z)) \vee (Neg(z,y) \wedge \neg S(z)))))]$. Here, the monadic second order variable S is the set of atoms made true by a satisfying assignment.

Example 8. We can define formula satisfiability with an MSO formula over vocabulary τ_{PL}. Such a formula can be obtained from the formula of Example 7 and the translation scheme Φ_T using Theorem 1.

5   Guarded Reductions

In this section, we define a family of guarded reductions and show that these reductions are treewidth preserving.

Definition 6 (Treewidth Preserving Reduction). *We say a reduction f from $\mathcal{L}$ to $\mathcal{K}$ is bounded treewidth preserving (or just treewidth preserving) if there is a computable function g such that, for every $\mathcal{A} \in \mathcal{L}$, $\mathrm{TW}(f(\mathcal{A})) \leq g(\mathrm{TW}(\mathcal{A}))$.*

5.1   Treewidth of CNF Formulas

Treewidth of CNF formulas is usually defined in terms of on one of two graphs associated CNF formula. The primal graph has a vertex for each atom, and an edge (u, v) iff u and v occur together in a clause (regardless of polarity). The incidence graph has a vertex for each atom and for each clause, and an edge (a, c) if atom a occurs in clause c (with either polarity). These induce two notions of treewidth for CNF formulas, the primal treewidth and incidence treewidth. Since the indcidence treewidth is at most one more than the primal treewidth (and sometimes much smaller), it is the more interesting measure.

For every propositional CNF formula Γ, the Gaifman graph of the τ_{CNF}-structure for Γ is the incidence graph of Γ. Therefore, in this paper the treewidth of a CNF formula means its incidence treewidth. (In some work the primal graph is called the Gaifman graph. This results from a different association of structures with CNF formulas).

An example of a treewidth preserving reduction to SAT is the usual reduction from 3-Colouring. Each vertex is mapped to 3 atoms (one for each colour). For each vertex there is a clause saying it must be coloured, and for each edge three clauses say the ends have distinct colours. A graph of treewidth w is mapped to a CNF formula of treewidth $3w$.

If $P \neq NP$ there are problems in NP which do not have treewidth preserving reductions to SAT. SAT is FPT for treewidth, so any problem with a treewidth preserving reduction to SAT must also be FPT for treewidth. However, there are problems that are NP-complete on trees (e.g. Call Scheduling [6] and Common Embedded Subtree [10]) or on bounded treewidth graphs (e.g. Edge Disjoint Paths [17], and Weighted Colouring [15]), and thus not FPT for treewidth.

5.2   Guarded FO Reductions

The primitive positive formulas are the smallest set of FO formulas containing all atoms, including those of the form $x = y$, and closed under conjunction and existential quantification.

Definition 7. *Let $\phi(\bar{x})$ be a FO formula. A FO formula γ is a* packed guard *for ϕ if it is a primitive positive formula of the form $\gamma_1 \wedge \ldots \wedge \gamma_m$, where each γ_i is either an atom or an existentially quantified atom, such that:*

1. *$free(\gamma) = free(\phi) = \bar{x}$;*
2. *Each pair y, z of distinct variables from $\bar{x}$ appears among the free variables of some γ_i in γ.*

The name is taken from the Packed Fragment of FO, introduced in [14], in which guards are of this form.

Definition 8 (Guarded Translation Scheme, Guarded Reduction). *A FO τ-σ translation scheme Φ is* guarded *if every formula in the scheme, except possibly the domain-defining formula ϕ_0, is a disjunction of formulas of the form $(\gamma(\bar{x}) \wedge \psi(\bar{x}))$, where γ is a packed guard for ψ. A* guarded reduction *is a reduction defined by a guarded translation scheme.*

The guards relativize quantification to the contents of instance relations. They ensure that, if Φ is a guarded translation scheme, every edge of the Gaifman graph of $\mathcal{B} = \overrightarrow{\Phi}(\mathcal{A})$ has a corresponding edge in the Gaifman graph of $\mathcal{A}$. To see this, consider a tuple $\bar{b}$ in a relation of $\mathcal{B}$. If $\bar{b}$ contributes an edge to the Gaifman graph, it contains at least two elements. These elements are constructed from tuples of elements from A. Any two of these elements had to co-occur in the relation defined by one of the atoms of a guard formula, and therefore has a corresponding edge in the Gaifman graph of $\mathcal{A}$.

5.3   Guarded Reductions Preserve Bounded Treewidth

To show guarded reductions preserve bounded treewidth we construct a small-width tree decomposition of $\overrightarrow{\Phi}(\mathcal{A})$ from a small-width decomposition of $\mathcal{A}$.

Theorem 4. *Let Φ be a guarded k-ary τ-σ translation scheme. If $\mathcal{A}$ is a τ-structure with $TW(\mathcal{A}) \leq w$ and $\mathcal{B} = \overrightarrow{\Phi}(\mathcal{A})$ then $TW(\mathcal{B}) \leq (w+1)^k$.*

Proof. Let $\mathcal{B} = \overrightarrow{\Phi}(\mathcal{A})$, and let $T_{\mathcal{A}} = (U, F, L_{\mathcal{A}})$ be a width w tree decomposition of $\mathcal{A}$. We will construct a tree decomposition $T_{\mathcal{B}} = (U, F, L_{\mathcal{B}})$ of $\mathcal{B}$ that is isomorphic to $T_{\mathcal{A}}$, but with a different labelling function (a.k.a. "bag" contents). Let **B** be the set of all tuples occurring in some relation of $\mathcal{B}$. Construct a total function $f : \mathbf{B} \rightarrow U$ as follows. Let $\bar{b} = (b_1, \ldots, b_r)$ be a tuple in a relation of $\mathcal{B}$. Then $\bar{b}$ is of the form $(\langle a_{1,1}, \ldots a_{1,k} \rangle, \ldots \langle a_{r,1}, \ldots a_{r,k} \rangle)$. By construction of the guards of Φ, each pair $(a_{i,j}, a_{i',j'})$ in $\bar{b}$ has a corresponding edge in the Gaifman graph of $\mathcal{A}$. Therefore, there is a clique in $G(\mathcal{A})$ containing all of the elements $a_{i,j}$ in $\bar{b}$. It follows that there must be a vertex $u \in U$ of $T_{\mathcal{A}}$ such that $\{a \mid a \text{ occurs in } \bar{b}\} \subset L_{\mathcal{A}}(u)$. Let $f(\bar{b})$ be such a u. Now define $L_{\mathcal{B}}$ in terms of f by $L_{\mathcal{B}}(u) = \cup\{\bar{b} \mid \bar{b} \in \mathbf{B} \text{ and } f(\bar{b}) = u\}$. The first condition for $T_{\mathcal{B}}$ to be a tree decomposition of $\mathcal{B}$ is now satisfied. We establish the second condition by modifying $T_{\mathcal{B}}$ as follows: If for some $b \in B$ and for two distinct tree nodes u,v we have that $b \in L_{\mathcal{B}}(u)$ and $b \in L_{\mathcal{B}}(v)$ but $b \notin L_{\mathcal{B}}(w)$, we add b to $L_{\mathcal{B}}(w)$. It remains to establish an upper bound on the size of bags (the sets $L_{\mathcal{B}}(u)$). Each element of $b \in B$ is a tuple $b = \langle a_1, \ldots a_k \rangle$ of k elements from A. By construction of $T_{\mathcal{B}}$, if $b \in L_{\mathcal{B}}(u)$ then for each $a_i \in b$, we have that $a_i \in L_{\mathcal{A}}(u)$. The bound is established by the number of elements in $L_{\mathcal{A}}(u)$ and the number of elements $b \in B$ that could be constructed from these. This number is at most $(w + 1)^k$, since an element of B is a k-tuple of elements from $L_A(u)$. $\qquad\square$

Preservation of bounded treewidth, follows immediately.

Theorem 5. *Let Γ be a guarded reduction from $\mathcal{K}$ to $\mathcal{L}$. Then there is a computable function f such that, for every $\mathcal{A} \in \mathcal{K}$*

$$\mathrm{TW}(\Gamma(\mathcal{A})) \leq f(\mathrm{TW}(\mathcal{A}))$$

6 Automatically Generating Reductions

We now consider how specifications induce reductions. We consider a problem specification to be a formula Ψ of the form $\exists \bar{R}\psi$, where $\bar{R}$ is a tuple of second order variables, and ψ is a FO sentence. If the problem defined is a class of τ-structures, then Ψ is a τ-formula, and ψ is a formula with vocabulary $(\tau, \bar{R})$. The decision problem is: given a τ-structure $\mathcal{A}$, decide if $\mathcal{A} \models \Psi$. The associated search problem is to find a witness for the existential SO variables, or, equivalently, to find a $\tau \cup \bar{R}$-structure $\mathcal{B} = (\mathcal{A}, \bar{\mathcal{R}}^{\mathcal{B}})$ that is an expansion of $\mathcal{A}$ to the vocabulary of ψ, and such that $\mathcal{B} \models \psi$.

We may regard $\Psi = \exists \bar{R}\psi$ as implicitly defining a reduction to SAT, based on the following four-step construction:

1. Given $\mathcal{A}$, construct a quantifier-free formula from ψ by rewriting each quantified subformula as a large conjunction or disjunction over elements of A;
2. Transform the resulting ground formula to CNF by Tseitin's method;
3. Evaluate any atoms over the vocabulary τ of $\mathcal{A}$, deleting any clauses that evaluate to true;
4. Replace each distinct ground FO atom with a distinct propositional atom.

For Step 1, we must introduce new constant symbols to denote elements of A in the ground formula. For Step 2, we must introduce new atoms corresponding to subformulas, to be used in the Tseitin construction. For the purpose of associating tree decompositions of the final propositional CNF formula with tree decompositions of $\mathcal{A}$, we will construct these "Tseitin" atoms as ground FO atoms using new relation symbols. Roughly speaking, the result is a ground formula Γ such that models of Γ correspond to the expansions of $\mathcal{A}$ that witness the SO existentials in Ψ.

We write $\tilde{A}$ for the set of (new) constant symbols $\tilde{A} = \{\tilde{a}|a \in A\}$. Then, the first step of our construction is defined by the following recursive function $\Gamma(\phi, \nu, A)$. Here ϕ is a FO formula, ν is a partial map from variables to domain elements, $\nu\langle x \to a\rangle$ is the valuation just like ν except that it maps x to a, and A is the domain.

$$\Gamma(\phi, \nu, A) = \begin{cases} \phi(\nu(\bar{x})) & \text{if } \phi \text{ is an atom } \phi(\bar{x}) \\ (\Gamma(\psi_1, \nu, A) \vee \Gamma(\psi_2, \nu, A)) & \text{if } \phi \text{ is } (\psi_1 \vee \psi_2) \\ (\Gamma(\psi_1, \nu, A) \wedge \Gamma(\psi_2, \nu, A)) & \text{if } \phi \text{ is } (\psi_1 \wedge \psi_2) \\ \neg\Gamma(\psi, \nu, A) & \text{if } \phi \text{ is } \neg\psi \\ (\bigwedge_{a \in A} \Gamma(\psi, \nu\langle x \to \tilde{a}\rangle, A)) & \text{if } \phi \text{ is } \forall x\psi \\ (\bigvee_{a \in A} \Gamma(\psi, \nu\langle x \to \tilde{a}\rangle, A)) & \text{if } \phi \text{ is } \exists x\psi \end{cases}$$

For step 2, associate with each non-atomic subformula η of ψ a new relation symbol P_η, with arity $|\text{free}(\eta)|$. Associate with each non-atomic subformula β of $\Gamma(\psi, A) = \Gamma(\psi, \emptyset, A)$ a ground atom $P\bar{a}$, where P is a relation symbol associated to the corresponding subformula of ψ, and $\bar{a} = \nu(\text{free}(\psi))$ with ν the substitution used in evaluating $\Gamma(\psi, \nu, A)$ in constructing $\Gamma(\psi, A)$. Now, apply Tseitin's reduction to CNF to the formula $\Gamma(\psi, A)$, using the atoms just defined as the Tseitin atoms corresponding to the non-atomic subformulas. Denote the resulting formula $\beta = \text{CNF}(\Gamma(\psi, A))$.

β is a ground FO formula in CNF form, over atoms with constant symbols from $\tilde{A}$, with each relation symbol either a vocabulary symbol of Ψ, a second order variable symbol of Ψ, or a Tseitin symbol as just introduced. Step 3 is to eliminate atoms over τ by replacing them with their truth values determined by $\mathcal{A}$. More precisely, we delete each clause that contains a true atom and delete all false atoms from remaining clauses.

Definition 9 (Grounding). *Let Ψ be a $\exists SO$ τ-formula $\exists\bar{R}\psi$ where ψ is a FO sentence. Let $\mathcal{A}$ be a τ-structure. We call a formula Γ a grounding of Ψ over $\mathcal{A}$ if it satisfies the following properties:*

1. *Γ is a ground formula for a vocabulary σ that includes $\tau, \bar{R}, \tilde{A}$;*
2. *If $\mathcal{A} \models \Psi$ then there is an expansion $\mathcal{B}$ of $\tilde{A}$ to σ such that $\mathcal{B} \models \Gamma$;*
3. *If $\mathcal{B}$ is a σ-structure that is an expansion of $(A, \tilde{A})$ and $\mathcal{B} \models \Gamma$ and $\mathcal{C}$ is the τ-reduct of $\mathcal{B}$, then $\mathcal{C} \models \Psi$.*

So $\Gamma = \text{CNF}(\Gamma(\psi, A))$ is a grounding. Of particular interest are certain proper subsets of $\text{CNF}(\Gamma(\psi, A))$ that also are groundings.

Definition 10 (Direct Grounding). *We call a formula Ψ a direct grounding of ψ over $\mathcal{A}$ if it satisfies the following:*

1. Ψ is a grounding of ψ over $\mathcal{A}$
2. For every clause C of Ψ, there is a clause C' of $\mathrm{CNF}(\Gamma(\psi, A))$ with $C \subseteq C'$.

Any subformula of ψ that contains no symbols from $\bar{R}$ can be directly evaluated over $\mathcal{A}$. This can be used to reduce the number of clauses included in a grounding of Ψ over $\mathcal{A}$. In particular, consider subformula $\psi(\bar{x})$ of ψ of the form $\psi_1(\bar{x}) \wedge \psi_2(\bar{x})$, and suppose that ψ_2 contains symbols from $\bar{R}$ but ψ_1 does not. Then, for each substitution ν for which $\psi_1(\nu(\bar{x}))$ evaluates to false, the clauses corresponding to $\Gamma(\psi_2, \nu, A)$ may be left out of Γ, and it will still be a direct grounding. A dual property holds for disjunctive subformulas. We say that a grounding that leaves out such clauses satisfies the *lazy generation property*. Practical grounding software has this property. In a direct recursive implementation of $\Gamma(\phi, \nu, A)$, lazy generation amounts to little more than lazy evaluation.

6.1 Direct CNF Grounding as a FO Transduction

We wish to show that, from a $\exists$SO problem specification $\Psi = \exists \bar{R} \psi$, we can (algorithmically) construct a FO reduction Δ from the class of models of Ψ to SAT. The image of $\mathcal{A}$ under Δ is a structure for vocabulary $\tau_{CNF} = (\mathrm{At}, \mathrm{Cl}, \mathrm{Pos}, \mathrm{Neg})$. So, the formulas of Δ will have certain elements in common with those of our transduction Φ_T from Sect. 4.

As before, the domain B of $\mathcal{B} = \Delta(\mathcal{A})$ has elements corresponding to atoms and clauses of the resultant CNF formula. It needs an element identified with each ground atom $P\bar{a}$, where P is an element of $\bar{R}$ or a Tseitin relation symbol corresponding to a subformula of Ψ, and $\bar{a}$ is a tuple of elements of A. Let r be the maximum number of free variables in a subformula of Ψ. Then our domain elements will be $k = r + 2$-tuples $\langle P, \bar{a}, C \rangle$, where P is a special constant symbol denoting a relation symbol, C is a special constant symbol denoting an atom or clause-type from {atom,topClause, orClause,andClause,notClause}, and $\bar{a}$ is a tuple from $(A \cup \{\_\})^w$. The special constant $\_$ is a place-holder letting us model a tuple $\bar{a}$ of arity less than w with a w-tuple.

For each subformula of ψ that is a disjunction, our ground CNF formula has one ternary clause for every instantiation of the free variables. For example, the disjunction $\phi(\bar{x}) = (\phi_1(\bar{x}) \vee \phi_2(\bar{x}))$ contributes a clause of the form $(\neg P_\phi \bar{a} \vee P_{\phi_1} \bar{a} \vee P_{\phi_2} \bar{a})$ for each instantiation $\bar{a}$ of its free variables. Such a clause contributes three pairs to relations of $\mathcal{B}$: one in $Neg^{\mathcal{B}}$ and two in $Pos^{\mathcal{B}}$, as in the propositional case. Supposing $\bar{x}$ to be of size 2, the two in Pos, for all instantiations, can be defined by a formula

$$\alpha(\langle p, x_1, \ldots, x_k, c_1 \rangle, \langle p, x_1, \ldots, x_k, c_2 \rangle) =$$
$$(p = P_\phi \ \wedge (x_1 = x_1) \wedge (x_2 = x_2) \wedge (x_3 = \_) \wedge \ldots \wedge (x_k = \_)$$
$$\wedge (c_1 = \mathrm{atom}) \wedge (c_2 = \mathrm{orClause})) \quad (1)$$

Similarly, for each subformula of the form $\exists x \phi$, we have $|A| + 1$ pairs in $\mathcal{B}$, one in $Neg^{\mathcal{B}}$ and the rest in $Pos^{\mathcal{B}}$. We proceed similarly for all connectives and for the atoms. The formula ϕ_P of Δ, that defines the relation $Pos^{\mathcal{B}}$ is defined by the disjunction of all formulas defining particular subsets of $Pos^{\mathcal{B}}$, and similarly for $Neg^{\mathcal{B}}$. From the complete construction, we obtain the following.

Theorem 6. *For every $\exists SO$ formula Ψ, there is a FO transduction Δ that is a reduction from Mod Ψ to SAT. In particular, $\Delta(\mathcal{A})$ is a direct grounding of Ψ over $\mathcal{A}$.*

7   Guarded Existential SO Specifications

As in other guarded logics, we assume FO quantifiers apply to blocks of variables, and that every formula of the form $\exists x \exists y \phi$ has been re-written as $\exists xy\, \phi$, and similarly for $\forall$.

Definition 11 (Guarded $\exists SO$). *Call an $\exists SO$ formula $\Psi = \exists \bar{R} \psi$ guarded if*

1. *In every subformula that is of the form $\exists \bar{x} \phi$ and that contains a non-monadic symbol from $\bar{R}$, ϕ is of the form $(\gamma(\bar{y}) \wedge \phi'(\bar{y}))$, where γ is a packed guard for ϕ', and $\bar{y} \supset \bar{x}$.*
2. *In every subformula that is of the form $\forall \bar{x} \phi$ and that contains a non-monadic symbol from $\bar{R}$, ϕ is of the form $(\gamma(\bar{y}) \rightarrow \phi'(\bar{y}))$, where γ is a packed guard for ϕ', and $\bar{y} \supset \bar{x}$.*

In practice, guarded specifications are very natural. For example, consider the Vertex Cover problem, in which we are given a graph $G = (V, E)$ and must find a set S (normally with some size bound) containing at least one end point of each edge. This property is naturally described with the guarded formula

$$\forall u, v (Euv \rightarrow (Su \vee Sv)).$$

However, for the Domatic Partition problem, which calls for a partition of vertices into sets $P_1, \ldots, P_k$, it does not seem that there is a guarded version (the only possible guard being E) of the property:

$$\forall v \exists i (Pvi \wedge \forall j (Pvj \rightarrow j = i)).$$

Theorem 7. *Let Ψ be a guarded $\exists SO$ formula with vocabulary τ, and Δ a reduction that implements a direct grounding of Ψ over $\mathcal{A}$ that satisfies the lazy evaluation property. Then for any τ-structure $\mathcal{A}$, we have that*

$$TW(\Delta(\mathcal{A})) \leq f(TW(\mathcal{A}))$$

where $f(x) = O((x + 1)^r)$, with r determined by Ψ.

The proof is quite similar to that of Theorem 4.

Proof (Sketch). Let T be $T_{\mathcal{A}} = (U, F, L_{\mathcal{A}})$ be a width w tree decomposition of $\mathcal{A}$, and Γ be the formula $\Gamma = CNF(\Gamma(\Psi, \mathcal{A}))$. We will construct a tree decomposition $T_{\mathcal{B}}$ for Γ that is isomorphic to $T_{\mathcal{A}}$, but has different bags. Consider any existentially quantified subformula $\psi'(\bar{y})$ of Ψ that has a symbol from $\bar{R}$. This formula is of the form $\exists x (\gamma(\bar{y}, x) \wedge \psi(\bar{y}, x))$, where γ is a packed guard for ψ. By the lazy generation property, no clauses corresponding to $\gamma(\nu(\bar{y}, x))$ are included in Γ, and for each tuple $\bar{a} = \nu(\bar{y}, x)$, clauses corresponding to $\psi(\nu(\bar{y}, x))$ are only included in Γ if $\Gamma(\nu(\bar{y}, x))$ evaluates to true. Consider the formula $\psi(\nu(\bar{y}, x))$. If it is quantifier-free, then all atoms in corresponding clauses of Γ are of the form $P\bar{a}$ where P is either in $\bar{R}$ or a Tseitin symbol corresponding to a subformula of ψ. Since $\bar{a}$ was "sanctioned" by the guard, we know that there is a bag in $T_{\mathcal{A}}$ containing all elements of $\bar{a}$. We put all the atoms from all the clauses into the corresponding bag in $T_{\mathcal{B}}$. If ψ has quantified subformulas, then we include as part of the current step the Tseitin atoms corresponding to those subformulas, but not the clauses corresponding to them. Following this, we have that the first of the tree decomposition properties is satisfied by $T_{\mathcal{B}}$. As in the proof of Theorem 4, we add to each bag the minimum collection of atoms to satisfy the second property. It then remains to bound the size of the bags. As in the proof of Theorem 4, it is sufficient to bound the number of elements in a bag of $T_{\mathcal{B}}$ in terms of the number in a related bag of $T_{\mathcal{A}}$. From $w + 1$ elements of A, in a bag $L_A(u)$ we can construct $\binom{w+1}{r}$ tuples. The number of relation symbols is bounded by $|\Psi|$, and there are 5 special constants (which were used to distinguish the syntactic types of subformulas) for the last element, so the number of elements in a $L_B(u)$ is at most

$$5|\Psi| \binom{w + 1}{r} = O((w + 1)^r)$$

which is polynomial in w because the constant 5, $|\Psi|$ and r are all fixed by Ψ. $\square$

7.1   Guarded FO Reductions from Guarded Specifications

To obtain a guarded FO reduction from a guarded $\exists$SO specification, we first construct an FO reduction according to the process in Sect. 6.1. We then modify it by conjoining a suitable guard to each disjunct of each formula defining the reduction. The appropriate guard for the formula defining elements of $N^{\mathcal{B}}$ or $P^{\mathcal{B}}$ corresponding to a subformula ϕ of the specification, is the guard for the least subformula of the specification ψ that is quantified and that contains ϕ as a subformula. Consider the formula 1 of Sect. 6.1. Suppose that the subformula of Ψ it addresses, $(\phi_1(x_1, x_2) \vee \phi_2(x_1, x_2))$, appears in a subformula of the form

$$\exists x_1((\gamma(x_1, x_2) \wedge (\phi_3(x_1) \wedge ((\phi_1(x_1, x_2) \vee \phi_2(x_1, x_2))))).$$

Then we add the guard $\gamma(x_1, x_2)$, to obtain the guarded formula

$$\alpha(\langle p, x_1, \ldots, x_k, c_1 \rangle, \langle p, x_1, \ldots, x_k, c_2 \rangle) = [\gamma(x_1, x_2) \wedge$$
$$(p = P_\phi \wedge (x_1 = x_1) \wedge (x_2 = x_2) \wedge (x_3 = \_) \wedge \ldots \wedge (x_k = \_)$$
$$\wedge (c_1 = \text{atom}) \wedge (c_2 = \text{orClause}))] \quad (2)$$

From the complete construction, we obtain the following.

Theorem 8. *For every guarded ∃SO formula Ψ, there is a guarded FO reduction Δ from Mod Ψ to SAT.*

8   Discussion

When writing specifications in constraint modelling or knowledge representation languages, it is common practice to write constraints in guarded form when this is easy. Our results demonstrate one possible benefit of this, and also suggest that making an effort to write guarded specifications might improve solving time. The potential speedup does not depend on instances being of bounded treewidth, since guarded reductions will preserve related sparseness properties for instance families that are somehowe "close to" having small treewidth. More importantly, our work takes a step toward understanding when reductions obtained from declarative problem specifications may preserve interesting structural instance properties.

We would like to also obtain necessary conditions for existence of treewidth preserving reductions. We conjecture that the class of problems with treewidth preserving reductions to SAT is strictly larger than the class of problems with guarded FO reductions to SAT. We also conjecture that guarded reductions preserve structural sparseness properties more general than bounded treewidth.

Related Work. Bliem et al. [2] demonstrated treewidth affecting ASP solver run-time, and introduced connection guarded ASP programs. These preserve bounded treewidth in grounding, but only if degree is also bounded. Bliem [1] defined guarded ASP programs and showed that grounding for these preserves bounded treewidth regardless of degree. The definition is quite restrictive, and we conjecture there are problems with no guarded ASP formulas but with guarded ∃SO definitions and guarded reductions to SAT. However, it is possible that with a carefully formulated use of defined predicates in guards this could be remedied. The paper [3] illustrates extending the features of the IDP system by using the system itself to compute transductions. Results in [8,18] indicate that there should be very efficient algorithms for grounding guarded specifications. The MSO transductions in [4] preserve treewidth but are restricted in a way that makes them too weak for our application.

Acknowledgements. Phokion Kolaitis suggested studying "good" reductions by via special classes such as FO reductions [11]. Marc Denecker suggested that guarded formulas should produce groundings with bounded treewidth [5]. This work was supported in part by an NSERC Discovery Grant.

References

1. Bliem, B.: ASP programs with groundings of small treewidth. In: Ferrarotti, F., Woltran, S. (eds.) FoIKS 2018. LNCS, vol. 10833, pp. 97–113. Springer, Cham (2018). https://doi.org/10.1007/978-3-319-90050-6_6

2. Bliem, B., Moldovan, M., Morak, M., Woltran, S.: The impact of treewidth on ASP grounding and solving. In: Proceedings of the Twenty-Sixth International Joint Conference on Artificial Intelligence, IJCAI 2017, Melbourne, Australia, 19–25 August 2017, pp. 852–858 (2017)
3. Bogaerts, B., Jansen, J., de Cat, B., Janssens, G., Bruynooghe, M., Denecker, M.: Bootstrapping inference in the IDP knowledge base system. New Gener. Comput. **34**(3), 193–220 (2016)
4. Courcelle, B., Engelfriet, J.: Graph Structure and Monadic Second-Order Logic -A Language-Theoretic Approach. Encyclopedia of Mathematics and Its Applications, vol. 138. Cambridge University Press, Cambridge (2012)
5. Denecker, M.: Personal communication (2015)
6. Erlebach, T., Jansen, K.: Call scheduling in trees, rings and meshes. In: 30th Annual Hawaii International Conference on System Sciences (HICSS-30), 7–10 January 1997, Maui, Hawaii, USA, p. 221 (1997)
7. Fagin, R.: Generalized first-order spectra and polynomial-time recognizablesets'. In: Proceedings of the SIAM-AMS, vol. 7 (1974)
8. Flum, J., Frick, M., Grohe, M.: Query evaluation via tree-decompositions. J. ACM **49**(6), 716–752 (2002)
9. Freuder, E.C.: A sufficient condition for backtrack-free search. J. ACM **29**(1), 24–32 (1982)
10. Kilpelainen, P., Mannila, H.: Ordered and unordered tree inclusion. SIAM J. Comput. **24**(2), 340–356 (1995)
11. Kolaitis, P.: Personal communication (2014)
12. Lovász, L., Gács, P.: Some remarks on generalized spectra. Math. Log. Q. **23**(36), 547–554 (1977)
13. Makowsky, J.A.: Algorithmic uses of the Feferman-Vaught theorem. Ann. Pure Appl. Log. **126**(1–3), 159–213 (2004)
14. Marx, M.: Tolerance logic. J. Log. Lang. Inform. **10**(3), 353–374 (2001)
15. McDiarmid, C., Reed, B.: Channel assignment on graphs of bounded treewidth. Discrete Math. **273**(1), 183–192 (2003). EuroComb 2001
16. Nethercote, N., Stuckey, P.J., Becket, R., Brand, S., Duck, G.J., Tack, G.: MiniZinc: towards a standard CP modelling language. In: Bessière, C. (ed.) CP 2007. LNCS, vol. 4741, pp. 529–543. Springer, Heidelberg (2007). https://doi.org/10.1007/978-3-540-74970-7_38
17. Nishizeki, T., Vygen, J., Zhou, X.: The edge-disjoint paths problem is NP-complete for series-parallel graphs. Discrete Appl. Math. **115**(1), 177–186 (2001). First Japanese-Hungarian Symposium for Discrete Mathematics and its Applications
18. Patterson, M., Liu, Y., Ternovska, E., Gupta, A.: Grounding for model expansion in k-guarded formulas with inductive definitions. In: Proceedings of the 20th International Joint Conference on Artificial Intelligence, IJCAI 2007, Hyderabad, India, 6–12 January 2007, pp. 161–166 (2007)
19. Samer, M., Szeider, S.: Constraint satisfaction with bounded treewidth revisited. J. Comput. Syst. Sci. **76**(2), 103–114 (2010)
20. Tseitin, G.S.: On the complexity of derivation in propositional calculus. In: Siekmann, J.H., Wrightson, G. (eds.) Automation of Reasoning. Symbolic Computation (Artificial Intelligence), pp. 466–483. Springer, Heidelberg (1983). https://doi.org/10.1007/978-3-642-81955-1_28

Automatic Streamlining for Constrained Optimisation

Patrick Spracklen[✉], Nguyen Dang, Özgür Akgün, and Ian Miguel

School of Computer Science, University of St Andrews, St Andrews, UK
{jlps,nttd,ozgur.akgun,ijm}@st-andrews.ac.uk

Abstract. Augmenting a base constraint model with additional constraints can strengthen the inferences made by a solver and therefore reduce search effort. We focus on the automatic addition of *streamliner* constraints, which trade completeness for potentially very significant reduction in search. Recently an automated approach has been proposed, which produces streamliners via a set of streamliner generation rules. This existing automated approach to streamliner generation has two key limitations. First, it outputs a single streamlined model. Second, the approach is limited to satisfaction problems. We remove both limitations by providing a method to produce automatically a *portfolio* of streamliners, each representing a different balance between three criteria: how aggressively the search space is reduced, the proportion of training instances for which the streamliner admitted at least one solution, and the average reduction in quality of the objective value versus the unstreamlined model. In support of our new method, we present an automated approach to training and test instance generation, and provide several approaches to the selection and application of the streamliners from the portfolio. Empirical results demonstrate drastic improvements both to the time required to find good solutions early and to prove optimality on three problem classes.

Keywords: Constraint programming · Streamliners

1 Introduction

An initial constraint model can be augmented through additional constraints. If well chosen, these constraints strengthen the inferences the solver can make and therefore reduce search. *Implied* constraints are inferred directly from the initial model and therefore do not alter the set of solutions to the model. Manual [15,16] and automated [7,9,17] approaches to generating implied constraints have been successful.

In contrast, *streamliner* constraints [20] (our focus herein) are not inferred from the initial model and often radically alter the set of solutions to the model in an attempt to focus effort on a highly restricted but promising portion of the search space. Streamliners trade the completeness offered by implied constraints

© Springer Nature Switzerland AG 2019
T. Schiex and S. de Givry (Eds.): CP 2019, LNCS 11802, pp. 366–383, 2019.
https://doi.org/10.1007/978-3-030-30048-7_22

for potentially much greater search reduction. They were originally derived manually by examining solutions of small instances of a problem class for patterns, which were used as the basis for streamliners [20, 22–24]. For example Gomes and Sellmann added a streamliner requiring a latin square structure when searching for diagonally ordered magic squares [20].

More recently, an automated approach has been proposed, which produces streamliners via a set of streamliner generation rules [32, 35] operating on the ESSENCE [12–14] specification of a problem class. Using training instances drawn from the problem class under consideration, streamliner candidates are evaluated automatically and the most promising ones are used to solve more difficult instances from the same problem class.

The existing automated approach to streamliner generation has two key limitations. First, it outputs a single streamlined model. If on a test instance this streamliner excludes all solutions the only remedy is to revert to the initial model. Second, the approach is limited to satisfaction problems. We remove both limitations by providing a method to produce automatically a *portfolio* of streamliners, each representing a different balance between three criteria: how aggressively the search space is reduced, the proportion of training instances for which the streamliner admitted at least one solution, and the average reduction in quality of the objective value versus an unstreamlined model.

In support of our new method, we present an automated approach to training and test instance generation, and provide several approaches to the selection and application of the streamliners from the portfolio. The result is the first automatic method to produce streamliners for optimisation problems and to offer alternatives if the most preferred streamliner is unsuccessful.

2   Candidate Streamliner Generation

As in [32], our approach proceeds from a specification of a problem class in the abstract constraint specification language ESSENCE [14], such as the SONET example in Fig. 1. An ESSENCE specification comprises the problem class parameters (given); the combinatorial objects to be found (find); the constraints the objects must satisfy (such that); identifiers declared (letting); and an optional objective function (min/maximising). The key feature of the language is support for abstract decision variables, such as multiset, relation and function, as well as *nested* types, such as the multiset of sets in Fig. 1.

The highly structured description of a problem an ESSENCE specification provides is better suited to streamliner generation than a lower level representation, such as a constraint modelling language like MiniZinc [27]. This is because nested types like multiset of sets must be represented as a constrained collection of more primitive variables, obscuring the structure that is useful to drive streamliner generation. We employ the same set of streamliner generation rules as [32], summarised in Table 1. High-order rules take another rule as an argument and lift its operation onto a decision variable with a nested domain such as the complex multi-set structure present in SONET. This allows for the generation of a rule such as enforcing that approximately half (with softness parameter) of the sets in the multiset only contain even numbers. Imposing extra structure in this manner can reduce search very considerably. Table 2 presents candidate streamliners automatically generated for the problem classes considered herein. Although rich, the set of ESSENCE type constructors is not exhaustive. Graph types, for example, are a work in progress [10]. At present, therefore, we might specify such a problem in terms of a set of pairs. The streamliner generator constraints would produce candidate streamliners based on this representation.

Using training instances drawn from the problem class under consideration, streamliner candidates are evaluated as follows. The CONJURE [1,3] automated modelling tool is used to refine the ESSENCE specification (including streamliner) into the solver-independent constraint modelling language ESSENCE PRIME, which SAVILE ROW [29] translates into input suitable for the constraint solver MINION [19].

```
1   $ SONET
2   given nnodes, nrings, capacity : int(1..)
3   letting Nodes be domain int(1..nnodes)
4   given demand : set of set (size 2) of Nodes
5
6   find network : mset (size nrings) of
7                  set (maxSize capacity) of Nodes
8   minimising sum ring in network . |ring|,
9   such that forAll pair in demand .
10    exists ring in network . pair subsetEq ring
11
12  $ Minimum Energy Broadcast
13  letting dNodes   be domain int(1..n_nodes)
14  letting dDepths  be domain int(1..n_nodes)
15  find parents: function (total)
16                   dNodes --> dNodes,
17       depths : function (total)
18                   dNodes --> dDepths
19
20  $ Progressive Party
21  letting Boat be domain int(1..n_boats)
22  find hosts: set (minSize 1) of Boat,
23       sched: set (size n_periods) of
24                  function (total) Boat --> Boat
```

Fig. 1. ESSENCE specifications for the three problem classes considered herein. Synchronous Optical Networking (SONET) [28] is given in full. For brevity, only the parameters and decision variable declarations (from which streamliners are generated) are shown for the Progressive Party Problem [33] and the Minimum Energy Broadcast Problem [6]

3   Searching for a Streamliner Portfolio

Candidate streamliners are often most effectively used in combination [20]. In an attempt to find a single "best" streamlined model, Spracklen et al. described a Monte Carlo Tree Search [5] (MCTS)-based algorithm to search the lattice of models where the root is the original ESSENCE specification and an edge represents the addition of a streamliner to the combination associated with the parent node.

This search had a single objective, average search effort reduction across a set of training instances, which generates only one streamlined model per problem class. This model tends to achieve a high search effort reduction, but has difficulty generalising across the problem class. Furthermore, it is designed only for satisfaction problems. The optimisation problems with which Spracklen et al. experimented were converted into satisfaction problems by bounding the objective and searching for a satisfying solution. This is a serious limitation since a candidate streamlined model may find a solution quickly, but of poor quality, and may exclude the set of optimal solutions entirely.

Table 1. The rules used to generate conjectures. Rows with a softness parameter specify a family of rules each member of which is defined by an integer parameter.

Class	Trigger domain	Name	Softness
First-order	`int`	odd{even}	No
		lower{upper}Half	No
	`function int --> int`	monotonicIncreasing{Decreasing}	No
		largest{smallest}First{Last}	No
	`function (X,X) --> X`	commutative	No
		associative	No
		non-commutative	No
	`partition from X`	quasi-regular	Yes
	`sequence`	montonicIncreasing{Decreasing}	No
		largest{smallest}First{Last}	No
Higher-order	`matrix/set of X`	all	No
		most	Yes
		half	No
		approxHalf	Yes
	`function X --> Y`	range	No
		defined	No
		pre{post}fix	Yes
		allBut	Yes
	`function (X,X) --> Y`	diagonal	No
	`partition from X`	parts	No
	`sequence`	range	No
		defined	No

Table 2. Sample streamliners generated for the three problem classes we consider (see Fig. 1 for their ESSENCE specifications). References to odd/even are with respect to the integer identifiers associated with entities such as nodes or boats. Streamliner Id is a unique reference given to a streamliner when generated through CONJURE; we shall refer to these examples in Sect. 8.1

Problem	Streamliner Id	Description
Sonet	6	Exactly half the nodes installed on each ring are odd
	13	Approx. half the nodes installed on each ring are odd
	15	Approx. half the nodes on each ring are from the lower half of the Nodes domain
	67	The objective variable is constrained to the lower half of its domain
MEB	18	Approx. half of the entries in the range of the parents function must be even
	41	The range of the depths function contains all odd entries
PPP	7	For half of the hosts the boats must be in the lower half of the Boats domain
	14	For approx. half of the hosts the Boats must be odd

To address these problems we adopt a multi-objective optimisation approach, where each point x in the search space X is associated with a d-dimensional (d is the number of objectives) reward vector r_x in R^d. Our three objectives allow us explicitly to balance considerations of solution quality against how aggressively the streamlined model reduces search:

1. **Applicability.** The proportion of training instances for which the streamlined model admits a solution.
2. **Search Reduction.** The mean reduction in time to prove optimality in comparison with an unstreamlined model.
3. **Optimality Gap.** The mean percentage difference between the optimal value found by the streamlined model and the true optimal value under the unstreamlined model.

All objectives are transformed such that they can be maximized. With these three objectives for each streamliner combination we define a partial ordering on R^d and so on X using the Pareto dominance test. Given $x, x\prime \in X$ with vectorial rewards $r_x = (r_1, \ldots, r_d)$ and $r_{x\prime} = (r_{1\prime}, \ldots, r_{d\prime})$ r_x dominates $r_{x\prime}$ iff r_i is greater than or equal to $r_{i\prime}$ for $i = 1 \ldots d$.

To search the lattice structure for a portfolio of Pareto optimal streamlined models we have adapted the *dominance-based multi-objective MCTS (MOMCTS-DOM)* algorithm [34]. This has four phases, as summarised below and in Fig. 2:

1. **Selection:** Starting at the root node, the Upper Confidence Bound applied to Trees (UCT) [5] policy is applied to traverse the explored part of the lattice until an unexpanded node is reached.
2. **Expansion:** Uniformly select and expand an admissible child
3. **Simulation:** The collection of streamliners associated with the expanded node are evaluated. The vectorial reward (Applicablity, Search Reduction, Optimality Gap) across the set of training instances is calculated and returned.
4. **BackPropagation:** The current portfolio; which contains the set of non dominated streamliner combinations found up to this point during search; is used to compute the Pareto dominance. The reward values of the Pareto dominance test are non stationary since they depend on the portfolio, which evolves during search. Hence, we use the cumulative discounted dominance (CDD) [34] reward mechanism during reward update. If the current vectorial reward is not dominated by any streamliner combination in the portfolio then the evaluated streamliner combination is added to the portfolio and a CDD reward of 1 is given, otherwise 0. Dominated streamliner combinations are removed from the portfolio. The result of the evaluation is propagated back up through all paths in the lattice to update CDD reward values, as shown in the figure.

4 Generating Diverse Training Instances

Our method relies on training instances from a given problem class to construct a high quality portfolio of streamlined models. Ideally these should be diverse,

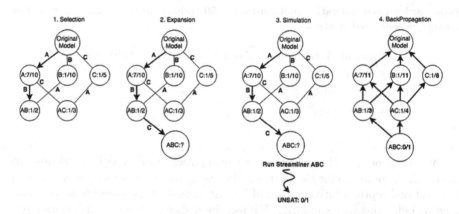

Fig. 2. MOMCTS-DOM operating on the streamliner lattice. A, B and C refer to single candidate streamliners generated from the original ESSENCE specification. As MOMCTS-DOM descends down through the lattice the streamliners are combined through the conjunction of the individual streamliners (AB, ABC). The nodes are labelled with *CDD reward value/times visited.*

otherwise the portfolio may be skewed towards instances of one type and so not generalise across the problem class. To ensure diversity, we employ an automated approach combining a per-class parameter generator and an algorithm configuration tool, described below.

For each problem class we wrote a simple instance generator that accepts a parameter setting and a random seed, and outputs a problem instance. At the moment the instance generator has to be manually created, and is the only part of the whole system that is not automated. However, this issue has been tackled in a recent work [2] within the same pipeline, which can be integrated into our system in the future. To keep the computational cost manageable, we require a set of relatively easy (but not trivial) instances for the training phase, which we define as solvable by MINION [19] on an unstreamlined model within a time limit of $[10, 300]$ s.

To find instances satisfying our criteria, the automatic algorithm configuration tool irace [25] is used. Parameters of each generator are tuned by irace with a performance measure guiding it towards regions of satisfiable instances within the required range of solving time. As the tuning procedure usually converges at certain regions of the search, multiple tunings with two settings of irace (the default and another that allows more exploration) are performed per problem class to obtain more diverse sets of instances.

There is an inherent tradeoff with the number of training instances used during search. If too few instances are used it diminishes the ability of the generated portfolios to generalise across the problem class, whereas a larger set reduces the iteration speed of MOMCTS to the point where it is ineffective in searching the streamliner lattice. Taking these considerations into account, for the experiments in this paper we have set the number of training instances to 50.

Table 3. Instance generation and clustering. 50 training instances are selected from among the generated clusters.

Problem	Total number of instances	Number of clusters
SONET	517	3
MEB	989	8
PPP	1264	8

We first generate a large instance set using irace. Table 3 (column 2) presents the results of doing so for the problem classes we consider in this paper. In order to select our representative subset of 50 instances, instance-specific features are used to judge instance similarity. We use the features proposed in [18] and generated by MINION. All features are normalised according to the z-score standardisation. GMeans clustering is used on the generated features to detect the number of instance clusters (see column 3 of Table 3). To build the training set instances are randomly selected from each cluster, with the number of instances taken from each weighted according to the relative size of each cluster.

The time limit for training instances, and the size of the training set are both parameters to our method, which will be investigated in future work.

5   Pruning the Streamliner Portfolio

As the number of objectives increases so, typically, does the size of the Pareto front, and hence the size of the generated streamliner portfolio. This is demonstrated in Table 4, which, in column 2, records the size of the streamliner portfolios generated through MOMCTS for our three problem classes. A large portfolio is cumbersome when considering streamliner selection and scheduling. We observed, however, that the streamlined models were not distributed evenly across the Pareto front. Therefore, GMeans clustering is used to identify the number of clusters present in the portfolio and a point from each cluster is then selected to form a representative subset of the full portfolio (see column 3 of Table 4).

Table 4. We prune an initially generated streamliner portfolio through GMeans clustering and select a representative point from each cluster.

Problem	Initial portfolio size	Pruned portfolio size
SONET	57	6
MEB	56	3
PPP	64	9

6   Selecting from the Streamliner Portfolio

Having constructed a streamliner portfolio for a particular problem class using MOMCTS and the set of training instances, for a given test instance the question arises as to which streamlined models from the portfolio should be used, in what

---

Algorithm 1. Lexicographic Streamliner Selection

---

procedure SELECTION(Portfolio P, Ordering, $Time_{total}$, Instance)
    $P \leftarrow sort(P, by = Ordering)$
    $Time_{Taken} \leftarrow 0$
    while $Time_{Taken} \leq Time_{Total}$ **do**
        Streamliner $\leftarrow$ P.next()
        Stats $\leftarrow$ Apply(Streamliner, Instance)
        if Stats→sat() **then**
            setBound(Instance, Stats.bound)   ▷ Set new bound on the instance
        end if
        $Time_{Taken} + = Stats.time$
    end while
end procedure

---

order, and according to what schedule. We consider both static lexicographic selection methods, which establish a priority order over our three objectives of Applicability, Search Reduction and Optimality Gap, and a dynamic method, which adjusts the selection based on the performance on the instance thus far.

6.1 Lexicographic Selection Methods

It is possible to order the streamlined models in a portfolio lexicographically by, for example, prioritising Applicability, then Search Reduction, and finally the Optimality Gap. Given three objectives, there are six such orderings to consider. Through preliminary testing it became apparent that only two of these orderings are effective, where the Applicability objective is prioritised. The other orderings trade Applicability for either Search Reduction or a better Optimality Gap. On more difficult test instances, significant search effort can be required to prove that an aggressive streamliner has rendered an instance unsatisfiable, which can lead to poor overall performance. Thus two lexicographic selection methods are used herein: {*Applicability First, Optimality Second, Reduction Third*} and {*Applicability First, Reduction Second, Optimality Third*}.

The selection process involves traversing the portfolio (using the defined ordering) for a given time period and applying each streamliner in turn to the given instance as shown in Algorithm 1. The schedule is static in that it only moves to the next streamlined model when the search space of the current one is exhausted. A key parameter is $Time_{total}$, which specifies the total budget in seconds for traversing the streamliner portfolio. In Sect. 8 for each selection method four different settings for this parameter are experimented with to explore its effect on overall performance.

6.2 UCB Streamliner Selection

During optimisation, typically a number of feasible solutions are discovered before the optimal objective value is found. This intermediate information can be used as an indicator of the performance of the streamlined model. For a given instance we have no prior knowledge of the suitability of a particular streamlined model and as such it is important to balance the time taken exploring the portfolio to identify the performance of each model while exploiting those that have already found solutions. Representing this as a multi-armed bandit problem allows us to employ well known regret-minimising algorithms to deal with the exploration/exploitation dilemma. The multi-armed bandit can be seen as a set of real distributions, each distribution being associated with the rewards delivered by one of the K levers. In our case this is the K streamlined models that comprise the portfolio. On each iteration a streamliner is selected to search the given instance and a reward is observed based upon the improvement to the objective value. The aim is at each iteration to apply the optimal streamliner, where optimality is defined as producing the largest increase/decrease in the value of the objective. The regret ρ after T rounds is defined as the expected difference between the reward sum associated with an optimal strategy and the

---

Algorithm 2. UCBSelection

---

procedure SELECTION(Portfolio, Ordering, $Time_{total}$, Instance)
   $Time_{taken} \leftarrow 0$
   UCBTimeLimit $\leftarrow 1$
   NumberOfIterations $\leftarrow 0$
   Map   ▷ Mapping from Streamliner to Process
   while $Time_{taken} \leq Time_{total}$ **do**
      Streamliner $\leftarrow$ UCTSelection(Portfolio)
      if Map[Streamliner].restart **then**
         Process $\leftarrow$ remodel(instance, streamliner)   ▷ Remodel with the new bound
         Map[Streamliner].process $\leftarrow$ Process
         Stats $\leftarrow$ run(Process, UCBTimeLimit)
      else
         Process $\leftarrow$ Map[Streamliner].process
         Stats $\leftarrow$ run(Process, UCBTimeLimit)   ▷ Continue running existing process
      end if
      Map[Streamliner].visits += 1
      NumberOfIterations += 1
      if Stats→sat() **then**
         Map[Streamliner].reward += 1
         setBound(Instance, Stats.bound)   ▷ Set new bound on the instance
         for $S \leftarrow Map$ **do**
            if S != Streamliner **then**
               Map[S].restart = True ▷ New Bound was found; restart all other processes
            end if
         end for
      end if
      $Time_{taken}$ += Stats.time
   end while
end procedure

---

sum of the collected rewards observed. The UCB1 [4] algorithm was chosen to solve the multi-armed bandit problem as first and foremost its regret grows logarithmically in line with the number of actions taken.

For each streamliner k we record the average reward x^k and the number of times k has been tried in the selection (n_j) out of a total of n iterations. On each iteration a streamliner is chosen that maximizes $x^k + \sqrt{2 \log(n)/n_j}$. The reward distributions for an individual streamliner are not fixed, so this is not a Stationary Multi-Armed Bandit problem. However, if a streamliner performs well, we expect it will continue performing well during search even if there is a slight variation in the mean reward. We have found that using UCB1 gives good results. Future work could investigate the use of Upper Confidence Bound policies for non-stationary bandit problems, such as the family of Exp3 algorithms [21,26].

When traversing the portfolio UCB performs incremental evaluation, it runs a streamliner for a set time, observes the results, and potentially moves on before the corresponding search space has been exhausted. When the streamliner is preempted it is necessary to pause the search in order to avoid repeating work if it is rescheduled at a later point. The only exception to this is whenever a new bound on the objective is discovered all of the streamliners from the portfolio, aside from the current streamliner, are restarted and remodeled with the new bound. There are two main benefits to doing this. Firstly, by restarting the streamliner has the newly constrained bound at the top of the search tree which allows it to make more informed decisions higher up without descending into unsatisfactory subtrees. Secondly, by remodeling it takes advantage of the toolchain (CONJURE and SAVILE ROW) which may be able to reformulate the model based upon this new information and produce reductions at the solver level. Algorithm 2 shows the UCBSelection process in detail.

7   Experimental Setting

We evaluate our automated streamlining approach on the three problem classes in Fig. 1. We selected these problems to give good coverage of the abstract domains available in ESSENCE, such as set, multi-set and function. Furthermore, SONET and Progressive Party have nested domains: multi-set of set and set of function respectively.

Our hypothesis is that a streamliner portfolio, generated automatically on a set of automatically generated training instances from a given problem class, can be employed to solve more difficult test instances to deliver substantial performance improvements relative to an unstreamlined model. Training instances were generated as per Sect. 2, with a time limit of $[10, 300]$ s. Test instances are generated using the same instance generator and the tuning tool irace but with a time limit of $(300, 3600]$ s. 50 instances are selected randomly to form the test set.

Care must be taken when considering the proof of optimality of our test instances. Although in solving a streamlined model the constraint solver may exhaust the search space this is not a proof that the current objective value is optimal. This is because streamliners are not necessarily sound, hence a streamlined model may exclude the set of optimal solutions. For this reason, after the streamliner portfolio has been run for its allotted time, we use the remainder of the time budget to run the unstreamlined model, starting from the best objective value found by the streamliner portfolio, to provide the optimality proof. The benefit of streamlining in this context is in finding high quality solutions much more quickly than the unstreamlined model.

All experiments were run on a cluster of 280 nodes, each with two 2.1 GHz, 18-core Intel Xeon E5-2695 processors. MOMCTS was run on a single core with a budget of 4 CPU days for each problem class. Results on 50 test instances under the unstreamlined and streamlined models are reported, where every test instance was run with three random seeds.

Source code, instance generators, datasets and detailed results are available at https://github.com/stacs-cp/CP2019-Streamlining.

8   Results

Table 5 summarises results on 50 test instances (3 runs/instance) for each of our three problem classes. We evaluate four different approaches: an unstreamlined model, and streamliner portfolios with UCB selection, lexicographic ordering {*Applicability First, Optimality Second, Reduction Third*} (denoted *opt-second*), and lexicographic ordering {*Applicability First, Reduction Second, Optimality Third*} (denoted *red-second*). For each streamliner selection method, a parameter is the amount of time allocated to the streamliner portfolio before handing over to the unstreamlined model to prove optimality. Four different values for this time budget were tested: 30, 60, 120 and 300 s.

Results in Table 5 are strongly positive. They show that all the streamliner portfolio approaches can not only find an optimal solution and prove optimality on more test instances than the unstreamlined model, but also vastly reduce the amount of time required for both tasks. In general, the UCB-30s variant has the best overall performance across the three problem classes, and provides consistently robust improvement over the unstreamlined model.

Figure 3 presents more details of how the streamliner approaches improve on the unstreamlined models on an instance basis. In these plots, we use the time-reduction ratio, a "normalised" version of the speed-up values reported in Table 5 for presentation: as the speed-up values can be arbitrarily large, many data points in the speed-up plots can appear in a very small range, making them difficult to distinguish. The reduction ratio, which is calculated as $1 - 1/speed\text{-}up$, is limited to at most one and can be easily scaled. For brevity, we only show in Fig. 3 results of the streamliner variants with the time limit of 30 s. Each data point corresponds to a pair of instances and random seeds. The plots show that the solving time of the test instances are well distributed across the x-axis, which is a good indication for the diversity of the test instance set. There are several cases where the unstreamlined model cannot find or prove optimality within the time budget and the streamliner can, which are represented by the data points on the rightmost side after the vertical red lines.

The MEB results demonstrate strong performance of all three streamliner approaches on all test instances. On SONET, UCB-30s clearly has better performance compared with the other two approaches, which aligns with the summary results in Table 5. While still strongly positive, on PPP the reduction provided by the streamliner approaches is not quite as strong as for the other two problem classes. There are a minority of cases where even the best streamliner approach, UCB-30s, cannot find or prove optimality within the time budget, as shown by the data points in the bottom-right corners.

Table 5 and Fig. 3 demonstrate that the time to prove optimality is very significantly reduced through the application of streamliners. This stems from their ability to find high quality feasible solutions quickly. Hence, once the time allocated to the streamlined models has elapsed, the unstreamlined model begins from an optimal or very high quality objective value, requiring much less effort to exhaust the search space.

Table 5. Summary results on 50 test instances (3 runs/instance) on three optimisation problem classes: MEB, PPP and SONET. The first column, mean #proved 1-h, represents the average number of instances solved within one hour. All streamliner portfolio variants significantly outperform the unstreamlined model by this simple measure. The remaining columns report results where each run is now given a maximum amount of 96 CPU-hours (as tuning and generation of test instances is performed on the basis of one seed, on the two other seeds it is possible for the unstreamlined model to time out at one CPU hour). They include the time to reach an optimal solution, the time to both reach an optimal solution and prove its optimality; and the corresponding speed-up ratios when compared to the unstreamlined model. For each measurement, we report the 10th percentile (p10), the median (p50), and the 90th percentile (p90). These values are reported as the mean can be skewed by outliers. In particular, if the optimal solution is not proved this results in a large time value (96 h = 345600 s) for that run. The percentiles avoid this situation and show a clearer overall trend.

	Strategy	mean #proved (1-hour)	Finding an optimal solution time(s)			Finding an optimal solution speed-up ratio			Finding and prove optimality time(s)			Finding and prove optimality speed-up ratio		
			p10	p50	p90	p10	p50	p90	p10	p50	p90	p10	p50	p90
MEB	unstreamlined	35	157.9	1185.2	13893.9				311.1	1976.2	16781.3			
	UCB-30s	**50**	6.1	8	**11.0**	14.2	158.2	1583	15.2	22.2	**176.7**	6.6	43.6	492.2
	UCB-60s	**50**	4.4	7.2	12	15	150.3	1552.2	16.1	24.6	188.7	6.9	35.6	521.9
	UCB-120s	**50**	4.5	7.8	12.1	14.9	158	1604	15.1	24.8	220.9	6.2	36.1	518.4
	UCB-300s	**50**	4.5	7.1	12.1	15	157.5	1605.4	14.9	24.9	345.1	5.2	32.1	416.6
	opt-second-30s	49.7	**4.1**	6.3	13.4	14.1	171.1	1701.5	11.6	22.9	221.3	7.3	44.6	605.9
	opt-second-60s	49.7	**4.1**	6.6	14.9	15.7	174.3	1833.4	11.7	22.5	199.6	7	45.3	625.4
	opt-second-120s	**50**	4.2	6.2	13.6	**19.9**	178.3	1776.7	11.7	21.8	181.6	7.3	46.5	594.9
	opt-second-300s	**50**	**4.1**	**6.1**	12.8	**19.9**	170.9	1865.8	**11.5**	21.8	176.9	7.5	**47.6**	**647.0**
	red-second-30s	49.7	**4.1**	6.7	13.6	14.1	156	1845.1	11.8	22.8	249.1	7.3	43	532.2
	red-second-60s	49.7	4.2	**6.1**	12.8	15.3	**187.0**	1878.3	11.8	**21.7**	198.7	7.3	45	646.3
	red-second-120s	**50**	**4.1**	6.2	12.6	16.9	177.4	**1903.5**	11.6	22.1	178.1	7.2	46.3	605.4
	red-second-300s	**50**	**4.1**	**6.1**	13.5	16.8	167.5	1891	11.7	22.3	178.8	**7.6**	47.5	625.1
PPP	unstreamlined	41.3	73.4	564.3	3123				313	1339.7	6908.1			
	UCB-30s	47.7	13	**73.7**	**1007.9**	**1.2**	**4.1**	52	49.2	**350.8**	1946.6	**1.0**	**3.0**	**29.3**
	UCB-60s	**48.3**	19.2	105.9	1078.7	0.9	2.9	28.7	86.1	428.8	2141.5	0.9	2.5	24.4
	UCB-120s	**48.3**	18.9	163.3	1129.7	0.7	2.5	31.8	135.5	449.6	**1936.2**	0.9	2.1	16.8
	UCB-300s	**48.3**	19	344.6	1311.3	0.4	1.6	30.1	323.9	646.3	2273.2	0.6	1.4	10.5
	opt-second-30s	46.7	8.3	105.1	1340.5	0.9	3.5	75.1	**44.1**	419.4	2592.5	0.9	2.4	26.2
	opt-second-60s	47	**8.1**	105.8	1444.2	0.8	3.4	75.2	73.7	453.5	2640.3	0.8	2.3	18.9
	opt-second-120s	47.3	8.9	142.9	1765.1	0.7	3.6	**76.5**	113.1	486	2716.7	0.8	1.9	17.6
	opt-second-300s	47.7	8.9	211	1349.3	0.5	3.1	72.4	110.8	599.1	2703.2	0.7	1.8	15.5
	red-second-30s	45	14.7	177.7	2344.7	0.7	2	18.9	73.3	626.2	3537.7	0.8	1.7	14.8
	red-second-60s	45.3	21.2	195.2	2341.6	0.6	2.1	15.6	96.1	643.2	3174.7	0.7	1.8	13.8
	red-second-120s	45.7	13.6	175.7	2384	0.6	2.1	17.5	136.5	591.5	3095	0.6	1.8	11.1
	red-second-300s	45.3	13.6	228	2731.5	0.6	1.9	16.8	157	657.6	3339.1	0.6	1.4	8.4
SONET	unstreamlined	43	539.5	1263.2	3820.3				574.4	1417.8	3954			
	UCB-30s	**50**	5	21.8	**121.9**	**10.3**	49.7	341.5	34	**42.3**	**174.0**	**6.6**	**23.4**	60.5
	UCB-60s	**50**	6.1	28	131.9	8.5	38.1	300.3	63.3	75.3	198.7	4.9	14.4	42.1
	UCB-120s	46	6	31.1	246.8	3.4	31.5	321.9	121.2	132.2	581.2	2.3	7.6	32.1
	UCB-300s	**50**	7	30.7	344.5	3.8	33.4	287.3	111.8	310.8	437.8	1.7	4.2	22.9
	opt-second-30s	49.3	3.5	9	1023.8	1.4	112.7	553.9	27.7	72.7	1023.2	1.4	19.2	**70.5**
	opt-second-60s	49.7	3.5	9	443.2	1.5	113.1	611.4	27.6	93.6	644	1.5	15.9	66.9
	opt-second-120s	49.3	3.3	**8.3**	455.6	1.3	117.3	**677.9**	**26.9**	120.7	701	1.3	14.6	68.6
	opt-second-300s	49.3	3.7	8.4	549	3.6	**121.0**	549.9	28	123.1	770.6	1.1	10.3	69.4
	red-second-30s	47.7	**3.0**	115.4	1749.6	0.8	10.6	483.5	27.7	227.3	2167.4	0.8	5.2	61.7
	red-second-60s	47.7	**3.0**	105.3	1760.9	0.8	14.2	530.9	28.1	185	2137.2	0.8	7.2	64.1
	red-second-120s	47.3	**3.0**	96.7	1532.5	0.8	16	506.3	28.3	157.8	2295.6	0.8	7.6	62.6
	red-second-300s	47.7	**3.0**	96	1451.4	0.9	18.2	533.8	27.1	221.4	1717.6	0.8	6.1	65.2

8.1   UCB Streamliner Selection: Discussion

In this section, we discuss the UCB approach for streamliner selection in more detail, as UCB-30s achieves the best overall performance across the three problem classes, both in terms of reduction to finding the optimal objective value and reduction to proving optimality. In contrast to the lexicographic methods, which only move on to the next streamlined model when the search space of the current one is exhausted, UCB benefits from its ability to sample the entire streamliner portfolio. After the initial exploration phase, where each streamliner is given its initial application, UCB then selects streamliners based upon the observed rewards. Its main advantage is the ability to balance the exploration and exploitation of the streamlined models in the portfolio.

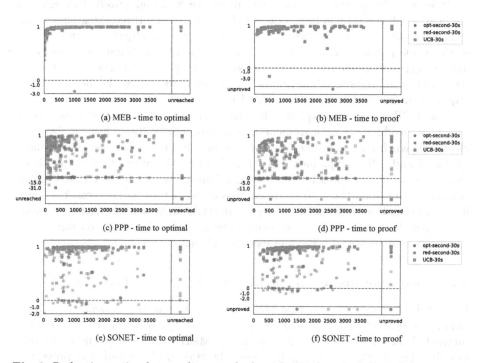

(a) MEB - time to optimal

(b) MEB - time to proof

(c) PPP - time to optimal

(d) PPP - time to proof

(e) SONET - time to optimal

(f) SONET - time to proof

Fig. 3. Reduction ratio of streamliner methods with 30 s for scheduling of the streamliner portfolio. Two reduction ratio values are reported: reduction in time to reach an optimal solution, and reduction in time to reach an optimal solution and prove its optimality. The x-axis represents the time required by the unstreamlined model. The y-axis shows the the reduction value. Each data point corresponds to a pair of (instance, random seed). These plots focus on the region within a 1-h time limit: all data points outside that ranges are shrunk into the same region. More specifically, runs where the (unstreamlined model) streamliner methods do not reach an optimal solution or does not prove optimality in one hour are separated by the red (vertical) horizontal lines. The reduction values, however, are still the true values calculated based on the 4-day CPU limit. As most data points lie within the range of $y \in [0, 1]$, the plot is rescaled so that this range is zoomed in for better visualisation.

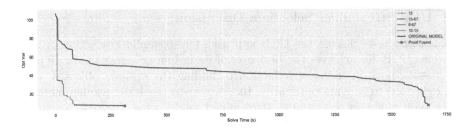

Fig. 4. Objective value progression from the unstreamlined model compared with its progression under the UCB selection method for a representative SONET instance.

It is not always the case that the objective is found purely through the application of one streamliner. For SONET, on average three streamliners are used across the 50 test instances to arrive at the optimal objective value. Access to the whole portfolio allows UCB to descend upon the optimal objective value more quickly and is one reason for its success. The application of several different streamliners at different time points can be used to reduce the bound of the objective in an effective manner as per Fig. 4.

The UCB algorithm exploits the streamliners that have previously been shown to produce an improvement in the objective value. This can be very clearly shown from Fig. 4 where for an instance from SONET the streamliners 13, 13–67 and 6–67[1] (explained in Table 2) improve the objective multiple times during the course of the selection process. This is due to the fact that UCB is continuing to exploit those streamliners as previously they had success. However, it is also crucial to continually explore the portfolio in an attempt to find streamliners that did not initially have success but may do after a certain number of iterations. Streamliner 13–15 is an example of such a case.

8.2 Time Allocated to the Streamliner Portfolio: Discussion

From Table 5 it can be seen that the $Time_{Total}$ parameter as defined in Algorithms 1 and 2 can have a large impact on the overall performance of the selection method. There is a general trend (excluding MEB which will be discussed separately) that as the $Time_{Total}$ increases the time both to find and prove the optimal objective value increases. This may seem puzzling initially: if using a $Time_{Total}$ of 30s reduces the time to find the optimal objective value to a certain extent, it might be expected that a $Time_{Total}$ of 300s will do equally well. However, there are two things to consider. First, streamliners from the portfolio are not guaranteed to preserve the optimal value and so there is the potential for an optimality gap between what the streamliners can find and the true optimal of the instance. Therefore, the true optimal is only found after the switch to the unstreamlined model occurs. Second, on average the streamliners converge

---

[1] 13–67, for example, indicates a streamlined model including both streamliner 13 and 67.

upon their optimal value in a very short period of time, 17 s, 7 s and 12 s for SONET, MEB and PPP respectively. By increasing the $Time_{Total}$ parameter it delays the point at which the switch occurs to the unstreamlined model which in turn delays the point at which the true optimal is found. However, for MEB the $Time_{Total}$ does not have a large impact on performance and this is due to the fact that the streamliners in the portfolio generally exhaust their search space very quickly. Hence, the whole portfolio can be traversed before $Time_{Total}$ is reached and so the time at which the switch to the unstreamlined model occurs is generally the same across all parameter settings.

The increase in time to prove optimality occurs as if the T_{total} parameter is set too large then when the optimal value is found at time T_{opt}, the whole duration from $T_{opt} \rightarrow T_{total}$ is spent proving the optimality of that solution in the streamlined subspaces. Since proving optimality with respect to the streamliners does not prove optimality on the unstreamlined model and so the whole time from $T_{opt} \rightarrow T_{total}$ is wasted.

9   Conclusion and Future Work

We have presented the first automated approach to generating streamliners automatically for optimisation problems, and for their selection and scheduling when employed on unseen instances. On three quite different problem classes the results are very encouraging, with vastly reduced effort both to find and to prove optimal objective values.

An important question we plan to investigate further is the applicability of our method to identify in which contexts our streamliner can and cannot help. In the context of optimisation the benefit of streamlining lies in the early identification of the optimal, or at least high quality, values for the objective. Where an unstreamlined model is able to identify the optimal value quickly, the benefit of streamlining will be limited. When considering satisfaction problems, however, streamlining can be used throughout the search and we will compare the portfolio approach developed herein with the single selection provided by the method presented in Spracklen et al. [32].

Furthermore, there are several methods for devising good search strategies for constrained optimisation problems. Recent research suggest using machine learning to design a promising search ordering [8], using solution density as a heuristic indicator [31] and a number of value ordering heuristics to find good solutions early [11,30]. Streamlining constraints can potentially be used in combination with the existing methods for devising good variable and value selection heuristics to achieve even better results.

Acknowledgements. This work is supported by UK EPSRC grant EP/P015638/1. It used the Cirrus UK National Tier-2 HPC Service at EPCC (http://www.cirrus.ac.uk) funded by the University of Edinburgh and EPSRC (EP/P020267/1).

References

1. Akgün, Ö.: Extensible automated constraint modelling via refinement of abstract problem specifications. Ph.D. thesis, University of St Andrews (2014)
2. Akgün, Ö., Dang, N., Miguel, I., Salamon, A.Z., Stone, C.: Instance generation via generator instances. In: Schiex, T., de Givry, S. (eds.) CP 2019. LNCS, vol. 11802, pp. 3–19. Springer, Cham (2019)
3. Akgün, Ö., Gent, I.P., Jefferson, C., Miguel, I., Nightingale, P.: Breaking conditional symmetry in automated constraint modelling with Conjure. In: ECAI, pp. 3–8 (2014)
4. Auer, P., Cesa-Bianchi, N., Fischer, P.: Finite-time analysis of the multiarmed bandit problem. Mach. Learn. **47**(2), 235–256 (2002). https://doi.org/10.1023/A: 1013689704352
5. Browne, C., et al.: A survey of Monte Carlo tree search methods. IEEE Trans. Comput. Intell. AI **4**, 1–43 (2012)
6. Burke, D.A., Brown, K.N.: CSPLib problem 048: minimum energy broadcast (MEB). http://www.csplib.org/Problems/prob048
7. Charnley, J., Colton, S., Miguel, I.: Automatic generation of implied constraints. In: ECAI, vol. 141, pp. 73–77 (2006)
8. Chu, G., Stuckey, P.J.: Learning value heuristics for constraint programming. In: Michel, L. (ed.) CPAIOR 2015. LNCS, vol. 9075, pp. 108–123. Springer, Cham (2015). https://doi.org/10.1007/978-3-319-18008-3_8
9. Colton, S., Miguel, I.: Constraint generation via automated theory formation. In: Walsh, T. (ed.) CP 2001. LNCS, vol. 2239, pp. 575–579. Springer, Heidelberg (2001). https://doi.org/10.1007/3-540-45578-7_42
10. Dunlop, F., Enright, J., Jefferson, C., McCreesh, C., Prosser, P., Trimble, J.: Expression of graph problems in a high level modelling language. In: Proceedings of the International Workshop on Graphs and Constraints (2018)
11. Fages, J.G., Prud'Homme, C.: Making the first solution good! In: 2017 IEEE 29th International Conference on Tools with Artificial Intelligence (ICTAI), pp. 1073–1077. IEEE (2017)
12. Frisch, A.M., Grum, M., Jefferson, C., Hernández, B.M., Miguel, I.: The essence of essence. Modelling and Reformulating Constraint Satisfaction Problems, p. 73 (2005)
13. Frisch, A.M., Grum, M., Jefferson, C., Hernández, B.M., Miguel, I.: The design of essence: a constraint language for specifying combinatorial problems. In: IJCAI, vol. 7, pp. 80–87 (2007)
14. Frisch, A.M., Harvey, W., Jefferson, C., Martínez-Hernández, B., Miguel, I.: Essence: a constraint language for specifying combinatorial problems. Constraints **13**(3), 268–306 (2008)
15. Frisch, A.M., Jefferson, C., Miguel, I.: Symmetry breaking as a prelude to implied constraints: a constraint modelling pattern. In: ECAI, vol. 16, p. 171 (2004)
16. Frisch, A.M., Miguel, I., Walsh, T.: Symmetry and implied constraints in the steel mill slab design problem. In: Proceedings of the CP01 Workshop on Modelling and Problem Formulation (2001)
17. Frisch, A.M., Miguel, I., Walsh, T.: CGRASS: a system for transforming constraint satisfaction problems. In: O'Sullivan, B. (ed.) CologNet 2002. LNCS, vol. 2627, pp. 15–30. Springer, Heidelberg (2003). https://doi.org/10.1007/3-540-36607-5_2
18. Gent, I.P., et al.: Learning when to use lazy learning in constraint solving. In: ECAI, pp. 873–878. Citeseer (2010)

19. Gent, I.P., Jefferson, C., Miguel, I.: Minion: a fast scalable constraint solver. In: ECAI, vol. 141, pp. 98–102 (2006)
20. Gomes, C., Sellmann, M.: Streamlined constraint reasoning. In: Wallace, M. (ed.) CP 2004. LNCS, vol. 3258, pp. 274–289. Springer, Heidelberg (2004). https://doi.org/10.1007/978-3-540-30201-8_22
21. Kocsis, L., Szepesvári, C.: Bandit based Monte-Carlo planning. In: Fürnkranz, J., Scheffer, T., Spiliopoulou, M. (eds.) ECML 2006. LNCS (LNAI), vol. 4212, pp. 282–293. Springer, Heidelberg (2006). https://doi.org/10.1007/11871842_29
22. Kouril, M., Franco, J.: Resolution tunnels for improved SAT solver performance. In: Bacchus, F., Walsh, T. (eds.) SAT 2005. LNCS, vol. 3569, pp. 143–157. Springer, Heidelberg (2005). https://doi.org/10.1007/11499107_11
23. Le Bras, R., Gomes, C.P., Selman, B.: On the Erdős Discrepancy Problem. In: O'Sullivan, B. (ed.) CP 2014. LNCS, vol. 8656, pp. 440–448. Springer, Cham (2014). https://doi.org/10.1007/978-3-319-10428-7_33
24. LeBras, R., Gomes, C.P., Selman, B.: Double-wheel graphs are graceful. In: IJCAI, pp. 587–593 (2013)
25. López-Ibáñez, M., Dubois-Lacoste, J., Cáceres, L.P., Birattari, M., Stützle, T.: The irace package: iterated racing for automatic algorithm configuration. Oper. Res. Perspectives 3, 43–58 (2016)
26. Munos, R.: From bandits to Monte-Carlo tree search: the optimistic principle applied to optimization and planning. FTML 7(1), 1–129 (2014). https://doi.org/10.1561/2200000038
27. Nethercote, N., Stuckey, P.J., Becket, R., Brand, S., Duck, G.J., Tack, G.: MiniZinc: towards a standard CP modelling language. In: Bessière, C. (ed.) CP 2007. LNCS, vol. 4741, pp. 529–543. Springer, Heidelberg (2007). https://doi.org/10.1007/978-3-540-74970-7_38
28. Nightingale, P.: CSPLib problem 056: synchronous optical networking (SONET) problem. http://www.csplib.org/Problems/prob056
29. Nightingale, P., Akgün, O., Gent, I.P., Jefferson, C., Miguel, I., Spracklen, P.: Automatically improving constraint models in Savile Row. Artif. Intell. 251, 35–61 (2017). https://doi.org/10.1016/j.artint.2017.07.001
30. Palmieri, A., Perez, G.: Objective as a feature for robust search strategies. In: Hooker, J. (ed.) CP 2018. LNCS, vol. 11008, pp. 328–344. Springer, Cham (2018). https://doi.org/10.1007/978-3-319-98334-9_22
31. Pesant, G.: Counting-based search for constraint optimization problems. In: Thirtieth AAAI Conference on Artificial Intelligence (2016)
32. Spracklen, P., Akgün, Ö., Miguel, I.: Automatic generation and selection of streamlined constraint models via Monte Carlo search on a model lattice. In: Hooker, J. (ed.) CP 2018. LNCS, vol. 11008, pp. 362–372. Springer, Cham (2018). https://doi.org/10.1007/978-3-319-98334-9_24
33. Walsh, T.: CSPLib problem 013: progressive party problem. http://www.csplib.org/Problems/prob013
34. Wang, W., Sebag, M.: Hypervolume indicator and dominance reward based multi-objective monte-carlo tree search. Mach. Learn. 92(2–3), 403–429 (2013)
35. Wetter, J., Akgün, Ö., Miguel, I.: Automatically generating streamlined constraint models with essence and conjure. In: Pesant, G. (ed.) CP 2015. LNCS, vol. 9255, pp. 480–496. Springer, Cham (2015). https://doi.org/10.1007/978-3-319-23219-5_34

Compiling Conditional Constraints

Peter J. Stuckey[1,2] and Guido Tack[1,2(✉)]

[1] Monash University, Melbourne, Australia
{peter.stuckey,guido.tack}@monash.edu
[2] Data61, CSIRO, Melbourne, Australia

Abstract. Conditionals are a core concept in all programming languages. They are also a natural and powerful mechanism for expressing complex constraints in constraint modelling languages. The behaviour of conditionals is complicated by undefinedness. In this paper we show how to most effectively translate conditional constraints for underlying solvers. We show that the simple translation into implications can be improved, at least in terms of reasoning strength, for both constraint programming and mixed integer programming solvers. Unit testing shows that the new translations are more efficient, but the benefits are not so clear on full models where the interaction with other features such as learning is more complicated.

Keywords: Constraint modelling · Conditional constraints · MiniZinc

1 Introduction

Conditional expressions are a core part of virtually any programming and modelling language. They provide a way to change behaviour of the program/model depending on some test. MiniZinc 1.6 [10] and earlier versions provided the conditional expression

```
if cond then thenexp else elseexp endif
```

restricted to the case that the expression *cond* could be evaluated at compile time to *true* or *false*. This is simple to handle, since the MiniZinc compiler can replace this expression by *thenexp* if *cond = true* and *elseexp* if *cond = false*. From MiniZinc 2.0 onwards the expression *cond* is no longer restricted to be known at compile time, it can be an expression involving decision variables whose truth will be determined during the execution of the solver. This extension is very useful, it makes the expression of many complex constraints much more natural. For example, the absolute value function can be simply expressed as

```
function var int: abs(var int: x) = if x >= 0 then x else -x endif;
```

The MiniZinc compiler must translate these conditional expressions into primitive constraints that are implemented by the solver. For example, Constraint

© Springer Nature Switzerland AG 2019
T. Schiex and S. de Givry (Eds.): CP 2019, LNCS 11802, pp. 384–400, 2019.
https://doi.org/10.1007/978-3-030-30048-7_23

Programming (CP) solvers may use *reification* or dedicated constraints for *logical connectives* [1,7,8] to link the truth of x >= 0 to the result of the function being x or -x, whereas for Mixed Integer Programming (MIP) solvers we would employ techniques such as *big-M* or *indicator* constraints (see e.g. [6,9]).

The effective translation of conditional expressions is the focus of this paper. But before we determine how to translate conditional expressions we must understand their *exact meaning*, including how they interact with *undefinedness*.

Undefinedness naturally arises in constraint models through the use of partial functions like x div y or, much more commonly, out of bounds array lookups a[i] where i takes a value outside the index set of a. In the remainder of the paper we will argue for the correct semantics of conditionals with undefinedness (Sect. 3), and then illustrate in Sect. 4 how to compile conditional constraints to a form that can be executed by solvers while respecting the correct semantics. We then show in Sect. 5 how we can improve this translation for constraint programming (CP) and mixed integer programming (MIP), and how we can often improve models created by non-experts, who use conditionals in a familiar procedural programming style (Sect. 6). Section 7 shows how the different translations compete in terms of solving efficiency, and Sect. 8 concludes the paper.

2   Model Translation

Modern modelling languages like MiniZinc [10], Essence [5], AMPL [3] and OPL [12] provide highly expressive ways of defining a constraint problem. But the underlying solvers only solve one form of problem, typically

$$\text{minimize } o \text{ subject to } \exists V. \bigwedge_{i=1}^{n} c_i$$

where V is a set of variables and each c_i is a *primitive constraint* understood by the solver. In some solvers the primitive constraints are very limited, e.g. SAT solvers only consider clauses, and MIP solvers only consider linear constraints.

The MiniZinc compiler translates high level models to models that only contain primitive constraints suitable for a given target solver. Solver-level models are represented in a language called FlatZinc, in a process called *flattening*. FlatZinc is a much richer low level language than used by SAT and MIP, in order to support all the primitive constraints natively supported by CP solvers. For this paper we will assume the solver supports the following primitives.

```
sum(i in S)(a[i]*x[i]) <= a0;                  % linear inequality
sum(i in S)(a[i]*x[i]) = a0;                   % linear equality
b <-> sum(i in S)(a[i]*x[i]) <= a0;            % reified linear inequality
b <-> sum(i in S)(a[i]*x[i]) = a0;             % reified linear equality
y1 = x[y2];                                    % element
b1 <-> y1 = x[y2];                             % reified element
[not] b1 \/ [not] b2 \/ ... \/ [not] bn;       % clause, [not] is optional
b1 <-> b2;                                     % Boolean equality
```

where x is an integer array, yn is an integer, a is a fixed integer array, a0 is a fixed integer, bn is a Boolean, and S is a fixed integer set. Note that we will often write these in slightly different syntactic form, e.g. b <-> y = 0 is a reified linear equality, and b1 /\ not b2 -> b3 is a clause.[1] Most CP solvers directly support a wide array of *global constraints* which also appear in FlatZinc. We shall introduce new global constraints as needed during the paper.

The reader may ask why we need *reified* versions of primitive constraints. This arises because in a complex model not all constraints that appear in the model must hold.

Example 1. Consider the model **constraint y <= 0 \/ x = a[y]**.
Neither y <= 0 nor x = a[y] must hold all the time (one of the two is sufficient). A correct flattening making use of reified primitive constraints is

```
constraint b1 \/ b2;
constraint b1 <-> y <= 0;
constraint b2 <-> x = a[y];
```

In a MiniZinc model each expression occurs in a *context*, which is the nearest enclosing Boolean expression. The *root context* is all Boolean expressions that can be syntactically determined to be *true* in any solution, that is the context of a top-level constraint or a top level conjunction. A *non-root context* is any other Boolean expression. When flattening a constraint, all total functional expressions in the constraint can be moved to the root context, while relations or partial functional expressions must be reified to maintain the meaning of the model.

3   Semantics of Conditionals

Any sufficiently complex modelling language has the ability to express undefined values, which we will represent as $\perp$, for example division by zero or array index out of bounds. The treatment of undefinedness in modelling languages is complicated. Whereas in a traditional programming language it would be handled by runtime exception or abort, in a relational language this is not correct, since the solver will be making decisions that may result in undefinedness. Frisch and Stuckey [4] considered three different semantics for the treatment of undefinedness in modelling languages: a three-valued *Kleene semantics* (agreeing with a usual logical interpretation of undefinedness, but requiring three valued logic), a two-valued *strict semantics* (essentially making any undefinedness cause the model to have no solution), and the two-valued *relational semantics* (that agrees with the relational interpretation discussed above). MiniZinc and other modelling languages such as ESSENCE PRIME [11] implement the relational semantics, since it most closely accords with a modeller's intuition and does not require introducing three truth values.

---

[1] Note that we use a simplified FlatZinc syntax, including support for reified element constraints, to improve readability.

The core abstract modelling language introduced in [4] did not consider conditional expressions. There are a few plausible interpretations of the desired interaction between conditionals and undefinedness. We invite the reader to consider the simple one-line constraint models given below (each sharing the same declarations of x, y and a) and determine what they believe to be the correct set of solutions for each:

```
var 0..2: x;
var 0..2: y;
array[1..2] of int: a = [0,2];
constraint if y = 0 then x = a[3] else x = a[y] endif;         %% (1)
constraint x = if y = 0 then a[3] else a[y] endif;             %% (2)
constraint x = if y = 0 then a[y] else a[1] endif;             %% (3)
constraint y = 0 \/ x = if y = 0 then a[3] else a[y] endif; %% (4)
```

The basic rule for the relational semantics is *"an undefined value causes the nearest containing Boolean expression to be false"*. That means when we find an expression with value $\bot$ we must propagate this upward through all enclosing expressions until we reach a Boolean context, where the $\bot$ becomes *false*. For instance, assume that the index set of a in Example 1 is 1..2, then a[y] is undefined for y=-1. The nearest containing Boolean context is x = a[y], which therefore becomes *false*. The overall disjunction would still be true, since its other disjunct y<=0 is true. If we consider what this means for conditional expressions, we can distinguish two cases.

- An undefined result occurs somewhere in the condition cond or it occurs in the thenexp or elseexp and the type of thenexp and elseexp is Boolean.
- An undefined results occurs in the thenexp or elseexp and their type is not Boolean.

In the first case, the undefinedness is captured by the subexpressions and the conditional does not directly deal with undefinedness. In the second case there are two possible solutions. The *eager* approach says that the entire conditional expression takes the value $\bot$ if either thenexp or elseexp takes the value $\bot$. The *lazy* approach says that the conditional expression takes the value $\bot$ iff the cond = true and thenexp =$\bot$ or cond = false and elseexp = $\bot$. Clearly the lazy approach reduces the effect of undefinedness. We would argue that it also accords with traditional programming intuitions, since in languages such as C, (b ? x = y / 0 : x = y;) does not raise an exception unless b is true. Hence in this paper we propose to adopt the *lazy* interpretation of the relational semantics for conditional expressions. Note that this accords with the *lazy* interpretation of array lookup which is considered in the original paper on relational semantics [4]. Hence for the examples above we find:

(1) The undefined expression a[3] causes the thenexp to be *false* requiring the condition $y = 0$ to be false, leaving solutions $x = 0 \land y = 1$ and $x = 2 \land y = 2$.
(2) The undefined expression a[3] causes the thenexp to be $\bot$ requiring the condition $y = 0$ to be false, or the whole expression evaluates to *false* leaving solutions $x = 0 \land y = 1$ and $x = 2 \land y = 2$. Note how the lazy approach gives the identical answers to the "equivalent" expression (1).

(3) The **thenexp** `a[y]` is only of interest when $y = 0$ (and hence if `a[y]` is $\perp$), so the else case must hold, leaving solutions $x = 0 \wedge y = 1$ and $x = 0 \wedge y = 2$.

(4) The then case is $\perp$ so the second disjunct evaluates to *false* when $y = 0$, leaving solutions $x = 0 \wedge y = 1, x = 2 \wedge y = 2$ and $x = 0 \wedge y = 0, x = 1 \wedge y = 0, x = 2 \wedge y = 0$.

MiniZinc supports an extended conditional expression of the form

```
if c1 then e1 elseif c2 then ... else ek endif
```

This is semantically equivalent to

```
if c1 then e1 else if c2 then ... else ek endif ... endif
```

To simplify presentation, we will assume an alternative syntax that uses two arrays of size k: `ite([c1,c2,...,true],[e1,e2,...,ek])` The c array are the *conditions*, the e array are the *results*. Note how the last condition $c_k = true$ for the else case. The (lazy) relational semantics applied to this expression requires that the conditional takes the value of expression e_i iff $c_i \wedge \bigwedge_{j=1}^{i-1} \neg c_j$ holds. Hence if e_i is $\perp$ then either the ith condition cannot hold or the nearest enclosing Boolean context will be *false*. This is the same semantics as if we treated the `ite` as a nested sequence of `if then else endif` expressions.

4   Translating Conditionals

In this section we examine how to translate the conditional expression `ite(c,e)` where c is an array of k Boolean expressions with the last one being *true*, and e is an array of k expressions. We usually introduce a new variable x to hold the value of this expression, `x = ite(c,e)`.

Boolean Result. In its most basic form, a conditional is a Boolean expression, i.e., x has type **par bool** or **var bool**. In this case the semantics can be defined in terms of simple expressions:[2]

```
x = forall (i in 1..k) (c[i] /\ not exists (c[1..i-1]) -> e[i])
```

If x is *true* (e.g. if the conditional appears directly as a constraint) this can be encoded using k simple clauses. Because Boolean values can never be $\perp$ (they simply become *false*) there is no difficulty with undefinedness here.

Non-Boolean Result. If the `e[i]` are not Boolean, a straightforward translation scheme would result in the Boolean expression:

```
forall (i in 1..k) (c[i] /\ not exists (c[1..i-1]) -> x = e[i])
```

If none of the expressions `e[i]` can be $\perp$ this is correct. But if case i is the selected case and `e[i]` is $\perp$, then this translation causes the entire model to have no solution. If the conditional expression occurs at the root context this is correct, but if it occurs in a non-root context, the desired semantics is that it enforces that the context is *false*.

---

[2] We use array slicing notation `c[1..u]` equivalent to `[ c[j] | j in 1..u ]`.

Example 2. Consider the translation of example (3). Since the expression appears in the root context we can use the simple translation:

```
constraint b1 <-> y = 0;               % Boolean for condition y = 0
constraint b2 <-> x = a[y];            % then expression
constraint b3 <-> x = a[1];            % else expression
constraint b1 -> b2;                   % then case
constraint true /\ not b1 -> b3;       % else case
```

The solutions, projecting onto x and y, are $x = 0 \wedge y = 1$ and $x = 0 \wedge y = 2$ as expected. Note the undefinedness of the expression a[y] is captured by the reified constraint b2 <-> x = a[y].   □

In order to deal with non-root contexts we introduce an explicit representation of undefinedness by associating with every term t appearing in a non-root context the Boolean def(t) which records whether it is defined (*true*) or undefined, i.e. def(t) = *false* iff t = ⊥. The translation will ensure that if def(t) = *true* then t takes its appropriate defined value, and if def(t) = *false* then t is free to take some value (which does not affect the satisfiability of other constraints). We can now define the full semantics of a conditional ite(c,e):

```
forall(i in 1..k)( def(x)    /\ c[i] /\ not exists(c[1..i-1]) -> x=e[i] )
forall(i in 1..k)( def(e[i]) /\ c[i] /\ not exists(c[1..i-1]) -> def(x) )
```

Let us look at the two parts of this definition in turn. The first part (line 1) states that if the overall expression is defined (def(x)), and alternative i is selected, then the expression takes the value of branch i. Due to the relational semantics of x=e[i], this also implies that e[i] is defined. The second part (line 2) states that if the value of the selected branch is defined, then the value of the overall conditional is defined. Note how if e[i] is ⊥ and the ith case is chosen, then the first part will enforce def(x) to be *false*, and there will be effectively no usage of the value of e[i].

Example 3. Let us consider the translation of example (4). Here the conditional expression is not in a root context. The resulting translated form is

```
constraint b1 \/ b2;                   % top level disjunction
constraint b1 <-> y = 0;               % Boolean for condition y = 0
constraint b2 <-> def(x);          % second disjunct holds iff x is defined
constraint b3 <-> x = a[3];            % then expression
constraint b4 <-> x = a[y];            % else expression
constraint def(x) /\ b1 -> b3;         % then case
constraint def(a[3]) /\ b1 -> def(x);
constraint def(x) /\ true /\ not b1 -> b4; % else case
constraint def(a[y]) /\ true /\ not b1 -> def(x);
constraint def(a[3]) <-> false;        % definedness of base
constraint def(a[y]) <-> y in 1..2;
```

The solutions, projecting onto x and y, are $x = 0 \wedge y = 1$, $x = 2 \wedge y = 2$, $x = 0 \wedge y = 0$, $x = 1 \wedge y = 0$, and $x = 2 \wedge y = 0$ as expected.   □

5   Improving the Translation of Conditionals

The direct translation of conditionals explored in the previous section is correct, but does not necessarily propagate very well. We can create better translations if we know more about the underlying solver we will be using.

5.1   Element Translation

We first restrict ourselves to the case that none of the expressions e can be $\perp$. In this case we can translate x = ite(c,e) as x = e[arg_max(c)]. Note that the arg_max function returns a variable constrained to be the least index which takes the maximum value. Since c[k] is always *true* this returns the first condition which is *true*. Thus this expression first calculates the first case j whose condition holds, and then uses an *element constraint* to equate x to the appropriate expression in e. Flattening simply yields x = e[j] /\ maximum_arg(c, j) where maximum_arg is the predicate version of the arg_max function. Note in the case of the simple conditional if b then e[1] else e[2] endif we can avoid the use of arg_max and simply use x = e [2 - b]. If b is *true* we obtain x = e[1] and otherwise x = e[2].

The advantage of the element translation is that we can get stronger propagation. A solver that has native support for the element constraint can apply a form of *constructive disjunction* [13]: it can reason about all of the cases together, and propagate any common information that all cases (disjuncts) share.

Example 4. Consider the translation of the expression
$$x = \text{if } c1 \text{ then } a \text{ elseif } c2 \text{ then } b \text{ else } c \text{ endif}$$
where variables $a, b, c \in \{5, 8, 11\}$. The implication form compiles to

```
constraint b1 <-> x = a;
constraint c1 -> b1;
constraint b2 <-> x = b;
constraint not c1 /\ c2 -> b2;
constraint b3 <-> x = c;
constraint not c1 /\ not c2 -> b3;
```

Initially no propagation is possible. But the element translation is

```
constraint maximum_arg([b1,b2,true],j);
constraint x = [a,b,c][j];
```

Immediately propagation determines that x can only take values $\{5, 8, 11\}$. If some other constraints then cause the value 5 to be removed from the domains of a, b and c, the element constraint will also remove it from x, while the implication form will not be able to perform this kind of reasoning.   □

We can show that the element translation is domain consistent when using domain consistent propagators for element and maximum_arg. Hence it is a strongest possible decomposition, and therefore must propagate at least as much as the implication decomposition.

Theorem 1. *The element translation of* **ite** *is domain consistent assuming domain propagation of the primitives.*[3]   □

In order to take into account definedness we can extend the translation by using another element constraint to determine the definedness of the result:

```
x = e [ arg_max(c) ] /\
def(x) = [ def(e[i]) | i in index_set(e) ] [ arg_max(c) ];
```

Suppose e[i] is undefined, then since def(e[i]) is *false* the resulting variable will be unconstrained, which will leave x unconstrained as well.

5.2 Domain Consistent Decomposition of maximum_arg

Note that the element translation relies on domain consistent propagation for the element and maximum_arg constraints to enforce domain consistency. Many CP solvers support domain propagation for element, but not necessarily for maximum_arg. We can enhance the standard decomposition of maximum_arg so that it enforces domain consistency.

The new decomposition is shown below. It tests for the special case that one of the elements in the array is known to be at the upper bound of all elements at compile time, i.e exists(j in 1..u)(lb([x[j]) = ubx). The decomposition introduces variables d such that d[i] is true iff x[k]=ubx for some k<=i.

```
predicate maximum_arg(array[int] of var int: x, var int: y) =
    let { int: l = min(index_set(x));
          int: u = max(index_set(x));
          int: ubx = ub_array(x);
        } in
    if exists(j in 1..u)(lb(x[j]) = ubx)        % max is known to be ubx
    then let { array[1..u] of var bool: d; } in
        x[y] = ubx /\                           % ith case must be equal to ub
        forall(j in 1..u)(x[j] = ubx -> y <= j) /\ % lower bound
        d[1] = (x[1] = ubx) /\
        forall(j in 1+1..u)(d[j] <-> (d[j-1] \/ (x[j] = ubx))) /\
        forall(j in 1..u)(not d[j] -> y >= j+1)     % upper bound
    else % otherwise use standard integer decomposition
        maximum_arg_int(x, y)
    endif;
```

Theorem 2. *The decomposition of* **maximum_arg** *above maintains domain consistency in the case where the maximum is known.*   □

5.3 Domain Consistent Propagator for maximum_arg

Given the key role that maximum_arg plays for translating conditionals it may be worth building a specialised propagator for the case where it applies: on an array

---

[3] Proofs of theorems not included in this paper can be found in the extended version at https://www.minizinc.org/pub/mzn_conditionals.pdf.

of Booleans where the maximum is known to be *true*. The following pseudocode implements a domain consistent propagator for this case. In a learning solver each propagation must be explained. The correct explanations are given in the parentheses following the *because* statements.

index_of_first_true(x, y)
    $ol := lb(y)$
    for $(j \in 1..ol - 1)$
        propagate $\neg x[j]$ *because* $(y \geq ol \rightarrow \neg x[j])$
    $l := ub(y) + 1$
    $u := lb(y)$
    for $(j \in dom(y))$
        if $(l > ub(y) \wedge ub(x[j]) = true)$ $l := j$
        if $(ub(x[j]) = false)$ propagate $y \neq j$ *because* $(\neg x[j] \rightarrow y \neq j)$
        if $(lb(x[j]) = true)$ $u := j$; **break**
    propagate $y \geq l$ *because* $(y \geq ol \wedge \bigwedge_{k \in ol..l-1} y \neq k \rightarrow y \geq l)$
    propagate $y \leq u$ *because* $(x[u] \rightarrow y \leq u)$
    if $(lb(y) = ub(y))$ propagate $x[lb(y)]$ *because* $(y = lb(y) \rightarrow x[lb(y)])$
    for $(j \in ol..lb(y) - 1)$
        propagate $\neg x[j]$ *because* $(y \geq lb(y) \rightarrow \neg x[j])$

The propagation algorithm above first ensures that all indexes lower than the lower bound of y cannot be true. Propagation does nothing if this is already the case. The propagator iterates through the remaining possible values of y. It finds the first element indexed by y that might be true and records this as l. If it finds index positions j where $x[j]$ is false, it propagates that y cannot take these values. If it finds an index position j where $x[j]$ is true, it records the position as u and breaks the loop. Once the loop is exited it sets the new lower bound as l and the new upper bound as u. If the *updated* lower bound and upper bound of y are equal we propagate that x at that position must be true. Finally we ensure that all x positions less than the new lower bound of y are false. Note that if $l = ub(y) + 1$ at the end of the loop this effectively propagates failure, similarly if $lb(y) = ub(y)$ and $x[lb(y)]$ is false, or $x[j], j < lb(y)$ is true.

Theorem 3. *The index_of_first_true(x, y) propagator enforces domain consistency for maximum_arg(x, y) when the max is known to be true.* □

5.4   Linear Translation

When the target solver to be used for conditional constraints is a MIP solver, the conditional constraints have to be linearised. The implication form of a conditional constraint is in fact not too difficult to linearise, however it may result in a weak linear relaxation. We now discuss how to improve over the

Example 5. Consider the linearisation of the implication form of the expression:
```
x = if c1 then 5 elseif c2 then 8 else 11 endif
```

Assuming the initial bounds on x are 0..100, the result is (see [2] for details):

```
constraint x >= 5*b1;              % b1 -> x >= 5
constraint x <= 5 + 95*(1-b1);     % b1 -> x <= 5
constraint x >= 8*b2;              % b2 -> x >= 8
constraint x <= 8 + 92*(1-b2);     % b2 -> x <= 8
constraint x >= 11*b3;             % b3 -> x >= 11
constraint x <= 11 + 89*(1-b3);    % b3 -> x <= 11
constraint b1 >= c1;               % c1 -> b1
constraint c1 + b2 >= c2;          % not c1 /\ c2 -> b2
constraint c1 + c2 + b3 >= 1;      % .. -> b3
```

This has a weak linear relaxation, x can take any value in the range 0..100. □

We can do better, at least in the case where all the expressions e are fixed, by translating $x = \mathtt{ite(c,e)}$ as follows:

```
int: k = length(c);
array[1..k] of var 0..1: b;
constraint forall(i in 1..k) (sum(c[1..i-1]) + b[i] >= c[i]);
constraint 1 = sum(b);       % redundant: exactly one case is true
constraint x = sum(i in 1..k)(b[i]*e[i]);
```

The case selector variables b are the same as in the implication form, but we add that exactly one of them is 1, and generate x as a linear combination of the cases. If all elements of the array e are fixed integer expressions then the last constraint is simply a linear equation. In the case of a conditional with just one condition we again have a much simpler translation: $x = \mathtt{b*e[1] + (1-b)*e[2]}$;

Example 6. The result of translating the expression of Example 5 is

```
constraint b1 >= c1;                        % c1 -> b1
constraint c1 + b2 >= c2;                    % not c1 /\ c2 -> b2
constraint c1 + c2 + b3 >= 1;                % .. -> b3
constraint b1 + b2 + b3 = 1;                 % exactly one case
constraint x = b1 * 5 + b2 * 8 + b3 * 11;    % definition of x
```

Bounds propagation can immediately determine from the last equation that x takes values in the range 0..24. A linear solver can immediately determine (using the last two constraints) that x takes values in the range 5..11. □

The example above shows that the linear translation can sometimes be much stronger than the linearisation of the implication translation. The following theorem establishes that it is indeed never weaker.

Theorem 4. *Assuming all values in e are constant, the linear relaxation of the linear translation of a conditional expression is no weaker than the linear relaxation of the implication translation.* □

Since the linear translation is not worthwhile unless all the e expressions are fixed we do not need to extend it to handle undefinedness.

6   Improving "Procedural" Code

Many non-expert users of MiniZinc make heavy use of the conditional statement, since they are familiar with it from procedural programming. This section shows that reformulating the procedural style can lead to improved propagation when the element translation is used.

Example 7. Consider the constraint

```
if c1 then x = 5 elseif c2 then x = 8 else x = 11 endif;
```

The resulting element translation is

```
constraint b1 <-> x = 5;
constraint b2 <-> x = 8;
constraint b3 <-> x = 11;
constraint maximum_arg([c1,c2,true],j);
constraint true = [b1,b2,b3][j];
```

Initially a CP solver will be able to propagate no information. But consider the equivalent expression

```
constraint x = if c1 then 5 elseif c2 then 8 else 11 endif;
```

discussed in Example 5. Using the element translation the solver will immediately propagate that the domain of x is $\{5, 8, 11\}$.

The linear translation of the original form above is the same as shown in Example 5. The linear relaxation of this system does not constrain x more than its original bounds 0..100. The linear relaxation of the translation of the equivalent expression discussed in Example 6 enforces that x is in the range 5..11. □

We can define a transformation for arbitrary conditional expressions that share at least some equational constraints for the same variable. This is a very common modelling pattern for inexperienced modellers. Consider the constraint

```
if x = 0 then y = 0 /\ z = 0
elseif x <= 1 then y = 1 /\ z = x
elseif x <= u then y = x /\ z = u
else x >= y + z endif
```

In most cases the conditional (conjunctively) defines values for y and z. To compile this efficiently we can in effect duplicate the conditional structure to give separate partial definitions of y, z and any leftover constraints.

```
y = if x=0 then 0 elseif x<=1 then 1 elseif x<=u then x else y endif /\
z = if x=0 then 0 elseif x<=1 then x elseif x<=u then u else z endif /\
if x = 0 then true elseif x<=1 then true elseif x<=u then true
else x >= y + z endif
```

When translating the result using the element translation we can reuse the maximum_arg computation for each part, arriving at

```
var 1..4: j;
constraint maximum_arg([x = 0, x <= 1, x <= u, true], j);
constraint y = [0,1,x,y][j];
constraint z = [0,x,u,z][j];
constraint true = [true,true,true,x >= y+z][j];
```

We can show that lifting the equalities out of an if-then-else expression can only improve propagation. The theorem assumes that each branch has an equality for variable x, if this is not the case we can add the equation x = x to give the right syntactic form.

Theorem 5. *Using the element translation for a conditional constraint of the form* $ite(c, [x = ex_1 \wedge r_1, x = ex_2 \wedge r_2, \ldots x = ex_m \wedge r_m])$ *and the equivalent constraint* $x = ite(c, [ex_1, ex_2, ..., ex_m]) \wedge ite(c, [r_1, r_2, \ldots, r_m]$, *then the second form propagates at least as much.*   □

7 Experimental Evaluation

This section presents an evaluation of the different approaches to compiling conditionals using an artificial unit test and a more natural model of a problem that uses conditionals.

All experiments were run on an Intel Core i7 processor at 4 GHz with 16 GB of RAM. The models were compiled with current development versions of MiniZinc (revision e1d9d10), Gecode (revision b3fceb0) and Chuffed (revision b74a2d7), CBC version 2.9.8/1.16.10 and CPLEX version 12.8.0. The solvers were run in single-core mode with a time out of 120 s. All models use fixed search for CP solvers so that the differences for traditional CP solvers only arise from differences in propagation, in learning solvers differences also arise from different learning. For the MIP solvers CBC and CPLEX, presolving was switched off to improve consistency of results.

7.1 Unit Tests

Let π be a sorted sequence of $4n$ distinct integers in the range $0 \ldots 100n$, with $\pi(i)$ denoting the ith integer in the sequence. We build a (procedural) constraint system consisting of two ite constraints (the ++ operator stands for array concatenation):

$$ite([x \leq \pi(4i - 3)|i \in 1..n - 1]\text{++}[\textbf{true}], [y = \pi(4i - 2)|i \in 1..n])$$
$$ite([y \leq \pi(4i - 1)|i \in 1..n - 1]\text{++}[\textbf{true}], [y = \pi(4i)|i \in 1..n - 1]\text{++}[0])$$

By definition the first constraint implies $y > x$ unless $x > \pi(4n - 7)$ then $y = \pi(4n - 2)$. The second constraint implies $x > y$ unless $y > \pi(4n - 5)$ then $x = 0$. Together these are unsatisfiable. We also consider the equivalent functional form

$$y = ite([x \leq \pi(4i - 3)|i \in 1..n - 1]\text{++}[\textbf{true}], [\pi(4i - 2)|i \in 1..n])$$
$$x = ite([y \leq \pi(4i - 1)|i \in 1..n - 1]\text{++}[\textbf{true}], [\pi(4i)|i \in 1..n - 1]\text{++}[0])$$

Table 1. Unit testing for the Gecode and Chuffed CP solvers.

n	Gecode					Chuffed				
	$\rightarrow$	$[]$	$[]_P$	$x = []$	$x = []_P$	$\rightarrow$	$[]$	$[]_P$	$x = []$	$x = []_P$
25	0.07	0.06	**0.05**	0.06	0.06	0.07	0.07	0.06	**0.05**	**0.05**
50	0.06	0.07	**0.05**	0.06	**0.05**	0.08	0.08	0.08	0.06	**0.05**
100	0.21	0.15	0.22	0.24	**0.05**	0.10	0.10	0.07	0.17	**0.05**
200	0.15	0.15	0.07	0.13	**0.05**	0.17	0.17	0.08	0.14	**0.05**
400	0.38	0.33	0.09	0.30	**0.06**	0.42	0.36	0.11	0.31	**0.07**
800	1.43	1.12	0.21	1.03	**0.08**	1.43	0.98	0.17	1.03	**0.09**
1600	5.11	3.08	0.24	3.24	**0.12**	5.40	3.68	0.27	3.39	**0.15**
3200	26.23	12.92	0.44	12.78	**0.19**	26.29	13.53	0.53	13.70	**0.23**
6400	—	67.39	3.78	72.43	**0.39**	—	60.54	1.97	60.64	**0.52**

Table 2. Unit testing for the CBC and CPLEX MIP solvers.

n	CBC			CPLEX		
	$\rightarrow$	$x = []$	$\sum$	$\rightarrow$	$x = []$	$\sum$
25	0.21	1.84	**0.06**	1.41	0.19	**0.05**
50	0.27	1.68	**0.08**	0.22	0.37	**0.07**
100	0.95	13.08	**0.17**	0.71	1.16	**0.12**
200	4.69	48.92	**0.38**	3.90	4.52	**0.30**
400	40.99	—	**1.27**	38.62	17.12	**1.01**
800	—	—	**5.34**	—	71.97	**3.94**
1600	—	—	**20.51**	—	—	**18.76**
3200	—	—	**94.75**	—	—	**70.37**

Table 1 compares the Gecode and Chuffed CP solvers on five different versions: $\rightarrow$ is the implication translation (which is identical for both forms); $[]$ is the element translation of the procedural form with domain consistent `maximum_arg` decomposition; $[]_P$ is the element translation of the procedural form with domain consistent `maximum_arg` propagator; $x = []$ is the element translation of the functional form with domain consistent `maximum_arg` decomposition; and $x = []_P$ is the element translation of the functional form with domain consistent `maximum_arg` propagator; A—indicates 120 s timeout reached. The results demonstrate the clear benefits of the mapping from procedural form to functional form, and the benefit of the element translation, in particular with a global propagator for `maximum_arg` over the implication form.

Table 2 shows the results of the MIP solvers of the functional form of the model with: $\rightarrow$ linearisation of the implication translation; $x = []$ linearisation of the element translation with decomposed `maximum_arg`; and $\sum$ the direct linearisation defined in Sect. 5.4. Note here we cannot control the search, so that there may be more variance in results due to other factors. The linearisation of the implication translation is generally better than the linearisation of the element translation, since the `maximum_arg` decomposition makes the whole thing much more complex. The direct linear translation is substantially better and much more scalable.

7.2 Fox-Geese-Corn

To study the effects of different translations on a more realistic model, we consider a generalisation of the classic *Fox-Geese-Corn* puzzle, where a farmer needs to transport goods from one side of a river to the other side in several trips. It contains several rules such as "When the farmer leaves some foxes and geese alone on one side of the river, and there are more foxes than geese, one fox dies in argument over geese, and no geese die; if there are no more foxes than geese, each fox eats a goose". The overall goal is to maximise the farmer's profit of goods successfully transported across.[4]

The most natural way of modelling this problem follows the structure of the specification, using conditionals to constrain state variables for each time point and type of object. We manually analysed models written by students and submitted as part of an online assignment for the Coursera course *Modeling Discrete Optimization*, which ran from 2015–2018. We have identified a number of different modelling approaches based on conditional expressions. Almost all students use the "procedural" syntax. In general, the constraints take the form such as `if cond then geese[i+1]=geese[i]-c elseif ... endif`, where `geese[i]` is the number of geese at step `i`, `cond` specifies the condition from the rules, and `c` is a constant (e.g. 1 in the case that one goose dies) or a variable (e.g. `fox[i]` in the case that each fox eats one goose). We manually performed the conversion into expression form such as `geese[i+1] = if cond then geese[i]-c elseif ... endif` and call this Model 1.

Several students described all cases explicitly in the conditions, not noticing that the case distinction is in fact exhaustive. For example, they might have written `if cond1 then geese[i+1]=geese[i]-c elseif not cond then geese[i+1]=geese[i] else true endif`, instead of the simpler `if cond then geese[i+1]=geese[i]-c else geese[i+1]=geese[i] endif`. The resulting expression form, which we call Model 2, would contain the less compact `geese[i+1]=if cond then geese[i]-c elseif not cond then geese[i] else geese[i+1] endif`.

Table 3 shows the results for the two models on ten instances, with the implication ($\rightarrow$) and element ($x = []_P$) encoding, and the additional linear encoding ($\sum$) for CPLEX. A number without superscript represents time in seconds for finding the optimal solution and proving optimality. For cases where search timed out, a number with superscript (e.g. 1.75^{120}) represents the time when the best solution was found (1.75 s) and the objective value of the best solution found (120). The last line in each table summarises the results for each solver: for a

---

[4] See https://github.com/minizinc/minizinc-benchmarks.

given solver, an encoding wins for an instance if it proves optimality faster than the other encoding(s), or if neither encoding proves optimality, if it finds a better solution, or the same objective but faster. A—represents that no solution was found within the 120 s timeout. Numbers in bold indicate the winning model (highest objective found in least amount of time) for each solver.

In more complex models the relative performances of the different encodings are less distinct. For Gecode the $x = []_P$ version continues to be superior, for Chuffed the $\rightarrow$ version is more competitive because it can learn on intermediate literals. Surprisingly for CPLEX the $x = []$ translation is usually better than the implication translation and overall best for Model 1. For Model 2 the Σ translation is strongest even though in this case the e arguments are not fixed.

Table 3. Fox-Geese-Corn

Model 1 (exhaustive conditional):

	Gecode		Chuffed		CPLEX		
	$\rightarrow$	$x = []_P$	$\rightarrow$	$x = []_P$	$\rightarrow$	$x = []$	Σ
1	**0.16**	**0.16**	0.10	**0.06**	1.86	1.74	**0.22**
2	0.53	**0.42**	**0.12**	0.13	0.92	0.76	**0.67**
3	1.75^{120}	$\mathbf{1.39^{120}}$	7.18^{148}	8.70^{148}	95.46^{112}	$\mathbf{50.57^{141}}$	$\mathbf{11.92^{132}}$
4	0.79	**0.64**	0.34	**0.25**	0.45	**0.28**	0.29
5	113.53^{38}	$\mathbf{102.16^{40}}$	97.97^{52}	$\mathbf{86.38^{52}}$	48.01^{40}	$\mathbf{105.14^{46}}$	86.35^{44}
6	116.91^{64}	$\mathbf{91.89^{64}}$	$\mathbf{90.29^{71}}$	103.45^{71}	$\mathbf{106.70^{71}}$	118.12^{71}	19.22^{69}
7	72.36	**61.35**	**1.50**	1.55	6.63	**1.73**	2.46
8	88.34	**75.29**	2.43	**2.27**	10.61	**4.13**	6.16
9	24.61^{72}	$\mathbf{19.39^{72}}$	84.65^{130}	$\mathbf{43.01^{130}}$	93.49^{116}	**57.06**	58.58^{128}
10	0.52^{140}	$\mathbf{0.40^{140}}$	81.76^{280}	$\mathbf{77.41^{280}}$	36.82^{219}	$\mathbf{116.24^{430}}$	115.81^{391}
Wins	0	9	4	6	1	7	2

Model 2 (additional **else true** case):

	Gecode		Chuffed		CPLEX		
	$\rightarrow$	$x = []_P$	$\rightarrow$	$x = []_P$	$\rightarrow$	$x = []$	Σ
1	0.16	**0.06**	0.10	**0.07**	1.76	**0.41**	**0.28**
2	0.13	**0.11**	0.10	**0.10**	0.90	**0.79**	0.96
3	$\mathbf{0.15^{75}}$	$\mathbf{0.15^{75}}$	$\mathbf{3.16^{148}}$	3.54^{148}	45.76^{142}	$\mathbf{15.16^{148}}$	82.25^{143}
4	0.14	**0.12**	**0.11**	0.12	0.53	0.60	**0.33**
5	0.21^{14}	$\mathbf{0.16^{14}}$	$\mathbf{39.62^{52}}$	57.61^{52}	109.03^{32}	$\mathbf{118.09^{47}}$	23.22^{24}
6	0.13^{21}	$\mathbf{0.10^{21}}$	$\mathbf{48.06^{71}}$	49.84^{71}	12.65^{67}	111.29^{65}	$\mathbf{14.13^{70}}$
7	3.09	**2.56**	**0.29**	0.32	**2.55**	2.77	2.69
8	1.90	**1.47**	0.24	**0.20**	**1.30**	1.37	1.33
9	0.15^{29}	$\mathbf{0.12^{29}}$	55.69^{130}	$\mathbf{54.68^{130}}$	64.47^{130}	51.36^{132}	$\mathbf{36.42^{132}}$
10	0.34^{112}	$\mathbf{0.25^{112}}$	$\mathbf{21.46^{350}}$	22.62^{350}	—	51.92^{419}	$\mathbf{75.52^{525}}$
Wins	0	9	6	3	2	3	5

8   Conclusion

Conditional expressions are one of the most expressive constructs that appear in modelling languages, allowing natural expression of complex constraints. They are also frequently used by beginner modellers used to thinking procedurally. In this paper we show how to translate conditional expressions to a form that underlying solvers can handle efficiently. Unit tests clearly demonstrate the improvements from new translations, though this is not so clear on more complex models, where other factors interact with the translations. The concepts presented here are available in MiniZinc 2.3.0, Gecode 6.2.1, and Chuffed 0.10.5.

Acknowledgements. We would like to thank the anonymous reviewers for their comments that helped improve this paper. This work was partly sponsored by the Australian Research Council grant DP180100151.

References

1. Bacchus, F., Walsh, T.: Propagating logical combinations of constraints. In: Kaelbling, L.P., Saffiotti, A. (eds.) IJCAI-05, Proceedings of the Nineteenth International Joint Conference on Artificial Intelligence, Edinburgh, Scotland, UK, 30 July - 5 August 2005, pp. 35–40 (2005)
2. Belov, G., Stuckey, P.J., Tack, G., Wallace, M.: Improved linearization of constraint programming models. In: Rueher, M. (ed.) CP 2016. LNCS, vol. 9892, pp. 49–65. Springer, Cham (2016). https://doi.org/10.1007/978-3-319-44953-1_4
3. Fourer, R., Kernighan, B.: AMPL: A Modeling Language for Mathematical Programming. Duxbury, Massachusetts (2002)
4. Frisch, A.M., Stuckey, P.J.: The proper treatment of undefinedness in constraint languages. In: Gent, I.P. (ed.) CP 2009. LNCS, vol. 5732, pp. 367–382. Springer, Heidelberg (2009). https://doi.org/10.1007/978-3-642-04244-7_30
5. Frisch, A.M., Harvey, W., Jefferson, C., Hernández, B.M., Miguel, I.: Essence: a constraint language for specifying combinatorial problems. Constraints 13(3), 268–306 (2008)
6. Hooker, J.N.: Integrated Methods for Optimization, 2nd edn. Springer, Heidelberg (2011). https://doi.org/10.1007/978-1-4614-1900-6
7. Jefferson, C., Moore, N.C.A., Nightingale, P., Petrie, K.E.: Implementing logical connectives in constraint programming. Artif. Intell. 174(16–17), 1407–1429 (2010)
8. Lhomme, O.: Arc-consistency filtering algorithms for logical combinations of constraints. In: Régin, J.-C., Rueher, M. (eds.) CPAIOR 2004. LNCS, vol. 3011, pp. 209–224. Springer, Heidelberg (2004). https://doi.org/10.1007/978-3-540-24664-0_15
9. McKinnon, K.I.M., Williams, H.P.: Constructing integer programming models by the predicate calculus. Ann. Oper. Res. 21(1), 227–245 (1989)
10. Nethercote, N., Stuckey, P.J., Becket, R., Brand, S., Duck, G.J., Tack, G.: MiniZinc: towards a standard CP modelling language. In: Bessière, C. (ed.) CP 2007. LNCS, vol. 4741, pp. 529–543. Springer, Heidelberg (2007). https://doi.org/10.1007/978-3-540-74970-7_38
11. Nightingale, P.: Savile Row, a constraint modelling assistant (2018). http://savilerow.cs.st-andrews.ac.uk/

12. Van Hentenrcyk, P.: The OPL Optimization Programming Language. MIT Press, Cambridge (1999)
13. Van Hentenryck, P., Saraswat, V., Deville, Y.: Design, implementation, and evaluation of the constraint language cc(FD). In: Podelski, A. (ed.) TCS School 1994. LNCS, vol. 910, pp. 293–316. Springer, Heidelberg (1995). https://doi.org/10.1007/3-540-59155-9_15

Training Binarized Neural Networks
Using MIP and CP

Rodrigo Toro Icarte[1,2(✉)], León Illanes[1], Margarita P. Castro[3],
Andre A. Cire[4], Sheila A. McIlraith[1,2], and J. Christopher Beck[3]

[1] Department of Computer Science, University of Toronto, Toronto, Canada
{rntoro,lillanes,sheila}@cs.toronto.edu
[2] Vector Institute, Toronto, Canada
[3] Department of Mechanical and Industrial Engineering, University of Toronto,
Toronto, Canada
{mpcastro,jcb}@mie.utoronto.ca
[4] Department of Management, University of Toronto Scarborough, Toronto, Canada
acire@utsc.utoronto.ca

Abstract. Binarized Neural Networks (BNNs) are an important class
of neural network characterized by weights and activations restricted to
the set $\{-1, +1\}$. BNNs provide simple compact descriptions and as such
have a wide range of applications in low-power devices. In this paper, we
investigate a model-based approach to training BNNs using constraint
programming (CP), mixed-integer programming (MIP), and CP/MIP
hybrids. We formulate the training problem as finding a set of weights
that correctly classify the training set instances while optimizing objective functions that have been proposed in the literature as proxies for generalizability. Our experimental results on the MNIST digit recognition
dataset suggest that—when training data is limited—the BNNs found by
our hybrid approach generalize better than those obtained from a state-
of-the-art gradient descent method. More broadly, this work enables the
analysis of neural network performance based on the availability of optimal solutions and optimality bounds.

Keywords: Binarized Neural Networks · Machine learning ·
Constraint programming · Mixed integer programming ·
Discrete optimization

1 Introduction

Deep learning is responsible for recent breakthroughs in image recognition, speech recognition, language translation, and artificial intelligence [7,18].
Roughly speaking, deep learning aims to find a set of weights for a neural network (NN) that maps training inputs to target outputs (e.g., English sentences
to their Spanish translations), a process known as *training*. The most notable
feature of deep learning is that NNs *generalize* when trained over large datasets,
i.e., they can map unseen inputs to their target outputs with high accuracy.

© Springer Nature Switzerland AG 2019
T. Schiex and S. de Givry (Eds.): CP 2019, LNCS 11802, pp. 401–417, 2019.
https://doi.org/10.1007/978-3-030-30048-7_24

Hubara et al. [9] recently showed that Binarized Neural Networks (BNNs)—NNs with weights and activations in $\{-1, +1\}$—have comparable test performance to standard NNs in two well-known image recognition datasets. This is a remarkable result because BNNs can be implemented using Boolean operations with low memory and energy consumption, enabling, for example, the application of deep learning in mobile devices. While training BNNs is a discrete optimization problem, it has not been addressed by model-based techniques such as mixed-integer programming (MIP) or constraint programming (CP). Instead, BNNs are trained using gradient descent (GD) methods over continuous weights which are binarized during the forward pass of the algorithm [17,20,26,33].

Model-based approaches have stronger convergence guarantees than GD and, as such, can potentially find better solutions given enough time and resources. There are two reasons, however, why a model-based approach—in particular MIP—may be disadvantageous, as stated by Gambella et al. [6]. First, it may not scale to large datasets since the size of the model depends on the size of the training set. Second, solutions with provably-optimal training error are likely to *overfit* the data, that is, they will classify the training examples effectively but will not generalize.

The main contribution of this paper is a collection of model-based training methods that explicitly address these issues. The key insights are (i) improving scalability by taking advantage of CP's ability to find BNNs that fit the training data and (ii) avoiding overfitting by optimizing well-known proxies for generalizability. Specifically, we propose MIP, CP, and CP/MIP hybrid approaches to train BNNs while optimizing objective functions based on two machine learning principles for generalization: *simplicity* and *robustness*.

We experimented over subsets of the widely-used MNIST dataset [19]. Our experiments focused on limited training data, a setting known as *few-shot learning* [32]. This setting is important in Machine Learning because collecting labeled data is expensive—or even impossible—in many important real-world applications, including healthcare [3,21]. Our results show that our hybrid methods scale significantly better than MIP (i.e., they solve problems with larger networks and more training data) and produced BNNs that generalize better than those trained with GD. In fact, our BNNs correctly classified up to 3 times more unseen examples than BNNs learned by GD on a few-shot learning regime. However, model-based approaches are still far from scaling at the level of GD and, hence, GD should be preferred when a large amount of data is available. Finally, since model-based approaches find provably-optimal solutions—GD does not—they allow for principled empirical comparisons between generalization proxies. Our results suggest that optimizing for robustness leads to better test performance than simplicity.

2   Problem Definition

BNNs are NNs with weights and activations restricted to the values -1 and $+1$. A BNN architecture is defined by the number of layers L and the set of

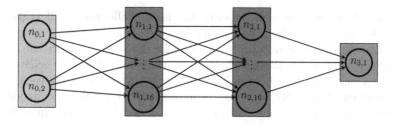

Fig. 1. A BNN with 2 inputs, two hidden layers with 16 neurons each, and 1 output neuron. We use the notation $n_{\ell j}$ to represent neuron j from layer ℓ.

neurons $\mathcal{N} = \langle N_0, \ldots, N_L \rangle$, where N_ℓ corresponds to the set of neurons in layer $\ell \in \{0, .., L\}$. For instance, Fig. 1 shows a BNN with two input neurons ($n_{0,1}$ and $n_{0,2}$), two hidden layers with 16 neurons each ($n_{1,1}$ to $n_{1,16}$ and $n_{2,1}$ to $n_{2,16}$), and one output neuron ($n_{3,1}$), i.e., its architecture is $\mathcal{N} = \langle N_0, N_1, N_2, N_3 \rangle$ with $|N_0| = 2$, $|N_1| = |N_2| = 16$, and $|N_3| = 1$. Every neuron $j \in N_\ell$ ($\ell \geq 1$) is connected to every neuron $i \in N_{\ell-1}$ by a weight $w_{i\ell j} \in \{-1, 0, 1\}$. Note that setting $w_{i\ell j} = 0$ is equivalent to removing the corresponding connection from the BNN. Given a value $\mathbf{x}$ for the input neurons, the *preactivation* $a_{\ell j}(\mathbf{x})$ of neuron $j \in N_\ell$ and its *activation* $n_{\ell j}(\mathbf{x})$ are, respectively,

$$a_{\ell j}(\mathbf{x}) = \sum_{i \in N_{\ell-1}} w_{i\ell j} \cdot n_{(\ell-1)i}(\mathbf{x}) \quad \text{and} \quad n_{\ell j}(\mathbf{x}) = \begin{cases} \mathbf{x}_j & \text{if } \ell = 0 \\ +1 & \text{if } \ell > 0,\, a_{\ell j}(\mathbf{x}) \geq 0 \\ -1 & \text{otherwise.} \end{cases}$$

The activations of all the neurons in a BNN are -1 or $+1$ except for the input neurons, which may take any real value. A weight assignment $\mathbf{W}$ to the network defines a function $f_{\mathbf{W}} \colon \mathbb{R}^{|N_0|} \to \{-1, 1\}^{|N_L|}$ that maps input vectors $\mathbf{x} \in \mathbb{R}^{|N_0|}$ to output vectors $\mathbf{y} \in \{-1, 1\}^{|N_L|}$, where $\mathbf{y}$ represents the neuron activations in the last layer. *Training* a BNN consists of finding a weight assignment that fits a training set $\mathcal{T} = \langle (\mathbf{x}^1, \mathbf{y}^1), \ldots, (\mathbf{x}^\tau, \mathbf{y}^\tau) \rangle$, i.e., finding $\mathbf{W}$ such that $f_{\mathbf{W}}(\mathbf{x}^k) = \mathbf{y}^k$ for all pairs $(\mathbf{x}^k, \mathbf{y}^k) \in \mathcal{T}$. The task of learning functions from input-output examples is known as *supervised learning*.

The goal of supervised learning is *generalization* [4]. A trained BNN is useful only if it can map unseen examples to their correct outputs (i.e., good test performance). Hence, a central problem is how to distinguish BNNs that generalize from those that overfit the training data. There are two main principles to avoid overfitting in machine learning (ML): simplicity and robustness.

The simplicity principle, also known as Occam's razor, suggests that we should prefer the simplest BNNs that fit the training set. For NNs, a natural measure of simplicity is the number of connections [23]. Our first optimization problem, therefore, looks for a BNN that fits the training data and minimizes the number of nonzero weights:

$$\min_{\mathbf{W}} \left\{ \sum_{w \in \mathbf{W}} |w| : f_{\mathbf{W}}(\mathbf{x}) = \mathbf{y},\ \forall (\mathbf{x}, \mathbf{y}) \in \mathcal{T},\ w \in \{-1, 0, 1\},\ \forall w \in \mathbf{W} \right\}. \quad \text{(min-weight)}$$

While the effectiveness of this principle has been challenged [4], it is the basis for most forms of regularizers used in modern deep learning [27].

In contrast, the robustness principle looks for BNNs that classify the training set correctly despite small perturbations to their weights. It is believed that deep NNs avoid overfitting because GD implicitly drives the exploration toward robust solutions [12,13,25]. One way of finding robust BNNs is by maximizing the *margins* of their neurons. Given a training set $\mathcal{T}$, the margin of neuron $n_{\ell j}$ is equal to the minimum absolute value of its preactivation $a_{\ell j}(\mathbf{x})$ for any $(\mathbf{x}, \mathbf{y}) \in \mathcal{T}$. Intuitively, neurons with larger margins require bigger changes on their inputs and weights to change their activation values. Recent work shows that margins are good predictors for the generalization of deep convolutional NNs [11]. Our second optimization problem searches for BNNs that fit the training data and have the maximum sum of neuron margins:

$$\max_{\mathbf{W}} \quad \sum_{\ell \in \{1..L\}} \sum_{j \in N_\ell} \min\{|a_{\ell j}(\mathbf{x})| : (\mathbf{x}, \mathbf{y}) \in \mathcal{T}\} \qquad \text{(max-margin)}$$

$$\text{s.t.} \quad f_{\mathbf{W}}(\mathbf{x}) = \mathbf{y} \qquad\qquad\qquad \forall (\mathbf{x}, \mathbf{y}) \in \mathcal{T}$$

$$\qquad\quad w \in \{-1, 0, 1\} \qquad\qquad\quad \forall w \in \mathbf{W}$$

We focus on these two criteria because they are well-supported by previous work. However, there are likely to be other objective functions worth studying and our models may be extended to do so. Additionally, our models assume that the training set has no incorrectly labeled training examples. Extensions to handle noise can be done by including slack variables as proposed in the Support Vector Machine literature [29].

3   Related Work

Unfortunately, BNNs cannot be trained using standard backpropagation because their weights are discrete. Hubara et al. [9] proposed using two sets of weights: $\mathbf{W}$ and $\mathbf{W}_b$, with $\mathbf{W}$ taking continuous values. When computing the activations, the weights $\mathbf{W}$ and activations a are projected to -1 or $+1$ using $\mathbf{W}_b = \text{sign}(\mathbf{W})$ and $a_b = \text{sign}(a)$. Then, the gradients are computed as usual except for the activation function. To backpropagate over $\text{sign}(a)$ they assume that its gradient is equal to 1 if $|a| \leq 1$ and is 0 otherwise. These gradients update $\mathbf{W}$ and then the process repeats. While most work on training BNNs follows this approach [20,26,31,33], there are a few gradient-based alternatives such as *Apprentice* [22] and *Self-Binarizing Networks* [17].

Other work has explored the use of model-based approaches, in particular MIP, in tasks related to NNs [6]. For example, MIP models have been proposed for NN verification [1] and for finding adversarial examples for NNs [5,30] and BNNs [15]. Given a pre-trained network and a target input, the problem of finding an adversarial example consists of discovering the smallest perturbation of the target input such that the output of the network changes. In particular,

Khalil et al. proposed a MIP model that, similarly to our work, uses big-M constraints to model the neuron activations [15]. They recognize those big-M constraints as the main bottleneck in scaling their approach and propose a heuristic method that finds adversarial examples by fixing different subsets of the activations over time. One of our hybrid CP/MIP models, HA, exploits a similar idea but for training BNNs. SAT models have also been used in the context of verifying properties over BNNs [2,24].

With regards to training BNNs, Khalil and Dilkina discussed the viability of using MIP models in an extended abstract at CPAIOR 2018 [14]. They report no specific results, but suggest that their MIP approach fails to scale. To the best of our knowledge, there is no other work on training BNNs using MIP or CP nor any applications of CP to BNNs.

4 Monolithic Models for Training BNNs

We now introduce CP and MIP models to train BNNs. The models receive the training set $T = \langle (\mathbf{x}^1, \mathbf{y}^1), \ldots, (\mathbf{x}^\tau, \mathbf{y}^\tau) \rangle$ and the network's architecture $\mathcal{N} = \langle N_0, \ldots, N_L \rangle$ as input. Our models use $T = \{1, \ldots, \tau\}$ as the set of training indices and $\mathcal{L} = \{1, \ldots, L\}$ as the set of layers.

4.1 Constraint Programming Models

Our CP models use the formalism and global constraints available in IBM ILOG CP Optimizer [10]. Let $w_{i\ell j} \in \{-1, 0, 1\}$ be a decision variable indicating the weight of the connection going from neuron $i \in N_{\ell-1}$ to $j \in N_\ell$. Let $n_{\ell j}^k$ be a CP expression representing the activation of neuron j in layer ℓ when the training instance $\mathbf{x}^k$ is fed to the BNN. Our model uses the vector notation $\mathbf{w}_{\ell j} = [w_{1\ell j}, \ldots, w_{|N_{\ell-1}|\ell j}]^\top$ and $\mathbf{n}_\ell^k = [n_{\ell 1}^k, \ldots, n_{\ell|N_\ell|}^k]^\top$. The constraints are:

$$n_{0j}^k = x_j^k \qquad\qquad\qquad\qquad \forall j \in N_0, k \in T \quad (1)$$

$$n_{\ell j}^k = 2\left(\texttt{scal\_prod}(\mathbf{w}_{\ell j}, \mathbf{n}_{\ell-1}^k) \geq 0\right) - 1 \qquad \forall \ell \in \mathcal{L} \setminus \{L\}, j \in N_\ell, k \in T \quad (2)$$

$$n_{Lj}^k = y_j^k \qquad\qquad\qquad\qquad \forall j \in N_L, k \in T \quad (3)$$

$$w_{i\ell j} \in \{-1, 0, 1\} \qquad\qquad\qquad \forall \ell \in \mathcal{L}, i \in N_{\ell-1}, j \in N_\ell \quad (4)$$

The first three constraints recursively define the neuron activations. Constraint (1) instantiates N_0 to be the same as the input vector for each training example. Constraint (2) defines the activations for the remaining layers, which depend on the variables $w_{i\ell j}$. This constraint uses a reified scalar product constraint, $\texttt{scal\_prod}(\mathbf{v}_1, \mathbf{v}_2) = \mathbf{v}_1^\top \cdot \mathbf{v}_2$, to compute the neuron activation. Constraint (3) matches the last neuron layer values to the output vector of each training example. Constraint (4) defines the variable domains.

Our CP models have identical sets of constraints but different objectives. Model $\texttt{CP}_w$ minimizes the total number of weights using the expression $\texttt{abs}(a) = |a|$ for absolute value, i.e.,

$$\min \sum_{\ell \in \mathcal{L}} \sum_{i \in N_{\ell-1}} \sum_{j \in N_\ell} \texttt{abs}(w_{i\ell j}), \quad \text{s.t. } (1)-(4), \tag{CP$_w$}$$

while model CP_m maximizes the sum of neuron margins, i.e.,

$$\max \sum_{\ell \in \mathcal{L}} \sum_{j \in N_\ell} \min \left(\{\texttt{abs}(\texttt{scal\_prod}(\mathbf{w}_{\ell j}, \mathbf{n}_{\ell-1}^k)) | \ k \in T\} \right), \quad \text{s.t. } (1)-(4). \tag{CP$_m$}$$

Each CP model has $O(W)$ decision variables and $O(|N_L| \cdot \tau)$ constraints, where W is the number of weights, $|N_L|$ is the number of output neurons, and τ is the size of the training set.

4.2   Mixed Integer Programming Models

The MIP and CP models share the same main decision variables. Variable $w_{i\ell j} \in \{-1, 0, 1\}$ indicates the weight of the connection from neuron $i \in N_{\ell-1}$ to neuron $j \in N_\ell$. Variable $u_{\ell j}^k \in \{0, 1\}$ models the activation of neuron $j \in N_\ell$ when the training instance $\mathbf{x}^k$ is fed to the BNN. Note that the actual neuron activation is $n_{\ell j}^k = 2u_{\ell j}^k - 1$ in this case. In addition, we use an auxiliary variable to model the non-linearities inside the BNN. Variable $c_{i\ell j}^k \in \mathbb{R}$ represents the multiplication of neuron activation $i \in N_{\ell-1}$ for a given $k \in T$ and weight $w_{i\ell j}$, i.e., $c_{i\ell j}^k = (2u_{(\ell-1)i}^k-1) \cdot w_{i\ell j}$. Lastly, we use sets $\mathcal{L}_2 = \{2, \dots, L\}$ and $\mathcal{L}^{L-1} = \{1, \dots, L-1\}$, and a small constant $\epsilon > 0$ to model strict inequalities.

Our minimum-weight MIP_w model introduces a binary variable $v_{i\ell j} \in \{0, 1\}$ to represent the absolute value of each weight $w_{i\ell j}$. Constraints (5) and (6) force the BNN output to be equal to target value y_j^k in the training set. Constraints (7) and (8) are implication constraints (which can be reformulated as big-M constraints) that define the activations. Constraint (9) sets the value of c_{i1j}^k for the input layer, while constraints (10) to (13) ensure that $c_{i\ell j}^k = (2u_{(\ell-1)i}^k - 1) \cdot w_{i\ell j}$. Constraint (14) defines the absolute values of each weight. Lastly, constraints (15) to (18) specify the domains of the variables.

$$\min \sum_{\ell \in \mathcal{L}} \sum_{i \in N_{\ell-1}} \sum_{j \in N_\ell} v_{i\ell j} \tag{MIP$_w$}$$

$$\text{s.t.} \quad \sum_{i \in N_{L-1}} c_{iLj}^k \geq 0 \qquad\qquad \forall j \in N_L, k \in T : y_j^k = 1 \tag{5}$$

$$\sum_{i \in N_{L-1}} c_{iLj}^k \leq -\epsilon \qquad\qquad \forall j \in N_L, k \in T : y_j^t = -1 \tag{6}$$

$$(u_{\ell j}^k = 1) \implies \left(\sum_{i \in N_{\ell-1}} c_{i\ell j}^k \geq 0 \right) \qquad \forall \ell \in \mathcal{L}^{L-1}, j \in N_\ell, k \in T \tag{7}$$

$$(u_{\ell j}^k = 0) \implies \left(\sum_{i \in N_{\ell-1}} c_{i\ell j}^k \leq -\epsilon \right) \qquad \forall \ell \in \mathcal{L}^{L-1}, j \in N_\ell, k \in T \tag{8}$$

$$c_{i1j}^k = x_i^k \cdot w_{i1j} \qquad\qquad \forall i \in N_0, j \in N_1, k \in T \tag{9}$$

$$c_{i\ell j}^k - w_{i\ell j} + 2u_{(\ell-1)i}^k \leq 2 \qquad \forall \ell \in \mathcal{L}_2, i \in N_{\ell-1}, j \in N_\ell, k \in T \tag{10}$$

$$c_{i\ell j}^k + w_{i\ell j} - 2u_{(\ell-1)i}^k \leq 0 \qquad \forall \ell \in \mathcal{L}_2, i \in N_{\ell-1}, j \in N_\ell, k \in T \quad (11)$$

$$c_{i\ell j}^k - w_{i\ell j} - 2u_{(\ell-1)i}^k \geq -2 \qquad \forall \ell \in \mathcal{L}_2, i \in N_{\ell-1}, j \in N_\ell, k \in T \quad (12)$$

$$c_{i\ell j}^k + w_{i\ell j} + 2u_{(\ell-1)i}^k \geq 0 \qquad \forall \ell \in \mathcal{L}_2, i \in N_{\ell-1}, j \in N_\ell, k \in T \quad (13)$$

$$-v_{i\ell j} \leq w_{i\ell j} \leq v_{i\ell j} \qquad \forall \ell \in \mathcal{L}, i \in N_{\ell-1}, j \in N_\ell \quad (14)$$

$$w_{i\ell j} \in \{-1, 0, 1\} \qquad \forall \ell \in \mathcal{L}, i \in N_{\ell-1}, j \in N_\ell \quad (15)$$

$$u_{\ell j}^k \in \{0, 1\} \qquad \forall \ell \in \mathcal{L}^{L-1}, j \in N_\ell, k \in T \quad (16)$$

$$c_{i\ell j}^k \in \mathbb{R} \qquad \forall \ell \in \mathcal{L}, i \in N_{\ell-1}, j \in N_\ell, k \in T \quad (17)$$

$$v_{i\ell j} \in \{0, 1\} \qquad \forall \ell \in \mathcal{L}, i \in N_{\ell-1}, j \in N_\ell \quad (18)$$

The maximum-margin $\mathtt{MIP}_m$ model introduces a variable $m_{\ell j} \in \mathbb{R}^+$ to represent the margin of each neuron $j \in N_\ell$. The set of constraints is similar to the previous model with the exception that it includes neuron margin variables in the neuron activation constraints (19)–(22).

$$\max \sum_{\ell \in \mathcal{L}} \sum_{j \in N_\ell} m_{\ell j} \qquad\qquad\qquad (\mathtt{MIP}_m)$$

s.t. (9)−(13), (15)−(17)

$$\sum_{i \in N_{L-1}} c_{iLj}^k \geq m_{Lj} \qquad \forall j \in N_L, k \in T : y_j^k = 1 \quad (19)$$

$$\sum_{i \in N_{L-1}} c_{iLj}^k \leq -\epsilon - m_{Lj} \qquad \forall j \in N_L, k \in T : y_j^t = -1 \quad (20)$$

$$(u_{\ell j}^k = 1) \implies \left(\sum_{i \in N_{\ell-1}} c_{i\ell j}^k \geq m_{\ell j} \right) \qquad \forall \ell \in \mathcal{L}^{L-1}, j \in N_\ell, k \in T \quad (21)$$

$$(u_{\ell j}^k = 0) \implies \left(\sum_{i \in N_{\ell-1}} c_{i\ell j}^k \leq -\epsilon - m_{\ell j} \right) \qquad \forall \ell \in \mathcal{L}^{L-1}, j \in N_\ell, k \in T \quad (22)$$

$$m_{lj} \geq 0 \qquad \forall \ell \in \mathcal{L}, j \in N_\ell \quad (23)$$

Note that each MIP model has $O(W + N \cdot \tau)$ integer decision variables and $O((W + N)\tau)$ constraints, where W is the number of weights, N is the total number of neurons, and τ is the size of the training set.

5 CP/MIP Hybrid Approaches

Our experimental results (Sect. 7.1) suggest that CP is good at finding a feasible set of weights, while MIP is good at optimizing them towards solutions that generalize better. This motivates our hybrid methods that find a first feasible solution using CP and then use a MIP model to optimize. We use a CP model without objective function, $\mathtt{CP}_f$, since it finds feasible solutions on a larger number of instances than $\mathtt{CP}_w$ and $\mathtt{CP}_m$ (Sect. 7.2).

We propose two alternatives to incorporate the CP solution into the MIP models. Our first hybrid model, HW, uses the CP solution as a warm-start for either $\mathtt{MIP}_w$ or $\mathtt{MIP}_m$. The second hybrid variant, $\mathtt{HA}$, fixes the activations of all the neurons in the MIP model and searches only over the weights. As a result, all the big-M constraints and variables $c_{i\ell j}^k$ are removed, albeit at the cost of potentially pruning optimal solutions.

Given a feasible set of weights $\hat{w}_{i\ell j}$ returned by $\mathtt{CP}_f$, $\mathtt{HA}$ computes the neuron activations for a training example $k \in T$ as $\hat{n}_{0i}^k = x_i^k$ for the input layer and $\hat{n}_{\ell i}^k = 2\left(\sum_{i \in N_{\ell-1}} \hat{n}_{(\ell-1)i}^k \hat{w}_{i\ell j} \geq 0\right) - 1$ for $\ell \in \mathcal{L}$. Then, the fixed activation models for min-weight $\mathtt{HA}_w$ and max-margin $\mathtt{HA}_m$ are as follows.

$$\min \sum_{\ell \in \mathcal{L}} \sum_{i \in N_{\ell-1}} \sum_{j \in N_\ell} v_{i\ell j} \qquad (\mathtt{HA}_w)$$

s.t. $(14), (15), (18)$

$$\sum_{i \in N_{\ell-1}} w_{i\ell j} \cdot \hat{n}_{(\ell-1)i}^k \geq 0 \qquad \forall \ell \in \mathcal{L}, j \in N_\ell, k \in T : \hat{n}_{\ell j}^k = 1 \qquad (24)$$

$$\sum_{i \in N_{\ell-1}} w_{i\ell j} \cdot \hat{n}_{(\ell-1)i}^k \leq -\epsilon \qquad \forall \ell \in \mathcal{L}, j \in N_\ell, k \in T : \hat{n}_{\ell j}^k = -1 \qquad (25)$$

$$\max \sum_{\ell \in \mathcal{L}} \sum_{j \in N_\ell} m_{\ell j} \qquad (\mathtt{HA}_m)$$

s.t. $(15), (23)$

$$\sum_{i \in N_{\ell-1}} w_{i\ell j} \cdot \hat{n}_{(\ell-1)i}^k \geq m_{\ell j} \qquad \forall \ell \in \mathcal{L}, j \in N_\ell, k \in T : \hat{n}_{\ell j}^k = 1 \qquad (26)$$

$$\sum_{i \in N_{\ell-1}} w_{i\ell j} \cdot \hat{n}_{(\ell-1)i}^k \leq -\epsilon - m_{\ell j} \qquad \forall \ell \in \mathcal{L}, j \in N_\ell, k \in T : \hat{n}_{\ell j}^k = -1 \qquad (27)$$

Hybrid methods are not necessary when the BNN has no hidden layers; in such scenarios, the implication constraints (7)–(8) and (21)–(22) are not needed and, as a result, the $\mathtt{HA}$ models reduce to our $\mathtt{MIP}$ models.

6   Gradient Descent Baselines

Current methods to train BNNs follow Hubara et al.'s GD-based algorithm [9] described in Sect. 3. This algorithm is a highly optimized local search method that starts from a random weight assignment and locally changes the weights towards minimizing a *Square Hinge loss* function. The Square Hinge loss function quadratically penalizes the errors on the training set. The most relevant hyperparameter is the *learning rate* that defines how much each weight is updated in every step.

Hubara et al.'s approach learns BNNs only with -1 and $+1$ weights, whereas our models also allow for zero-value weights. To make a fair comparison, we also extended Hubara et al.'s approach to work with zero-value weights. Instead of learning one binary weight per connection we learn two, w_b^1 and w_b^2. The final weight for the connection is the average between those two values, i.e, $w_b = (w_b^1 + w_b^2)/2 \in \{-1, 0, 1\}$. Our experiments report the performance of both the original approach $\mathtt{GD}_b$ and our extension $\mathtt{GD}_t$.

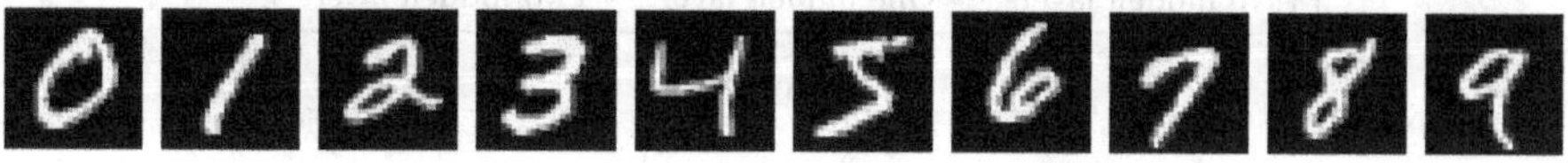

Fig. 2. The first 10 examples from the MNIST dataset.

7 Experimental Evaluation for Few-Shot Learning

We tested our models over subsets of the MNIST dataset [19], which consists of 70,000 labeled images of handwritten digits. Each image has 28×28 gray-scale pixels with values between 0 to 255. Every example has a label representing the digit that appears on the input image, i.e., its class. Figure 2 shows 10 examples from the MNIST training set.

To emulate the conditions of a few-shot learning scenario, we limited the training set size to a range varying from 1 to 10 examples per class. We sampled 10 problem instances for each class and trained BNNs with $28 \times 28 = 784$ input neurons and 10 output neurons (one per class). If the image label is i, then the ith output neuron should be active ($y_i = 1$) and the rest inactive ($y_j = -1$ for all $j \neq i$). Each BNN has 0, 1, or 2 hidden layers with 16 neurons each.

We compare our models using three metrics. The first two metrics correspond to the number of instances solved (i.e., finding a weight assignment that fits the training data) and the quality of those solutions w.r.t. the objective functions. The third metric compares the test performance over the 10,000 test instances from MNIST. We use the *all-good* metric that evaluates the percentage of instances where the value of the 10 output neurons is correct. As such, the expected performance of a BNN with random weights is 0.098%.

Approaches. We use Gurobi 8.1 [8] to solve the MIP models and IBM ILOG CP Optimizer 12.8 [10] for the CP models. For MIP, the implications are formulated using Gurobi's special construct. The GD baselines were solved using Tensorflow 1.9.0 and Adam optimizer [16]. We evaluated the following approaches:[1]

- CP_w and CP_m: min-weight and max-margin CP models, respectively.
- MIP_w and MIP_m: min-weight and max-margin MIP models, respectively.
- HW_w and HW_m: min-weight and max-margin warm-start hybrid models.
- HA_w and HA_m: min-weight and max-margin fixed-activation hybrid models.
- GD_b and GD_t: Hubara et al.'s approach [9] and our extension for zero-weights.

As the GD baselines find different solutions depending on their starting point and learning rate, we tested four common learning rates ($10^{-3}, \ldots, 10^{-6}$) starting from 5 independently sampled BNNs for each problem instance. We defined the performance of each learning rate to be the average performance across its five starting points. Our experimental results report the performance of the *best* learning rate for each problem. This is an upper bound on the GD performance

---

[1] Our source code is publicly available at https://bitbucket.org/RToroIcarte/bnn.

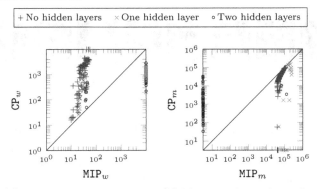

(a) Min-weight optimization   (b) Max-margin optimization

Fig. 3. Solution quality comparison between CP and MIP.

that assumes the existence of an oracle that can predict the best learning rate for each sampled training instance and BNN architecture.

Each approach was run with a 2-h time limit using one thread on an Intel Xeon E5-2680 2.70 GHz processor with 96 GB of RAM. This time limit is long enough for GD_b and GD_t to converge in most of our experiments.

Data Preprocessing. An input neuron is considered *dead* if its value is the same in the entire training set. As those neurons add no new information to discriminate the correct output for a given input, they can be removed without losing correctness. We exploit this structure in our models (and baselines) by fixing the value of every weight connected to a dead input neuron to zero.

7.1   Solution Quality Comparison Between MIP and CP

We now compare the efficiency of our monolithic models for finding high-quality BNNs that fit the training data. Figure 3 compares the quality of the solutions found by the MIP and CP models for the min-weight and max-margin objectives. A point (x, y) in the plots corresponds to a single instance where x and y represent the objective value obtained using MIP and CP, respectively. Points that appear along the vertical (resp. horizontal) axes correspond to instances where MIP (resp. CP) timed out before finding any feasible solution. For the min-weight objective, points above the diagonal represent instances where the MIP model found better solutions. The inverse is true for the max-margin graph.

The results show that MIP struggled to find feasible solutions when using one and two hidden layers. This is mainly explained by the large number of big-M constraints and variables that both MIP_w and MIP_m have in those cases. In contrast, CP found feasible solutions for most of the problem instances.

Table 1. Number of instances where a feasible solution was found.

	One hidden layer										Two hidden layers										
$|\mathcal{T}|$	10	20	30	40	50	60	70	80	90	100	10	20	30	40	50	60	70	80	90	100	**Total**
GD_b	9.4	0.2	0	0	0	0	0	0	0	0	5.6	0	0	0	0	0	0	0	0	0	15.2
GD_t	9.6	5.6	0.4	0	0	0	0	0	0	0	9.2	8.4	5.2	6.2	4.2	2.2	0	0	0	0	51
MIP_m	10	3	2	1	0	1	0	0	0	0	2	0	0	0	0	0	0	0	0	0	19
MIP_w	10	7	0	0	0	0	0	0	0	0	9	0	0	0	0	0	0	0	0	0	26
CP_m	10	6	6	3	3	3	0	0	0	0	10	10	10	10	10	9	6	5	2	0	103
CP_w	10	10	10	10	10	10	10	10	10	8	10	10	10	8	4	7	2	0	0	0	149
CP_f	10	10	9	8	8	7	3	3	1	0	10	10	10	10	10	10	10	10	8	6	**153**

When both methods found feasible solutions, the graphs suggest that MIP was better at finding high-quality solutions. With both objectives, MIP consistently found equal or better quality solutions than CP. In fact, MIP found proven optimal solutions for 68 out of 300 instances when minimizing weights and 15 out of 300 when maximizing margins, while CP never found and proved optimal solutions.

7.2   Comparison Between Hybrid Methods

When hidden layers are used, our hybrid methods find a first feasible solution using a CP model without an objective function, CP_f, and give it to a MIP model to optimize. To find feasible solutions, we could have instead used MIP, GD, CP_w, or CP_m. However, CP_f tends to finds more feasible solutions than the other methods. This is well-supported by Table 1, which shows the number of instances where a feasible solution was found (for each method) in under 2 h.

To compare the solution quality of our model-based approaches, we analyze the optimality gaps across different network architectures and training examples. The gap computation uses the best dual bound found by any approach. Figure 4 shows the average optimality gaps obtained for the min-weight and max-margin criteria using the monolithic and hybrid methods. We omit the gap lines for HA and HW for the experiments with no hidden layer since our hybrid methods are not needed in this case (see Sect. 5).

These results suggest that the hybrid methods exhibit the best characteristics of the CP and MIP models. HW and HA scaled to larger training sets and network architectures in a manner similar to CP—significantly outperforming MIP in this metric—while obtaining high quality solutions that are comparable to those produced by MIP. In addition, HA consistently outperformed HW when maximizing the margins and found similar quality solutions for the min-weight criteria.

7.3   Test Performance Comparison with GD

Figure 5 compares the test performance of our methods and the best performing GD baseline. A data point represents the performance of a BNN for each training set and network architecture. Points below the diagonal represent instances

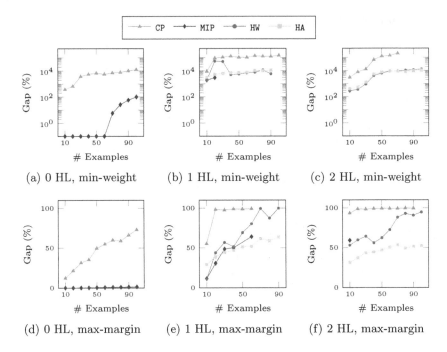

Fig. 4. Optimality gap comparison with different number of hidden layers (HL) and size of the training set (# Examples).

where our approaches outperformed GD. These results show that MIP outperformed GD methods when it found a solution, and that the hybrid methods found many more solutions than MIP while maintaining a similar test performance profile. In particular, HA_m has a remarkable performance in comparison with GD in these experiments. For instance, with 2 hidden layers and 100 training examples, HA_m correctly classifies up to $5,612$ of $10,000$ unseen examples while GD predicted the true class in at most $1,563$ cases. Note that a BNN with random weights is expected to correctly predict less than 10 examples.

To better represent these results, Fig. 6 shows the number of instances where model-based approaches have a strictly better test performance than the best GD baseline. When limited data was used, the hybrid approaches consistently outperformed GD. However, as the training sets get larger, some of our models timed out before finding feasible solutions. In contrast, GD always returned a solution. Such solution might not fit all the training data but it can still be evaluated on the test set. Hence, GD is superior with large training sets.

It is equally important to consider by how much the solutions found by our models outperform the solutions found by GD. Figure 7 displays the average test performance across the instances solved by each approach. Under this metric, the clear winner is HA_m (which reduces to MIP_m when no hidden layers are used) as it largely outperformed GD and CP while scaling better than MIP.

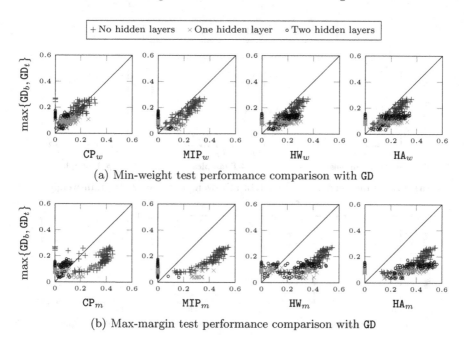

(a) Min-weight test performance comparison with GD

(b) Max-margin test performance comparison with GD

Fig. 5. Test performance comparison between the best GD and model-based methods with different number of hidden layers (HL) and optimization criteria.

7.4   Discussion

Our experiments demonstrate the merits of model-based approaches—in particular, MIP and CP—to train BNNs. When data is scarce, these methods can find solutions that generalize better than the solutions found by GD. This is a notable result that opens many opportunities for future work. In particular, there are three interesting questions that arise from our experimental evaluation.

What are the advantages and limitations of model-based approaches?
The main advantage of training BNNs using model-based approaches is in finding solutions that generalize better using fewer examples. Consider the results on Fig. 7(e) and (f). They show that our hybrid models need only 10 examples to find solutions that generalize better than the ones found by GD using 100 examples. That being said, their main limitation is scalability. We expect that more sophisticated model-based approaches, such as decompositions and specialized CP propagators, will push the boundary of problems that can be solved. We also believe that model-based approaches will become a new tool for ML researchers, as they allow for principled empirical comparisons of generalization criteria based on provable bounds.

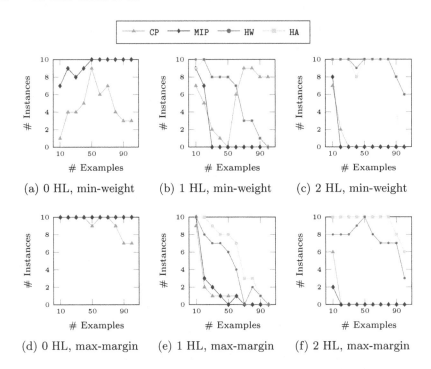

(a) 0 HL, min-weight   (b) 1 HL, min-weight   (c) 2 HL, min-weight

(d) 0 HL, max-margin   (e) 1 HL, max-margin   (f) 2 HL, max-margin

Fig. 6. Number of instances where model-based approaches have better test performance than $\max\{\mathtt{GD}_b, \mathtt{GD}_t\}$.

Are min-weight and max-margin the best proxies for generalization?
Our results suggest that both min-weight and max-margin are good proxies for generalization. In fact, given two BNNs that perfectly fit the training data, our models can accurately predict which one generalizes better. Through a pairwise comparison of all the perfect-fit BNNs generated for each instance in our experiments, we saw that the BNN with bigger margin generalized better over 85% of the time. The min-weight criteria is not as good at predicting generalization, but still does a reasonable job: over 79% of the time, the BNN with fewer nonzero weights generalized better. However, it would be very surprising if there are no other criteria that could better predict generalization. Looking for such criteria is a promising future work direction.

Why are deep BNNs not generalizing better than shallow ones?
A major insight from the deep learning literature is that adding hidden layers improves generalization [28]. Surprisingly, this was not the case in our experiments. A possible explanation is that our training sets are not big enough to justify the use of hidden layers. This a reasonable hypothesis, especially considering that the test performance of GD methods also decreased when adding more hidden layers (see Fig. 7). However, it does not explain why adding hidden layers improves generalization when using 10 training examples for $\mathtt{HW}_m$ and $\mathtt{HA}_m$.

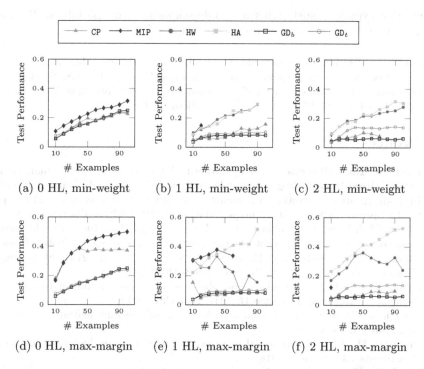

Fig. 7. Test performance comparison for all our methods in BNNs with different number of hidden layers (HL) using the two optimization criteria.

Another possible explanation is that we are not finding close-to-optimal solutions when using hidden layers (see Fig. 4). Hence, while the test performance reaches its full potential for the case with no hidden layers, there is room for improvement for BNNs with hidden layers.

8   Concluding Remarks

Our work examines the use of MIP and CP to train BNNs. We formulate the training problem as finding a BNN that perfectly fits the training set while optimizing two proxies for generalizability. When solving this problem, we note that CP is good at finding feasible solutions and MIP is good at optimizing them. Hence, we propose two CP/MIP hybrids that exploit the strengths of CP and MIP. With limited training data, our hybrid approaches found BNNs that generalized better than the ones found by GD. In contrast, GD scaled better, making it more appealing when large training sets are available.

This work opens many opportunities for future work at the intersection between ML and OR. From an ML perspective, model-based approaches allow for principled empirical comparisons between proxies for generalization and seem effective for few-shot learning. From an OR perspective, training BNNs is a challenging combinatorial optimization problem with interesting structure. We believe that exploiting such structure via decompositions or specialized CP propagators presents a promising direction for future work.

Acknowledgements. We would like to thank Toryn Klassen, Maayan Shvo, and Ethan Waldie for their help running experiments and Kyle Booth, Arik Senderovich, and the anonymous reviewers for helpful comments. We gratefully acknowledge funding from CONICYT (Becas Chile), NSERC, and Microsoft Research.

References

1. Anderson, Ross, Huchette, Joey, Tjandraatmadja, Christian, Vielma, Juan Pablo: Strong Mixed-Integer Programming Formulations for Trained Neural Networks. In: Lodi, Andrea, Nagarajan, Viswanath (eds.) IPCO 2019. LNCS, vol. 11480, pp. 27–42. Springer, Cham (2019). https://doi.org/10.1007/978-3-030-17953-3_3
2. Cheng, Chih-Hong, Nührenberg, Georg, Huang, Chung-Hao, Ruess, Harald: Verification of Binarized Neural Networks via Inter-neuron Factoring. In: Piskac, Ruzica, Rümmer, Philipp (eds.) VSTTE 2018. LNCS, vol. 11294, pp. 279–290. Springer, Cham (2018). https://doi.org/10.1007/978-3-030-03592-1_16
3. Ching, T., et al.: Opportunities and obstacles for deep learning in biology and medicine. J. Roy. Soc. Interface **15**(141), 20170387 (2018)
4. Domingos, P.: A few useful things to know about machine learning. Commun. ACM **55**(10), 78–87 (2012)
5. Fischetti, M., Jo, J.: Deep neural networks and mixed integer linear optimization. Constraints **23**, 296–309 (2018)
6. Gambella, C., Ghaddar, B., Naoum-Sawaya, J.: Optimization models for machine learning: a survey. arXiv preprint arXiv:1901.05331 (2019)
7. Goodfellow, I., Bengio, Y., Courville, A.: Deep Learning. MIT Press, Cambridge (2016)
8. Gurobi Optimization, LLC: Gurobi Optimizer Reference Manual (2018). http://www.gurobi.com
9. Hubara, I., Courbariaux, M., Soudry, D., El-Yaniv, R., Bengio, Y.: Binarized neural networks. In: Proceedings of the 29th Conference on Advances in Neural Information Processing Systems (NIPS), pp. 4107–4115 (2016)
10. IBM: ILOG CP Optimizer 12.8 Manual (2018)
11. Jiang, Y., Krishnan, D., Mobahi, H., Bengio, S.: Predicting the generalization gap in deep networks with margin distributions. In: Proceedings of the 7th International Conference on Learning Representations (ICLR) (2019)
12. Kawaguchi, K., Kaelbling, L.P., Bengio, Y.: Generalization in deep learning. arXiv preprint arXiv:1710.05468 (2017)
13. Keskar, N.S., Mudigere, D., Nocedal, J., Smelyanskiy, M., Tang, P.T.P.: On large-batch training for deep learning: generalization gap and sharp minima. In: Proceedings of the 5th International Conference on Learning Representations (ICLR) (2017)
14. Khalil, E.B., Dilkina, B.: Training binary neural networks with combinatorial algorithms. In: Extended abstract at the 15th International Conference on the Integration of Constraint Programming, Artificial Intelligence, and Operations Research (CPAIOR) (2018)
15. Khalil, E.B., Gupta, A., Dilkina, B.: Combinatorial attacks on binarized neural networks. In: Proceedings of the 7th International Conference on Learning Representations (ICLR) (2019)

16. Kingma, D.P., Ba, J.: Adam: a method for stochastic optimization. In: Proceedings of the 3rd International Conference on Learning Representations (ICLR) (2015)
17. Lahoud, F., Achanta, R., Márquez-Neila, P., Süsstrunk, S.: Self-binarizing networks. arXiv preprint arXiv:1902.00730 (2019)
18. LeCun, Y., Bengio, Y., Hinton, G.: Deep learning. Nature **521**(7553), 436–444 (2015)
19. LeCun, Y., Cortes, C., Burges, C.J.: The MNIST database of handwritten digits (1998). http://yann.lecun.com/exdb/mnist
20. Li, F., Zhang, B., Liu, B.: Ternary weight networks. arXiv preprint arXiv:1605.04711 (2016)
21. Miotto, R., Wang, F., Wang, S., Jiang, X., Dudley, J.T.: Deep learning for healthcare: review, opportunities and challenges. Brief. Bioinform. **19**(6), 1236–1246 (2017)
22. Mishra, A., Marr, D.: Apprentice: using knowledge distillation techniques to improve low-precision network accuracy. In: Proceedings of the 6th International Conference on Learning Representations (ICLR) (2018)
23. Moody, J.E.: The effective number of parameters: an analysis of generalization and regularization in nonlinear learning systems. In: Proceedings of the 4th Conference on Advances in Neural Information Processing Systems (NIPS), pp. 847–854 (1991)
24. Narodytska, N.: Formal analysis of deep binarized neural networks. In: Proceedings of the 27th International Joint Conference on Artificial Intelligence (IJCAI), pp. 5692–5696 (2018)
25. Neyshabur, B., Bhojanapalli, S., McAllester, D., Srebro, N.: Exploring generalization in deep learning. In: Proceedings of the 30th Conference on Advances in Neural Information Processing Systems (NIPS), pp. 5947–5956 (2017)
26. Rastegari, Mohammad, Ordonez, Vicente, Redmon, Joseph, Farhadi, Ali: XNOR-Net: ImageNet Classification Using Binary Convolutional Neural Networks. In: Leibe, Bastian, Matas, Jiri, Sebe, Nicu, Welling, Max (eds.) ECCV 2016. LNCS, vol. 9908, pp. 525–542. Springer, Cham (2016). https://doi.org/10.1007/978-3-319-46493-0_32
27. Schmidhuber, J.: Deep learning in neural networks: an overview. Neural Netw. **61**, 85–117 (2015)
28. Simonyan, K., Zisserman, A.: Very deep convolutional networks for large-scale image recognition. In: Proceedings of the 3rd International Conference on Learning Representations (ICLR) (2015)
29. Suykens, J.A., Vandewalle, J.: Least squares support vector machine classifiers. Neural Process. Lett. **9**(3), 293–300 (1999)
30. Tjeng, V., Xiao, K., Tedrake, R.: Evaluating robustness of neural networks with mixed integer programming. In: Proceedings of the 7th International Conference on Learning Representations (ICLR) (2019)
31. Umuroglu, Y., et al.: FINN: a framework for fast, scalable binarized neural network inference. In: Proceedings of the 25th International Symposium on Field-Programmable Gate Arrays (FPGA), pp. 65–74 (2017)
32. Vanschoren, J.: Meta-learning: a survey. arXiv preprint arXiv:1810.03548 (2018)
33. Wan, D., et al.: TBN: convolutional neural network with ternary inputs and binary weights. In: Proceedings of the 15th European Conference on Computer Vision (ECCV), pp. 315–332 (2018)

Application Track

Models for Radiation Therapy Patient Scheduling

Sara Frimodig[1,2]([⊠])[iD] and Christian Schulte[1]([⊠])[iD]

[1] KTH Royal Institute of Technology, Stockholm, Sweden
{sarhal,cschulte}@kth.se
[2] RaySearch Laboratories, Stockholm, Sweden

Abstract. In Europe, around half of all patients diagnosed with cancer are treated with radiation therapy. To reduce waiting times, optimizing the use of linear accelerators for treatment is crucial. This paper introduces an Integer Programming (IP) and two Constraint Programming (CP) models for the non-block radiotherapy patient scheduling problem. Patients are scheduled considering priority, pattern, duration, and start day of their treatment. The models include expected future patient arrivals. Treatment time of the day is included in the models as time windows which enable more realistic objectives and constraints. The models are thoroughly evaluated for multiple different scenarios, altering: planning day, machine availability, arrival rates, patient backlog, and the number of time windows in a day. The results demonstrate that the CP models find feasible solutions earlier, while the IP model reaches optimality considerably faster.

1 Introduction

Radiation therapy (RT), chemotherapy, and surgery are the most commonly used cancer therapies worldwide. In RT, machines called linear accelerators (LINACs) deliver beams of radiation to the tumor in order to kill malignant tumor cells.

A long waiting time between when a patient is ready for RT and when the treatment starts has a negative effect on its outcome due to for example tumor growth, psychological distress of the patient, and progressed symptoms [8, 12, 14, 22, 30]. Hence, many cancer institutes worldwide have adopted waiting time targets that determine the maximum waiting time before treatment starts.

The intent of RT treatments is either curative or palliative, where the latter mainly aims to provide pain relief. Furthermore, cancer patients are generally divided into three urgency levels depending on the site of the cancer, treatment intent, and the size and progress of the tumor. The waiting time targets depend on the patient's urgency level.

RT treatments are generally divided into a number of fractions that are delivered once a day and together sum up to the planned radiation dose. The duration of the fractions vary between patients due to for example treatment technique and tumor complexity [16]. There are also many uncertainties in the RT process, including for example patient inflow and unexpected machine failures.

© Springer Nature Switzerland AG 2019
T. Schiex and S. de Givry (Eds.): CP 2019, LNCS 11802, pp. 421–437, 2019.
https://doi.org/10.1007/978-3-030-30048-7_25

The scheduling of RT patients on LINACs can be divided into block or non-block systems [10]. Block scheduling systems divide days into slots of equal duration, whereas non-block systems allow for different treatment durations. Block systems are more widely used, but have severe drawbacks since there is no way to control the variability of treatment time, which can generate costs related to machine underutilization, staff overtime, and patient waiting time.

Scheduling patients is mostly done manually and is a considerable challenge for RT clinics. Designing more efficient appointment schedules would be of great significance and could potentially save lives. In order to improve the scheduling of radiotherapy patients, this paper makes the following contributions:

- Two CP models and one IP model of the non-block RT patient scheduling problem are introduced that take expected future patients into account. To the best of our knowledge, these are the very first CP models as well as the first IP model to include expected future patient arrivals.
- The treatment time of the day is included in the models which for the first time supports objectives and constraints on treatment time of the day.
- The models capture real-world constraints such as non-consecutive treatment days, different treatment durations, allowed start days, and patient priorities.
- The models are evaluated and compared using a patient arrival model and several experiments based on data from a European cancer clinic.

Plan of the Paper. Section 2 discusses related work. Section 3 presents the setup of the problem. Section 4 describes the models, followed by a description of the search heuristics used for the CP models in Sect. 5. The models are evaluated in Sect. 6. Section 7 presents conclusions and potential extensions.

2   Related Work

Related work on optimal scheduling in health care has mainly focused on nurse scheduling (see [5]), outpatient assignment (see [7]), and surgery scheduling (see [21]). The RT patient scheduling problem shares some characteristics with these problems, but has particular attributes that make it difficult to apply the models and methods proposed in the literature.

Scheduling of RT patients is a relatively young field with limited literature. In 2016, a review on the literature using operations research for resource planning in RT was published [31]. The authors found 12 papers addressing the problem of scheduling patients on LINACs, where the first ones are published in 2006.

In [18], the authors show that RT patient scheduling can be seen as a special case of a dynamic job-shop problem. They review different exact and metaheuristic methods suitable for solving the problem. A heuristic that schedules patients forward from the first feasible start date (ASAP) is developed in [25]. A local search heuristic that outperforms the ASAP approach is developed in [26].

The first use of IP for optimization of RT appointments is presented in [9,10]. Another IP model for non-block scheduling is presented in [17]. A limitation of these papers is that they do not consider all the constraints present in RT scheduling, such as for example treatments on non-consecutive days and LINAC eligibility. In [6], the authors develop an IP model that includes more realistic constraints, but still using a myopic scheduling policy, i.e., not taking future patient arrivals into account.

Using a block scheduling strategy, [27] presents a method for advance RT patient scheduling, where appointments are scheduled in advance of the service date with future demand still unknown. A Markov decision process (MDP) and approximate dynamic programming are used to solve the problem, and they achieve very good results. In [13], the authors use the same problem setup but also include patient cancellations using simulation-based solution methods. For these approaches, time of the day for the treatments is impossible to include.

A hybrid combining stochastic and online optimization is presented in [19]. The authors use a block-scheduling strategy to schedule curative patients at the same time every day and require that patients leave the center with their appointment, which calls for short computation times. This is different from earlier published methods that all schedule multiple patients in a batch.

CP has been used in RT treatment planning [2] and in chemotherapy patient scheduling [15]. Scheduling is a field where CP has shown to be effective, see for example [3]. A comprehensive review of operations research methods for optimization in radiation oncology is presented in [11], where it is stated that CP has not yet had a significant impact on medical physics.

3   Radiation Therapy Patient Scheduling

This section introduces the RT patient scheduling problem, the assumptions made in this paper, and some fundamental modeling aspects.

Time. Radiation therapy clinics have different routines for scheduling patients on LINACs. Some gather patients into a *batch* and schedule them once or several times a day, while others immediately schedule a single patient. This paper focuses on batch scheduling and assumes that the scheduling is done at the end of each day taking patients from previous days into account. As previously stated, RT clinics can be divided into two categories; those who use *block* and *non-block* scheduling systems. In order to be able to control the variability of treatment time, this paper uses a *non-block* scheduling strategy.

A day is divided into *time windows*. A time window is typically 1.5–4 h while a treatment takes 10–45 min. Patients are assigned to windows instead of specific start times as this leads to simpler and more efficient models while maintaining an adequate level of detail from a clinical perspective.

Patients. A physician assigns a *priority* to each patient based on urgency and treatment intent (palliative or curative). It is assumed that there are three priority groups, and therefore three waiting time targets: 2 days for priority A (the highest), 14 days for priority B, and 28 days for priority C patients.

A patient is assigned to a *treatment protocol*, which states the *fractionation scheme* (that is, how many days the patient is to be treated and with which frequency) and the *duration* of each treatment. Different protocols have different allowed *start days*, which enforces that fewer patients are scheduled on weekends. Some protocols also specify that treatment must start on a *certain time of the day*. In this paper, the protocols used are from a large cancer center in Europe.

The scheduled times are communicated to the patient at most one week before the start date or immediately for priority A patients. All fractions are communicated and cannot be re-planned, as this is the collaboration clinic's approach. The schedule can change until being communicated: booking decisions are postponed to the next day if patients are scheduled more than a week away.

When creating a patient schedule in practice, the booking administrator needs to make sure that there is room in the schedule for more urgent future patients. In most cases, this is done by leaving some empty time on each machine. In the models, the *expected future patient arrivals* are included to predict the expected utilization of resources. Only the expected future patients who have a waiting time target shorter than the maximum waiting time target of current patients are included. This is as patients with longer waiting time targets have little or no effect on the current schedule.

An overall *arrival rate* can be extracted from historical data for each clinic, as well as the proportion of arrivals for each priority group. This paper uses the same proportions between the priorities as [19]: 31% are priority A, 19% are priority B, and 50% are priority C. The proportions can easily be adjusted to a particular clinic. In the models, a separate priority group D is created for expected future patients of priority A, since the actual priority A patients should have higher priority than expected future priority A patients. These patients are also treated differently in the search heuristics for the CP models, see Sect. 5. Each arriving patient is randomly assigned to a treatment protocol.

Machines. The radiation is delivered on LINACs. As a rule, larger centers have multiple machine types used for different sorts of treatment and multiple identical machines to have a redundancy in case of machine failures. In small centers, there may be a few identical machines that only serve some treatment types, while more complex cases are sent to larger centers.

This paper assumes that there are *multiple machines* but only *one machine type*. The machines are exchangeable in that a patient can be scheduled on any machine each day. This scenario is a realistic way of decomposing the multiple machine problem, since the clinics may consider separate scheduling tasks for each machine type. Instead of M machines with W windows each, this is modeled as having one machine with MW windows. Thus, if for example $M = 3$, $W = 4$, then if a patient is scheduled in window 1–4, this corresponds to machine 1, window 5–8 is machine 2, and window 9–12 is machine 3. An alternative would be to model this as one machine with W windows and multiply each window length by M, but this would be a relaxation of the actual problem.

Using multiple machines represents a real-world setting and also allows for having a higher arrival rate. If there were only one machine available, the arrival rate would be very low and dividing very few patients into three different priority groups would not give good statistics for the expected future patient arrivals.

Objective. In this paper, the main objective is to minimize a weighted sum of the violations of the target dates, where the weights reflect that it is worse to violate the target date for a patient with higher priority. The secondary objective is to schedule patients at approximately the same time each day. Some patients may still work or study during treatment and hence prefer mornings or late afternoons. More importantly, the biological effects of the radiation is calculated on having 24 h between each fraction, however, in most cases it is allowed to deviate from this and it is thus an objective rather than a hard constraint. The second objective is the reason why time windows are used in the models; scheduling a patient in the same time window every day ensures that the treatment is delivered at approximately the same time every day.

4   Models

Three models are developed to capture the RT patient scheduling problem; a scheduling-based CP model, a packing-based CP model, and an IP model. These are designed to capture the same real-world constraints and objectives. Using the set $\mathbb{B} := \{0, 1\}$, the inputs to these models are:

$\mathcal{P} = \{1, \ldots, P\}$	set of all patients, $P \in \mathbb{N}$
$\mathcal{D} = \{1, \ldots, D\}$	set of days in the planning horizon, $D \in \mathbb{N}$
$\mathcal{W} = \{1, \ldots, W\}$	set of time windows in a day, $W \in \mathbb{N}$
$w_L = T_s/W$	the window length, where $T_s \in \mathbb{N}$ is the number of time slots during a day with the chosen discretization
$\mathcal{T}_w = \{t_1, t_2, \ldots, t_W\}$	set of times when each window starts counting from the beginning of day 1, where $t_1 = 1, \ldots, t_i = (i-1)w_L + 1$
$\mathcal{M} = \{1, \ldots, M\}$	set of machines, $M \in \mathbb{N}$
$dur_p \in \mathbb{N}$	duration of a fraction for patient p in time slots
$L_p \in \{13, \ldots, 47\}$	schedule length for patient p in days
$\mathcal{F} = \{1, \ldots, \max(L_p)\}$	set of all treatment days
$FS_p \in \mathbb{B}^{L_p}$	a vector holding the fractionation schedule for patient p, where ones represent treatment and zeros pause days
$S \in \mathbb{B}^D \times \mathbb{B}^{T_s}$	a matrix holding the partially occupied schedule, where $S_{d,t_s} = 1$ iff time slot t_s on day d is occupied
$\mathcal{A}_p \in \{1, \ldots, 7\}$	the set of allowed start days for patient p
$c_p \in \mathbb{N}$	penalty for missing the waiting time target for each priority
$d_{L,p} \in \mathbb{N}$	day limit, i.e., the waiting time target for patient p

4.1   Scheduling-Based CP Model

Variables. The basic decision variables of the model are as follows:

$start_{p,f} \in \mathcal{T}_w$	the start time for the window patient $p \in \mathcal{P}$ is scheduled in during treatment day $f \in \mathcal{F}$
$window_{p,f} \in \{0, \ldots, W\}$	the window patient $p \in \mathcal{P}$ is scheduled in during treatment day $f \in \mathcal{F}$, where window 0 represents no treatment
$fraction_{p,d} \in \{0, \ldots, L_p\}$	fraction that is delivered to patient $p \in \mathcal{P}$ on day $d \in \mathcal{D}$, where fraction 0 represents no treatment

There following variables are derived from the basic variables:

$day_{p,f} \in \mathcal{D}$ day patient $p \in \mathcal{P}$ is treated with fraction $f \in \mathcal{F}$

$start\_day_p = day_{p,1}$ start day for patient $p \in \mathcal{P}$

Constraints. In the scheduling-based CP model, the `cumulative` constraint [1] is used to ensure that no two treatments overlap:

$$\texttt{cumulative}([\langle start_{p,f}, w_L, dur_p\rangle | p \in \mathcal{P}, f \in \mathcal{F}], w_L), \tag{1}$$

where w_L is the window length. The partially occupied input schedule is included as patients with fixed window, start time, and duration to ensure that no new patients are scheduled in a window that is already full. The constraint is used "backwards", setting the duration of each treatment equal to the window length w_L and the resource requirement as the duration of the treatment dur_p.

The variables $day_{p,f}$ and $fraction_{p,d}$ are dual to each other:

$$fraction_{p,d} = f, \text{ where } d = day_{p,f} \quad \forall p \in \mathcal{P}, f \in \mathcal{F}, d \in \mathcal{D}, \tag{2}$$

while $fraction_{p,d} = 0$ if patient p is before the start or after the end of treatment:

$$(d < start\_day_p) \vee (d \geq start\_day_p + L_p) \rightarrow fraction_{p,d} = 0 \quad \forall p \in \mathcal{P}, d \in \mathcal{D}. \tag{3}$$

The day $day_{p,f}$ for fraction $f \in \mathcal{F}$ for patient $p \in \mathcal{P}$ is given by the time when the fraction starts, $start_{p,f}$, divided by the number of time slots T_s. To avoid division, this is expressed as: $(day_{p,f} - 1)T_s + 1 \leq start_{p,f} \leq day_{p,f}T_s$.

The days are connected to each other by the constraint:

$$day_{p,f+1} = day_{p,f} + 1 \quad \forall p \in \mathcal{P}, f \in \mathcal{F},$$

which means that if treatment day f is on day d, then $f + 1$ is on $d + 1$.

Next, connect $window_{p,f}$ to $start_{p,f}$. The vector FS_p is the fractionation schedule for patient p and is input to the problem given by a protocol. $FS_{p,f} = 1$ corresponds to treatment day f being active for patient p, i.e., treatment is delivered that day. Thus, if the input $FS_{p,f}$ is indeed one, this gives

$$start_{p,f} = (day_{p,f} - 1)T_s + (window_{p,f} - 1)w_L + 1, \tag{4}$$

and for all other $p \in \mathcal{P}, f \in \mathcal{F}$, set $window_{p,f} = 0$. This constraint states that the start time for the window patient p is scheduled during fraction f is equal to the start time of that day, plus the start time of the window on that day.

Bounds are also given for when each fraction can start earliest and latest. For example, the patient's second fraction cannot be on the first day, and similarly, the patient's first fraction cannot be on the last day:

$$f \leq day_{p,f} \leq D - (L_p - f) \quad \forall f \in \mathcal{F}. \tag{5}$$

Similar constraints limit the start day for each patient to be at latest L_p days from the end of the planning horizon. Patients have to start treatment on an allowed start day: $start\_day_p \in \mathcal{A}_p \quad \forall p \in \mathcal{P}$.

To break some dominance, patients of the same priority and treatment protocol are sorted by their waiting time target. A constraint enforces that an earlier target patient always starts their treatment before a later target patient.

Objective Function. The first objective is to start each treatment within the waiting time targets. The target violation is measured as the number of days that the patient misses their treatment target date with

$$target\_violation_p = \max(0, start\_day_p - d_{L,p}) \quad \forall p \in \mathcal{P}, \qquad (6)$$

where $d_{L,p}$ is the day limit for when treatment should start.

The second objective is to schedule patients in the same window each day. Therefore, a penalty is added each time the window is switched. Since, for example, 3 machines with 4 windows is modeled as 1 machine with 12 windows, there is no penalty for moving from window 1 to 5 or 9, as they are the same but on different machines. An array m_w maps model windows to real machine windows. Only the active treatment days $\mathcal{F}_a$, when $window_{p,f} \neq 0$, are considered:

$$window\_diff_p = \sum_{f \in \mathcal{F}_a} (m\_w[window_{p,f}] \neq m\_w[window_{p,f+1}]) \quad \forall p \in \mathcal{P} \qquad (7)$$

$$\text{where } m\_w = [1, 2, 3, 4, 1, 2, 3, 4, 1, 2, 3, 4]$$

The two parts of the objective function given by (6) and (7) are combined into a weighted sum. Each entry of $window\_diff_p$ is multiplied by 2 to make it equal to the IP formulation, see Sect. 4.3. The total objective function is then:

$$\sum_{p \in \mathcal{P}} (100 c_p target\_violation_p + 2 window\_diff_p), \qquad (8)$$

where c_p are weights for each priority group used to capture that it is worse to violate the target for higher prioritized patients. The weight 100 reflects that waiting time targets are more important than minimizing window switches.

4.2   Packing-Based CP Model

In the packing-based CP model, $fraction_{p,d} = 0$ if patient p has not started or has finished treatment on day d is expressed by the **regular** constraint [23], replacing (3). For example, if the treatment length is 10 days, the regular expression is $r = 0^* \cdot 1 \cdot 2 \cdot \ldots \cdot 8 \cdot 9 \cdot 10 \cdot 0^*$ (where $\cdot$ is concatenation) in:

$$\textbf{regular}([fraction_{p,d}|d \in \mathcal{D}], r) \quad \forall p \in \mathcal{P}. \qquad (9)$$

The variables $start_{p,d}$ are removed as the constraints (1) and (4) are removed. Instead, regular expressions on the time windows define the treatment patterns, where window 0 corresponds to no treatment. The example regular expression $r = 0 \cdot ([1, 2, 3, 4]^5 \cdot 0^2)^6$ states that $f = 0$ gives $window_{p,0} = 0$, then $window_{p,f} \in \{1, \ldots, 4\}$ for five days, followed by two pause days where $window_{p,f} = 0$, and then repeating this pattern six times, leading to:

$$\textbf{regular}([window_{p,f}|f \in \{0, \ldots, L_p\}], r) \quad \forall p \in \mathcal{P}. \qquad (10)$$

The `bin_packing` constraint [29] ensures that the patients fit in each window:

$$\texttt{bin\_packing}([\infty, w_L, w_L, w_L, w_L], [window_{p,f}|p \in \mathcal{P}], [dur_p|p \in \mathcal{P}]) \quad \forall d \in \mathcal{D}, \tag{11}$$

where constraint (2) connects d to f. The capacity of window 0 is infinite and the other windows have the window length w_L as capacity. The partially occupied input schedule is also included as patients with fixed windows.

4.3 IP Model

Variables. The basic variables of the IP model are (again using $\mathbb{B} = \{0, 1\}$):

$s_{p,d} \in \mathbb{B}$ $\quad s_{p,d} = 1$ iff patient $p \in \mathcal{P}$ starts treatment on day $d \in \mathcal{D}$
$q_{p,d,f} \in \mathbb{B}$ $\quad q_{p,d,f} = 1$ iff patient $p \in \mathcal{P}$ has their f-th treatment day on $d \in \mathcal{D}, f \in \mathcal{F}$
$x_{p,d,w} \in \mathbb{B}$ $\quad x_{p,d,w} = 1$ iff patient $p \in \mathcal{P}$ is scheduled in window $w \in \mathcal{W}$ on day $d \in \mathcal{D}$

There are also variables that are derived from the basic variables:

$u_{p,d} \in \mathbb{B}$ $\quad u_{p,d} = 1$ iff patient $p \in \mathcal{P}$ has an active treatment day on $d \in \mathcal{D}$
$y_{p,d,w} \in \mathbb{B}$ $\quad$ help variable for the objective function where $y_{p,d,w} = 1$ if patient $p \in \mathcal{P}$ is scheduled in window $w \in \mathcal{W}$ on day $d \in \mathcal{D}$, and if patient p is not scheduled in window w on day d, $y_{p,d,w} \in \mathbb{B}$

Constraints. Patient $p \in \mathcal{P}$ will be treated for L_p days. Thus, the last day to start treatment is $daylimit_p = D - L_p + 1$. The treatment should start exactly one time on an allowed start day given by $\mathcal{A}_p$:

$$\sum_{d \subseteq \mathcal{A}_p}^{daylimit_p} s_{p,d} = 1, \quad \forall p \in \mathcal{P}.$$

For the CP models, constraint (5) states which fractions that can be delivered on which days. In the IP model, this is enforced by setting $q_{p,d,f} = 0$ for all $p \in \mathcal{P}, d \in \mathcal{D}, f \notin \mathcal{F}_{p,d}$ where $\mathcal{F}_{p,d} := \{\max(0, d - (D - L_p)), \ldots, \min(d, L_p)\}$. Using $\hat{\mathcal{F}}_{p,d}$ to denote $\mathcal{F}_{p,d}$ with the last element excluded, the constraint:

$$q_{p,d,f} = q_{p,d+1,f+1} \quad \forall p \in \mathcal{P}, d \in \{1, \ldots, D-1\}, f \in \hat{\mathcal{F}}_{p,d}$$

enforces that all treatment days are scheduled after each other.

The following constraints state that the f-th treatment day can only happen once and that patient p is scheduled at most once every day d:

$$\sum_{d \in \mathcal{D}} q_{p,d,f} \leq 1 \quad \forall p \in \mathcal{P}, f \in \mathcal{F}_{p,d} \qquad \sum_{f \in \mathcal{F}_{p,d}} q_{p,d,f} \leq 1 \qquad \forall p \in \mathcal{P}, d \in \mathcal{D}.$$

In the CP models, this is enforced by constraints (4) or (9).

The first treatment day $f = 1$ is given on the start day for each patient:

$$q_{p,d,1} = s_{p,d} \quad \forall p \in \mathcal{P}, d \in \mathcal{D}.$$

In the fractionation schedule for patient p, an active treatment day f gives $FS_{p,f} = 1$. A variable $u_{p,d}$ is introduced so that $u_{p,d} = 1$ iff patient p is during treatment on day d ($q_{p,d,f} = 1$) and has an active treatment day ($FS_{p,f} = 1$) and zero otherwise, thus, it controls if d is an active day or not for patient p:

$$u_{p,d} = \sum_{f \in \mathcal{F}_{p,d}} (q_{p,d,f} FS_{p,f}) \quad \forall p \in \mathcal{P}, d \in \mathcal{D}.$$

Each patient is scheduled in exactly one time window on active treatment days, and not in any window on off-days:

$$\sum_{w \in \mathcal{W}} x_{p,d,w} = u_{p,d} \quad \forall p \in \mathcal{P}, d \in \mathcal{D}.$$

In the CP models, this constraint is expressed by constraints (4) or (10).

In order to make sure that all treatments fit within each time window, $u_{p,d}$ is used to keep track of if the patient has an active day or not, together with $x_{p,d,w}$, which is one iff patient p is scheduled in window w on day d:

$$\sum_{p \in \mathcal{P}} x_{p,d,w} u_{p,d} \, dur_p + \sum_{t_s \subseteq w} S_{d,t_s} \leq w_L \quad \forall d \in \mathcal{D}, w \in \mathcal{W}.$$

S is the input schedule, where an element is one iff time slot t_s on day d is occupied. Thus, the sum of the duration of all patients in window w plus the previously occupied slots in that window must be less than or equal the window length w_L. In the CP models, this is enforced by constraints (1) or (11). A major difference is that the number of constraints in the IP model grows with the number of time windows, which is not the case in the CP models.

Objective Function. The objective is the same as in the CP model. A penalty is added for the time by which the target is missed:

$$f_{1,p} = \sum_{d=d_{L,p}}^{D} s_{p,d}(d - d_{L,p}) \quad \forall p \in \mathcal{P},$$

where $d_{L,p}$ corresponds to the waiting target in days for patient p and $s_{p,d} = 1$ on the start day. This constraint corresponds to the CP objective function (6).

The other objective is to schedule the patient in the same time window each day. To do this, a help variable $y_{p,d,w} \in \mathbb{B}$ is introduced so that $y_{p,d,w}$ is one when $x_{p,d,w}$ is one, and the sum of $y_{p,d,w}$'s is one on all days:

$$y_{p,d,w} \geq x_{p,d,w} \quad \forall p \in \mathcal{P}, d \in \mathcal{D}, w \in \mathcal{W}$$

$$\sum_{w \in \mathcal{W}} y_{p,d,w} = 1 \quad \forall p \in \mathcal{P}, d \in \mathcal{D}.$$

As for the CP models, the problem with 3 machines with 4 windows each is modeled as having 1 machine with 12 windows, and hence there should be no penalty for switching from window 1 to 5 or 9. Introduce $\mathcal{W}_m = \{1, \ldots, W_m\}$ where W_m is the number of windows on each machine. Each window switch is penalized in the second objective function, where we here assume 3 machines:

$$f_{2,p} = \sum_{d \in \hat{\mathcal{D}}} \sum_{w \in \mathcal{W}_m} \left| \sum_{i = \{w, w+W_m, w+2W_m\}} (y_{p,d,i} - y_{p,d+1,i}) \right| \quad \forall p \in \mathcal{P}. \tag{12}$$

Since $f_{2,p}$ sums both $0 - 1$ and $1 - 0$ for each switch, there is a penalty of 2 for every switch. To make the CP objective function equivalent, the factor 2 difference is adjusted for in (8).

New variables avoid absolute values, as they would render the model non-linear: $z_{p,d,w} = \sum_{i = \{w, w+W_m, w+2W_m\}} (y_{p,d,i} - y_{p,d+1,i})$ for $w \in \mathcal{W}_m$. $z_{p,d,w}$ is divided into a positive and negative part; $z_{p,d,w} = z^+_{p,d,w} - z^-_{p,d,w}$:

$$f_{2,p} = \sum_{d \in \hat{\mathcal{D}}} \sum_{w \in \mathcal{W}_m} z^+_{p,d,w} + z^-_{p,d,w} \quad \forall p \in \mathcal{P}$$

$$z^+_{p,d,w} \geq 0, z^-_{p,d,w} \geq 0 \quad \forall p \in \mathcal{P}, d \in \mathcal{D}, w \in \mathcal{W}_m. \tag{13}$$

The two formulations (12) and (13) are equivalent in a minimization setting.

In total, the objective function is equivalent to the CP objective function (8):

$$\sum_{p \in \mathcal{P}} (100 c_p f_{1,p} + f_{2,p}).$$

5   CP Search

When solving the CP models, the search is conducted in the following order:

1. Assign priority A patients randomly to a start day as early as possible.
2. Assign priority C and D patients to a start day randomly by choosing the $start\_day_p$ variable with smallest domain size over weighted degree [4].
3. Assign priority B patients to their earliest possible start day.
4. Assign the number of window switches $window\_diff_p$ as small as possible for all patients sorted by their duration, with longest duration assigned first.

For easy instances, with few patients to schedule, it is possible to construct deterministic search heuristics for the CP models that perform much better than the random search strategies described above. However, for more difficult instances these heuristics fail to even find a solution within reasonable time. This is the reason for including randomization.

When solving a minimization problem in CP, branch-and-bound tree search is used, which follows the same branch until it has failed. However, using a restart strategy causes search to restart from the top node whenever search finds a solution or after a specific number of failures defined by the restart strategy. In this case, using a restart strategy yields better results because the problem is somewhat under-constrained; it is easy to find feasible solutions and relatively few failures occur. Therefore, the Luby restart strategy [20] multiplied by a factor of 100 is used. The interval is a result of testing many different intervals on different problem setups. Doing restarts this often can be compared to approximating a Large Neighborhood Search (LNS) [28] (see also Sect. 7).

6   Experiments and Results

Multiple experiments are run to evaluate and compare the three models. The experiments are run on a Windows 10 machine with an Intel® Core™ i9-7940X X-series processor and 64 GB of RAM. The IP model is solved using the MIP solver of CPLEX 12.8 in the Python API with default parameters. The CP models use MiniZinc 2.2.2 and are solved with Gecode 6.1.0. Other solvers have been tested, such as the lazy clause solver Chuffed, but Gecode gave the best overall results on the tested problem instances.

A simulation engine is built with Python 3.6. In this engine, the first day starts from an empty schedule and patients to be scheduled are assumed to arrive according to a Poisson process. For each simulated day, a patient schedule is created using the previously described models, and the patients are fixed to the schedule if they have a start day within a week (since this is the limit assumed to communicate the schedules to the patients). The schedule from the previous day is used as input for the next day in the simulation, together with the backlog of yet unscheduled patients. For these patients, the waiting time target is adjusted by one day since it is counted from the day the patient is ready for treatment.

The simulation engine is used to generate problem benchmarks, that each have a partially occupied schedule, a patient backlog, and the number of expected patients arriving per day (as discussed in Sect. 3). Occupation in a schedule is measured as the average occupation of the first two weeks in that schedule. This is not a perfect measure; if the first week is completely booked and the second is completely free, some urgent patients will not be able to meet their target dates, although occupancy is 50% in total for these weeks.

Scheduling patients on three different machines, 16 different benchmarks that are grouped into three categories are summarized in Table 1. The categories capture the following aspects:

(a) The average number of patients arriving each day.
(w) Which weekday the schedule is created. Note that a benchmark w-7 with Sunday as planning day is omitted as it is identical to benchmark l-1.
(l) The load in the input, i.e., the amount of partial occupation in the input schedule and the size of the backlog. These are closely related, since an almost empty schedule does not come with a large backlog, and vice versa. Benchmarks l-3 and l-4 examine scalability and do not represent realistic scenarios, as patients would be transferred to other clinics.

The weights in the objective function are chosen as $c_1 = 10$, $c_2 = 3$, $c_3 = 1$, corresponding to priority group A to C, and are the same as in [24]. Priority D

Table 1. Setup of the benchmarks, 3 machines

Benchmark		Occupation (%) (4/6 windows)	Expected number of arriving patients	Number of patients in backlog (4/6 windows)	Number of patients including future arrivals (4/6 windows)	Planning day
1	a-1	18.5/10.2	4	5/8	49/52	Sunday
2	a-2	18.5/10.2	6	5/8	77/80	Sunday
3	a-3	18.5/10.2	8	5/8	93/96	Sunday
4	a-4	53.9/55.0	4	8/8	52/52	Tuesday
5	a-5	53.9/55.0	6	8/8	80/80	Tuesday
6	a-6	53.9/55.0	8	8/8	96/96	Tuesday
7	w-1	53.9/55.0	5	8/8	79/79	Monday
8	w-2	53.9/55.0	5	8/8	79/79	Tuesday
9	w-3	53.9/55.0	5	8/8	79/79	Wednesday
10	w-4	53.9/55.0	5	8/8	79/79	Thursday
11	w-5	53.9/55.0	5	8/8	79/79	Friday
12	w-6	53.9/55.0	5	8/8	79/79	Saturday
13	l-1	53.9/55.0	5	8/8	79/79	Sunday
14	l-2	60.9/61.8	5	16/14	87/85	Sunday
15	l-3	69.8/64.8	5	24/34	90/100	Friday
16	l-4	73.3/69.5	5	45/48	111/114	Thursday

is for expected future patients of priority A and has weight $c_4 = 5$. The day is divided into 4 or 6 time windows. The timeout is set to 6 h.

The performance of the models is measured as the objective function value as a function of runtime for the benchmarks, for both 4 and 6 windows.

Average Patient Arrival. The results for benchmarks a-1 to a-6 are shown in Fig. 1. They show that except for the 6 window case in benchmark a-4, the IP model reaches optimality considerably faster, and the packing-based CP model outperforms the scheduling-based CP model on all instances. When the arrival rate is low, as in benchmarks a-1 and a-4, the IP and CP models have similar performance. The CP models are however more sensitive to an increase in the average number of patients arriving each day, while the time to reach optimality in the 4 window IP model does not change significantly. Increasing the number of time windows from 4 to 6 makes the IP model slower in all cases. When the partial occupation is low (benchmarks a-1 to a-3), the packing-based CP model is not slower with more windows, which is not the case when the partial occupation is higher (benchmarks a-4 to a-6).

Weekday. The results when varying the weekday the schedule is created are presented in Fig. 2. Some treatment protocols state that treatment can be initiated only on certain days, but this does not have a large effect on runtime. An observation is that for the 6-window case, although the IP model reaches optimality faster, the packing-based CP model has a better objective value initially. Benchmarks w-1 to w-6 show that if the time limit is short, the packing-based CP model performs better than the IP model.

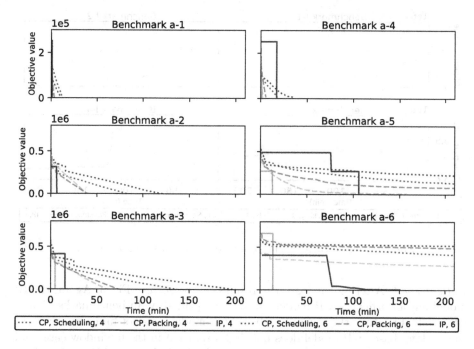

Fig. 1. Results with varying arrival rate, 4 windows and 6 windows

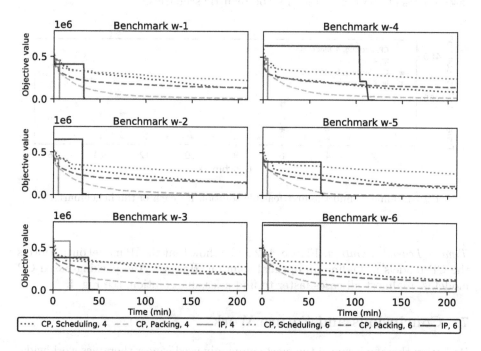

Fig. 2. Results with varying planning day, 4 windows and 6 windows

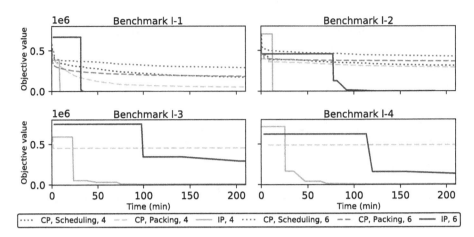

Fig. 3. Results when varying the load, 4 windows and 6 windows

Patient Load. Altering the load, the results can be seen in Fig. 3. Again, the IP model outperforms the CP models in finding optimality. In benchmarks I-3 and I-4, in 6 h of runtime the scheduling-based CP model does not find any solutions, the packing-based CP model does not find a solution in the 6-window case, and the IP model does not reach optimality in the 6-window case. However, these cases represent too heavy a load to be realistic scenarios.

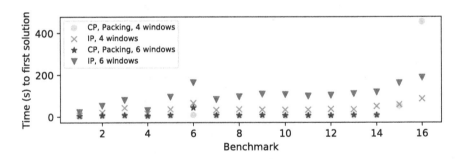

Fig. 4. Time to find the first feasible solution for each of the benchmarks

Time to Feasible Solution. Figures 1, 2 and 3 show that the IP model outperforms the CP models in finding an optimal solution. However, the packing-based CP model is in almost all cases faster to find a feasible solution, see Fig. 4.

7   Conclusions and Future Work

Radiation therapy is one of the most commonly used cancer therapies worldwide. The waiting times can be reduced optimizing the use of LINACs for treatment.

This paper introduces three different models that capture the non-block radiotherapy patient scheduling problem; one IP model, a scheduling-based CP model, and a packing-based CP model. The patients have different priority levels, treatment patterns, treatment durations, and start days for treatment. The expected future patient arrivals are included in the models to predict future resource utilization.

The models are evaluated for multiple different scenarios. The results show that the packing-based CP model outperforms the scheduling-based CP model on all problem instances. In general, the CP models find feasible solutions faster than the IP model, however, the IP model reaches optimality considerably faster. All models are sensitive to an increase in the number of time windows per day.

Future Work. To make the models more realistic, overtime on the machines and cancellation of treatments should be included in the models. Cancellation of treatments is common in RT, and allowing for overtime on the machines is important since it is often used in practice to reduce waiting times. Both aspects are bound to increase the complexity of the models, and overtime is likely to make the IP model nonlinear. Another future direction is to extend the models to include multiple machine types, which also increases the models' complexity.

For the models to be able to handle these complexities, a potential future extension is to take advantage of the strengths of each model in an IP/CP hybrid. Another option would be to use some decomposition method on the IP model, for example Benders decomposition. To better explore the search space, an option is to use Large Neighborhood Search (LNS) when solving the CP models.

A potential improvement of the stochastic aspect of the models is to use scenario-based probabilities instead of expected values when accounting for patients arriving in the future.

The models have so far been compared with each other. The next step would be to see how the models perform over time, using the simulation engine and comparing to both a myopic scheduling strategy (not taking future patient arrivals into account) and to historical data from a clinic.

Acknowledgments. The authors thank Per Enqvist at KTH and Kjell Eriksson at RaySearch Laboratories for a fruitful collaboration. The authors are grateful for insightful discussions about the CP models with Mats Carlsson and Peter J. Stuckey and constructive comments from the anonymous reviewers.

References

1. Aggoun, A., Beldiceanu, N.: Extending chip in order to solve complex scheduling and placement problems. Math. Comput. Modell. **17**(7), 57–73 (1993)
2. Baatar, D., Boland, N., Brand, S., Stuckey, P.J.: CP and IP approaches to cancer radiotherapy delivery optimization. Constraints **16**, 173–194 (2011)
3. Barták, R., Salido, M., Rossi, F.: New trends in constraint satisfaction, planning, and scheduling: a survey. Knowl. Eng. Rev. **25**, 249–279 (2010)

4. Boussemart, F., Hemery, F., Lecoutre, C., Sais, L.: Boosting systematic search by weighting constraints. In: de Mántaras, R.L., Saitta, L. (eds.) Sixteenth European Conference on Artificial Intelligence, pp. 146–150. IOS Press, Valencia (2004)

5. Burke, E.K., De Causmaecker, P., Berghe, G.V., Van Landeghem, H.: The state of the art of nurse rostering. J. Sched. **7**(6), 441–499 (2004)

6. Burke, E.K., Leite-Rocha, P., Petrovic, S.: An integer linear programming model for the radiotherapy treatment scheduling problem. arXiv e-prints arXiv:1103.3391 (2011)

7. Cayirli, T., Veral, E.: Outpatient scheduling in health care: a review of literature. Prod. Oper. Manage. **12**(4), 519–549 (2009)

8. Chen, Z., King, W., Pearcey, R., Kerba, M., Mackillop, W.J.: The relationship between waiting time for radiotherapy and clinical outcomes: a systematic review of the literature. Radiother. Oncol. **87**(1), 3–16 (2008)

9. Conforti, D., Guerriero, F., Guido, R.: Optimization models for radiotherapy patient scheduling. 4OR **6**(3), 263–278 (2008)

10. Conforti, D., Guerriero, F., Guido, R.: Non-block scheduling with priority for radiotherapy treatments. Eur. J. Oper. Res. **201**(1), 289–296 (2010)

11. Ehrgott, M., Holder, A.: Operations research methods for optimization in radiation oncology. J. Radiat. Oncol. Inform. **6**(1), 1–41 (2014)

12. Fortin, A., Bairati, I., Albert, M., Moore, L., Allard, J., Couture, C.: Effect of treatment delay on outcome of patients with early-stage head-and-neck carcinoma receiving radical radiotherapy. Int. J. Radiat. Oncol. Biol. Phys. **52**(4), 929–936 (2002)

13. Gocgun, Y.: Simulation-based approximate policy iteration for dynamic patient scheduling for radiation therapy. Health Care Manage. Sci. **21**(3), 317–325 (2018)

14. Gomez, D.R., et al.: Time to treatment as a quality metric in lung cancer: staging studies, time to treatment, and patient survival. Radiother. Oncol. **115**(2), 257–263 (2015)

15. Hahn-Goldberg, S., Beck, J.C., Carter, M.W., Trudeau, M., Sousa, P., Beattie, K.: Solving the chemotherapy outpatient scheduling problem with constraint programming. J. Appl. Oper. Res. **6**(3), 135–144 (2014)

16. Halperin, E.C., Wazer, D.E., Brady, L.W., Perez, C.A.: Perez and Brady's Principles and Practice of Radiation Oncology, 6th edn. Lippincott Williams and Wilkins, Philadelphia (2013)

17. Jacquemin, Y., Marcon, E., Pommier, P.: A pattern-based approach of radiotherapy scheduling. In: IFAC Proceedings Volumes, vol. 44, pp. 6945–6950 (2011)

18. Kapamara, T., Sheibani, K., Haas, O.C.L., Reeves, C., Petrovic, D.: A review of scheduling problems in radiotherapy. In: Proceedings of the International Control Systems Engineering Conference on Systems Engineering (ICSE 2006), pp. 201–207 (2006). https://onlinelibrary.wiley.com/doi/abs/10.1111/j.1937-5956.2011.01221.x

19. Legrain, A., Fortin, M.A., Lahrichi, N., Rousseau, L.M.: Online stochastic optimization of radiotherapy patient scheduling. Health Care Manage. Sci. **18**, 110–123 (2015)

20. Luby, M., Sinclair, A., Zuckerman, D.: Optimal speedup of Las Vegas algorithms. Inform. Process. Lett. **47**, 173–180 (1993)

21. May, J.H., Spangler, W.E., Strum, D.P., Vargas, L.G.: The surgical scheduling problem: current research and future opportunities. Prod. Oper. Manage. **20**, 392–405 (2011)

22. O'Rourke, N., Edwards, R.: Lung cancer treatment waiting times and tumour growth. Clin. Oncol. **12**(3), 141–144 (2000)

23. Pesant, G.: A regular language membership constraint for finite sequences of variables. In: Wallace [32], pp. 482–495
24. Petrovic, D., Castro, E., Petrovic, S., Kapamara, T.: Radiotherapy scheduling. In: Uyar, A., Ozcan, E., Urquhart, N. (eds.) Automated Scheduling and Planning. Studies in Computational Intelligence, vol. 505, pp. 155–189. Springer, Heidelberg (2013). https://doi.org/10.1007/978-3-642-39304-4_7
25. Petrovic, S., Leung, W., Song, X., Sundar, S.: Algorithms for radiotherapy treatment booking. In: Qu, R. (ed.) 25th Workshop of the UK Planning and Scheduling Special Interest Group, pp. 105–112, April 2006
26. Riff, M.C., Cares, J.P., Neveu, B.: RASON: a new approach to the scheduling radiotherapy problem that considers the current waiting times. Expert Syst. Appl. **64**, 287–295 (2016)
27. Sauré, A., Patrick, J., Tyldesley, S., Puterman, M.L.: Dynamic multi-appointment patient scheduling for radiation therapy. Eur. J. Oper. Res. **223**(2), 573–584 (2012)
28. Shaw, P.: Using constraint programming and local search methods to solve vehicle routing problems. In: Maher, M., Puget, J.-F. (eds.) CP 1998. LNCS, vol. 1520, pp. 417–431. Springer, Heidelberg (1998). https://doi.org/10.1007/3-540-49481-2_30
29. Shaw, P.: A constraint for bin packing. In: Wallace [32], pp. 648–662
30. Van Harten, M.C., Hoebers, F.J., Kross, K.W., Van Werkhoven, E.D., Van Den Brekel, M.W., Van Dijk, B.A.: Determinants of treatment waiting times for head and neck cancer in the netherlands and their relation to survival. Oral Oncol. **51**(3), 272–278 (2015)
31. Vieira, B., Hans, E.W., Van Vliet-Vroegindeweij, C., Van De Kamer, J., Van-Harten, W.: Operations research for resource planning and - use in radiotherapy: a literature review. BMC Med. Inform. Decis. Making **16**(149) (2016)
32. Wallace, M. (ed.): CP 2004. LNCS, vol. 3258. Springer, Heidelberg (2004). https://doi.org/10.1007/b100482

Constraint Programming-Based Job Dispatching for Modern HPC Applications

Cristian Galleguillos[1,2(✉)], Zeynep Kiziltan[2], Alina Sîrbu[3], and Ozalp Babaoglu[2]

[1] Pontificia Universidad Católica de Valparaíso, Valparaíso, Chile
`cristian.galleguillos.m@mail.pucv.cl`
[2] University of Bologna, Bologna, Italy
{`zeynep.kiziltan,ozalp.babaoglu`}`@unibo.it`
[3] University of Pisa, Pisa, Italy
`alina.sirbu@unipi.it`

Abstract. HPC systems are increasingly being used for big data analytics and predictive model building that employ many short jobs. In these application scenarios, HPC job dispatchers need to process large numbers of short jobs quickly and make decisions on-line while ensuring high Quality-of-Service (QoS) levels and meet demanding timing requirements. Constraint Programming (CP) is an effective approach for tackling job dispatching problems. Yet, the state-of-the-art CP-based job dispatchers are unable to satisfy the challenges of on-line dispatching and take advantage of job duration predictions. These limitations jeopardize achieving high QoS levels, and consequently impede the adoption of CP-based dispatchers in HPC systems. We propose a class of CP-based dispatchers that are more suitable for HPC systems running modern applications. The new dispatchers are able to reduce the time required for generating on-line dispatching decisions significantly, and are able to make effective use of job duration predictions to decrease waiting times and job slowdowns, especially for workloads dominated by short jobs.

1 Introduction

Easy access to massive data sets, data analytics tools and High-Performance Computing (HPC) have been fueling the trend towards *data-driven* computational scientific discovery [3], with big-data processing frameworks such as Hadoop and Spark increasingly integrated with HPC systems [2, 17, 31, 34]. Workloads of HPC systems engaged in data-driven analytics tend to be a mix of many short jobs (<1 h) with fewer longer jobs [32]. Hence, HPC job dispatchers need to rapidly process a large number of short jobs in making on-line decisions so as to minimize both waiting times and *slowdown* (the ratio between the total job duration including waiting time and the actual job duration during runtime). These measures of Quality-of-Service (QoS) are particularly important when HPC systems

© Springer Nature Switzerland AG 2019
T. Schiex and S. de Givry (Eds.): CP 2019, LNCS 11802, pp. 438–455, 2019.
https://doi.org/10.1007/978-3-030-30048-7_26

are used to provide real-time services, such as big-data visualization [29,33,39], where response times are critical for acceptable user experience.

While the on-line job dispatching problem in HPC systems is NP-hard [6], it can be formulated as a job scheduling and resource allocation problem for which Constraint Programming (CP) has produced good results [4]. The first CP-based HPC dispatcher with job waiting times as a measure of QoS was introduced in [5] and shown to obtain better solutions compared to a Priority Rule-Based (PRB) dispatcher [10,21], which is widely adopted in commercial HPC workload management systems such as Altair PBS Professional [1] and SLURM Workload Manager [35]. The dispatcher was later embedded as a plug-in within the software framework of PBS professional [9]. Subsequently, another CP-based dispatcher with a similar measure of QoS with the additional feature of limiting system power consumption was presented in [7,8] and proved to outperform a PRB dispatcher on the instances with tight power capping values.

Despite the potential of these CP-based job dispatchers, certain limitations hinder their adoption for modern HPC systems. As reported in [9], the first dispatcher is not resilient to *heavy workloads*—workloads where resource requests greatly exceed available resources. The time spent by this dispatcher in generating a dispatching decision increases dramatically as more jobs requiring high system utilization arrive to the system. The second CP-based HPC dispatcher was initially employed in off-line mode [8], and later also in on-line mode [7] but on workloads of maximum 1000 jobs submitted in a time window of half an hour. A more realistic scenario where jobs arrive continuously and many of them end up waiting in a queue due to unavailable computational resources increases greatly the difficulty of generating dispatching decisions. Our experimental results confirm that both dispatchers are not resilient to heavy workloads that are present in real datasets, which is undesirable in the quest for fast response times.

Another limitation is related to the actual runtime duration of a job on a specific HPC system which is not known before it is executed and yet is crucial for generating dispatching decisions to guarantee high QoS levels. Dispatchers often use the expected job duration, which is the maximum time a job is allowed to execute on the system. In the above mentioned dispatchers, the expected duration is the default value assigned by the system, which is typically the default wall-time of the queue where the job is submitted, unless the job owner supplied her own expected duration. Even in the latter case, however, users tend to use the maximum wall-time and user estimations are acknowledged to be overestimated in general [13,16,27]. A dispatcher that relies on overestimated durations is likely to schedule fewer jobs than possible at dispatching time, and consequently, is likely to cause unnecessary delays. Prediction of actual runtime durations using simple heuristics or more sophisticated machine learning techniques is an active area of research [15,19,20,38]. Recent studies show that the use of job duration predictions when generating dispatching decisions can substantially improve QoS levels in backfilling-based dispatchers [15,19,20,37].

Our contribution is a class of novel CP-based dispatchers that are more suitable for HPC systems running modern applications. We build on [5,8] and redesign their main components. First, we revisit their model and search control mechanism so as to make them resilient to heavy workloads and applicable to on-line dispatching. Second, we study the use of job duration prediction, instead of the expected duration, when generating dispatching decisions. We discuss why naively replacing the expected duration with a predicted duration may be ineffective, if not detrimental for QoS. Consequently, we adapt the model and search algorithm of our dispatchers to the use of job duration predictions to obtain high QoS levels in terms of job waiting times and slowdown. We conduct a simulation study on a workload trace collected from an HPC system containing large numbers of short jobs. We use predictions with different accuracy, underestimation and overestimation rates on the dataset. Our results demonstrate that with our approach, the CP-based dispatchers can: (i) significantly reduce the time required to generate dispatching decisions; and (ii) benefit from good job duration predictions and considerably decrease the waiting times and the slowdown of the jobs, especially for workloads dominated by short to medium jobs.

The rest of the paper is organized as follows. In Sect. 2, we introduce the on-line job dispatching problem in HPC systems and give an overview of the CP-based dispatchers introduced in [5,8]. In Sects. 3 and 4, we describe our approach. In Sects. 5 and 6, we detail our experimental study and present our results. We discuss the related work in Sect. 7 and conclude in Sect. 8.

2   Formal Background

2.1   On-line Job Dispatching Problem in HPC Systems

A user request in an HPC system consists of the execution of a computational application over the system resources. Such a request is referred to as *job* and the set of all jobs is known as *workload*. Each job in the workload is associated to a name, required resources (cores, memory, etc.) to run the corresponding application, and its *expected duration* which is the maximum time it is allowed to execute on the system. An HPC system typically receives multiple jobs simultaneously from different users, placing them in a *queue* together with the other waiting jobs (if there are any). The time interval during which a job remains in the queue until its execution time is known as *waiting time*. At a given *dispatching time*, a job *dispatcher decides* when the jobs waiting in the queue can start executing and on which resources they can execute. The goal is to dispatch in the best possible way according a measure of QoS, such as by reducing the waiting times or the slowdown of the jobs, which is directly perceived by the HPC users. During execution, a job exceeding its expected duration is killed.

Formally, *on-line dispatching* in an HPC system takes place at a specific time t for (a subset of) the queued jobs Q. A typical HPC system is composed of N nodes, with each node $n \in N$ having a capacity $cap_{n,r}$ for each of its resource type $r \in R$, giving the total amount of available resource. Each job $i \in Q$ has the arrival time $q_i \leq t$ to the queue, which is unknown before the arrival, and a demand $req_{i,r}$ giving the amount of resources required from r. The *on-line dispatching problem* at time t consists in *scheduling* each job i by assigning it a start time $s_i \geq t$, and *allocating* i to the requested resources during its expected duration d_i, such that the capacity constraints are satisfied: at any time in the schedule, the capacity $cap_{n,r}$ of a resource r is not exceeded by the total demand $req_{i,r}$ of the jobs i allocated on it, taking into account the presence of jobs already in execution. A typical objective is to minimize the sum of the waiting times $s_i - q_i$. Once the problem is solved, only the jobs with $s_i = t$ are dispatched. The remaining jobs with $s_i > t$ are queued again with their original q_i. It is the workload management system software that decides the dispatching time t and the subsequent dispatching times.

A solution to the problem (i.e., a *dispatching decision*) is obtained according to a policy using the current system status, such as the queued jobs, the running jobs and the availability of the resources. A sub-optimal solution could cause exceptional delays in the queue, hurting the QoS. While a (near-)optimal solution is a critical requirement in HPC systems, the on-line job dispatching problem is an NP-hard problem [6] and thus needs to be addressed with a dedicated approach. In [5,8], the first CP-based dispatchers for HPC systems are developed and tested on a workload trace collected from the Eurora system [11].

2.2 CP-Based Dispatchers for HPC Systems

In the first dispatcher [5], the entire dispatching problem is modelled and solved using a CP solver. The second dispatcher [8] instead relies on a hybrid method. While the scheduling problem is modelled and solved in a CP solver, the allocation problem is solved separately using a heuristic search algorithm. We will refer to them as PCP and HCP, respectively, to mean the use of a Pure CP and a Hybrid CP method in their dispatching algorithms.

Scheduling. In both PCP and HCP, the scheduling problem is modeled with Conditional Interval Variables (CIVs) [25]. A CIV $\tau_i \in \tau$ represents a job i and defines the time interval during which i runs. At a certain dispatching time t, there may already be jobs in execution which were previously scheduled and allocated. We refer to such jobs as running jobs. The scheduling model considers in the τ variables both the running jobs and the queued jobs in Q. The properties $s(\tau_i)$ and $d(\tau_i)$ correspond respectively to the start time and the duration of the job i. Since the actual runtime duration d_i^r of a running or queued job i is unknown at the modeling time, PCP and HCP rely on an estimation and use the expected duration d_i for $d(\tau_i)$. Thus we have $d(\tau_i) = d_i$ for the queued jobs and $d(\tau_i) = s(\tau_i) + d_i - t$ for the running jobs. While the start time of the running jobs have already been decided, the queued jobs have $s(\tau_i) \in [t, eoh]$, where eoh is the end of the worst-case makespan calculated as $t + \sum_{\tau_i} d(\tau_i)$. Expected durations d_i are supplied by the users. In the absence of this information, the dispatchers use the default wall-time of the queue. It is important to note that even user-supplied values tend to be equal to the wall-time of the queue, which is indeed the maximum allowed value for d_i. We will refer to the use of such d_i to define $d(\tau_i)$ as the *wall-time approach*.

Unlike PCP, HCP searches for a start time for the first m jobs in Q (referred to as $\bar{Q}$). The remaining jobs in $Q \setminus \bar{Q}$ are still in the model, but they are postponed to the end of the makespan by fixing their start time as $s(\tau_i) = eoh - d(\tau_i)$. The capacity constraints in PCP are enforced via a cumulative constraint for all $n \in N$ and for all $r \in R$, ensuring that at any given time in the makespan the total $req_{i,r}$ of the jobs i using r does not exceed $cap_{n,r}$. In HCP, resources of the same type across all nodes are considered as a pool of resources, hence the cumulative constraints are posted for each $r \in R$ with the total capacity $Cap_r^T = \sum_{n \in N} cap_{n,r}$. Any infeasibility that may be introduced due to this modelling choice is fixed during the allocation phase. HCP considers also power as a resource type, allowing to restrict the total power consumption of the jobs. We here omit this feature as it is not relevant to our study.

We consider the objective function which minimizes the sum of the waiting times of the jobs. In PCP it is formalized as $\sum_{\tau_i} \max(0, \frac{s(\tau_i) - q_i - ewt_i}{ewt_i})$. It is a weighted sum so as to give priority to the jobs that stay in the queue longer than their ewt_i. The ewt_i value is the average waiting time of the queue where i is submitted, and is obtained by analyzing the Eurora workload data which was collected by the PBS dispatcher [22]. In the objective function of HCP, the weights are slightly different, giving priority to the jobs of the queues with lower expected waiting times: $\sum_{\tau_i} \frac{\max(ewt_i)}{ewt_i} * (s(\tau_i) - q_i)$. We will explain later how the corresponding scheduling models are solved by PCP and HCP.

Allocation. In PCP, the allocation problem is modelled via an alternative constraint [25] for all $\tau_i \in \tau$, which ensures that the requested resources $req_{i,r}$ are satisfied by selecting a subset of the alternative possibilities for allocating i. An alternative possibility is an optional CIV, which may or may not be present in the allocation decision, and represents an individual allocation to the resources of a given node n. Instead in HCP, it is solved by a PRB algorithm for the jobs which have $s(\tau_i) = t$ after the scheduling model is solved. This heuristic algorithm iteratively tries to allocate each scheduled job using the best-fit allocation strategy. The jobs are chosen based on their priority. The jobs that have been waiting the longest at time t have the highest priority. Such a priority is calculated in line with the priority of the jobs in the objective function: $\frac{\max(ewt_i)}{ewt_i} * (t - q_i)$. As a tie breaker, job demand is used, which is the job's resource requirements multiplied by job duration $d(\tau_i)$. Hence, among the high priority jobs, those that have requested fewer resources and have shorter durations have further priority. Since the scheduling decision may contain some inconsistencies due to considering the resources of the same type as a pool, a job may not be allocated, in which case it is postponed to the next dispatching time.

Search. To solve the scheduling and the allocation model altogether, PCP uses the self-adapting large neighborhood search algorithm [24] which is the default search available in the solver where PCP is implemented [26]. HCP instead uses a custom search algorithm derived from the schedule-or-postpone algorithm [30] to solve the scheduling model. The criteria used to select a job among all the available ones at each decision node follows the priority rule used in the PRB allocation algorithm, thus preferring the jobs that can start first and whose priority are highest. Note that the priorities are calculated once statically at the dispatching time t before search starts. Due to problem complexity, search in both PCP and HCP is bounded by a time limit δ. Thus, the best solution found within the limit is the dispatching/scheduling decision. If, however, no solution is found within the limit, the search is restarted with an increased time limit $2 * \delta$. This procedure continues while no solution is found and $\delta \geq \delta_{max}$, where δ_{max} is the maximum time available to generate a decision.

3   Resiliency to Heavy Workloads

In this section, we *reduce the model size and improve the search control* of the dispatchers in an effort to make the dispatchers resilient to heavy workloads and applicable to on-line dispatching.

At a dispatching time t, PCP searches for a solution for all the jobs in Q which can be very time consuming when many jobs are waiting. While this problem is tackled in HCP by searching for a solution for the jobs in $\bar{Q}$ and postponing the remaining jobs in $Q \setminus \bar{Q}$ to the end of the makespan, there raises another issue: when many jobs are postponed in the same way, they are likely to overlap and create excess demand for the system resources at a given time in the schedule. It may therefore be not be possible to find a feasible solution that satisfies the resource constraints, consequently the entire Q may be postponed to the next dispatching time $t + 1$. To address this problem, we remove the remaining jobs in $Q \setminus \bar{Q}$ from the model and place them in the queue with their original q_i.

During the typical operation of an HPC system, job submission by users has a stochastic nature and actual runtime durations are known only when jobs terminate. Additionally, at a dispatching time t, only the jobs with $s(\tau_i) = t$ are dispatched. Thus, it is not fruitful to generate a dispatching decision for the entire schedule makespan $[t, eoh]$. We therefore remove from the model all the jobs requiring more amount of resources than available at time t and queue them again with their original q_i. In addition to reducing the model size in terms of decision variables, we also eliminate the unnecessary variables and constraints in the model of a given problem instance. Specifically, for a given resource type r (in a node n), if none of the jobs in the model require it, we remove the corresponding cumulative constraint from the model. Moreover, in PCP, if there is no availability to allocate i in the system resources, we remove i and its corresponding alternative constraint from the model, and queue it again with its original q_i. Note that removing jobs from the model and putting them back in the queue does not cause any starvation problem. As we will argue in Sect. 4 and confirm experimentally in Sect. 6, their priority grow with their slowdown and eventually they are all dispatched.

During search for a solution, both solvers of PCP and HCP use a time limit δ to interrupt the search and return the best solution found. If, no solution is found within the limit, the search is restarted with an increased time limit $2 * \delta$. In the latter case, the dispatchers cannot distinguish an unsatisfiable problem instance from a difficult instance that is not solved yet. This has the consequence of searching for a solution again and again for an instance known to be unsatisfiable. To address this problem, we add the solver state to the search control. Consequently, if the solver proved unsatisfiability, this will be known when the search is interrupted by the time limit, and the subsequent restart will be avoided by placing the jobs in the queue for the next dispatching time. Finally, we avoid a restart if the solution quality did not change after k consecutive restarts.

In the following, we refer to the versions of PCP and HCP whose model and search control are built as described here as PCP$_1$ and HCP$_1$.

4   Incorporation of Job Duration Prediction

A straightforward way to incorporate the duration prediction d_i^d of a job i into our dispatchers is to use it for defining the duration $d(\tau_i)$ as $d(\tau_i) = d_i^d$ for the queued jobs and $d(\tau_i) = s(\tau_i) + d_i^d - t$ for the running jobs, without any other changes to the dispatchers. In this section, we argue that this naive use may be ineffective, if not worsen the QoS, thus we adapt the model and search algorithm of both dispatchers to the use of job duration predictions in order to obtain high QoS levels in terms of job waiting times and slowdown.

A duration prediction d_i^d of a job i may be perfectly accurate ($d_i^d = d_i^r$), underestimated ($d_i^d < d_i^r$), or overestimated ($d_i^d > d_i^r$). If a running job i is underestimated, at a certain dispatching time t, we will have $s(\tau_i) + d_i^d < t$ and thus $d(\tau_i) = s(\tau_i) + d_i^d - t < 0$. That is, the duration of a running job will have a negative value even if the job is still running. A negative $d(\tau_i)$ for a running job directly affects the calculation of the makespan $\sum_{\tau_i} d(\tau_i)$ of the queued jobs. With a reduced makespan, it may not be possible to find a schedule and/or allocation for the queued jobs, consequently they may all be postponed to the next dispatching time $t + 1$, worsening the QoS. If instead, a running job is overestimated at t, we will surely have $s(\tau_i) + d_i^d > t$ and $d(\tau_i) > 0$, thus the makespan will not be shorter than necessary.

To *address the problem of duration prediction underestimation*, we extend the duration $d(\tau_i)$ of a running job i which has $d(\tau_i) < 0$ at time t. Specifically, we redefine it as $d(\tau_i) = 1$, assuming that the job i needs at least one more unit of time as of t. This value is necessary and sufficient. It is the minimum value necessary to prevent a feasible problem instance from turning into an unfeasible one, as the makespan will be large enough to fit all the queued jobs in a schedule. To show that it is sufficient, we remind that at t, only the jobs for which the dispatcher decides that $s(\tau_i) = t$ are dispatched (the remaining are queued again). The allocation decision made for such jobs is valid until the next dispatching time $t + 1$ and is not affected by the actual runtime durations of the running jobs even if they are underestimated. By using the minimum possible value for the duration of the underestimated running jobs, we keep the search space size compact. Our initial experiments confirm that higher values of $d(\tau_i)$ make the problem more difficult. In the following, we refer to this version of PCP$_1$ and HCP$_1$ as PCP$_2$ and HCP$_2$.

Table 1. Dispatcher versions.

Enhancement	PCP_1	HCP_1	PCP_2	HCP_2	PCP_3	HCP_3
Reduced model size, improved search control	✓	✓	✓	✓	✓	✓
Addressing duration prediction underestimation			✓	✓	✓	✓
Job durations in the obj. function and search					✓	✓

Even if the job duration prediction is accurate, resulting in $d_i^d \sim d_i^r$ for all jobs, the dispatchers may still not be able to exploit them fruitfully for targeting low job waiting time $s(\tau_i) - q_i$ and slowdown $(s(\tau_i) - q_i + d_i^r)/d_i^r$. As we saw in Sect. 2.2, both dispatchers assign a priority to the jobs that should not wait long. Then the jobs with higher priority are forced to be scheduled first via the objective function, as well as in the custom search of the scheduling problem and in the heuristic search of the allocation problem of the HCP dispatcher. However, job duration $d(\tau_i)$ is ignored in the priority. It is used only as a tie breaker among the jobs having the same priority during the search of the scheduling and the allocation problems of HCP. The priority instead focuses on a relation between the current waiting time $t - q_i$ of the job i and its expected waiting time ewt_i. The problem is that ewt_i is not a job specific feature that can be decided on-line at the time of dispatching. It is a feature of the queue where the job is submitted and is calculated offline. Such a value may not be informative on the current job submission status so as to generate a dispatching decision of high quality.

We tackle this limitation by involving *job durations in the objective function and in the search* of the scheduling and allocation, via the use of job slowdown as job priority. Thus, the new objective function and the priority of a job i at a dispatching time t become $\sum_{\tau_i} \frac{s(\tau_i) - q_i + d(\tau_i)}{d(\tau_i)}$ and $(t - q_i + d(\tau_i))/d(\tau_i)$, respectively. This is the normalization of the job waiting time, which has a higher value for jobs waiting more than their duration than for jobs waiting less than their duration. We foresee the following benefits. First, since it gives priority to short jobs, the dispatcher will aim at lowering both the total job waiting times and the total job slowdown, as required by modern HPC applications. Our experimental results in Sect. 6 show that by giving priority to short jobs, we never penalize the medium and long jobs. Second, it prioritizes the jobs based on a job specific feature $d(\tau_i)$ which can be calculated on-line and which can reflect better the current job submission status. Finally, integrating $d(\tau_i)$ in the objective function and search of the dispatchers paves the way to exploit job duration predictions.

In the following, we refer to the versions of PCP_2 and HCP_2 whose model and search algorithms are adapted as described here as PCP_3 and HCP_3. Table 1 summarizes all the dispatcher versions. We note that, similar to HCP, the HCP_3 dispatcher uses the job priorities in the custom search of the scheduling problem and in the heuristic search of the allocation problem, and calculates the priorities once statically at the dispatching time t before search starts. Our initial experiments revealed that updating them dynamically during search is not beneficial. As we described in Sect. 2.2, the search of PCP relies on the default search of the underlying solver and does not exploit priorities. We observed in our initial experiments that the custom search of the scheduling model in HCP is valuable also for PCP to solve the entire scheduling and allocation problem, hence we adopt that kind of search and exploit priorities also in PCP$_3$.

5 Experimental Study

To evaluate the significance of our approach, we conducted an experimental study, by simulating on-line job submission to an HPC system.

HPC System and its Workload Dataset. Our study is based on a workload trace collected from the Eurora system [11], with (the portions of) which the original CP-based dispatchers were tested. [5,7–9]. We repeated the same study using another workload trace collected from the Gaia system [14] and obtained similar results which we omit in the paper due to space restrictions. The Eurora system was hosted by CINECA [12], the largest Italian datacenter. Eurora occupied the first place in the Green500 list of June 2013, and was in production until August 2015. It consisted of 64 nodes, each equipped with 2 8-core GPUs, 16 GB of RAM memory, and 2 accelerators: GPUs and MICs. The workload, collected by the PBS dispatcher between March 2014 and August 2015, consists of logs for over 400,000 jobs submitted to one of its four queues, including job duration and detailed resource usage. The workload is dominated by short jobs (under 1 h), making up 93.14% of all jobs, while the remaining 6.10% are medium jobs (between 1 and 5 h) and 0.75% are long jobs (over 5 h).

Job Duration Prediction. To derive job durations, we used three prediction methods with varying accuracy levels, and underestimation and overestimation rates: (i) the wall-time approach, (ii) a data-driven prediction heuristic [19] which is simple to implement and has a low overhead, and as a baseline (iii) the actual runtime (real) durations. In [19], the authors have applied the heuristic prediction to the Eurora dataset. The mean absolute error (MAE) of the heuristic and the wall-time approach with respect to the real duration were shown to be 40 mins and 225 mins, respectively. The heuristic prediction shows thus an improvement of 82% over the wall-time approach. In Fig. 1, we show the empirical cumulative distribution function (ECDF) of the prediction accuracy $A = d_i^r/d(\tau_i)$, the ratio between the real and the predicted duration of a job, of all the three methods. The empirical ECDF shows the proportion of scores that are less than or equal to each score of A on Eurora. When $A = 1$, the duration $d(\tau_i)$ matches the real duration d_i^r. We have underestimation when $A > 1$, overestimation when $A < 1$. In theory, we should not have underestimation with the wall-time approach because in a real system a job is killed if it takes longer than its d_i. However, a system requires extra time after a job is killed or completed to bring the resources on-line again and this extra time is reflected to the dataset.

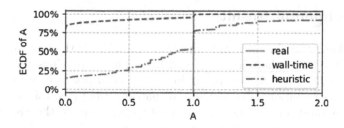

Fig. 1. The distribution of the accuracy of the three prediction methods.

Therefore, in some cases we have $A > 1$ in Fig. 1. We have $0.75 \leq A \leq 1.25$ for about 50% of the workload with the heuristic, and for less than 10% with the wall-time approach. On the other hand, the heuristic introduces considerable underestimation. The exact under and overestimation rates are 3.6% and 96.3% for the wall-time and 25.8% and 53.7% for the heuristic, respectively.

Experimental Setup. We used the open-source discrete event simulator AccaSim [18] to simulate the Eurora system with its workload dataset. Each job submission is simulated by using its available data, for instance, the owner, the requested resources, and the real duration, the execution command or the name of the application executed. AccaSim uses the real duration to simulate the job execution during its entire duration. Therefore job duration prediction errors do not affect the running time of the jobs with respect to the real workload data. The dispatchers under study are implemented using the AccaSim directives to allow them to generate the dispatching decisions during the system simulation.

With the heuristic prediction, as opposed to calculating the predictions off-line as in [19], we calculate them on-line during the simulation and update the knowledge base upon job terminations. The accuracy of the heuristic thus depends on the generated dispatching decisions. As a CP modelling and solving toolkit, we used Google OR-Tools[1] version 6.7 and ported it to Python 3.6 to implement the dispatchers in AccaSim. The PCP, PCP$_1$, and PCP$_2$ dispatchers use the default search algorithm of OR-Tools for CIVs, which is the schedule-or-postpone algorithm. As explained in Sects. 2.2 and 4, all the other dispatchers use the custom search derived from schedule-or-postpone. In terms of the dispatcher parameters, we set $\delta = 1s$, $k = 2$, and $\delta_{max} = 16s$ to small values to keep the dispatcher overhead low. We keep $m = 100$ as in HCP. Both dispatchers need in some of their versions the estimated waiting time $ewt_Q = \frac{1}{|Q|} \sum_{i \in Q} s_i - q_i$ of each queue Q in the system. These values were calculated for the Eurora workload in [5,8] and reused here. All experiments were performed on a dedicated server with a 16-core Intel Xeon CPU and 8 GB of RAM, running Linux Ubuntu 16.04. The source code of the CP-based dispatchers is available at https://git.io/fjia1.

---

[1] https://developers.google.com/optimization/.

6   Experimental Results

In this section, we report our experimental results. While the best and the final versions of the dispatchers are PCP_3 and HCP_3, all the previous versions (PCP, PCP_1, PCP_2, HCP, HCP_1, HCP_2) appear in the experiments in order to evaluate each of our contributions. To refer to a dispatcher using a certain job duration prediction method, we append -W, -D or -R to the name of the dispatcher for the Wall-time approach, the Data-driven heuristic and the Real duration, resp.

Table 2. Times and problem sizes.

Dispatcher	PCP	PCP_1-W	PCP_3-R	PCP_3-D	HCP	HCP_1-W	HCP_3-R	HCP_3-D
Avg. disp. time [ms]	∞	743	692	701	1,014	703	523	575
Total pred. time [s]	-	-	-	289	-	-	-	308
Total sim. time [s]	∞	262,436	261,985	262,764	374,788	245,663	201,223	215,814
Avg. # of intervals	-	145	94	115	379	100	51	63
Avg. # of req. res	-	853	142	584	6,267	1,292	258	571
Avg. # of avl. res	-	1,476	1,471	1,473	1,487	1,477	1,473	1,474

6.1   Dispatcher Performance and Problem Size

We first assess the impact of reducing the model size and improving the search control of the dispatchers for resiliency to heavy workloads. Following the original dispatchers PCP and HCP, we use the wall-time approach in PCP_1 and HCP_1 for job duration prediction, and compare the performance of and the problem size in PCP and PCP_1-W, as well as HCP and HCP_1-W. We report in Table 2 the mean CPU time spent in generating a dispatching decision over all dispatcher invocations, including the time for modeling the dispatching problem instance and searching for a solution. We also report the total simulation time from the first job submission until the last job completion, and the average problem size: number of intervals, number of requested resources, number of available resources.

PCP crashes before the completion of the entire workload, demonstrating that it is not resilient to heavy workloads. We therefore underline the improvement reached by the PCP_1-W dispatcher which is now able to process the workload. Compared to HCP, the HCP_1-W dispatcher reduces the total time by around 34% and reduces the problem size and time required for dispatching significantly. These results demonstrate that our approach has significantly better performance, making the dispatchers applicable to heavy workloads and paving the way to the use of CP-based dispatchers for HPC on-line dispatching.

6.2   Quality of the Dispatching Decisions

Next, we evaluate the value of adapting the model and search algorithm of the dispatchers to the use of job duration predictions by comparing the quality of the decisions made by PCP, PCP_1, PCP_2, PCP_3, as well as by HCP, HCP_1, HCP_2, HCP_3.

Since we are aiming at reducing both the slowdown and waiting time of jobs, we consider both of these metrics. We first study the effectiveness of PCP, PCP_1, PCP_2, and PCP_3 with each job duration prediction method. Then, we analyze HCP, HCP_1, HCP_2, and HCP_3. We show the results of PCP_2 and HCP_2 only in conjunction with the data-driven heuristic. This is because on our workload the heuristic has a considerable underestimation rate while the other prediction methods have negligible or no underestimation, so the behaviour was very similar to PCP_1 and HCP_1. We also compare the various dispatchers with the performance of PBS in the original system, by calculating the slowdown and waiting time from the workload data.

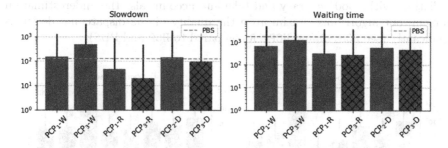

Fig. 2. Average and error bars showing one standard deviation of slowdown and waiting times [s] using the PCP dispatchers.

PCP Results. Fig. 2 shows the slowdown and waiting times obtained by various versions of the dispatchers, compared to PBS. PCP is missing from the plot due to the fact that it is not able to process the workload, hence we consider PCP_1-W as a baseline, which is the enhancement most similar to the original algorithm. Additionally, we do not report the results of PCP_1-D because the simulation was too heavy and did not terminate in more than two weeks, so we interrupted it. We believe the long simulation time is due to the fact that PCP_1-D does not deal with underestimation, so it tends to use the maximum time limit for the instances in which jobs are underestimated, generating long queues.

A first observation is that, our best dispatcher coupled with the best duration predictor (PCP_3-R) and the heuristic predictor (PCP_3-D) always outperform PBS. PCP_3-W has lower performance compared to PCP_1-W. This is probably because the wall-time approach has a high overestimation rate, which is not beneficial when the dispatcher involves job durations in dispatching decisions. However, if we look at the dispatchers using real durations, we observe a significant increase in performance compared to PCP_1-W but also when moving from PCP_1-R to PCP_3-R. The reduction in the slowdown and waiting time from PCP_1-R to PCP_3-R reach up to 58% and 13%. This is due to the accuracy of the prediction method which does not present any underestimation nor overestimation. This proves that our approach is essential when a good quality prediction is available.

On a more realistic prediction, the results confirm that great care needs to be taken when integrating predictions. A straightforward integration of the predictions in previous algorithms is not helpful at all: PCP_1-D takes too long. By handling underestimation as in PCP_2-D, we are able to improve the results compared to PCP_1-W. Further improvement is observed when moving to PCP_3-D, demonstrating again the benefits of including predictions, albeit imperfect, into the model and search algorithm. Specifically, we observe 37% and 29% reduction in the average slowdown and the average waiting time.

HCP Results. Figure 3 shows the performance of HCP, HCP_1, HCP_2 and HCP_3 compared to PBS. Unlike the PCP case, here the original dispatcher HCP is able to process the entire workload so we can compare our results directly with the state-of-the-art method, besides PBS. We observe that in general, if we include predictions with good accuracy and take into account also the underestimation problem, our algorithms can improve the quality of the dispatching decisions significantly (see HCP_3-D and HCP_3-R compared to HCP and PBS).

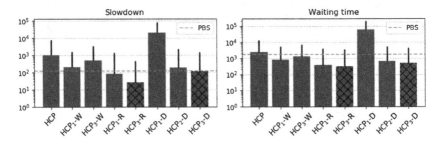

Fig. 3. Average and error bars showing one standard deviation of slowdown and waiting times [s] using the HCP dispatchers.

In more detail, we observe that simply moving from HCP to HCP_1-W, with an approach aimed at reducing the CPU time for dispatching, we also improve the quality of the solutions. HCP_3-W does not improve HCP_1-W, since the accuracy of predictions using wall-time is rather low. We observe the most significant improvements over HCP with HCP_3-R, proving again the importance of our approach when a good quality prediction is available. The decreased performance of HCP_1-D compared to all other algorithms confirms again that naively including predictions can be detrimental. The gains obtained by HCP_2-D with respect to HCP_1-D support again the need of dealing with underestimated jobs. We note that, while HCP_1-D performs worse than the original HCP, HCP_2-D becomes better than HCP and HCP_3-D further improves HCP_2-D, demonstrating again the benefits of including predictions, albeit imperfect, into the model and search algorithm.

Discussion. We conclude that suitable incorporation of job duration predictions in PCP and HCP, such as PCP_3 and HCP_3, can lead to significantly higher levels of QoS especially for workloads dominated by short jobs. To benefit from this potential, durations should rely on predictions with acceptable levels of accuracy, going beyond the standard wall-time approach. The quality of the decisions generated by PCP$_1$-W and HCP$_1$-W is much worse than PCP$_3$-R and HCP$_3$-R. On the other hand, PCP$_3$-D and HCP$_3$-D offer valid alternatives to PCP$_1$-W and HCP$_1$-W with further reductions in problem size (as reported in Table 2) and with QoS measures closer to those of PCP$_3$-R and HCP$_3$-R. Table 2 shows also the time cost of this gain. While PCP$_3$-D and HCP$_3$-D come each with a cost of prediction, the total simulation times of PCP$_3$-D and PCP$_1$-W are similar, and HCP$_3$-D reduces notably the time with respect to HCP$_1$-W. The fact that the new dispatchers give priority to short jobs does not penalize the medium and long jobs, as can be witnessed in Fig. 4. Finally, our approach does not affect the system utilization. We did not observe any major differences between the various dispatchers (results not shown due to space limitations). This is probably because all the dispatchers are using the best-fit allocation strategy.

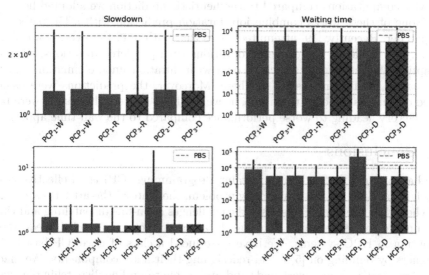

Fig. 4. Average and error bars showing one standard deviation of slowdown and waiting times [s] on medium and long jobs using all the dispatchers.

7   Related Work

Job duration prediction has been used to optimize job dispatchers. In [20], a simple linear model can improve the slowdown of backfilling techniques by 28%. On an IBM Blue Gene/P machine, adjusted user estimates were able to improve up to 20% the performance of the dispatchers favoring short jobs [37], while a predictive heuristic was shown to double the performance of a backfilling-based dispatcher [38]. The heuristic prediction that we employ here is similar to [38], however it considers more complex job profiles. When the prediction underestimates the job duration, [38] apply a correction step, to keep the job alive, similar to the adjustment that we make. However, they adjust the duration to define a new backfilling window whereas in our approach it is intended for defining valid CP models. Recently, machine learning methods were applied to predict job duration [23,36], including metadata such as job names as features. In fact, the heuristic method we employ also relies on job metadata, however it is much simpler, being an heuristic that does not require model training. Neither of the methods is integrated within a dispatcher for testing. An adaptive on-line machine learning method based on state space models is used in [28] to predict job duration. The authors show that their predictions allow for reducing waiting times by 25% in backfilling-based dispatchers.

Underestimation of job duration is a problem that appears often in the literature, since it negatively affects dispatcher performance, more than overestimation. Recently, [15] proposed a predictive method based on a censored regression model, which could minimize underestimation. Although promising, it requires heavier computations compared to the heuristic prediction we adopted here.

None of these works combine job duration prediction with a CP-based job dispatcher. Recently, [19] attempted to do that with HCP. However, it was done naively by replacing the expected durations with predicted durations, without adapting the model and search to deal with duration underestimation and to the use of predictions, as we did here. Moreover, the predictions were calculated off-line, as opposed to on-line, as we did here. Indeed, the results were not satisfactory, leading to worse performance compared to the wall-time approach.

8   Conclusions

We have argued that, while Constraint Programming (CP) is an effective approach in tackling the job dispatching problem, the-state-of-the-art CP-based dispatchers [5,8] are unable to satisfy the challenges of on-line dispatching and they are unable to take advantage of job duration predictions, which impede their adoption in HPC systems. We have introduced a class of novel CP-based dispatchers by building on [5,8] and redesigning their main components. We made them resilient to heavy workloads and applicable to on-line dispatching, as well as adapted them to the use of job duration predictions to obtain high QoS levels in terms of job waiting times and slowdown. We evaluated the significance of our approach on a workload trace collected from an HPC system, using predictions with different accuracy and underestimation and overestimation rates on the dataset. The experimental results are excellent. Compared to the original dispatchers, the time spent by the new dispatchers in generating decisions on a heavy workload is significantly reduced. Moreover, the new dispatchers can benefit from job duration predictions and generate decisions of higher QoS levels on a workload dominated by short jobs. The new dispatchers are thus more suitable for HPC systems running modern applications that employ short jobs. To benefit from this potential, the durations should rely on predictions with acceptable levels of accuracy, going beyond the standard wall-time approach. While the heuristic prediction considered in the paper is not the best, we have shown that it is a valid alternative to the wall-time approach, despite its simplicity.

In future work, we will include the allocation problem in the search of the new PCP dispatcher, which currently focuses only on the scheduling problem.

We also plan to test the dispatchers with other, more sophisticated, duration prediction methods, as well as to integrate dedicated allocation strategies in the dispatchers so as to enhance system utilization.

Acknowledgements. We thank A. Bartolini, L. Benini, M. Milano, M. Lombardi and the SCAI group at Cineca for providing the Eurora data, and A. Borghesi and T. Bridi for sharing the original implementations of the dispatchers. We thank the IT Center of the University of Pisa and M. Marzolla for providing computing resources. C. Galleguillos has been supported by Postgraduate Grant PUCV 2018. A. Sîrbu has been partially funded by the SoBigData EU project (grant agreement 654024).

References

1. Altair: Altair PBS professional (2019). http://www.pbsworks.com
2. Anderson, M.J., et al.: Bridging the gap between HPC and big data frameworks. PVLDB **10**(8), 901–912 (2017)
3. Ashby, S., et al.: The opportunities and challenges of exascale computing-summary report of the advanced scientific computing advisory committee (ASCAC) subcommittee. US Department of Energy Office of Science, pp. 1–77 (2010)
4. Baptiste, P., Laborie, P., Pape, C.L., Nuijten, W.: Chapter 22 - constraint-based scheduling and planning. In: Handbook of Constraint Programming, Foundations of Artificial Intelligence, vol. 2, pp. 761–799. Elsevier (2006)
5. Bartolini, A., Borghesi, A., Bridi, T., Lombardi, M., Milano, M.: Proactive workload dispatching on the EURORA supercomputer. In: O'Sullivan, B. (ed.) CP 2014. LNCS, vol. 8656, pp. 765–780. Springer, Cham (2014). https://doi.org/10.1007/978-3-319-10428-7_55
6. Blazewicz, J., Lenstra, J.K., Kan, A.H.G.R.: Scheduling subject to resource constraints: classification and complexity. Discrete Appl. Math. **5**(1), 11–24 (1983)
7. Borghesi, A., Bartolini, A., Lombardi, M., Milano, M., Benini, L.: Scheduling-based power capping in high performance computing systems. Sustain. Comput. Inf. Syst. **19**, 1–13 (2018)
8. Borghesi, A., Collina, F., Lombardi, M., Milano, M., Benini, L.: Power capping in high performance computing systems. In: Pesant, G. (ed.) CP 2015. LNCS, vol. 9255, pp. 524–540. Springer, Cham (2015). https://doi.org/10.1007/978-3-319-23219-5_37
9. Bridi, T., Bartolini, A., Lombardi, M., Milano, M., Benini, L.: A constraint programming scheduler for heterogeneous high-performance computing machines. IEEE Trans. Parallel Distrib. Syst. **27**(10), 2781–2794 (2016)
10. Buddhakulsomsiri, J., Kim, D.S.: Priority rule-based heuristic for multi-mode resource-constrained project scheduling problems with resource vacations and activity splitting. Eur. J. Oper. Res. **178**(2), 374–390 (2007)
11. Cavazzoni, C.: EURORA: a European architecture toward exascale. In: Proceedings of Future HPC Systems - The Challenges of Power-Constrained Performance, pp. 1–4. ACM (2012)
12. CINECA: The Italian Interuniversitary Consortium for High Performance Computing (2019). https://www.cineca.it/
13. Cirne, W., Berman, F.: A comprehensive model of the supercomputer workload. In: Proceedings of the Fourth Annual IEEE International Workshop on Workload Characterization, pp. 140–148, December 2001

14. Emeras, J., Varrette, S., Guzek, M., Bouvry, P.: EVALIX: classification and prediction of job resource consumption on HPC platforms. In: Desai, N., Cirne, W. (eds.) JSSPP 2015-2016. LNCS, vol. 10353, pp. 102–122. Springer, Cham (2017). https://doi.org/10.1007/978-3-319-61756-5_6

15. Fan, Y., Rich, P., Allcock, W.E., Papka, M.E., Lan, Z.: Trade-off between prediction accuracy and underestimation rate in job runtime estimates. In: Proceedings of IEEE International Conference on Cluster Computing, CLUSTER 2017, pp. 530–540. IEEE Computer Society (2017)

16. Feitelson, D.G., Weil, A.M.: Utilization and predictability in scheduling the IBM SP2 with backfilling. In: Proceedings of First Merged International Parallel Processing Symposium and Symposium on Parallel and Distributed Processing, IPPS/SPDP 1998, pp. 542–546 (1998)

17. Fox, G., Qiu, J., Jha, S., Ekanayake, S., Kamburugamuve, S.: Big data, simulations and HPC convergence. In: Rabl, T., Nambiar, R., Baru, C., Bhandarkar, M., Poess, M., Pyne, S. (eds.) WBDB -2015. LNCS, vol. 10044, pp. 3–17. Springer, Cham (2016). https://doi.org/10.1007/978-3-319-49748-8_1

18. Galleguillos, C., Kiziltan, Z., Netti, A., Soto, R.: AccaSim: a customizable workload management simulator for job dispatching research in HPC systems. Cluster Computing (2019)

19. Galleguillos, C., Sîrbu, A., Kiziltan, Z., Babaoglu, O., Borghesi, A., Bridi, T.: Data-driven job dispatching in HPC systems. In: Nicosia, G., Pardalos, P., Giuffrida, G., Umeton, R. (eds.) MOD 2017. LNCS, vol. 10710, pp. 449–461. Springer, Cham (2018). https://doi.org/10.1007/978-3-319-72926-8_37

20. Gaussier, É., Glesser, D., Reis, V., Trystram, D.: Improving backfilling by using machine learning to predict running times. In: Proceedings of International Conference for High Performance Computing, Networking, Storage and Analysis, SC 2015, pp. 1–10. ACM (2015)

21. Haupt, R.: A survey of priority rule-based scheduling. Operat. Res. Spektrum **11**(1), 3–16 (1989)

22. Henderson, R.L.: Job scheduling under the Portable Batch System. In: Feitelson, D.G., Rudolph, L. (eds.) JSSPP 1995. LNCS, vol. 949, pp. 279–294. Springer, Heidelberg (1995). https://doi.org/10.1007/3-540-60153-8_34

23. Wyatt II, M.R., Herbein, S., Gamblin, T., Moody, A., Ahn, D.H., Taufer, M.: PRIONN: predicting runtime and IO using neural networks. In: Proceedings of 47th International Conference on Parallel Processing, ICPP 2018, pp. 1–12. ACM (2018)

24. Laborie, P., Godard, D.: Self-adapting large neighborhood search: application to single-mode scheduling problems. In: Proceedings of 3rd Multidisciplinary International Conference on Scheduling: Theory and Applications, MISTA 2007, pp. 276–284 (2007)

25. Laborie, P., Rogerie, J.: Reasoning with conditional time-intervals. In: Proceedings of Twenty-First International Florida Artificial Intelligence Research Society Conference, FLAIRS 2008, pp. 555–560. AAAI Press (2008)

26. Laborie, P., Rogerie, J., Shaw, P., Vilím, P.: IBM ILOG CP Optimizer for scheduling. Constraints **23**(2), 210–250 (2018)

27. Bailey Lee, C., Schwartzman, Y., Hardy, J., Snavely, A.: Are user runtime estimates inherently inaccurate? In: Feitelson, D.G., Rudolph, L., Schwiegelshohn, U. (eds.) JSSPP 2004. LNCS, vol. 3277, pp. 253–263. Springer, Heidelberg (2005). https://doi.org/10.1007/11407522_14

28. Naghshnejad, M., Singhal, M.: Adaptive online runtime prediction to improve HPC applications latency in cloud. In: Proceedings of 11th IEEE International Conference on Cloud Computing, CLOUD 2018, pp. 762–769. IEEE Computer Society (2018)

29. Nonaka, J., Sakamoto, N., Shimizu, T., Fujita, M., Ono, K., Koyamada, K.: Distributed particle-based rendering framework for large data visualization on HPC environments. In: 2017 International Conference on High Performance Computing Simulation (HPCS), pp. 300–307 (2017)

30. Pape, C.L., Couronne, P., Vergamini, D., Gosselin, V.: Time-versus-capacity compromises in project scheduling. AISB Q. **91**, 19–31 (1995)

31. Qiu, J., Jha, S., Luckow, A., Fox, G.C.: Towards HPC-ABDS: an initial high-performance big data stack. Building Robust Big Data Ecosystem ISO/IEC JTC **1**, 18–21 (2014)

32. Reuther, A., et al.: Scalable system scheduling for HPC and big data. J. Parallel Distrib. Comput. **111**, 76–92 (2018)

33. Rückemann, C.: Using parallel multicore and HPC systems for dynamical visualisation. In: 2009 International Conference on Advanced Geographic Information Systems Web Services, pp. 13–18 (2009)

34. Singh, D., Reddy, C.K.: A survey on platforms for big data analytics. J. Big Data **2**(1), 8 (2015)

35. SLURM: SLURM workload manager (2019). http://slurm.schedmd.com

36. Soysal, M., Berghoff, M., Streit, A.: Analysis of job metadata for enhanced wall time prediction. In: Klusáček, D., Cirne, W., Desai, N., et al. (eds.) JSSPP 2018. LNCS, vol. 11332, pp. 1–14. Springer, Cham (2018). https://doi.org/10.1007/978-3-030-10632-4_1

37. Tang, W., Desai, N., Buettner, D., Lan, Z.: Analyzing and adjusting user runtime estimates to improve job scheduling on the Blue Gene/P. In: Proceedings of 24th IEEE International Symposium on Parallel and Distributed Processing, IPDPS 2010, pp. 1–11. IEEE (2010)

38. Tsafrir, D., Etsion, Y., Feitelson, D.G.: Backfilling using system-generated predictions rather than user runtime estimates. IEEE Trans. Parallel Distrib. Syst. **18**(6), 789–803 (2007)

39. Vivodtzev, F., Bertron, I.: Remote visualization of large scale fast dynamic simulations in a HPC context. In: Proceedings of 4th IEEE Symposium on Large Data Analysis and Visualization, LDAV 2014, pp. 121–122. IEEE (2014)

Scheduling of Mobile Robots Using Constraint Programming

Stanislav Murín$^{(\boxtimes)}$ and Hana Rudová

Faculty of Informatics, Masaryk University, Brno, Czech Republic
murin.stanislav@mail.muni.cz, hanka@fi.muni.cz

Abstract. Mobile robots in flexible manufacturing systems can transport components for jobs between machines as well as process jobs on selected machines. While the job shop problem with transportation resources allows encapsulating of transportation, this work concentrates on the extended version of the problem, including the processing by mobile robots. We propose a novel constraint programming model for this problem where the crucial part of the model lies in a proper inclusion of the transportation. We have implemented it in the Optimization Programming Language using the CP Optimizer, and compare it with the existing mixed integer programming solver. While both approaches are capable of solving the problem optimally, a new constraint programming approach works more efficiently, and it can compute solutions in more than an order of magnitude faster. Given that, the results of more realistic data instances are delivered in real-time, which is very important in a smart factory.

Keywords: Scheduling · Constraint programming · Mobile robot · Flexible manufacturing system · Transportation · IBM ILOG CPLEX Optimization Studio

1 Introduction

The concept of the smart factory was defined by the Industry 4.0 [18] project recently. There is an emphasis on automation, data exchange, and flexible manufacturing technologies. A flexible manufacturing system consists of the work machines such as automated CNC machines connected by a material handling system and the central control computer. The material handling system can be realized by conveyors or automatic guided vehicles (AGV). AGVs are used to transport materials between machines [8].

Traditional job shop scheduling problems have been extended to work with the AGVs for transportation of material whenever the job changes from one machine to another. This class of problems is called the job shop scheduling with transportation (JSPT) as discussed by Nouri *et al.* [21] who reviewed various approaches applied to this problem. The classical work of Bilge and Ulusoy [8] formulated a nonlinear mixed integer programming model which was intractable

© Springer Nature Switzerland AG 2019
T. Schiex and S. de Givry (Eds.): CP 2019, LNCS 11802, pp. 456–471, 2019.
https://doi.org/10.1007/978-3-030-30048-7_27

due to the size and nonlinearity. To handle that, they applied an iterative heuristic approach to solve the problem. Later on, many metaheuristic and heuristic approaches have been studied to solve this problem [21].

Recent approach [11] extended the JSPT to allow for the inclusion of mobile robots who can perform various value-added tasks on machines as well as transportation of material between machines. They proposed an exact approach using linear mixed integer programming and solved the corresponding extension of benchmark problems from [8] optimally. These problems have varying difficulty, some of them can be solved by the mixed integer programming (MIP) approach within few seconds while computations of others took more than ten hours. To deliver solutions in a short time (within several seconds), the hybrid heuristic based on genetic algorithms was also proposed and implemented in [11]. Very latest work studied solution to this problem using adaptive large neighborhood search [12], and it is aimed to obtain solutions in real-time. Our paper also concentrates on this problem while proposing a different exact approach represented by constraint programming (CP).

In our approach, we use IBM ILOG CP Optimizer and its Optimization Programming Language [2,17]. Our problem is a combination of scheduling [6,24] and vehicle routing [16,25]. To solve the problem, we need to assign mobile robots to each transportation and processing where the robot is needed, as well as assign starting time to each transportation and processing. The transportation includes pickup of the job components and its delivery to the consequent machine where the job is processed.

Since we are not aware of any CP approach to our problem, we explored similarities with other close problems. There are relations to pickup and delivery problems [22,23], as well as dial-a-ride problems [10] where their vehicles correspond to mobile robots and pickup and delivery requests between origins and destinations can be seen as our transportations between the origin and destination machine. Berbeglia *et al.* proposed the first exact algorithm to check the feasibility of dial-a-ride problems using CP [7]. Liu *et al.* [19] approached the senior transportation problem using CP, MIP, logic-based Benders decompositions as well as constructive heuristic and found CP being the best approach. A similar problem of the patient transportation [9] was recently solved by CP efficiently. All these CP approaches [7,9,19] consider activities for pickup and delivery separately, corresponding to the fact that they may be interleaved among each other. However, this is in contrast to our approach, where each pickup is followed by the corresponding delivery. We show that the model with separate variables for pickup and delivery is not effective enough, and it must be replaced by a model where pickup and delivery activities are replaced by one transportation activity.

Other related problems are represented by scheduling with setup times or costs [4] where sequence-dependent setup times may correspond to our transportations between machines processed for consequent operations. Taking into account CP, various approaches solved the job shop scheduling problem with sequence-dependent setup times integrating setups with the objective function and search [5,14,15]. Recently, the propagation procedure, including transition times into the classic filtering algorithm for unary resources [13] was proposed to solve this problem. This is also the approach taken by CP Optimizer, where non-overlapping constraint with transition times is available [17]. While this is an interesting concept, it cannot be directly applied in our case since our transportation activities must be related to two different locations. To handle that, we propose a more complex modeling approach with both non-overlapping constraints as well as transportation activities.

Let us summarize the contributions of our work.

1. We introduce the first CP model for scheduling with mobile robots.
2. A novel approach to handle complex transportations using non-overlapping constraints is proposed.
3. A new CP approach is in more than an order of magnitude faster in computing of optimal solution than the earlier MIP approach [11].
4. For smart factories, it is important that real-time computation (within a second) is achieved for more realistic data instances.

The next section describes our scheduling problem with mobile robots. Section 3 presents the detail CP model with transportation activities and the alternative approach with the pickup and delivery activities. The section concludes by the discussion of the search options. Section 4 introduces data instances, explores different versions of our model and search, and compares our approach with the MIP solver taken from [11]. The final section summarizes the results and presents some ideas for future work.

2   Problem Description

We have m machines where n jobs are processed. Each job is composed of the set of operations each processed on a different machine. Typically, jobs are not processed on all machines, so the number of operations per job differs. For each job, there is a specific order of the operations, *i.e.*, we know in advance the order of processing of operations on machines for each job. Operations of one job cannot overlap, and only one operation (and job) can be processed on one machine at any time.

Our problem is related to the combination of the job shop scheduling problem and vehicle routing, which is called the job shop scheduling with transportation resources [21]. It means that there are vehicles, *e.g.*, automated guided vehicles (AGV), which are used for transportation of the components between every two consequent operations of one job [8]. The AGVs are identical, and the travel times between machines are specified (they include loading and unloading times) by the distance matrix. There are sufficient input and output buffer space at each machine where components of the job or robots can wait. There is also a special loading/unloading "dummy" machine (L/U station) where all AGVs start, and all materials for jobs are available. So, the first transportation for each job starts at the L/U station and the last one ends there.

In this paper, we study a recently proposed problem where some operations need processing by mobile robots [11]. Mobile robots are identical and perform both transportations as well as the processing of selected operations. Each robot can perform at most one activity (processing or transportation) at a time.

Example 1. A sample problem with the schedule is shown in Fig. 1. There are two robots, three jobs, three machines and the L/U station where the processing of all jobs starts. Routes for both robots are also depicted with the numbered transportations and processings.

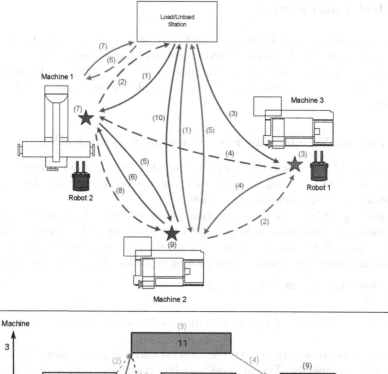

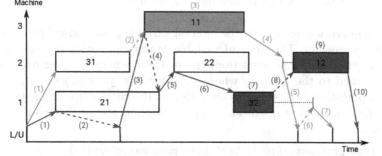

Fig. 1. Layout of FMS with mobile robots, and the schedule (based on [11]).

For instance, we can see that the third job is started by transportation (1) using the blue robot. Consequently, operation 31 is processed on the machine 2.

After some delay, the red robot transports components of the third job to the machine 1 by (6) which is followed by the processing of the operation 32 using the same red robot (7). Finally, job components are transported by the blue robot (7) to the L/U station, again after some delay.

In the robot routes, we can see the processing by robots (colored boxes) and the transportation of the components (so-called loaded trips; solid lines). There are also dashed lines showing transportations of robots without any materials (so-called empty trips) which are necessary to move the robot to the machine where it is needed, *e.g.*, the transportation (4) between the machines 3 and 1.

3   Model and Search

We will describe particular components of the model, starting from the base variables and constraints followed by more sophisticated concepts. Also, we would like to relate similar ideas together to make an understanding of the model easier. The final part of this section will discuss explored possibilities of the search method.

3.1   Interval Variables for Processing and Transportation

Interval variables are used to encapsulate both processing of operations as well as transportation of job components between machines. These variables represent processing and transportation activities.

A pair `<j,o>` identifies an operation processed as an o-th operation of the job j based on predefined ordering. We have a `Set` of such tuples to represent all operations. Note that operations of each job are organized in `Set` such that ordering of operations represents the processing order of operations on machines.

```
tuple pair {int j; int o;}
{pair} Set;
```

Using the set of tuples allows to handle variable number of operations per job easily.

For each operation `jo=<j,o>`, the interval variable `processing[jo]` is available for its processing. The interval variable `transport[jo]` represents transportation of the job components from the machine where the previous operation is processed to the machine where the operation `jo` is processed (we will write: transportation from previous operation to the current operation `jo`). The transportation to the first operation starts at the L/U station (which is further index by 0).

```
dvar interval processing[jo in Set] size processingSize[jo];
dvar interval transport[jo in Set] size transportSize[jo];
```

The size of the activity corresponds either to the processing time of the operation `jo` given by `processingSize[jo]` or to the travel time `transportSize[jo]` between the machines where the operation `jo` and its previous operation are processed.

Example 2. The operation 21 in Fig. 1 is represented by the processing activity `processing[<2,1>]`. The transportation (5) from the operation 21 to the operation 22 is represented by the transportation activity `transportation[<2,2>]`.

3.2   Temporal Constraints

The activities for processing and transportation of one job are related by the precedence constraints. The transportation activity `transport[jo]` from the previous operation to the operation `jo` must precede the processing activity `processing[jo]`.

```
forall (jo in Set)
   endBeforeStart(transport[jo], processing[jo]);
```

At the same time, the transportation activity `transport[jo]` must be preceeded by the processing activity from the previous operation. The previous operation of `jo` in the `Set` can be obtained using `prev(Set,jo)` function.

```
forall (jo in Set: jo.o != 1)
   endBeforeStart(processing[prev(Set,jo)], transport[jo]);
```

Note that the first transportation from L/U station to the next machine (`jo.o=1`) is not proceeded by any processing.

3.3   Non-overlapping of Operations on Machines

The operations processed in the same machine cannot overlap. This can be achieved by the inclusion of the sequence variable

```
dvar sequence machinePlan[i in 1..m] in
     all (jo in Set: operation[jo].machine == i) processing[jo];
```

for each of the `m` machines. The structure `operation[jo].machine` stores the machine for each operation `jo`. Consequently, the `noOverlap` constraint is posted for all machines.

```
forall(i in 1..m) noOverlap(machinePlan[i]);
```

3.4   Alternative Interval Variables

There are interval variables `rbtTransport[r][jo]` and `rbtProcessing[r][jo]` which ensure selection of one proper robot `r` for all transportation activities and for the processing activities which requires a robot for processing.

```
dvar interval rbtTransport [r in 1..rb][jo in Set] optional
     size transportSize[jo];
dvar interval rbtProcessing [r in 1..rb][jo in Set] optional
     size processingSize[jo];
```

The keyword `optional` corresponds to the fact that (at most) one robot activity will be selected by the `alternative` constraints to choose among the `rb` robots.

The following `alternative` constraint means that one robot transportation activity `rbtTransport[r][jo]` is selected (is present) for each transportation activity `transport[jo]`.

```
forall (jo in Set)
   alternative(transport[jo], all (r in 1..rb) rbtTransport[r][jo]);
```

The other `alternative` constraint is posted for all processing activities where the robot is working on the operation together with the machine (the data structure `operation[jo].robotRequired` stores information about needed robots). Again one robot processing activity `rbtProcessing[r][jo]` will be present for each processing activity `processing[jo]` requiring a robot.

```
forall (jo in Set: operation[jo].robotRequired == 1)
   alternative(processing[jo], all (r in 1..rb) rbtProcessing[r][jo]);
```

In addition, all robot processing activities `rbtProcessing[r][jo]` are set not be present by the `presenceOf` constraint when the robot is not needed.

3.5   Non-overlapping of Activities for Robots

Finally, we have all important elements of the model to propose how to handle non-overlapping of activities for each robot. Firstly, we define the sequence variable `rbtRoute[r]` for each robot `r` which includes all activities for the robot `r`. There are robot processing activities `rbtProcessing[r][jo]` for all operations `jo` where a robot is required. In addition, there are the robot transportation activities `rbtTransport[r][jo]`. Of course, the present activities are involved in the final sequence only.

```
dvar sequence rbtRoute[r in 1..rb]
   in append(all (jo in Set) rbtProcessing[r][jo],
             all (jo in Set) rbtTransport[r][jo])
   types append(all (jo in Set) processingType[jo],
                all (jo in Set) transportType[jo]);
```

As you can see, the sequence variables have defined their type which allows to handle the *empty* transportations of the robots between machines when their next processing or the origin of the next transportation is scheduled on a different machine than the robot is placed.

Example 3. The red robot in Fig. 1 performs the transportation from the L/U station to the operation 11 (to the machine 3) by the robot transportation activity `rbtTransport[red][<1,1>]` denoted by (3). Consequently, the red robot needs to go the machine 1 by the transportation (4). At the machine 1, the red robot needs to pick up the job components for the job 2 and transport them to the machine 2 by (5) and `rbtTransport[red][<2,2>]`. The transportation (4) represents the empty trip (and it is not represented by any activity).

For each empty trip, *e.g.* (8), we need to make sure that there is enough time to perform transportation from the origin activity (rbtProcessing[red][<3,2>] at machine 1) to the destination activity (rbtProcessing[red][<1,2>] at machine 2). This is completed by the noOverlap constraint with the distanceMatrix.

```
forall(r in 1..rb) noOverlap(rbtRoute[r], distanceMatrix);
```

This constraint ensures non-overlapping of all activities for each robot r as well as handling of travel times for their empty trips. The distance matrix is a set of tuples encoding the distances (travel times) between every two types.

```
tuple Triplet { int type1; int type2; int distance; }
{Triplet} distanceMatrix;
```

In vehicle routing problems [16,17], each activity represents processing at some location. Locations have defined their distances by the distance matrix. When we assign the corresponding location as a type to each activity, the noOverlap constraint enforces the minimal travel time between every two consequent activities from the different locations.

As mentioned in Sect. 1, the concept in vehicle routing problems cannot be directly implemented in our case, because our transportation activities are related to two different machines/locations. The transportation activity starting at the machine i1 and ending at the machine i2 cannot simply be related to one of the machines. Instead, we propose to have types associated with the two machines i1,i2. Next, the distance matrix needs to define distances for each quadruple i1,i2 and i3,i4 corresponding to the travel time between i2 and i3. Note that the distance between the type i1,i2 and i2,i3 is zero. Of course, the processing activities still resides on one machine which results in the type i,i for the machine i. It means that the space complexity for the distance matrix corresponds to $\mathcal{O}(m^4)$.

Example 4. The robot transportation activity rbtTransport[red][<1,1>] (3) has the type 03 because it corresponds to the transportation from the L/U station to the machine 3. The activity rbtTransport[red][<2,2>] (5) has the type 12 corresponding to the transportation from the machine 1 to the machine 2. The empty trip (4) has secured the minimal travel time given by the distance between the types 03 and 12 which is given by the travel time between the machines 3 and 1.

The robot processing activity rbtProcessing[red][<3,2>] for the operation 32 by (7) has the type 11 as it is processed on the machine 1.

3.6 Different Approach with Pickup and Delivery Activities

The initial solution approach for our problem was based on interval variables for both pickup and delivery, *i.e.*, there are interval variables pickup and delivery as well as rbtPickup and rbtDelivery. Both pickup and delivery variables are time points, *i.e.*, their size equals to zero.

Example 5. The transportation (5) in Fig. 1 from the operation 21 to the operation 22 is represented by the pickup activity pickup[<2,2>] and the delivery activity delivery[<2,2>].

There are new constraints related to the fact that the same robot must complete pickup and delivery for one transportation. It means that their presence variables must have the same value. Also, the pickup robot activity directly precedes the delivery robot activity for the corresponding transportation. This is achieved by the prev constraint.

```
forall (r in 1..rb, jo in Set) {
  presenceOf(rbtPickup[r][jo]) == presenceOf(rbtDelivery[r][jo]);
  prev(rbtRoute[r],rbtPickup[r][jo],rbtDelivery[r][jo]); }
```

For one transportation, the ending time of the pickup activity must be separated from the starting time of the delivery activity just by the travel time between machines. Travel times are stored in the structure travelTimes[i1][i2] for each two machines i1, i2 including the L/U station. Note that this data structure is also used to construct the distance matrix.

```
forall (jo in Set: jo.o != 1)
  endAtStart(pickup[jo], delivery[jo],
    travelTimes[operation[prev(Set,jo)].machine][operation[jo].machine]);
```

Since we need to include traveling from the L/U station to the first machine for each job, we also have the following constraints.

```
forall (jo in Set: jo.o == 1)
  endAtStart(pickup[jo], delivery[jo],
    travelTimes[0][operation[jo].machine]);
```

Finally, there is a slightly different implementation of the precedence constraints from Sect. 3.2. In the first constraint, we consider precedence between delivery (instead of transportation) and processing activities. The second constraint implements precedence between processing and pickup (again replacing transportation) activities.

3.7 Objective Function

The objective of the problem is to minimize the completion time of all activities, *i.e.*, the makespan. In our case, we minimize the maximal completion time of the last processing activity of each job (the data structure nbOperations stores the number of operations per each job). It means that we do not consider the last transportation to the L/U station which is in correspondence with earlier works [8,11].

```
minimize max(jo in Set: jo.o == nbOperations[jo.j])
          endOf(processing[jo]);
```

3.8   Search

CP Optimizer [17] employs a combination of Large Neighbourhood Search and Failure-directed Search [26] by default. Other options are introduced by multi-point search and depth-first search [1,17]. In the experimental part, we will explore the differences among these search algorithms. Based on our analysis, the default search is significantly better than other options (see Sect. 4.2).

In all cases, the search process is performed to find an optimal solution as well as prove the optimality of the found solution. Certainly, proving the optimality takes much longer time than the computation of the first optimal solution.

Another possibility of how to tune the search in CP Optimizer can be intro-duced by consideration of search phases together with grouping variables and their specific ordering. However, this does not make much sense for our problem. We have also explored various built-in variable and value ordering, but none of them appeared to have a positive effect.

To conclude, we use the default search setting of CP Optimizer.

4   Experiments

In this section, we describe the experimental evaluation. Our approach is imple-mented using the academic distribution of IBM ILOG CPLEX Optimization Studio 12.8 in OPL[1]. We compare it with the MIP implementation in OPL [11] using the same version and setting. We use random seeds to diversify the solu-tion approach, which plays a significant role in statistical comparison in both CP and CPLEX engines [3,17]. To allow for that, 30 runs are completed for each experiment. The experiments are run on a computer with Intel Xeon Gold 6130, 16 GB RAM using a single thread.

4.1   Data Instances

We use data instances from [11] available from the website[2]. These data instances extend JSPT data from [8] by introducing the processing by robots on some operations.

There are 82 data instances with 4 machine layouts and different t/p ratios (travel time/processing time). There are 4 machines (plus the L/U station) in each layout, 5–8 jobs, 13–21 operations, and 2 robots. The first group of 40 data instances has relatively high t/p ratio ($t/p > 0.25$), while the other 42 data instances have it relatively low ($t/p < 0.25$). The names given to the instances start with prefix "EX" followed by the number of the job set and the layout. The names of the instances in the second group include the additional 0, 1 digit implying that the processing time is doubled or tripled, respectively. In the second group, travel times are halves (*e.g.*, "EX541" corresponds to the job

---

[1] The source code, including data instances, is available from https://github.com/StanislavMurin/Scheduling-of-Mobile-Robots-using-Constraint-Programming.

[2] https://sites.google.com/site/schedulingmobilerobots/.

Table 1. Performance of various setting in the CP approach (in seconds).

	EX110		EX210		EX11		EX22		EX63	
	Time	S.D.	Time	S.D.	Time	S.D.	Time	S.D.	Time	S.D.
Default	0.010	0.003	0.028	0.015	0.745	0.142	0.447	0.133	0.564	0.077
DFS	0.006	0.005	2.443	0.939	0.473	0.018	2.567	1.235	16.763	0.844
MultiPoint	0.018	0.007	0.136	0.083	–	–	–	–	–	–

set 5 with tripled processing time and the layout 4 with half travel time). For the second group, it results in a more realistic data set, since the robot takes a long time for processing in comparison with its travel time [20].

4.2   Setting

In this section, we compare different versions of our model as well as built-in search methods.

Model Setting. In Table 1, the experiments exploring efficiency of our model using the default CP Optimizer setting are presented in the first line denoted *Default*. We have also tried different levels of propagation as available within CP Optimizer, but none of them has a positive effect.

The alternative model replaces transportation activities by the pickup and delivery activities as described in Sect. 3.6. Performance of this model was very weak, resulting in 96.794 ± 5.086 s even for the easiest EX110 problem. This confirmed our expectation after some trial experiments that this model cannot be used at all.

Search Setting. In Table 1, the results for the default search (*Default*), the depth-first search (*DFS*), and the multi-point search (*Multi-point*) are given.

While the best performance of the default search is not clear based on the instance EX110 and EX11 where the *DFS* is almost two times faster, it becomes decided on other instances. The depth-first search was significantly worse for other problems. The speed-up of the default search was 84.25, 5.74, and 29.72 for EX210, EX22, and EX63, respectively. It confirmed our preliminary experiments where the depth-first search was very slow on many problems. The multi-point search was not even able to find a solution within 12 h for three problems, even though there are still rather easy data instances.

Given the initial experiments, we can conclude that the default setting of CP Optimizer is the best, and it is further used to perform experiments on all data instances.

4.3   Results and Comparison with MIP

We compare the results of our CP solver with the results of the MIP solver from [11]. Table 2 includes the computational times (*Time*) with the standard deviations (*S.D.*) for both approaches together with the speed-up (*Speed.*) of the CP approach. The left columns show results for 40 data instances with $t/p > 0.25$, and the right columns for 42 instances with $t/p < 0.25$. Different levels of the blue color are used to emphasize differences in computational times. Darker colors show shorter computational times while lighter colors demonstrate higher times. Similarly, red colors show differences in speed-up. Here darker colors demonstrate better results of our CP solver.

The results for the first set of 40 data instances were mostly computed within 10 s (15 instances within a second, other 17 instances in less than 10 s). The three most demanding instances EX71, EX74, and EX104 needed 23.5, 14.3, and 4.8 min, respectively. The remaining 5 instances required the computational time between $10-100$ s (23.85 s on average). For the MIP solver, the instances EX71, EX74, and EX104 are very hard, and the results were computed within 12 h in 3 out of 90 cases (values in the column for speed-up specify the number of solved instances). Other instances were also much harder for MIP than for CP. Only 5 instances suffice with the computational time less than ten seconds. For 12 other instances, solutions were computed within a hundred seconds, and 11 more instances needed $100-1,000$ s. The last 9 instances required more than a thousand seconds. We cannot compare speed-up for EX71, EX74, and EX104 instances which were hard for both solvers, because the MIP solver has not mostly computed solutions. The CP solver was 1,000, $100-1,000$, and less than 100 faster for 3, 18, and 16 problems, respectively. The smallest speed-up was 8.8, and it was 38.1 on average for 16 problems with the speed-up smaller than a hundred.

We can observe the correlation between the computational time and the job set or the layout complexity. The layouts 1 and 4 appear to be generally more difficult for both solvers, having average computational times of 6.5 and 1,975.8 s for the layout 1, and 8.1 and 963.0 s for the layout 4 for CP and MIP, respectively. Layout 2 and 3 are generally easier with the average computational times of 1.0 and 217.6 s for the layout 2, and 1.4 and 261.2 s for the layout 3 for CP and MIP, respectively. Please note, that job sets 7 and 10 were excluded from average computation over layouts, because of unfinished MIP tests in layout 1 and 4.

The results for the second set of instances is in the right part of Table 2. We can see that computational times of the CP approach are always smaller than one second, which is very important because these problems represent more realistic data sets as discussed before. Computational times rather rely on the given job set which is given by smaller importance of travel times in this set of problems, since the travel times were halved and processing times were doubled or tripled. The problems from the job sets 9, and 10 need the highest computational time, mostly higher than 0.5 s. The job sets 4 and 6 with the exclusion of EX420 require between 0.1 and 0.6 s. 27 remaining instances suffice with solution time smaller than 0.05 s on average. The MIP solver needs more than 9 s for the job sets 8–10

468   S. Murín and H. Rudová

Table 2. Performance of CP and MIP approach (computational times in seconds).

	$t/p > 0.25$						$t/p < 0.25$				
No.	CP		MIP			No.	CP		MIP		
	Time	S.D.	Time	S.D.	Speed.		Time	S.D.	Time	S.D.	Speed.
EX11	0.74	0.14	21.09	14.00	28.3	EX110	0.010	0.003	0.418	0.097	40.4
EX21	6.40	0.98	813.09	307.91	127.0	EX210	0.028	0.015	5.052	1.084	178.3
EX31	1.45	0.35	151.71	54.37	104.8	EX310	0.011	0.004	2.401	1.293	218.3
EX41	29.64	6.11	11,062.71	4,671.52	373.3	EX410	0.381	0.093	10.043	2.625	26.4
EX51	2.09	0.35	103.41	28.65	49.4	EX510	0.009	0.003	0.451	0.113	52.0
EX61	7.81	1.92	2,144.12	819.86	274.5	EX610	0.295	0.092	9.954	12.807	33.7
EX71	1,410.05	1,098.22	–	–	0/30	EX710	0.014	0.005	2.318	1.471	161.7
EX81	0.03	0.05	39.11	28.99	1,448.5	EX810	0.011	0.003	43.181	30.684	3,810.1
EX91	4.51	0.70	1,471.48	1,134.73	326.0	EX910	0.496	0.131	10.945	2.517	22.1
EX101	40.30	4.86	11,034.87	3,868.56	273.8	EX1010	0.484	0.072	52.802	69.282	109.1
EX12	0.20	0.07	2.15	0.76	10.5	EX120	0.008	0.005	0.333	0.071	41.6
EX22	0.45	0.13	12.47	5.74	27.9	EX220	0.045	0.026	3.975	0.905	87.7
EX32	0.35	0.04	5.52	0.97	15.6	EX320	0.009	0.005	1.217	0.592	130.4
EX42	4.93	0.64	1,439.60	614.47	291.8	EX420	0.023	0.004	7.103	1.787	313.4
EX52	0.38	0.09	3.57	1.11	9.4	EX520	0.010	0.005	0.449	0.147	44.9
EX62	0.39	0.06	38.15	28.02	98.7	EX620	0.213	0.112	7.555	2.816	35.5
EX72	1.65	0.43	269.45	145.27	162.9	EX720	0.012	0.005	4.875	1.840	417.9
EX82	0.01	0.01	72.78	63.20	7,529.4	EX820	0.012	0.006	27.567	21.253	2,362.9
EX92	0.95	0.22	166.36	79.32	174.7	EX920	0.575	0.157	9.306	3.044	16.2
EX102	2.11	0.87	555.15	184.64	262.9	EX1020	0.443	0.261	42.312	29.385	95.5
EX13	0.45	0.14	3.98	1.11	8.8	EX130	0.008	0.004	0.361	0.058	43.3
EX23	1.17	0.50	65.59	37.82	56.0	EX230	0.013	0.006	4.091	0.805	322.9
EX33	0.34	0.08	5.17	1.11	15.3	EX330	0.010	0.003	1.171	0.475	113.4
EX43	6.32	0.85	1,481.41	673.00	234.5	EX430	0.157	0.075	4.488	1.168	28.5
EX53	0.63	0.11	14.61	7.14	23.3	EX530	0.009	0.006	0.367	0.098	42.4
EX63	0.56	0.08	86.11	52.28	152.6	EX630	0.338	0.171	6.646	1.705	19.7
EX73	5.18	1.25	661.64	341.92	127.8	EX730	0.009	0.005	4.340	1.602	500.8
EX83	0.01	0.00	76.68	83.95	8,216.0	EX830	0.012	0.005	26.695	18.926	2,288.1
EX93	1.69	0.29	356.05	144.02	210.2	EX930	0.482	0.145	9.748	3.511	20.2
EX103	3.53	0.45	948.83	366.85	269.1	EX1030	0.389	0.130	41.732	22.047	107.4
EX14	1.23	0.19	22.03	15.27	17.9	EX140	0.013	0.005	0.440	0.146	33.0
EX24	8.86	1.54	610.69	241.69	68.9	EX241	0.013	0.005	3.407	0.974	269.0
EX34	3.80	0.64	205.19	87.97	54.0	EX340	0.012	0.004	2.282	1.213	185.0
						EX341	0.015	0.005	1.637	0.637	109.1
EX44	18.11	3.03	2,316.20	994.28	127.9	EX441	0.274	0.045	7.826	1.841	28.6
EX54	1.77	0.29	84.69	29.51	47.9	EX541	0.008	0.004	0.464	0.100	55.6
EX64	21.06	3.46	1,651.60	772.60	78.4	EX640	0.508	0.502	8.311	2.380	16.4
EX74	858.75	1,761.51	–	–	1/30	EX740	0.011	0.005	4.541	1.930	412.8
						EX741	0.011	0.004	4.690	1.645	413.8
EX84	0.11	0.06	67.76	55.66	637.3	EX840	0.009	0.005	25.581	20.276	2,951.7
EX94	10.17	1.50	2,745.50	1,645.57	269.9	EX940	0.547	0.211	13.448	3.775	24.6
EX104	286.20	43.52	–	–	2/30	EX1040	0.509	0.048	57.863	45.987	113.8

(together with EX410 and EX610). Less than 1 s is needed for the job sets 1 and 5. The remaining 20 instances require 1–9 s. This results in a tremendous speed-up for the job set 8 being more than 1,000. The speed-up in two orders of magnitude was achieved for the job set 7, 10 and some problems from 2–4 job sets. Finally, 21 other instances were running 35.6 faster on average, and the smallest speed-up was 16.2. It is good to see very good results across all the instances for the CP solver because such running times allow applying the solver even with the real-time demands.

As we can see, the performance of the CP engine is much better in comparison to the CPLEX engine using their standard setting. It has been shown that the CP approach is better than the MIP approach in more than an order of magnitude, being even in three orders of magnitude faster for some of the problems.

5   Conclusion

We have proposed a novel CP approach for scheduling with mobile robots. It is an important recent problem which needs to be handled in smart factories nowadays. The approach is very effective given our new proposal with non-overlapping constraints, including transportation activities. For more realistic data instances (42 problems), all solutions can be computed within one second, which allows real-time computation needed by a smart factory. In this case, our exact solver can replace even a hybrid heuristic solver proposed in [11] to compute solutions fast. Data instances from the second harder data set (40 problems) can be mostly solved within ten seconds, less than a quarter of them has higher computational demands. When we compare these results with the earlier MIP approach, there is a significant speed-up in one, sometimes two, or even three orders of magnitude. To conclude, this makes up a nice example of the constraint programming application.

In the future, we would like to extend further our problems based on real life. Interesting characteristics can be studied by consideration of an uncertain and dynamic environment where changes are happening due to the uncertain behavior of mobile robots or the existence of new jobs. Certainly, our interests lie in further improvements to the current model and search. Last but not least, we would like to study other combinations of scheduling and vehicle routing problems where non-overlapping constraints with transportation activities can be helpful.

Acknowledgements. We would like to thank the anonymous reviewers for their careful reading of our paper and their insightful comments and suggestions.

Access to computing and storage facilities owned by parties and projects contributing to the National Grid Infrastructure MetaCentrum provided under the program "Projects of Large Research, Development, and Innovations Infrastructures" (CESNET LM2015042), is greatly appreciated.

References

1. IBM ILOG CPLEX Optimization studio CP Optimizer user's manual, version 12 release 8. IBM Corporation (2017)
2. IBM ILOG CPLEX Optimization studio OPL language user's manual version 12 release 8. IBM Corporation (2017)
3. Achterberg, T.: Random seeds. IBM Community CPLEX Optimizer Forum, IBM Corporation (2013)
4. Allahverdi, A., Ng, C., Cheng, T., Kovalyov, M.Y.: A survey of scheduling problems with setup times or costs. Eur. J. Oper. Res. **187**(3), 985–1032 (2008)
5. Artigues, C., Belmokhtar, S., Feillet, D.: A new exact solution algorithm for the job shop problem with sequence-dependent setup times. In: Régin, J.-C., Rueher, M. (eds.) CPAIOR 2004. LNCS, vol. 3011, pp. 37–49. Springer, Heidelberg (2004). https://doi.org/10.1007/978-3-540-24664-0_3
6. Baptiste, P., Laborie, P., Le Pape, C., Nuijten, W.: Chapter 22 constraint-based scheduling and planning. In: Rossi, F., van Beek, P., Walsh, T. (eds.) Handbook of Constraint Programming, Foundations of Artificial Intelligence, vol. 2, pp. 761–799. Elsevier, Amsterdam (2006)
7. Berbeglia, G., Pesant, G., Rousseau, L.M.: Checking the feasibility ofdial-a-ride instances using constraint programming. Transp. Sci. **45**, 399–412 (2011)
8. Bilge, U., Ulusoy, G.: A time window approach to simultaneous scheduling of machines and material handling system in an FMS. Oper. Res. **43**(6), 1058–1070 (1995)
9. Cappart, Q., Thomas, C., Schaus, P., Rousseau, L.-M.: A constraint programming approach for solving patient transportation problems. In: Hooker, J. (ed.) CP 2018. LNCS, vol. 11008, pp. 490–506. Springer, Cham (2018). https://doi.org/10.1007/978-3-319-98334-9_32
10. Cordeau, J.F., Laporte, G.: The dial-a-ride problem: models and algorithms. Ann. Oper. Res. **153**(1), 29–46 (2007)
11. Dang, Q.V., Nguyen, C.T., Rudová, H.: Scheduling of mobile robots for transportation and manufacturing tasks. J. Heuristics **25**(2), 175–213 (2019)
12. Dang, Q.V., Rudová, H., Nguyen, C.T.: Adaptive large neighborhood search for scheduling of mobile robots. In: Proceedings of ACM GECCO Conference, pp. 224–232. ACM (2019)
13. Dejemeppe, C., Van Cauwelaert, S., Schaus, P.: The unary resource with transition times. In: Pesant, G. (ed.) CP 2015. LNCS, vol. 9255, pp. 89–104. Springer, Cham (2015). https://doi.org/10.1007/978-3-319-23219-5_7
14. Focacci, F., Laborie, P., Nuijten, W.: Solving scheduling problems with setup times and alternative resources. In: Artificial Intelligence for Planning and Scheduling (AIPS), pp. 92–101 (2000)
15. Grimes, D., Hebrard, E.: Job shop scheduling with setup times and maximal time-lags: a simple constraint programming approach. In: Lodi, A., Milano, M., Toth, P. (eds.) CPAIOR 2010. LNCS, vol. 6140, pp. 147–161. Springer, Heidelberg (2010). https://doi.org/10.1007/978-3-642-13520-0_19
16. Kilby, P., Shaw, P.: Chapter 23 vehicle routing. In: Rossi, F., van Beek, P., Walsh, T. (eds.) Handbook of Constraint Programming, Foundations of Artificial Intelligence, vol. 2, pp. 801–836. Elsevier, New York (2006)
17. Laborie, P., Rogerie, J., Shaw, P., Vilím, P.: IBM ILOG CP optimizer for scheduling. Constraints **23**(2), 210–250 (2018)

18. Lasi, H., Fettke, P., Kemper, H.G., Feld, T., Hoffmann, M.: Industry 4.0. Bus. Inf. Syst. Eng. **6**(4), 239–242 (2014)
19. Liu, C., Aleman, D.M., Beck, J.C.: Modelling and solving the senior transportation problem. In: van-Hoeve, W.-J. (ed.) CPAIOR 2018. LNCS, vol. 10848, pp. 412–428. Springer, Cham (2018). https://doi.org/10.1007/978-3-319-93031-2_30
20. Madsen, O., et al.: Integration of mobile manipulators in an industrial production. Ind. Robot Int. J. Robot. Res. Appl. **42**(1), 11–18 (2015)
21. Nouri, H.E., Driss, O.B., Ghédira, K.: A classification schema for the job shop scheduling problem with transportation resources: state-of-the-art review. In: Silhavy, R., Senkerik, R., Oplatkova, Z.K., Silhavy, P., Prokopova, Z. (eds.) Artificial Intelligence Perspectives in Intelligent Systems. AISC, vol. 464, pp. 1–11. Springer, Cham (2016). https://doi.org/10.1007/978-3-319-33625-1_1
22. Parragh, S.N., Doerner, K.F., Hartl, R.F.: A survey on pickup and delivery problems, Part I: transportation between customers and depot. Journal für Betriebswirtschaft **58**(1), 21–51 (2008)
23. Parragh, S.N., Doerner, K.F., Hartl, R.F.: A survey on pickup and delivery problems, Part II: transportation between pickup and delivery locations. Journal für Betriebswirtschaft **58**(2), 81–117 (2008)
24. Pinedo, M.L.: Planning and Scheduling in Manufacturing and Services, 2nd edn. Springer, New York (2009)
25. Toth, P., Vigo, D.: Vehicle Routing: Problems, Methods, and Applications. Society for Industrial and Applied Mathematics, 2 edn. (2014)
26. Vilím, P., Laborie, P., Shaw, P.: Failure-directed search for constraint-based scheduling. In: Michel, L. (ed.) CPAIOR 2015. LNCS, vol. 9075, pp. 437–453. Springer, Cham (2015). https://doi.org/10.1007/978-3-319-18008-3_30

Decomposition and Cut Generation Strategies for Solving Multi-Robot Deployment Problems

Adriana Pacheco[✉], Cédric Pralet, and Stéphanie Roussel

ONERA/DTIS, Université de Toulouse, 31055 Toulouse, France
{adriana.pacheco,cedric.pralet,stephanie.roussel}@onera.fr

Abstract. In this paper, we consider a multi-robot deployment problem involving a set of robots which must realize observation tasks at different locations and navigate through a shared network of waypoints. To solve this problem, we develop a two-level approach which alternates between (a) quickly obtaining high-level schedules based on a coarse grain CP model which approximates navigation tasks as setup times between observations, and (b) generating more accurate schedules based on a fine grain CP model which takes into account all resource usage conflicts during traversals of the shared network. The low-level layer also contains an explanation module able to generate constraints holding on high-level decision variables. These constraints (or cuts) account for interferences found in the low-level solutions and which the high-level scheduler should take into account to minimize the makespan. The proposed variants of the cut generation strategy are incomplete, the aim being to obtain good quality solutions in a short time, and they differ in the way they allow to diversify search. Experiments show the efficiency of this approach and the complementarity of the cut generation schemes proposed.

Keywords: Multi-Robot Missions · Constraint-based scheduling · Problem decomposition

1 Problem Description

We consider a Multi-Robot Deployment (MRD) problem where a fleet of robots must perform, as quickly as possible, a set of observations on specific areas of a field. Each candidate observation must be allocated to a robot, and for each robot the sequence of its observation tasks must be scheduled. Between two successive observations, a robot must also navigate through a network composed of waypoints and links, as illustrated in Fig. 1. Several candidate paths can be considered to navigate between pairs of observation areas, and each alternative path can be broken down into successive movements through links and waypoints. One specificity is that the network is shared between all robots. To avoid collisions during traversals of the network, or at least to reduce the need to deal with collision situations online, we consider that each link and each waypoint

© Springer Nature Switzerland AG 2019
T. Schiex and S. de Givry (Eds.): CP 2019, LNCS 11802, pp. 472–487, 2019.
https://doi.org/10.1007/978-3-030-30048-7_28

can be occupied by at most one robot at a time (disjunctive resources). A finer version could be used by considering a non-unit capacity, but this would require handling cumulative resources which are left for future work. The only locations which are considered as sharable are those associated with observation areas (see Fig. 1 again). The rationale for this assumption is that the robots have a smaller (or even null) speed when performing observations at these locations, so the online management of collisions is easier and less hazardous in this case. According to the time frame during which each move monopolizes the shared link and waypoint resources, two different approaches can be distinguished:

- *Minimum Handover.* In this first approach, a robot r successively consumes the network resources involved in a path, as illustrated in Fig. 2 for a transition between observations o and o' successively using link l_1, waypoint wp, and link l_2. One specificity though is that there exists a positive time lapse, called the *handover duration*, during which a robot switches between network resources (move between a link and a waypoint for example). During each handover period, the robot moves to the next resource on its path, but must also remain "connected" to the previous one, to prevent robots from "jumping" between resources. The goal is to forbid inconsistent solutions where two robots would instantaneously cross each other over the network (*e.g.* solutions where at a given time t, one robot instantaneously moves for link l to waypoint w while another one instantaneously moves from w to l).
- *Path Isolation.* In this second approach, for a transition of robot r between two observations o and o', all path resources used during the transition are reserved for the whole move duration, as illustrated in Fig. 3. The path monopolization starts just before the robot leaves observation o and ends just after the robot arrives to observation o'.

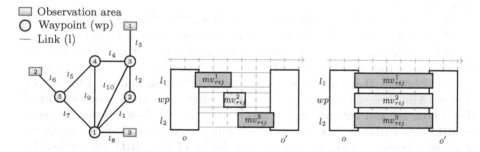

Fig. 1. Observation field   **Fig. 2.** Minimum handover   **Fig. 3.** Path isolation

The two approaches differ in terms of robustness and in terms of required synchronization between the robots at execution time, where duration of robot moves can be shorter or longer that expected. With the minimum handover strategy, the usage of resources is more finely optimized but there is a need to synchronize the robots at each basic move. On the contrary, the path isolation strategy takes more margins to get a collision-free deployment, but it only requires synchronizing the robots at the start and at the end of the global moves between observation tasks, that is when the speed of robots is low. The path isolation approach is inspired by works dealing with inter-core interferences due to shared hardware resources in multi-core processors [4, 16], to temporally isolate hard real-time applications [12].

In the mission specifications, robots cannot perform more than one observation at a time, and they must transfer observation data in real time to a mission center, using a specific emission frequency. To avoid communication interferences, two robots that use the same frequency cannot transfer observation data simultaneously. Redundancy is also useful in this kind of application to be robust to robot failures at execution time, therefore each observation area must be observed by several different robots at times spaced by a defined lapse. Last, each robot is initially located at a given location and must come back at the end of the mission to a predefined goal location.

The rest of the paper is organized as follows. Section 2 describes a first possible approach for dealing with the MRD problem based on a global CP model. Section 3 describes a two-layer approach which decomposes the global problem into a coarse-grain scheduling layer L1 used to compute sequences of observations, and a fine-grain scheduling layer L2 responsible for detailing navigation paths through the shared network of waypoints. Section 4 then introduces an iterative process of interaction between layers L1 and L2, based on four different *cut generation strategies* that allow L2 to provide L1 with new relevant constraints. This strategy can be seen as a kind of Logic-Based Benders Decomposition [5] (more details later on the relationship with LBDD). Section 5 presents the results of the decomposed approach and shows the complementarity of the cut generation techniques introduced. Finally, Sect. 6 concludes and gives some perspectives.

Note that the MRD application has been widely addressed to tackle real-world problems related to situation awareness issues, such as cooperative sensing using air-ground teams [3], disaster response [13], and exploration-rescue systems in hostile environments [14]. The aim of this paper is to propose more generic solutions that can be applied to such industrial and academic applications. Also, one of our goals is to exploit existing CP solvers to the best of their potential, and not to compare them with other resolution techniques such as MILP [7], PDDL [15] or Greedy algorithms. Finally, the proposed approach seeks to decrease the computational complexity of the MRD application, by considering the navigation interferences of a detailed solution plan. In most of the existing references, these interferences are not taken into consideration at a planning stage, and the use of anti-collision mechanisms is supposed at execution time.

2   A First Global Constraint-Based Scheduling Approach

Input Data. A first possible approach to solve the MRD problem is to develop a global CP model covering all specifications of the mission. To define such a model, we consider the following input data:

- a set of *frequencies* $\mathcal{F}$ available for communicating observation data in real time to the mission center;
- a set of *robots* $\mathcal{R}$; for each robot $r \in \mathcal{R}$, $freq_r \in \mathcal{F}$ is the (unique) frequency used by r to emit data during observations;
- a set of *observation areas* $\mathcal{A}$, corresponding to areas of the field that must be observed; for each area $a \in \mathcal{A}$, $duObs_a \in \mathbb{N}$ denotes the duration required to observe a.
- a number *NobsPerArea* of observations required over each observation area; all observations of a given area must be performed by distinct robots for redundancy issues;
- a set of *observations* $\mathcal{O}$ to be performed, which contains as many elements as the number of (a, k) pairs in $\mathcal{A} \times [1..NobsPerArea]$; for each observation $o \in \mathcal{O}$, $ar_o \in \mathcal{A}$ denotes the area associated with o;
- for each robot $r \in \mathcal{R}$, two specific observations denoted by α and β which represent virtual observations that must be performed at the beginning and at the end of the plan of r respectively; fictitious observations α and β allow us to model the initial and goal locations of r;
- a set of connected shared *waypoints* $\mathcal{W}$; for each robot $r \in \mathcal{R}$ and each waypoint $w \in \mathcal{W}$, $duMv_{r,w} \in \mathbb{N}$ is the minimum duration spent in w during a navigation of r through w; the observation areas are not considered as shared waypoints;
- a set of *links* $\mathcal{L}$, which correspond to direct connections between adjacent waypoints or between an observation area and a waypoint; for each robot $r \in \mathcal{R}$ and each link $l \in \mathcal{L}$, $duMv_{r,l} \in \mathbb{N}$ is the minimum duration required by r to traverse link l;
- a minimum temporal distance d between two observations of the same area;
- a *temporal horizon* $H \in \mathbb{N}$ available for the whole mission.

Constraint-Based Scheduling Model. From the previous input data, we can define a constrained-based scheduling model. For space limitation reasons, and because this global approach does not scale well compared to the decomposed approach detailed later, we only give the main lines of the model.

Basically, the global CP model is built upon so-called *interval variables* used in CP Optimizer. If $[Ts, Te]$ denotes the time frame available for realizing the associated task, each interval variable *itv* is characterized by a start value $start(itv) \in [Ts, Te]$, an end value $end(itv) \in [Ts, Te]$, and a presence $pres(itv) \in \{0, 1\}$ expressing whether the task is present in the solution schedule. The global CP model involves intervals representing observation activities and navigation activities through the shared network. In particular, it contains a huge number of optional intervals $mv_{r,o,o',p,q}$ to model the move of robot r on the

qth network resource (link or waypoint) of the pth path available to travel from observation o to observation o'. To boost constraint propagation, the model also contains a coarse-grain no overlap constraint taking into account the minimum temporal distance between observation tasks. Also, a so-called *sequence variable* seq_r is associated with each robot $r \in \mathcal{R}$. The value of this variable corresponds to a total ordering of all observation tasks realized by r. The CP model contains constraints over such sequences to ensure that the start and end locations of a path chosen between two observations o, o' is consistent with the locations of o and o'. Finally, a solution is optimal if it minimizes the makespan C_{max}, which corresponds to the maximum end time of the fictitious last observation tasks realized by the robots to reach their goal locations.

3   Decomposition into a Two-Layer Scheduling Problem

To decrease the computational complexity, the MRD problem can be split into two parts: (1) one part which decides on the successive observations realized by each robot based on a coarse grain model of navigation operations (so-called layer L1), and (2) one part responsible for detailing the navigations of robots within the network of shared waypoints (so-called layer L2).

3.1   Coarse-Grain Scheduling Model: Layer L1

In the high-level scheduling model of layer L1, the navigation between two given observations o and o' is abstracted in a very coarse way as a simple integer setup time required between the end of the realization of o and the start of the realization of o'. In other words, it considers the robots as disjunctive resources with setup times, as in Sequence Dependent Setup Time Job Shop Scheduling Problems [8].

Inputs. In addition to some inputs already mentioned in Sect. 2 (the set of *frequencies*, the set of *robots*, the set of *observation areas*, the set of *observations* to be performed, and the *temporal horizon*), an additional input of the high-level MRD scheduling problem considers:

- a constant setup time $setup_{r,o,o'} \in \mathbb{N}$ for each robot $r \in \mathcal{R}$ and each pair of observations (o, o') successively realized by r; this setup time represents the minimum duration required for r between the end of o and the start of o'; it is obtained from the length of the shortest path available to go from the location of o to the location of o'; this length is computed in polynomial time in preprocessing, and it corresponds to an optimistic evaluation assuming that r is alone over the network during its navigation from o to o'.

We also define $setup_{r,\alpha,o}$ as the shortest duration required to move from the initial location of r to observation o, and $setup_{r,o,\beta}$ as the shortest duration required to move from observation o to the goal location of r. For each robot $r \in \mathcal{R}$, the associated function to obtain the setup time between each pair of observations, is denoted $setup_r$.

Scheduling Problem. To define a CP model for layer L1, we use several scheduling constructs available in the CP Optimizer tool. More precisely, we introduce the following decision variables:

- for each observation $o \in \mathcal{O}$, one *interval variable* obs_o which must be placed during time frame $[0, H]$ and whose duration is $duObs_{ar_o}$, that is the observation duration of the area associated with o;
- for each observation $o \in \mathcal{O}$ and each robot $r \in \mathcal{R}$, one *optional* interval variable $obs_{o,r}$ used to represent the realization of observation o by robot r;
- for each robot $r \in \mathcal{R}$, two (non-optional) interval variables $obs_{\alpha,r}$ and $obs_{\beta,r}$ representing fictitious observations that r must realize at the beginning and end of its plan respectively; we recall that these fictitious observations allow us to model the initial and goal locations of the robots; interval $obs_{\alpha,r}$ has a null duration and must be placed at time 0, and interval $obs_{\beta,r}$ has a null duration and must be placed during time frame $[0, H]$;
- for each robot $r \in \mathcal{R}$, one *sequence variable* $\mathbf{seq}_r$ which represents an ordering over all present intervals associated with r, *i.e.* over all present intervals in set $\{obs_{o,r} \mid o \in \mathcal{O} \cup \{\alpha, \beta\}\}$.

The set of decision variables of the high-level model is therefore

$$V^1 = (\cup_{o \in \mathcal{O}}\{obs_o\}) \cup (\cup_{o \in \mathcal{O} \cup \{\alpha,\beta\}, r \in \mathcal{R}}\{obs_{o,r}\}) \cup (\cup_{r \in \mathcal{R}}\{\mathbf{seq}_r\}) \qquad (1)$$

Constraints 2 to 7 are imposed over these variables. Constraint (2) imposes that the first and last observations in a robot sequence must correspond to the initial and final fictitious observations. Constraints (3) uses the *alternative* constraint of CP Optimizer and expresses that each observation is realized by a unique robot. Constraint (4) imposes that each robot can realize at most one observation for a given area (redundancy requirement). Constraint (5) expresses that observation tasks using the same frequency cannot overlap. Constraint (6) expresses that observations of a given area cannot overlap, taking into account the minimum delay d defined in the input data. Constraint (7) enforces that observation tasks using the same robot must not overlap, taking into account the approximated setup durations required to move from one observation to the next for each robot. Symmetry breaking constraints could also be added.

$$\forall r \in \mathcal{R},\ first(\mathbf{seq}_r, obs_{\alpha,r}) \wedge last(\mathbf{seq}_r, obs_{\beta,r}) \qquad (2)$$

$$\forall o \in \mathcal{O},\ alternative(obs_o, \{obs_{o,r} \mid r \in \mathcal{R}\}) \qquad (3)$$

$$\forall a \in \mathcal{A},\ \forall r \in \mathcal{R},\ \sum_{o \in \mathcal{O} \mid ar_o = a} pres(obs_{o,r}) \leq 1 \qquad (4)$$

$$\forall f \in \mathcal{F},\ noOverlap(\{obs_{o,r} \mid o \in \mathcal{O},\ r \in \mathcal{R},\ freq_r = f\}) \qquad (5)$$

$$\forall a \in \mathcal{A},\ noOverlap(\{obs_{o,r} \mid o \in \mathcal{O},\ r \in \mathcal{R},\ ar_o = a\}, d) \qquad (6)$$

$$\forall r \in \mathcal{R},\ noOverlap(\{obs_{o,r} \mid o \in \mathcal{O} \cup \{\alpha, \beta\}\}, setup_r) \qquad (7)$$

The objective is to minimize the makespan, defined as the time at which each robot reaches its goal position:

$$\underset{r \in \mathcal{R}}{\text{minimize}\ \max}\ end(obs_{\beta,r}) \qquad (8)$$

Output. A high-level *solution schedule* σ^1 for layer L1 is an assignment of the values of all the decision variables in V^1 that satisfies all the problem constraints. A solution schedule σ^1 is said to be optimal if it minimizes the makespan.

3.2 Fine-Grain Scheduling Model: Layer L2

The low-level scheduling model of layer L2 takes into consideration all the navigation paths available in the waypoint graph representing the observation field (see Fig. 1), to detail the routing of robots in the shared network and manage navigation conflicts. This decision layer only considers the navigations between the observation tasks which are present in the coarse-grain solution σ^1 produced by L1 (much less navigation options compared to the global CP model).

Inputs. In addition to the inputs mentioned in Sect. 2 for the MRD problem, the additional inputs of the low-level multi-robot scheduling problem are:

- the high-level solution schedule σ^1 produced by layer L1; in this solution, we keep for each robot r the value of sequence **seq**$_r$, which defines the successive observations planned for r; to have flexibility in L2, we do not keep the exact dates found by layer L1 for present intervals in σ^1; in the following, to make some notations easier, we denote by Tr the set of all triples (r, o, o') such that in solution σ^1, observation o' is realized just after observation o for r;
- for each robot r and each pair of successive observations (o, o') realized by r (*i.e.* $(r, o, o') \in Tr$), a set of candidate paths $P_{r,o,o'}$ which can be used by r to go from o to o'; this set contains all paths whose length is not longer than the duration between the end of $obs^{L1}_{o,r}$ and the start of $obs^{L1}_{o',r}$ in the plan generated by layer L1; each path p in $P_{r,o,o'}$ is specified by the sequence $[p_1, \ldots, p_Q]$ of successive network resources (waypoints or links) traversed by the path ($p_q \in \mathcal{W} \cup \mathcal{L}$ for every $q \in [1..Q]$); for instance, in Fig. 1, several paths are available to move from observation area 3 to observation area 1; according to the solution obtained in σ^1, Fig. 4 details the setup operations between observation tasks $O_{3,2}$ and $O_{1,2}$ performed by robot 2, which can use either path $[l_8, wp_1, l_1, wp_2, l_2, wp_3, l_3]$ or path $[l_8, wp_1, l_{10}, wp_3, l_3]$.

Scheduling Problem. In the scheduling problem built for layer L2, the detailed routing between observation tasks must be defined. For space limitation reasons, we give the model only for the minimum handover case (Fig. 2). The model for the path isolation case (Fig. 3) is a bit simpler. For each robot $r \in \mathcal{R}$ and each observation o realized by r, the CP model contains one interval variable $obs_{o,r}$ as in L1. For each robot $r \in \mathcal{R}$ and each pair of observations o, o' successively realized by r in the solution found by L1, we also consider the following variables:

- one (mandatory) interval variable $mv_{r,o,o'}$ representing the global move of r from o to o';
- for each candidate path $p \in P_{r,o,o'}$, one optional interval variable $mv_{r,o,o',p}$ representing a move along path p;

– for each candidate path $p = [p_1, \ldots, p_Q] \in P_{r,o,o'}$ and each index $q \in [1..Q]$, one optional interval variable $mv_{r,o,o',p,q}$ representing the usage of the qth network resource of path p.

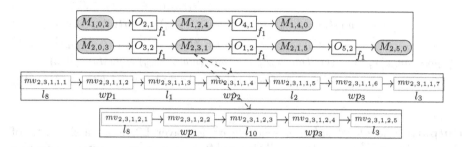

Fig. 4. Possible global move decompositions for layer L2

Together, these interval variables make up the set of decision variables V^2 of layer L2. Fine-grain constraints associated with the minimum handover case are given below. Constraint (9) forbids the temporal overlapping of tasks that use the same link or waypoint. Constraint (10) ensures that exactly one path is used between each pair of successive observations. Constraint (11) expresses that this path spans all its elementary moves. Constraint (12) states that the presences of elementary moves must be consistent with the presences of the selected paths. Constraints (13) and (14) define the start and end times of the moves from and to the first and last fictitious observations respectively. Constraints (15) to (17) enforce a handover period between the successive intervals involved in a chosen navigation path. Constraint (18) defines the minimum duration of each elementary move interval, taking into account the handover period. We consider an inequality here since in the minimum handover configuration, a robot is allowed to wait on a link or a waypoint. Constraint (19) forbids the temporal overlapping of tasks that use the same frequency (the ordering of observations over frequency resources is not transferred from L1 to L2 to keep more flexibility in L2). Finally, the goal is still to minimize the makespan (same expression as in Eq. (8)).

$$\forall \gamma \in \mathcal{W} \cup \mathcal{L}, \; noOverlap(\{mv_{r,o,o',p,q} \mid \tag{9}$$
$$((r,o,o') \in Tr) \wedge (p \in P_{r,o,o'}) \wedge (q \in [1..|p|]) \wedge (p_q = \gamma)\})$$
$$\forall (r,o,o') \in Tr, \; alternative(mv_{r,o,o'}, \{mv_{r,o,o',p} \mid p \in P_{r,o,o'}\}) \tag{10}$$
$$\forall (r,o,o') \in Tr, \forall p \in P_{r,o,o'}, \; span(mv_{r,o,o',p}, \{mv_{r,o,o',p,q} \mid q \in [1..|p|]\}) \tag{11}$$
$$\forall (r,o,o') \in Tr, \; \forall p \in P_{r,o,o'}, \; \forall q \in [1..|p|], \tag{12}$$
$$pres(mv_{r,o,o',p}) = pres(mv_{r,o,o',p,q})$$
$$\forall (r,\alpha,o') \in Tr, \; endAtStart(obs_{\alpha,r}, mv_{r,\alpha,o'}) \tag{13}$$
$$\forall (r,o,\beta) \in Tr, \; endAtStart(mv_{r,o,\beta}, obs_{\beta,r}) \tag{14}$$

$$\forall (r, o, o') \in \mathit{Tr}, \ \forall p \in P_{r,o,o'}, \ \forall q \in [2..|p|], \quad (15)$$
$$pres(mv_{r,o,o',p}) \rightarrow (start(mv_{r,o,o',p,q}) = end(mv_{r,o,o',p,q-1}) - 1)$$
$$\forall (r, o, o') \in \mathit{Tr}, \ \forall p \in P_{r,o,o'}, \quad (16)$$
$$pres(mv_{r,o,o',p}) \rightarrow (start(mv_{r,o,o',p,1}) = end(obs_{o,r}) - 1)$$
$$\forall (r, o, o') \in \mathit{Tr}, \ \forall p \in P_{r,o,o'}, \quad (17)$$
$$pres(mv_{r,o,o',p}) \rightarrow (end(mv_{r,o,o',p,|p|}) = start(obs_{o',r}) + 1)$$
$$\forall (r, o, o') \in \mathit{Tr}, \ \forall p \in P_{r,o,o'}, \ \forall q \in [1..|p|], \quad (18)$$
$$pres(mv_{r,o,o',p,q}) \rightarrow (end(mv_{r,o,o',p,q}) - start(mv_{r,o,o',p,q}) \geq duMv_{r,p_q} + 2)$$
$$\forall f \in \mathcal{F}, \ noOverlap(\{obs_{o,r} \mid o \in \mathcal{O}, \ r \in \mathcal{R}, \ freq_r = f\}) \quad (19)$$

Output. A low-level *solution schedule* σ^2 for layer L2, is an assignment of all variables in V^2 that satisfies all the problem constraints. It corresponds to a solution of the global MRD problem. A solution schedule σ^2 is said to be optimal if it minimizes the makespan (end time of the fictitious last observation tasks performed by the robots).

4   Iteration Resolution and Cut Generation Strategies

4.1   Iterative Resolution Approach

When using a top-down approach such as the one described in the previous section, the highest quality solutions may be missed since high-level decisions are computed from a coarse-grain model. This is why we use an iterative resolution strategy related to Logic-Based Benders Decompositions (LBBD), where a master solver iteratively proposes solutions to a slave solver which generates new constraints called *cuts*. Iterations between the master and the slave solvers are realized until convergence or until a maximum CPU time is reached. In our case, layer L1 first transfers to layer L2 the sequence of tasks realized by each robot. Then, layer L2 obtains a consistent solution schedule σ^2 for the low-level scheduling problem. In L2, we introduce an explanation module which detects interferences between tasks consuming the shared network resources. As shown in Fig. 5, this explanation module synthesizes cuts which are sent as a feedback to L1.

Compared to standard LBDD, one specificity of the technique proposed is that, as shown later, the explanation module generates cuts that are not necessarily valid in the sense that they might prune optimal solutions. The purpose of these cuts, which could be called *heuristic cuts*, is not to converge towards an optimal solution, but to speed the search for good solutions by forbidding in a coarse way some observation sequence patterns which might lead to interferences on detailed navigation activities. These patterns can be more or less precise and the generated cuts range from cuts usable to intensify search around the best known solution to cuts usable for exploring completely different regions of the search space. We emphasize again that the purpose of this process is not to obtain an optimal solution for the global problem but to get good quality solutions within a short computation time, which is more crucial than finding optimality in most MRD problems. Last, to perform several iterations between L1 and L2, we do not solve each problem in L1 or L2 to optimality. Instead, each run of L1 and L2 has a maximum allocated CPU time, which depends on the problem instance considered.

The approach proposed also differs from a strategy introduced in a previous work [10]. First, from an application point of view, [10] considers a simpler problem where the only disjunctive network resources are the links, and where the robots can freely cross each other at waypoints, in contrast to the approach followed in our "continuous" minimum handover strategy. The latter makes the model of layer L2 more complex but has the advantage of being more collision-safe. Also, [10] does not consider the path isolation configuration, which makes the moves of robots more constrained but which can be useful for robustness reasons. From a technical point of view, [10] also considers a two-layer approach but without any generation of cuts. Instead, the feedback from L2 to L1 corresponds to a simple update of the abstract setup durations of L1 by formula $setup_{r,o,o'} \leftarrow (1 - \mu) \cdot setup_{r,o,o'} + \mu \cdot du$, where μ corresponds to a learning rate and du corresponds to the duration of transition $o \rightarrow o'$ obtained for robot r in layer L2. Doing so, layer L1 learns a setup duration model from L2 and is close to work on surrogate models for optimization [6]. On the opposite, our approach exploits more detailed information (see Sect. 4.2) and is closer to LBDD. Other works already addressed similar real-world applications using two-stage decompositions involving CP models (Decomposed-CP), such as the deployment of multiple robots to assist the residents in a retirement home environment [15]. This last work also involves robots moving through the environment and a number of tasks that potentially increases with the number of robots and locations, and it also sets the value of certain decision variables for the sub-problem (layer L2), using the solution of the master problem (layer L1). Their Decomposed-CP approach may not find the optimal solution, but one distinctive feature is that no feedback loop from L2 to L1 is used. On this point, the authors of [15] state that it's not straightforward to determine whether their problem structure allows for a decomposition such as LBBD to be implemented.

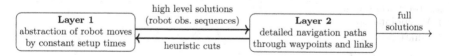

Fig. 5. Interactions between the decision layers

4.2   Cut Generation in the Explanation Module of Layer L2

The explanation module of L2 returns information about interferences found in the low-level solution and that have a negative impact on makespan minimization. These interferences are detected by examining conflicts related to the usage of network resources during path traversals. More precisely, for a given robot r_1, if the duration required to traverse a path between two successive observations o_1, o_1' is strictly greater than the duration obtained in the solution of L1, this means that there is a resource precedence constraint which creates an interference at some point during the transition from o_1 to o_1'. The goal of the explanation module of L2 is to detect interferences related to network resource usages in the obtained sequence from σ^2. For instance, Fig. 6 illustrates a scenario where two robots are in conflict for using waypointq wp_1 and wp_2, and link l_2 to traverse the paths needed to perform the sequences of observations shown in Figs. 7 and 8 (handover duration not represented). In this case, the duration of the transition is longer for robot r_1 since it must wait for some network resources to be released by r_2. The explanations of these longer transition durations are depicted in red in Figs. 7 and 8. In the general case, the explanation module of L2 detects through critical path analysis all triples (r_2, o_2, o_2') such that there is a transition from observation o_2 to observation o_2' for robot r_2 and such that at some point between o_1 and o_1', robot r_1 waits for a network resource to be released by r_2 during its transition from o_2 to o_2'. In terms of scheduling, we identify the critical resource precedence constraints associated with the network resources. In the end, each interference produced by the explanation module is defined by a 6-tuple $(r_1, o_1, o_1', r_2, o_2, o_2')$. In the following, the set of all interferences synthesized from low-level solution σ^2 is denoted by $\mathbf{Itf}(\sigma^2)$.

Four categories of cuts that can be generated through the explanation module are introduced below, by increasing order of refinement. In the following, we respectively denote by x^{L1} and x^{L2} the variables manipulated by layers L1 and L2. For instance, $obs_{o,r}^{L1}$ denotes the observation interval of o by robot r manipulated by L1, while $obs_{o,r}^{L2}$ denotes the observation interval $obs_{o,r}$ manipulated by L2 for representing the same task. Moreover, to get more concise expressions, we denote by $next_{r,o,o'}^{L1} \in \{0,1\}$ the variable taking value 1 if interval $obs_{o,r}^{L1}$ is the predecessor of interval $obs_{o',r}^{L1}$ in the sequence of intervals $\mathbf{seq}_r^{L1}$ associated with robot r in layer L1. Also, $\sigma^2(start(obs_{o,r}^{L2}))$ and $\sigma^2(end(obs_{o,r}^{L2}))$, denote respectively the start and the end date of $obs_{o,r}^{L2}$ in the low-level solution σ^2.

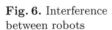

Observation area

Waypoint (wp)

Link (l)

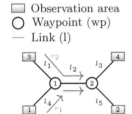

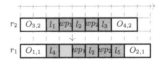

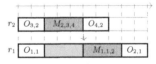

Fig. 6. Interference between robots

Fig. 7. Minimum handover interferences

Fig. 8. Path isolation interferences

Broad Cuts: Setup Times (Cuts C1). From the set of interferences $\mathbf{Itf}(\sigma^2)$, temporal constraints holding on high-level decision variables can be added to the scheduling problem of layer L1. A first possible approach is to return the following cuts:

$$\forall(r_1, o_1, o_1', r_2, o_2, o_2') \in \mathbf{Itf}(\sigma^2), \quad (20)$$
$$start(obs_{o_1',r_1}^{L1}) - end(obs_{o_1,r_1}^{L1}) \geq \sigma^2(start(obs_{o_1',r_1}^{L2})) - \sigma^2(end(obs_{o_1,r_1}^{L2}))$$

Such cuts are not valid since the initial abstract setup duration between o_1 and o_1' for r_1 (the setup duration considered by L1) could be met by updating the sequences of observations realized by other robots. However, these cuts can allow to quickly diversify search by penalizing, at the level of L1, a transition $o_1 \to o_1'$ which *might* lead to an interference at the level of network resources.

Note that cuts C1 are equivalent to use the solution σ^2 from layer L2 to update the inputs of layer L1 (the coarse-grain duration of the setup operations between locations for each robot). Remember that $setup_{r,o,o'} \in \mathbb{N}$ corresponds to the high-level approximation of the duration required by r to move from the location of observation o to the location of observation o' over all possible paths of the waypoint network. The previous cuts amount to update $setup_{r,o,o'}$ by:

$$\forall(r_1, o_1, o_1', r_2, o_2, o_2') \in \mathbf{Itf}(\sigma^2), \quad (21)$$
$$setup_{r_1,o_1,o_1'} \leftarrow \mathbf{max}(setup_{r_1,o_1,o_1'}, \ \sigma^2(start(obs_{o_1',r_1}^{L2})) - \sigma^2(end(obs_{o_1,r_1}^{L2})))$$

Moderate Cuts: Setup Times and Sequencing (Cuts C2). In contrast to the previous cut generation strategy, we can consider another category of cuts which take into consideration the precise transitions creating the interference in σ^2. Such cuts are defined by:

$$\forall(r_1, o_1, o_1', r_2, o_2, o_2') \in \mathbf{Itf}(\sigma^2), \quad (22)$$
$$(next_{r_1,o_1,o_1'}^{L1} \wedge next_{r_2,o_2,o_2'}^{L1}) \rightarrow$$
$$(start(obs_{o_1',r_1}^{L1}) - end(obs_{o_1,r_1}^{L1}) \geq \sigma^2(start(obs_{o_1',r_1}^{L2})) - \sigma^2(end(obs_{o_1,r_1}^{L2})))$$

and can be added to the scheduling problem of layer L1. These cuts impose longer setup times in the high-level approximation whenever the successive observations involved in the interferences are successive again in a new sequence considered by L1. Cuts of type C2 are weaker than cuts of type C1, meaning that C2 prunes less solutions than C1.

Refined Cuts: Setup Times, Sequencing, and Temporal Positioning (Cuts C3). More refined cuts coming from the solution analysis of layer L2 can be sent to layer L1. Unlike the previous cut generation strategies, these new cuts consider the time frame during which the setup tasks between observations are performed. Basically, they add high-level constraints which impose longer coarse-grain setup times only in case of temporal overlapping between the transitions involved in the interference. More precisely, let $overlap^{L1}(r_1, o_1, o_1', r_2, o_2, o_2')$ denote an expression taking value true when transitions $o_1 \rightarrow o_1'$ and $o_2 \rightarrow o_2'$ overlap in time, that is:

$$overlap^{L1}_{r_1,o_1,o_1',r_2,o_2,o_2'} = \qquad (23)$$
$$(end(obs^{L1}_{o_1,r_1}) > start(obs^{L1}_{o_2',r_2})) \wedge (end(obs^{L1}_{o_2,r_2}) > start(obs^{L1}_{o_1',r_1})))$$

The detailed cuts are then given by:

$$\forall (r_1, o_1, o_1', r_2, o_2, o_2') \in \mathbf{Itf}(\sigma^2), \quad (24)$$
$$(next^{L1}_{r_1,o_1,o_1'} \wedge next^{L1}_{r_2,o_2,o_2'} \wedge overlap^{L1}_{r_1,o_1,o_1',r_2,o_2,o_2'}) \rightarrow$$
$$(start(obs^{L1}_{o_1',r_1}) - end(obs^{L1}_{o_1,r_1}) \geq \sigma^2(start(obs^{L2}_{o_1',r_1})) - \sigma^2(end(obs^{L2}_{o_1,r_1})))$$

which means that if transitions $o_1 \rightarrow o_1'$ and $o_2 \rightarrow o_2'$ appear again in a solution for L1 and if these transitions overlap in time, then a higher setup time must be used at the level of L1. Cuts of type C3 are weaker than cuts of type C2, meaning that C3 prunes less solutions than C2.

Valid Global Cut (Cut C4). A quite simple valid cut consists in forbidding the entire sequence obtained for L1 at the previous step. This cut is defined by:

$$\neg \left[\bigwedge_{(r,o,o') \in Tr} next^{L1}_{r,o,o'} \right] \qquad (25)$$

It can be added to the scheduling problem of L1 as a global scheduling constraint. This cut will only force to seek for a different high-level solution, bypassing the synthesized information about the interferences found.

5   Experiments

Benchmarks. The two-layer approach and the four cut generation strategies proposed were evaluated over several MRD problem instances containing from 1 to 15 observation areas, connected through a network of shared links and waypoints. Several randomly generated observation scenarios were tested, considering from 1 to 3 frequencies available to transfer observation data, and from 2 or 3 homogeneous robots available to carry out the observations. The fields generated are regular grids of size $N \times M$ containing waypoints which are connected to their 4 adjacent neighbors. Random fields such as the one in Fig. 1, and other grid configurations were also tested, leading to the same experimental conclusion. Observation areas are randomly positioned so as to be connected to one waypoint of the grid, and for most observation pairs o, o' there are several navigation paths of minimum length from o to o'. Each area requires observations from 1 to 3 robots (redundancy). The generated instances were all tested on IBM ILOG CP Optimizer 12.5 on an Intel Xeon CPU E5-1603, 2.80 GHz 8 GB RAM, setting an adequate number of iterations depending on the problem size and on cpuMax. Experiments were performed for both the minimum handover and path isolation configurations, to test the algorithms on instances which are more or less constrained in terms of usage of the shared network.

Representative results of the tested configurations are given in Figs. 9 and 10, where two different robots must observe each observation area. For nearly all problem instances, the four proposed strategies for the two-layer approach achieve better makespan results than the global CP approach, in a significantly shorter computation time. They provide good quality solutions in just a few seconds, even for the largest instances for which the global CP approach is not able to reach any solution with a CPU time of 30 min. For the smallest instances, most of the strategies of the two-layer approach manage to find the optimal solution, but without proving its optimality. As shown in Table 1, the results also demonstrate that over the set of benchmarks tested, there is not a single winner among the four cut generation strategies for the two-layer approach. One explanation is that for some instances, it may be more advantageous to diversify the exploration of the search space by generating moderate cuts (strategy C2) or coarse-grain cuts updating the entire set of setup times (strategy C1), while for other instances it may be more convenient to explore a search space not so far from the current problem by generating fine-grain cuts (strategies C3 and C4). In other words, there is a kind of exploration/exploitation trade-off depending on the instance, leading to a disparity in the number of added cuts and in the elapsed time until the best solution is found, averaging between 1 and 2 min for the different strategies. To take advantage of all cuts, the next step would be to define a portfolio solver exploiting the different kinds of cuts, the goal being to outperform each individual cut generation strategy. Portfolio approaches combine different solvers to get a globally better one, and their efficiency was already shown in the CP field [1,2,9].

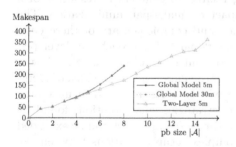

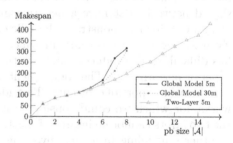

Fig. 9. Minimum handover   **Fig. 10.** Path isolation

Table 1. Makespan found along with the number of cuts added until the best solution is found (in brackets) for different sizes of the set of observation areas $\mathcal{A}$ and for both configurations (Minimum Handover and Path Isolation); results are given for cut generation strategies C1 to C4, with 5-min time limit (cpuMax), and for the global CP model, with 5 and 30-min time limits

	Minimum Handover						Path Isolation					
	Cut Strategies				Global		Cut Strategies				Global	
$\|\mathcal{A}\|$	C1	C2	C3	C4	5m	30m	C1	C2	C3	C4	5m	30m.
1	42 [0]	42 [0]	42 [0]	42 [0]	42	42	58 [0]	58 [0]	58 [0]	58 [0]	58	58
2	52 [0]	52 [0]	52 [0]	52 [0]	52	52	96 [0]	85 [2]	96 [0]	96 [0]	85	85
3	76 [2]	77 [0]	76 [2]	76 [1]	76	76	97 [0]	97 [0]	97 [0]	97 [0]	97	97
4	93 [0]	93 [0]	93 [0]	93 [0]	96	93	111 [4]	111 [8]	178 [0]	178 [0]	111	111
5	115 [3]	115 [3]	115 [3]	115 [1]	121	118	130 [4]	130 [11]	129 [6]	130 [7]	134	134
6	133 [13]	133 [5]	133 [8]	133 [0]	154	154	150 [11]	153 [10]	153 [10]	150 [13]	168	156
7	156 [37]	162 [8]	156 [31]	159 [11]	195	195	174 [11]	174 [4]	177 [19]	187 [17]	269	211
8	185 [17]	173 [42]	185 [34]	182 [1]	240	240	212 [6]	212 [21]	198 [18]	208 [15]	316	304
9	205 [38]	204 [13]	211 [9]	205 [8]	-	-	240 [16]	237 [24]	234 [10]	238 [15]	-	-
10	235 [41]	235 [6]	235 [9]	235 [10]	-	-	288 [5]	270 [13]	268 [24]	252 [9]	-	-
11	256 [40]	260 [55]	255 [25]	257 [7]	-	-	306 [14]	290 [20]	317 [12]	309 [7]	-	-
12	296 [0]	286 [51]	296 [0]	291 [3]	-	-	326 [13]	325 [26]	335 [10]	343 [5]	-	-
13	312 [33]	306 [12]	320 [13]	312 [7]	-	-	354 [4]	374 [16]	380 [7]	364 [5]	-	-
14	326 [5]	333 [3]	332 [3]	313 [4]	-	-	389 [0]	376 [11]	389 [0]	389 [0]	-	-
15	373 [11]	362 [11]	362 [7]	377 [4]	-	-	430 [4]	430 [16]	432 [9]	430 [4]	-	-

6   Conclusion

We proposed four strategies to generate cuts in a two-layer approach for solving Multi-Robot Deployment Problems. The generated cuts account for the interferences found in the low-level solutions, related to conflicts in resources of a shared network that have a negative impact on makespan minimization. The results obtained demonstrate the efficiency and complementary of these cuts. Even for large size problems, in which the global CP approach we developed has difficulties to produce a first solution, the cut generation strategies show a superior performance. The complementary of the cuts leads to the idea of merging them in a portfolio of cuts. This idea will be further refined in upcoming works, for which we could consider restart strategies when solutions found by the two layers cannot be improved. Last, the proposed approach can be extended to other scheduling problems involving complex setup operations between the main tasks. An example of such problems is the placement of embedded functions on a many-core processor [11], where the functions placed on the different cores interact through data exchanges over a shared network. Similarly, logistic in warehouses involves object transfers between locations and requires the utilization of shared resources whose activities must also be scheduled.

References

1. Amadini, R., Gabbrielli, M., Mauro, J.: An extensive evaluation of portfolio approaches for constraint satisfaction problems. Int. J. Interact. Multimed. Artif. Intell. **3**(7), 81–86 (2016)
2. Amadini, R., Gabbrielli, M., Mauro, J.: SUNNY-CP and the Minizinc challenge. CoRR (2017)
3. Chaimowicz, L., et al.: Deploying air-ground multi-robot teams in urban environments. In: Parker, L.E., Schneider, F.E., Schultz, A.C. (eds.) Multi-Robot Systems. From Swarms to Intelligent Automata Volume III, pp. 223–234. Springer, Dordrecht (2005). https://doi.org/10.1007/1-4020-3389-3_18
4. Girbal, S., Jean, X., Rhun, J.L., Pérez, D.G., Gatti, M.: Deterministic platform software for hard real-time systems using multi-core cots. In: 2015 IEEE/AIAA 34th Digital Avionics Systems Conference (DASC) (2015)
5. Hooker, J., Ottosson, G.: Logic-based Benders decomposition. Math. Program. **96**(1), 33–60 (2013)
6. Vu Khac, K., D'Ambrosio, C., Hamadi, Y., Liberti, L.: Surrogate-based methods for black-box optimization: surrogate-based methods for black-box optimization. Int. Trans. Oper. Res. **24**, 393–424 (2016)
7. Koes, M., R. Nourbakhsh, I., Sycara, K.: Heterogeneous multirobot coordination with spatial and temporal constraints. In: International Conference on Artificial Intelligence (AAAI), pp. 1292–1297 (2005)
8. Oddi, A., Rasconi, R., Cesta, A., Smith, S.F.: Applying iterative flattening search to the job shop scheduling problem with alternative resources and sequence dependent setup times. In: Proceedings of the Workshop on Constraint Satisfaction Techniques for Planning and Scheduling Problems, pp. 15–22 (2011)
9. O'Mahony, E., Hebrard, E., Holland, A., Nugent, C., O'Sullivan, B.: Using case-based reasoning in an algorithm portfolio for constraint solving. In: Irish Conference on Artificial Intelligence and Cognitive Science (2008)
10. Pacheco, A., Pralet, C., Roussel, S.: Constraint-based scheduling with complex setup operations: an iterative two-layer approach. In: International Joint Conference on Artificial Intelligence (IJCAI) (2019)
11. Perret, Q., Maurère, P., Noulard, E., Pagetti, C., Sainrat, P., Triquet, B.: Mapping hard real-time applications on many-core processors. In: Proceedings of the 24th International Conference on Real-Time Networks and Systems, pp. 235–244. ACM (2016)
12. Perret, Q., Maurère, P., Noulard, E., Pagetti, C., Sainrat, P., Triquet, B.: Temporal isolation of hard real-time applications on many-core processors. In: RTAS: Real-Time Embedded Technology & Applications Symposium (2016)
13. Ren, Q., Man, K.L., Lim, E.G., Lee, J., Kim, K.K.: Cooperation of multi robots for disaster rescue. In: 2017 International SoC Design Conference (ISOCC), pp. 133–134 (2017)
14. Sugiyama, H., Tsujioka, T., Murata, M.: Real-time exploration of a multi-robot rescue system in disaster areas. Adv. Robot. **27**, 1313–1323 (2013)
15. Tran, T.T., Vaquero, T.S., Nejat, G., Beck, J.C.: Robots in retirement homes: applying off-the-shelf planning and scheduling to a team of assistive robots. J. Artif. Intell. Res. **58**, 523–590 (2017)
16. Wilhelm, R., Reineke, J.: Embedded systems: many cores - many problems. In: 7th IEEE International Symposium on Industrial Embedded Systems, pp. 176–180 (2012)

Multi-agent and Parallel CP Track

An Improved GPU-Based SAT
Model Counter

Johannes K. Fichte[1]([✉])([iD]), Markus Hecher[2,3]([iD]), and Markus Zisser[2]

[1] TU Dresden, Dresden, Germany
johannes.fichte@tu-dresden.de
[2] TU Wien, Vienna, Austria
{markus.hecher,markus.zisser}@tuwien.ac.at
[3] University of Potsdam, Potsdam, Germany
hecher@uni-potsdam.de

Abstract. In this paper, we present and evaluate a new parallel propositional model counter, called gpusat2, which is based on dynamic programming (DP) on tree decompositions using log-counters. gpusat2 extends its predecessor by a novel architecture for DP that includes using customized tree decompositions, storing solutions to parts of the input instance during the computation variably in arrays or binary search trees, and compressing solution parts. In addition, we avoid data transfer between the RAM and the VRAM whenever possible and employ extended preprocessing by means of state-of-the-art preprocessors for propositional model counting. Our novel architecture allows gpusat2 to be competitive with modern model counters when we also take preprocessing into consideration. As a side result, we observe that state-of-the-art preprocessors allow to produce tree decompositions of significantly smaller width.

Keywords: Propositional model counting · Dynamic programming · Parameterized algorithmics · Bounded treewidth

1 Introduction

The *model counting problem* (#SAT) asks to compute the number of solutions of a propositional formula. A natural generalization of #SAT is weighted model counting (WMC), where formulas are extended by weights. Both #SAT and WMC are special cases of the weighted constraint satisfaction problem [30,39]. Nonetheless, they can already be used to solve a variety of applications to real-world questions in modern society, reasoning, and combinatorics [8,12,13,38,42]. Both #SAT and WMC are known to be complete for the class #P [4,35].

Our system gpusat2 is available under GPL3 license at github.com/daajoe/GPUSAT. A preliminary version has been presented at the workshop Pragmatics of SAT 2019. The work has been supported by the Austrian Science Fund (FWF), Grants Y698 and P26696, and the German Science Fund (DFG), Grant HO 1294/11-1.

ⓒ Springer Nature Switzerland AG 2019
T. Schiex and S. de Givry (Eds.): CP 2019, LNCS 11802, pp. 491–509, 2019.
https://doi.org/10.1007/978-3-030-30048-7_29

In this paper, we consider both problems from the practical perspective. We present and evaluate a new version of a parallel model counter, called gpusat2, which is based on *dynamic programming (DP)* on tree decompositions (TDs) [36]. The idea of solving #SAT decomposing graph representations of the formula and applying DP on them is in fact quite well-known [36] and has earlier already been introduced for the constraint satisfaction problem (CSP) by Kask et al. [24]. Its underlying ideas are as follows. A TD of a propositional formula F is defined on a graph representation of F and formalizes a certain static relationship of the variables of F among each other. The decomposition then gives rise to an evaluation order and to sets of variables, which define which variables have to be evaluated together when solving the given formula. Intuitively, the width of a TD indicates how many variables have to be considered exhaustively together during the computation. Our previous solver gpusat1 already implements DP-based weighted model counting and model counting using uniform weights on a GPU [21]. Prior to this, Fioretto et al. [22] presented an approach and implementation to compute one solution in weighted CSP, which could also be extended to solve the sum-of-products problem[1]. Here, we focus on an efficient computation and implementation of #SAT solving by introducing a novel architecture in our solver gpusat2. We focus on the so-called primal graph as graph representation, even though the incidence graph [36] theoretically allows for smaller width (off by one), mainly because simplicity of algorithms on the primal graph often outweighs the benefits of potential smaller width [15,21]. Our solver implements the principle of parallel programming of single instructions on multiple threads (SIMT) on a GPU. Therefore, we parallelize by executing the computation of variables that have to be considered exhaustively together on multiple threads, since the computation of an assignment to these variables is independent of other assignments during DP.

Contribution. For our new solver gpusat2, we implement a variety of techniques and introduce an innovative architecture for DP. (i) We employ extended preprocessing [26,27]. (ii) We use customized TDs [2]. (iii) We split the DP computation if we cannot ensure that all resulting data (as well as previously computed data) fit into the VRAM. (iv) We implement width dependent data structures and compression for storing counts during the computation, i.e., arrays or binary search trees. (v) We store the model count during the computation by floating *log-counters*, which increases the accuracy and applicability of our solver to instances with very high solution count. Storing values by the log of the value is a common technique in the domain of probabilistic inference. In addition, we avoid data transfer between the RAM and the VRAM whenever possible. Finally, we present experimental work, where we compare the runtime of our system with state-of-the-art model counters and observe a competitive behavior. In fact, gpusat2 solves about 200 #SAT instances more than its predecessor if we also take preprocessing for both solvers into account. As a side result, we observe that state-of-the-art preprocessors allow to produce TDs of significantly smaller width.

---

[1] The sum-of-product problem is often also referred to as weighted counting, partition function, or probability of evidence.

2   Preliminaries

Propositional Satisfiability. A literal is a propositional variable x or its negation $\neg x$. A *clause* is a finite set of literals, interpreted as the disjunction of these
literals. A *(CNF) formula* is a finite set of clauses, interpreted as a conjunction
of the clauses. Let F be a formula. A *sub-formula* S of F consists of subsets of
clauses of F. For a clause $c \in F$, $\text{var}(c)$ consists of all variables that occur in c and
$\text{var}(F) := \bigcup_{c \in F} \text{var}(c)$. A *(partial) assignment* is a mapping $\sigma : \text{var}(F) \to \{0,1\}$.
The formula $F(\sigma)$ *under assignment* σ is obtained by removing all clauses c from
F that contain a literal set to 1 by σ and removing from the remaining clauses
all literals set to 0 by σ. An assignment σ is *satisfying* if $F(\sigma) = \emptyset$. The problem
#SAT asks to output the number of satisfying assignments of a formula. This
problem can be generalized with weights of literals by assigning weights between
0 and 1 to each literal and taking the sum of weights for satisfying assignments.

Listing 1: Algorithm $\mathsf{K}(t, \chi_t, F_t, \langle \rho_1, \dots, \rho_\ell \rangle, R)$ for Step 3 (DP) and nice TDs.

In: Node t, bag χ_t, clauses F_t, $\langle \rho_1, \dots \rho_\ell \rangle$ is the sequence of tables for child
     nodes $\langle t_1, \dots, t_\ell \rangle$ of t, set $R \subseteq 2^{\chi_t \to \{0,1\}}$ of assignments. **Out:** Local Storage

1 **if** type$(t) = \textit{leaf}$ **then** $\rho_t := \{ \langle \emptyset, 1 \rangle \mid \emptyset \in R \}$
2 **else if** type$(t) = \textit{intr}$, *and* $a \in \chi_t$ *is introduced* **then**
3 $\quad \mid \quad \rho_t := \{ \langle \beta, c \rangle \qquad\qquad \mid \langle \alpha, c \rangle \in \rho_1, \beta \in \{ \alpha^+_{a \mapsto 0}, \alpha^+_{a \mapsto 1} \}, F_t(\beta) = \emptyset, \beta \in R \}$
4 **else if** type$(t) = \textit{rem}$, *and* $a \notin \chi_t$ *is removed* **then**
5 $\quad \mid \quad \rho_t := \{ \langle \alpha^-_a, \Sigma_{\langle \beta, c \rangle \in \rho_1 : \alpha^-_a = \beta^-_a} c \rangle \qquad \mid \langle \alpha, \cdot \rangle \in \rho_1, \alpha^-_a \in R \}$
6 **else if** type$(t) = \textit{join}$ **then**
7 $\quad \mid \quad \rho_t := \{ \langle \alpha, c_1 \cdot c_2 \rangle \qquad \mid \langle \alpha, c_1 \rangle \in \rho_1, \langle \alpha, c_2 \rangle \in \rho_2, \alpha \in R \}$
8 **return** ρ_t

$\alpha^-_e := \alpha \setminus \{ e \mapsto 0, e \mapsto 1 \}$, $\alpha^+_{e \mapsto b} := \alpha \cup \{ e \mapsto b \}$.

Tree Decomposition and Treewidth. A *tree decomposition (TD)* of a given graph G
is a pair $\mathcal{T} = (T, \chi)$ where T is a rooted tree and χ is a mapping which assigns
to each node $t \in V(T)$ a set $\chi(t) \subseteq V(G)$, called *bag*, such that: (i) $V(G) = \bigcup_{t \in V(T)} \chi(t)$ and $E(G) \subseteq \{ \{u, v\} \mid t \in V(T), \{u, v\} \subseteq \chi(t) \}$; and (ii) for each
$r, s, t \in V(T)$, such that s lies on the path from r to t, we have $\chi(r) \cap \chi(t) \subseteq \chi(s)$. We let width$(\mathcal{T}) := \max_{t \in V(T)} |\chi(t)| - 1$. For a node $t \in V(T)$, we say
that type(t) is *leaf* if t has no children and $\chi(t) = \emptyset$; *join* if t has children t'
and t'' with $t' \neq t''$ and $\chi(t) = \chi(t') = \chi(t'')$; *intr* ("introduce") if t has a
single child t', $\chi(t') \subseteq \chi(t)$ and $|\chi(t)| = |\chi(t')| + 1$; *rem* ("removal") if t has a
single child t', $\chi(t') \supseteq \chi(t)$ and $|\chi(t')| = |\chi(t)| + 1$. If for every node $t \in N$,
type$(t) \in \{ \textit{leaf}, \textit{join}, \textit{intr}, \textit{rem} \}$, then the TD is called *nice*. The *treewidth* tw(G)
of G is the minimum width$(\mathcal{T})$ over all TDs $\mathcal{T}$ of G. The *primal graph* P_F [36]
of a formula F has as vertices its variables and two variables are joined by an
edge if they occur together in a clause of F. For brevity, we refer by *treewidth
of a formula* to the treewidth of its primal graph. For a given node $t \in T$ of
the primal graph P_F, we let $F_t := \{ c \mid c \in F, \text{var}(c) \subseteq \chi(t) \}$ be the clauses
entirely covered by $\chi(t)$. The formula $F_{\leq s}$ denotes the union over all F_t for all
descendant nodes $t \in V(T)$ of s. In other words, $F_{\leq s}$ is the sub-formula of F

that contains all clauses that have been entirely covered by a bag $\chi(s)$ for t and any of its descendant nodes.

Dynamic Programming on TDs. A solver based on *dynamic programming (DP)* for formulas evaluates the input formula F in parts along a given TD of the primal graph P_F. For each node t of the TD, results are usually stored in a *local storage* ρ_t. The approach works in four steps as follows:

1. Construct the primal graph P_F of the input formula F.
2. Heuristically compute a tree decomposition $\mathcal{T} = (T, \chi)$ of the primal graph P_F.
3. DP: Traverse the nodes in $V(T)$ in post-order O.
   At every node $t \in O$, run an algorithm K that takes as input t, $\chi(t)$, the sub-formula F_t and previously computed results of its children and stores the results in ρ_t, which in turn is used by the algorithm at the parent (if exists).
4. Print the (weighted) model count by interpreting the result ρ_n, which has been computed for the root n of T.

Algorithm K in Step 3 for nice TDs is depicted in Listing 1, cf., [21,36]. Let therefore parameter R ("range") of K be a set of assignments, i.e., $R \subseteq 2^{\chi(t) \to \{0,1\}}$. We assume $R = 2^{\chi(t) \to \{0,1\}}$ for sequential DP. Then, algorithm K stores in ρ_t pairs of the form $\langle \alpha, c \rangle$ consisting of an assignment $\alpha : \chi(t) \to \{0,1\}$ together with a counter c. Each pair $\langle \alpha, c \rangle$ indicates that there are c many satisfying assignments restricted to $\chi(t)$ of $F_{\leq t}$. These pairs are carefully maintained for all the different types of nodes of nice TDs in Listing 1. For details, we refer to the literature [21,36].

While in theory we often prefer nice TDs, due to simpler cases distinctions, in an actual implementation of K one handles also interleaved cases. Note that a very compact way to represent ρ_t is simply by taking a sequence of model counts c for the sub-formula $F_{\leq t}$ without explicitly storing α for a fixed ordering of the considered assignments. Each counter in the sequence is entirely independent of another counter in the sequence as each counter in ρ_t depends only on results previously computed at the children. This allows directly to parallelize the computation of the counts in ρ_t. Since we have $2^{|\chi(t)|}$ many assignments at each node t, for which we can compute the (potentially zero) counters by the very same operations, we can immediately parallelize the operations on the GPU [21] by employing a *single instruction on multiple threads (SIMT)* computation model. More detailed, K in Step 3 refers to a program that can be executed by a GPU thread, taking a small set of instructions but multiple input data, e.g., different ranges R. Then, for each node t one can compute ρ_t for $|2^{\chi(t)}|$ many (singular) ranges in parallel. Such a procedure is also called *(GPU-)kernel* for this hardware architecture. The simplest possible data structure for ρ_t is an *array* that just contains the counts, where an assignment is addressed by the memory address (index) of an entry in the array. However, this data structure has to be allocated on the *video RAM (VRAM)* prior to running the kernel on the GPU. There, one has to take care of out-of-memory issues caused by huge space requirements.

3   An Improved GPU-Based DP Architecture

In this section, we build upon the idea above and present an innovative architecture for parallel dynamic programming on the GPU, which is outlined in Fig. 1. Novel parts of the architecture are the *preprocessors*, tree *decomposition selection* heuristics (customized TDs), generalization to allow for adaptable, more advanced *data structures*, *caching* intermediate results on the GPU, and the idea of *compressing* counters for assignments. Note that the architecture is independent from the underlying data structures, i.e., in Step 2 we refer by ρ_t to a storage for data in the RAM, which can be an array or another data structure. Analogously, ι_t also denotes a storage, that caches results in the VRAM for GPU computations. In the following, we discuss the novel steps of the architecture, whereas details on data structures are presented in Sect. 4.

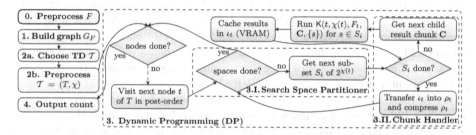

Fig. 1. Architecture of our DP-based solver for parallel execution. Yellow colored boxes indicate tasks that are required as initial step for the DP-run or to finally read the model count from the computed results. The parts framed by a dashed box illustrate the DP-part. Boxes colored in red indicate computations that run on the CPU. Boxes colored in blue indicate computations that are executed on the GPU (with waiting CPU).

Step 0: Instance Preprocessing. Before we decompose our instance, we *simplify the formula F* by a preprocessor for formulas [26,27]. There, we preserve the number of satisfying assignments and potentially decrease the treewidth of F. For weighted model counting, vivification and literal elimination can be applied [27].

Step 2: Tree Decomposition Computation. In Step 2.I, we heuristically compute a tree decomposition for the dynamic programming. Various recent literature suggests [3,7,23] that tree decompositions for practical solving require in addition to "small" width other criteria to speed up the performance of a solver. Such tree decompositions are frequently called *customized tree decompositions*. We compute m different tree decompositions via heuristics [2] and then select among the m decompositions one decomposition according to a selection criterion. In the implementation, we use the library *htd* version 1.2 with default settings [2] where $m = 30$. The selection criterion is as follows. We first try to minimize the width. Then, if several decompositions of the same width are found, we select the decomposition with the smallest maximal cardinality $v(\mathcal{T})$ of the intersection of bags of any node with its children, i.e., $v(\mathcal{T}) = \max\{|\chi_t \cap (\chi_{t_1} \cup \chi_{t_2} \cup \ldots \cup \chi_{t_\ell})| \mid t \in V(T), t_1, t_2, \ldots, t_\ell$ are children of t in $T\}$ where $\mathcal{T} = (T, \chi)$. The idea of the selection is to balance the *trade-off* between runtime and space requirements in the worst-case as outlined in earlier work [23]. In that way, we first improve on the worst-case *runtime (and VRAM consumption)* and then on the number of IO operations required to copy data between RAM and VRAM. After the selection of tree decomposition $\mathcal{T}$, we *preprocess* $\mathcal{T}$ (Step 2.II). There, we combine nodes to obtain bags of size s, which is the largest number such that on the chosen hardware 2^s GPU threads can still run in parallel. This reduces the overhead of copying onto VRAM, and GPU thread allocation.

Step 3: Dynamic Programming

The more involved architecture of Step 3 consists of multiple parts as follows.

Step 3.I: Search Space Partitioning. As described in the preliminaries, the DP proceeds by traversing a tree decomposition in post-order. At each node, we consider assignments restricted to the variables in $\chi(t)$ and its corresponding counters. Overall we can have at most $2^{|\chi(t)|}$ assignments ("local search space" at a node). Thus, the number of assignments can simply be too large to even store just one counter per assignment in the VRAM. In practice, we would expect that plenty of these assignments result in a counter that is zero and hence we could actually avoid the out-of-memory issue as data can be compressed. However, on the VRAM we have to allocate memory prior to the computation and hence it would require to detect the point where we run out-of-memory then to copy the data back to the RAM resulting in turn in an unutilized GPU. To avoid this situation, we simply split in Step 3.I all possible assignments that are considered together at once on the GPU into several disjoint subsets $S_1, S_2, \ldots, S_k$ of $2^{|\chi(t)|}$, which we call *search space partitioning*. On these grounds, we do the solution space splitting before the GPU kernels are even executed to ensure that no out-of-memory issue occurs. Splitting is independent of the actually used data structure and can, e.g., be used if we store counters in an array similar to gpusat1.

Step 3.II: Splitting Input Result from Children and Compression. In the next step, we systematically process each set S_i for $1 \leq i \leq k$. Therefore, we consider the assignments in S_i and the corresponding counters for the children, i.e., the counters and corresponding assignments at the children which we need to compute the counter for an assignment at the currently considered node t. Since we have both to copy these relevant assignments of the children onto the VRAM and still allocate enough VRAM for S_i, we might run into the situation that both would not fit into the VRAM. Hence, we need to split for the counts and its corresponding assignments in S_i the relevant results in $\rho_{t_1}, \ldots, \rho_{t_\ell}$ computed at the child nodes $t_1, \ldots, t_\ell$ of t into subsets $C_1 \subseteq \rho_{t_1}, \ldots, C_\ell \subseteq \rho_{t_\ell}$. We call a tuple $C = \langle C_1, \ldots, C_\ell \rangle$ of these subsets *chunk*. Then, the *chunk handler* systematically takes each tuple C relevant for S_i and executes a kernel in a GPU thread for each element in S_i using C. Subsequently, the resulting counts are summed up accordingly and kept inside cache storage ι_t on the VRAM. This allows to reduce the number of IO operations between RAM and VRAM for tree decompositions of larger width. Finally, if all chunks are processed for S_i, the memory region ι_t is merged into the RAM ρ_t for node t. There, depending on the data structure, it can be beneficial to merge and *compress* resulting solutions obtained for two different solution spaces S_i and S_j. For this task, we use a *(support) kernel* on the GPU, where one can merge two regions, say $\iota_t^{(i)}$ and $\iota_t^{(j)}$, into memory ρ_t, with the same idea for parallelization as in K, in parallel. In turn, this might decrease the number of child chunks needed at the parent of t. In other words, for the parent node of t, one might prevent splitting results from children.

4   Data Representation

We implemented our solver gpusat2 based on the architecture presented in the previous section. In this section, we describe advancements in the implementation of the solver such as *data structures* optimized for GPUs, and improved accuracy in form of *log-counters*.

4.1   Binary Search Tree on the GPU

A naive approach to store counters on the VRAM ι_t is simply to exhaustively consider all possible assignments in $2^{\chi(t)}$ and store for each assignment a counter, even if zero, in an array. In order to compactly store assignments at a node t in the VRAM, we propose a new data structure, which is in a broader sense a *binary search tree (BST)* for assignments on a very low level architecture. The binary search tree contains only assignments to F_t whose corresponding counter is non-zero, i.e., only counters for assignments that can be extended[2] to satisfying assignments of $F_{\leq t}$. The BST data structure allows us to allocate memory on the VRAM in advance, which is required by OpenCL 1.2, as kernels itself are not allowed to allocate memory on the VRAM during the execution. Further, it allows us to avoid of synchronization between threads.

Internally, a BST consists of a continuous sequence of *cells* that are implemented as 64-bit unsigned integers. We have three types of cells, namely *empty cells*, *value cells*, and *index cells*. An empty cell contains a zero whereas a value cell contains an integer greater than zero. For value cells the 64-bit integer corresponds to a counter that is internally actually interpreted as a double floating point type. We discuss details in the next paragraph. Index cells have either one or two successors in the tree and refer to a value or index cell. For an index cell, the *lower 32 bits* of the integer represent the *index* to the next cell, where a corresponding variable is set to false. Symmetrically, the *upper 32 bits* form an index to the cell if the variable is set to true. Note that either the lower or upper bits can be zero indicating that the respective index is empty. In Example 1 we illustrate BST memory allocation including the three types of cells by a simple example where an assignment is inserted into a non-empty previously BST.

---

[2] Extending an assignment can be done by recursively considering previously computed assignments at the children that correspond to an assignment at the node.

(index)	(var)	cell low	high
0	x	ε	1
1	y	3	2
2	-	**23**	
3	-	**42**	
4	-	ε	
...	...	...	

(index)	(var)	cell low	high
0	x	4	1
1	y	3	2
2	-	**23**	
3	-	**42**	
4	y	ε	5
5	-	**1**	
6		ε	
...	...	...	

Fig. 2. Initial BST $\mathcal{B}$ (left) and BST $\mathcal{B}'$ (right), which was obtained after inserting assignment $\{x \mapsto 0, y \mapsto 1\}$. Value cells are depicted in bold. Both empty cells and empty indices are indicated by the symbol ε. Note: We store *only* cells ("low", "high").

Example 1. Figure 2 (left) illustrates a binary search tree $\mathcal{B}$, where value cells are depicted in **bold** face. Both empty cells and empty indices are represented by the symbol "ε".In the BST $\mathcal{B}$ we assume $x < y$. In Fig. 2 (right), we can see the BST $\mathcal{B}'$ obtained by inserting the assignment $\alpha := \{x \mapsto 0, y \mapsto 1\}$ into $\mathcal{B}$. In order to insert α, we recursively search for α in $\mathcal{B}$ by traversing $\mathcal{B}$ according to the variable order, beginning at start index 0. Then, depending on the assignment of the variable at index 0 in α, we continue searching using the next variable at the respective index. As soon as an empty index is found, new index cells for the remaining variables are subsequently inserted, followed by an inserted value cell of value 1. As a result, the search for α in $\mathcal{B}$ stops at the lower 32 bits of the index cell at index 0. In turn, these bits refer to a new index cell for y at index 4, whose upper 32 bits point to a new value cell at index 5. ∎

Note that we need some *fixed order on the variables* in $\chi(t)$, to distinguish index and value cells in order to search, insert, update, and delete counters for a given assignment over the variables. The binary search tree enables us to address $2^{32} - 1$ many 64-bit integers, which can be changed to relative indices (offsets) if more address space is required. Further, for a given number b of variables, the tree requires in the worst-case at most $2^{b+1} - 1$ many 64-bit integers, since there are at most 2^b many value cells (all assignments have non-zero counters) and $2^b - 1$ index cells (perfectly balanced BST) needed.

Obviously, our data structure has to be manipulated by several GPU threads in parallel. In contrast to the array data structure, where each GPU thread has unique access to one entry, the BST has to *prevent race conditions* between threads. Our strategy is to allow atomic write access only to non-empty cells. To this end, we keep track of the number of non-empty cells of the tree, and assign indices of non-empty cells in ascending order. Then, we need (i) to atomically reserve free (empty) indices and (ii) to run only synchronized updates on existing value cells. In the actual implementation this is efficiently done by *atomic operations* for 32-bit and 64-bit data types provided by the OpenCL framework [32]. In Case (i) we rely on atomic_cmpxchg for inserting into the index, but only if it is empty (atomic operation). In Case (ii) we use atomic_add for concurrently updating counters.

Balancing Between BST and Arrays. It is easy to see that the BST data structure introduces an overhead in the computation. For instances of small width where all possible partial truth assignments (and the counts even if 0) easily fit into the VRAM using an array, the BST might not pay off. Hence, one can also design an implementation that uses the array up to a threshold and then switches to BST as a data structure. We implemented this option into our solver and call the resulting variant gpusat2(A+B). Further, whenever gpusat2 has to combine BSTs for a node with more than one child node, we use a support kernel for transforming BSTs into arrays. Kernel K then combines the arrays into a BST.

4.2   Accuracy of Large-Scale Counters

In order to be able to apply our solver to instances with very high solution count while still preserving a high accuracy, we store the model count during the dynamic programming by the log of the value. This technique is common in the domain of probabilistic inference and also known as *log-counter*.

In more detail, we described in the previous paragraph that one could take 64-bit floating point numbers to store the values of a counters. While IEEE 64-bit floating point numbers allow to represent only values below 10^{308} [1], counters can have a significantly higher value. Therefore, we need an extended data type for each counter c. We store the value of c in relation to 2^e for a 64-bit integer e. We start with significant digit 1 before the decimal point where $c = 1.x \cdot 2^e$ and chosen e accordingly. Then, we store $1.x$ and e to reconstruct the value of c. In the implementation, we start with $e = 0$ and increase it dynamically during the computation at a node t whenever necessary. We normalize the largest counter c at a node t as described above and all other counters computed at node t are represented with the same exponent e, i.e., we need only one exponent per node. We call the resulting e the *largest exponent for t*. In that way, we remain highest accuracy while still being able to represent high values. The largest exponent is carefully maintained on the GPU during computation of ρ_t in kernel K and passed along to parent nodes. For the largest exponents at a node, which has more than one child (join node), we may have to combine counters with respect to different largest exponents for child nodes of t. In more details, algorithm K additionally takes the sum e of the largest exponents e_1, e_2, ... of its children as parameter, which represent the counters. Exponent e is then also used to represent counters at t, however, it might increase during the computation and therefore K returns an updated exponent after the computation.

5   Experiments

We conducted a series of experiments using several benchmark sets for model counting and weighted model counting. Benchmark sets [17] and our results [18] are publicly available and also on github at daajoe/gpusatspsexperiments.

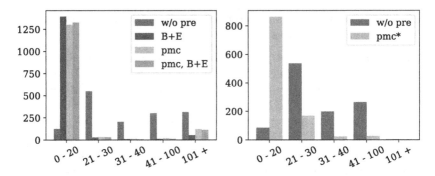

Fig. 3. Width distribution of #SAT instances (left) before and after preprocessing (using both B+E and pmc). Width distribution of WMC instances (right) before and after preprocessing using pmc*. Results are based on the primal treewidth and presented in intervals. X-axis labels the intervals, y-axis labels the number of instances.

Measure, Setup, and Resource Enforcements. As we use different types of hardware in our experiments and other natural measures such as power consumption cannot be recorded with current hardware, we compare wall clock time and number of timeouts. In the time we include, if applicable, *preprocessing time* as well as *decomposition time* for computing 30 decompositions with a random seed and decomposition selection time. However, we avoid IO access on the CPU solvers whenever possible, i.e., we load instances into the RAM before we start solving. For parallel CPU solvers we allow access to 12 or 24 physical cores on machines where hyperthreading was disabled. We set a timeout of 900 s and limited available RAM to 14 GB per instance and solver.

Benchmark Instances. We considered a selection of overall 1494 instances from various publicly available benchmark sets for model counting consisting of fre/meel benchmarks[3] (1480 instances), and c2d benchmarks[4] (14 instances). For WMC, we used the overall 1091 instances from the Cachet benchmark set[5].

Benchmarked Solvers. In our experimental work, we present results for the most recent versions of publicly available #SAT solvers, namely, c2d 2.20 [10], d4 1.0 [28], DSHARP 1.0 [31], miniC2D 1.0.0 [33], cnf2eadt 1.0 [25], bdd_minisat_all 1.0.2 [41], and sdd 2.0 [11] (based on knowledge compilation techniques). We also considered rather recent approximate solvers ApproxMC2, ApproxMC3 [6] and sts 1.0 [14], as well as CDCL-based solvers Cachet 1.21 [37], sharpCDCL[6], and sharpSAT 13.02 [40]. Finally, we also included multi-core solvers gpusat 1.0 [21], as well as countAntom [5] on 12 physical CPU cores, which performed better than on 24 cores. Note that we benchmarked additional solvers, which we omitted from the presentation here and where we

---

[3] See: tinyurl.com/countingbenchmarks.
[4] See: reasoning.cs.ucla.edu/c2d.
[5] See: cs.rochester.edu/u/kautz/Cachet.
[6] See: tools.computational-logic.org.

Table 1. Overview on upper bounds of the primal treewidth for considered #SAT and WMC benchmarks before and after preprocessing. vMdn median of variables, cMdn median of clauses, t[s] Mdn of the decomposition runtime in seconds, maximum runtime t[s] Max, median Mdn and percentiles of upper bounds on treewidth, and min/max/mdn of the width improvement after preprocessing. Negative values indicate worse results.

prob	pre	vMdn	cMdn	t[s] Mdn	to	t[s] Mdn	pre	to	Mdn	50%	80%	90%	95%	Min	Max	Mdn
#SAT	w/o pre	637	810	0.07	6		n/a	n/a	31	31	166	378	922	n/a	n/a	n/a
	pmc, B+E	231	350	0.02	6	0.06	192		3	3	**17**	201	**823**	−72	**755**	22
	pmc	**231**	189	0.03	6	**0.03**	103		3	4	19	228	**823**	−1839	547	23
	B+E	**231**	**185**	**0.02**	6	0.04	189		3	3	18	**192**	**823**	−2	633	23
WMC	w/o pre	200	519	0.04	0		n/a	n/a	28	28	40	43	54	n/a	n/a	n/a
	pmc*	200	**300**	**0.03**	0	**0.03**	0		11	11	20	25	30	0	330	16

placed results online in our result data repository. For WMC, we considered the following solvers:sts, gpusat1, gpusat2, *miniC2D*, *Cachet*, *d4*, and d-DNNF reasoner 0.4.180625 (on top of d4 as underlying knowledge compiler). All experiments were conducted with default solver options. For solver gpusat2, we also benchmarked variant gpusat2(A+B) where we used 30 as threshold above which we apply the BST.

Benchmark Machines. The non-GPU solvers were executed on a cluster of 9 nodes. Each node is equipped with two Intel Xeon E5-2650 CPUs consisting of 12 physical cores each at 2.2 GHz clock speed and 256 GB RAM. The results were gathered on Ubuntu 16.04.1 LTS machines with disabled hyperthreading on kernel 4.4.0-139, which is already a post-Spectre and post-Meltdown kernel[7]. For gpusat1 and gpusat2 we used a machine equipped with a consumer GPU: Intel Core i3-3245 CPU operating at 3.4 GHz, 16 GB RAM, and one Sapphire Pulse ITX Radeon RX 570 GPU running at 1.24 GHz with 32 compute units, 2048 shader units, and 4GB VRAM using driver amdgpu-pro-18.30-641594 and OpenCL 1.2. The system operated on Ubuntu 18.04.1 LTS with kernel 4.15.0-34.

5.1   Results

First, we present how existing preprocessors for #SAT and equivalence-preserving preprocessors for WMC influence the treewidth on the considered instances.

Treewidth Analysis. We computed upper bounds on the primal treewidth for our benchmarks before and after preprocessing and state them in intervals. For model-count preserving preprocessing we explored both B+E Apr2016 [26] and pmc 1.1 [27]. For WMC, we used pmc with documented options −vivification −eliminateLit −litImplied −iterate = 10 to preserve all the models, which we refer to by *pmc**. In this experiment, we used different timeouts. We set the timeout of the preprocessors to 900 s and allowed further 1800 s for the decomposer

---

[7] Details on spectre and meltdown: spectreattack.com.

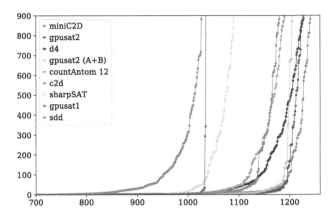

Fig. 4. Runtime for the top 5 sequential and all parallel solvers over all the #SAT instances with pmc preprocessor. The x-axis refers to the number of instances and the y-axis depicts the runtime sorted in ascending order for each solver individually.

to get a detailed picture of treewidth upper bounds. Figure 3 (left) presents the width distribution of number of instances (y-axis) and their corresponding upper bounds (x-axis) for primal treewidth, both before and after preprocessing using B+E, pmc, and both preprocessors in combination (first pmc, then B+E) for #SAT. Table 1 (top) provides statistics on the benchmarks combined, including runtime of the preprocessor, runtime of the decomposer to obtain a decomposition, upper bounds on primal treewidth, and its improvements before and after preprocessing. Further, the table also lists the median of the widths of the obtained decompositions and their percentiles, which is the treewidth upper bound a given percentage of the instances have. Interestingly, overall we have that a majority of the instances after preprocessing has width below 20. In more details, more than 80% of the #SAT instances have primal treewidth below 19 after preprocessing, whereas 90% of the instances have treewidth below 192 for B+E. With pmc we observed a corner case where the primal treewidth upper bound increased by 1839, however, on average we observed a mean improvement on the upper bound of slightly above 23. The best improvement among the widths of all our instances was achieved with the combination of pmc and B+E where we improved the width by 755. Overall, both B+E and pmc managed to *drastically reduce* the widths, the decomposer ran below 0.1 s in median. Interestingly, even the upper bounds on the treewidth of the WMC instances reduced with pmc* as depicted in Fig. 3 (right). In more detail, after preprocessing 95% of the instances have primal treewidth below 30, c.f., Table 1 (bottom).

Solving Performance Analysis. Figure 4 illustrates the top five sequential solvers, and all parallel counting solvers with preprocessor pmc in a cactus-like plot. Table 2 presents detailed runtime results for #SAT with preprocessors pmc, B+E, and without preprocessing, respectively. Since the solver sts produced results that varied from the correct result on average more than the value of the correct result, we excluded it from the presented results. If we disallow preprocessing, gpusat2 and gpusat1 perform only slightly better in the overall standing of the solvers. But gpusat2 solves 42 instances more and requires about 10 h less of wallclock time. Further, we can observe, that the variant gpusat2($A+B$) performs particular well, mainly since for instances below width 30, the BST compression seems relatively expensive compared to the array data structure. Interestingly, when considering the results on preprocessing in Table 2 (top, mid) and Fig. 4 we observe that the architectural improvements pay off quite well. gpusat2 can solve the vast majority of the instances and ranks second place. If one uses the B+E preprocessor shown in Table 2 (mid), gpusat2 solves even more instances as well as the other solvers. Still, it ranks fifth solving only 26 instances less than the best solver and 10 less than the third best solver and solves the most instances having width below 30.

While we focus on #SAT with our implementation, we also conducted the experiments with WMC in order to compare our solver with gpusat1 in the setting for which it was designed. Table 3 (top) lists results of the top five best solvers capable of solving WMC on our instances. Compared to gpusat1, our solver gpusat2 shows an improvement when the width of the instance is between 21 and 40, in more detail gpusat2 solves 44 instances more. After preprocessing with pmc*, one can observe that the majority of the instances has width below 20, c.f., Table 3 (bottom). As a result, gpusat2 does not provide significant improvement over gpusat1 there apart from small runtime improvements.

Currently, we are unable to measure the speed-up of the implementations in terms of the used cores, mainly due to the fact that we aimed for an implementation that is close to gpusat1 so that the improvements are actually from the architectural changes and not just from the different framework or drivers. Note that OpenCL does not support disabling certain cores on the GPU. We also benchmarked gpusat2 on an Nvidia GPU, whose runtimes are quite similar. We also provide preliminary data online with the experiments; which are however not conclusive yet. However, we ran into bugs, which seems to be attributed to the OpenCL1.2 Nvidia driver. Therefore, we aim as future work for a new implementation in CUDA [9].

6 Related Work and Conclusion

Related Work. Fioretto et al. [22] introduced a solver for outputting a solution to the weighted CSP problem using a GPU. Their technique is effectively a version of dynamic programming on tree decompositions also known as bucket-elimination, which they limited to an incomplete elimination by introducing shortcuts and discarding non-optimal solutions in order to speed up the computation for the problem of outputting just one solution. While the underlying idea of a dynamic programming based solver still exists in our solver, gpusat2 is very different when just taking a slightly more detailed look. We approach the *counting question – not just outputting one solution*, which disallows certain simplifications. We can neither apply an incomplete bucket-elimination technique (mini-bucket elimination) nor discard non-optimal solutions. But then,

Table 2. Number of #SAT instances (grouped by treewidth upper bound intervals) solved by sum of the top five sequential and all parallel counting solvers with preprocessor pmc (top), B+E (mid), and without preprocessing (bottom). time [h] is the cumulated wall clock time in hours, where unsolved instances are counted as 900 s.

	Solver	0–20	21–30	31–40	41–50	51–60	>60	best	unique	$\sum$	time [h]
pmc preprocessing	miniC2D	1193	29	**10**	2	1	7	13	0	**1242**	**68.77**
	gpusat2	**1196**	32	1	0	0	0	250	**8**	1229	71.27
	d4	1163	20	**10**	2	**4**	28	52	1	1227	76.86
	gpusat2($A+B$)	1187	18	1	0	0	0	120	7	1206	74.56
	countAntom 12	1141	18	**10**	5	**4**	13	101	0	1191	84.39
	c2d	1124	31	**10**	3	3	10	20	0	1181	84.41
	sharpSAT	1029	16	**10**	2	**4**	**30**	**253**	1	1091	106.88
	gpusat1	1020	16	0	0	0	0	106	7	1036	114.86
	sdd	1014	4	7	1	0	2	0	0	1028	124.23
	Solver	0–20	21–30	31–40	41–50	51–60	>60	best	unique	$\sum$	time [h]
B+E preprocessing	c2d	1199	24	**9**	0	2	23	14	0	**1257**	**63.46**
	miniC2D	1203	**27**	8	0	2	12	8	0	1252	64.92
	d4	1182	15	**9**	1	**3**	31	79	1	1241	69.32
	countAntom 12	1177	14	8	0	2	**34**	100	0	1235	69.79
	gpusat2	**1204**	26	1	0	0	0	**150**	3	1231	68.15
	gpusat2($A+B$)	1201	21	1	0	0	0	67	3	1223	70.39
	sdd	1106	11	4	**1**	1	4	0	0	1127	100.48
	gpusat1	1037	16	0	0	0	0	87	3	1053	110.87
	bdd_minisat_all	926	6	3	**1**	1	0	101	0	937	140.59
	Solver	0–20	21–30	31–40	41–50	51–60	>60	best	unique	$\sum$	time [h]
Without preprocessing	countAntom 12	118	511	139	**175**	**21**	181	318	15	**1145**	**96.64**
	d4	124	514	148	162	**21**	168	69	15	1137	104.94
	c2d	119	525	**165**	161	18	120	48	15	1108	110.53
	miniC2D	122	514	128	149	9	62	0	0	984	141.22
	sharpSAT	100	467	124	156	12	123	**390**	4	982	135.41
	gpusat2($A+B$)	**125**	**539**	96	138	0	0	94	**19**	898	151.16
	gpusat2	**125**	523	96	138	0	0	78	17	882	155.43
	gpusat1	**125**	524	67	140	0	0	82	9	856	162.03
	cachet	99	430	71	152	8	57	3	0	817	176.26
	Solver	0–20	21–30	31–40	41–50	51–60	>60	best	unique	$\sum$	time [h]

we consider the binary case, which allows us to introduce various simplifications including the way we store the data enabling us to save memory and to avoid copying data. Also, we would like to point out that bucket-elimination is used in the decomposer htd to compute just the tree decomposition. In that way, our architecture is quite general as it completely separates the decomposition and the actual computation part resulting in a framework that can also be used for other problems. Moreover, we use more sophisticated data

Table 3. Number of WMC instances solved (with)out preprocessing. time [h] is the cumulated wall clock time in hours, where unsolved instances count as 900 s.

	Solver	0–20	21–30	31–40	41–50	51–60	>60	best	unique	$\sum$	time [h]
With pmc*	miniC2D	858	**164**	**6**	**0**	**0**	3	13	8	**1031**	21.29
	gpusat1	**866**	158	0	0	0	0	**348**	4	1024	18.03
	gpusat2(A+B)	**866**	156	0	0	0	0	343	4	1022	**17.86**
	gpusat2	**866**	138	0	0	0	0	299	4	1004	22.43
	d4	810	106	0	0	0	0	46	0	916	55.36
	cachet	617	128	1	0	0	3	106	1	749	93.65
Without pre	d4	82	501	**142**	**156**	**10**	**19**	111	**24**	**910**	**53.97**
	miniC2D	84	517	134	152	3	4	19	7	894	59.69
	gpusat2(A+B)	86	**527**	98	138	0	0	167	19	849	64.40
	gpusat2	86	511	98	138	0	0	131	7	833	68.61
	gpusat1	86	513	68	140	0	0	**182**	10	807	73.78
	cachet	60	447	100	145	2	9	118	1	763	89.80

structures and split data whenever the data does not fit into the VRAM of the GPU. Finally, we balance between small width during the computation and not too small width as we want to employ the full computational power of the parallelization with the GPU. In the past, a variety of model counters and weighted model counters have been implemented based on several different techniques. We listed them in details in Sect. 5. However, here we want to highlight a few differences between our technique and knowledge compilation-based techniques as well as distributed computing. The solver d4 [28], which implements a knowledge compilation-based approach, employs heuristics to compute decompositions of an underlying hypergraph, namely the dual hypergraph, and uses this during the computation. Note that the following relationships are known for treewidth (i.e., the width of a tree decomposition of smallest width) of an arbitrary formula F, $\mathrm{inctw}(F) \leq \mathrm{dualtw}(F)+1$ and $\mathrm{inctw}(F) \leq \mathrm{primtw}(F)+1$, where inctw refers to the treewidth of the incidence graph, dualtw of the dual graph, and primtw of the primal graph. However, there is no such relationship between the treewidth of the primal and dual graph. We are currently unaware of how these theoretical results generalize to hypergraphs. Experimentally, it is easy to verify that a decomposition of the dual graph is often not useful in our context as it provides only decompositions of large width. When we consider parallel solving, a few words on distributed counting are in order. In fact, the model counter DMC [29] is intended for parallel computation on a cluster of computers using the message passing model (MPI). However, this distributed computation requires a separate setup of the cluster and exclusive access to multiple nodes. We focus on parallel counting with a shared memory model. For details, we refer to the difference between parallel and distributed computation [34].

Conclusion. We presented an improved OpenCL-based solver gpusat2 for solving #SAT and WMC. Compared to the weighted model counter gpusat1 that uses the GPU, our solver gpusat2 implements adapted memory management, specialized data structures on the GPU, improved data type precision handling, and an initial approach to use customized TDs. We carried out rigorous experimental work, including establishing upper bounds for treewidth after preprocessing of commonly used benchmarks and comparing to most recent solvers.

Future Work. Our results give rise to several research questions. Since established preprocessors are mainly suited for #SAT, we are interested in additional preprocessing methods for weighted model counting (WMC) that reduce the treewidth or at least allow us to compute TDs of smaller width. It would also be interesting whether GPU-based techniques can successfully be used within knowledge compilation-based model counters. An interesting research direction is to study whether efficient data representation techniques can be combined with dynamic programming in order to lift our solver to counting in WCSP [22]. Further, we are also interested in extending this work to projected model counting [16,19,20].

References

1. IEEE standard for floating-point arithmetic: IEEE Std 754–2008, pp. 1–70 (2008)
2. Abseher, M., Musliu, N., Woltran, S.: htd – a free, open-source framework for (customized) tree decompositions and beyond. In: Salvagnin, D., Lombardi, M. (eds.) CPAIOR 2017. LNCS, vol. 10335, pp. 376–386. Springer, Cham (2017). https://doi.org/10.1007/978-3-319-59776-8_30
3. Abseher, M., Musliu, N., Woltran, S.: Improving the efficiency of dynamic programming on tree decompositions via machine learning. J. Artif. Intell. Res. **58**, 829–858 (2017)
4. Bacchus, F., Dalmao, S., Pitassi, T.: Algorithms and complexity results for #SAT and Bayesian inference. In: Chekuri, C.S., Micciancio, D. (eds.) Proceedings of the 44th IEEE Symposium on Foundations of Computer Science (FOCS 2003), pp. 340–351. IEEE Computer Society, Cambridge, October 2003
5. Burchard, J., Schubert, T., Becker, B.: Laissez-faire caching for parallel #SAT solving. In: Heule, M., Weaver, S. (eds.) SAT 2015. LNCS, vol. 9340, pp. 46–61. Springer, Cham (2015). https://doi.org/10.1007/978-3-319-24318-4_5
6. Chakraborty, S., Fremont, D.J., Meel, K.S., Seshia, S.A., Vardi, M.Y.: Distribution-aware sampling and weighted model counting for SAT. In: Brodley, C.E., Stone, P. (eds.) Proceedings of the 28th AAAI Conference on Artificial Intelligence (AAAI 2014), pp. 1722–1730. The AAAI Press, Québec City (2014)
7. Charwat, G., Woltran, S.: Expansion-based QBF solving on tree decompositions. In: RCRA@AI*IA. CEUR Workshop Proceedings, vol. 2011, pp. 16–26. CEUR-WS.org (2017)
8. Choi, A., Van den Broeck, G., Darwiche, A.: Tractable learning for structured probability spaces: a case study in learning preference distributions. In: Yang, Q. (ed.) Proceedings of 24th International Joint Conference on Artificial Intelligence (IJCAI 2015). The AAAI Press (2015)
9. Cook, S.: CUDA Programming: A Developer's Guide to Parallel Computing with GPUs, 1st edn. Morgan Kaufmann Publishers Inc., San Francisco (2013)

10. Darwiche, A.: New advances in compiling CNF to decomposable negation normal form. In: López De Mántaras, R., Saitta, L. (eds.) Proceedings of the 16th European Conference on Artificial Intelligence (ECAI 2004), pp. 318–322. IOS Press, Valencia (2004)

11. Darwiche, A.: SDD: a new canonical representation of propositional knowledge bases. In: Walsh, T. (ed.) Proceedings of the 22nd International Joint Conference on Artificial Intelligence (IJCAI 2011), pp. 819–826. AAAI Press/IJCAI, Barcelona, Catalonia, Spain, July 2011

12. Domshlak, C., Hoffmann, J.: Probabilistic planning via heuristic forward search and weighted model counting. J. Artif. Intell. Res. **30**, 565–620 (2007)

13. Dueñas-Osorio, L., Meel, K.S., Paredes, R., Vardi, M.Y.: Counting-based reliability estimation for power-transmission grids. In: Singh, S.P., Markovitch, S. (eds.) Proceedings of the Thirty-First AAAI Conference on Artificial Intelligence (AAAI 2017), pp. 4488–4494. The AAAI Press, San Francisco, February 2017

14. Ermon, S., Gomes, C.P., Selman, B.: Uniform solution sampling using a constraint solver as an oracle. In: de Freitas, N., Murphy, K. (eds.) Proceedings of the 28th Conference on Uncertainty in Artificial Intelligence (UAI 2012), pp. 255–264. AUAI Press, Catalina, August 2012

15. Fichte, J.K., Hecher, M., Morak, M., Woltran, S.: Answer set solving with bounded treewidth revisited. In: Balduccini, M., Janhunen, T. (eds.) LPNMR 2017. LNCS (LNAI), vol. 10377, pp. 132–145. Springer, Cham (2017). https://doi.org/10.1007/978-3-319-61660-5_13

16. Fichte, J.K., Hecher, M., Morak, M., Woltran, S.: Exploiting treewidth for projected model counting and its limits. In: Beyersdorff, O., Wintersteiger, C.M. (eds.) SAT 2018. LNCS, vol. 10929, pp. 165–184. Springer, Cham (2018). https://doi.org/10.1007/978-3-319-94144-8_11

17. Fichte, J.K., Hecher, M., Woltran, S., Zisser, M.: A Benchmark Collection of #SAT Instances and Tree Decompositions (Benchmark Set), June 2018. https://doi.org/10.5281/zenodo.1299752

18. Fichte, J.K., Hecher, M., Zisser, M.: Analyzed Benchmarks and Raw Data on Experiments for gpusat2 (Dataset), July 2019. https://doi.org/10.5281/zenodo.3337727

19. Fichte, J.K., Hecher, M.: Treewidth and counting projected answer sets. In: Balduccini, M., Lierler, Y., Woltran, S. (eds.) LPNMR 2019. LNCS, vol. 11481, pp. 105–119. Springer, Cham (2019). https://doi.org/10.1007/978-3-030-20528-7_9

20. Fichte, J.K., Hecher, M., Meier, A.: Counting complexity for reasoning in abstract argumentation. In: Hentenryck, P.V., Zhou, Z.H. (eds.) Proceedings of the 33rd AAAI Conference on Artificial Intelligence (AAA 2019). Honolulu, Hawaii, USA (2018). https://arxiv.org/abs/1811.11501

21. Fichte, J.K., Hecher, M., Woltran, S., Zisser, M.: Weighted model counting on the GPU by exploiting small treewidth. In: Azar, Y., Bast, H., Herman, G. (eds.) Proceedings of the 26th Annual European Symposium on Algorithms (ESA 2018). Leibniz International Proceedings in Informatics (LIPIcs), vol. 112, pp. 28:1–28:16. Dagstuhl Publishing (2018)

22. Fioretto, F., Pontelli, E., Yeoh, W., Dechter, R.: Accelerating exact and approximate inference for (distributed) discrete optimization with GPUs. Constraints **23**(1), 1–23 (2018)

23. Jégou, P., Ndiaye, S.N., Terrioux, C.: Computing and exploiting tree-decompositions for solving constraint networks. In: van Beek, P. (ed.) CP 2005. LNCS, vol. 3709, pp. 777–781. Springer, Heidelberg (2005). https://doi.org/10.1007/11564751_63

24. Kask, K., Dechter, R., Larrosa, J., Cozman, F.: Bucket-tree elimination for automated reasoning. Technical report, Donald Bren School of Information and Computer Sciences University of California at Irvine (2001). https://www.ics.uci.edu/~dechter/publications/
25. Koriche, F., Lagniez, J.M., Marquis, P., Thomas, S.: Knowledge compilation for model counting: affine decision trees. In: Rossi, F., Thrun, S. (eds.) Proceedings of the 23rd International Joint Conference on Artificial Intelligence (IJCAI 2013). The AAAI Press, Beijing, August 2013
26. Lagniez, J., Lonca, E., Marquis, P.: Improving model counting by leveraging definability. In: Kambhampati, S. (ed.) Proceedings of 25th International Joint Conference on Artificial Intelligence (IJCAI 2016), pp. 751–757. The AAAI Press, New York, July 2016
27. Lagniez, J., Marquis, P.: Preprocessing for propositional model counting. In: Brodley, C.E., Stone, P. (eds.) Proceedings of the 28th AAAI Conference on Artificial Intelligence (AAAI 2014), pp. 2688–2694. The AAAI Press, Québec City (2014)
28. Lagniez, J.M., Marquis, P.: An improved decision-DDNF compiler. In: Sierra, C. (ed.) Proceedings of the 26th International Joint Conference on Artificial Intelligence (IJCAI 2017), pp. 667–673. The AAAI Press, Melbourne (2017)
29. Lagniez, J.M., Marquis, P., Szczepanski, N.: DMC: a distributed model counter. In: Proceedings of the Twenty-Seventh International Joint Conference on Artificial Intelligence, IJCAI 2018, pp. 1331–1338. The AAAI Press, August 2018
30. Larrosa, J.: Node and arc consistency in weighted CSP. In: Dechter, R., Kearns, M., Sutton, R.S. (eds.) Proceedings of the 18th AAAI Conference on Artificial Intelligence (AAAI 2002), pp. 48–53. The AAAI Press, Edmonton, July 2002
31. Muise, C., McIlraith, S.A., Beck, J.C., Hsu, E.I.: DSHARP: fast d-DNNF compilation with sharpSAT. In: Kosseim, L., Inkpen, D. (eds.) AI 2012. LNCS (LNAI), vol. 7310, pp. 356–361. Springer, Heidelberg (2012). https://doi.org/10.1007/978-3-642-30353-1_36
32. Munshi, A., Gaster, B., Mattson, T.G., Fung, J., Ginsburg, D.: OpenCL Programming Guide, 1st edn. Addison-Wesley, Boston (2011)
33. Oztok, U., Darwiche, A.: A top-down compiler for sentential decision diagrams. In: Yang, Q., Wooldridge, M. (eds.) Proceedings of the 24th International Joint Conference on Artificial Intelligence (IJCAI 2015), pp. 3141–3148. The AAAI Press (2015)
34. Raynal, M.: Parallel computing vs. distributed computing: a great confusion? (position paper). In: Hunold, S., et al. (eds.) Euro-Par 2015. LNCS, vol. 9523, pp. 41–53. Springer, Cham (2015). https://doi.org/10.1007/978-3-319-27308-2_4
35. Roth, D.: On the hardness of approximate reasoning. Artif. Intell. **82**(1–2), 273–302 (1996)
36. Samer, M., Szeider, S.: Algorithms for propositional model counting. J. Discrete Algorithms **8**(1), 50–64 (2010)
37. Sang, T., Bacchus, F., Beame, P., Kautz, H., Pitassi, T.: Combining component caching and clause learning for effective model counting. In: Hoos, H.H., Mitchell, D.G. (eds.) Online Proceedings of the 7th International Conference on Theory and Applications of Satisfiability Testing (SAT 2004), Vancouver, BC, Canada (2004)
38. Sang, T., Beame, P., Kautz, H.: Performing Bayesian inference by weighted model counting. In: Veloso, M.M., Kambhampati, S. (eds.) Proceedings of the 29th National Conference on Artificial Intelligence (AAAI 2005). The AAAI Press (2005)
39. Shapiro, L.G., Haralick, R.M.: Structural descriptions and inexact matching. IEEE Trans. Pattern Anal. Mach. Intell. **PAMI-3**(5), 504–519 (1981)

40. Thurley, M.: sharpSAT – counting models with advanced component caching and implicit BCP. In: Biere, A., Gomes, C.P. (eds.) SAT 2006. LNCS, vol. 4121, pp. 424–429. Springer, Heidelberg (2006). https://doi.org/10.1007/11814948_38
41. Toda, T., Soh, T.: Implementing efficient all solutions SAT solvers. ACM J. Exp. Algorithmics **21**, 44 (2015). 1.12, Special Issue SEA 2014, Regular Papers and Special Issue ALENEX 2013
42. Xue, Y., Choi, A., Darwiche, A.: Basing decisions on sentences in decision diagrams. In: Hoffmann, J., Selman, B. (eds.) Proceedings of the 26th AAAI Conference on Artificial Intelligence (AAAI 2012). The AAAI Press, Toronto (2012)

Reducing Bias in Preference Aggregation for Multiagent Soft Constraint Problems

Alexander Schiendorfer[(✉)] and Wolfgang Reif

Institute for Software & Systems Engineering, University of Augsburg,
Augsburg, Germany
{schiendorfer,reif}@isse.de

Abstract. Most distributed constraint optimization problems assume
the overall objective function to be the "utilitarian social welfare", i.e.,
a sum of several utility functions, belonging to different agents. This
also holds for the most popular soft constraint formalisms, cost function networks and weighted constraints. While, in theory, this model is
sound, it is susceptible to manipulation and resulting bias in practice.
Even without malevolent intentions, bias can result from the way orderings over solutions are transformed into numerical values or normalized.
Alternatively, preferences can be aggregated directly using the tools of
social choice theory to discourage manipulations and practically reduce
unwanted bias. Several common voting functions can be implemented
on top of constraint modeling languages through incremental search and
suitable improvement predicates. We demonstrate that our approach, in
particular Condorcet voting, can undo bias which is shown on two real-
life-inspired case studies using the soft constraint extension MiniBrass
on top of MiniZinc.

Keywords: Soft constraints · Distributed constraint optimization ·
Social choice · Modeling languages · MiniZinc

1 Motivation

Many real-life problems such as coordinating a fleet of mobile sensors [31] or
scheduling devices in smart grids and homes [13] have recently been reduced
to distributed constraint optimization problems (DCOP) that involve multiple
agents. Similarly, more mundane tasks such as assigning seminar topics or agreeing on a shared meal plan are problems that (logically) involve several agents,
even if solved centrally. By far the most popular way of aggregating agents'
preferences (including the cited examples) is to assume them to be specified as
(numerical) cost/utility functions that are summed up [12]. This is equivalent to
weighted constraint satisfaction problems (WCSP), the most common class of
soft constraint problems [18]. While this may be tolerable if other measures constrain the utilities, this commonly accepted notion of "social welfare" is prone
to unfairness and bias in practice, especially if the utilities are unconstrained

© Springer Nature Switzerland AG 2019
T. Schiex and S. de Givry (Eds.): CP 2019, LNCS 11802, pp. 510–526, 2019.
https://doi.org/10.1007/978-3-030-30048-7_30

$X = \{f, s\}, \qquad D_f = D_s = \{\text{curry}, \text{chili}, \text{stew}\}$
$\text{sols}(\text{CSP}) = \{(f \mapsto \text{stew}, s \mapsto \text{stew}), (f \mapsto \text{stew}, s \mapsto \text{curry}), (f \mapsto \text{curry}, s \mapsto \text{chili})\}$

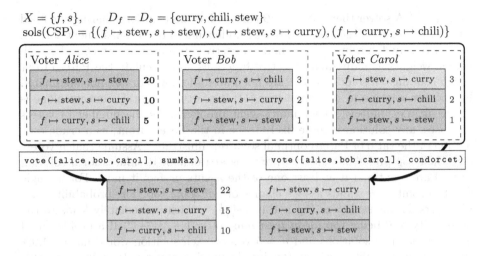

Fig. 1. Social welfare functions can reduce weight-induced bias. Agents pick meals for Friday f and Saturday s, with three solutions in sols(CSP) due to hard constraints. *Alice* submits manipulated weights which puts $(f \mapsto \text{stew}, s \mapsto \text{stew})$ to the top whereas the other agents like it least. Condorcet voting mitigates that by ranking $(f \mapsto \text{stew}, s \mapsto \text{curry})$ first since it wins both pairwise majority contests against the other solutions.

and only known at runtime (agents can just outbid each other). The left part (sumMax) of Fig. 1 illustrates this rather naïvely by allowing an agent *Alice* to vote with higher weights and manipulate the group decision in her favor.

There are, broadly speaking, three approaches to the problem:

(i) We normalize a single objective function to be less biased.
(ii) We address the problem as a multi-objective optimization problem looking for the Pareto front.
(iii) We devise *ordinal* operators that operate on the preference relations.

Regarding (i), of course, we would try to take care of such blunt manipulations as those in Fig. 1 in real-life problems. There are still more subtle ways how biased weights can emerge in preference specifications: Assume, e.g., that students rank seminar topics. Unless we make every student rank *every* topic, we are forced to introduce bias: consider student A stating six preferences whereas student B only specifies three. How should we relate a violation of A's top choice to one of B's top preference? The only fact we can safely deduce is which outcomes A and B prefer in isolation. Naïvely modeled as WCSPs, A's top choice is weighted six whereas B's top choice gets a weight of three. Summing them up clearly favors solutions that please A. Alternatively, we could allocate a fixed budget of q to every agent that proportionally distributes q. For instance, A could split 21 points as $\langle 6, 5, 4, 3, 2, 1 \rangle$ whereas B could split the same 21 points as $\langle 8, 7, 6 \rangle$.[1] A solver then caters to B since A has more options sharing the fixed budget. Either way, the model is biased, independent of the subsequent solvers.

Conversely, following (ii), for multi-objective optimization, we consider each voters' weights as individual objective functions and calculate the Pareto front which contains all solutions that are not Pareto-dominated, i.e., not dominated by another solution in *all* dimensions [11]. Unfortunately, this concept alone is too weak (i.e., indecisive) for multi-agent problems. As the number of voters increases, any solution is more likely to be in the Pareto front, as a crude estimation of the ratio of Pareto-optimal solutions illustrates: Assume that n voters pick their top-preference out of m options at random. It suffices for a solution to be Pareto-optimal if at least one of the agents prefers it most. For a single agent, a solution θ stands a $\frac{m-1}{m}$ chance of *not being* top. The probability that *all* voters do not rank θ as top is thus $(\frac{m-1}{m})^n$ which immediately leads to the probability of θ being top for *at least one* voter: $1 - (\frac{m-1}{m})^n$. If a problem had $m = 100$ possible solutions and $n = 5$ voters, each solution would have a 4.9% chance of being Pareto-optimal. Raising the number of voters to 20 increases this chance to 18.2%, and with 40 voters, every solution already has a 33.1% chance of ending up Pareto-optimal. When facing such a large number of Pareto-optimal solutions, our problem is to be more selective within the Pareto-front – we would still insist on *at least* choosing a solution within the Pareto front since otherwise *all* agents agree that another one is better for them.

In terms of purely ordinal operators (alternative iii), Pareto and lexicographic combinations are the canonical combinations of preference relations [1] that do not need numeric utilities. Yet, we already discussed shortcomings of Pareto orderings and the lexicographic combination is a too strict form of preference aggregation. This is where *social choice theory* [3] comes into play. Rooted in electoral systems[2], it discusses how to amalgamate a group's preference relations. On the right side of Fig. 1, we see that voting based on only ordinal information (here, Condorcet's method that prefers an option to another if a majority favors it) can lead to fairer decisions. Voting over *solutions* to a constraint satisfaction problem corresponds to traversing the search space effectively (e.g., by constraint propagation and search heuristics). Therefore, we implement our approach with modeling languages that compile to a variety of algorithmically efficient solvers.

Our contribution in this paper is thus **to make voting methods such as Condorcet's amenable to soft constraint optimization on the modeling language level.** We extend the open source soft constraint modeling language MiniBrass [24] built on top of MiniZinc [20]. In contrast to other approaches (see Sect. 2.3) this allows agents to vote on the *solution level* instead of the individual *variable level*. Our key insight is that some voting methods can be conveniently mapped to branch-and-bound search.

---

[1] The same logic obviously applies to the normalized case of $q = 1$ where, e.g., B would get $[1/2, 1/3, 1/6]$ and A's top choice only gets a weight of 0.28.

[2] For instance, the Schulze method [27] is a Condorcet-based method used for elections in open source organizations such as Ubuntu, Debian, or the Wikimedia Foundation.

2   Preliminaries

Our approach combines (soft) constraint programming and social choice theory. In essence, soft constraint programming orders solutions using an overall valuation (not necessarily numerical) in a (partially) ordered set that results from individual valuations using a combination operation in an algebraic structure. Social choice theory deals with aggregating preference relations over outcomes to a single relation. Hence, these two ideas naturally complement each other.

2.1   Constraint Optimization and Soft Constraints

As usual, a *constraint (satisfaction) problem* $\text{CSP} = (X, D, C)$ is described by a set of (decision) variables X, their associated family of domains $D = (D_x)_{x \in X}$ of possible values, and a set of (hard) constraints C that restrict valid assignments. For a CSP, an assignment θ over scope X is a function from X to D such that each variable x maps to a value in D_x. The set of all assignments is written as $[X \to D]$. A (hard) constraint $c \in C$ is a function $c : [X \to D] \to \mathbb{B}$ where $\theta \models c$ expresses that θ satisfies c. For solving by inference, i.e., reducing valid domain items by logical implications (so-called *constraint propagation*), *global constraints* offer dedicated filtering algorithms [4]. Consequently, an assignment θ is a solution if $\theta \models c$ holds for all $c \in C$. We write $\theta \in \text{sols(CSP)}$.

We move from satisfaction to *constraint optimization problems* (COP) by adding an objective function $f : [X \to D] \to P$ where $(P, \leq_P)$ is a partial order, i.e., $\leq_P$ is a reflexive, antisymmetric, and transitive relation over P. Elements of P are called *satisfaction degrees*. Without loss of generality, we interpret $m <_P n$ as satisfaction degree m being strictly *worse* than n and restrict our attention to *maximization problems* regarding P. Consequently, θ_1 is worse than θ_2 if $f(\theta_1) <_P f(\theta_2)$ which results in a partial quasi-ordering over solutions since multiple solutions may map to the same satisfaction degree and anti-symmetry does not hold. A solution θ is *optimal* with respect to a COP if for all solutions θ' it holds either that $f(\theta') \leq_P f(\theta)$ or $f(\theta') \parallel_P f(\theta)$, expressing incomparability with respect to $\leq_P$.

Soft constraint problems are specialized COPs where each soft constraint s_i maps $[X \to D]$ to an algebraic structure $(M, \cdot_M, \varepsilon_M, \leq_M)$, i.e., a partially-ordered, commutative monoid called a *partial valuation structure* (PVS) [15] which subsumes several specific soft constraint formalisms such as WCSP, cost function networks, or fuzzy constraints [18]. The combination operator $\cdot_M$ aggregates all soft constraints' satisfaction degrees, ε_M denotes maximal satisfaction and is neutral with respect to the $\cdot_M$ operator. In terms of COPs, the overall objective $f : [X \to D] \to (M, \leq_M)$ is defined by $f(\theta) = \Pi_M\{s_i(\theta) \mid s_i \in S\}$ for a set of soft constraints S, also written as $S(\theta)$. We use PVS for soft constraint problems since they are more general than c-semirings [5] or total valuation structures [25]. Therefore, they are used as basic building block in MiniBrass [24]. Since in this paper we only care about aggregating several agents' overall satisfaction degrees, we do not rely on the precise properties of the algebraic structure that are described in [15,24]. Given two PVS M and N, the most natural combination is the Cartesian product $M \times N$ that orders elements according to a Pareto ordering:

$$(m,n) \leq_{M \times N} (m',n') \Leftrightarrow m \leq_M m' \wedge n \leq_N n'$$

The Pareto-ordering leads to a "fair" but not decisive aggregation of several PVS, as we discussed in the introductory section.

By contrast, the ordering of the lexicographic combination is defined as

$$(m,n) \leq_{M \ltimes N} (m',n') \Leftrightarrow (m <_M m') \vee (m = m' \wedge n \leq_N n')$$

It allows us to express strictly hierarchical relationships between PVS to distinguish, e.g., organizational from individual goals.

In terms of software implementations, most existing constraint solvers offer an API to model constraint problems in imperative code. For higher layers of abstraction, there have been several proposals for domain-specific languages, including MiniZinc [20] or Essence [14]. Due to its popularity (see, e.g., the annual MiniZinc challenge [29]) and language features, we favor the former in this paper. MiniZinc is a high-level constraint modeling language understood by many constraint, MIP, or SAT solvers:

```
array[1..3] of var 1..3: x;              % decision variables
constraint forall (i in 1..2) (x[i] <= x[i+1]);   % constraints
solve satisfy; % minimize sum(x) / maximize sum(x)  % objectives
```

In its default version, MiniZinc allows for a rich variety of global constraints but only limited capacity for optimization objectives. Only totally ordered numeric objectives are supported. To increase generality, incremental search is needed that facilitates adding and retracting new constraints during traversal of the search tree. MiniSearch [22] enables such customizable search on the solution level and MiniZinc itself has extensions for large neighborhood search where some variables are fixed to stay unchanged (using added constraints), thereby defining a large neighborhood [10]. MiniBrass [24] "softens" MiniZinc to incorporate PVS-based soft constraints. Agents specify their preferences as PVS and aggregate them using lexicographic or Pareto combinations, as Fig. 2 shows.

2.2 Social Choice Theory

Formally, the field of social choice theory is concerned with aggregating preference relations [3]. For a (usually finite and small) set of outcomes (or candidates) O, we call $\mathcal{Q}_O$, $\mathcal{P}_O$, and $\mathcal{T}_O$ the sets of quasi, partial, and total orders over O, respectively. Most often in social choice, quasi-orders (total, transitive but allowing for ties at the cost of anti-symmetry) are used whereas SCSPs lead to partial quasi-orders over solutions. We denote a set of agents (or voters) as $N = \{1, \ldots, n\}$. Then, a *preference profile* $[\preceq] = (\preceq_i)_{i \in N} \in \mathcal{P}_O^n$ (or $\mathcal{T}_O^n$, etc.) is a tuple containing a preference relation $\preceq_i$ for every agent $i \in N$ where, again, $o \preceq_i o'$ indicates that outcome o is *worse* than o'. Voting methods then map

```
PVS: alice = new                          PVS: bob = new
  WeightedCsp("alice") {                     ConstraintPreferences("bob") {
  scons c1:'f = stew /\ s = stew' :: w('20'); scons c1: 'f = curry /\ s = chili';
  scons c2:'f = stew /\ s = curry':: w('10'); scons c2: 'f = stew /\ s = curry';
  scons c3:'f = curry /\ s = chili':: w('5'); scons c3: 'f = stew /\ s = stew';
};                                          crEdges : '[| mbr.c3, mbr.c2 |
% PVS: carol = new FuzzyCsp("carol") {                  mbr.c2, mbr.c1 |]';
                                          };
```

```
solve alice lex (bob pareto carol);
```

Fig. 2. Example problem in MiniBrass, slightly adapted from Fig. 1. Constraint preferences require ordinal information only (`mbr.c3, mbr.c2` denotes that c_3 is less important than c_2). Other PVS types (fuzzy) could be used as well.

a preference profile either to a single winning outcome or to a full preference relation over O. Both tasks strive to represent the agents' joint wishes. A *social welfare function* W maps a preference profile $[\preceq]$ of n agents to a preference relation, formally written as $W : \mathcal{P}_O^n \to \mathcal{P}_O$. By contrast, a *social choice function* $C : \mathcal{P}_O^n \to O$ only returns a winner [28]. The problems are strongly related since we convert a social welfare function to a social choice function by picking a top option, or conversely, repeatedly call a social choice function to obtain a full ordering as a social welfare function.

To list a few examples, the *majority voting* rule builds a welfare ordering by ranking all outcomes according to the number of top occurrences they achieve. *Borda voting* asks every agent to assign a score from 0 (least desirable) to $|O| - 1$ (most desirable) and adds up the scores of all agents. *Condorcet voting* fixes an ordering over O, say $[o_1, \ldots, o_m]$, and performs pairwise competitions to determine the welfare ordering. That is, agents vote for o_1 or o_2 and the winner (according to a majority) challenges o_3, and so on. A *Condorcet winner* is an outcome that wins all pairwise competitions. There can however be cycles such that no proper ordering emerges (e.g., transitivity is violated) [3]. Approval voting, on the other hand, lets agents only partition the set of outcomes into "approved" and "disapproved" and ranks outcomes by their number of approvals.

Besides these examples of voting methods, social choice theory offers several general impossibility results based on axiomatic characterizations of welfare functions, most notably Arrow's famous theorem [2]: It states that for at least three outcomes and two voters, no welfare function can simultaneously be Pareto-efficient (PE, all agents preferring o over o' must imply $o \succ_W o'$), independent of irrelevant alternatives (IIA, the relative ordering of o and o' does not change when agents change their preferences with respect to other outcomes), and non-dictatorial (ND, no single agent gets to determine the welfare ordering).

2.3 Related Work

There have already been efforts to combine soft constraints and voting. Most notably, the first algorithm to solve a problem specified with n c-semirings (precisely, fuzzy constraints) is *sequential voting* [7–9]: Agents vote sequentially over each variable's assigned domain value using social choice functions, according to a pre-defined ordering. The authors investigate voting-theoretical properties of their method. Specifically, the relationship between voting axioms assumed for *local* voting rules (i.e., voting over a single variable's assignment) to the global *solution* level was investigated. For instance, it is a necessary but not sufficient condition that all local rules be IIA in order for the global rule to be IIA as well. Also, if a single local rule is non-dictatorial, the global rule is non-dictatorial. Although their approach is mathematically appealing, in practice it suffers from the fine granularity that agents have to vote on: domain items for a single variable. This leads to myopic and overly optimistic estimations: Assuming, for example, that the combinations "(fish, white wine)" and "(meat, red wine)" are acceptable for an agent A with a slight preference for fish. Then A would place his or her bet on "white wine" although A might end up with the least desirable option "(meat, white wine)" since decisions cannot be retracted. Moreover, sequential voting may choose Pareto inefficient solutions deterministically even if the local rules are Pareto efficient (an example is provided in [23, Chap. 9]). Therefore, we propose to vote over solutions instead of individual variables' values.

Conversely, DCOP research is most prominently concerned with distributed settings and algorithms that operate across computational nodes, such as, e.g., the ADOPT algorithm [30]. Netzer and Meisels extended the classical sum-of-costs DCOP model (that, again, is equivalent to WCSP in soft constraints) to "distributed social constraint optimization problems" where social welfare functions replace summation [21]. Still, their approach calculates a single score for each assignment based on the agents' individual (numeric) valuations instead of a preference relation and can thus suffer from the bias problems shown in our introduction. Moreover, they assume some form of commensurability of utilities in the sense that an operator such as "maximize the unhappiest agent's value" is meaningful – if agents operate on distinct (esp. non-numerical) ordering relations, this is not obvious. In this paper, we abstract from the underlying distribution of the computational nodes or the specific distributed optimization algorithms and focus on adequate, unbiased *models* of "how to aggregate multiple agents' preferences" – as the first step towards more distributed solutions. Our experimental evaluations are thus conducted in a centralized setting.

Outcomes resulting from strategic interactions among several self-interested agents is central to game theory. Morgenstern and von Neumann introduced the foundations of numeric utility functions for ordinal preference relations in [19]. However, it is hard to consider bias resulting from such utility functions when multiple independent agents are involved. Mechanism design adds the strategic component of truth-telling to social choice situations [28]. We do not yet address such questions other than disincentivizing manipulation with weights.

The problem of bias reduction, in particular, has not yet been addressed with voting methods. Moreover, the existing proposals come with specialized implementations and are not readily available to end-users. Since our approach employs state-of-the-art constraint modeling languages, we expect it to inherit their benefits in terms of efficient solvers, user-friendliness, and flexibility. It is the first proposal in terms of applicable software and systems to offer access to social choice functions and discrete optimization.

3 Implementation

Our goal is to combine soft constraints and voting on the solution level. Formally, assume a CSP $= (X, D, C)$ and a set of agents $N = \{1, \dots, n\}$ along with sets of soft constraints $(S_j)_{j \in N}$ mapping to PVS $(M_j)_{j \in N}$. Then the set of outcomes O corresponds to sols$([X \to D])$ and the preference profile $[\preceq] = (\preceq_j)_{j \in N}$ results from applying soft constraints: $\theta_1 \preceq_j \theta_2 \Leftrightarrow S_j(\theta_1) \leq_{M_j} S_j(\theta_2)$. Ideally, we can obtain a full social welfare ordering $W([\preceq])$, but we also settle for choice functions that return a single solution $C([\preceq]) \in [X \to D]$ as the group's favorite.

To implement this form of optimization for modeling languages, we first revisit how incremental search proceeds in a branch-and-bound fashion that systematically explores the full search space (we discuss extensions to local search later). For instance, in MiniSearch [22], we could write this as follows:

```
function ann: maximize_bab(var int: obj) =
  repeat(
      if next() then commit() /\
        post(obj > sol(obj))
      else break endif );
```

Any time the solver returns a solution, we can formulate new constraints to be propagated based on the current solution's values, e.g., to bound the objective. For instance, upon finding a solution with objective value obj = 17, we add a constraint obj > 17 to the constraint problem to find the next solution. If the resulting problem becomes unsatisfiable, we have found an optimal solution. More generally, that logic extends to arbitrary constraints for improvement which we refer to as getBetter predicates. MiniBrass [24] already generates these (hidden from the end-user) for atomic or complex objectives, including lexicographic and Pareto combinations. For instance, assuming two PVS M and N, solving for their Pareto combination leads to the following predicate:

```
predicate getBetterPareto(var M: overall_M, var N: overall_N) =
  post( % both agents' PVS find the current solution worse or equal
    is_worse_or_equal_M(sol(overall_M), overall_M ) ) /\
    is_worse_or_equal_N(sol(overall_N), overall_N ) ) /\
    sol(overall_M) != overall_M \/ sol(overall_N) != overall_N ) );
```

Our goal is to align voting methods with this optimization principle by generating getBetter constraints for them. Indeed, some voting methods are better suited for this task than others. For example, Borda voting would need to enumerate all, say, k solutions, rank the best solution with $k-1$, the next best with $k-2$ and so forth, for every agent. A priori, it is hard to guess the relative position any solution in the search space has as well as the size of $k = |\text{sols(CSP)}|$. By contrast, a variation of Condorcet's method is more "local" in the search space as it only requires pairwise comparisons, which we can exploit for optimization.

3.1   Condorcet Voting

Recall from Sect. 2.2 that canonical Condorcet voting proceeds by fixing an ordering of the outcomes O, comparing two adjacent options with respect to the pairwise majority, and keeping the winner. With the set of outcomes being the search space, ordering them in advance is, of course, infeasible. Nevertheless, we can tweak the method slightly for constraint-based optimization in MiniBrass:

1. Find the next solution (call it θ).
2. Impose a constraint that enforces that the *next* solution θ' must be preferred by a *majority* of voters and search for the next solution.
3. Repeat until no such solution can be found (or a cycle is detected).

In MiniZinc-style pseudocode, this predicate is generated as follows (M_i refers to the specific PVS element type of PVS i):

```
predicate getBetterCondorcet(array[1..N] of var M_i : overall) =
  % M_1 represents the PVS of agent i
  post(
    % # agents that find the current solution worse than the next
    sum(i in 1..N) (bool2int(is_worse_i(sol(overall[i]), overall[i] ) ) )
    >
    % # agents that find the next solution worse than the current one
    sum(i in 1..N) (bool2int(is_worse_i(overall[i], sol(overall[i]) ) ) )    );
```

In fact, this is weakening the Pareto condition that *all* agents have to accept or prefer a new solution. If a solution is indeed a Condorcet-winner, this method will find it. If there is a Condorcet-cycle, there is no guarantee with respect to the outcome since the moment of termination depends on the ordering of the solutions as returned from the solver. However, such a cycle is easily detected by inspecting the trace of solutions. The only guarantee we can give upon termination is that there is no *unseen* solution that a majority of agents would prefer to the current one – which arguably still makes for a reasonable social choice function. The complexity of generating the above predicate is, analogously to Pareto and lexicographic combinations, hidden from MiniBrass end-users. For example, they would rewrite the `solve` expression in Fig. 2 as follows:

```
solve vote([alice, bob, carol], condorcet);
```

On a technical side-note, this approach requires the `getBetter` predicates to be *reifiable*, i.e., allow additional boolean variables to take their truth values.

3.2   Approval and Majority-Tops Voting

Arrow's theorem provides a hint for another suitable voting method to consider. Instead of allowing agents to order solutions arbitrarily, we can have them partition the search space into "acceptable" and "unacceptable" solutions, and search for a solution that is approved by the highest number of agents. Hence, we implement "approval voting" which turns out to have beneficial voting-theoretical properties: Approval voting is a non-dictatorial social welfare function that satisfies PE and IIA – due to the restriction to only two options, approved or disapproved;

Lemma 1. *With only approval or disapproval at hand, approval voting satisfies PE, IIA, and is non-dictatorial [28, p. 267].*

Arrow's theorem applies for votings with $|O| \geq 3$ and does therefore not apply in this restricted setting. Despite the restricted generality of having only two levels of satisfaction, a variety of use cases fall into that category (e.g., students proposing a personalized set of acceptable exam dates). Moreover, generating an optimization predicate (or even a numeric objective in this case) is straightforward. For a given assignment θ, we count the number of approving agents. Once we see a solution, we must impose a constraint that more agents approve of the next one. Since for approval voting, we know that the type of every agent's PVS type must be boolean, the `getBetter` predicate is simplified:

```
predicate getBetterApproval(array[1..N] of var bool: overallAgents) =
  post(
    sum(i in 1..N) (bool2int(overallAgents[i])) >
      sum(i in 1..N) (bool2int(sol(overallAgents[i])))  );
```

This results in a social welfare ordering over solutions. MiniBrass end-users would again only write

```
solve vote([alice, bob, carol], approval);
```

where MiniBrass would ensure that each submitted PVS for approval voting is indeed boolean-valued.

For more general orderings, approval voting is insufficient except if valuations are, e.g., thresholded. The most canonical threshold value imaginable is to approve a solution only if it evaluates to the top value ε_M in the corresponding PVS (e.g., 0 in WCSP, 1.0 for fuzzy CSP, or $\emptyset$ for a violation-set-based formalism). Solutions are then ranked according to the number of top-values. Since this is an adaption of the majority rule that asks for an outcome to count the number of agents that place it on top of their ranking, we call this variant *majority-tops*. We would then just count the number of agents that get all their wishes satisfied. In MiniBrass, we write, e.g.:

```
solve vote([alice, bob, carol], majorityTops);
```

To sum up, our proposed voting methods in MiniBrass encompass `condorcet`, `majorityTops`, `approval`, and (for numeric objectives) also `sumMin` and `sumMax`. The latter two sum up the overall valuations analogously to conventional WCSP or DCOP formulations.

3.3 Voting with Local Search

We want to conclude our proposed implementation with a word of caution. Larger problem instances can be prohibitive for systematic and complete search space traversal and require heuristic local search approaches such as large-neighborhood search. Arrow's theorem then proves practically relevant. More specifically, a social welfare function W violating IIA can lead to unexpected results: If n voters choose over a subset of the available solutions $\Theta \subseteq [X \rightarrow D]$, e.g., a neighborhood $\mathcal{N}(\theta) \subseteq [X \rightarrow D]$ of a solution θ, the local ordering obtained by applying W to the profile over Θ can be different from the ordering over Θ when the whole search space $[X \rightarrow D]$ is present. In practice, this means that even though the agents agree to switch from θ_1 to $\theta_2 \in \mathcal{N}(\theta_1)$ when voting over the neighborhood's options, they would rank θ_1 better than θ_2 if *all* solutions were up for election. Hence, an IIA welfare function such as approval voting is a more sensible choice for local search than Condorcet voting.

4   Experimental Evaluation

For our evaluation, we investigate two problems that are inspired by real-life decision situations faced at a typical university research group.

The first one, **lunch selection**, serves as our initial proof of concept and consists of deciding a shared meal plan during a research retreat. We assume a given set of prospective dishes and that a fixed-cardinality subset of those has to be selected for a week. Each voter has preferences concerning the presence or absence of certain dishes in the final selection.

The second one, **mentor matching**, is more complex than lunch selection and involves assigning students to industry mentoring partners where, for simplicity, we only consider students' preferences regarding companies. There are cardinality constraints on how many students each company can supervise.

In the real-life counterparts to the experimental setting proposed in this paper, agents were only asked to provide a partial order in MiniBrass and can leave out options. Here, we force them to submit a total order in both cases for better comparability. Some of the agents are allowed to cheat, i.e., to *amplify* their weights to gain an unfair advantage in DCOP/WCSP-style (summation-based) optimization. This *emulates* other less-obvious ways that introduce weight bias. Our goal is *to test if Condorcet voting can undo this artificial amplification*. Still, for the example problems, we could easily apply numerical normalization to undo the amplification effects. In more general settings involving non-numerical orderings, this would no longer be an option.

As a result of this forced total ordering over desirable outcomes, both instances give rise to an interpretable unit of satisfaction over all agents. This allows for measuring how well the proposed voting methods mitigate unfair preference specifications that we expect to influence the WCSP approach heavily. To quantify this in a controlled fashion, we also calculate a ground-truth baseline distribution of satisfaction values emerging from truthful, non-amplified weights from amplifying agents. We investigate several preference aggregation strategies (ignoring approval voting and majority-tops due to the ranked setting):

- **WCSP Unbiased:** Classical weight-based summed optimization with amplification deactivated
- **WCSP Biased:** Classical weight-based summed optimization with amplification activated
- **Condorcet:** asking for pairwise majority improvements
- **Pareto:** searching for Pareto-improvements

We hypothesize that the satisfaction distribution resulting from unbiased WCSP is similar to that of Condorcet voting whereas biased WCSP would strongly prefer the amplifier group. For both problems, we create 200 random instances (i.e., synthesized preferences). Since we observed qualitatively the same behavior for various parameter settings (e.g., the size of the restricted set of dishes or the number of students and companies) in both problems, we present the results for a fixed setting. For every instance, we pick a random subset of voters to become amplifiers, according to varying ratios. The allowed amplification factor is proportional to the number of agents, analogously to Fig. 1. All four aggregation methods are applied to the same 200 instances, to ensure comparability.

We solve lunch selection using Google OR-Tools 7.0 (CP solver) [16] and mentor matching using Chuffed [6], since we found these to be performing best for the respective problems. Each presented experiment runs on a machine having 4 Intel Xeon CPU 3.20 GHz cores and 14.7 GB RAM on 64 bit Ubuntu 16.04.[3]

4.1   Lunch Selection

First, we consider the results obtained from the lunch selection experiment. Given the set of available dishes F, upon deciding the lunches in $L \subseteq F$, we can determine the satisfaction values per agents as follows: We assign one unit of satisfaction for every meal that agent i likes that is in L and, equivalently, one unit for every disliked meal which is not in L.[4] Agents specify their wishes as a ranking over F (not only approval sets), with the last positions corresponding to disliked dishes.

Figure 3 presents the average satisfaction degrees per group (amplifiers/non-amplifiers) obtained for this problem, once with 50% and once with 25% amplifiers. Figure 3(a) shows the results of equally splitting the agents into the groups of amplifiers and non-amplifiers. We can see that (as expected) the unbiased WCSP leads to an equal distribution of satisfaction degrees in both groups, whereas amplification clearly and unfairly treats amplifiers significantly better (plotted in Fig. 3(c), and also revealed by a student t-test at $\alpha = 10^{-3}$). However, that effect can almost entirely be reversed by using Condorcet voting instead that ends up having insignificantly varying satisfaction distributions (cf. Fig. 3(d)). This confirms the intuition presented in Fig. 1.

Besides, searching for Pareto improvements leads to significantly worse satisfaction degrees than Condorcet for both groups. This confirms our intuition that Pareto combinations alone are too indecisive, i.e., the society cannot make many improvement steps in the search tree since *all* agents need to agree on a better solution as opposed to Condorcet's method that only requires a *majority*

---

[3] The raw result data and experimental code can be found online at https://github.com/isse-augsburg/minibrass/tree/master/evaluation/minibrass-voting-experiment.

[4] In the experiments, $n = 12$ agents vote over a set L with $|L| = 4$ chosen from $|F| = 7$ available objects. This leaves us with only $\binom{7}{4} = 35$ solutions. Nevertheless, the bias reduction effects are already apparent in this small example.

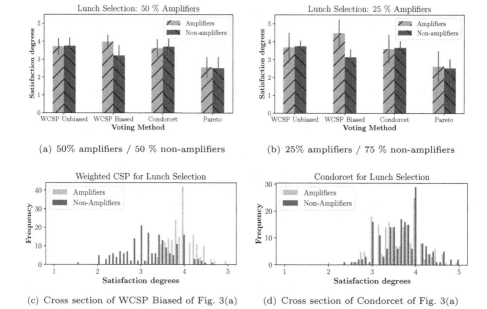

(a) 50% amplifiers / 50 % non-amplifiers   (b) 25% amplifiers / 75 % non-amplifiers

(c) Cross section of WCSP Biased of Fig. 3(a)   (d) Cross section of Condorcet of Fig. 3(a)

Fig. 3. Lunch selection: Average satisfaction values per group over 200 instances. In the unbiased case, the amplification factor was deactivated.

of them to do so. For a smaller set of amplifiers (25% instead of 50%), Fig. 3(b) presents qualitatively similar results. The amplifiers gain even more benefit in biased WCSP since there are fewer other amplifiers to split satisfaction degrees.

4.2   Mentor Matching

In mentor matching, we assign students to their mentoring companies. This gives us a natural measure of satisfaction, i.e., the rank of the assigned company, with $r_i = j$ denoting that agent i gets their j^{th} preference and $r_i = 1$ meaning top satisfaction. The minimal and maximal number of students a company can supervise are constrained which makes supervision a scarce resource.[5] The students' preferences are not arbitrary, but some companies had a higher probability of appearing higher than others – which corresponds to our real-life experiences.

Figure 4 shows overall results similar to the lunch selection case (we converted the students' achieved ranks to satisfaction degrees to have axes consistent with the previous example by subtracting ranks from a constant value). Condorcet voting mitigates the bias introduced by amplified weights, although the resulting average values are slightly lower than in the unbiased case. Interestingly,

---

[5] Here, $n = 18$ agents vote over assignments to six companies, with each company supervising at least two and at most three students. There are at most $6^{18} \approx 1.015^{14}$ solutions to explore, not accounting for the cardinality constraints.

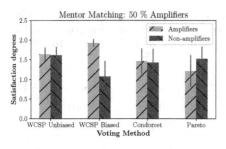
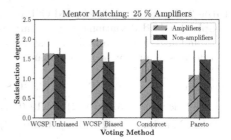

(a) 50% amplifiers / 50 % non-amplifiers   (b) 25% amplifiers / 75 % non-amplifiers

Fig. 4. Mentor matching: Average satisfaction values per group over 200 instances. In the unbiased case, the amplification factor was deactivated.

Table 1. An illustrative example for skewed satisfaction resulting from Condorcet voting. Average value and sample standard deviation are denoted by $\bar{r}$ and σ_r.

Voting method	r_1	r_2	r_3	r_4	r_5	r_6	r_7	r_8	r_9	r_{10}	r_{11}	r_{12}	r_{13}	r_{14}	r_{15}	r_{16}	r_{17}	r_{18}	$\bar{r}$	σ_r
WCSP (Unbiased)	2	2	1	1	1	1	1	1	2	2	1	3	1	1	1	2	1	1	1.39	0.59
Condorcet	1	2	1	1	1	4	1	1	1	2	1	1	1	5	1	2	1	1	1.56	1.12

the average disadvantage of non-amplifiers is stronger in the 50% case depicted in Fig. 4(a) than in the 25% case shown in Fig. 4(b) whereas the average satisfaction of the amplifiers remains close to optimal. Pareto voting (unexpectedly) discriminates *against* amplifiers. This interesting result is due to the solver's default search strategy that favors smaller domains (i.e., the non-amplifiers' cost variables). Since Pareto search does not dive deeply into the search tree, we end up close to these biased solutions.[6]

We also note that, compared to lunch selection, the standard deviations of Condorcet voting are higher. Closer investigation of that issue reveals a weakness apparent in Condorcet voting that we exemplified with a case (ID 26 in the online results) in Table 1: While unbiased WCSP results in a rather fair allocation that never pairs an agent with a company worse than their third preference, Condorcet's method offers a solution that results in the fourth or even fifth preference for two students. Both assignments have similar average satisfaction degrees over all students, but we might consider Condorcet's result less fair. It is, however, a logical consequence of the focus on majority improvements: If all but two agents agree that (here) a mentoring assignment is better, it gets picked – even if this means substantial deterioration in satisfaction for the two agents. It does not make a difference *how strong* the dissatisfaction is.

To confirm this suspicion, in Table 2, we also show the distribution of sample standard deviations of satisfaction degrees per assignment as a rough measure of

---

[6] We confirmed that a different solver (Gecode [26]) is more balanced but still keep these unexpected results in the paper to highlight issues with Pareto optimization.

Table 2. Comparison of weighted CSP optimization (unbiased) and Condorcet voting. Sample standard deviations (S-STD) are metrics to measure each instance's "unfairness" and aggregated over all 200 instances. Analogously, SAT corresponds to satisfaction degrees as presented in Figs. 3 and 4.

Problem	WCSP SAT	WCSP S-STD	Condorcet SAT	Condorcet S-STD
Lunch Selection	3.72 (0.22)	1.19 (0.24)	3.64 (0.28)	1.19 (0.23)
Mentor Matching	1.62 (0.12)	**0.6** (0.14)	1.45 (0.18)	**1.02** (0.28)

unfairness (σ_r in Table 1). While this effect does not show up in lunch selection (due to the smaller search space), we can observe for mentor matching, regardless of how many amplifiers were present (therefore we only present the 50% amplifiers results). The average sample standard deviation of 0.6 for unbiased WCSP optimization is significantly lower than the value of 1.02 that Condorcet reaches. A student t-test at $\alpha = 10^{-3}$ confirms this for both amplifier ratios.[7]

5   Conclusion and Future Work

We presented an extension to conventional soft constraint optimization problems that allows to aggregate preferences using voting methods instead of either the usual numeric operations such as summation or Pareto/lexicographic combinations. This extension was able to "correct" the bias introduced by amplified weight specifications on two real-life-inspired problems to almost the level of unbiased specifications using Condorcet voting. Our evaluation revealed, however, that pure Condorcet voting is susceptible to producing unbalanced assignments at the expense of few agents due to its focus on the majority.

Using only ordinal information gives us no "metric" sense of different levels of dissatisfaction among a group of agents – we cannot relate discomfort a from agent A to discomfort b of another agent B. Therefore, a potential direction for research is to focus on methods that can produce more balanced assignments. It might still be necessary, for that matter, to introduce a numeric scale and allowing some form of transferable utility/budget or conversion rates. We hope to provide more variety in this type-constrained setting.

Additionally, we expect better fairness guarantees in repeated optimization settings (e.g., rostering problems over many weeks) as well as an "iterative deepening" approach that first votes over coarser classes of problems (using diverse solution search [17]) and subsequently refines those choices. Finally, we intend to implement our model formulation also in actual DCOP solvers that mostly need to be able to understand MiniZinc models. On a somewhat similar note, the scalability of the voting approach remains to be tested for larger models, perhaps

---

[7] Using the student t-test for significance is justified by observing that the sample standard deviations of satisfaction degrees follow a normal distribution according to a Shapiro-Wilk test at $\alpha = 10^{-3}$.

involving large neighborhood search. We expect many real-life problems close to end-users to benefit strongly from fairer voting methods than simple summation.

References

1. Andréka, H., Ryan, M., Schobbens, P.Y.: Operators and laws for combining preference relations. J. Log. Comput. **12**(1), 13–53 (2002)
2. Arrow, K.J.: Social Choice and Individual Values. Yale University Press, London (1951)
3. Arrow, K.J., Sen, A., Suzumura, K.: Handbook of Social Choice and Welfare, vol. 2. Elsevier, Amsterdam (2010)
4. Beldiceanu, N., Carlsson, M., Demassey, S., Petit, T.: Global constraint catalogue: past present and future. Constraints **12**(1), 21–62 (2007)
5. Bistarelli, S., Montanari, U., Rossi, F.: Semiring-based constraint satisfaction and optimization. J. ACM **44**(2), 201–236 (1997)
6. Chu, G.: Improving combinatorial optimization. Ph.d. thesis, University of Melbourne (2011)
7. Cornelio, C., Pini, M.S., Rossi, F., Venable, K.B.: Multi-agent soft constraint aggregation via sequential voting: theoretical and experimental results. Auton. Agent. Multi-Agent Syst. **33**(1–2), 159–191 (2019)
8. Dalla Pozza, G., Pini, M.S., Rossi, F., Venable, K.B.: Multi-agent soft constraint aggregation via sequential voting. In: Proceedings of the 22nd International Joint Conference on Artificial Intelligence (IJCAI 2011), pp. 172–177 (2011)
9. Dalla Pozza, G., Rossi, F., Venable, K.B.: Multi-agent soft constraint aggregation: a sequential approach. In: Proceedings of the 3rd International Conference on Agents and Artificial Intelligence (ICAART 2011), vol. 11 (2010)
10. Dekker, J.J., de la Banda, M.G., Schutt, A., Stuckey, P.J., Tack, G.: Solver-independent large neighbourhood search. In: Hooker, J. (ed.) CP 2018. LNCS, vol. 11008, pp. 81–98. Springer, Cham (2018). https://doi.org/10.1007/978-3-319-98334-9_6
11. Ehrgott, M.: Multicriteria Optimization, vol. 491. Springer Science & Business Media, New York (2005)
12. Fioretto, F., Pontelli, E., Yeoh, W.: Distributed constraint optimization problems and applications: a survey. J. Artif. Intell. Res. **61**, 623–698 (2018)
13. Fioretto, F., Yeoh, W., Pontelli, E., Ma, Y., Ranade, S.J.: A Distributed Constraint Optimization (DCOP) approach to the economic dispatch with demand response. In: Proceedings of the 16th International Conference on Autonomous Agents and Multiagent Systems (AAMAS 2017), pp. 999–1007. International Foundation for Autonomous Agents and Multiagent Systems (2017)
14. Frisch, A.M., Harvey, W., Jefferson, C., Martínez-Hernández, B., Miguel, I.: Essence: a constraint language for specifying combinatorial problems. Constraints **13**(3), 268–306 (2008)
15. Gadducci, F., Hölzl, M., Monreale, G.V., Wirsing, M.: Soft constraints for lexicographic orders. In: Castro, F., Gelbukh, A., González, M. (eds.) MICAI 2013. LNCS (LNAI), vol. 8265, pp. 68–79. Springer, Heidelberg (2013). https://doi.org/10.1007/978-3-642-45114-0_6
16. Google Optimization Tools (2017). https://developers.google.com/optimization. Accessed 29 June 2017

17. Hebrard, E., Hnich, B., O'Sullivan, B., Walsh, T.: Finding diverse and similar solutions in constraint programming. In: Proceedings of the 23rd National Conference Artificial Intelligence (AAAI 2005), vol. 5, pp. 372–377 (2005)
18. Meseguer, P., Rossi, F., Schiex, T.: Soft constraints. In: Rossi, F., van Beek, P., Walsh, T. (eds.) Handbook of Constraint Programming, chap. 9. Elsevier, Amsterdam (2006)
19. Morgenstern, O., von Neumann, J.: Theory of games and economic behavior (1944)
20. Nethercote, N., Stuckey, P.J., Becket, R., Brand, S., Duck, G.J., Tack, G.: MiniZinc: towards a standard CP modelling language. In: Bessière, C. (ed.) CP 2007. LNCS, vol. 4741, pp. 529–543. Springer, Heidelberg (2007). https://doi.org/10.1007/978-3-540-74970-7_38
21. Netzer, A., Meisels, A.: SOCIAL DCOP - social choice in distributed constraints optimization. In: Brazier, F.M.T., Nieuwenhuis, K., Pavlin, G., Warnier, M., Badica, C. (eds.) Intelligent Distributed Computing V. SCI, vol. 382, pp. 35–47. Springer, Heidelberg (2011). https://doi.org/10.1007/978-3-642-24013-3_5
22. Rendl, A., Guns, T., Stuckey, P.J., Tack, G.: MiniSearch: a solver-independent meta-search language for MiniZinc. In: Pesant, G. (ed.) CP 2015. LNCS, vol. 9255, pp. 376–392. Springer, Cham (2015). https://doi.org/10.1007/978-3-319-23219-5_27
23. Schiendorfer, A.: Soft Constraints in MiniBrass: Foundations and Applications. Dissertation, Universität Augsburg (2019)
24. Schiendorfer, A., Knapp, A., Anders, G., Reif, W.: MiniBrass: soft constraints for MiniZinc. Constraints 23, 403–450 (2018)
25. Schiex, T., Fargier, H., Verfaillie, G.: Valued constraint satisfaction problems: hard and easy problems. In: Proceedings of the 14th International Joint Conference on Artificial Intelligence (IJCAI 1995), vol. 1, pp. 631–639. Morgan Kaufmann (1995)
26. Schulte, C., Lagerkvist, M.Z., Tack, G.: Gecode: generic constraint development environment. In: INFORMS Annual Meeting (2006)
27. Schulze, M.: A new monotonic, clone-independent, reversal symmetric, and condorcet-consistent single-winner election method. Soc. Choice Welfare 36(2), 267–303 (2011)
28. Shoham, Y., Leyton-Brown, K.: Multiagent Systems: Algorithmic, Game-theoretic, and Logical Foundations. Cambridge University Press, Cambridge (2008)
29. Stuckey, P.J., Feydy, T., Schutt, A., Tack, G., Fischer, J.: The MiniZinc challenge 2008–2013. AI Mag. 35(2), 55–60 (2014)
30. Yeoh, W., Felner, A., Koenig, S.: BnB-ADOPT: an asynchronous branch-and-bound DCOP algorithm. J. Artif. Intell. Res. 38, 85–133 (2010)
31. Zivan, R., Yedidsion, H., Okamoto, S., Glinton, R., Sycara, K.: Distributed constraint optimization for teams of mobile sensing agents. Auton. Agent. Multi-Agent Syst. 29(3), 495–536 (2015)

Testing and Verification Track

A Cube Distribution Approach to QBF Solving and Certificate Minimization

Li-Cheng Chen[1] and Jie-Hong R. Jiang[1,2(✉)]

[1] Graduate Institute of Electronics Engineering, National Taiwan University,
Taipei 10617, Taiwan
hank12322@gmail.com
[2] Department of Electrical Engineering, National Taiwan University,
Taipei 10617, Taiwan
jhjiang@ntu.edu.tw

Abstract. Quantified Boolean Formulas (QBFs) are powerful expressions to naturally and concisely encode many decision problems in computer science, such as robotic planning, hardware/software synthesis and verification, among others. Their effective solving and certificate (in terms of model and countermodel) generation play crucial roles to enable practical applications. In this work, we give a new view on QBF solving and certificate generation by the *cube distribution* interpretation. It provides a largely increased flexibility for QBF reasoning and allows compact certificate derivation with don't cares. Through this interpretation, we develop a QBF solver based on the prior clause selection framework. Experimental results demonstrate the superiority of our solver in both solving performance and certificate size compared to other state-of-the-art solvers with certificate generation ability.

Keywords: Quantified Boolean formula · Certificate · Cube distribution

1 Introduction

Quantified Boolean formulas (QBFs) are powerful logic expressions to compactly encode various decision problems in, e.g., robotic planning [20], ontology reasoning [13], formal verification [6], design debugging [23], circuit synthesis [12], program synthesis [22], engineering change order [5], and so on. The universal and existential quantifiers of QBF provide succinct descriptive power, but raise the reasoning complexity to PSPACE complete. The broad application and computation challenge of QBF attract much research attention in recent years.

QBF solving involves two important tasks. One is to determine the truth or falsity of a QBF; the other is to compute the corresponding model or countermodel certificate. However, the importance of the latter task is often overlooked. In fact, the model and countermodel certificates are key to many applications. For example, they may correspond to a solution plan to a planning task, a circuit or program fragment under synthesis, a rectification solution to an erroneous

© Springer Nature Switzerland AG 2019
T. Schiex and S. de Givry (Eds.): CP 2019, LNCS 11802, pp. 529–546, 2019.
https://doi.org/10.1007/978-3-030-30048-7_31

design, etc. Not only the derivation, but also the quality of certificate is of significant practical relevance. In this work, we focus on QBF solving with compact certificate derivation.

Among the vast efforts on QBF research, there are a few state-of-the-art solvers capable of generating certificates, especially in the form of Skolem and Herbrand functions. The representatives include: DepQBF [14], a solver based on search with clause learning, whose Q-resolution proofs can be converted to Skolem/Herbrand certificates [1]; CAQE [18,24], a solver based on clausal abstraction; CADET [17,19], a solver based on incremental determinization, which however is restricted to QBFs with two quantification levels; QuAbS [9], a non-CNF solver based on counterexample guided abstraction and refinement. Apart form their certificate generation abilities, the issue of deriving compact certificates remains a challenge, especially for QBFs with more than two quantification levels.

We note that while preprocessing is an important technique for effective QBF solving, there are QBF preprocessors supporting partial certificate generation [10,25]. Although the partial certificates may be combined with the certificates generated by certifying QBF solvers on preprocessed formulas to provide full certificates, they are currently limited to Skolem certificates for true QBFs only.

In this work, we present a *cube distribution* interpretation of QBF solving. Essentially this new view provides flexibility for both QBF reasoning and certificate construction. Based on the cube distribution principle, we develop a solver for general QBFs with multiple quantification levels under the clause selection framework [11]. Experiments show the superiority of our method in both solving performance and certificate size compared to other state-of-the-art certificate generating solvers dealing with general QBFs.

2   Preliminaries

For the brevity of a Boolean expression, the Boolean connective of conjunction $\wedge$ is sometimes omitted; the disjunction $\vee$ is also denoted by the symbol "$+$"; the negation $\neg$ is also denoted by an overline. The appearance of a Boolean variable x in a Boolean expression can be in the form of a *positive literal* x or a *negative literal* $\neg x$. We denote the variable corresponding to a literal l by $var(l)$. A *clause* is a disjunction of a set of literals; a *cube* is a conjunction of a set of literals. We alternatively specify a clause or cube by a set of literals. A Boolean formula is in the *conjunctive normal form* (CNF) if it is expressed as a conjunction of a set of clauses. We alternatively specify a CNF formula by a set of clauses.

A Boolean function f over variables X, with $|X| = n$, corresponds to a mapping $f : \mathbb{B}^n \to \mathbb{B} \cup \{\perp\}$, where "$\perp$" denotes the *don't care* value. Let $[\![X]\!]$ denote the set of valuations/assignments on variables X. The *onset*, denoted f^+, of function f is the set of input assignments $\subseteq [\![X]\!]$ that make f valuate to TRUE; its *offset*, denoted f^-, is the set of input assignments that make f valuate to FALSE; the *don't care set*, denoted $f^\perp$, is the set of input assignments that make f valuate to $\perp$. If $f^\perp$ is empty, f is *completely specified*; otherwise, f is *incompletely specified*.

A *quantified Boolean formula* (QBF) Φ over variables $X = X_1 \cup X_2 \cup \ldots \cup X_n$, where $X_i \neq \emptyset$ and $X_i \cap X_j = \emptyset$ for $i \neq j$ in the *prenex conjunctive normal form* (PCNF) can be expressed as

$$Q_1 X_1, \ldots, Q_n X_n . \varphi, \tag{1}$$

where $Q_1 X_1 \cdots Q_n X_n$, called the *prefix* and denoted Φ_{pfx}, satisfies $Q_i \in \{\exists, \forall\}$ and $Q_i \neq Q_{i+1}$, and φ, called the *matrix* and denoted Φ_{mtx}, is a quantifier-free CNF formula in terms of variables X. A variable $x \in X_i$ is said to be at *quantification level* i. The set X of variables of Φ can be partitioned into *existential variables* $X_\exists = \{x \in X_i \mid Q_i = \exists\}$ and *universal variables* $X_\forall = \{x \in X_i \mid Q_i = \forall\}$. A literal l is called an *existential literal* and a *universal literal* if $var(l)$ is in $X_\exists$ and $X_\forall$, respectively.

Note that for QBF Φ in the PCNF form, if $Q_n = \forall$, then Φ can be equivalently simplified with all the appearances of variables X_n in Φ being removed. Also, in PCNF form, the *forall-reduction* rule, which removes universal literals from a clause $C_i \in \varphi$ if their quantification level is the largest in C_i, can be applied to simplify Φ. To simplify our discussion, we assume $X_n = \exists$ and φ has been simplified by the forall-reduction rule.

The QBF Φ is TRUE if and only if there exists a set of *Skolem functions* [21], one f_x for each existential variable $x \in X_i$ referring only to the universal variables $y \in X_j$ with $j < i$, such that substituting x with f_x in Φ_{mtx} makes Φ_{mtx} a tautology. By duality [1], the QBF Φ is FALSE if and only if there exists a set of *Herbrand functions*, one f_x for each universal variable $x \in X_i$ referring only to the existential variables $y \in X_j$ with $j < i$, such that substituting x with f_x in Φ_{mtx} makes Φ_{mtx} unsatisfiable. That is, the Skolem functions and Herbrand functions serve as the model and countermodel of Φ, respectively. In this work, we are concerned with Skolem and Herbrand function derivation as the certificate of the truth or falsity of a QBF.

2.1 Clause Selection

Clause selection [11], or *clausal abstraction* [18], is a QBF solving technique to track the clause satisfaction status and to facilitate learning with abstract variables. Given a QBF Φ of Eq. (1), the subclause of a clause $C_i \in \varphi$ consisting of literals $l \in C_i$ with $var(l) \in X_j$ with $j \bowtie k$ for some k is denoted as $C_i^{\bowtie k}$ for $\bowtie \in \{=, <, \leq, >, \geq\}$. In particular, we abbreviate $C_i^{=k}$ as C_i^k in the sequel for simplicity. For clause $C_i \in \varphi$, we can decompose it into n subclauses C_i^1, C_i^2, $\ldots$, C_i^n, and write $C_i = C_i^1 + C_i^2 + \cdots + C_i^n$.

Let $c_{i,k}^j$ denote the kth literal of clause C_i^j. If the literal $c_{i,k}^j$ of clause C_i^j is TRUE under some assignment, we say C_i is *satisfied* or *deselected* by $c_{i,k}^j$. Since $C_i = C_i^1 + \cdots + C_i^n$, any $c_{i,k}^j$ in C_i^j valuated to TRUE under some variable assignment makes C_i be deselected. As the variables of a QBF are assigned in the order of quantification level, under some assignment once C_i is deselected at quantification level j, then C_i remains deselected at quantification levels greater than j regardless of the valuations of literals in $C_i^{>j}$. By introducing a fresh variable, called *selector*, s_i^j and letting $s_i^j \equiv \neg C_i^{\leq j}$, the selector tracks whether C_i has been deselected before or at level j.

3   Motivation and Intuition

To motivate our approach to QBF solving, consider the formula

$$\Gamma_1 = \forall x_1, x_2 \exists y_1, y_2, y_3.$$
$$(x_1 + y_1)(\overline{y_1} + \overline{y_2})(x_2 + y_2 + y_3)(\overline{x_1} + \overline{x_2} + \overline{y_3})(x_2 + y_1 + \overline{y_3})(\overline{x_2} + y_1 + y_3).$$

Let Γ_k be the formula of duplicating Γ_1 for k copies with the variables being renamed such that there are no common variables between different copies. It can be verified that Γ_k is satisfiable.

However, even with preprocessing[1], some of the state-of-the-art solvers, such as Qesto, CAQE, and DepQBF, may have difficulty solving Γ_k. Moreover, the generated certificate circuits of Skolem functions can be rather large.

In contrast, our method, as to be presented, requires one single query to a satisfiability (SAT) solver to solve Γ_k and generates simple certificate circuits. The way our method works in solving Γ_k can be explained by considering only Γ_1 as illustrated by Fig. 1, where each circle node represents the cube $B_i = \neg C_i^1$, which is the condition of falsifying the universal literals in clause C_i, each square node represents either a positive or negative literal of an existential variable, and distributor $d_{i,j}$, which connects nodes B_i and $c_{i,j}$ (the jth existential literal of clause C_i), signifies existential literal $c_{i,j}$ is in clause C_i.

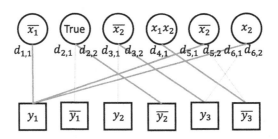

Fig. 1. QBF solving as cube distribution.

Essentially, the QBF solving of Γ_1 can be interpreted as a process of *cube distribution*. That is, for each cube B_i, at least one, $d_{i,j}$ say, of its distributors

---

[1] In fact, bloqqer [3] cannot directly solve the formula with all options being enabled, except for three options: covered clause elimination, variable elimination, and universal expansion. Disabling these three options is reasonable in that (1) covered clause elimination often considerably increases the size of the Skolem functions as discussed in [8], (2) variable elimination in principle can solve all QBFs, and (3) universal expansion in principle reduces all QBFs to SAT.

must be chosen such that the cube B_i is assigned/distributed to the onset (resp. offset) of the Skolem function of existential variable $y_m = var(c_{i,j})$ for literal $y_m \in C_i$ (resp. literal $\neg y_m \in C_i$). Moreover, in general, for two cubes B_i and $B_{i'}$ with $B_i \wedge B_{i'} \neq$ FALSE, then the distribution of them must be constrained such that $d_{i,j}$ and $d_{i',j'}$ cannot be chosen simultaneously if the two edges connect to complementary literals, i.e., one to y_m and the other to $\neg y_m$. In the case of Γ_1, if we choose $d_{1,1}$, $d_{2,2}$, $d_{3,2}$, $d_{4,1}$, $d_{5,1}$, and $d_{6,1}$ (as indicated by the thick edges in Fig. 1), all the six cubes $B_1 = \neg x_1$, $B_2 =$ TRUE, $B_3 = \neg x_2$, $B_4 = x_1 x_2$, $B_5 = \neg x_2$, and $B_6 = x_2$ can be successfully distributed. Therefore Γ_1 is satisfiable.

The above cube distribution encoding can be translated into a CNF formula for SAT solving. The formula is merely in terms of the distribution variables $d_{i,j}$ such that $d_{i,j} =$ TRUE if and only if B_i is distributed to the onset/offset (depending on the appearance of positive/negative literal of variable y_m in C_i) of Skolem function of y_m. For $d_{i,j} =$ TRUE, we refer to it by saying "*cube B_i is distributed to variable y_m*" for simplicity in the sequel. Note that if all the cubes can be distributed, then the QBF is TRUE. However the converse does not hold because only non-disjoint cubes need to be distributed simultaneously. Essentially a satisfying assignment to the CNF formula provides sufficient information for Skolem function construction as we can accordingly distribute cubes to the onset f_y^+ and offset f_y^- of the Skolem function f_y of each existential variable y. If $f_y^+ \vee f_y^-$ is not a tautology, then the don't care set $f_y^\perp$ of the Skolem function f_y is non-empty and can be used to minimize the circuit of f_y. In the case of solving Γ_1, we have $f_{y_1}^+ = \neg x_1 + \neg x_2 + x_2 =$ TRUE, $f_{y_1}^- =$ FALSE, $f_{y_1}^\perp =$ FALSE, $f_{y_2}^+ =$ FALSE, $f_{y_2}^- =$ TRUE, $f_{y_2}^\perp =$ FALSE, $f_{y_3}^+ = \neg x_2$, $f_{y_3}^- = x_1 x_2$, and $f_{y_3}^\perp = \neg x_1 x_2$, where the sets are represented with characteristic functions, e.g., the characteristic function $\neg x_1$ represents the set of assignments $(x_1, x_2) = \{($FALSE, FALSE$), ($FALSE, TRUE$)\}$. Exploiting the don't cares for certificate minimization, we get Skolem functions $f_{y_1} =$ TRUE, $f_{y_2} =$ FALSE, $f_{y_3} = \neg x_2$.

We remark that cube distribution break (to some extend) the conventional 2QBF solving loop that first assigns universal variables and then checks for matching assignments to the existential variables. It exhibits distinct capability of simultaneously matching multiple (possibly conflicting) universal assignments to existential variables.

Extending clause selection to a new principle, we formally define *cube distribution* as follows.

Definition 1 (Cube Distribution). *Given a QBF $\Phi = Q_1 X_1, \ldots, Q_n X_n.\varphi$ and an assignment A to variables $\bigcup_i X_i$, the cube $B^j = \neg C^{<j}$ of a clause $C \in \varphi$ at level j is said* distributed *to literal l (or variable $var(l)$) under A if both B^j and l are satisfied under A.*

4   Cube Distribution Interpretation

Consider the QBF $\Phi = \forall X_1, \exists Y_1, \ldots, \forall X_k, \exists Y_k.\varphi$. Assume, except for X_1, the variable sets are non-empty. Let C_i^j be the subclause of $C_i \in \varphi$ with literals

$\{l \in C_i \mid var(l) \in X_j \cup Y_j\}$; $C_i^{\geq j}$ be $\{l \in C_i \mid var(l) \in \bigcup_{m=j}^{k}(X_m \cup Y_m)\}$; $C_i^{j\forall}$ be $\{l \in C_i \mid var(l) \in X_j\}$; $C_i^{j\exists}$ be $\{l \in C_i \mid var(l) \in Y_j\}$. Let cube B_i^j be $\neg C_i^{j\forall}$. Note that if $X_1 = \emptyset$, then $B_i^1 = \text{TRUE}$. Observe that if cube B_i^1 is distributed into the onset (if literal $y \in C_i^{1\exists}$) or offset (if literal $\neg y \in C_i^{1\exists}$) of the Skolem function f_y of variable $y \in Y_1$, then when variable y being substituted with Skolem function f_y, the clause C_i^1 will always be satisfied, i.e., a tautology, regardless of the valuations of variables X_1. Given a distribution D_A on a set of cubes $\{B_i^1 \mid C_i \in A\}$ with respect to a set of clauses $A \subseteq \varphi$, we define the *induced QBF* by distribution D_A on Φ as $\forall X_2, \exists Y_2, \ldots, \forall X_k, \exists Y_k.\varphi|_{D_A}$, with $\varphi|_{D_A} = \bigwedge_{C_i \in \varphi} C_i|_{D_A}$, where $C_i|_{D_A} = \text{TRUE}$ if $C_i \in A$ and $C_i|_{D_A} = C_i^{\geq 2}$ if $C_i \notin A$.

With the above observation and definition, the cube distribution interpretation of QBF solving can be formally stated in the following theorem.

Theorem 1. *The QBF* $\Phi = \forall X_1, \exists Y_1, \ldots, \forall X_k, \exists Y_k.\varphi$ *is* TRUE *if, and only if, for every maximal set of clauses* $\varphi' \subseteq \varphi$ *with non-disjoint cubes, i.e.,* $(\bigwedge_{C_i \in \varphi'} B_i^1) \neq$ FALSE, *there exists a set of clauses* $A \subseteq \varphi'$ *and a distribution* D_A *on cubes* $\{B_i^1 \mid C_i \in A\}$ *such that the induced QBF* $\Phi' = \forall X_2, \exists Y_2, \ldots, \forall X_k, \exists Y_k.\varphi'|_{D_A}$ *is* TRUE.

Proof. ($\Leftarrow$) If a maximal set $\mathcal{B} = \{B_i^1 \mid C_i \in \varphi'\}$ of cubes with a common minterm m can be distributed at the outermost quantification level, the cube distribution defines legitimate Skolem functions of existential variables Y_1 over the minterms common to all cubes in $\mathcal{B}$. Note that for a minterm, the maximal cube set covering it is unique. In this case, $A = \varphi'$ and Φ' must be TRUE. Hence Φ is TRUE under the assignment m over X_1 with the constructed Skolem functions.

On the other hand, assume some of the cubes of $\mathcal{B}$ cannot be distributed, i.e., $A \subset \varphi'$, and the induced QBF Φ' is TRUE. The existence of a model of Skolem functions for Φ' guarantees that for each $C_i \in \varphi' \backslash A$ the cube B_i^j can be distributed at some level $j \geq 2$. Moreover, we let the distribution of cubes $\{B_i^1 \mid C_i \in A\}$ define Skolem functions of existential variables Y_1 over the minterms m common to all cubes in $\mathcal{B}$. Combining the defined Skolem functions, because every cubes in $\mathcal{B}$ is either distributed at level 1 or another larger level, Φ is TRUE under the assignment m over X_1 with the constructed Skolem functions. Because QBF Φ is TRUE under every minterm assignment of X_1, QBF Φ is TRUE.

($\Rightarrow$) For QBF Φ to be TRUE, there exists a legitimate set of Skolem functions. By the valuations of the Skolem functions under each minterm m of the universal variables X_1, we let the maximal set of clauses φ' for cube distribution be the set of clauses selected under m. Then A corresponds to the clauses in φ' deselected by the Y_1 assignment determined by the Skolem function valuation under m. In addition, the Skolem functions of Φ under the assignment m on X_1 form a model to the induced QBF Φ'. Hence Φ' is TRUE. Therefore the theorem holds.

Note that, by recursion, Theorem 1 provides a cube distribution procedure for solving QBFs of arbitrary number of quantification levels.

5 QBF Solving with Cube Distribution

Instead of resorting to recursive reasoning as suggested by Theorem 1, we seek for a non-recursive procedure. Based on the cube distribution principle, we develop a QBF solver on top of the clause selection framework of Qesto [11], whose main procedure is sketched in Fig. 2. In the following, we will emphasize the differences while referring the reader to [11] for background details. In the pseudo code of Figs. 2, 3, 4 and 5, the highlighted lines are those that differ from Qesto.

In the main procedure $QbfSolve$ of Fig. 2, we assume the QBF Φ, the model (satisfying assignment) M^j returned by the SAT solver, the winning condition W^j, Skolem/Herbrand functions f_x^+ and f_x^- are global in scope, and can be accessed by other function calls. In line 01, the winning condition W^j, $j = 1, \ldots, n$, and the onset f_x^+ and offset f_x^- of Skolem/Herbrand functions of $x \in X$ are initialized to FALSE. In line 02, the selector constraint formula α^j, which maintains the clause selection status of quantification level j, is initialized by $InitConstraint$. In line 03, the backtrack level index btlev and model M^0 are initialized. The subsequent while-loop iterates until the truth or falsity of Φ is determined.

If the iteration is at level $j = n+1$, the learned clause C_R in line 12 indicates a loss condition of the universal player. If the iteration is at level $j \neq n + 1$, line 06 collects the set A of selector literals of the previous level SAT model. In line 07, α^j is solved with A being imposed as the unit assumption to the SAT solver. If α^j is satisfiable under A, the procedure proceeds to the next quantification level. Otherwise, the final conflict clause C_R returned by the SAT solver is analyzed in $Analyze_\exists$ in line 14 if $Q_j = \exists$ or in $Analyze_\forall$ in line 19 if $Q_j = \forall$ to obtain the corresponding backtrack level and blocking clause. If btlev $= -1$, the falsity and Herbrand functions are returned in line 17, or the truth and Skolem functions are returned in line 22. Otherwise, the selector constraint is strengthened in line 23 and new quantification level is updated in line 24.

Besides the certificate initialization step in line 01, and warp-up steps in lines 16 and 21, $QbfSolve$ differs from Qesto mainly in the procedures $InitConstraint$, $Analyze_\exists$ and $Analyze_\forall$ as we detail below.

Procedure $InitConstraint$ of Fig. 3 differs from Qesto in the construction of α^j for $Q_j = \exists$ in lines 04 and 05. In the formula of α^j, the distributor variable $d_{i,k}^j$ is a fresh new variable introduced for every existential literal $c_{i,k}^j$ in φ. Variable $d_{i,k}^j$ valuates to TRUE if and only if the cube $\neg C_i^{<j}$ is to be distributed to (the Skolem function of) variable $var(c_{i,k}^j)$. Note that in our notation, two literals $c_{i,k}^j$ and $c_{i',k'}^j$ may refer to the same literal. However, their corresponding distributor variables $d_{i,k}^j$ and $d_{i',k'}^j$ are distinct. This distinctness makes QBF reasoning more flexible and provides freedom for certificate construction.

In Qesto, α^j for $Q_j = \exists$ is constructed as

$$\bigwedge_{C_i \in \varphi} (\neg s_i^j \equiv (\neg s_i^{j-1} \vee C_i^j)) \tag{2}$$

QbfSolve
input: QBF $\Phi = Q_1X_1, Q_2X_2, \ldots, Q_nX_n.\varphi$
output: the truth or falsity of Φ (and certificate)
begin
```
01    InitCert();
02    for 1 ≤ j ≤ n do α^j ← InitConstraint(j);
03    j ← 1; btlev ← −1; M^0 ← {s_i^0 | C_i ∈ φ};
04    while TRUE
05        if j ≠ n + 1
06            A ← {l ∈ M^{j−1} | var(l) = s_i^{j−1}, C_i ∈ φ};
07            (satResult, M^j, C_R) ← SatSolve(α^j, A);
08            if satResult = SAT
09                j ← j + 1;
10                continue;
11        else
12            C_R ← {s_i^n | C_i ∈ φ};
13        if Q_j = ∃
14            (btlev, C_B) ← Analyze_∃(j, C_R);
15            if btlev = −1
16                F_H ← EndCert(∃);
17                return (FALSE, F_H);
18        else
19            (btlev, C_B) ← Analye_∀(j, C_R);
20            if btlev = −1
21                F_S ← EndCert(∀);
22                return (TRUE, F_S);
23        α^{btlev} ← α^{btlev} ∧ C_B;
24        j ← btlev;
```
end

Fig. 2. Procedure: Solve QBF.

in line 04 of Fig. 3, and without the line 05 step. Essentially, it is the two lines 04 and 05 of code in Fig. 3 that fulfill the cube distribution principle in a non-recursive way. To see the connection and difference between Qesto and our method, Eq. (2) can be rewritten as

$$\bigwedge_{C_i \in \varphi} (\neg s_i^j \equiv (\neg s_i^{j-1} \vee \bigvee_{c_{i,k}^j \in C_i^j} (c_{i,k}^j \wedge s_i^{j-1}))). \qquad (3)$$

By the definition of the selector variables, it is logically equivalent to

$$\bigwedge_{C_i \in \varphi} (\neg s_i^j \equiv (\neg s_i^{j-1} \vee \bigvee_{c_{i,k}^j \in C_i^j} (c_{i,k}^j \wedge \neg C_i^{<j}))). \qquad (4)$$

By comparing Eq. (4) to the right-hand side formula of line 04 of Fig. 3 and the definition of cube distribution, the correctness of our construction can be established.

Procedure $Analyze_\exists$ of Fig. 4 differs from Qesto in lines 02, 04 and 05. In line 02, unlike Qesto, the distributor literals are collected instead of existential literals. In line 05, on-the-fly certificate construction is invoked while there is no certificate construction in Qesto.

Procedure $Analyze_\forall$ of Fig. 5 is similar to Qesto. The difference is that because distributor literals are used to replace the original existential literals, clause deselection at existential levels is controlled by distributor variables. In lines 01 and 04, $\Upsilon_i^{j,k}$ is a predicate denoting whether there is a distributor $d_{i,t}^h$, for some t, and $j \leq h \leq k$, deselecting C_i. In line 05, there is an additional certificate construction step.

6 Certificate Generation

Our on-the-fly certificate construction procedures $AnalyzeCert_\exists$ and $Analyze\text{-}Cert_\forall$ are invoked in $QbfSolve$, and are shown in Figs. 6 and 7, respectively.

InitConstraint
input: level j, QBF Φ
output: selector constraint α^j
begin
01   **if** $Q_j = \forall$
02        $\alpha^j \leftarrow (\bigwedge_{C_i \in \varphi} (\neg s_i^j \equiv (\neg s_i^{j-1} \vee C_i^j))) \wedge \bigvee_{C_i \in \varphi} s_i^j;$
03   **if** $Q_j = \exists$
04        $\alpha^j \leftarrow \bigwedge_{C_i \in \varphi} (\neg s_i^j \equiv (\neg s_i^{j-1} \vee \bigvee_{c_{ik}^j \in C_i^j} d_{i,k}^j));$
05        $\alpha^j \leftarrow \alpha^j \wedge \bigwedge_{\substack{(\neg C_i^{<j} \wedge \neg C_{i'}^{<j}) \neq \text{FALSE}, \\ c_{i,k}^j = \neg c_{i',k'}^j}} (\neg d_{i,k}^j \vee \neg d_{i',k'}^j);$
06   **for** $C_i \in \varphi$
07   **if** j is the largest level in C_i
08                $\alpha^j \leftarrow \alpha^j \wedge \neg s_i^j;$
09   **return** α^j;
end

Fig. 3. Procedure: Initialize selector constraint.

For Herbrand function construction, $AnalyzeCert_\exists$ of Fig. 6 is invoked when a conflict occurs in an existential level in $QbfSolve$. Given a set of conflicting clauses R, backtrack level btlev, and level qlev where the conflict occurs, we known that if all the clauses in R are selected at btlev (the winning condition for the universal player characterized by λ in line 01), the existential player has no way to make the matrix true at level qlev under the current universal assignment (the winning move collected by L in line 02). In lines 08 and 10, the newly explored winning condition, characterized by $\neg W^j \wedge \lambda$, is added to the either the onset and offset of the Herbrand function of a variable x according to the winning strategy collected in L. In line 11, the so-far explored winning condition W^j is updated.

For Skolem function construction, procedure $AnalyzeCert_\forall$ of Fig. 7 is similar to procedure $AnalyzeCert_\exists$ with some minor differences. First, the distribution information is collected in lines 02 and 03. Second, the Skolem functions are updated in lines 09 and 11 by adding $\neg W^j \wedge \lambda \wedge \neg C_i^{<j}$, instead of $\neg W^j \wedge \lambda$. The additional conjunction of $\neg C_i^{<j}$ makes the onset and offset smaller, and increases the don't case set. The increased flexibility can be exploited for certificate minimization.

Note that in $QbfSolve$, procedure $InitCert$ initializes the characteristic functions f_x^+, f_x^-, and W^j to FALSE for all $x \in X$ and $j = 1, \ldots, n$; procedure $EndCert$ collects Skolem functions for the existential variables for a TRUE QBF or Herbrand functions for the universal variables for a FALSE QBF.

7   Experimental Results

The proposed cube distribution algorithm, named Cued, was implemented in the C++ language under the framework of Qesto [11] and used MiniSat 2.2 [7] as the SAT engine. The experiments were conducted on a Linux machine with Intel Xeon E5-2620 v4 2.10 GHz CPU and 125 GB RAM. The benchmarks were taken from QBFEVAL'16[2] (Note that because no Herbrand functions currently can be obtained from preprocessing [8], we thus focus on non-preprocessed formulas. The results on preprocessed formulas are omitted due to space limitation).

Analyze$_\exists$
input: level j, conflict clause C_R
output: btlev, blocking clause
begin
01    $R \leftarrow \{C_i \in \varphi \mid \neg s_i^{j-1} \in C_R\}$;
02    $D \leftarrow \{d_{i,k}^h \mid \neg d_{i,k}^h \in M^h, C_i \in R, 1 \leq h < j\}$;
03   **if** $D = \emptyset$ **then return** $(-1, \text{FALSE})$;
04   btlev $\leftarrow \max\limits_{d_{i,k}^h \in D} h$;
05    $AnalyzeCert_\exists(R, \text{btlev}, j)$;
06   **return** (btlev, $\{\neg s_i^{\text{btlev}} \mid C_i \in R\}$);
end

Fig. 4. Procedure: Analyze $\exists$-Loss.

---

[2] As QBFEVAL'16 contains more benchmark instances than QBFEVAL'17 and QBFEVAL'18, we took QBFEVAL'16 benchmarks for our experimental study.

Analyze$_\forall$
input: level j, conflict clause C_R
output: btlev, blocking clause
begin
01    $R \leftarrow \{C_i \in \varphi \mid \Upsilon_i^{1,j} = \text{FALSE}, s_i^{j-1} \in C_R\}$;
02   **if** $R = \emptyset$ **then return** $(-1, \text{FALSE})$;
03   btlev $\leftarrow \underset{\neg s_i^k \in M^k, C_i \in R, Q^k = \forall, k < j}{\max} k$;
04    $R \leftarrow \{C_i \in \varphi \mid \Upsilon_i^{\text{btlev},j} = \text{FALSE}, s_i^{j-1} \in C_R\}$;
05   *AnalyzeCert$_\forall$*(R, btlev, j);
06   **return** $(\text{btlev}, \{s_i^{\text{btlev}} \mid C_i \in R\})$;
end

Fig. 5. Procedure: Analyze $\forall$-Loss.

We compared our solver Cued with Qesto and two other state-of-the-art certifying solvers for general QBFs in PCNF, including CAQE [18,24] (with PicoSAT 965 [2] as the underlying SAT solver) and DepQBF [15,16]. As the newest versions of CAQE and DepQBF have been enhanced by several techniques that disallow the generation of certificates. We therefore disabled the options in CAQE and DepQBF that prevent certificate generation. Specifically, for CAQE, we disabled strong UNSAT refinement and miniscoping in CAQE v2; for DepQBF, we applied traditional QCDCL option, simple dependency manager, and no generalized axioms in DepQBF v6.03. We denote the so-configured CAQE and DepQBF as CAQE-c and DepQBF-c, respectively. Note that although the performance of CAQE-c and DepQBF-c may not be as good as their counterparts under default settings, turning off the modern options could make it more transparent to compare the baselines.

To evaluate the solver performance, a time limit of 600 s was imposed for solving each instance. Figure 8 compares Cued, Qesto, CAQE-c, DepQBF-c in terms of their performance of solving the 825 PCNF track formulas. In the plot, the x-axis shows the number of solved instances, and the y-axis shows the runtime in seconds with the instances sorted in an ascending order according to their solving time. In measuring the solving time, certificate generation computation was turned off for all the solvers. In total, Cued, Qesto, CAQE-c, DepQBF-c solved 432, 354, 316, and 368 instances, respectively. Evidently, Cued outperformed Qesto, CAQE-c, and DepQBF-c.

To evaluate the certificate quality, we compared the Skolem/Herbrand function circuits computed from Cued, CAQE-c, and DepQBF-c. Note that because Qesto has no certificate generation option, it is not in our main comparison. (Nevertheless, we implemented a certificate generation version of Qesto. The results can be found in Supplements, where the superiority of Cued to Qesto can be seen.) A time limit of 300 s was imposed for solving (with certificate computation being turned on). The certificates obtained by the three solvers were converted to and-inverter graphs (AIGs) and further minimized by

AnalyzeCert$_\exists$
input: conflict clauses R, btlev, qlev
output: void (Herbrand func and W^j modified through side effect)
begin
01    $\lambda \leftarrow \bigwedge\limits_{C_i \in R} \neg C_i^{\leq \text{btlev}}$;
02    $L \leftarrow \{l \mid l \in M^j, var(l) \in X_j, \text{btlev} < j < \text{qlev},$
                $Q_j = \forall, \neg l \in C_i, C_i \in R\}$;
03   **for** btlev $< j <$ qlev, $Q_j = \forall$
04        $L^j \leftarrow \{l \mid var(l) \in X_j, l \in L\}$;
05   **for** $l \in L^j$
06            $x \leftarrow var(l)$;
07   **if** l is positive literal
08                $f_x^+ \leftarrow f_x^+ \vee \neg W^j \wedge \lambda$;
09   **else**
10                $f_x^- \leftarrow f_x^- \vee \neg W^j \wedge \lambda$;
11        $W^j \leftarrow W^j \vee \lambda$;
end

Fig. 6. Procedure: Construct Herbrand Function.

AnalyzeCert$_\forall$
input: conflict clauses R, btlev, qlev
output: void (Skolem func and W^j modified through side effect)
begin
01    $\lambda \leftarrow \bigwedge\limits_{C_i \in R} C_i^{\leq \text{btlev}}$;
02    $D \leftarrow \{d_{i,k}^j \mid \text{literal } d_{i,k}^j \in M^j, \text{btlev} < j < \text{qlev}, Q_j = \exists\}$;
03    $D' \leftarrow \{\text{pick one literal } d_{i,k}^j \in D \text{ (if any) for each } C_i \notin R\}$;
04   **for** btlev$< j <$qlev, $Q_j = \exists$
05        $D^j \leftarrow \{d_{i,k}^j \mid d_{i,k}^j \in D'\}$;
06   **for** $d_{ik}^j \in D^j$
07            $x \leftarrow var(c_{i,k}^j)$;
08   **if** $c_{i,k}^j$ is positive literal
09                $f_x^+ \leftarrow f_x^+ \vee \neg W^j \wedge \lambda \wedge \neg C_i^{<j}$;
10   **else**
11                $f_x^- \leftarrow f_x^- \vee \neg W^j \wedge \lambda \wedge \neg C_i^{<j}$;
12        $W^j \leftarrow W^j \vee \lambda$;
end

Fig. 7. Procedure: Construct Skolem Function.

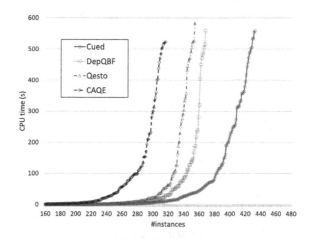

Fig. 8. Solver performance evaluation.

Berkeley ABC logic synthesis tool [4] using commands `dc2` and `dfraig`. A time limit of 300 s was imposed for synthesis. The certificate quality is measured by AIG size, i.e., the number of AND gates.

For the PCNF track formulas, Cued, CAQE-c, and DepQBF-c solved 380, 262, 354 cases, respectively, with certificates being generated. Among the generated certificates, only 339 cases of Cued, 255 of CAQE-c, and 303 of DepQBF-c were successfully synthesized by ABC. The remaining 41 cases of Cued, 7 of CAQE-c, and 51 of DepQBF-c are too large for ABC to handle within the time limit or out of the memory limit of 5 GB; their sizes were counted in terms of the original non-synthesized AIG nodes. On the other hand, for an unsolved instance of a solver, its certificate size is counted as infinity.

Figure 9 shows the results, where CAQE-c and Cued are compared in (a), and DepQBF-c and Cued are compared in (b). For true and false QBF instances, they can be distinguished with the Skolem and Herbrand spots, respectively. Because the x- and y-axes are in a logarithmic scale, size 0 is shifted to 1 for better visualization. Moreover, for the unsolved or memory-exploded instances, their AIG sizes are counted as 100,000,000 (much larger than other actual AIG sizes) in the plot.

When CAQE-c and Cued are compared, it can be observed that Cued tends to produce smaller certificates, especially there are more spots outside of the upper 10x line than the lower 10x line. Among the 251 (113 true/138 false) instances solved by both solvers, there are 101 (50/51) instances that Cued achieved smaller sizes, and there are 65 (27/38) instances that CAQE-c achieved smaller sizes. Moreover, there are 117 (54/63) instances solved by Cued but not CAQE-c, and there are 11 (3/8) instances solved by CAQE-c but not Cued.

When DepQBF-c and Cued are compared, the spots are more scattered. Among the 273 (131/142) instances solved by both solvers, there are 111 (57/54) instances that Cued achieved smaller sizes, and there are 84 (38/46) instances

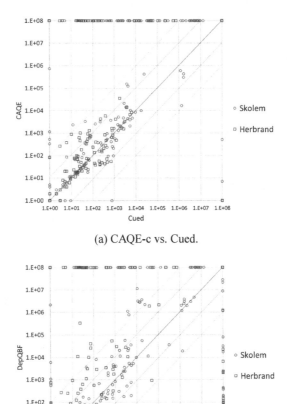

(a) CAQE-c vs. Cued.

(b) DepQBF-c vs. Cued.

Fig. 9. Certificate size comparison on PCNF track instances.

that DepQBF-c achieved smaller sizes. Moreover, there are 95 (36/59) instances
solved by Cued but not DepQBF-c, and there are 55 (11/44) instances solved
by DepQBF-c but not Cued.

For the 2QBF track formulas, Cued, CAQE-c, and DepQBF-c solved 80, 45,
42 cases, respectively, with certificates being generated. Among the generated
certificates, only 65 cases of Cued, 37 of CAQE-c, and 31 of DepQBF-c were
successfully synthesized by ABC. The remaining 15 cases of Cued, 8 of CAQE-c,
and 11 of DepQBF-c are too large for ABC to handle within the time or memory
limit; their sizes were counted in terms of the original non-synthesized AIG
nodes. On the other hand, for an unsolved instance of a solver, its certificate size
is counted as 100,000,000.

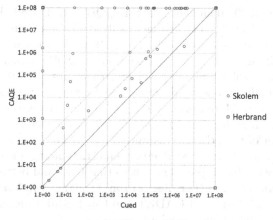

(a) CAQE-c vs. Cued.

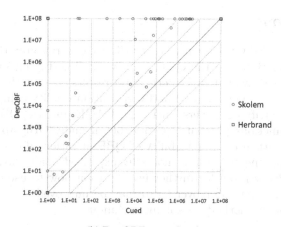

(b) DepQBF-c vs. Cued.

Fig. 10. Certificate size comparison on 2QBF track instances.

Figure 10 shows the results. Evidently, Cued dominates both CAQE-c and DepQBF-c. Note that for the 2QBF case, because the universal variables are in the first quantification level, the Herbrand functions are either constant TRUE or FALSE and thus the size of a Herbrand circuit, if generated successfully, is always 0.

When CAQE-c and Cued are compared, among the 44 (25/19) instances solved by both solvers, there are 18 (18/0) instances that Cued achieved smaller sizes, and there is 1 (1/0) instance that CAQE-c achieved the smaller size. Moreover, there are 32 (23/9) instances solved by Cued but not CAQE-c, and there is 1 (0/1) instance solved by CAQE-c but not Cued.

When DepQBF-c and Cued are compared, among the 34 (20/14) instances solved by both solvers, there are 19 (19/0) instances that Cued achieved smaller sizes, and there is 0 (0/0) instance that DepQBF-c achieved the smaller size. Moreover, there are 42 (28/14) instances solved by Cued but not DepQBF-c, and there is 0 (0/0) instance solved by DepQBF-c but not Cued.

When CADET and Cued are compared on the 2QBF instances, among the 33 (31/2) instances solved by both solvers, there are 7 (7/0) instances that Cued achieved smaller sizes, while there are 8 (8/0) instances that CADET achieved smaller sizes. Moreover, there are 43 (17/26) instances solved by Cued but not CADET, while there are 141 (93/48) instance solved by CADET but not Cued. As a matter of fact, the 149 instances that CADET did better than Cued come from the `irqlkeapclte`, `terminator`, `RankinfFunctions` and `wmiforward` families. On the other hand, the 50 instances that Cued did better than CADET come from the `MutexP`, `Qshifter`, `Reduction-finding`, and `Sorting_networks` families. This fact suggests that CADET and Cued exhibit distinct capability in solving some of the benchmark families.

8   Conclusions

This work has tackled QBF solving and certificate generation in the new view of cube distribution. Based on the Qesto framework, a certifying solver Cued has been developed and evaluated. Experimental results suggest the superiority of Cued in both solving performance and certification quality compared to other state-of-the-art certifying solvers, including CAQE and DepQBF. As Skolem and Herbrand functions play an essential role in many QBF applications, our results may strengthen the applicability of QBF in various domains. For future work, in the experiments we have not fully exploited don't cares in synthesis, there remains plenty room for further investigation how to use don't cares for certificate minimization. Moreover Cued may be improved through integration with other reasoning methods or solving techniques.

Acknowledgment. The authors thank Mikolas Janota for providing the Qesto source code. This work was supported in part by the Ministry of Science and Technology of Taiwan under grants 105-2221-E-002-196-MY3, 105-2923-E-002-016-MY3, and 108-2221-E-002-144-MY3.

References

1. Balabanov, V., Jiang, J.H.R.: Unified QBF certification and its applications. Formal Methods Syst. Des. **41**, 45–65 (2012)
2. Biere, A.: PicoSAT essentials. J. Satisfiability Boolean Model. Comput. **4**, 75–97 (2008)
3. Biere, A., Lonsing, F., Seidl, M.: Blocked clause elimination for QBF. In: Bjørner, N., Sofronie-Stokkermans, V. (eds.) CADE 2011. LNCS (LNAI), vol. 6803, pp. 101–115. Springer, Heidelberg (2011). https://doi.org/10.1007/978-3-642-22438-6_10
4. Brayton, R., Mischenko, A.: ABC: an academic industrial-strength verification tool. In: Touili, T., Cook, B., Jackson, P. (eds.) CAV 2010. LNCS, vol. 6174, pp. 24–40. Springer, Heidelberg (2010). https://doi.org/10.1007/978-3-642-14295-6_5

5. Dao, A.Q., et al.: Efficient computation of eco patch functions. In: Design Automation Conference (DAC), pp. 51:1–51:6 (2018)
6. Dershowitz, N., Hanna, Z., Katz, J.: Bounded model checking with QBF. In: Bacchus, F., Walsh, T. (eds.) SAT 2005. LNCS, vol. 3569, pp. 408–414. Springer, Heidelberg (2005). https://doi.org/10.1007/11499107_32
7. Een, N., Mishchenko, A., Sörensson, N.: Applying logic synthesis for speeding up SAT. In: Marques-Silva, J., Sakallah, K.A. (eds.) SAT 2007. LNCS, vol. 4501, pp. 272–286. Springer, Heidelberg (2007). https://doi.org/10.1007/978-3-540-72788-0_26
8. Fazekas, K., Heule, M.J.H., Seidl, M., Biere, A.: Skolem function continuation for quantified boolean formulas. In: Gabmeyer, S., Johnsen, E.B. (eds.) TAP 2017. LNCS, vol. 10375, pp. 129–138. Springer, Cham (2017). https://doi.org/10.1007/978-3-319-61467-0_8
9. Hecking-Harbusch, J., Tentrup, L.: Solving QBF by abstraction. In: International Symposium on Games, Automata, Logics, and Formal Verification (GandALF), pp. 88–102 (2018)
10. Heule, M.J.H., Seidl, M., Biere, A.: Solution validation and extraction for QBF preprocessing. J. Automated Reasoning 58(1), 97–125 (2017)
11. Janota, M., Marques-Silva, J.: Solving QBF by clause selection. In: International Joint Conference on Artificial Intelligence (IJCAI), pp. 325–331 (2015)
12. Jiang, J.H.R., Lin, H.P., Hung, W.L.: Interpolating functions from large Boolean relations. In: International Conference on Computer-Aided Design (ICCAD), pp. 779–784 (2009)
13. Kontchakov, R., et al.: Minimal module extraction from DL-lite ontologies using QBF solvers. In: International Joint Conference on Artificial Intelligence (IJCAI), pp. 836–841. AAAI Press (2009)
14. Lonsing, F., Biere, A.: DepQBF: a dependency-aware QBF solver (system description). J. Satisfiability Boolean Model. Comput. 7, 71–76 (2010)
15. Lonsing, F., Egly, U.: Depqbf 6.0: a search-based QBF solver beyond traditional QCDCL. In: International Conference on Automated Deduction (CADE), pp. 371–384 (2017)
16. Niemetz, A., Preiner, M., Lonsing, F., Seidl, M., Biere, A.: Resolution-based certificate extraction for QBF - (tool presentation). In: Cimatti, A., Sebastiani, R. (eds.) SAT 2012. LNCS, vol. 7317, pp. 430–435. Springer, Heidelberg (2012). https://doi.org/10.1007/978-3-642-31612-8_33
17. Rabe, M.N., Seshia, S.A.: Incremental determinization. In: Creignou, N., Le Berre, D. (eds.) SAT 2016. LNCS, vol. 9710, pp. 375–392. Springer, Cham (2016). https://doi.org/10.1007/978-3-319-40970-2_23
18. Rabe, M.N., Tentrup, L.: CAQE: a certifying QBF solver. In: International Conference on Formal Methods in Computer-aided Design (FMCAD), pp. 136–143 (2015)
19. Rabe, M.N., Tentrup, L., Rasmussen, C., Seshia, S.A.: Understanding and extending incremental determinization for 2QBF. In: International Conference on Computer Aided Verification (CAV), pp. 256–274 (2018)
20. Rintanen, J.: Asymptotically optimal encodings of conformant planning in QBF. In: National Conference on Artificial Intelligence (AAAI), pp. 1045–1050 (2007)
21. Skolem, T.: Über die mathematische Logik. Norsk Mat. Tidsskrift 106, 125–142 (1928). (Translation in From Frege to Gödel, A Source Book in Mathematical Logic, J. van Heijenoort, Harvard Univ. Press, 1967)

22. Solar-Lezama, A., Tancau, L., Bodik, R., Seshia, S., Saraswat, V.: Combinatorial sketching for finite programs. In: International Conference on Architectural Support for Programming Languages and Operating Systems (ASPLOS), pp. 404–415 (2006)
23. Staber, S., Bloem, R.: Fault localization and correction with QBF. In: International Conference on Theory and Applications of Satisfiability Testing (SAT), pp. 355–368 (2007)
24. Tentrup, L.: On expansion and resolution in CEGAR based QBF solving. In: International Conference on Computer Aided Verification (CAV), pp. 475–494 (2017)
25. Wimmer, R., Reimer, S., Marin, P., Becker, B.: HQSpre – an effective preprocessor for QBF and DQBF. In: Legay, A., Margaria, T. (eds.) TACAS 2017. LNCS, vol. 10205, pp. 373–390. Springer, Heidelberg (2017). https://doi.org/10.1007/978-3-662-54577-5_21

Functional Synthesis with Examples

Grigory Fedyukovich$^{(\boxtimes)}$ and Aarti Gupta

Princeton University, Princeton, USA
{grigoryf,aartig}@cs.princeton.edu

Abstract. Functional synthesis (FS) aims at generating an implementation from a declarative specification over sets of designated input and output variables. Traditionally, FS tasks are formulated as ∀∃-formulas, where input variables are universally quantified and output variables are existentially quantified. State-of-the-art approaches to FS proceed by eliminating existential quantifiers and extracting Skolem functions, which are then turned into implementations. Related applications benefit from having concise (i.e., compact and comprehensive) Skolem functions. In this paper, we present an approach for extracting concise Skolem functions for FS tasks specified as *examples*, i.e., tuples of concrete values of integer variables. Our approach builds a decision tree from *relationships* between inputs and outputs and *preconditions* that classify all examples into subsets that share the same input-output relationship. We also present an extension that is applied to *hybrid* FS tasks, which are formulated in part by examples and in part by arbitrary declarative specifications. Our approach is implemented on top of a functional synthesizer AE-VAL and evaluated on a set of reactive synthesis benchmarks enhanced with examples. Solutions produced by our tool are an order of magnitude smaller than ones produced by the baseline AE-VAL.

1 Introduction

One way to ensure the absence of bugs in programs is to replace a human developer with a machine that leverages automated decision procedures and theorem provers to develop programs that are correct-by-construction. But the task of automatically synthesizing programs from given specifications is notoriously hard and often depends crucially on the way specifications are formulated. Furthermore, it is often tedious to formulate a specification precisely and completely, such that it adequately represents the targeted intent. For humans, it is usually easier to provide a set of examples, such as tuples of concrete values for input and output variables. The task of the automated synthesizer is to generate an implementation, which produces given outputs for given inputs. In addition, the synthesizers should envision as much as possible which input-output tuples could appear in the actual programs, and implementations should be general enough to cover such cases. Finally, a specification may also include arbitrary additional requirements, and the resulting implementation should be consistent with both input-output examples and constraints at the same time.

T. Schiex and S. de Givry (Eds.): CP 2019, LNCS 11802, pp. 547–564, 2019.
https://doi.org/10.1007/978-3-030-30048-7_32

Many different techniques have been studied under the general umbrella of program synthesis, with a wide range in kinds of specifications and search techniques [1–3, 26–29, 31]. Typically, Functional Synthesis (FS) requires a declarative *relational specification* which connects input and output variables. Programming by Examples (PBE) requires a set of input-output tuples consisting of *concrete values* of variables. Although both FS and PBE have been developed successfully in many domains, there is a relative lack of unifying efforts that take advantages of them together to provide a general solution.

A classic formulation of an FS task is via checking the validity of a $\forall\exists$-formula, in which inputs are universally quantified, and outputs are existentially quantified. The validity of this formula guarantees realizability of the synthesis task, and a witnessing Skolem function can be turned into an implementation. In [11], Skolem functions are generated while lazily eliminating quantifiers in (and proving the validity of) $\forall\exists$-formulas in linear integer arithmetic (LIA). The generated solutions are represented in the form of decision trees, where decision nodes denote formulas over inputs called *preconditions*, and leaves denote equalities of outputs with terms over inputs called *local Skolem terms*. This method can be applied in a straightforward manner to a PBE task too: in the corresponding decision tree, the preconditions would be represented by (conjunctions of) equalities over inputs and their values, and the local Skolem terms would simply be the corresponding values of the output variables.

To obtain concise Skolem terms, decision trees can be compacted by potentially merging decision nodes that could share the same leaves. In the context of PBE, this idea is in general inapplicable because the terms in the leaves are always constants. To apply any compaction, there should be a way to replace these constants by terms over inputs. For LIA, this can be done by discovering linear equations over input and output variables.

The challenge is that not all given examples would be classified by a single linear equation. Thus, in our approach, we first *partition* the set of examples into subsets, such that all examples within each subset share the same linear relationship. Clearly, such a partitioning is not unique, and we target deriving a small number of subsets. Another criterion we consider is that all examples within each subset should be classified concisely by some precondition over inputs. In particular, a precondition that simply disjoins all equalities between inputs and concrete values would be too bulky (growing linearly with the size of the subset). Instead, we seek an opportunity to replace it by some inequality or conjunction of inequalities. These criteria lead to compact decision trees.

One key novelty in our approach is a completely automated procedure to discover compact preconditions and local Skolem terms for PBE tasks in LIA. Existing synthesis approaches, e.g., those based on enumerative search [26, 27, 31], require the user to additionally supply formal grammars (or templates) that specify a pool of candidate formulas and terms. They search for suitable candidates from the grammars and iteratively test them on given examples. While this general capability is useful for rich grammars and specifications, for LIA it is possible to completely automate these steps. In particular, our approach does not require any extra input from the user and automatically *infers* candidates for local Skolem terms directly from data using canonical equations in linear arithmetic.

The candidate preconditions are also inferred automatically from data – they specify ranges of input values. To find suitable candidates, we pose queries over certain ranges in the ∀∃-form which intuitively say "for all inputs within the candidate range, there exists an output value which is consistent with the candidate assignment and all given examples". By counting how many examples are covered by each candidate that passes the ∀∃-test, we create a ranking of candidates and pick those with the highest rank.

We also extended the approach to *hybrid* PBE and FS tasks, formulated in part by using input-output examples and in part by using arbitrary input-output relational constraints. To solve such problems, the formula describing an FS-part of the task is simply added to our ∀∃-test and is taken into account when filtering suitable candidate ranges and the corresponding local Skolem terms.

Our implementation on top of the AE-VAL [11] tool has been evaluated on a range of reactive synthesis benchmarks enhanced with examples. The discovered solutions are an order of magnitude smaller than straightforward Skolem terms and less sensitive to the number of examples.

2   Running Example

Table 1 gives a set of examples by means of integer values of input variables x_1, x_2, and x_3, and an output variable y. Each row represents a transition from concrete inputs to the concrete output. Some examples are *incomplete*, i.e., a subset of input values is not given (denoted "."). For instance, the first row specifies input values for only x_1 and x_3; and it should be interpreted as "if both x_1 and x_3 are equal to one, then y should be equal to one as well".

Our goal is to find (1) symbolic linear relationships among given values in each input-output tuple, and (2) preconditions that uniquely determine equivalence classes

Table 1. Input output tuples.

x_1	x_2	x_3	y
1	·	1	1
0	·	2	0
·	1	3	2
·	2	4	4
2	4	5	6
2	0	6	2

of these relationships. For instance, for the first two rows, it is true that $y = x_1$. Precondition $1 \leq x_3 \leq 2$ uniquely determines the first two rows, in a sense that for the remaining four rows, it does not hold. Similarly, for the next two rows, $y = 2 \cdot x_2$ under precondition $3 \leq x_3 \leq 4$; and for the last two rows, $y = x_1 + x_2$ under precondition $5 \leq x_3 \leq 6$. Combining preconditions and relationships, we can formally describe how y can be computed from x_1, x_2, and x_3:

$$y = ite(1 \leq x_3 \leq 2, x_1, ite(3 \leq x_3 \leq 4, 2 \cdot x_2, x_1 + x_2))$$

In fact, such a *decision tree* is not unique for values in the table. A more compact one can be found by our algorithm:

$$y = ite(2 \leq x_3 \leq 5, 2 \cdot x_3 - 4, x_1)$$

In the rest of the paper, we show how such a solution can be discovered automatically.

3   Background and Notation

A many-sorted first-order theory consists of disjoint sets of sorts $\mathcal{S}$, function symbols $\mathcal{F}$ and predicate symbols $\mathcal{P}$. A set of *terms* is defined recursively as follows:

$$term ::= f(term, \ldots, term) \mid const \mid var$$

where $f \in \mathcal{F}$, *const* is an application of some $v \in \mathcal{F}$ of zero arity, and *var* is a variable uniquely associated with a sort in $\mathcal{S}$. A set of quantifier-free *formulas* is built recursively using the usual grammar:

$$formula ::= true \mid false \mid p(term, \ldots, term) \mid Bvar \mid$$
$$\neg formula \mid formula \wedge formula \mid formula \vee formula$$

where *true* and *false* are Boolean constants, $p \in \mathcal{P}$, and *Bvar* is a variable associated with sort *Bool*.

In this paper, we consider the theory Linear Integer Arithmetic (LIA). In LIA, $\mathcal{C} \stackrel{\text{def}}{=} \{\mathbb{Z}, Bool\}$, $\mathcal{F} \stackrel{\text{def}}{=} \{+, \cdot, div\}$, and $\mathcal{P} \stackrel{\text{def}}{=} \{=, >, <, \geq, \leq, \neq\}$. We define *ite* as a shortcut for *if-then-else*, i.e., $ite(x, y, z) \stackrel{\text{def}}{=} (x \wedge y) \vee (\neg x \wedge z)$.

Formula φ is called satisfiable if there exists an interpretation m, called a model, of each element (i.e., a variable, a function or a predicate symbol), under which φ evaluates to *true*; otherwise φ is called unsatisfiable. If every model of φ is also a model of ψ, then we write $\varphi \implies \psi$. A formula φ is called *valid* if $true \implies \varphi$.

For existentially-quantified formulas of the form $\exists y . \psi(\vec{x}, y)$, the validity requires that each interpretation for variables in $\vec{x}$ and each function and predicate symbol in ψ can be *extended* to a model of $\psi(\vec{x}, y)$. For a valid formula $\exists y . \psi(\vec{x}, y)$, a term $sk_y(\vec{x})$ is called a *Skolem term*, if $\psi(\vec{x}, sk_y(\vec{x}))$ is valid.

In the paper, we assume that all free variables $\vec{x}$ are implicitly universally quantified. For simplicity, we omit the arguments and simply write φ when the arguments are clear from the context.

Extracting Skolem Terms. Our work is built on top of a lazy quantifier-elimination method for checking validity and performing synthesis called AE-VAL [11,12]. It generates a structured synthesis solution in the form of a decision tree. Its main procedure is based on deriving a sequence of Model-Based Projections (MBPs) [19] to lazily decompose the overall problem, where each model is used to derive a precondition that captures an arbitrary subspace on the $\vec{x}$ variables and a Skolem term for the $\vec{y}$ variables. Unlike other prior work [21], AE-VAL does not require converting the formula into Disjunctive Normal Form (DNF), which often leads to larger and redundant solutions. AE-VAL also uses minimization and compaction procedures for on-the-fly compaction of the generated synthesis solution. In particular, it derives Skolem terms that can be re-used across multiple preconditions for a single output and shares the preconditions in a common decision tree across multiple outputs in a program. This is done by identifying theory terms that can be shared both within and across outputs.

However, AE-VAL handles relational specifications (FS tasks) only, and it is not designed to handle input-output examples (PBE tasks) properly. When given concrete input-output examples, it would generate an implementation in the form of a decision tree with the depth equal to the number of examples (as described in more detail in the next section).

4   Synthesis by Examples

We formalize the case when all examples are complete and defer the case of partially defined examples till the next section.

Definition 1. *Let* $\vec{x} = \langle x_1, \ldots, x_n \rangle$ *be a vector of input variables and* E *be a set of* m *examples, where each* $\vec{e} \in E$ *is a vector of integers and* $\vec{e}$ *has* $n + 1$ *components. For an output variable* y, *vectors* $\vec{x}$ *and* $\vec{e} \in E$ *are connected through an* example-formula ζ:

$$\zeta(e, \vec{x}, y) \stackrel{\text{def}}{=} \bigwedge_{1 \le i \le n} (\vec{x}[i] = \vec{e}[i]) \implies y = \vec{e}[n+1]$$

We assume consistency among all examples in E, i.e., that the following formula is valid:

$$\forall \vec{x} . \exists y . \bigwedge_{\vec{e} \in E} \zeta(\vec{e}, \vec{x}, y) \tag{1}$$

Note that the formula could only be invalid if there are two vectors $\vec{e}_1, \vec{e}_2 \in E$, such that:

$$\vec{e}_1[n+1] \ne \vec{e}_2[n+1] \wedge \forall i . 0 \le i \le n \implies \vec{e}_1[i] = \vec{e}_2[i]$$

A Skolem term for y in (1) can be derived in the form of a nested *ite*-block of depth m as shown below:

$$ite\left(\bigwedge_{1 \le i \le n} (\vec{x}[i] = \vec{e}_1[i]), \vec{e}_1[n+1], ite\left(\bigwedge_{1 \le i \le n} (\vec{x}[i] = \vec{e}_2[i]), \vec{e}_2[n+1], \ldots, 0 \right) \right) \tag{2}$$

where each $\vec{e}_i \in E$ identifies the i-th level of the decision tree, and the last *else*-branch represents the case when none of examples match current values of $\vec{x}$, thus an arbitrary value (e.g., 0 as in (2)) can be assigned to y.

We wish to generate a Skolem term for y as a decision tree with a smaller depth. That is, among the space of terms of form (3), we wish to identify the one with a (preferably) minimal number of *ite*-blocks.

Definition 2. *A* Skolem term *for an example-formula* (1) *is called* generalized *if it has the following form:*

$$ite\left(pre[1](\vec{x}), sk[1](\vec{x}), ite\left(pre[2](\vec{x}), sk[2](\vec{x}), \ldots, 0 \right) \right) \tag{3}$$

where the vector pre *collects formulas over* $\vec{x}$ *(called* preconditions*), and the vector* sk *collects terms over* $\vec{x}$ *(called* local Skolems*). Each pair* $\langle pre_F, sk_F \rangle$ *corresponds to a subset of examples* $F \subseteq E$, *such that* (4) *and* (5) *hold:*

$$\forall \vec{e} \in F . \bigwedge_{1 \le i \le n} (\vec{x}[i] = \vec{e}[i]) \implies pre_F(\vec{x}) \tag{4}$$

$$\forall \vec{e} \in F . \bigwedge_{1 \le i \le n} (\vec{x}[i] = \vec{e}[i]) \wedge y = sk_F(\vec{x}) \implies y = \vec{e}[n+1] \tag{5}$$

We present an algorithm that partitions the given set E into disjoint subsets, which give rise to vectors *pre* and *sk*. An overview of the proposed algorithm is shown in Algorithm 1. The key insight is to identify each subset $F \subseteq E$ by inferring a precondition pre_F and a local Skolem term sk_F from pairs of examples $\langle \vec{e}_1, \vec{e}_2 \rangle \in E \times E$. Once a subset F, such that (5) holds, is discovered, it is straightforward to generate pre_F: The algorithm relies on helper procedures to discover a candidate precondition (line 4) and a candidate term for each pair (line 5). These procedures, applied to all pairs of examples, produce a set of candidate preconditions and a set of candidate terms. However, there is no guarantee that a precondition and a term, which suit all given examples, could be discovered. But we can often find some precondition and some term that will suit *many* examples, which will constitute the desired subset F. In order to identify it, our algorithm filters bad preconditions and terms and ranks successful ones. In the rest of this section, we outline a particular instantiation of subroutines of Algorithm 1 for LIA[1].

Method GETRANGE. To define a range of values of variables $\vec{x}$ between $\vec{e}_1$ and $\vec{e}_2$ we introduce a function M:

$$M(\vec{e}_1, \vec{e}_2, i) \stackrel{\text{def}}{=} \begin{cases} \vec{e}_1[i] \le \vec{x}[i] \wedge \vec{x}[i] \le \vec{e}_2[i], \text{ if } \vec{e}_1[i] \le \vec{e}_2[i] \\ \vec{e}_2[i] \le \vec{x}[i] \wedge \vec{x}[i] \le \vec{e}_1[i], \text{ otherwise} \end{cases}$$

Then, formula γ representing a range between $\vec{e}_1$ and $\vec{e}_2$ is simply computed as:

$$\gamma \stackrel{\text{def}}{=} \bigwedge_{1 \le i \le n} M(\vec{e}_1, \vec{e}_2, i) \tag{6}$$

Method CONNECT. Relationships between variables $\vec{x}$ and y are determined by a *canonical equation of a line* and two vectors of their values, $\vec{e}_1$ and $\vec{e}_2$:

$$\frac{\vec{x}[1] - \vec{e}_1[1]}{\vec{e}_2[2] - \vec{e}_1[1]} = \cdots = \frac{\vec{x}[n] - \vec{e}_1[n]}{\vec{e}_2[n] - \vec{e}_1[n]} = \frac{y - \vec{e}_1[n+1]}{\vec{e}_2[n+1] - \vec{e}_1[n+1]} \tag{7}$$

---

[1] With the required support for quantifier elimination, it can be immediately adapted to rational arithmetic, nonlinear arithmetic, and bitvectors. But to achieve more compact solutions, these algorithms could benefit from additional adjustments in method CONNECT which are left for future work.

Algorithm 1: GETBESTCLASS($\vec{x}, y, E$)

Input: $\vec{x}, y, E$
Output: F, pre_F, sk_F, s.t. (4) and (5) hold

1 $Cands \leftarrow \varnothing$;
2 $R \leftarrow \lambda\xi\,.\,\varnothing$;
3 **for** $\langle\vec{e}_1, \vec{e}_2\rangle \in E \times E$ **do**
4 $\quad$ $\gamma \leftarrow$ GETRANGE($\vec{e}_1, \vec{e}_2$);
5 $\quad$ $X \leftarrow$ CONNECT($\vec{e}_1, \vec{e}_2$);
6 $\quad$ **for** $\xi \in X$ **do**
7 $\quad\quad$ **if** SANITYTEST(γ, ξ) **then**
8 $\quad\quad\quad$ $\xi \leftarrow$ LOCALSKOLEM(γ, ξ);
9 $\quad\quad\quad$ $Cands \leftarrow Cands \cup \{\xi\}$;
10 $\quad\quad\quad$ $R(\xi) \leftarrow R(\xi) \cup \{\gamma\}$;
11 **for** $\xi \in Cands$ **do**
12 $\quad$ $R(\xi) \leftarrow R(\xi) \cup$ GENERALIZE($R(\xi)$);
13 $\quad$ **for** $\gamma \in R(\xi)$ **do** RANK(γ, ξ, E);
14 **return** LARGEST(E, γ, ξ);

Algorithm 2: PBE($\vec{x}, y, E$)

Input: $\vec{x}, y, E$
Output: Skolem term sk for y in (1)

1 **if** $E = \varnothing$ **then return** PICKANY($\mathbb{Z}$);
2 $F, pre_F, sk_F \leftarrow$ GETBESTCLASS($\vec{x}, y, E$);
3 **if** $F = \varnothing$ **then**
4 $\quad$ $\vec{e} \leftarrow$ PICKANY(E);
5 $\quad$ $F \leftarrow \{\vec{e}\}$;
6 $\quad$ $pre_F \leftarrow \bigwedge_{1 \leq i \leq n} (\vec{x}[i] = \vec{e}[i])$;
7 $\quad$ $sk_F \leftarrow (y = \vec{e}[n+1])$;
8 **return** $ite(pre_F, sk_F, \text{PBE}(\vec{x}, y, E \setminus F))$;

It gives rise to various possible equalities connecting components of $\vec{x}$ and y. In particular, any two equalities of form $(\vec{x}[i] - \vec{e}_1[i]) \cdot (\vec{e}_2[n+1] - \vec{e}_1[n+1]) = (\vec{e}_2[i] - \vec{e}_1[i]) \cdot (y - \vec{e}_1[n+1])$, where $1 \leq i \leq n$, can be summed (or subtracted) side-by-side.

Example 1. Recall our set of input-output tuples from Sect. 2. Suppose, in the first loop of Algorithm 1, we are considering the first two tuples (i.e., rows in Table 1): $\zeta_1 \overset{\text{def}}{=} (x_1 = 1 \wedge x_3 = 1) \implies (y = 1)$ and $\zeta_2 \overset{\text{def}}{=} (x_1 = 0 \wedge x_3 = 2) \implies (y = 0)$. The GETRANGE method produces $\gamma_{1,2} \overset{\text{def}}{=} 0 \leq x_1 \leq 1 \wedge 1 \leq x_3 \leq 2$. The CONNECT method produces equalities $X_{1,2} = \{y = x_1, \, y = 2 - x_3$, and $2 \cdot y = x_1 - x_3 + 2\}$ (the last one is produced by summing left and right sides of the first two equalities).

Methods SANITYTEST *and* LOCALSKOLEM. Let γ be a range-formula over $\vec{x}$, and ξ be a formula over $\vec{x}$ and y. We filter a set of pairs $\langle \gamma, \xi \rangle$ based on the following criterion:

$$\forall \vec{x} . \gamma(\vec{x}) \implies \exists y . \xi(\vec{x}, y) \tag{8}$$

If formula (8) is valid, a Skolem term for y exists and can be extracted, e.g., using the AE-VAL algorithm.

Example 2. Recall ζ_1 and ζ_2 produced from our input-output tuples (see Sect. 2) in Example 1. For each $\xi \in X_{1,2}$ and $\gamma_{1,2}$, we pose a query of form (8):

$$\forall x_1, x_2, x_3 . 0 \leq x_1 \leq 1 \wedge 1 \leq x_3 \leq 2 \implies \exists y . y = x_1 \qquad \text{valid}$$
$$\forall x_1, x_2, x_3 . 0 \leq x_1 \leq 1 \wedge 1 \leq x_3 \leq 2 \implies \exists y . y = 2 - x_3 \qquad \text{valid}$$
$$\forall x_1, x_2, x_3 . 0 \leq x_1 \leq 1 \wedge 1 \leq x_3 \leq 2 \implies \exists y . 2 \cdot y = x_1 - x_3 + 2 \qquad \text{invalid}$$

The results for these queries are shown on the right. Since the last query is invalid, we thus proceed with the other candidates $y = x_1$ and $y = 2 - x_3$ only.

Note that if for some $\xi \in X$, the coefficient for y is 1, then any query of form (8) is valid (and a Skolem function for y is ξ itself).

Method GENERALIZE. Given a set of factored preconditions of ξ (recall (6)), we fix a variable $x \in \vec{x}$ and identify factors over x across all preconditions. Then, we iteratively prune this set of formulas by applying the following rule:

$$\frac{\alpha_1 \leq x \wedge x \leq \alpha_2 \quad \alpha_3 \leq x \wedge x \leq \alpha_4}{min(\alpha_1, \alpha_3) \leq x \leq max(\alpha_2, \alpha_4)} \text{ if } \alpha_3 \leq \alpha_2 \wedge \alpha_1 \leq \alpha_4$$

Repeating this operation yields a new formula over x. Repeating this for all $x \in \vec{x}$ and conjoining the resulting formulas gives us a new range-formula for ξ.

Note that this new formula is an over-approximation of the disjunction of the original preconditions for ξ. By using these preconditions for all candidate Skolem terms, we face a trade-off between the depth of the resulting decision tree and the syntactic size of preconditions. That is, some of the over-approximated preconditions could be too coarse, and thus filtered away (see method RANK of the algorithm). But if an over-approximated precondition has not been filtered, it is likely to be more compact and general.

Example 3. Let $\gamma_{1,6} \overset{\text{def}}{=} 1 \leq x_1 \leq 2 \wedge 1 \leq x_3 \leq 6$ and $X_{1,6} = \{y = x_1\}$.[2] Following Examples 1 and 2, $y = x_1$ is also associated with $\gamma_{1,2} = 0 \leq x_1 \leq 1 \wedge 1 \leq x_3 \leq 2$. Thus, our generalization produces $\gamma_{1,2,6} \overset{\text{def}}{=} 0 \leq x_1 \leq 2 \wedge 1 \leq x_3 \leq 6$.

Methods RANK *and* LARGEST. These two methods identify the best formula (or a combination of formulas) among the candidates. We evaluate a precondition γ and a suitable candidate local Skolem term ξ on all examples $\vec{e} \in E$. In particular, we identify a subset of examples, for which implication (9) holds (denoted $F(\gamma)$) and a subset of examples, for which implication (10) does not hold (denoted $G(\xi)$).

$$\bigwedge_{1 \leq i \leq n} (\vec{x}[i] = \vec{e}[i]) \implies \gamma(\vec{x}) \tag{9}$$

$$\xi(\vec{x}) \wedge \bigwedge_{1 \leq i \leq n} (\vec{x}[i] = \vec{e}[i]) \implies y = \vec{e}[n+1] \tag{10}$$

Cardinalities of $F(\gamma)$ and $G(\xi)$ give a ranking to each $\langle \gamma, \xi \rangle$. If $G(\xi)$ is nonempty, then the ranking is zero. Otherwise, the ranking is $|F(\gamma)|$.

Example 4. To rank precondition $\gamma_{1,2,6}$ for a candidate $y = x_1$ generated in Example 3, we enumerate all input-output tuples from Table 1 and test implications (9) and (10). It appears that set $G(y = x_1)$ is nonempty since for the fifth tuple (10) is invalid:

$$y = x_1 \wedge x_1 = 2 \wedge x_2 = 4 \wedge x_3 = 5 \not\implies y = 6$$

Another precondition $\gamma_{2,3,4,5} \overset{\text{def}}{=} 2 \leq x_3 \leq 5$ for candidate $y = 2 \cdot x_3 - 4$ (computed similarly) gets ranking 4 since set $G(y = 2 \cdot x_3 - 4)$ is empty, and $F(\gamma_{2,3,4,5})$ consists of four examples.

Since ranking explicitly checks partially generated functions w.r.t. specifications, our solutions are correct by construction. More formally, it is represented by the following lemma.

Lemma 1. *Any pair of formulas $\langle \gamma, \xi \rangle$ with a non-zero ranking can be used to extract the outer ite-block of the Skolem term.*

For getting a candidate formula with the best coverage, we select the formula with a higher (and non-zero) ranking. It intuitively corresponds to the largest subset of examples that can be described by a single precondition and a single local Skolem term.

---

[2] We refer the reader to Sect. 5 that describes a process of learning from partial examples.

Algorithm 2 describes an algorithm to construct a decision tree recursively. It starts with the full set of given examples E, and uses Algorithm 1 to identify the largest subset $F \subseteq E$, elements of which share the same precondition and local Skolem term (to be used at one level of the decision tree). In the case when F is empty, it is enough to pick any element of E and create a precondition and a local Skolem term in a straightforward way. Then, all elements of F are excluded from E, and the algorithm recurses. It converges when E is empty, and for this (the deepest) level of the decision tree, we can pick any local Skolem term (e.g., an integer constant) with no precondition.

5   Synthesis by Partial Examples

In this section, we present a generalization of the synthesis by examples algorithm (described in Sect. 4) that relies on *subvectors* of examples.

Definition 3. *Let $\vec{x}$ be a vector containing n components and s be an injective function to $\{1, \ldots, n\}$. A subvector of $\vec{x}$ (denoted $\vec{x}_{|s}$) is a vector, such that for all i, $\vec{x}_{|s}[i] = \vec{x}[s(i)]$.*

Intuitively, $\vec{x}_{|s}$ is produced by *removing* components from $\vec{x}$ and preserving the order of the remaining components. We naturally extend this definition to sets of vectors, i.e., $E_{|s} \stackrel{\text{def}}{=} \{\vec{e}_{|s} \mid \vec{e} \in E\}$.

The algorithms from Sect. 4 can be used for subvectors of input variables and sets of subvectors of examples. In particular, let s be an injective function to $\{1, \ldots, n\}$, we can apply Algorithm 1 to $\vec{x}_{|s}$ and $E_{|s}$, if the following formula is valid:

$$\forall \vec{x}_{|s} . \exists y . \bigwedge_{\vec{e} \in E} \zeta(\vec{e}_{|s}, \vec{x}_{|s}, y) \tag{11}$$

There are two main advantages for doing this. First, it may give us more concise and general solutions (which are expressible using fewer variables). Second, while extracting subvectors, we shrink the set of examples, which lowers the cost of the synthesis procedure.

Thus, the whole procedure can be supplied with a preprocessing, during which various mappings s are considered and formulas of form (11) are checked for validity. The mapping s with the smallest domain size can be then used for synthesis by examples. The speed of the entire procedure could then be improved, but the effectiveness of the resulting solution could worsen.

Example 5. Recall Example 1, let s be a function with $dom(s) = \{1\}$ and $img(s) = \{3\}$. Then $E_{|s}$ is constructed from E by keeping the values of x_3. Formula (11) is compiled as follows:

$$\forall x_3, \exists y . (x_3 = 1 \implies y = 1) \wedge (x_3 = 2 \implies y = 0) \wedge (x_3 = 3 \implies y = 2)$$
$$(x_3 = 4 \implies y = 4) \wedge (x_3 = 5 \implies y = 6) \wedge (x_3 = 6 \implies y = 2)$$

The formula above is valid, and Algorithm 2 can be applied to extract the following Skolem term:

$$y = ite(2 \leq x_3 \leq 5, 2 \cdot x_3 - 4, ite(x_3 = 1, 1, 2))$$

Note that this Skolem term is not optimal (the one provided in Sect. 2 has a fewer nested *ite*-blocks). A heuristic in the rest of the section aims at discovering a more effective solution.

Some of examples could be defined only partially, i.e., using a sequence of injective functions $s_1, \ldots, s_m$ to $\{1, \ldots, n\}$ that gives rise to sequences $\vec{x}_{|s_1}, \ldots, \vec{x}_{|s_m}$ and $E_1, \ldots, E_m$. For each s_i, examples from E_i use values of $\vec{x}_{|s_i}$ and y.

The task is to extract a Skolem term for the given *valid* formula:

$$\forall \vec{x} . \exists y . \bigwedge_{1 \leq i \leq m} \bigwedge_{\vec{e} \in E_i} \zeta(\vec{e}, \vec{x}_{|s_i}, y) \tag{12}$$

Algorithm 3 shows an adaptation of Algorithm 2 applicable to the union of sets of all examples $E \overset{\text{def}}{=} E_1 \cup \ldots \cup E_m$. It iteratively produces subvectors of all examples and finds such a subset of them, which gives the valid example-formula (line 3). Then, it applies Algorithm 1 to detect a level of the decision tree (line 4) and shrinks the set of examples accordingly (lines 10–13). Similarly to Algorithm 2, the algorithm recurses until the entire decision tree is constructed (line 3).

Theorem 1. *If* $\bigcap_{1 \leq i \leq m} img(s_i) \neq \varnothing$, *then Algorithm 3 returns a Skolem term for* (12).

Example 6. In the first iteration, Algorithm 3 considers function s from Example 5. As a result, it extracts four input-output tuples (recall Example 4). In the second iteration, Algorithm 3 takes as input just two remaining tuples and considers function s', such that $dom(s') = \{1\}$ and $img(s') = \{1\}$. It appears that $\gamma_{1,6}$ and the $y = x_1$ are considered again (recall Example 1). But in this case (as opposed to Example 4), their ranking is computed with respect to only two input-output tuples, thus resulting in $F(\gamma_{1,6}) = \varnothing$ and $|G(y = x_1)| = 2$. This concludes the search, and the final Skolem term gets composed from two nested *ite*-blocks (i.e., exactly as provided in Sect. 2).

Algorithm 3: PARTIALEXSPBE$(\vec{x}, y, \{\langle s_i, E_i \rangle\}_{1 \leq i \leq n})$

Input: $\vec{x}, y$: variables, $\{\langle s_i, E_i \rangle\}_{1 \leq i \leq n}$: set of pairs of functions and sets of
partial examples, $E = \bigcup\limits_{1 \leq i \leq n} E_i$

Output: Skolem term sk for y in (12)

1   **if** $E = \varnothing$ **then return** PICKANY$(\mathbb{Z})$;

2   let s be such that $\forall s_j, img(s) \subseteq img(s_j)$;
3   $E' \leftarrow$ GETVALIDSUBSET$(E_{|s})$;
4   $F, pre_F, sk_F \leftarrow$ GETBESTCLASS$(\vec{x}_{|s}, y, E')$;
5   **if** $F = \varnothing$ **then**
6     $\vec{e} \leftarrow$ PICKANY(E');
7     $F \leftarrow \{\vec{e}\}$;
8     $pre_F \leftarrow \bigwedge\limits_{i \in img(s)} (\vec{x}[i] = \vec{e}[i])$;
9     $sk_F \leftarrow (y = \vec{e}[n+1])$;
10   $Rem \leftarrow \varnothing$;
11   **for** $1 \leq i \leq n$ **do**
12     $E_i \leftarrow \{e \in E_i \mid \vec{e}_{|s} \notin E'\}$;
13   **if** $E_i \neq \varnothing$ **then** $Rem \leftarrow Rem \cup \langle s_i, E_i \rangle$;
14   **return** $ite(pre_F, sk_F, \text{PARTIALEXSPBE}(\vec{x}, y, Rem))$;

---

6   Hybrid Synthesis: PBE + FS

Suppose we are given an additional requirement $\psi(\vec{x}, y)$ for the input and output variables. Note that ψ may be a *partial* specification, i.e., it may impose necessary but not sufficient conditions for correctness. The goal is to discover a Skolem term for (13):

$$\forall \vec{x} . \exists y . \bigwedge_{\vec{e} \in E} \zeta(\vec{e}, \vec{x}, y) \wedge \psi(\vec{x}, y) \tag{13}$$

If ψ is consistent with the set of examples, then a Skolem term for (13) can be discovered by the procedure from Sect. 4 with the following differences:

– *Default local Skolem term in Algorithm* 2 (line 1).
   The random choice is replaced with a Skolem term for y in formula $\forall \vec{x} . \exists y . \psi(\vec{x}, y)$ (i.e., solve a standard functional synthesis task without examples). In our implementation, we use AE-VAL.
– *Criteria (8) and (10) for methods* SANITYTEST *and* RANK, *respectively.* We check the validity of formulas, respectively (14) and (15), enhanced with ψ.

$$\forall \vec{x} . \gamma(\vec{x}) \implies \exists y . \xi(\vec{x}, y) \wedge \psi(\vec{x}, y) \tag{14}$$

$$\xi(\vec{x}) \wedge \bigwedge_{1 \leq i \leq n} (\vec{x}[i] = \vec{e}[i]) \implies y = \vec{e}[n+1] \wedge \psi(\vec{x}, y) \tag{15}$$

– *Extra criterion for method* GENERALIZE. We perform an extra sanity check (14) for each over-approximated precondition.

Theorem 2. *With these adjustments, the output of Algorithm 2 is a Skolem term for* (13).

Example 7. Consider a synthesis task consisting of (1) values specified in Table 1, and (2) an additional requirement $y \geq x_1 \vee y \geq x_2$. Thus, the entire formula is as follows:

$$\forall x_1, x_2, x_3 . \exists y . y \geq x_1 \vee y \geq x_2 \wedge$$
$$(x_1 = 1 \wedge x_3 = 1 \implies y = 1) \wedge (x_1 = 0 \wedge x_3 = 2 \implies y = 0) \wedge$$
$$(x_2 = 1 \wedge x_3 = 3 \implies y = 2) \wedge (x_2 = 2 \wedge x_3 = 4 \implies y = 4) \wedge$$
$$(x_1 = 2 \wedge x_2 = 4 \wedge x_3 = 5 \implies y = 6) \wedge (x_1 = 2 \wedge x_2 = 0 \wedge x_3 = 6 \implies y = 2)$$

This is a suitable task for AE-VAL, but it would return a Skolem term as a decision tree with six levels. In contrast, our algorithm produces a Skolem term for y with just three levels:

$$ite\Big(4 \leq x_3 \leq 6 \wedge 0 \leq x_2 \leq 4, x_2 + 2,$$
$$ite\big(1 \leq x_3 \leq 2 \wedge 0 \leq x_1 \leq 1, x_1, ite(x_2 = 1 \wedge x_3 = 3, 2, x_1)\big)\Big)$$

The deepest decision, x_1, is a Skolem term for y in formula $\forall x_1, x_2, x_3 . \exists y . y \geq x_1 \vee y \geq x_2$. The preconditions and relationships identified in Example 6 are not suitable.

7   Evaluation

We implemented our synthesis algorithm on top of the AE-VAL [11] tool [3] which uses the Z3 SMT solver [7]. To compare our implementation with state-of-the-art tools, we considered the "plain" AE-VAL, CVC4 [23], EUSOLVER [3] and DRYADSYNTH [15]. None of them supports discovery of relationships among data tuples: AE-VAL, CVC4, and DRYADSYNTH return a *straightforward Skolem*, i.e., a formula of form (2) with nested *ite*-blocks of the highest depth m; and EUSOLVER has frontend issues. The timings for discovery of a straightforward Skolem are usually small even for a large number of examples. Since we do not consider a straightforward Skolem an acceptable solution for our class of tasks, we do not present a detailed evaluation report for the competing tools. Instead, we focus on details of our AE-VAL-PBE and AE-VAL.

We considered 59 benchmarks from various Assume-Guarantee contracts written in the Lustre programming language [17]. These are the relational specifications derived mainly from industrial projects, such as a Quad-Redundant Flight Control System, a Microwave model, a Generic Patient Controlled Analgesia infusion pump, a Cinderella-Stepmother game, and several tricky hand-written examples. The depths of solutions for these original benchmarks, generated by AE-VAL, range from 1 to 8 (median is 3, geometric mean is 2.3).

---

[3] The source code and benchmarks are available at https://github.com/grigoryfedyuko vich/aeval.

The specifications of the system were enhanced by the designer with sets of examples that describe some additional features of the desired implementations (thus, the Skolem terms generated for the original specification might no longer be valid for the corresponding enhanced specifications). We considered 32 unique examples to enhance each benchmark. The depths of straightforward solutions, generated by AE-VAL for these benchmarks, range from 1 to 106 (median is 37, geometric mean is 32). In contrast, the depths of the solutions by our AE-VAL-PBE for these benchmarks are an order of magnitude smaller, i.e., they range from 1 to 17 (median is 5, geometric mean is 5.5). Thus, the AE-VAL-PBE was shown to be *more effective when computing compact solutions*: the ratio between depths ranges from 1 to 24, median is 6.8, geometric mean is 5.8. The synthesis time for producing the default local Skolem terms, as well as the straightforward decision trees was negligible.

Effect of number of examples. A common characteristic exhibited by the "plain" AE-VAL when enhancing relational specifications with examples is the growth of the resulting decision trees. Intuitively, the more examples are given, the larger solutions are generated. In this subsection we show that such a scenario is uncommon for our approach.

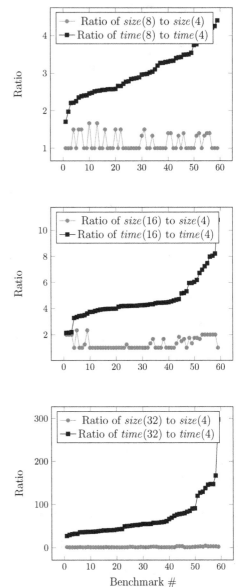

Fig. 1. Stability of our solutions.

We performed three additional experiments, in which we kept respectively 16, 8, and 4 given examples out of the original 32 and repeated our synthesis procedure. Although the computation of a decision tree for fewer examples is less resource-demanding, the precision of solutions remained roughly the same.

For 16 examples, the median depth of the decision tree is 4.9 (geometric mean is 5). For 8 examples, the median depth is 4.4 (geometric mean is 4), and for 4 examples, the median depth is 3.9 (geometric mean is 4).

We refer to this feature of our algorithm as *stability*. More statistics are shown in Fig. 1 on these three experiments. For every $i \in \{8, 16, 32\}$ and for each benchmark, we computed a ratio of the decision-tree depth for i examples to the depth of the decision tree for 4 examples (shown grey). Then, we compared the two for the runtime (shown black). Intuitively, the two graphs in each plot show the growths of the solution size and the synthesis time, respectively, when increasing the number of examples.

Clearly, for most of our benchmarks, the resulting solutions have the same depths, and thus do not significantly differ from each other. For a few benchmarks, however, we witnessed certain anomalies with the solving time, which we believe can be explained by the greediness of the algorithm and a large number of computed candidate relationships. In the future, we would like our procedure to invest effort in optimizing this better.

8   Related Work

Our work is broadly related to automated synthesis as well as verification techniques that utilize decision procedures.

Synthesis Techniques. Many successful instances of the general synthesis framework are based on enumerative search, where a user-provided grammar is used to constrain the space of candidate programs, along with checking correctness with respect to a specification. These include techniques that collect input-output examples lazily, by querying the specification [2,3]. In contrast, our approach deals with input-output examples only if they are explicitly given. More importantly, our technique does not require any additional templates or grammar. In this respect, our technique is closer to functional synthesis approaches [11,21,23] that directly formulate the synthesis tasks as quantified formulas to be solved by decision procedures. However, deriving compact implementations continues to be a challenge and provides the motivation for the new ideas developed in this paper. A compaction algorithm, employed by [3], proceeds by repairing the decision trees guided by new examples. In contrast, our approach performs a *global search*: a compact decision tree is constructed at once, by taking into account all available examples. In other words, our algorithm never revisits the upper levels of already constructed decision trees and never asks for more examples.

Another class of techniques has been successfully used for synthesis of programs *only* by examples, e.g., string and other transformations in spreadsheets [14,26,27,31]. These often require a domain-specific grammar or some type specifications to constrain the search for programs. Since a set of examples is often incomplete in practice, some generalization in dealing with examples is useful, e.g., via interaction with the user [9] or by using machine learning techniques [5,8,25]. We are inspired by the success of these techniques and the relative ease with which users can provide examples. However, our focus is strictly on numerical domains only, and we have experimented with applications in the area of reactive synthesis [17]. As mentioned earlier, a straightforward application of existing functional synthesis techniques on such input-output examples results in large implementations. Our motivation is to find smaller implementations. We expect that our completely automated technique could be potentially used as a submodule within a broader synthesis framework targeting a richer domain.

Table constraints, which express the combinations of values of variables that are allowed or forbidden, are widely used in Constraint Programming. Several heuristics to compress tables have been proposed [6,16,18,30,32]. While the table compression task can be seen as a generalization of our PBE task, none of these approaches proceeds further and generates an implementation from the compressed tables.

Verification Techniques. Our technique for finding symbolic linear relationships among examples is similar to techniques [10,13,20,22,24] for synthesis of invariants in program verification. In particular, these techniques can generate formulas from concrete values of program variables while discovering inductive invariants of loops. In this line of work, various feasible paths are obtained using execution or symbolic execution to generate data with values of all variables. An invariant requires generating a relation over all program variables that transit through the loop. Functional synthesis tasks, such as the one we are solving, aim at embedding a function into this relation, thus requiring more work.

9 Conclusions

We have presented a novel approach to synthesis that leverages PBE specifications and uses an FS framework for LIA. Our approach discovers preconditions and local Skolem terms by iterative partitioning of the set of examples into subsets. Each subset is described using detected relationships over inputs and outputs, which are directly used in the resulting implementations. The approach is easily extendable to deal with hybrid tasks, which are formulated in part by examples and in part by FS specifications. Our implementation on top of AE-VAL exhibits a promising performance on a set of reactive synthesis benchmarks enhanced with examples. Decision trees produced by our tool are an order of magnitude smaller than ones produced by the "plain" AE-VAL. In the future, we would like to extend this approach to other theories, such as arrays, strings, and algebraic data types, as well as to adopt more advanced ordering criteria and strategies for solution counting [4].

Acknowledgments. This work was supported in part by NSF Grant 1525936. Any opinions, findings, and conclusions expressed herein are those of the authors and do not necessarily reflect those of the NSF.

References

1. Alur, R., et al.: Syntax-guided synthesis. In: FMCAD, pp. 1–17. IEEE (2013)
2. Alur, R., Černý, P., Radhakrishna, A.: Synthesis through unification. In: Kroening, D., Păsăreanu, C.S. (eds.) CAV 2015. LNCS, vol. 9207, pp. 163–179. Springer, Cham (2015). https://doi.org/10.1007/978-3-319-21668-3_10
3. Alur, R., Radhakrishna, A., Udupa, A.: Scaling enumerative program synthesis via divide and conquer. In: Legay, A., Margaria, T. (eds.) TACAS 2017. LNCS, vol. 10205, pp. 319–336. Springer, Heidelberg (2017). https://doi.org/10.1007/978-3-662-54577-5_18
4. Beldiceanu, N., Simonis, H.: A constraint seeker: finding and ranking global constraints from examples. In: Lee, J. (ed.) CP 2011. LNCS, vol. 6876, pp. 12–26. Springer, Heidelberg (2011). https://doi.org/10.1007/978-3-642-23786-7_4
5. Bhatia, S., Kohli, P., Singh, R.: Neuro-symbolic program corrector for introductory programming assignments. In: ICSE, pp. 60–70. ACM (2018)
6. Charlier, B.L., Khong, M.T., Lecoutre, C., Deville, Y.: Automatic synthesis of smart table constraints by abstraction of table constraints. In: IJCAI, pp. 681–687. ijcai.org (2017)
7. de Moura, L., Bjørner, N.: Z3: an efficient SMT solver. In: Ramakrishnan, C.R., Rehof, J. (eds.) TACAS 2008. LNCS, vol. 4963, pp. 337–340. Springer, Heidelberg (2008). https://doi.org/10.1007/978-3-540-78800-3_24
8. Devlin, J., Uesato, J., Bhupatiraju, S., Singh, R., Mohamed, A., Kohli, P.: Robust-Fill: Neural program learning under noisy I/O. In: ICML, vol. 70, pp. 990–998. PMLR 2017
9. Drachsler-Cohen, D., Shoham, S., Yahav, E.: Synthesis with abstract examples. In: Majumdar, R., Kunčak, V. (eds.) CAV 2017. LNCS, vol. 10426, pp. 254–278. Springer, Cham (2017). https://doi.org/10.1007/978-3-319-63387-9_13
10. Ernst, M.D., Czeisler, A., Griswold, W.G., Notkin, D.: Quickly detecting relevant program invariants. In: ICSE, pp. 449–458. ACM (2000)
11. Fedyukovich, G., Gurfinkel, A., Gupta, A.: Lazy but effective functional synthesis. In: Enea, C., Piskac, R. (eds.) VMCAI 2019. LNCS, vol. 11388, pp. 92–113. Springer, Cham (2019). https://doi.org/10.1007/978-3-030-11245-5_5
12. Fedyukovich, G., Gurfinkel, A., Sharygina, N.: Automated discovery of simulation between programs. In: Davis, M., Fehnker, A., McIver, A., Voronkov, A. (eds.) LPAR 2015. LNCS, vol. 9450, pp. 606–621. Springer, Heidelberg (2015). https://doi.org/10.1007/978-3-662-48899-7_42
13. Fedyukovich, G., Prabhu, S., Madhukar, K., Gupta, A.: Solving constrained horn clauses using syntax and data. In: FMCAD, pp. 170–178. ACM (2018)
14. Gulwani, S.: Automating string processing in spreadsheets using input-output examples. In: POPL, pp. 317–330. ACM (2011)
15. Huang, K., Qiu, X., Wang, Y.: DRYADSYNTH: a concolic SyGuS solver (2019). https://github.rcac.purdue.edu/cap/DryadSynth
16. Jefferson, C., Nightingale, P.: Extending simple tabular reduction with short supports. In: IJCAI, pp. 573–579. IJCAI/AAAI (2013)
17. Katis, A., Fedyukovich, G., Guo, H., Gacek, A., Backes, J., Gurfinkel, A., Whalen, M.W.: Validity-guided synthesis of reactive systems from assume-guarantee contracts. In: Beyer, D., Huisman, M. (eds.) TACAS 2018. LNCS, vol. 10806, pp. 176–193. Springer, Cham (2018). https://doi.org/10.1007/978-3-319-89963-3_10
18. Katsirelos, G., Walsh, T.: A compression algorithm for large arity extensional constraints. In: Bessière, C. (ed.) CP 2007. LNCS, vol. 4741, pp. 379–393. Springer, Heidelberg (2007). https://doi.org/10.1007/978-3-540-74970-7_28

19. Komuravelli, A., Gurfinkel, A., Chaki, S.: SMT-based model checking for recursive programs. In: Biere, A., Bloem, R. (eds.) CAV 2014. LNCS, vol. 8559, pp. 17–34. Springer, Cham (2014). https://doi.org/10.1007/978-3-319-08867-9_2
20. Krishna, S., Puhrsch, C., Wies, T.: Learning invariants using decision trees. CoRR abs/1501.04725 (2015)
21. Kuncak, V., Mayer, M., Piskac, R., Suter, P.: Functional synthesis for linear arithmetic and sets. STTT 15(5–6), 455–474 (2013)
22. Padhi, S., Sharma, R., Millstein, T.D.: Data-driven precondition inference with learned features. In: PLDI, pp. 42–56. ACM (2016)
23. Reynolds, A., Deters, M., Kuncak, V., Tinelli, C., Barrett, C.: Counterexample-guided quantifier instantiation for synthesis in SMT. In: Kroening, D., Păsăreanu, C.S. (eds.) CAV 2015. LNCS, vol. 9207, pp. 198–216. Springer, Cham (2015). https://doi.org/10.1007/978-3-319-21668-3_12
24. Sharma, R., Gupta, S., Hariharan, B., Aiken, A., Liang, P., Nori, A.V.: A data driven approach for algebraic loop invariants. In: Felleisen, M., Gardner, P. (eds.) ESOP 2013. LNCS, vol. 7792, pp. 574–592. Springer, Heidelberg (2013). https://doi.org/10.1007/978-3-642-37036-6_31
25. Singh, R.: BlinkFill: semi-supervised programming by example for syntactic string transformations. PVLDB 9(10), 816–827 (2016)
26. Singh, R., Gulwani, S.: Synthesizing number transformations from input-output examples. In: Madhusudan, P., Seshia, S.A. (eds.) CAV 2012. LNCS, vol. 7358, pp. 634–651. Springer, Heidelberg (2012). https://doi.org/10.1007/978-3-642-31424-7_44
27. Singh, R., Gulwani, S.: Predicting a correct program in programming by example. In: Kroening, D., Păsăreanu, C.S. (eds.) CAV 2015. LNCS, vol. 9206, pp. 398–414. Springer, Cham (2015). https://doi.org/10.1007/978-3-319-21690-4_23
28. Solar-Lezama, A., Tancau, L., Bodík, R., Seshia, S.A., Saraswat, V.A.: Combinatorial sketching for finite programs. In: ASPLOS, pp. 404–415. ACM (2006)
29. Torlak, E., Bodík, R.: A lightweight symbolic virtual machine for solver-aided host languages. In: PLDI, pp. 530–541. ACM (2014)
30. Verhaeghe, H., Lecoutre, C., Deville, Y., Schaus, P.: Extending compact-table to basic smart tables. In: Beck, J.C. (ed.) CP 2017. LNCS, vol. 10416, pp. 297–307. Springer, Cham (2017). https://doi.org/10.1007/978-3-319-66158-2_19
31. Wang, X., Dillig, I., Singh, R.: Program synthesis using abstraction refinement. PACMPL 2(POPL), 1–30 (2018)
32. Xia, W., Yap, R.H.C.: Optimizing STR algorithms with tuple compression. In: Schulte, C. (ed.) CP 2013. LNCS, vol. 8124, pp. 724–732. Springer, Heidelberg (2013). https://doi.org/10.1007/978-3-642-40627-0_53

SolverCheck: Declarative Testing
of Constraints

Xavier Gillard[(✉)] [iD], Pierre Schaus[(✉)] [iD], and Yves Deville[(✉)]

Université Catholique de Louvain, Ottignies-Louvain-la-Neuve, BE, Belgium
{xavier.gillard,pierre.schaus,yves.deville}@uclouvain.be

Abstract. This paper introduces SolverCheck, a property-based testing (PBT) library specifically designed to test CP solvers. In particular, SolverCheck provides a declarative language to express a propagator's expected behavior and test it automatically. That language is easily extended with new constraints and flexible enough to precisely describe a propagator's consistency. Experiments carried out using Choco [41], JaCoP [27] and MiniCP [35] revealed the presence of numerous nontrivial bugs, no matter how carefully the test suites of these solvers have been engineered. Beyond the remarkable effectiveness of our technique to assess the correctness and robustness of a solver, our experiments also demonstrated the practical usability of SolverCheck to test actual CP-solvers.

Introduction

Constraint Programming (CP) owes much of its success to the declarative aspect of its models and the expressiveness of its constraints. Obviously, CP wouldn't have been the achievement we all know if it weren't for the efficiency of the propagators that have been devised over the years to enforce some degree of consistency for the constraints enlisted in the catalog [5]. E.g. *alldiff* [42], *regular* [40], *element* [24]. Nevertheless, the success of the tools developed in our community remains fragile as results of a solver might all be invalidated by a bogus implementation of one single propagator. As it turns out, the algorithms and data structures involved in those propagators are quite advanced and sometimes rely on state-restoration mechanisms. This is why, ensuring the correctness and robustness of their implementation is crucial to the success of CP as a whole. However, checking the correctness of a propagator by focusing solely on the absence of solution removal is far from enough. Indeed, in order to be able to tackle real world problems, it is essential that a solver be both correct and efficient. In practice, the efficiency of a propagator results from a balance between the strength of the enforced consistency and the complexity of the algorithm used to implement it. Hence, being able to test the consistency level imposed by a propagator becomes a necessity. Else, should the consistency

Massart and Rombouts have worked on a preliminary version of this work for their MSc. thesis which we supervised. They presented it at the CP-2018 Doctoral Programme [33].

ⓒ Springer Nature Switzerland AG 2019
T. Schiex and S. de Givry (Eds.): CP 2019, LNCS 11802, pp. 565–582, 2019.
https://doi.org/10.1007/978-3-030-30048-7_33

be weaker than announced, some problem instances might become intractable and that intractability could hardly be analyzed or reasoned about.

In that context, we propose SolverCheck: an open-source property-based testing (PBT) library inspired by QuickCheck [13] for Haskell. It has been specifically designed and engineered to improve the quality of the tests used to validate CP solvers. In practice, SolverCheck makes it easy to both test the *correctness* of the propagators and to test the level of consistency enforced by the latter. Moreover, SolverCheck aims at being an extensible framework. Therefore, it comes with simple interfaces through which a user can easily describe the relation imposed by a new constraint. Concretely, this relation is described using a *Checker*, a predicate deciding whether or not a tuple belongs to the constraint relation. Similarly, the consistency level that can be tested need not necessarily be one of the classical consistency level (DC, BC(D), BC(Z), RC, FC) [6] as SolverCheck permits the definition of custom mixed consistencies matching the exact expected behavior of some given propagator. Additionally, SolverCheck is able to perform *dynamic checking* and hence to explicitly test the correctness of the state-restoration mechanisms involved in the targeted propagators.

Our contribution with this paper is the following: we propose SolverCheck as a DSL and tool to help improve the quality and robustness of JVM-based CP solvers. Given the very implementation-minded nature of the CP community, we hope that it can benefit the whole community and foster further innovation in the same way as Minizinc [36], XCSP3 [7], CPViz [44], Essence [23], etc. have in the past.

Outline. The rest of our paper is organized as follows: Sects. 1 and 2 present the background material necessary to understand the purpose and methodology applied in SolverCheck. Then, Sect. 3 briefly presents other related lines of research and how these relate to our work. After that, Sect. 4 introduces the various capabilities of SolverCheck through a simple yet illustrative example. Section 5 gives some more details relative to the implementation of our tool. Finally, Sect. 6 reports on the experiments that were made to validate the effectiveness and practical usability of SolverCheck before conclusions are drawn in Sect. 7.

1   Property-Based Testing

SolverCheck adopts the so-called *property-based testing* paradigm which tackles the weaknesses of the classical *example-based testing* methodology. All the open-source solvers that we are aware of, in particular Gecode [46], Choco [41], JaCop [27], Or-tools [38], OscaR [39], and MiniCP [35], maintain a test suite to test the solver at the granularity of the constraints. The test suites of most of the solvers[1] follow the classical example-based approach.

---

[1] Gecode, and likewise Choco for some of its propagators, are a notable exception which is covered in the related work section.

As the name suggests, example-based testing relies on a tester to describe concrete situations (example, with actual variables and domain instantiation) supposedly representative of a class of errors. By combining many such examples, the tester creates a broad test suite covering a large number of potential problems. However, we point out two weaknesses of this approach. First, example-based unit tests are expensive to write and to maintain. Manually finding interesting instances to test is no easy task. It requires some expertise and intuition. Also, test code is often treated as a second class citizen: the quality standards applied to that fraction of the code are less stringent than for the rest of the code base. Therefore, it results that the code composing the test suites is often crippled with duplicate fragments. Moreover, the hard-coded instances fail to clearly communicate the intent regarding what important property is being tested with a given example. For instance, the objective of testing a global constraint's consistency level does not shine from any given test example. Add to that picture the fact that example-based tests often opt for an all imperative coding style, and the original goal of the test becomes difficult to grasp. Meanwhile, example-based testing does not offer any means to improve on that floor or to test that kind of property in a generic way.

Property-based testing (PBT) addresses those weaknesses by a combination of *fuzzing* [45] and *formal specification*. Doing so, PBT changes the role of the test engineer. With PBT he must express the general *properties* that must hold for all executions of a given software rather than manually crafting lots of *test cases* (example-based testing). These properties are expressed in a high-level declarative language which abstracts away the details of actual test cases. As the name suggests, this method is *test-based*. Hence, it is inherently incomplete. Nevertheless, moving the burden of actual test case generation from the human tester to an automated tool makes PBT a remarkably effective approach to identifying bugs in practice.

2 Mixed Consistency

In order to solve a CSP, filters are used to reduce the search space. A filter applied on a constraint aims at establishing some consistency property of this constraint by removing some values in the domain of its variables, without removing any solutions. We thus hereby only consider filters for *domain-based* consistencies. That is filters reducing the domain of variables.

In particular, we would like to set the focus on filters where different consistencies are mixed in a constraint. The idea of mixed consistency is to maintain different levels of consistency on the different variables of a constraint. The concept of mixed consistency has been introduced in [17] to handle graph and set variables. It is also used in [29,31].

A *constraint* c over the variables $(x_1, \ldots, x_n)$ (its scope) is a relation over the values of the variables. Like any relation, a constraint c can either be defined *in extension* as the set of all the n-tuples belonging to c. Or it can be defined *in comprehension* using a *checker* $c(v_1, \ldots, v_n)$ stating if $\langle v_1, \ldots, v_n \rangle$ belongs to c.

The *domain* of a variable x is a finite set of discrete values $D(x) \subset \mathbb{Z}$. It inherits the usual properties of proper finite subsets of $\mathbb{Z}$. In particular, it is either empty or it has a minimum (noted $lb(D(x))$) and a maximum (noted $ub(D(x))$). We denote D the set of tuples $D(x_1) \times \ldots \times D(x_n)$. A tuple $\tau = \langle v_1, \ldots, v_n \rangle$ is said to be valid if $\tau \in D$. The element v_i of the tuple is denoted τ_i.

A *partial assignment* is a mapping associating a domain to each variable.

The idea of support is central in the notion of consistency. Intuitively, a support of a value of a variable is a valid tuple, involving this value and satisfying the constraint. The definition of support can also be extended by considering sets larger than the actual variables domains. For instance, one can consider all the integer values between the bounds of the domain. We define $D^{\mathbb{Z}}(x) = \{v \in \mathbb{Z} \mid lb(D(x)) \leqslant v \leqslant ub(D(x)\}$.

Definition 1. *A* support *on c of (x_i, v) in D is a tuple $\tau \in D$ such that $\tau \in c$ and $\tau_i = v$. A* bound(Z) support *on c of (x_i, v) in D is a tuple $\tau \in D^{\mathbb{Z}}$ such that $\tau \in c$ and $\tau_i = v$.*

Different classical levels of consistency can now be defined. Each consistency, however, focuses on a single variable of the constraint. This will allow to later combine them in a mixed consistency.

Definition 2. *(DC) A constraint c is* domain consistent *on x_i with respect to D iff $\forall v \in D(x_i)$, there exists a support on c of (x_i, v) in D.*

Definition 3. *(RC) A constraint c is* range consistent *on x_i with respect to D iff $\forall v \in D(x_i)$, there exists a bound(Z) support on c of (x_i, v) in D.*

Definition 4. *(BC(D)) A constraint c is* bound(D) consistent *on x_i with respect to D iff $(x_i, lb(D(x_i)))$ and $(x_i, ub(D(x_i))$ have a support on c in D.*

Definition 5. *(BC(Z)) A constraint c is* bound(Z) consistent *on x_i with respect to D iff $(x_i, lb(D(x_i)))$ and $(x_i, ub(D(x_i))$ have a bound(Z) support on c in D.*

Definition 6. *(FC) A constraint c is* forward checking consistent *on x_i with respect to D iff when forall $j \neq i$ $D(x_j)$ is a singleton, then c is domain consistent on x_i with respect to D.*

Mixed consistency can now be defined with a consistency level associated to each variable of the constraint.

Definition 7. *Let $\Phi = \langle \phi_1, \ldots, \phi_n \rangle$ with $\phi_i \in \{DC, RC, BC(D), BC(Z), FC\}$. The constraint c is Φ* mixed consistent *with respect to D iff c is ϕ_i consistent on x_i with respect to D.*

When Φ is a constant tuple, the above definition reduces to the standard definition of domain consistency or to the other standard levels of consistencies.

Given a consistency ϕ and a constraint c, we associate a filter $\phi_c(x, D)$ yielding a domain D' such that $D' \subseteq D$, $c \cap D = c \cap D'$ (same solutions) and c is ϕ_i consistent on x with respect to D'. A filter $\Phi_c(D)$ is also associated to a tuple of consistency Φ. It yields a domain D' such that $D' \subseteq D$, $c \cap D = c \cap D'$ (same solutions) and c is Φ mixed consistent with respect to D'.

Example 1. Given an array A of integer values, and two variables x, y, the *element(A,x,y)* constraint [24] is satisfied whenever $A[x] = y$. It is not uncommon for CP-solvers to implement a filter achieving the mixed consistency $(RC, BC(D))$ on the two variables (x, y). This kind of filter ensures that all values in the domain of x have a bound(Z) support, and that $lb(y)$ and $ub(y)$ have a support.

Algorithm 1 is a basic implementation of a filter, parameterized with a tuple of filters, achieving mixed consistency. This algorithm repeatedly reduces the domain of each individual variable x_i using its associated filter until a fixed point is reached. Assuming all the filters on the variables are correct, this algorithm yields a domain D' such that $D' \subseteq D$, $c \cap D = c \cap D'$ (same solutions) and c is Φ mixed consistent with respect to D'.

Algorithm 1: Filter achieving mixed consistency

```
1  Filter Φc(⟨φ1,...,φn⟩, D)
2      fixedpoint ← False ;
3      while !fixedpoint do
4          fixedpoint ← True;
5          foreach xi ∈ scope(c) do
6              D'xi ← φi(xi, D) ;
7              if D'xi.isEmpty() then return Fail ;
8              fixedpoint ← fixedpoint ∧ D(xi) = D'xi ;
9              D(xi) ← D'xi;
10     return D ;
```

Generally speaking, a filter of a constraint modifies domains. A filter f_c should be contracting $(f_c(D) \subseteq D)$ and idempotent, that is a repeated application of the filter does not further reduce the domain $(f_c(f_c(D)) = f_c(D))$. In [43], Schulte and Tack have shown that weak monotony is the minimal necessary condition that any filter must fulfill in order to guarantee the soundness and the completeness of constraint propagation. A filter f_c is weakly monotonic iff $\forall D, \forall v \in D : f_c(\{v\}) \subseteq f_c(D)$. A correct filter for some constraint c is thus necessarily weakly monotonic and contracting. The corollary of this property is that a correct filter behaves as the checker applied to a singleton domain (i.e. an assignment).

Two filters f_1, f_2 of a given constraint can be compared thanks to their relative *strength*. A filter f_1 is said to be *stronger* than f_2 (noted $f_1 \sqsubseteq f_2$) iff $\forall D : f_1(D) \subseteq f_2(D)$. Similarly, f_1 is said to be *weaker* than f_2 ($f_1 \sqsupseteq f_2$) iff f_2 is stronger than f_1. Finally, f_1 and f_2 are *equivalent* whenever both $f_1 \sqsubseteq f_2$ and $f_1 \sqsupseteq f_2$ hold.

This paper aims at comparing filters. Therefore, we will say that a filter realizes a given consistency ϕ if it does not remove any further values than the ones required per the consistency definition. That is, we say that a filter realizes the consistency ϕ iff it is the *weakest filter* (removing as few values as possibly can) complying with the definition of ϕ. Any other filter also realizing ϕ that removes additional (non-solution) values is therefore *stronger than* ϕ.

Example 2. It is clear from Definitions 2 and 4 that whenever a filter f realizes DC, it also realizes BC(D). However, f possibly removes more values than a hypothetical filter g that enforces BC(D) but not DC. Hence, we have $f \sqsubseteq g$. Thus, f is not the *weakest* filter realizing BC(D). Therefore, in this paper, we would not say that f is equivalent to BC(D) – although it realizes that consistency. Instead, we would say that f is *stronger than* BC(D).

3   Related Work

The purpose of our research differs from the line of work started in the late '80s [14,15,20,28,34]. Indeed, that rich body of investigations aimed at verifying whether the *CP program* (today, one would rather talk about CP *model* instead) was correct. SolverCheck, on the other hand, aims at testing the implementation of a CP solver, which is a different concern by large. It also differs from the research embodied in FocalTest [11] which uses CP to define *smart generators* for PBT. Instead, SolverCheck provides a PBT library to assess the correctness and robustness of CP solvers.

Even though the properties to be tested are formally specified, SolverCheck is a *testing library*, not a formal verification tool. That distinction typically makes it simpler to use. Indeed, despite the many advances in the domain, proof-checkers for general purpose languages either require some human guidance, do not support all language constructs [1,4], or are currently unable to deal with programs as large and complex as modern CP-solvers [21,22,26]. Similarly, as of today, formally certified CP solvers [12,19] are nowhere close to the state of refinement and efficiency of state-of-the-art solvers. For instance, these rely on (efficient but suboptimal) OCaml code extracted from Coq [47] and only support constraints of arity greater than 3 through a decomposition into equivalent binary constraints (using the hidden variable encoding) [18].

Recently, the SAT/SMT/ASP/QBF communities have undertaken a line of work that closely relates to ours [3,8,9,37]. Just like SolverCheck, these techniques also apply fuzzing in order to ensure the quality of the tools they develop. However, that body of work ignores the specifics of a CP solver. In particular, they disregard consistency related issues (mixed or not). Meanwhile, as explained earlier, this is one of the essential aspects of the reasoning and development of a CP solver.

As it has already been mentioned, Gecode [46] and Choco [41] adopt an original test strategy which allows them to test the consistency (DC, BC(D)) imposed by some of their propagators[2]. Their approach, albeit elegant and efficient, is unable to deal with mixed consistencies (eg. that of the *element* [24] constraint).

---

[2] Actually, both solvers adopt a slightly different approach, but this is not relevant for our matter as they are based on the same idea. For the full details, see http://bit.ly/cst-gecode and http://bit.ly/cst-choco.

Last year, Akgün et al. proposed at the CP conference an interesting approach based on metamorphic testing [2] to test the implementation of a solver. Their goal, as well as their initial intuition is the same as those behind SolverCheck. Both target the testing of propagators implemented in actual CP solvers, and both rely on having two distinct implementation of each filter. However, their approach relies on the *table* propagator from the target solver and requires the test-engineer to provide a table with all the solutions of the tested constraint (the authors of [2] provide a python function to help alleviate that burden). SolverCheck uses a different approach: it automatically derives a naive alternate implementation of the propagator which is completely independent from the target solver. Moreover, SolverCheck sets the focus on mixed consistencies, which is not the case of [2]. Additionally, the approach used in SolverCheck makes it easy to test properties that do not depend on a specific consistency level such as "stronger filtering". This kind of comparison is particularly well suited to compare the filtering for NP-hard constraints such as bin-packing [16].

4 What SolverCheck Has to Offer

We will use the example reproduced in Listing 1.1 as a starting point. The latter is actually the verbatim copy of a property we specified when writing a test suite for JaCoP.

Listing 1.1: Example: JaCoP LexOrder($\leq$) must enforce GAC.

```
 1  @Test
 2  public void statelessLexLE() {
 3    assertThat(
 4      forAll(listOf("x", jDom())).assertThat(x ->
 5      forAll(listOf("y", jDom())).assertThat(y ->
 6      a(statelessJacopLexOrder(false))
 7       .isEquivalentTo(arcConsistent(lexLE(x.size(), y.size())))
 8       .forThePartialAssignment(x, y)
 9    )));
10  }
11
12  // Generate a domain respecting JaCoP's documented limits
13  public GenDomainBuilder jDom() {
14    return domain().withValuesBetween(
15      IntDomain.MinInt,
16      IntDomain.MaxInt);
17  }
18  // Discriminate solutions from non-solutions
19  public Checker lexLE(int x_sz, int y_sz) {
20    return assignment -> {
21      var xs = assignment.subList(0, x_sz);
22      var ys = assignment.subList(x_sz, x_sz+y_sz);
23      for (int i=0; i < min(x_sz, y_sz); i++) {
24        if (xs.get(i) < ys.get(i)) return true;
25        if (xs.get(i) > ys.get(i)) return false;
26      }
27      return x_sz <= y_sz;
28    };
29  }
30  // Adapter to expose the actual constraint as a SolverCheck Filter
31  private Filter statelessJacopLexOrder(final boolean lt) {
32    return partialAssignment -> {
33      Store store = new Store();
```

```
34    IntVar[][] vars = componentsToVars(store, partialAssignment);
35    store.impose(new LexOrder(vars[0], vars[1], lt));
36    if (!store.consistency()) {
37      return PartialAssignment.error(partialAssignment.size());
38    } else {
39      return vars2Partial(vars);
40    }
41  };
42 }
```

4.1 Declarative Testing

The declarative aspect of the test code reproduced in Listing 1.1 is obvious. No mention is ever made in the code about any concrete test case. Instead, that code snippet uses a declarative style close to that of a domain-specific-language to express a *property*, a *specification* of what the code should do. The details of the actual tests that are used to validate the implementation are left to the system. Assuming a basic knowledge of Java, it is clear from Listing 1.1 that any reader – familiar with SolverCheck or not – will grasp the expressed property (lines 3–9). In our example, it states that for any two given lists x and y of variables, the filtering of the domains imposed by the actual LexOrder constraint from JaCoP should strictly enforce domain consistency.

All the other functions declared in the example are actually utility methods: jDom() (lines 13–17) provides a means to generate pseudo-random domains [3] having their values in the range of values accepted by JaCoP. The lexLE() method (lines 19–29) returns a Checker for the Lex constraint. That is, it returns a predicate deciding whether or not an assignment belongs to the constraint relation. The Checker API is SolverCheck's mechanism to test constraints that are not built in the framework. The statelessJacopLexOrder() method (lines 31–42) adapts the actual constraint from JaCoP (LexOrder) and exposes it as a Filter that SolverCheck can interact with.

4.2 Consistency

Despite its apparent simplicity, the example from Listing 1.1 is a good illustration of the flexibility provided by SolverCheck. It shows how to parameterize the consistency level used to test a given propagator. It would only take a change of line 8 in the example to modify the property expressed in Listing 1.1 and let it state that the propagator should enforce BC(Z) rather than DC. For that purpose, the only change required would be to replace arcConsistent(lexLE(x.size(), y.size())))) by boundZConsistent(lexLE(x.size(), y.size())))).

Because solvers developers tend to be pragmatic people who favor general case efficiency over the compliance to pure mathematical consistency definitions, it is often the case that discrepancies exist between the implemented artifacts and the theoretical framework. To cope with that reality, SolverCheck offers facilities to express that a filtering should be stronger than (isStrongerThan(·)),

---

[3] sets of pseudo-random int.

weaker than (isWeakerThan($\cdot$)) or equivalent to (isEquivalentTo($\cdot$)) a given consistency level. This is illustrated by line 7 in our illustrating example. However, a relative positioning wrt a "standard" consistency level might be deemed too weak. This is why SolverCheck also supports the definition of custom mixed consistencies. The example of Listing 1.2 illustrates how the exact mixed consistency of a propagator is specified with SolverCheck (line 8). That example shows that for any array A of integer and pair of variables x and y, MiniCP's $element(A, x, y) \equiv A[x] = y$ constraint does not comply with any of the standard consistencies. Instead, the property states that each value in the domain of x should have a support in y whereas only the upper and lower bounds of $D(y)$ should have a support in x. Additionally, this example illustrates (line 3) how a time limit can be set to check a property.

Listing 1.2: A[x] = y has a mixed consistency

```
1  @Test
2  public void elementIsHybridConsistent() {
3    given(TIMELIMIT, TimeUnit.SECONDS).assertThat(
4      forAll(listOf("A", integer())).assertThat(A ->
5      forAll(domain("x")).assertThat(x ->
6      forAll(domain("y")).assertThat(y ->
7        a(minicpElement1D(A))
8          .isEquivalentTo(hybrid(element(A), rangeDomain(), bcDDomain()))
9          .forThePartialAssignment(x, y)
10     ))));
11 }
```

4.3 Extensibility

The example from Listing 1.1 also illustrates how SolverCheck's capabilities can be extended to support constraints that were not initially foreseen[4]. To that end, it suffices to implement a new **Checker** for the desired constraint. That is a predicate on assignment which is true iff the assignment belongs to the constraint relation.

On top of the assertions meant to test the strength of a propagator, SolverCheck provides several extension points making it possible to check virtually any property of the tested filter. For instance, in the snippet a(tested).is(property), the method is($\cdot$) will accept any predicate on partial assignments for its property argument. In particular, this is how the checks isContracting(), isIdempotent() and isWeaklyMonotonic() have been implemented in the library.

4.4 Dynamic Checking

Because there are many cases where existing solvers implement the filtering of their constraints as *incremental propagators*, they do not exactly fit the ideal of *pure* filtering functions having no side effects. Indeed, propagators hold and

---

[4] SolverCheck comes with built-in checkers for the usual constraints alldiff, element, gcc, etc.

manipulate some internal state in order to deliver an efficient filtering in practice. But the efficiency gains often come at the expense of an increased risk of error. In order to detect the bugs caused by this internal state handling, SolverCheck proposes two operating modes.

Static Checking. Pseudo random test cases are fed to the filter. Then, the library tests if the outcome of *one application* of the filter delivers the expected result. This corresponds to the way of using SolverCheck which has been presented until now.

Dynamic Checking. The tested solver searches through the state-space, making branch decisions and backtracking, in conditions similar to those of an actual CSP solving.

Algorithm 2 describes how dynamic checks operate. This algorithm accepts five parameters: two stateful filters *trusted* and *tested* matching the interface of Listing 1.3, a property *prop* that must hold of all executions, a natural number N and a pseudo-randomly generated partial assignment *pa*. As opposed to static checking, dynamic checking considers the partial assignment *pa* as the root of a search tree and explores a fraction of that search tree with a series of N dives. That is, it dives N times in the search tree until a leaf (assignment or error) is reached (lines 7–12). At that moment, the library rolls back a few decisions it made when diving (lines 13–15). Then it starts the exploration of a new branch. At each visited node of the search tree, the current states of *tested* and *trusted* are compared to check whether the property is verified (line 12).

Algorithm 2: Dynamic checking, the *dive* algorithm

---

1 **Dive** *(trusted, tested, prop, N, pa)*
2 | trusted.initialize(pa); tested.initialize(pa);
3 | **if** *not prop.holds(trusted, tested)* **then** **fail**;
4 | **for** *N times* **do**
5 | | CheckBranch(trusted, tested, prop, pa);
6 | | BackJump(trusted, tested);

7 **CheckBranch** *(trusted, tested, prop, pa)*
8 | **while** *neither trusted nor tested reached a leaf* **do**
9 | | trusted.saveState(); tested.saveState();
10 | | decision ← RandomDecision(pa);
11 | | trusted.branchOn(decision); tested.branchOn(decision);
12 | | **if** *not prop.holds(trusted, tested)* **then** **fail**;

13 **BackJump** *(trusted, tested)*
14 | **while** *not trusted.isAtRootLevel() and RandomBool()* **do**
15 | | trusted.restoreState(); tested.restoreState();

Writing Dynamic Checks, in Practice. As is made clear by the previous paragraphs, using the dynamic checking mode requires slightly more work from the human tester. Indeed, rather than writing a stateless `Filter` adapter similar to the one shown in Listing 1.1 (lines 31–42), the tester must write an adapter matching the `StatefulFilter` interface (Listing 1.3).

The `branchOn()` method stands for the addition of usual branching conditions such as $\neq$, $<$, $>$. The `pushState()` and `popState()` methods adopt the terminology used in trail-based solvers where these methods are at the heart of the backtracking mechanism (see [35] for further information on that matter).

For the rest, thanks to the declarative nature of SolverCheck, the code remains almost identical to what is required in the static case.

Listing 1.3: Interface of a Stateful Filter in SolverCheck

```
1  public interface StatefulFilter {
2      void setup(PartialAssignment initialDomains);
3      void pushState();
4      void popState();
5      void branchOn(int variable, Operator op, int value);
6      PartialAssignment currentState();
7  }
```

5   Implementing SolverCheck

SolverCheck posits that the *implementation* of current CP solvers have become incredibly efficient at the expense of an increased code complexity. Therefore, they no longer fit in Hoare's *"obviously no deficiencies"* category [25]. The idea behind SolverCheck is then fairly simple: the library tests the behavior of an actual (complicated) CP filter by simply comparing it with that of a (generated) implementation that is so outright simple that it is straightforward for anyone to *trust* that second implementation to be correct. In SolverCheck parlance, the filters obtained from a generated implementation are called *trusted filters*.

5.1   Deriving Trusted Filters

As explained in Sect. 4.3, SolverCheck trusted filters rely on a `Checker` to decide whether or not an assignment belongs to the tested constraint. On that basis, a naive but easily trusted filter implementation immediately follows from the definitions of Section 2. For that matter, one needs to distinguish *uniform* consistencies (DC, BC(D), BC(Z), RC, FC) from *mixed* consistencies.

Uniform Consistencies. Given a checker c, a consistency γ and the partial assignment $pa = (D(x_1), \ldots, D(x_n))$ the trusted filter $\phi_{c,\gamma}$ proceeds as follows. It starts by computing the set $D_{sol} = \{\tau \in D^* \mid c(\tau)\}$ of solutions, where D^* stands for either D (when $\gamma \in \{DC, BC(D)\}$) or $D^{\mathbb{Z}}$ (when $\gamma \in \{RC, BC(Z)\}$). Then it computes a partial assignment pa' associating the domain $D'(x_i) =$

$\bigcup_{\tau \in D_{sol}} \tau_i$ to each variable x_i. Finally, it uses pa' to filter the original domains in pa according to the rules of γ. This is the final output of $\phi_{c,\gamma}$.

Deriving a trusted filter implementation for the uniform FC consistency is trivial: when the domains of all but one variable are singletons, the filter behaves as in the DC case. Otherwise, the original partial assignment is returned.

Mixed Consistencies. The trusted filters derived for mixed consistencies are a direct implementation of Algorithm 1. There the filters ϕ_i used to filter the domain of each variable x_i simply consist in the application of a uniform trusted filter which is then projected over x_i.

5.2   Generation of Pseudo-random Test Cases

In order to check that a property holds, SolverCheck generates pseudo-random test cases which are fed to both the trusted and tested filters. Because it is widely accepted among software engineers that extreme values often exhibit extreme behaviors, it was decided that SolverCheck would not use a uniform random distribution to generate its test inputs. Instead, it draws its values from a multi-modal distribution – the modes being the *usually problematic* values (zero, min and max). Doing so, it introduces a bias on the values occurring in the generated test cases.

6   Experimental Results

We conducted a series of experiments, all of which are based on three solvers[5]: Choco [41], JaCoP [27] and MiniCP [35]. These solvers have been chosen because, on the one hand, they run on the JVM which is our target platform; and on the other hand, because they have been carefully developed by domain experts. Among the large panel of possible constraints, we picked seven that were deemed representative of the kind of constraint typically available in a CP solver.

For each of the selected constraint, we present two distinct experiments. The first one aimed at evaluating the effectiveness of our library when it comes to detecting bugs in an actual solver. In practice, this experiment consisted in a phase of *exploratory testing* during which we went through the documentation of solvers/constraints and wrote specifications matching the documented behavior for each of the tested artifacts. The goal of our second experiment was to assess the usability of our library in practice. That is, we wanted to make sure SolverCheck could actually be used and be useful in a continuous integration setup. To that end, we measured the time it took for our exploratory tests to complete as well as the code coverage they achieved.

---

[5] Experiments were also realized using AbsCon [30]. However, even though we highlighted some defects in this solver, we chose not to report on the outcome of these experiments because we are still discussing some of our findings with the maintainer of that solver.

6.1   Exploratory Testing

We observed five different kind of outcomes during this experiment and summarize our findings in Table 1. The first possibility occurs when SolverCheck wasn't able to detect any mismatch between the tested propagators and their documented behavior (✔ in Table 1). An other possible result is observed when a propagator prunes more values than announced but never removes any solution (↑). The defective cases are split in three categories: the cases where a propagator was weaker than announced (↓), the cases where it provided an incorrect answer (✖) and those when an undesired behavior happened at runtime (⚡). Among others, this covers program crashes (cast errors, memory exhaustion, ...) and infinite loops. All of our findings have been reported to, and accepted by the solvers maintainers. As of today, the vast majority of the findings we reported have been fixed.

As shown per Table 1, SolverCheck was remarkably efficient at identifying discrepancies between the actual and documented behavior of implemented propagators. And that, even though all these propagators had already been carefully tested by their authors. The biased pseudo-random input generation used to produce "extreme" values naturally led to the identification of several over- and under-flows issues that are often counterintuitive for a human being. Table 1 shows however that it was far from the only type of error identified by our tool.

Table 1: Findings of the exploratory testing phase

Solver	Alldiff	Element	Table	Sum	GCC	Lex	Regular
Choco	↓✖	↓	⚡↓	↓✖	✔	⚡	✔
JaCoP	✔	↑	✖	✔	✔	↓	⚡
MiniCP	⚡	✔	↓✖	⚡✖	N/A	N/A	N/A

Errors in the State Management. The exploratory testing outdid our expectations wrt stateless issues detection. As a consequence, the stateful issues detection potential of our library remained unknown. Therefore, we conducted a variation of our exploratory testing experiment and designed it so as to specifically assess that potential. In practice, we manually introduced bugs in the state management of the stateful constraints. To that end, we replaced some *reversible integer* with its primitive counterpart in the source code. Then we used dynamic checking to test the properties of the targeted constraints. For all the seeded bugs, SolverCheck correctly reported a trace where the bug expressed itself.

Similarly, we also checked whether SolverCheck would identify bugs after we altered the implementation of the reversible structure to let it discard some modification from the trail. Again, SolverCheck reported a violation trace for all the cases that have been tested.

6.2 Effectiveness in Practice

The plots from Fig. 1 provide a good illustration of the behavior we observed when SolverCheck is used to test properties with a varying number of variables[6]. In particular, Fig. 1a plots the time it takes to test that the DC consistent propagator of each solver actually enforces DC with an increasing number of variables. While the exact duration of these tests is of little interest, the trend it indicates is informative. On the one hand, it clearly indicates an exponential duration blowup becoming significant beyond 12 variables. On the other hand, it also shows that dynamic checking consistently requires a longer amount of time to complete than a corresponding static check. The extreme similarity between the two groups of static and dynamic curves (the same graph with logarithmic y-axis does not show any major difference either) indicates that the test-completion time is dominated by SolverCheck rather than by the tested solver.

These observations had to be expected and directly stem from the algorithms implemented in our library. The need for our trusted filter to explicitly compute the set D_{sol} based on a filtering of D^* is sufficient to explain the exponential blowup in its own right. Similarly, the repeated application of filter operations (as per Algorithm 2) during dynamic checking explains why dynamic tests require longer to complete. Despite being expected and logically understood, both observations clearly highlight a limitation in the capabilities of our library: it does not scale and is not efficient when there are lots of variables to be considered (or when they have large domains).

That conclusion should nevertheless be contrasted by the information shown on Fig. 1b. It plots the line coverage of the tested propagators as measured during the tests whose runtime are plotted in Fig. 1a. From there, we first observe that the coverage stabilizes very quickly. As soon as three variables are considered, the coverage reaches a state where it marginally increases, if at all. We also

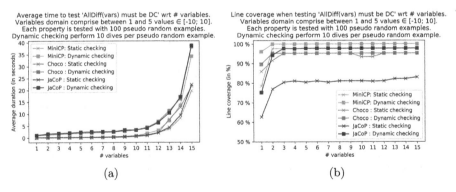

Fig. 1: Measuring the efficiency of SolverCheck for testing AllDiff (DC)

---

[6] The plots and observations to be made when it is the domain size that varies for a fixed number of variables are substantially the same as the case presented here (varying number of variables, fixed domain size). These are therefore omitted.

observe that in most cases, the tests cover about 95% of the lines. This is quite a high score, and is way above the usual 70–80% target from the industry. Finally, we observe that dynamic checking either improves or equates the static checking line coverage for all tests. The gap observed between the static and dynamic line coverage of JaCoP illustrates one of the benefits of dynamic checking. That strategy is able to exercise all parts of the propagators, including the ones related to state restoration of incremental propagators.

7   Conclusions and Future Work

In this paper we introduced SolverCheck, an open-source property-based testing library to effectively check the correctness of the propagators of any JVM-based solver. We showed how the library can be used to declaratively specify the properties which must hold for a constraint, and presented the two modes in which the tests can be operated.

Furthermore, we demonstrated the practical effectiveness of SolverCheck through an experimental study based on Choco, JaCoP and MiniCP. These results are promising as they show that our library has been able to identify bugs in the aforementioned solvers and provide counterexamples for each of the witnessed property violations. Besides that, we showed that SolverCheck is successful at its intended purpose. It can easily be integrated in the test suite of any JVM-based solver to produce a high quality set of tests (good coverage) that is easy to maintain. Moreover, given that SolverCheck allows a tester to control every aspect of how tests are generated, we are also confident that SolverCheck can be an integral part of the quality assurance process of any solver. In particular, checking properties with our library can seamlessly be integrated with the continuous integration of any JVM-based solver.

We envision several extensions of this work in the future. In particular, we believe that our library can be adapted and extended to cope with the specifics of scheduling constraints. For instance, it could be extended to generate trusted filters matching the filtering of an edge-finding propagator [10,32,48]. Also, it could be extended to target different classes of bugs. So far, SolverCheck is very good at finding value-related bugs like over/under-flows and logical errors in the propagation. We think that it would be interesting to leverage the features of SolverCheck to target aliasing issues which are also a common source of bugs in solvers supporting views. Beyond that, our library could benefit from the use of checkers that operate directly on partial assignments. With these, a trusted filter would not necessarily need to always test all assignments. Other possible extensions include microbenchmarking and the ability to test solvers outside of the JVM world through language-agnostic tests using MiniZinc [36] or XCSP3 [7].

References

1. Ahrendt, W., Beckert, B., Bubel, R., Hähnle, R., Schmitt, P.H., Ulbrich, M. (eds.): Deductive Software Verification - The KeY Book - From Theory to Practice. Lecture Notes in Computer Science, vol. 10001. Springer, Cham (2016). https://doi.org/10.1007/978-3-319-49812-6

2. Akgün, Ö., Gent, I.P., Jefferson, C., Miguel, I., Nightingale, P.: Metamorphic testing of constraint solvers. In: Hooker, J. (ed.) CP 2018. LNCS, vol. 11008, pp. 727–736. Springer, Cham (2018). https://doi.org/10.1007/978-3-319-98334-9_46

3. Artho, C., Biere, A., Seidl, M.: Model-based testing for verification back-ends. In: Veanes, M., Viganò, L. (eds.) TAP 2013. LNCS, vol. 7942, pp. 39–55. Springer, Heidelberg (2013). https://doi.org/10.1007/978-3-642-38916-0_3

4. Barnes, J.: SPARK: The Proven Approach to High Integrity Software. Altran Praxis, UK (2012). http://www.altran.co.uk

5. Beldiceanu, N., Carlsson, M., Rampon, J.X.: Global constraint catalog, 2nd edn, July 2010. Tech. Rep. Swedish Institute of Computer Science (2010)

6. Bessiere, C.: Constraint propagation. In: Rossi, F., van Beek, P., Walsh, T. (eds.) Handbook of Constraint Programming, Foundations of Artificial Intelligence, vol. 2, pp. 29–83. Elsevier, Amsterdam (2006)

7. Boussemart, F., Lecoutre, C., Piette, C.: XCSP3: An integrated format for benchmarking combinatorial constrained problems. arXiv preprint arXiv:1611.03398 (2016)

8. Brummayer, R., Järvisalo, M.: Testing and debugging techniques for answer set solver development. TPLP 10(4–6), 741–758 (2010)

9. Brummayer, R., Lonsing, F., Biere, A.: Automated testing and debugging of SAT and QBF solvers. In: Strichman, O., Szeider, S. (eds.) SAT 2010. LNCS, vol. 6175, pp. 44–57. Springer, Heidelberg (2010). https://doi.org/10.1007/978-3-642-14186-7_6

10. Carlier, J., Pinson, E.: Adjustment of heads and tails for the job-shop problem. Eur. J. Oper. Res. 78(2), 146–161 (1994). Project Management and Scheduling

11. Carlier, M., Dubois, C., Gotlieb, A.: FocalTest: a constraint programming approach for property-based testing. In: Cordeiro, J., Virvou, M., Shishkov, B. (eds.) ICSOFT 2010. CCIS, vol. 170, pp. 140–155. Springer, Heidelberg (2013). https://doi.org/10.1007/978-3-642-29578-2_9

12. Carlier, M., Dubois, C., Gotlieb, A.: A certified constraint solver over finite domains. In: Giannakopoulou, D., Méry, D. (eds.) FM 2012. LNCS, vol. 7436, pp. 116–131. Springer, Heidelberg (2012). https://doi.org/10.1007/978-3-642-32759-9_12

13. Claessen, K., Hughes, J.: Quickcheck: a lightweight tool for random testing of haskell programs. In: Odersky, M., Wadler, P. (eds.) Proceedings of the Fifth ACM SIGPLAN International Conference on Functional Programming (ICFP 2000), Montreal, Canada, 18–21 September 2000, pp. 268–279. ACM (2000)

14. Dahmen, M.: A debugger for constraints in prolog. Tech. Rep. ECRC-91-11. ECRC (1991)

15. Debruyune, R., Fekete, J.D., Jussien, N., Ghoniem, M.: Proposition de format concret pour des traces générées par des solveurs de contraintes réalisation rntl oadymppac 2.2. 2.1 (2001). http://pauillac.inria.fr/~contraintes/OADymPPaC/Public/d2.2.2.1.pdf

16. Derval, G., Régin, J., Schaus, P.: Improved filtering for the bin-packing with cardinality constraint. Constraints 23(3), 251–271 (2018)

17. Dooms, G., Deville, Y., Dupont, P.: CP(Graph): introducing a graph computation domain in constraint programming. In: van Beek, P. (ed.) CP 2005. LNCS, vol. 3709, pp. 211–225. Springer, Heidelberg (2005). https://doi.org/10.1007/11564751_18

18. Dubois, C.: Formally Verified Decomposition of Non-binary Constraints into Equivalent Binary Constraints. In: Magaud, N., Dargaye, Z. (eds.) Journées Francophones des Langages Applicatifs, January 2019. JFLA2019, Les Rousses, France (2019)

19. Dubois, C., Gotlieb, A.: Solveurs cp (fd) vérifiés formellement. In: Journées Francophones de Programmation par Contraintes (JFPC 2013), pp. 115–118. aix-en-provence, France (2013)

20. Ducassé, M.: Opium$^+$, a meta-debugger for prolog. In: ECAI, pp. 272–277 (1988)

21. Filliâtre, J.-C., Marché, C.: The Why/Krakatoa/Caduceus platform for deductive program verification. In: Damm, W., Hermanns, H. (eds.) CAV 2007. LNCS, vol. 4590, pp. 173–177. Springer, Heidelberg (2007). https://doi.org/10.1007/978-3-540-73368-3_21

22. Filliâtre, J.-C., Paskevich, A.: Why3 — where programs meet provers. In: Felleisen, M., Gardner, P. (eds.) ESOP 2013. LNCS, vol. 7792, pp. 125–128. Springer, Heidelberg (2013). https://doi.org/10.1007/978-3-642-37036-6_8

23. Frisch, A.M., Grum, M., Jefferson, C., Hernández, B.M., Miguel, I.: The design of ESSENCE: a constraint language for specifying combinatorial problems. In: Veloso, M.M. (ed.) IJCAI 2007, Proceedings of the 20th International Joint Conference on Artificial Intelligence, Hyderabad, India, 6–12 January 2007, pp. 80–87 (2007)

24. Hentenryck, P.V., Carillon, J.: Generality versus specificity: an experience with AI and OR techniques. In: Shrobe, H.E., Mitchell, T.M., Smith, R.G. (eds.) Proceedings of the 7th National Conference on Artificial Intelligence, St. Paul, MN, USA, 21–26 August 1988, pp. 660–664. AAAI Press/The MIT Press (1988)

25. Hoare, C.A.R.: The emperor's old clothes. Commun. ACM **24**(2), 75–83 (1981)

26. Kirchner, F., Kosmatov, N., Prevosto, V., Signoles, J., Yakobowski, B.: Frama-C: a software analysis perspective. Formal Asp. Comput. **27**(3), 573–609 (2015)

27. Kuchcinski, K., Szymanek, R.: JaCoP-java constraint programming solver. In: CP Solvers: Modeling, Applications, Integration, and Standardization, co-located with the 19th International Conference on Principles and Practice of Constraint Programming (2013)

28. Lazaar, N., Gotlieb, A., Lebbah, Y.: A CP framework for testing CP. Constraints **17**(2), 123–147 (2012)

29. Lecoutre, C., Prosser, P.: Maintaining singleton arc consistency. In: Proceedings of the 3rd International Workshop on Constraint Propagation and Implementation (CPAI 2006) held with CP 2006, pp. 47–61. Springer (2006)

30. Lecoutre, C., Tabary, S.: Abscon 112: towards more robustness. In: 3rd International Constraint Solver Competition (CSC 2008), Sydney, Australia, pp. 41–48 (2008). https://hal.archives-ouvertes.fr/hal-00870841

31. Lesaint, D., Mehta, D., O'Sullivan, B., Quesada, L., Wilson, N.: Solving a telecommunications feature subscription configuration problem. In: Stuckey, P.J. (ed.) CP 2008. LNCS, vol. 5202, pp. 67–81. Springer, Heidelberg (2008). https://doi.org/10.1007/978-3-540-85958-1_5

32. Martin, P., Shmoys, D.B.: A new approach to computing optimal schedules for the job-shop scheduling problem. In: Cunningham, W.H., McCormick, S.T., Queyranne, M. (eds.) IPCO 1996. LNCS, vol. 1084, pp. 389–403. Springer, Heidelberg (1996). https://doi.org/10.1007/3-540-61310-2_29

33. Massart, A., Rombouts, V., Schaus, P.: Testing global constraints. CoRR abs/1807.03975 (2018). http://arxiv.org/abs/1807.03975
34. Meier, M.: Debugging constraint programs. In: Montanari, U., Rossi, F. (eds.) CP 1995. LNCS, vol. 976, pp. 204–221. Springer, Heidelberg (1995). https://doi.org/10.1007/3-540-60299-2_13
35. Michel, L., Schaus, P., Van Hentenryck, P.: MiniCP: a lightweight solver for constraint programming (2018). www.minicp.org
36. Nethercote, N., Stuckey, P.J., Becket, R., Brand, S., Duck, G.J., Tack, G.: MiniZinc: towards a standard CP modelling language. In: Bessière, C. (ed.) CP 2007. LNCS, vol. 4741, pp. 529–543. Springer, Heidelberg (2007). https://doi.org/10.1007/978-3-540-74970-7_38
37. Niemetz, A., Preiner, M., Biere, A.: Model-based API testing for SMT solvers. In: Proceedings of the 15th International Workshop on Satisfiability Modulo Theories, p. 10. SMT (2017)
38. van Omme, N., Perron, L., Furnon, V.: OR-Tools user's manual. Google Inc. (2014). https://developers.google.com/optimization/
39. OscaR Team: OscaR: Scala in OR (2012). https://bitbucket.org/oscarlib/oscar
40. Pesant, G.: A regular language membership constraint for finite sequences of variables. In: Wallace, M. (ed.) CP 2004. LNCS, vol. 3258, pp. 482–495. Springer, Heidelberg (2004). https://doi.org/10.1007/978-3-540-30201-8_36
41. Prud'homme, C., Fages, J.G., Lorca, X.: Choco Documentation. TASC - LS2N CNRS UMR 6241, COSLING S.A.S. (2017). http://www.choco-solver.org
42. Régin, J.: A filtering algorithm for constraints of difference in CSPs. In: Hayes-Roth, B., Korf, R.E. (eds.) Proceedings of the 12th National Conference on Artificial Intelligence, Seattle, WA, USA, vol. 1, pp. 362–367, 31 July – 4 August 1994. AAAI Press/The MIT Press (1994)
43. Schulte, C., Tack, G.: Weakly monotonic propagators. In: Gent, I.P. (ed.) CP 2009. LNCS, vol. 5732, pp. 723–730. Springer, Heidelberg (2009). https://doi.org/10.1007/978-3-642-04244-7_56
44. Simonis, H., Davern, P., Feldman, J., Mehta, D., Quesada, L., Carlsson, M.: A generic visualization platform for CP. In: Cohen, D. (ed.) CP 2010. LNCS, vol. 6308, pp. 460–474. Springer, Heidelberg (2010). https://doi.org/10.1007/978-3-642-15396-9_37
45. Sutton, M.: Fuzzing: Brute Force Vulnerability Discovery. Addison-Wesley Professional, Reston (2007)
46. Team, G.: Gecode: generic constraint development environment (2008). https://www.gecode.org/
47. Coq development team, T.: The coq proof assistant (2019). https://coq.inria.fr/
48. Vilím, P.: Timetable edge finding filtering algorithm for discrete cumulative resources. In: Achterberg, T., Beck, J.C. (eds.) CPAIOR 2011. LNCS, vol. 6697, pp. 230–245. Springer, Heidelberg (2011). https://doi.org/10.1007/978-3-642-21311-3_22

Encodings for Enumeration-Based Program Synthesis

Pedro Orvalho[1,2], Miguel Terra-Neves[1,2], Miguel Ventura[2], Ruben Martins[3(✉)], and Vasco Manquinho[1]

[1] INESC-ID, Instituto Superior Técnico, Universidade de Lisboa, Lisbon, Portugal
{pmorvalho,vmm}@sat.inesc-id.pt
[2] OutSystems, Lisbon, Portugal
{miguel.neves,miguel.ventura}@outsystems.com
[3] Carnegie Mellon University, Pittsburgh, USA
rubenm@cs.cmu.edu

Abstract. Program synthesis is the problem of finding a program that satisfies a given specification. Most program synthesizers are based on enumerating program candidates that satisfy the specification. Recently, several new tools for program synthesis have been proposed where Satisfiability Modulo Theories (SMT) solvers are used to prune the search space by discarding programs that do not satisfy the specification.

The size of current tree-based SMT encodings for program synthesis grows exponentially with the size of the program. In this paper, a new compact line-based encoding is proposed that allows a faster enumeration of the program space. Experimental results on a large set of query synthesis problem instances show that using the new encoding results in a more effective tool that is able to synthesize larger programs.

Keywords: Program synthesis · Satisfiability Modulo Theories · Enumerative search · SQL

1 Introduction

The goal of program synthesis is to automatically generate programs that satisfy a given high-level specification. Once considered a utopian dream, the recent advances in program synthesis are making this approach more practical and have shown that it can be useful to both end-users and programmers. A common approach is to use input-output examples as specifications. Even though these specifications are incomplete, i.e. a program may satisfy the specification but may not be the program that the user desires, these are easy to create and can be used to solve many real-world applications. This approach is known as programming-by-example (PBE) and has been used to automate tedious tasks in a plethora of applications, such as string manipulations in spreadsheets [10, 15], list transformations [2,9], table reshaping [7], code completion [14], helping programmers to use libraries [8], and SQL queries [18–20]. Program synthesis is

© Springer Nature Switzerland AG 2019
T. Schiex and S. de Givry (Eds.): CP 2019, LNCS 11802, pp. 583–599, 2019.
https://doi.org/10.1007/978-3-030-30048-7_34

Fig. 1. Enumeration-based program synthesis

not merely an academic research topic since it is also transitioning into industry. Microsoft's FlashFill [10] is the most successful application of program synthesis by Microsoft for string manipulation and it is integrated into Microsoft Excel. Other companies are also starting to look for applications of program synthesis to their products, namely OutSystems [1] and query synthesis.

Even though there are many approaches to program synthesis, the most common one is to perform an enumerative search over the space of programs that satisfy the specifications. Figure 1 shows the high-level architecture of enumeration-based program synthesizers. They take as input the specification that describes the intention of the user (e.g., input-output examples) and a domain-specific language (DSL) that defines the search space. Program synthesizers typically enumerate programs in increasing order of the number of DSL components. For each candidate program $\mathcal{P}$, they check if $\mathcal{P}$ satisfies the specifications. If this is the case, then the desired program was found. Otherwise, the program synthesizer learns a reason for failure and enumerates the next candidate program.

Recent approaches combine enumerative search with deduction with the goal of performing early pruning of infeasible programs [7], or to learn from past failed candidate programs in order to prune all equivalent infeasible programs [6].

Suppose that a user wants to synthesize an SQL query using examples. In particular, given tables *supplier* and *parts* with the schema "supplier(id: integer, sname: string, address: string)" and "parts(id: integer, pname: string, color: string)", the user wants to find the *names* of parts, *pnames*, for which there is some supplier. [1] This could be accomplished with the following SQL query:

```
SELECT pname
FROM parts, supplier
WHERE parts.id = supplier.id
```

To enumerate the space of programs that satisfy the specifications, program synthesizers must first construct an underlying representation of the feasible space. Figure 2 shows the typical tree representation used by program synthesizers (e.g., [3,6,7]), for the above query example. Each node can be a library component or a terminal symbol. Program synthesizers can then traverse the space of possible candidates by enumerating all possible trees of a given depth. However, for approaches that rely on logical deduction, the space of feasible programs must be encoded a priori by using either a Boolean Satisfiability (SAT)

---

[1] This corresponds to exercise 5.2.1 from a classic textbook on databases [13].

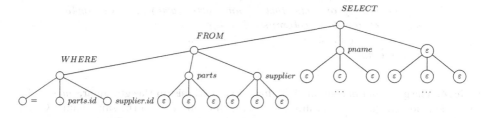

Fig. 2. Tree-based representation of the search space

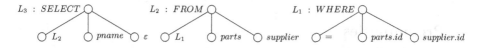

Fig. 3. Line-based representation of the search space

or Satisfiability Modulo Theory (SMT) encoding [6,7]. A common approach to encode all feasible programs is to represent them using a k-tree, where each node has exactly k children and k is the largest number of parameters of the functions in our library of components. Figure 2 shows an example of a 3-tree where each node has 3 children. A complete program corresponds to assigning a label to each node. Components that may have less than 3 parameters (e.g., SELECT), will have the empty label *empty* ε assigned to their unused children.

A large downside of a k-tree representation is the exponential growth of the size of the tree with respect to its depth. Since the encoding's complexity depends on the number of nodes, this makes it intractable to enumerate the search space of candidate programs using an SMT encoding.

In this paper, we propose a new line representation illustrated in Fig. 3, where we represent each line with its own subtree and add additional constraints to connect the multiple subtrees. For the above SQL query, we would only need 12 nodes using a line-based representation instead of the 3-tree representation's 40 nodes. When considering programs with 10 lines of code and $k = 4$, the line-based representation only needs 50 nodes instead of the 1,398,101 nodes required by the tree-based representation.

To summarize, this paper makes the following contributions:

- We formalize how to encode the traditional tree-based representation of a program into SMT which has an exponential growth with respect to the number of lines of a program.
- We propose a new compact SMT encoding based on a line representation of programs that grows linearly with the number of lines of a program.
- We integrate the line-based encoding into a program synthesizer and empirically evaluate our approach using SQL benchmarks. Experimental results show that the line-based encoding significantly outperforms the tree-based encoding and allows program synthesizers to more effectively enumerate the search space and synthesize larger programs.

$$table \quad \rightarrow select\_from(cols,\ table) \mid join(table,\ table) \mid parts \mid supplier$$
$$cols \quad \rightarrow column(col) \mid columns(col,\ cols)$$
$$col \quad \rightarrow pname \mid sname \mid id \mid color \mid address \mid *$$
$$empty \rightarrow empty$$

Fig. 4. The grammar of a simple DSL for query synthesis; in this grammar, *table* is the start symbol. All joins are natural joins between columns with the same name. Given as input the tables *supplier* and *parts*, with the schema "supplier(id: integer, sname: string, address: string)" and "parts(id: integer, pname: string, color: string)".

2 Preliminaries

The Satisfiability Modulo Theories (SMT) problem is a generalization of the well-known Propositional Satisfiability (SAT) problem. Given a decidable first-order theory $\mathcal{T}$, a $\mathcal{T}$-atom is a ground atomic formula in $\mathcal{T}$. A $\mathcal{T}$-literal is either a $\mathcal{T}$-atom t or its complement $\neg t$. A $\mathcal{T}$-formula is similar to a propositional formula, but a $\mathcal{T}$-formula is composed of $\mathcal{T}$-literals instead of propositional literals. Given a $\mathcal{T}$-formula ϕ, the SMT problem consists of deciding if there exists a total assignment over the variables of ϕ such that ϕ is satisfied. Depending on the theory $\mathcal{T}$, the variables can be of type integer, real, Boolean, among others.

Program synthesizers search the space of programs described by a given domain-specific language (DSL). The syntax of the DSL is described by a context-free grammar G. In particular, G is a tuple (Σ, R, S), where Σ represents the set of symbols, productions, and start symbol, respectively. Each symbol $\sigma \in \Sigma$ corresponds to built-in DSL constructs (e.g., *select_from, join*), constants, variables or inputs of the system. Each production rule $p \in R$ has the form $p = (A \rightarrow \sigma(A_1, \ldots, A_m))$, where $\sigma \in \Sigma$ is a DSL construct and $A_1, \ldots, A_m \in \Sigma$ are symbols for the arguments of σ.

Example 1. Consider a DSL D in Fig. 4, and suppose that a user wants to solve the query presented in Sect. 1, i.e. she wants to find all the names of *parts* for which there is some supplier. The desired query from D is the following *select_from(column(pname), join(parts, supplier))*. This query uses three production rules $p_1 = select\_from$, $p_2 = column$, and $p_3 = join$; the column *pname*; and input tables *parts* and *supplier*.

3 Tree-Based Encoding

This section describes the tree-based encoding used on several state-of-the-art synthesizers to perform program enumeration. Given a DSL, program synthesis frameworks search for a program that is consistent with the input-output examples provided by the user. For the search process to be complete, these frameworks use a structure capable of representing every possible program up to some given depth of n. Let k be the greatest arity among DSL constructs. For programs with $n - 1$ production rules, synthesizers adopt a tree structure of depth n, referred to

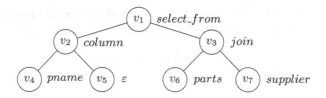

Fig. 5. k-tree representation of the query presented in Example 1

as k-tree, where each node has exactly k children. Figure 5 illustrates a 2-tree that can be used to represent the query presented in Example 1.

In order to perform program enumeration using the tree representation, program synthesizers encode the tree as an SMT formula such that a solution of the SMT formula encodes a complete program by assigning a symbol to each node.

A detailed description of the SMT model follows. First, the encoding variables are introduced. Next, the constraints of the SMT model are presented.

3.1   Encoding Variables

Let s be the length of the DSL's set of symbols, $s = |\Sigma|$. Let $id : \Sigma \to \mathbb{N}_0$ be a function that maps each symbol to a unique non-negative integer in a one-to-one mapping. As a result, this function provides a unique identifier (integer value between 0 and s) to each symbol in Σ. In our encoding, we assume that the empty production symbol (ε) is mapped to 0 (i.e. $id(\varepsilon) = 0$).

Consider the encoding for a program with a k-tree of depth n. Assume each node in the k-tree is assigned a unique index. Let N be the set of all k-tree nodes indexes such that $N = I \cup L$ where I denotes the set of internal node indexes and L denotes the set of leaf node indexes. Let $C(i)$ denote the set of child indexes of node $i \in N$. Clearly, if i is a leaf node ($i \in L$), then $C(i) = \emptyset$.

In our encoding we define the following variables:

- $V = \{v_i : 1 \le i \le |N|\}$: each variable v_i denotes the symbol identifier in node i of the k-tree;
- $B = \{b_i : 1 \le i \le |N|\}$: each variable b_i is a Boolean variable that denotes if node i is associated to a production symbol (true) or a terminal symbol (false).

3.2   Constraints

Let D be a DSL, $Prod(D)$ denotes the set of production rules in D and $Term(D)$ the set of terminal symbols in D. Furthermore, let $Types(D)$ denotes the set of types used in D and $Type(s)$ the type of symbol $s \in Prod(D) \cup Term(D)$. If $s \in Prod(D)$, then $Type(s)$ denotes the return type of production rule s.

To ensure that every program enumerated is well-typed the following constraints must be satisfied.

Leaf Nodes. The leaf nodes can only be assigned to terminal symbols because they have no children. Therefore, we define the following constraint:

$$\forall i \in L : \bigvee_{p \in Term(D)} v_i = id(p) \tag{1}$$

Example 2. Given the DSL D from Fig. 4, the set of terminal symbols is $Term(D) = \{parts, supplier, pname, sname, id, color, address, *, \varepsilon\}$ and the set of leaves is $L = \{4, 5, 6, 7\}$. Each leaf node in L must be assigned to a symbol in $Term(D)$. Hence, each leaf $i \in L$ must satisfy: $v_i = id(parts) \lor v_i = id(supplier) \lor v_i = id(pname) \lor v_i = id(sname) \lor v_i = id(id) \lor v_i = id(color) \lor v_i = id(address) \lor v_i = id(*) \lor v_i = id(\varepsilon)$.

Internal Nodes. If a production rule p is assigned to an internal node, then the type of its children nodes must match the types of parameters of p. Let $Type(p, j)$ denote the type of parameter j of production rule $p \in Prod(D)$. If $j > arity(p)$, then $Type(p, j) = empty$. If p is a terminal symbol, $p \in Term(D)$, then for every j, $Type(p, j) = empty$.

Let $\Sigma(Type(p, j))$ represent the subset of symbols in Σ of type $Type(p, j)$.

$$\forall i \in I, \ j \in C(i), \ p \in \Sigma \ :$$
$$v_i = id(p) \Rightarrow \bigvee_{t \in \Sigma(Type(p,j))} v_j = id(t) \tag{2}$$

With constraint (2), all the programs generated will be well-typed since each node is only assigned to a production rule if its children have the correct type.

Example 3. Consider again the query in Example 1. If the production *select_from* is assigned to the program's root, v_1, then $\Sigma(Type(select\_from, 1)) = \{column, columns\}$ and $\Sigma(Type(select\_from, 2)) = \{select\_from, join, parts, supplier\}$. The following constraint must be satisfied: $v_1 = id(select\_from) \Rightarrow (v_2 = id(column) \lor v_2 = id(columns))$ and $v_1 = id(select\_from) \Rightarrow (v_3 = id(select\_from) \lor v_3 = id(join) \lor v_3 = id(parts) \lor v_3 = id(supplier))$.

Output. Let t be the output type. Furthermore, consider that the program root identifier is 1. Then, v_1 must be assigned to a symbol that is consistent with the output type t. Hence, the following constraint must be satisfied.

$$\bigvee_{s \in \Sigma(t)} v_1 = id(s) \tag{3}$$

Input. Let IN be the set of symbols provided by the user as input. In order to guarantee that all generated programs use all the inputs provided by the user, the following constraint is added:

$$\forall p \in IN : \bigvee_{i \in N} v_i = id(p) \tag{4}$$

$$L_1 : ret_1 \leftarrow column(pname)$$
$$L_2 : ret_2 \leftarrow join(parts, supplier)$$
$$L_3 : ret_3 \leftarrow select\_from(ret_1, ret_2)$$

Fig. 6. Line-representation of the query from Example 1

Note that this is not required for the encoding's correction. Nevertheless, we are only interested in enumerating programs that use all inputs given by the user.

Exactly $n-1$ Production Rules. Finally, we are interested in enumerating programs using Exactly $n-1$ production rules by adding the following constraints:

$$\forall i \in N : b_i = 1 \iff \bigvee_{p \in Prod(D)} v_i = id(p) \tag{5}$$

$$\left(\sum_{i \in N} b_i \right) = n - 1 \tag{6}$$

With constraints (5) and (6), we guarantee that given a k-tree of depth n, each enumerated program will have exactly $n-1$ production rules. State-of-the-art program synthesizers iteratively search for programs in increasing depth. Thus, constraint (6) allows pruning the search space, in order to avoid enumerating repeated programs in future iterations of depth greater than n.

Encoding Complexity. Let k be the greatest arity between DSL constructs and let n denote the number of productions (lines of code) in a program. In terms of nodes complexity, the number of nodes increases exponentially with the number of productions, as follows: $\frac{k^{n+1}-1}{k-1}$.

4 Line-Based Encoding

In this section, we propose a new encoding to represent programs. Our goal is to represent a program as a sequence of lines where each line represents an operation in the DSL. Instead of using a single k-tree to represent a program, each line is represented as a tree with a depth of 1.

Consider the program in Fig. 6. One can represent this program as three trees of depth 1 as shown in Fig. 7. Note that the result of the program is the value returned by the third tree. Observe that ret_i is a new symbol that represents the return value of line i.

4.1 Encoding Variables

Recall that D denotes a DSL, $Prod(D)$ the set of production rules in D and $Term(D)$ the set of terminal symbols in D. Furthermore, $Types(D)$ denotes the set of types used in D and $Type(s)$ the type of symbol $s \in Prod(D) \cup Term(D)$. If $s \in Prod(D)$, then $Type(s)$ denotes the return type of production rule s.

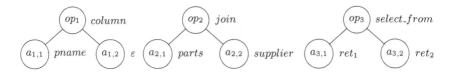

Fig. 7. Each tree represents a production rule. The first tree represents line 1, the second tree represents line 2 and the third tree represents line 3. ret_1 (resp. ret_2) denotes the value returned in line 1 (resp. line 2).

Consider the encoding for a program with n lines where the maximum arity of the operators is k, then we have the following variables:

- $O = \{op_i : 1 \le i \le n\}$: each variable op_i denotes the production rule used in line i;
- $T = \{t_i : 1 \le i \le n\}$: each variable t_i denotes the return type of line i;
- $A = \{a_{ij} : 1 \le i \le n, 1 \le j \le k\}$: each variable a_{ij} denotes the symbol corresponding to argument j in line i.

4.2   Constraints

Besides the production rules $Prod(D)$ and terminal symbols $Term(D)$, we define one return symbol for each line in the program. Let $Ret = \{ret_i : 1 \le i \le n\}$ denote the set of return symbols in the program.

In our encoding, we define a different non-negative identifier for each symbol. Here, we extend the id function to also consider the symbols that represent the return value of each line. Let $Symbols = Prod(D) \cup Term(D) \cup Ret$ define the set of all symbols used in the program. Finally, let $id : Symbols \to \mathbb{N}_0$ and $tid : Types(D) \to \mathbb{N}_0$ be one-to-one mappings of symbols and types, respectively, to non-negative integer values.

Operations. First, the operations in each line must be production rules. Hence, we have the following set of constraints:

$$\forall 1 \le i \le n : \bigvee_{p \in Prod(D)} (op_i = id(p)) \tag{7}$$

The operation symbol used in each line implies the line's return type.

$$\forall 1 \le i \le n, p \in Prod(D) : (op_i = id(p)) \Rightarrow (t_i = tid(Type(p))) \tag{8}$$

Given a sequence of operations, the arguments of operation i must either be terminal symbols or return symbols from previous operations. Hence, we have:

$$\forall 1 \le i \le n, 1 \le j \le k : \bigvee_{s \in Term(D) \cup \{ret_r : r < i\}} (a_{ij} = id(s)) \tag{9}$$

Arguments. The arguments for a given operation i must have the same types as the parameters of the production rule used in the operation. Let $Type(p, j)$

denote the type of parameter j of production rule $p \in Prod(D)$. If $j > arity(p)$, then $Type(p, j) = empty$. Hence, we have the following constraints when a return symbol is used as an argument of an operation:

$$\forall 1 \leq i \leq n, p \in Prod(D), 1 \leq j \leq arity(p), 1 \leq r < i :$$
$$((op_i = id(p)) \wedge (a_{ij} = id(ret_r))) \Rightarrow (t_r = tid(Type(p, j))) \tag{10}$$

A given terminal symbol $t \in Term(D)$ cannot be used as argument j of an operation i if it does not have the correct type:

$$\forall 1 \leq i \leq n, p \in Prod(D), 1 \leq j \leq arity(p),$$
$$s \in \{t \in Term(D) : Type(t) \neq Type(p, j)\} : \tag{11}$$
$$(op_i = id(p)) \Rightarrow (a_{ij} \neq id(s))$$

Since the arity of a given operation i can be smaller than k, we must also have that the arguments above the production's arity must be assigned to the empty symbol ε:

$$\forall 1 \leq i \leq n, p \in Prod(D), arity(p) < j \leq k :$$
$$(op_i = id(p)) \Rightarrow (a_{ij} = id(\varepsilon)) \tag{12}$$

Output. Let $Type(output)$ denote the type of the program's output and let $P_O \subseteq Prod(D)$ be the subset of production rules with return type equal to $Type(output)$, i.e., $P_O = \{p \in Prod(D) : Type(p) = Type(output)\}$. The following constraint ensures that the program's output (last line, n^{th}) has the desired type.

$$\bigvee_{p \in P_O} (op_n = id(p)) \tag{13}$$

Input. Let IN be the set of symbols provided by the user as input. In order to guarantee that all generated programs use all the inputs, the following constraint is used:

$$\forall s \in IN : \bigvee_{1 \leq i \leq n} \bigvee_{1 \leq j \leq k} (a_{ij} = id(s)) \tag{14}$$

Lines Used Exactly Once. A feature of this new encoding is that the result of a given operation can be used more than once. Notice that in the tree-based encoding, one would have to reproduce the same operations in a different branch of the tree. In order to compare the two types of enumeration, tree-based and line-based, we can add a set of constraints restricting the usage of each operation's result to only one usage. Clearly, the following constraints are not necessary to the encoding's correction.

$$\forall ret_r \in Ret(D) : \left(\sum_{r < i \leq n, 1 \leq j \leq k} (a_{ij} = id(ret_r)) \right) = 1 \tag{15}$$

$$
\begin{aligned}
L_1 &: ret_1 \leftarrow column(pname) \\
L_2 &: ret_2 \leftarrow join(parts,\ supplier) \\
L_3 &: ret_3 \leftarrow select\_from(ret_1,\ ret_2)
\end{aligned}
\qquad
\begin{aligned}
L_1 &: ret_1 \leftarrow join(parts,\ supplier) \\
L_2 &: ret_2 \leftarrow column(pname) \\
L_3 &: ret_3 \leftarrow select\_from(ret_2,\ ret_1)
\end{aligned}
$$

Fig. 8. Two line representations of the program from Example 1

Encoding Complexity. Let k be the greatest arity between DSL constructs and let n denote the number of productions (lines of code) in a program. In terms of nodes complexity, we can observe a drastic difference between both types of enumeration, tree-based and line-based. In tree-based enumeration, the number of nodes increases exponentially with the number of productions. On the other hand, the number of nodes used by line-based enumeration increases linearly, $(k+1) \times n$, because the enumerator uses n trees, with $k+1$ nodes each, to represent a program with n production rules.

4.3   Symmetric Programs

In line-based encoding, the number of solutions of the SMT formula is larger than the number of solutions in the corresponding tree-based encoding. There are two main reasons for this difference: (1) the line-based encoding can use the return value of the same line of code more than once, and (2) the same program can have more than one representation, i.e. symmetric programs.

Regarding reason (1), with constraint (15), we guarantee that the return value of each line is used exactly once. Concerning reason (2), in the line-based encoding, some programs can be represented with different sequences of lines. However, in the tree-based encoding, as a result of the single tree representation, the arguments of each production rule will always come from the same branch.

Example 4. Consider the DSL in Fig. 4 and the program *select_from (column (pname), join(parts, supplier))* from Example 1. In tree-based encoding this program has a single representation shown in Fig. 5. However, for the same program, line-based encoding has two possible representations shown in Fig. 8.

In order for the line-based process to enumerate the same number of solutions than the tree-based enumeration, it is necessary to find the symmetries in the line-based encoding and block them. Otherwise, symmetric programs as the one in Fig. 8 will be enumerated and the synthesizer will have to check both programs. Therefore, if we have a solution α of line-based SMT formula and the synthesizer verifies that the corresponding program is not consistent with the input-output examples, then all solutions that encode symmetric programs in relation to α can be blocked.

A simple way to find these symmetries is through a directed acyclic graph of dependencies, where a vertex is defined for each program line, and edges correspond to the line dependencies in the program. Let v_i and v_j denote the vertexes in the graph corresponding to lines i and j with $i < j$. If the return value of line i is used as argument in line j, then a directed edge (v_i, v_j) must be added to the graph. After building the graph, one can enumerate all possible topological orders of vertexes in the dependency graph. Next, each program associated with a topological order is blocked in the SMT formula.

Example 5. Consider the program from Example 1. Line 3 (L_3) depends on line 1 (L_1) and line 2 (L_2). Therefore, lines 1 and 2 must occur before line 3. However, the order of lines 1 and 2 can be changed. Hence, two solutions would be blocked corresponding to permutations $L_1 - L_2 - L_3$ and $L_2 - L_1 - L_3$ of the program.

5   Experimental Results

In order to evaluate the new line-based encoding, we integrated our proposal in the Trinity [4] synthesis framework. By default, Trinity uses tree-based enumeration to search for programs and uses the Z3 SMT solver [5] with the theory of Linear Integer Arithmetic in the enumeration process. Trinity, like most PBE state-of-the-art synthesizers, takes as input a DSL, a set of examples, and any constants or aggregate functions (e.g., mean) that the query may need. Trinity starts by searching for programs with 1 production rule and iteratively increases this bound until a program that satisfies all input-output examples is found.

All of the experiments presented in this section were conducted on an Intel(R) Xeon(R) with E5-2630 v2 2.60 GHz CPUs, using a memory limit of 64 GB and a time limit of 3,600 s. The goal of our evaluation was to answer the following questions:

Q1. How does line-based enumeration compare against tree-based enumeration in terms of encoding complexity? (Sect. 5.2)

Q2. How does line-based enumeration compare against tree-based enumeration in general? (Sect. 5.2)

Q3. How does line-based enumeration compare against tree-based enumeration for programs with more than three lines of code? (Sect. 5.2)

Q4. What is the performance impact of breaking symmetries in line-based enumeration? (Sect. 5.3)

5.1   SQL Benchmarks

We designed a DSL for SQL that can solve classic SQL queries from a database textbook [13]. These benchmarks were previously used by well-known SQL synthesizers [7,18,20]. We started with an initial set of 23 SQL benchmarks (corresponding to Sects. 5.1.1 and 5.1.2 of the database textbook [13]) and created variants of these benchmarks until a total of 55 benchmarks.

Since we want to study the performance of each encoding with respect to the size of the synthesized query, for each of these benchmarks, we generate six different SMT formulas to search for programs that use exactly n production rules from our DSL, for a total of 330 benchmarks (55 × 6). The SMT formulas differ in the number of productions that their programs must have, and it simulates the search performed by a program synthesizer until a solution is found.

Table 1. Number of tree nodes, variables and mean number of constraints used by each approach for a given program's size.

| Lines of code | Encoding | | | | | |
| | Tree-based | | | Line-based | | |
	Nodes	Variables	Constraints	Nodes	Variables	Constraints
1	5	10	379	5	6	44
2	21	42	1,703	10	16	118
3	85	170	6,999	15	30	224
4	341	682	28,183	20	48	362
5	1,365	2,730	112,919	25	70	532
6	5,461	10,922	451,863	30	96	734

Table 2. Number of solved benchmarks by each approach.

Lines of code	1	2	3	4	5	6	Total	% Solved	% Solved (LOC >= 4)
# Tests	55	55	55	55	55	55	330		
Tree-based	55	55	54	34	18	2	218	66.06%	32.73%
Line-based	55	55	54	49	48	39	300	90.91%	82.42%

5.2   Comparison Between Line-Based and Tree-Based Encodings

Encoding Complexity. As presented in Sects. 3 and 4, the number of nodes used by line-based enumeration increases linearly. On the other hand, in tree-based enumeration the number of nodes increases exponentially with the number of productions. The number of variables and constraints used by each type of enumeration varies with the number of nodes. Table 1 shows the number of nodes, variables and the mean number of constraints used by each type of enumeration on the 330 SQL benchmarks. The number of nodes and variables are always the same for a given program's size. The number of constraints varies with the DSL since each benchmark may use different constants and aggregate functions.

Performance. Table 2 shows the number of solved benchmarks by each encoding for a given number of lines in our DSL. The performance for both encodings is similar for programs with three or fewer lines of code. However, when the program size increases, the difference between these approaches becomes clear. The last line of Table 2, shows the percentage of solved benchmarks by each approach with more than three lines of code. The tree-based encoding only solves 33% of the benchmarks while line-based encoding solves 82%.

In terms of time spent in each benchmark, Fig. 9 shows two plots, a cactus plot in Fig. 9a and a scatter plot in Fig. 9b. The cactus plot shows the cumulative synthesis time (y-axis) against the number of benchmarks solved (x-axis). Each point in the scatter plot represents a benchmark where the x-value (resp. y-value) is the time spent by the line-based (resp. tree-based) enumerator on

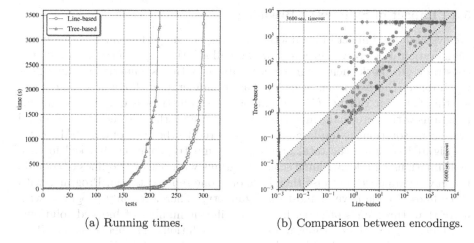

(a) Running times.   (b) Comparison between encodings.

Fig. 9. Tree-based vs Line-based Enumerators.

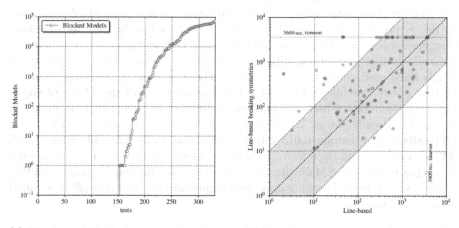

(a) Number of blocked symmetric solu-   (b) Runtime comparison without consid-
tions per benchmark.   ering the overhead to find symmetries.

Fig. 10. Impact of breaking symmetries.

a given benchmark. Both plots, in Fig. 9, support the results shown in Table 2. Additionally, the plots show that tree-based enumeration is, in general, significantly slower than line-based enumeration.

These differences in time and number of benchmarks solved, in particular for the instances with more than 3 lines, can be justified by the exponential number of variables and constraints required by tree-based enumeration.

5.3 Impact of Symmetry Breaking

We evaluate the impact of symmetry breaking on the performance of line-based enumeration. For every solution α, we find all solutions symmetric to α and add constraints to block them. Our experiments show that symmetry breaking does not improve the performance of line-based enumeration. Possible explanations for this behavior can be: (1) the number of symmetries is only significant for programs with more than three lines, and (2) the overhead to find and block all symmetric solutions is too large when compared to the time of each SMT call.

Figure 10a shows the total number of symmetric solutions blocked in each benchmark using a logarithmic scale. Programs with one or two lines of code do not have symmetries because they have only one representation. Programs with three lines of code have at most one symmetry. Therefore, only programs with more than three lines of code, have a significant number of blocked solutions, i.e., blocked more than a thousand symmetric solutions (117 benchmarks). If we only look at these 117 benchmarks, we observe that not breaking symmetries solves 87 benchmarks, while breaking symmetries only solves 68 benchmarks.

Since breaking symmetries is ineffective even when a large number of symmetries is present, we analyzed the current overhead of finding and blocking symmetric solutions. For each solution, we spend on average 0.091 s to find and block all symmetric solutions. Figure 10b shows, per benchmark, the time spent by the line-based enumerator with and without symmetry breaking, ignoring the time spent searching for and blocking symmetric solutions. This plot shows that, even if symmetry breaking was free, it does not improve the performance of the line-based enumerator. We observed that, without symmetry breaking, each SMT call takes on average 0.015 s. If we add symmetry breaking predicates, each SMT call doubles its time to 0.030 s, on average. Since our enumeration relies on solving many easy SMT calls, we concluded that the search space reduction enabled by symmetry breaking does not compensate for the extra effort required to break symmetries.

6 Related Work

Program synthesis has been widely used to synthesize queries using input-output examples [4,7,17,18,20] or natural language [19]. Approaches for query synthesis vary from using decision trees with fixed templates [17,20], to abstract representations of queries that can potentially satisfy the input-output examples [18], and to use SMT-based over-approximations to prune the search space [7].

Tree representations of program search spaces are commonly used in modern program synthesis applications. For example, Bonsai [3] is a validation framework for type systems that uses such representations to synthesize syntactically incorrect programs wrongly accepted by the type checker. State-of-the-art program synthesizers based on enumeration [4,6,7] also make use of tree-based SMT encodings in order to prune the search space by checking if it is possible to extend a given partial program to a complete program which satisfies the input-output examples. The encodings studied in this paper can improve the enumeration of program synthesizers based on SMT encodings [4,6,7]. These encodings are domain agnostic and can be used in other domains besides SQL (e.g., lists, strings, tables, etc.)˙with expected improvements of the same order of magnitude.

Alternatively, the synthesis problem can be directly encoded into SMT using quantified formulas [11,12,16]. Brahma [11,12] is an oracle-guided synthesizer that considers an SMT encoding with some similarities to the line-based encoding proposed in Sect. 4. However, there are some fundamental differences: (1) The program specification is generic and must be satisfied for all possible program inputs. Therefore, Brahma essentially solves a single universally quantified SMT formula in order to synthesize a program. (2) SMT specifications must capture the complete semantics of the respective components, while state-of-the-art enumeration-based synthesizers typically deal with specifications that over-approximate the components' behavior. (3) The authors focus on bit-vector manipulation and do not consider arguments of different types. Synudic [16] extends Brahma with additional restrictions on the structure of the program to be synthesized. This allows users to either search for all programs that satisfy the functional requirements or to narrow the search space to programs that satisfy a given template. Even though there are some similarities between our encoding and a purely SMT-based approach [11,12,16], we only need to encode a formula where each solution represents a well-typed program. The SMT encoding abstracts the semantics of the operators and is simpler than previous approaches that encode the entire synthesis problem as an SMT problem. Moreover, an enumeration-based approach makes thousands of SMT calls (each in the order of milliseconds), while the SMT encodings from previous approaches [11,12,16] typically solve one large quantified formula that can take a very long time.

7 Conclusions

In recent years, new platforms for software development have been made available where users with little programming skills are able to create and modify software applications. These tools are able to hide many aspects of programming, but some coding experience is still needed for some operations. Programming-by-example is making programming more accessible by allowing users to create incomplete specifications through input-output examples of these operations and automatically synthesize the desired program.

Currently, the most common approach to program synthesis is to perform an enumerative search on the space of programs and find one that satisfies the specifications. In this paper, we propose a new compact SMT encoding for program enumeration where each program is represented as a sequence of lines. Experimental results on synthesis of SQL queries show that the proposed line-based encoding allows a faster enumeration of programs when compared to the usual tree-based encoding. Moreover, while the tree-based encoding does not scale beyond a small number of operations, the new line-based encoding allows to find programs with a larger sequence of operations.

Acknowledgments. This work was supported by national funds through FCT with references UID/CEC/50021/2019, PTDC/CCI-COM/31198/2017 and DS-AIPA/AI/0044/2018, and by NSF award #1762363 and the CMU-Portugal program with reference CMU/AIR/0022/2017.

References

1. OutSystems (2019). https://www.outsystems.com. Accessed 15 July 2019
2. Balog, M., Gaunt, A.L., Brockschmidt, M., Nowozin, S., Tarlow, D.: Deepcoder: learning to write programs. In: Proceedings International Conference on Learning Representations. OpenReview (2017)
3. Chandra, K., Bodík, R.: Bonsai: synthesis-based reasoning for type systems. PACM Program. Lang. **2**(POPL), 1–34 (2018)
4. Chen, J., Martins, R., Chen, Y., Feng, Y., Dillig, I.: Trinity: an extensible synthesis framework for data science. PVLDB **12**(12), 1914–1917 (2019)
5. de Moura, L., Bjørner, N.: Z3: an efficient SMT solver. In: Ramakrishnan, C.R., Rehof, J. (eds.) TACAS 2008. LNCS, vol. 4963, pp. 337–340. Springer, Heidelberg (2008). https://doi.org/10.1007/978-3-540-78800-3_24
6. Feng, Y., Martins, R., Bastani, O., Dillig, I.: Program synthesis using conflict-driven learning. In: Proceedings Conference on Programming Language Design and Implementation, pp. 420–435. ACM (2018)
7. Feng, Y., Martins, R., Van Geffen, J., Dillig, I., Chaudhuri, S.: Component-based synthesis of table consolidation and transformation tasks from examples. In: Proceedings Conference on Programming Language Design and Implementation, pp. 422–436. ACM (2017)
8. Feng, Y., Martins, R., Wang, Y., Dillig, I., Reps, T.: Component-Based Synthesis for Complex APIs. In: Proceedings Symposium on Principles of Programming Languages, pp. 599–612. ACM (2017)
9. Feser, J.K., Chaudhuri, S., Dillig, I.: Synthesizing data structure transformations from input-output examples. In: Proceedings Conference on Programming Language Design and Implementation, pp. 229–239. ACM (2015)
10. Gulwani, S.: Automating string processing in spreadsheets using input-output examples. In: Proceedings Symposium on Principles of Programming Languages, pp. 317–330. ACM (2011)
11. Gulwani, S., Jha, S., Tiwari, A., Venkatesan, R.: Synthesis of loop-free programs. In: Proceedings Conference on Programming Language Design and Implementation, pp. 62–73. ACM (2011)
12. Jha, S., Gulwani, S., Seshia, S.A., Tiwari, A.: Oracle-guided component-based program synthesis. In: Proceedings International Conference on Software Engineering, pp. 215–224. ACM (2010)
13. Ramakrishnan, R., Gehrke, J.: Database management systems. McGraw Hill, New York (2000)
14. Raychev, V., Vechev, M., Yahav, E.: Code completion with statistical language models. In: Proceedings Conference on Programming Language Design and Implementation, pp. 419–428. ACM (2014)
15. Singh, R., Gulwani, S.: Transforming spreadsheet data types using examples. In: Proceedings Symposium on Principles of Programming Languagesm, pp. 343–356. ACM (2016)

16. Tiwari, A., Gascón, A., Dutertre, B.: Program synthesis using dual interpretation. In: Felty, A.P., Middeldorp, A. (eds.) CADE 2015. LNCS (LNAI), vol. 9195, pp. 482–497. Springer, Cham (2015). https://doi.org/10.1007/978-3-319-21401-6_33
17. Tran, Q.T., Chan, C., Parthasarathy, S.: Query by output. In: Proceedings International Conference on Management of Data, pp. 535–548. ACM (2009)
18. Wang, C., Cheung, A., Bodik, R.: Synthesizing highly expressive SQL queries from input-output examples. In: Proceedings Conference on Programming Language Design and Implementation, pp. 452–466. ACM (2017)
19. Yaghmazadeh, N., Wang, Y., Dillig, I., Dillig, T.: SQLizer: query synthesis from natural language. In: Proceedings International Conference on Object-Oriented Programming, Systems, Languages, and Applications, pp. 63:1–63:26. ACM (2017)
20. Zhang, S., Sun, Y.: Automatically synthesizing SQL queries from input-output examples. In: Proceedings International Conference on Automated Software Engineering, pp. 224–234. IEEE (2013)

Lemma Synthesis for Automating Induction over Algebraic Data Types

Weikun Yang$^{(\boxtimes)}$, Grigory Fedyukovich⦿, and Aarti Gupta

Princeton University, Princeton, NJ 08544, USA
{weikuny,grigoryf,aartig}@cs.princeton.edu

Abstract. In this paper we introduce a new approach for proving quantified theorems over inductively defined data-types. We present an automated prover that searches for a sequence of simplifications and transformations to prove the validity of a given theorem, and in the absence of required lemmas, attempts to synthesize supporting lemmas based on terms and expressions witnessed during the search for a proof. The search for lemma candidates is guided by a user-specified template, along with many automated filtering mechanisms. Validity of generated lemmas is checked recursively by our prover, supported by an off-the-shelf SMT solver. We have implemented our prover called ADTIND and show that it is able to solve many problems on which a state-of-the-art prover fails.

1 Introduction

Program verification tasks are often encoded as queries to solvers for Satisfiability Modulo Theories (SMT). Modern solvers, such as Z3 [26] and CVC4 [3], are efficient and scalable mainly on quantifier-free queries. Formulas with universally quantified formulas, which could be obtained from programs with algebraic data types (ADT), are still challenging. While quantifier-instantiation strategies [14, 18, 25] and superposition-based theorem proving [10, 23] are effective in some cases, a native support for inductive reasoning is needed to handle the full range of problems. Inductive reasoning over universally quantified formulas has been partially implemented in CVC4, in particular, using a conjecture-generation feature [30]. However, CVC4 often generates too many unrelated conjectures and does not utilize a problem-specific information.

Automating induction over ADTs has also been the target for many theorem provers. Tools such as IsaPlanner [11], ACL2 [6], Zeno [31], and HipSpec [8] can make use of induction when proving goals, with varying capabilities of automatic lemma discovery based on rippling [5] and generalization. However, the heuristics for lemma discovery are baked into the prover as fixed rules that target a limited space. These rule-based approaches are often ineffective when the form of the required lemmas is significantly different from expressions encountered during the proof attempt. There is no automated support for exploring a larger search space for candidates, and the user has to manually guide the overall search.

© Springer Nature Switzerland AG 2019
T. Schiex and S. de Givry (Eds.): CP 2019, LNCS 11802, pp. 600–617, 2019.
https://doi.org/10.1007/978-3-030-30048-7_35

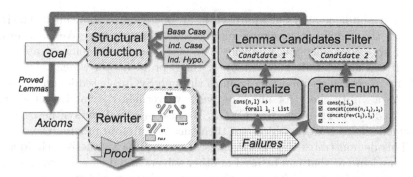

Fig. 1. ADTIND workflow.

A major challenge with both SMT-based tools and induction provers is that they often fail to produce crucial lemmas from which actual proofs follow. Technically this is due to the failure of cut-elimination in inductive theories [20], making the problem of finding proofs for many instances undecidable.

Our approach to automatic lemma discovery is inspired by the framework of syntax-guided synthesis (SyGuS) [1], applicable to program synthesis. To discover a program that meets a specification, SyGuS-based approaches take as additional input a formal grammar that defines the search space for the program. This framework has been successfully used in many applications and there exist dedicated SyGuS solvers that target various domains [2]. For example, SyGuS-based techniques have been used for program verification via generation of invariants [12,16,17] and termination arguments [13]. Although SyGuS has also inspired various SMT solver efforts [27–29] (described later in related work), to the best of our knowledge, none of these tackle automatic generation of lemmas for proofs by induction over ADTs.

An overview of our proposed framework is shown in Fig. 1. It is built on top of an automated theorem prover based on inductive reasoning. The prover decomposes given theorems into the base-case and inductive-case subgoals and uses a backtracking rewriter that sequentially simplifies each of the subgoals toward *true*. When the prover is unable to succeed, our approach first *generalizes* the partially-rewritten formula (as done in prior efforts) by replacing certain concrete subterms in a formula with fresh variables and attempts to prove its validity from scratch. If successfully proved, such a lemma can then be used to help prove the original subgoal.

However, generalization can discover only a limited number of lemmas. Therefore, we also perform a *SyGuS-based lemma enumeration* driven by *templates*, i.e., formulas with unknowns potentially provided by the user. Our key contribution is an algorithm that instantiates these unknowns with terms generated from syntactic elements obtained automatically from the formulas encountered during the proof search. Thus, the formal grammars, provided as input to SyGuS in our case are automatically generated, goal-directed, and in many cases small. Furthermore, we contribute a set of built-in grammar templates and techniques to effectively filter out invalid formulas produced by the enumeration.

We have implemented our approach in an open-source tool called ADTIND, including the inductive reasoning module and the rewriter. We have evaluated the tool on challenging problems with ADTs and have demonstrated that ADTIND can successfully solve many of these problems on which CVC4 [30] failed, by discovering supporting lemmas through SyGuS-based lemma enumeration. As a sanity check for our tool, we have verified the validity of lemmas synthesized by ADTIND by using CVC4. We also provided our synthesized lemmas as axioms to CVC4, which can then succeed often in proving the original goal. This demonstrates the effectiveness of our lemma synthesis techniques in generating lemmas that can be used in other solvers and different environments, not just in combination with a rewriter that we have used here to implement our ideas.

In summary, this paper makes the following contributions:

- an algorithm for automating proofs by induction over ADTs, where lemmas are synthesized by term enumeration guided by user-specified templates.
- an optimized enumeration process to propose lemma candidates, taking into account the formulas encountered during the proof search. This process includes filtering to reduce overhead of considering candidates that are invalid, or lemmas that may be valid but are less likely to be useful in a proof. These lemma synthesis techniques could be potentially integrated with other solvers.
- an implementation of our lemma synthesis procedures along with an inductive reasoning module and a rewriter that work together as a theorem prover ADTIND. We demonstrate its effectiveness in handling many challenging problems that cannot be solved by a state-of-the-art automated solver.

2   Preliminaries

A many-sorted first-order theory is defined as a tuple $\langle \mathcal{S}, \mathcal{F}, \mathcal{P} \rangle$, where $\mathcal{S}$ is a set of sorts, $\mathcal{F}$ is a set of function symbols, and $\mathcal{P}$ is a set of predicate symbols, including equality. A formula φ is called satisfiable if there exists a model where φ evaluates to $true$. If every model of φ is also a model of ψ, then we write $\varphi \implies \psi$. A formula φ is called $valid$ if $true \implies \varphi$.

An algebraic data type (ADT) is a tuple $\langle s, C \rangle$, where $s \in \mathcal{S}$ is a sort and C is a set of uninterpreted functions (called $constructors$), such that each $c \in C$ has some type $A \to s$. If for some s, A is s-free, we say that c is a $base$ constructor (otherwise, an $inductive$ constructor).

In this paper, we consider universally-quantified formulas over ADTs and uninterpreted functions. For proving validity of a formula $\forall x.\varphi(x)$, where variable x has sort s, we follow the well-known principle of $structural\ induction$:

Lemma 1. *Given an ADT $\langle s, \{bc_s : s, ic_s : \underbrace{s \times \ldots \times s}_{n} \to s\} \rangle$ and a formula φ,*

if the following two formulas (base case and inductive case) are valid:

$$\varphi(bc_s) \qquad and \qquad \forall x_1, \ldots, x_n. \Big(\bigwedge_{1 \leq i \leq n} \varphi(x_i) \Big) \implies \varphi(ic_s(x_1, \ldots, x_n))$$

then $\forall x.\varphi(x)$ is valid.

Lemma 1 is easily generalizable for ADTs with other constructor types. For instance, an inductive constructor *cons* of a single-linked list has an arity two (i.e., it takes an additional integer i as argument). Thus, to prove the inductive step, the validity of the following formula should be determined:

$$\forall x.\varphi(x) \implies \forall i.\varphi(cons(i, x))$$

We are interested in determining the validity of a universally-quantified formula $\forall x.\varphi(x)$, where φ may itself consist of universally quantified formulas:

$$\forall x.\left(\forall y.\psi(y)\right) \wedge \ldots \wedge \left(\forall z.\gamma(z)\right) \implies \theta(x) \tag{2.1}$$

We call the innermost universally-quantified formulas on the left side of the implication (2.1) *assumptions*. An assumption is called an *axiom* if it is not implied by any combination of other assumptions, and a *lemma* otherwise. We also assume that neither axioms nor lemmas have appearances of the variable x. The formula on the right side of the implication, which is the only formula over x, is called a *goal*.

A proof of a valid formula of the form (2.1) is derived by structural induction. In both the base and inductive cases, quantifier-free instances $Q(x)$ of axioms and lemmas are produced and used to (sequentially) rewrite the goal until it is simplified to *true*. In particular, we consider the following two simple proof rules (other rules on inequalities and user-proved predicates could also be added):

$$\frac{Q(x) \implies goal(x)}{true} \text{ [apply]} \qquad \frac{Q(x) \equiv (P(x) = R(x))}{goal[P \mapsto R](x)} \text{ [rewrite]}$$

If a (possibly transformed) goal cannot be further rewritten or simplified by the given axioms or lemmas, it is called a *failure* formula (if clear from the context, we drop "formula" and simply call it a failure). Clearly, when a goal is supplied by a larger number of assumptions, there is a wider room for possible simplifications, and thus a prover has more chances to succeed. We thus contribute a method that discovers new assumptions and enlarges the search space. An important condition for soundness of this method is that such newly introduced lemmas should themselves be derivable from given assumptions.

3   Motivating Example

Consider an ADT *Queue* defined as a tuple of two lists (the first one being the front, and the second one being the back but in reverse): $queue : list \times list \to Queue$. And some useful functions include: *concat* which concatenates two lists together, *len* which computes the length of a list, *qlen* which computes the length of a queue, *qpush* which appends one element to a queue, and finally *amrt* that balances the queue by concatenating the two lists together when the second list becomes longer than the first one. The given axioms and goal are shown below.

Axioms: definition of $concat, len, qlen, qpush, amrt$

$$\forall l.\ concat(nil, l) = l$$
$$\forall l_1, l_2, n.\ concat(cons(n, l_1), l_2) = cons(n, concat(l_1, l_2)) \tag{3.1}$$

$$\forall l.\ len(nil) = 0$$
$$\forall l, n.\ len(cons(n, l)) = 1 + len(l) \tag{3.2}$$

$$\forall l_1, l_2.\ qlen(queue(l_1, l_2)) = len(l_1) + len(l_2) \tag{3.3}$$

$$\forall l_1, l_2, n.\ qpush(queue(l_1, l_2), n) = amrt(l_1, cons(n, l_2)) \tag{3.4}$$

$$\forall l_1, l_2.amrt(l_1, l_2) = \begin{cases} queue(l_1, l_2) & \text{if } len(l_1) \geq len(l_2) \\ queue(concat(l_1, rev(l_2)), nil) & \text{Otherwise} \end{cases}$$
$$\tag{3.5}$$

Goal: prove that length of queue increases by 1 after $qpush$

$$\forall l_1, l_2.\ qlen(qpush(queue(l_1, l_2), n)) = 1 + qlen(queue(l_1, l_2)) \tag{3.6}$$

Our approach performs several rewriting steps of applying function definitions, then the base case of induction on variable l_1 leads to the following formula:

$$\forall l_2, n.\ 1 + len(l_2) = len(concat(rev(l_2), cons(n, nil))) \tag{3.7}$$

The formula (3.7) cannot be further simplified with existing axioms, and this constitutes a failure. Before moving on to the inductive case for l_1, we first try to apply generalization as follows:

– We could replace nil with new list variable l_3 on the right hand side (RHS), but there is no place to introduce l_3 on the left hand side (LHS).
– We could also replace $cons(n, nil)$ with new list variable l_3, again there is no corresponding replacement on the LHS.
– Function applications are possible candidates as well, e.g., replacing $rev(l_2)$ with new list variable l_3, yet again it cannot be applied on the LHS.

As seen above, generalization does not give us any suitable candidates. However, useful lemmas could still be generated from ingredients occurring in the failure. In this case, we apply the last two generalization rules (shown above) to the RHS of (3.7), and then automatically construct a formal grammar using one of the user-provided templates (to be explained in detail in Sect. 4.2). In particular, the last line in (3.8) below is regarded as a template, where the undefined symbols (shown as $\langle ??? \rangle$) have to be filled automatically with suitable terms. This is done by enumeration of integer-typed terms with the function symbols $len, concat$ (as well as constructors nil and $cons$), the variables l_3, l_4, and integer constants such as 0, 1, etc.

$$\forall l_2, n.\ 1 + len(l_2) = len(concat(rev(l_2), cons(n, nil)))$$

$$\downarrow \text{ replace } rev \text{ and } cons \text{ with new variables}$$

$$\forall l_2, n.\ 1 + len(l_2) = len(concat(l_3, l_4)) \tag{3.8}$$

$$\downarrow \text{ create template for term enumeration}$$

$$\forall l_3, l_4.\ \langle ??? \rangle + \langle ??? \rangle = len(concat(l_3, l_4))$$

For instance, our approach (explained in detail in Sect. 4) is able to discover the following lemma:

$$\forall l_3, l_4.\ len(concat(l_3, l_4)) = len(l_3) + len(l_4) \tag{3.9}$$

The lemma is proven valid by induction (i.e., a recursive invocation of our method), and then it can be used to prove the original goal.

4   Lemma Synthesis

In this section, we describe our key contributions on automated lemma synthesis. Algorithm 1 shows the top level procedure SOLVEWITHINDUCTION which applies structural induction to create the base-case and inductive-step subgoals for the rewriter to prove. If any subgoal cannot be proved with existing assumptions, the algorithm invokes GENERALIZE and ENUMERATELEMMAS to produce lemma candidates based on failures found in REWRITE. These procedures are further described in the following sections.

- REWRITE: a backtracking engine that attempts to rewrite a given goal towards *true*, using the provided assumptions (including discovered lemmas, as described in Sect. 2). For practical reasons, our implementation uses maximum limits on the depth of the recursive proof search and on the number of rewriting attempts using the same transformation. When a subgoal is not proved, the main output of this engine is a set of *failures*, i.e., formulas obtained during the search, to which no further rewriting rule can be applied (within the given limits). Our algorithm can utilize an external library of proven theorems while a set of heuristics must be developed to efficiently traverse a large search space, which is outside the scope of this work.
- GENERALIZE: an engine, further described in Sect. 4.1, which takes the failures discovered by REWRITE and applies transformations to replace concrete values by universally quantified variables in order to produce lemma candidates that may support proving the original goal.
- ENUMERATELEMMAS: a SyGuS-based lemma synthesis engine, further described in Sect. 4.2, which proposes a larger variety of lemma candidates than generalization. This incorporates more aggressive *mutation* of failures than generalization, to make the lemmas goal-oriented. This method is configurable by the choice of grammars, which can be guided by the user. In our implementation, we include grammars tailored to the most common applications appearing in practice in our benchmark examples.

For simplicity of the presentation, our pseudo-code in Algorithm 1 assumes only one quantified ADT variable, which is used to generate the base-case and inductive-step subgoals. However, our implementation also supports multiple quantifiers and nested induction. For both the subgoals, the proving strategy is to find a sequence of rewriting attempts using the set of assumptions. For the inductive case, the inductive hypotheses are also included in the set of assumptions. In the case of nested induction (omitted from the pseudo-code), all assumptions from the outer-induction are inherited by the inner-induction.

If the algorithm falls short in rewriting any of the subgoals using the existing assumptions, it attempts to synthesize new lemmas by (1) applying GENERALIZE to failures, and (2) identifying suitable terms from failures for applying SyGuS (inside ENUMERATELEMMAS). Generated this way, a lemma candidate needs to be checked for validity which is performed by calling Algorithm 1 recursively.

Algorithm 1: SOLVEWITHINDUCTION($Goal$, $Assumptions$)

Input: $Goal$: quantified formula to be proved, $Assumptions$: set of formulas
Output: $result \in \{\text{QED}, \text{UNKNOWN}\}$

1 **for** $subgoal \in \{baseCase(goal), indStep(goal)\}$ **do**
2   **if** $indStep$ **then** $Assumptions \leftarrow Assumptions \cup \{indHypo\}$
3      $result, failures \leftarrow$ REWRITE($subgoal$, $Assumptions$)
4   **if** $result$ **then continue**
5      $candidates \leftarrow$
         MAP(GENERALIZE, $failures$) $\cup$ ENUMERATELEMMAS($failures$)
6   **for each** $\psi \in candidates$ **do**
7   **if** SOLVEWITHINDUCTION(ψ, $Assumptions$) = QED **then**
8              $result \leftarrow$ SOLVEWITHINDUCTION($subgoal$, $Assumptions \cup \{\psi\}$)
9   **if** $result$ = QED **then break**
10   **if** $baseCase$ **and** $result$ = UNKNOWN **then return** UNKNOWN

11 **return** $result$

4.1 Lemma Synthesis by Generalization

The approach of generalizing a failure is widely applied among induction solvers such as IsaPlanner [11], ACL2 [6], and Zeno [31], based on the observation that proving a formula that applies to some specific value is often more difficult than proving a more general version. In our setting, we replace suitable subterms of the formula with fresh quantified variables, effectively weakening the formula.

Algorithm 2 shows the pseudocode of our generalization procedure that, given a formula φ, outputs a lemma candidate ψ. It starts by gathering common subterms in φ (e.g., when φ is an equality, it is possible that the same terms occur on both its sides). Then, it replaces occurrences of subterms by fresh variables and universally quantifies them. In our implementation, we prefer to generalize applications of inductive constructors first. If no lemma was discovered, we proceed to generalizing uninterpreted functions, and our last choice is to generalize base constructors.

Algorithm 2: GENERALIZE(φ)

Input: φ: formula to be generalized
Output: ψ: generalized formula

1 **while** $\exists t \in terms(\varphi)$, which occurs in φ twice **do**
2   let v be such that $v \notin vars(\varphi)$
3      $\psi \leftarrow \forall v.\varphi[t \mapsto v]$
4 **return** ψ

Algorithm 3: ENUMERATELEMMAS($Failures$)

Input: $Failures$ in the proof search
Output: $Candidates$ formulas

1 $\Phi \leftarrow terms(Failures)$
2 **while** $|Candidates| <$ THRESHOLD **do**
3      $\varphi \leftarrow$ LARGEST(Φ)
4      $G \leftarrow$ CREATEGRAMMAR($functions(\varphi), predicates(\varphi), vars(\varphi)$)
5   **for each** $\psi \in G$ **do**
6          $\psi \leftarrow \forall vars(\varphi).\psi$
7   **if** $\neg$REFUTED(ψ) **then** $Candidates \leftarrow Candidates \cup \{\psi\}$
8      $\Phi \leftarrow \Phi \setminus \{\varphi\}$
9 **return** $Candidates$

4.2 SyGuS-Based Lemma Synthesis

Applying generalization alone may not yield the desired supporting lemma at times. Algorithm 3 shows our SyGuS-style approach for synthesis of lemma candidates from formal grammars. These formal grammars are themselves generated on-the-fly by our procedure. Specifically, in each iteration of the outer loop, the algorithm picks a term which occurs in some failure, and then uses its parse tree to extract function and predicate symbols to construct a formal grammar. This grammar is then used to generate the desired candidate lemmas automatically. Our key contribution is the grammar construction algorithm (outlined in Sect. 4.3) that uses these function and predicate symbols in combination with user-provided templates. We also provide a set of built-in templates that have worked well on our practical benchmarks.

Finally, in Sect. 4.4, we describe how to enumerate lemma candidates (up to a certain size) from the grammar, and how to filter likely successful candidates for the original proof goal in Algorithm 1. These candidates must be proven correct first, as shown in Sect. 4.5.

4.3   Automatic Construction of Grammars

Although our algorithm does not depend on any particular grammar for lemma generation, it is practically important to consider grammars that are relevant for the failures (one or many), so that the generated lemmas have a higher likelihood of success in proving the original goal. Therefore, we focus on various elements (e.g., uninterpreted functions and predicates) that can be extracted from the parse trees of failures, to automate the process of grammar creation. Elements that do not appear in the failure are not considered to save efforts.

At the same time, a user might specify some *higher-level templates* that provide additional guidance for this process. Essentially, a higher-level template provided by a user can be viewed as a *partially defined grammar* that involves a set of *undefined nonterminals* (i.e., where the corresponding rules are still undefined). Our algorithm automatically constructs missing rules for these nonterminals by using the syntactic patterns obtained from failures, thus constructing fully-defined grammars. These fully-defined grammars are then used for automatically generating candidate lemmas.

To additionally optimize this process, our grammar construction algorithm focuses on individual subterms occurring in failures. Our particular strategy is to pick the largest subterm (referred to as φ in the pseudo-code and later in the text), but other heuristics could be used here as well.

Furthermore, we identified three useful higher-level templates that have been applied to solve our benchmarks[1]. These templates are in the form of an equality, they use undefined nonterminals (shown as $\langle ??? \rangle$), and interestingly, two of them have occurrences of φ on the left side of the equality:

$$\varphi = \langle ??? \rangle + \langle ??? \rangle \tag{4.1}$$

$$\varphi = \langle ??? \rangle \tag{4.2}$$

$$\langle ??? \rangle = \langle ??? \rangle \tag{4.3}$$

The first template is chosen when φ has an integer type, and the second one is chosen for all algebraic data types. Lemma candidates generated from the first two templates inherit information from the failure, having the subterm φ on one side. The third template is chosen as a last resort, when no valid lemmas are discovered after using the first two (as explained in Sects. 4.4 and 4.5).

After choosing one of these templates, our algorithm defines the rules for nonterminals $\langle ??? \rangle$, based on the variables, uninterpreted functions, and predicates occurring in φ.

Additionally, we identified two higher-level templates, applicable when the same function occurs in a failure multiple times. Intuitively, they correspond to the commutativity and the associativity of certain uninterpreted functions. After such functions are determined by a syntactic analysis of a failure, they immediately give instantiations of nonterminals $\langle ??? \rangle$ in templates (4.4) and (4.5).

---

[1] These templates are referred to as *built-in* templates, which need not be specified by the user. Furthermore, our current implementation automatically chooses a built-in template based on φ.

$$\langle ??? \rangle (a, b) = \langle ??? \rangle (b, a) \tag{4.4}$$

$$\langle ??? \rangle (a, \langle ??? \rangle (b, c)) = \langle ??? \rangle (\langle ??? \rangle (a, b), c) \tag{4.5}$$

Returning back to our motivating example, for failure (3.7), both sides of the equality have integer type. Thus, our algorithm chooses template (4.1), and we use the right side of (3.7) as φ, since it is the larger (more complex) expression. This allows more information from the failure to be retained, thereby enabling the enumerated lemma candidates to be goal-directed. The following grammar is then automatically extracted from φ:

$$\langle \mathbf{int - term} \rangle ::= n \mid len(\langle \mathbf{list - term} \rangle)$$

$$\langle \mathbf{list - term} \rangle ::= nil \mid l_2 \mid cons(\langle \mathbf{int - term} \rangle, \langle \mathbf{list - term} \rangle) \mid \tag{4.6}$$

$$concat(\langle \mathbf{list - term} \rangle, \langle \mathbf{list - term} \rangle) \mid rev(\langle \mathbf{list - term} \rangle)$$

Note that this grammar is recursive and relatively large in scope. For performance reasons, we try to reduce the grammar. We do this by heuristically generalizing φ first, where we replace function applications by fresh variables (i.e., similar to the strategy in Sect. 4.1), as shown in (3.8). The generalized φ gives rise to the following grammar:

$$\langle \mathbf{int - term} \rangle ::= len(\langle \mathbf{list - term} \rangle) \tag{4.7}$$

$$\langle \mathbf{list - term} \rangle ::= l_3 \mid l_4 \mid concat(\langle \mathbf{list - term} \rangle, \langle \mathbf{list - term} \rangle)$$

Finally, the resulting production rules are embedded into the chosen template to generate a complete grammar, where $\langle ??? \rangle$ is instantiated by $\langle \mathbf{int - term} \rangle$.

4.4   Producing Terms from Grammar

Given a grammar, constructed as shown in the previous subsection, our algorithm enumerates various candidate lemmas and checks their validity. Since larger candidate lemmas are typically more expensive to deal with, our algorithm starts by enumerating small formulas with terms upto some size. We define the *size* of an expression as the height of its parse tree. For example, variables and base constructors of data types, such as x, y, nil have size 1, while $cons(1, nil)$ and $rev(x)$ have size 2. By Ψ_k, we denote the set of expressions of size k, and by $\Psi_k[ty]$, we denote the set of expressions of size k that has type ty.

Given Ψ_k, it is straight-forward to enumerate expressions of size $k + 1$: for each function (including inductive constructors) f with m parameters typed $ty_1, \ldots, ty_m$, we first enumerate expressions $\tau_1, \ldots, \tau_m$ from sets $\Psi_k[ty_1], \ldots \Psi_k[ty_m]$, respectively, and second, we create a new expression $f(\tau_1, \ldots, \tau_m)$ which is inserted into Ψ_{k+1}. This process is repeated iteratively until we reach the desired size limit.

Our algorithm enforces the following two constraints on the generated candidate formulas. First, it checks the generated formula for *non-triviality*: there should be no application of a function on only base constructors of an ADT, e.g., *concat*(*nil*, *nil*). Such candidates are usually invalid for any non-trivial instantiation of a universally quantified variable. Second, a generated lemma candidate should cover as many variables occurring in the subterm φ from which we derived the templates (4.1) – (4.3) as possible. In our experience, prioritizing candidates with full coverage leads to significant performance gains.

To further reduce the number of candidates, we leverage symmetry of operators in a template (e.g., commutativity of integer addition) whenever possible.

4.5   Filtering by Refutation

We apply an additional filtering step on lemma candidates where we search for inexpensive counterexamples to validity. Given a candidate lemma, our algorithm instantiates quantified variables with concrete values, creates quantifier-free expressions, and repeatedly simplifies them by applying assumptions. In addition to the rules mentioned in Sect. 2, we also apply the following *refutation* rule:

$$\frac{goal(x) \implies false}{false} \text{ [apply}^r\text{]}$$

In our implementation, we limit the number of refutation attempts for each candidate and the complexity of the concrete instantiations. The concrete values of variables are produced by applying constructors of ADTs repeatedly.

For example, formula (4.8) is one of the possible candidates based on the template in (3.8). This lemma is shown invalid by instantiating l_3 and l_4 with concrete lists $cons(1, cons(2, nil))$ and $cons(3, nil)$, and then applying the given axioms to the resulting quantifier-free expression, as shown below.

$$\forall l_3, l_4.\, len(cons(len(l_3), nil)) + len(l_4) = len(concat(l_3, l_4))$$

$\quad\downarrow$ `instantiate quantified variables`

$$len(cons(len(cons(1, cons(2, nil))), nil)) + len(cons(3, nil))$$
$$= len(concat(cons(1, cons(2, nil)), cons(3, nil)))$$

$\quad\downarrow$ `apply axiom (3.2)`

$$1 + 1 = len(concat(cons(1, cons(2, nil)), cons(3, nil)))$$

$\quad\downarrow$ `apply axiom (3.1)`

$$1 + 1 = len(cons(1, cons(2, cons(3, nil))))$$

$\quad\downarrow$ `apply axiom (3.2)`

$$1 + 1 = 1 + 1 + 1 \text{ (False)}$$

(4.8)

If a lemma candidate passes (some number of) refutation tests, then a new instance of SOLVEWITHINDUCTION is created in an attempt to prove its validity. This recursive nature of our procedure allows proving lemma candidates that may further require discovering new supporting lemmas. However, creating a subgoal to prove a lemma candidate is a fairly expensive procedure. Therefore, we would like the filtering to be aggressive, to minimize the number of lemmas to be proved. Although the refutation tests are relatively cheap to perform, too many tests may result in wasted effort and delay lemma application in proving the original goal. Thus, we must strike a balance between testing for refutations and proof attempts. In our implementation, we perform three refutation tests by default (and the user can optionally set the number of such tests).

5   Implementation and Evaluation

We have implemented our algorithm in a prototype tool named ADTIND on top of Z3 [26]. Our backtracking REWRITE procedure uses the "apply" and "rewrite" proof rules repeatedly to simplify the goals and invokes Z3 to determine the validity of quantifier-free expressions encountered during such rewriting. Our implementation allows the user to specify the maximal depth of the backtracking search (15 steps by default); it also avoids divergence by limiting the consecutive applications of the same rewrite rules.

Our lemma synthesis procedures are also configurable. In GENERALIZE, the user can adjust the aggressiveness of generalization, opting to replace smaller or larger terms in failures (recall Sect. 4.1). In ENUMERATELEMMAS, the user sets a larger limit on sizes of enumerated terms to explore a larger space of lemma candidates. The number of refutation attempts is also configurable (3 times by default).

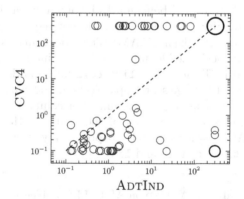

Fig. 2. Evaluation comparison (sec × sec): points above the diagonal represent runtimes for benchmarks on which ADTIND outperformed CVC4; points on the boundaries represent timeouts. The diameter of a circle represents the number of overlapping circles.

ADTIND has been evaluated on benchmarks from the CLAM [20] suite[2] consisting of 86 quantified theorems over common operations of natural numbers,

---

[2] The source code of ADTIND and benchmarks are available at: github.com/wky/aeval/tree/adt-ind.

lists and other data types. We have compared ADTIND against the CVC4 SMT solver (v1.7, which supports induction and subgoal generation). We used a timeout of 300 seconds. The scatter plot in Fig. 2 summarizes the results. In total, ADTIND proved 62, and CVC4 proved 47 benchmarks. These numbers include respectively 22 and 7 theorems proven only by the corresponding tool and not by the other (shown in Fig. 2 as crosses on the top horizontal line and as crosses clustered around the bottom right corner, respectively). Interestingly, there are not many cases when CVC4 takes a significant amount of time before delivering a successful result, i.e., it either terminates in less than a second or diverges. This is possibly due to an inability to discover a meaningful lemma candidate for these benchmarks. In contrast, ADTIND is often able to enumerate useful lemmas, but sometimes it requires a number of iterations (see, e.g., crosses on the bottom horizontal line). We hope to improve runtime performance of ADTIND in the future by adopting certain successful optimizations and heuristics from CVC4.

Of the 62 theorems proven by ADTIND, 29 did not require extra lemmas, 12 were proven with lemmas discovered through generalization, and 21 were proven with lemmas discovered through SyGuS. For the SyGuS-generated 21 theorems, ADTIND created on average 171 lemma candidates. In our experiments, 93% of processed candidates were refuted by tests, leaving only a small number of lemmas to be validated by the more expensive SOLVEWITHINDUCTION.

Experiment over ADTs and LIA. To fully demonstrate the power of SMT solvers, we considered several additional benchmarks involving linear integer arithmetic (LIA). With this capability, the benchmarks do not require specifying assumptions over natural numbers (like in CLAM). These benchmarks also motivate the usefulness of having a specialized lemma template for integers shown in Sect. 4.3. The results are listed in Table 1. The `list_rev2` benchmark took more solving time than others due to the large search space (about 500 lemma candidates were rejected before a sufficient lemma was found). For comparison, CVC4 failed to prove `list_rev` and exceeded a timeout of 300 s on the other 8 problems. The interactive prover ACL2 was only able to prove only 2 out of 9 problems in Table 1, namely `list_rev_concat` and `list_rev_len`.

Table 1. ADTIND on ADT+LIA problems.

Goal	ADTIND result	Goal	ADTIND result
`list_rev`	Proved, 10.6 s	`list_rev2_len`	Proved, 0.95 s
`list_rev_concat`	Proved, 2.52 s	`queue_push`	Proved, 33.9 s
`list_rev2_concat`	Proved, 1.95 s	`queue_len`	Proved, 7.6 s
`list_rev2`	Proved, 1 m 59 s	`tree_insert_all`	Proved, 1.9 s
`list_rev_len`	Proved, 2.30 s		

Future Work. There are other categories of theorems, mainly in the Leon [4] and TIP [7] suites, that require more advanced techniques for automated proving. Many theorems require non-trivial case-splitting transformations (some implemented in Why3 [15]) for *if-then-else* blocks in given axioms, which is not fully supported in our prototype yet. However, our lemma synthesis algorithms can be used in combination with a different solver or environment that handles case-splitting and other forms of goal decomposition. Thus, our tool can focus on producing lemma candidates without being dependent on current capabilities of our prototype rewriting engine.

Of the theorems that we cannot prove in the TIP set, many are mathematically challenging (e.g., Fermat's Last Theorem), involve high-order functions, contain sortedness properties, or require some form of pumping lemma to solve (e.g., proving equivalence of regular languages). These instances are currently outside the scope of our work.

6   Related Work

There is a wide range of approaches for proving quantified theorems defined on algebraic data types. These include SMT-based inductive reasoning in tools such as Dafny [24] and CVC4 [30]; Horn Clause solvers [33]; generic theorem provers such as ACL2 [6], and induction provers such as CLAM [20], IsaPlanner [22] and HipSpec [8]. The main issue with these tools is that even with the help of built-in heuristics such as rippling [5] and generalization of failures, they still require human interaction to discover necessary lemmas to complete a proof end-to-end. Our proposal to use term enumeration for lemma discovery (after failure of generalization) as a SyGuS-style synthesis task leverages information available at proof failures and explores a much larger space of possible lemma candidates. As shown in our evaluation in Sect. 5, this was enough to eliminate the need for human input in many practical cases.

On lemma discovery within induction provers, machine learning techniques have also been attempted in works such as ACL2(ml) [19] and Multi-Waterfall [21]. ACL2(ml) uses statistical machine learning algorithms to extract features present in the proof goal, and uses that to find similar patterns in a library of proven theorems to in order to suggest new lemmas. Multi-Waterfall runs multiple strategies in parallel, while a machine learning module trained by previous proofs in a library is used to select lemmas candidates based on their likelihood of advancing the current proof. The machine learning components in these tools typically require a sufficiently large set of proven theorems to learn from, whereas our tool uses term enumeration that does not depend on an external library.

Specifically, CVC4 [30] supports induction natively to solve quantified SMT queries with custom data types. The tool implements Skolemization with inductive strengthening to prove conjectures, and uses enumeration to find adequate subgoals (inspired by QuickSpec [9]). CVC4 employs filtering of candidates based on activation of function symbols, canonicity of terms and counterexamples, which is roughly analogous to our filtering techniques. However, our lemma candidates arise from grammars that combine user-provided (or built-in) templates with elements from failures in rewriting proof attempts and seem to have a better chance of proving the original goal.

ACL2 (a Boyer-Moore prover) is based on rewriting of terms and a number of induction heuristics. The tool identifies "key checkpoints" as subgoals to prove on its way to prove the outer theorem, and has rules to perform generalization similar to our approach described in Sect. 4.1. However, ACL2 does not have the ability to enumerate lemma candidates, although users can provide their own proof tactics or plug-ins to this theorem prover.

On the lemma synthesis front, the SLS framework [32] employs different techniques to automatically generate and validate lemmas, but within an interactive theorem prover environment. For symbolic heap verification using separation logic, the tool generates lemma templates with the heap structures from the goal entailment, and proposes unknown relations as constraints over the templates' variables, which are later solved to discover the desired lemmas. We were unable to experimentally compare with SLS because it works in an interactive theorem prover environment, targets a distinct type of problems (proof entailments in separation logic) and requires a different input format which is prohibitive for us to translate to.

Among SyGuS applications for solving quantified formulas, in another recent effort with CVC4 [28], a user can provide a grammar and a correctness specification to a synthesis task, whose goal is to find *rewrite rules* that transform and simplify SMT queries. The similarity here is that our tool also uses a SyGuS-style user-provided template to search for supporting lemmas, which will be used just like rewrite rules. However, the purpose of their technique is primarily goal-agnostic simplification, and it does not track information such as failures in proof search. More importantly, their grammars are not generated automatically from problem instances, but are fixed by the user. Another recent effort [27] uses SyGuS to synthesize invertibility conditions under which quantified bit-vector problems can be converted to quantifier-free problems, to be solved by an SMT solver. However, the purpose and specific techniques are different from our approach.

Finally, SyGuS was recently applied to verification of program safety and termination in the FREQHORN framework [12,13]. These works exploit the syntax of given programs to automatically generate grammar, from which the candidates for inductive invariants and ranking functions are produced. While their main insight is similar to ours, their approach does not support ADTs and hardly exploits any failures. In the future, we believe that our tool could be integrated to FREQHORN and help verify programs which are currently out of its scope.

7   Conclusions and Future Work

We have presented a new approach for automating induction over algebraic data-types that uses lemma synthesis based on automatic grammar generation and term enumeration guided by user-specified templates. Our prover ADTIND incorporates these ideas in a rewriting engine built on top of Z3. We demonstrated that it successfully solves many challenging problem instances that a state-of-the-art prover failed to solve.

So far, the proof goals in the examples that we considered (i.e., *List*, *Queue*, *Tree*) are mostly in the form of equalities. We intend to apply our ideas to support inequalities and other relations that demand non-trivial inductive reasoning and lemma discovery. Incorporating our lemma synthesis procedures into other theorem proving frameworks (such as CVC4) would allow us to leverage existing heuristics and proof tactics to deliver results on more complex problems. Also we will consider additional criteria for usefulness of lemma candidates to better filter the large number of candidates in certain benchmarks.

Acknowledgments. This work is supported in part by NSF Grant 1525936.

References

1. Alur, R., et al.: Syntax-guided synthesis. In: FMCAD, pp. 1–17. IEEE (2013)
2. Alur, R., Fisman, D., Singh, R., Solar-Lezama, A.: SyGuS-Comp 2017: results and analysis (2017). http://sygus.seas.upenn.edu/
3. Barrett, C., et al.: CVC4. In: Gopalakrishnan, G., Qadeer, S. (eds.) CAV 2011. LNCS, vol. 6806, pp. 171–177. Springer, Heidelberg (2011). https://doi.org/10.1007/978-3-642-22110-1_14
4. Blanc, R., Kuncak, V., Kneuss, E., Suter, P.: An overview of the leon verification system: verification by translation to recursive functions. In: Proceedings of the 4th Workshop on Scala. SCALA 2013, pp. 1:1–1:10. ACM, New York, NY, USA (2013). https://doi.org/10.1145/2489837.2489838
5. Bundy, A., Stevens, A., van Harmelen, F., Ireland, A., Smaill, A.: Rippling: a heuristic for guiding inductive proofs. Artif. Intell. **62**(2), 185–253 (1993)
6. Chamarthi, H.R., Dillinger, P., Manolios, P., Vroon, D.: The ACL2 sedan theorem proving system. In: Abdulla, P.A., Leino, K.R.M. (eds.) TACAS 2011. LNCS, vol. 6605, pp. 291–295. Springer, Heidelberg (2011). https://doi.org/10.1007/978-3-642-19835-9_27
7. Claessen, K., Johansson, M., Rosén, D., Smallbone, N.: TIP: tons of inductive problems. In: Kerber, M., Carette, J., Kaliszyk, C., Rabe, F., Sorge, V. (eds.) CICM 2015. LNCS (LNAI), vol. 9150, pp. 333–337. Springer, Cham (2015). https://doi.org/10.1007/978-3-319-20615-8_23
8. Claessen, K., Johansson, M., Smallbone, N.: HipSpec: Automating inductive proofs of program properties. In: Workshop on Automated Theory eXploration: ATX 2012 (2012)
9. Claessen, K., Smallbone, N., Hughes, J.: QUICKSPEC: guessing formal specifications using testing. In: Fraser, G., Gargantini, A. (eds.) TAP 2010. LNCS, vol. 6143, pp. 6–21. Springer, Heidelberg (2010). https://doi.org/10.1007/978-3-642-13977-2_3
10. Cruanes, S.: Superposition with structural induction. In: Dixon, C., Finger, M. (eds.) FroCoS 2017. LNCS (LNAI), vol. 10483, pp. 172–188. Springer, Cham (2017). https://doi.org/10.1007/978-3-319-66167-4_10
11. Dixon, L., Fleuriot, J.: IsaPlanner: a prototype proof planner in isabelle. In: Baader, F. (ed.) CADE 2003. LNCS (LNAI), vol. 2741, pp. 279–283. Springer, Heidelberg (2003). https://doi.org/10.1007/978-3-540-45085-6_22

12. Fedyukovich, G., Kaufman, S., Bodík, R.: Sampling invariants from frequency distributions. In: FMCAD, pp. 100–107. IEEE (2017)
13. Fedyukovich, G., Zhang, Y., Gupta, A.: Syntax-guided termination analysis. In: Chockler, H., Weissenbacher, G. (eds.) CAV 2018. LNCS, vol. 10981, pp. 124–143. Springer, Cham (2018). https://doi.org/10.1007/978-3-319-96145-3_7
14. Feldman, Y.M.Y., Padon, O., Immerman, N., Sagiv, M., Shoham, S.: Bounded quantifier instantiation for checking inductive invariants. In: Legay, A., Margaria, T. (eds.) TACAS 2017. LNCS, vol. 10205, pp. 76–95. Springer, Heidelberg (2017). https://doi.org/10.1007/978-3-662-54577-5_5
15. Filliâtre, J.-C., Paskevich, A.: Why3 — where programs meet provers. In: Felleisen, M., Gardner, P. (eds.) ESOP 2013. LNCS, vol. 7792, pp. 125–128. Springer, Heidelberg (2013). https://doi.org/10.1007/978-3-642-37036-6_8
16. Garg, P., Löding, C., Madhusudan, P., Neider, D.: ICE: a robust framework for learning invariants. In: Biere, A., Bloem, R. (eds.) CAV 2014. LNCS, vol. 8559, pp. 69–87. Springer, Cham (2014). https://doi.org/10.1007/978-3-319-08867-9_5
17. Garg, P., Neider, D., Madhusudan, P., Roth, D.: Learning invariants using decision trees and implication counterexamples. In: POPL, pp. 499–512. ACM (2016)
18. Ge, Y., Barrett, C., Tinelli, C.: Solving quantified verification conditions using satisfiability modulo theories. In: Pfenning, F. (ed.) CADE 2007. LNCS (LNAI), vol. 4603, pp. 167–182. Springer, Heidelberg (2007). https://doi.org/10.1007/978-3-540-73595-3_12
19. Heras, J., Komendantskaya, E.: Acl2(ml): Machine-learning for ACL2. In: Proceedings Twelfth International Workshop on the ACL2 Theorem Prover and its Applications, Vienna, Austria, 12–13th July 2014, pp. 61–75 (2014)
20. Ireland, A., Bundy, A.: Productive use of failure in inductive proof. J. Autom. Reasoning **16**, 79–111 (1996)
21. Jiang, Y., Papapanagiotou, P., Fleuriot, J.: Machine learning for inductive theorem proving. In: Fleuriot, J., Wang, D., Calmet, J. (eds.) AISC 2018. LNCS (LNAI), vol. 11110, pp. 87–103. Springer, Cham (2018). https://doi.org/10.1007/978-3-319-99957-9_6
22. Johansson, M., Dixon, L., Bundy, A.: Case-analysis for rippling and inductive proof. In: Kaufmann, M., Paulson, L.C. (eds.) ITP 2010. LNCS, vol. 6172, pp. 291–306. Springer, Heidelberg (2010). https://doi.org/10.1007/978-3-642-14052-5_21
23. Kersani, A., Peltier, N.: Combining superposition and induction: a practical realization. In: Fontaine, P., Ringeissen, C., Schmidt, R.A. (eds.) FroCoS 2013. LNCS (LNAI), vol. 8152, pp. 7–22. Springer, Heidelberg (2013). https://doi.org/10.1007/978-3-642-40885-4_2
24. Leino, K.R.M.: Automating induction with an SMT solver. In: Kuncak, V., Rybalchenko, A. (eds.) VMCAI 2012. LNCS, vol. 7148, pp. 315–331. Springer, Heidelberg (2012). https://doi.org/10.1007/978-3-642-27940-9_21
25. de Moura, L., Bjørner, N.: Efficient E-matching for SMT solvers. In: Pfenning, F. (ed.) CADE 2007. LNCS (LNAI), vol. 4603, pp. 183–198. Springer, Heidelberg (2007). https://doi.org/10.1007/978-3-540-73595-3_13
26. de Moura, L., Bjørner, N.: Z3: an efficient SMT solver. In: Ramakrishnan, C.R., Rehof, J. (eds.) TACAS 2008. LNCS, vol. 4963, pp. 337–340. Springer, Heidelberg (2008). https://doi.org/10.1007/978-3-540-78800-3_24

27. Niemetz, A., Preiner, M., Reynolds, A., Barrett, C., Tinelli, C.: Solving quantified bit-vectors using invertibility conditions. In: Chockler, H., Weissenbacher, G. (eds.) CAV 2018. LNCS, vol. 10982, pp. 236–255. Springer, Cham (2018). https://doi.org/10.1007/978-3-319-96142-2_16

28. Reynolds, A., et al.: Rewrites for SMT solvers using syntax-guided enumeration. In: SMT Workshop (2018)

29. Reynolds, A., Deters, M., Kuncak, V., Tinelli, C., Barrett, C.: Counterexample-guided quantifier instantiation for synthesis in SMT. In: Kroening, D., Păsăreanu, C.S. (eds.) CAV 2015. LNCS, vol. 9207, pp. 198–216. Springer, Cham (2015). https://doi.org/10.1007/978-3-319-21668-3_12

30. Reynolds, A., Kuncak, V.: Induction for SMT solvers. In: D'Souza, D., Lal, A., Larsen, K.G. (eds.) VMCAI 2015. LNCS, vol. 8931, pp. 80–98. Springer, Heidelberg (2015). https://doi.org/10.1007/978-3-662-46081-8_5

31. Sonnex, W., Drossopoulou, S., Eisenbach, S.: Zeno: an automated prover for properties of recursive data structures. In: Flanagan, C., König, B. (eds.) TACAS 2012. LNCS, vol. 7214, pp. 407–421. Springer, Heidelberg (2012). https://doi.org/10.1007/978-3-642-28756-5_28

32. Ta, Q., Le, T.C., Khoo, S., Chin, W.: Automated lemma synthesis in symbolic-heap separation logic. PACMPL 2(POPL), 9:1–9:29 (2018)

33. Unno, H., Torii, S., Sakamoto, H.: Automating induction for solving horn clauses. In: Majumdar, R., Kunčak, V. (eds.) CAV 2017. LNCS, vol. 10427, pp. 571–591. Springer, Cham (2017). https://doi.org/10.1007/978-3-319-63390-9_30

CP and Data Science Track

Modeling Pattern Set Mining Using Boolean Circuits

John O.R. Aoga(✉) ⓘ, Siegfried Nijssen ⓘ, and Pierre Schaus ⓘ

ICTEAM/INGI, UCLouvain, Ottignies-Louvain-la-Neuve, Belgium
{john.aoga,siegfried.nijssen,pierre.schaus}@uclouvain.be

Abstract. Researchers in machine learning and data mining are increasingly getting used to modeling machine learning and data mining problems as parameter learning problems over network structures. However, this is not yet the case for several pattern set mining problems, such as concept learning, rule list learning, conceptual clustering, and Boolean matrix factorization. In this paper, we propose a new modeling language that allows modeling these problems. The key idea in this modeling language is that pattern set mining problems are modeled as discrete parameter learning problems over *Boolean circuits*. To solve the resulting optimisation problems, we show that standard optimization techniques from the constraint programming literature can be used, including mixed integer programming solvers and a local search algorithm. Our experiments on various standard machine learning datasets demonstrate that this approach, despite its genericity, permits learning high quality models.

1 Introduction

A revolution is taking place in artificial intelligence, driven to a significant degree by *deep learning* toolkits for learning neural networks [26]. As a result, researchers and practitioners in machine learning and data mining are increasingly getting used to modeling and solving problems using the modeling languages offered in these toolkits. The key idea underlying these languages is that machine learning amounts to learning the parameters of a network structure that transforms inputs into predicted outputs.

However powerful these toolkits may be, they rely on a key underlying assumption: the functions applied to the inputs are continuous and differentiable in the parameters. This enables the use of gradient descent to identify values for the parameters.

Some problems in data mining and machine learning are however not continuous in nature. Good examples of such problems can be found in pattern *set* mining. Examples of pattern set mining problems include learning rule-based classifiers, conceptual clustering, and Boolean matrix factorization. In each of these problems, the task is not to find values for a set of continuous parameters, but to identify some discrete *patterns*.

The first author is supported by the FRIA-FNRS (Fonds pour la Formation à la Recherche dans l'Industrie et dans l'Agriculture, Belgium).

T. Schiex and S. de Givry (Eds.): CP 2019, LNCS 11802, pp. 621–638, 2019.
https://doi.org/10.1007/978-3-030-30048-7_36

This raises the question as to whether a generic framework exists that allows formalizing and solving these types of problems using a modeling language similar to the language that deep learning researchers are familiar with. This is the challenge that we address in this paper.

Generic modeling languages for pattern mining already exist in the literature. Well-studied is the *constraint programming for pattern mining* (CP4PM) framework [13,25]. However, this framework has two weaknesses. First, the modeling language is different from the network-based modeling language used in deep learning toolkits: pattern mining problems are formalized by modeling these problems using Boolean variables and constraints on these variables. Second, most of the studies on CP4PM focus on *pattern mining* instead of pattern *set* mining [14]. There is an important difference between pattern mining and pattern *set* mining. In pattern mining, one is interested in finding *all* patterns occurring in a dataset that satisfy a given set of constraints. The most well-known example of such a constraint is the requirement that a pattern is *frequent*. In pattern *set* mining, however, we are not interested in finding all patterns, but we are interested in finding a small set of patterns that together solve a well-defined data mining problem well, such as a classification task. Given the large number of frequent patterns that can typically be found in many datasets, in recent years, it has been argued that pattern set mining is the more relevant problem.

Only a limited number of studies have explored the extension of CP4PM to pattern set mining problems [12,14]. In these studies, a modeling language was proposed in which pattern set mining problems are formalized as constraint optimization problems. A solution strategy was proposed based on the use of a constraint programming solver. Unfortunately, this approach could, in practice, only be applied to relatively small datasets. Moreover, the modeling language proposed was rather different than the one currently used in machine learning toolkits.

We aim to address these weaknesses in this work. In this paper, we propose a new framework for **modeling pattern set mining problems**. Compared to existing frameworks, this framework has the following advantages:

- the modeling language that we propose is inspired by that of frameworks for deep learning: in our language, we represent learning problems as parameter learning problems on networks; however, instead of continuous functions, in the internal nodes of our networks we use Boolean operators, effectively turning the networks into *Boolean circuits*;
- the framework supports the generic use of a number of different solvers. In particular, in this work we will show how Mixed Integer Programming (MIP) and local search algorithms can be used to solve pattern set mining problems in a generic manner;
- as a consequence of the support of different solvers, the framework allows for finding larger pattern sets on larger datasets.

We introduce a domain specific language to ease the description of the network, its parameters, and the loss function to be computed on the data. Our experiments demonstrate the practicability and flexibility on standard benchmarks.

2   Related Work

A number of different modeling languages for pattern mining problems have already been proposed in the literature.

An important class of methods is based on the use of constraints to model pattern mining problems. The key idea in these approaches is to model a data mining problem as a *Constraint satisfaction problem (CSP)* and use a SAT-, CP- or MIP-solver for solving it. Several data mining tasks were studied, but most results were obtained for itemset mining [6,13,18,25] and sequence mining [3, 16,21]. However, these approaches are *solver- and task-dependent*, and do not solve pattern *set* mining problems.

Most related to this work is the work of De Raedt [8] and Guns et al. [14] on modeling pattern set mining problems as constraint satisfaction or optimization problems. Contrary to these earlier works the framework that we propose is solver independent and uses a modeling language familiar to machine learning researchers.

A modeling language for pattern mining that is solver independent is the *MiningZinc* [11] language. However, it does not address pattern set mining problems either.

The first modeling languages that were proposed in the data mining literature are those that use an SQL-like notation [19]. Also, these languages did not study pattern set mining problems and do not use a notation based on networks.

3   Pattern Set Mining Problems

In this section we will introduce the pattern set mining problems that are the focus of this work. We limit our attention to Boolean data that may or may not be supervised in nature. Let $\mathcal{I} = \{1, \ldots, m\}$ be the sets of items (features) and $\mathcal{T} = \{1, \ldots, n\}$ be the set of transaction (observation) identifiers. An *unsupervised* database can equivalently be seen as a set $\mathcal{D} = \{(t, T_t) \mid t \in \mathcal{T}, T_t \subseteq \mathcal{I}\}$ or as a matrix such that $\mathcal{D}_{ti} \in \{0, 1\}$ where $t \in \mathcal{T}$ and $i \in \mathcal{I}$. In a *supervised* database, we associate with every transaction a Boolean label, i.e. $\mathcal{D} = \{(t, T_t, a_t) \mid t \in \mathcal{T}, T_t \subseteq \mathcal{I}, a_t \in \{0, 1\}\}$.

Example 1. As an example consider the database in Fig. 1a. Ignoring the class label, this database can be represented as a set $\mathcal{D} = \{(1, \{1, 2, 3, 5\}), (2, \{1, 2, 4, 5\}), (3, \{1, 3, 4, 5\}), (4, \{1, 3, 4\})\}$.

Several pattern set mining problems have been proposed on such Boolean data. We will first consider supervised settings. In **Concept Learning** [1] the aim is to discover a set of itemsets that characterizes the positive examples in a supervised dataset as well as possible.

Definition 1 (Concept Learning). *Given a Boolean supervised dataset $\mathcal{D}$,* **find** *a set of itemsets $C \subseteq 2^{\mathcal{I}}$, also referred to as concepts, **such that** $|C| = k$ and $error(\cup_{I \in C} cover(I))$ is minimal.*

Here, we define the cover and the error as follows:

- $cover(I) = \{t \mid (t, T_t) \in \mathcal{D} \wedge I \subseteq T_t\}$, the set of transactions that contain a given itemset;
- $error(T) = |\{t \in \mathcal{T} \mid (a_t = 1 \wedge t \notin T) \vee (a_t = 0 \wedge t \in T)\}|$, the number of examples not characterized correctly.

Example 2. Consider the concepts $C = \big\{\{1,2,3\}, \{3,4\}, \{4,5\}\big\}$; these are highlighted in Fig. 1a. Observation 2 is misclassified, so the error is 1.

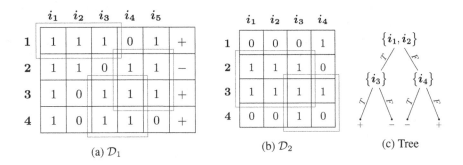

(a) $\mathcal{D}_1$ (b) $\mathcal{D}_2$ (c) Tree

Fig. 1. Itemset databases showing **(a)** Itemsets and its covers and **(b)** Tiles. **(c)** An example of itemset-based tree.

In **Rule Learning** [5,20], we treat the patterns as rules that can also predict a negative class label. Restricting ourselves to the Boolean context, we can define the problem of learning Rule Lists as follows.

Definition 2 (Rule List Learning). *Given a Boolean supervised dataset $\mathcal{D}$, **find** a list of k rules $\mathcal{R} = \big\langle I^{(r)} \to l^{(r)} \big\rangle_{r=1}^{k}$, where $I^{(r)}$ is an itemset and $l^{(r)} \in \{0, 1\}$ is a class label, **such that** $error(cover(\mathcal{R}))$ is minimal, and $I^{(k)} = \emptyset$, i.e., the kth rule serves as a default rule.*

Here, we define the cover of a rule list as the set of transactions for which the rule list predicts the positive class label, $cover(\mathcal{R}) = \big\{t \in \mathcal{T} \mid \exists(I^{(r)} \to 1) \in \mathcal{R} : I^{(r)} \subseteq T_t \wedge \neg(\exists r' < r : (I^{(r')} \to l^{(r')}) \in \mathcal{R} \wedge I^{(r')} \subseteq T_t)\big\}$.

Example 3. For the database of our running example (eg. 1), the rule list $\big\langle (\{1,2,3\} \to 1), (\{4,5\} \to 0), (\{3,4\} \to 1), (\emptyset \to 1) \big\rangle$ would have an error of 1: example 3 would be misclassified, as the first rule that applies to this example is the rule $\{4,5\} \to 0$).

Itemset-Based Decision Trees [10] are a generalization of rule lists. Essentially, they are decision trees in which each node tests for the presence of an itemset. A transaction (observation) will be put in the left-hand branch of an internal node v if the itemset $I^{(v)}$ is present, and in the right-hand branch if the itemset is absent. A *complete* decision tree is a tree that is filled completely till the lowest level.

Definition 3 (Learning Pattern-Based Decision Trees). *Given a Boolean supervised dataset $\mathcal{D}$, **find** a set of k itemsets $\{I^{(r)}\}$, corresponding to internal nodes of a complete decision tree, and labels in $\{0,1\}$ for each leaf of the tree, such that the error of the predictions at the leaves of the tree is minimal.*

Example 4. Fig. 1c is an example of a pattern-based tree obtained from $\mathcal{D}_1$ with 2 as error because examples 3 & 4 are misclassified.

We will consider a number of pattern set mining settings on unsupervised data next. **Conceptual Clustering** [9] aims to cluster examples while also finding descriptions for these clusters. One possible definition of this problem is the following.

Definition 4 (Conceptual Clustering). *Given a Boolean unsupervised dataset $\mathcal{D}$, **find** k itemsets C, such that $\left|\{t \in \mathcal{T} : |\{I \in C : I \subseteq T_t\}| \neq 1\}\right|$ is minimal; hence, the number of transactions of the given dataset not covered by exactly one itemset is minimal.*

Example 5. In our running example (eg. 1) each itemset now describes a cluster (the target attribute of $\mathcal{D}_1$ is not taken into account). These clusters cover respectively transactions $\{1\}$, $\{3,4\}$ and $\{2,3\}$; the $\{3\}$ is an overlapping transaction.

Boolean Matrix Factorization [17] aims to describe the 1s in a database using two Boolean matrices, which can be seen as matrices describing itemsets and their occurrences.

Definition 5 (Boolean Matrix Factorization). *Given a Boolean database $\mathcal{D}$, **find** a Boolean matrix A of size $n \times k$ and a Boolean matrix B of size $k \times m$, such that $error(A \circ B, \mathcal{D})$ is minimal.*

Here $\circ$ is the Boolean matrix product, in which the matrix product is redefined such that $1 + 1 = 1$, and *error* is a function that calculates the number of cells in the two given matrices that mismatch.

Example 6. Fig.1b shows an example of two rectangles of which the rows and columns are identified using Boolean matrices A and B. The error is 2: cells $(1,4)$ and $(4,4)$ are not described correctly in this matrix decomposition.

In the next section, we show how these problems can be reformulated as parameter learning problems over *Boolean circuits*.

4   Reformulating Pattern Set Mining as Parameter Learning in Logical Circuits

In this section we will define the problem of *parameter learning in Boolean circuits*; subsequently, we will show that the learning problems identified in the previous section can all be cast as such parameter learning problems. We first define *Boolean circuits*.

Definition 6. *A Boolean circuit C is a directed acyclic graph $G(V, E)$ in which each node $v \in V$ of in-degree zero represents an input variable, only one node has an out-degree of zero, and each internal node represents a logical gate and is labeled with an operator from the set $\{\wedge, \vee, \neg\}$.*

Logical circuits can be seen as the graphical representation of a Boolean formula. Each internal node v with label $\wedge$ corresponds to an expression $v = v_1 \wedge \cdots \wedge v_n$, where $v_1, \ldots, v_n$ are v's children. For any assignment to the input variables, the Boolean circuit calculates a Boolean value for the output variable. Hence, we can see the Boolean circuit as a function from the Boolean input variables to $\{0, 1\}$.

We can define the parameter learning problem for a given Boolean circuit as follows.

Definition 7. *Given (1) a Boolean circuit C, (2) a partition of the input variables into two sets X and W, where W represents the parameters of the circuit, and (3) a Boolean supervised dataset $\mathcal{D}$ over $|X|$ items, the parameter learning problem is the problem of finding an assignment to the variables W such that*

$$\sum_{(t, T_t, a_t) \in \mathcal{D}} |C(W, T_t) - a_t| \tag{1}$$

is minimized, i.e., C fits a_t well. Here we assume that in passing T_t as a parameter to C, we set all variables in X to **True** *that are included in T_t.*

Below, we will show that the pattern set mining problems presented earlier can be represented as parameter learning problems for Boolean circuits, for well chosen architectures for C and, in some cases, representations of the input data.

However, before doing so, we will introduce some additional notation. We found that modeling the pattern set mining problems at the level of basic Boolean circuits is cumbersome. To simplify our modeling task, we will use an approach that is common in Boolean circuit design: we will add additional gates to our notation that can be seen as a shorthand notation for underlying, larger circuits.

Our first additional gate is $\vargnothing$, which operates on two lists of inputs $\boldsymbol{w} = w_1, \ldots, w_n$ (each w_i corresponding to a variable in W) and $\boldsymbol{v} = v_1, \ldots, v_n$ (each v_i corresponding to a variable in V), and which can be understood as a shorthand notation for

$$\vargnothing(\boldsymbol{w}, \boldsymbol{v}) \equiv (w_1 \wedge v_1) \vee (w_2 \wedge v_2) \vee \cdots \vee (w_n \wedge v_n).$$

The idea is that the parameters $w_1, \ldots, w_n$ indicate which of the inputs of the $\vee$ should be taken into account.

Similarly, we define $\varobslash$ as a shorthand notation for

$$\varobslash(\boldsymbol{w}, \boldsymbol{v}) \equiv (\neg w_1 \vee v_1) \wedge (\neg w_2 \vee v_2) \wedge \cdots \wedge (\neg w_n \vee v_n),$$

where the parameters $w_1, \ldots, w_n$ indicate which of the inputs of the $\wedge$ should be taken into account.

In this paper, we will use a graphical notation for Boolean circuits. We will illustrate this notation first on the problem of learning rule lists.

Fig. 2 (left) shows the Boolean circuit for learning rule lists using the shorthand notation, for a dataset with 3 items and a rule list of at most two rules, plus the default rule. In the shorthand notation, we use an $\wedge$ symbol (respectively, $\vee$ symbol) with dotted incoming edges to represent the $\oslash$ (respectively, $\oslash$) gate. The dotted edges indicate that we have a parameter for each such edge, indicating whether or not to take into account that edge. Hence, in this diagram the input variables of the circuit representing parameters are not explicitly included. We can distinguish three types of layers in this circuit:

Layer 1 represents the problem of selecting which items are included in each rule; the output nodes of this layer can be seen as indicators for the absence or presence of a rule in a data instance;

Layer 3 represents the problem of selecting the class label for the two rules, as well as the class label for the default rule;

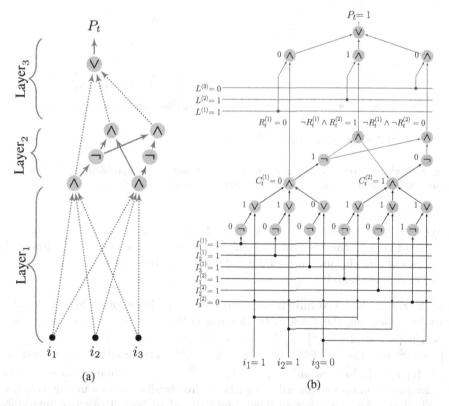

Fig. 2. Architecture of Boolean circuit for rule learning with $m = 3$ items and $k = 3$ rule list size: **(a)** general representation **(b)** full representation with decision variables (dotted arrows represent optional decisions and solid arrows mandatory decisions).

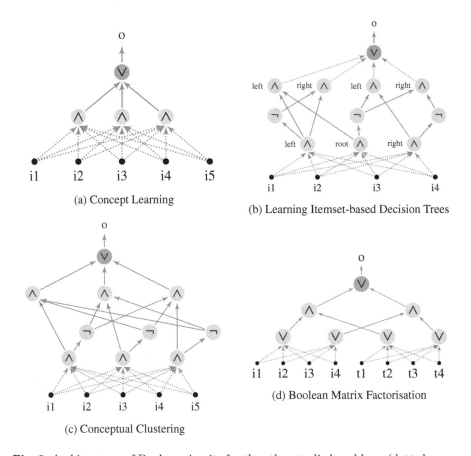

Fig. 3. Architectures of Boolean circuits for the other studied problems (dotted arrows represent optional decisions and solid arrows mandatory decisions).

Layer 2 expresses the order dependencies between the rules; it ensures that the second rule in the rule list will only be used for prediction if the first rule does not match, and the default rule will only be used if the previous two rules did not match.

Fig. 2 (right) show the full circuit representation with explicit binary decision variables encoding the rules to be discovered that can be retrieved as follows: rule 1 is $\{i \mid i \in \{1,\dots,3\} \wedge I_i^{(1)} = 1\} \to L^{(1)}$, rule 2 is $\{i \mid i \in \{1,\dots,3\} \wedge I_i^{(2)} = 1\} \to L^{(2)}$ and the last (default rule) is $\emptyset \to L^{(3)}$. The inputs of the circuit are $X = \{i_1, i_2, i_3\}$. The variables $I_1^{(1)} \dots I_3^{(2)}, L^{(1)} \dots L^{(3)}$ are shared among all the transactions to impose that all transactions are classified with a unique rule list.

Similarly, for the other problems introduced in Sect. 3, we can introduce Boolean circuits to formalize these mining and learning problems. The architecture of these circuits is illustrated in Fig. 3, for small examples.

For **concept learning** and **itemset-based decision trees** the data used for training is identical to that used for rule list learning. The difference between concept learning and rule list learning is that in concept learning the head of the rules is fixed and there is no order between the rules. The decision tree learning problem illustrated is for decision trees composed of 3 itemsets: one for the root and one for each of the children of the root. In the Boolean circuit every child of the root corresponds to one of the leaves of the decision tree. The extension towards perfect trees of larger depth is relatively straightforward.

Conceptual clustering in an unsupervised problem setting; however, in our parameter learning setting we need to provide a label for every training instance. We address this by giving every instance in the training data the label 1. The idea in this circuit is that we predict 1 for an example if there is exactly one itemset that matches it; if no itemset, or two or more itemsets match it, we predict 0. As a result, the error score for this circuit corresponds to counting the number of examples not exactly in one cluster.

Boolean matrix factorization is an unsupervised setting as well. The input data that we give to the Boolean circuit is different here. Given the original data $\mathcal{D}_{ti}$, we create a new dataset as follows:

$$\Big\{\big((t,i),\{t,i\},\mathcal{D}_{ti}\big) \mid t \in \mathcal{T}, i \in \mathcal{I}\Big\},$$

that is, we create a dataset in which each entry of the original matrix is an example; every new example consists of two items: one representing the original item, the other the original transaction identifier. In the lowest layer of the Boolean circuit, both transaction sets and itemsets are identified; the layer on top represents the Boolean matrix product. It can be shown that this model is equivalent to the original learning problem a well.

5   Generic Solving Framework

The question arises now how to solve these parameter learning problems. The benefit of our approach is that it allows for the use of alternative solvers. In this paper, we will consider two such approaches: one is the use of a greedy algorithm; the other is the use of Mixed Integer Programming (MIP) solvers.

5.1   Solving Using Greedy Algorithm

Greedy algorithms are among the most scalable algorithms for the pattern set mining algorithms studied in this work. Indeed, for rule learning tasks these are the most common type of algorithm, as in practice, the solution found by such algorithms is already of decent quality. The parameters of the Boolean circuit that minimize the error (1) can also be found greedily.

Algorithm 1 shows a greedy algorithm. This algorithm receives the Boolean circuit (together with its partition of inputs into two sets W and X) and the database $\mathcal{D}$. We will represent the values of the parameters by listing the subset

---

Algorithm 1: $Greedy(C, \mathcal{D})$

---

1 **Method** error (W: **Assignment to the parameters,** C: **Logical Circuit,** $\mathcal{D}$: **list of Transactions**)

2   $\lfloor$ **return** $\sum_{(t,T_t,a_t)\in\mathcal{D}} |C(W, T_t) - a_t|$

3 $W \leftarrow \emptyset$ minErr $\leftarrow$ error$(W, \mathcal{D})$

4 **do**

5   $\quad$ $w^* \leftarrow \underset{i\in\mathcal{P}\backslash W}{\mathrm{argmin}}\, \mathrm{error}(W \cup \{i\}, C, \mathcal{D})$

6   $\quad$ **if** error$(W \cup \{w^*\}, C, \mathcal{D}) <$ minErr **then** $W \leftarrow W \cup \{w^*\}$ $\triangleright$ *Add w^* to W* ;

7   $\quad$ **else break**;

8 **while** $\mathcal{P} \backslash W \neq \emptyset \wedge$ minErr > 0; $\triangleright$ *Stop if no more decision or* minErr $= 0$

9 **return** W

---

of parameters $\mathcal{P} = \{1, 2, \ldots, |W|\}$ that take the value 1. By abuse of notation, we hence treat a vector of assignments to the variables in W as a subset of parameters $\mathcal{P}$. The algorithm starts with an empty set of parameters W and then iteratively identifies the parameter for which a flip from the value 0 to the value 1 minimizes the error (Line 5). Once a local optimal parameter is found, W is updated accordingly in line 6. The process repeated until either *(i)* a solution better than W cannot be found (line 7) or *(ii)* all variables have been fixed to the value 1, or *(iii)* the minimum error is 0.

The set $\mathcal{P}$ represents all the dotted edges in our graphical representation of the learning problems. An empty list W indicates that no edge is selected. Taking a decision (include or not) corresponds to flipping the parameter value.

While we could apply this greedy algorithm on all parameters of the circuit, we perform an optimization when the root node of the circuit consists of a $\oslash$ node, such as in **rule learning** and **itemset-based decision trees**, and for every transaction only one of the children takes the value **True**. For such circuits it can be shown that the optimal choice for the children of the root can easily be calculated from the choices below those children. Hence, we do not perform a greedy search over this set of children.

For some problems, such as **conceptual clustering** and **Boolean matrix factorization**, we found that it can be beneficial to start from an assignment that puts all variables in W at the value 1. This can be emulated by putting a $\neg$ node between every parameter and the nodes that it is connected to.

5.2   Solving Using MIP

The key idea in this approach is to map the parameter learning problem to an optimization problem defined on integer variables and linear constraints. An additional integer variable is introduced for each node in the circuit and each gate is modeled using a set of linear constraints as follows:

- if a node y is an $\wedge$-*gate* of x_i nodes:

$$y = x_1 \wedge x_2 \wedge \cdots \wedge x_m \equiv \begin{cases} y \leq x_i & \forall i \in [1, m] \\ y \geq \Sigma_i x_i - (m - 1) \\ y \geq 0 \end{cases}$$

- if a node y is an $\vee$-*gate* of x_i nodes:

$$y = x_1 \vee x_2 \vee \cdots \vee x_m \equiv \begin{cases} y \geq x_i & \forall i \in [1, m] \\ y \leq \Sigma_i x_i \\ y \leq 1 \end{cases}$$

- finally, if a node y is a $\neg$-*gate* of x node:

$$y = \neg x \equiv y = 1 - x$$

We make a copy of the circuit for every example in the training data, fixing the corresponding inputs of the circuit to the values in the training example; the parameters are variables that the MIP solver will search over. For every training example, we will include the output v of the circuit in the optimization criterion, using v if the expected output is 0 and $(1 - v)$ if the expected output is 1. We minimize this error.

Algorithm 2: DSL to solve a rule learning problem with the architecture of Fig.2a and $\mathcal{D}$ in MIP and Greedy algorithm

---

1 ▷ *Building the network*
2 $L_1 \leftarrow \text{InputLayer}(m = 3)$
3 $L_2 \leftarrow AndSelectionLayer(L_1, k = 3)$
4 $L_3 \leftarrow NotLayer(L_2)$
5 $L_4 \leftarrow Layer(L_2[1], And(L_3[1], L_2[2]), And(L_3[1], L_3[2]))$
6 $L_5 \leftarrow OrSelectionLayer(L_4)$
7 $N \leftarrow L_5.network()$
8 ▷ *Load inputs and parameters*
9 $X \leftarrow getDB(\mathcal{D})\quad y \leftarrow getAttr(\mathcal{D})\quad \hat{y} \leftarrow L_5[0]$
10 $obj \leftarrow 1 - y.\hat{y} - (1 - y)(1 - \hat{y})$ ▷ *Objective function*
11 ▷ *Defining the procedures*
12 $greedy \leftarrow X$ into N using *Greedy.solver* minimizing obj
13 $stats \leftarrow greedy.run()$
14 $mip \leftarrow X$ into N using *MIP.solver* minimizing obj
15 $stats \leftarrow mip.run()$

Note that this gives a generic approach for solving the mining and learning problems discussed earlier using MIP solvers. There is already a literature on modeling some of these individual problems in MIP (see [4,22–24,27]); our approach provides a more general approach to modeling such problems.

6   Unified Modeling and Solving Language

In our vision, creating parameter learning problems for Boolean circuits can be seen as a programming task. The main benefit of our framework is that its modeling language is very similar to that of Deep Learning toolkits, and hence familiar to machine learning researchers.

Algorithm 2 is an example of an implementation of the rule list learning problem based on the architecture of Fig. 2a. From line 2 to 7 the Boolean circuit is defined by using *macro-functions* such as *AndSelectionLayer* which represents all the operations of Layer 1 in Fig. 2b. At line 10, we define the error function and at lines 12 and 14 the different solvers, which are launched in lines 13 and 15.

7   Learning Classifiers Based on Soft Rules

While we focused our modeling framework on Boolean parameters only, of interest can also be combinations of Boolean parameters with other types of parameters. We will study one first possible such combination here and will leave other combinations as future work.

In traditional learning, the interpretation of the conditions in a rule is typically *conjunctive*: all conditions in a rule need to be met. However, this conjunctive interpretation of rules can also be considered a limitation. More freedom would be allowed in rule-based classifiers in which we require to match a certain number of conditions, but not necessarily all.

We can model such problems by adding a *parameterized soft gate* to our network. The new *parameterized soft gate* requires that at least α inputs should

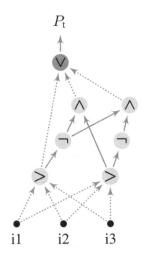

Fig. 4. Learning soft rule lists

be true, where α is also a parameter that needs to be learned. More formally, this gate operates on three inputs, and has the following semantics: $\ominus(\boldsymbol{v}, \boldsymbol{w}, \alpha) \equiv \sum_i v_i w_i \geq \alpha$, where $v_1, \ldots, v_n$ indicate Boolean input nodes, and $w_1, \ldots, w_n$ are Boolean indicators representing whether or not to take into account the ith input. Note that $\ominus(\boldsymbol{v}, \boldsymbol{w}, 1) = \oslash(\boldsymbol{v}, \boldsymbol{w})$: the gate generalizes the $\oslash$-gate. Furthermore, $\ominus(\boldsymbol{v}, \boldsymbol{w}, n) \equiv v_1 \wedge \cdots \wedge v_n$.

We can use this gate to define a rule learning problem in which each rule does specify not only conditions, but also the minimum number of conditions that need to be satisfied in order for the rule to apply. This is illustrated in Fig. 4. This model has a higher level of expressivity than traditional rule learning models: by fixing the parameters α to a sufficiently high level, we can still enforce that all conditions need to be satisfied for a rule to apply.

We can learn the parameters of such gates both using MIP and using greedy algorithms. In MIP, we exploit the fact that $v_i w_i \equiv v_i \wedge w_i$. Let $u_i \equiv v_i \wedge w_i$; then we implement the gate by first calculating the u_is using the representation for $\wedge$ discussed earlier and then using these u_is as follows:

$$y = \ominus(\boldsymbol{v}, \boldsymbol{w}, \alpha) = \begin{cases} \sum_{i=1}^{n} u_i - \alpha + (1 - y)n & \geq 0 \\ \sum_{i=1}^{n} u_i - \alpha - yn + 1 \leq 0. \end{cases}$$

Note that also here we copy the circuit for each example in the data; the u_is are calculated for each example separately. However, the parameters α and w_i are shared among all examples.

8   Experiments

In this section, we evaluate our framework from three perspectives: (i) the predictive power of the classifiers learned compared to other classifiers, (ii) the sensitivity of the pattern sets identified w.r.t. the variation of parameters (like k), (iii) the efficiency of the framework (using CPU time). All experiments were run in the JVM with maximum memory set to 8 GB on PCs with Intel Core i5 64bit processor (2.7 GHz) and 16 GB of RAM running MAC OS 10.13.3. Execution time is limited to one hour.

Datasets and Existing Classifiers. We use data from the CP4IM[1] repository. Statistics of these datasets are reported in Table 1a. We compare with the following methods: (i) Popular tree-based and neural network-based classifiers such as Random Forests (RF), decision trees (C4.5) and neural networks (NN) from the *scikit-learn* library (using default settings); (ii) a rule-based learner: Probabilistic Rule List (PRL) [2] and a k-pattern set miner [12,14] (KPATT) for the concept learning task, in which the concept learning problem is solved as a global optimization problem in a CP solver.

---

[1] https://dtai.cs.kuleuven.be/CP4IM/datasets/.

We denote our approaches by $X4Yz$, where $X \in \{MIP, G\}$ is the solving strategy and can be either MIP or G(greedy), $Y \in \{CL, RL, PDT, BMF, CC\}$ represents the pattern set mining problem solved and can be CL (concept learning), RL (rule learning), PDT (Pattern-based Decision Trees), BMF (Boolean Matrix factorization) or CC (conceptual clustering). We use $z \in \{+\alpha, -\alpha\}$ to indicate whether or not the soft gates are used (see Sect. 7).

Comparison of Model Quality. Table 1e shows the average accuracy evaluated using *stratified 10-fold cross-validation* on test data. Note that in some cases, within the time allocated the MIP solver could not prove optimality. We use the best pattern set found within the allocated amount of time. In these experiments, the number of patterns is fixed arbitrarily to $k = 5$.

RF and Neural Networks perform better than the rule-based methods, but our introduction of soft rules (Sect. 7) seems to improve the accuracy of rule-based methods in most cases over both the training and the test tests. The performance on test data of the greedy algorithm is sometimes worse and sometimes better than that of the MIP-based algorithm, although on concept learning the performance of the greedy algorithm is not satisfactory. For many cases where optimality could not be proven within the allocated time, the optimality gap is less than 10%.

Running Time Comparison. The greedy approach is generally very efficient and outputs results in a few seconds, so we only report the execution time of the MIP approach in Table 1. As one can see, in many cases, the timeout is reached with a gap from the optimum smaller than 10%, with a few exceptions (41%, 29%). Our approaches outperform KPATT on small databases and on (relatively) large databases KPATT fails to find a solution within the allocated amount of time.

MIP approaches are highly dependent on the number of variables and the number of constraints, which are $\mathcal{O}(k \times |\mathcal{T}| \times |\mathcal{I}|)$ and $\mathcal{O}(k \times |\mathcal{T}| \times (|\mathcal{I}| + k))$ respectively. However, the MIP pre-solving is able to drastically reduce the number of variables and constraints in some cases. For example, the initial numbers of variables $(424, 945)$ and constraints $(710, 632)$ in the Audi dataset were reduced by more than 97%.

Sensitivity to the Parameter k. Figure 5 shows the accuracy on training and test sets, the gap, and the execution time by varying k for the Hepa database; results on other datasets are similar. As can be seen, increasing k improves the accuracy on the training set. On the test set the outcome depends strongly either on whether optimality has been proven or not (as evidenced by the gap-plot) or overfitting on the training set.

Table 1. Experiments for pattern set mining problems over several datasets with $k = 5$ ("-" means the process stopped before a solution was found, either due to a *out of memory/timeout exception* in the pre-solving step of the MIP solver, or before the CP solver used in KPATT found a solution).

methods	Audi.	Aust.	HeCl.	Hepa.	KrKp.	Lymp.	Mush.	PrTu.	Soyb.	Spli.	TTT.	Vote	Zoo
a) Dataset Features													
$\|T\|$	216	653	296	137	3196	148	8124	336	630	3190	958	435	101
$\frac{\|\{t\in T\|a_t=1\}\|}{T}$	0.26	0.55	0.54	0.81	0.52	0.55	0.52	0.24	0.15	0.52	0.65	0.61	0.41
$\|\mathcal{I}\|$	148	125	95	68	74	68	119	31	50	287	27	48	36
b) Accuracies over training sets													
MIP4CL-α	1.0	**0.92**	0.89	0.98	0.87	0.97	0.55	0.89	0.97	-	**0.90**	0.98	**1.0**
MIP4CL+α	1.0	0.91	**1.0**	**1.0**	0.93	**1.0**	**1.0**	**0.91**	**1.0**	0.85	0.77	**1.0**	**1.0**
G4CL	0.73	0.45	0.46	0.19	0.47	0.46	0.49	0.76	0.86	0.48	0.35	0.37	0.59
KPATT	1.0	-	-	-	-	-	-	0.89	0.97	-	0.82	-	**1.0**
MIP4RL-α	1.0	-	-	1.0	-	0.95	-	**0.89**	0.96	-	0.81	0.98	1.0
MIP4RL+α	1.0	-	0.97	1.0	-	**1.0**	**1.0**	0.87	**0.98**	-	**0.83**	**1.0**	1.0
G4RL	0.98	**0.86**	0.83	0.89	**0.94**	0.86	0.98	0.84	0.86	**0.84**	0.70	0.96	1.0
MIP4PDT-α	1.0	-	0.89	1.0	-	0.96	-	0.89	0.94	-	-	0.97	1.0
MIP4PDT+α	1.0	-	**1.0**	1.0	-	**1.0**	**1.0**	-	**0.99**	-	-	**1.0**	1.0
G4PDT	0.99	**0.86**	0.82	0.89	**0.94**	0.88	**1.0**	0.84	0.86	**0.84**	**0.76**	0.96	1.0
c) Gap (%) for MIP over training sets													
MIP4CL-α	*	0.08	0.12	0.02	0.12	0.03	0.41	0.07	0.03	-	0.11	0.02	*
MIP4CL+α	*	0.10	*	*	0.07	*	*	0.05	*	0.16	0.29	*	*
MIP4RL-α	*	-	-	*	-	0.05	-	0.08	0.05	-	0.24	0.02	*
MIP4RL+α	*	-	0.03	*	-	*	*	0.10	0.02	-	0.20	*	*
MIP4PDT-α	*	-	0.12	*	-	0.04	-	0.08	0.06	-	-	0.03	*
MIP4PDT+α	*	-	*	*	-	*	*	-	0.01	-	-	*	*
d) Running time (in second) - TO$\equiv$ Timeout													
MIP4CL-α	26.09	TO	TO	TO	TO	TO	TO	TO	TO	-	TO	TO	1.65
MIP4CL+α	5.81	TO	2682.90	1.99	TO	2.50	251	TO	915.09	TO	TO	17.32	0.66
KPATT	20	-	-	-	-	-	-	TO	1730.30	-	TO	-	3.29
MIP4RL-α	45.45	-	-	31.65	-	TO	-	TO	TO	-	TO	TO	1.70
MIP4RL+α	7.73	-	TO	9.00	-	5.72	1103.95	TO	TO	-	TO	146.90	0.69
MIP4PDT-α	51.61	-	TO	3038.61	-	TO	-	TO	TO	-	-	TO	2.26
MIP4PDT+α	12.25	-	757.20	9.27	-	9.90	TO	-	TO	-	-	1265.56	1.05
MIP4CC-α	42.04	244.73	46.53	7.58	1072.34	1.91	-	16.77	123.81	-	153.56	**5.71**	2.57
MIP4CC+α	11.63	3.34	9.86	2.27	504.26	2.32	1118.64	7.26	1.62	443.68	22.56	7.38	**0.60**
MIP4BMF-α	-	-	-	556.20	-	697.97	-	1454.71	-	-	-	-	127.15
MIP4BMF+α	-	-	-	542.93	-	65.20	-	1446.05	-	-	-	-	123.20
e) Accuracies over test sets													
MIP4CL-α	0.87	**0.85**	**0.77**	**0.79**	0.86	0.69	0.55	**0.83**	0.98	-	0.94	0.98	1.0
MIP4CL+α	**0.91**	0.76	0.70	**0.79**	**0.89**	**0.75**	**1.0**	0.8	0.89	**0.82**	0.75	0.93	1.0
G4CL	0.77	0.48	0.42	0.21	0.51	0.38	0.45	0.71	0.84	0.47	0.34	0.49	0.64
MIP4RL-α	0.87	-	-	0.79	-	0.62	-	**0.89**	**0.95**	-	0.86	**1.0**	1.0
MIP4RL+α	0.87	-	**0.8**	**0.93**	-	0.62	**1.0**	0.8	0.89	-	0.81	0.98	1.0
G4RL	**1.0**	**0.94**	0.65	0.79	**0.95**	**0.81**	0.99	0.76	0.84	**0.85**	0.72	0.96	1.0
MIP4PDT-α	0.83	-	0.8	**0.86**	-	0.75	-	**0.77**	0.92	-	-	0.95	1.0
MIP4PDT+α	0.91	-	**0.83**	**0.86**	-	0.69	**1.0**	-	**0.95**	-	-	0.93	1.0
G4PDT	**1.0**	**0.94**	0.65	0.79	**0.95**	**0.81**	**1.0**	0.76	0.84	**0.85**	**0.77**	0.96	1.0
PRL	0.87	-	-	0.71	-	0.62	-	0.26	0.83	-	-	0.64	1.0
C4.5	0.94	0.81	0.75	0.72	**0.98**	0.84	0.97	0.75	0.91	0.93	0.82	0.95	0.97
RF	0.95	0.83	0.76	**0.82**	0.96	0.85	0.97	**0.79**	0.95	0.94	**0.87**	0.96	1.0
NN	**1.0**	0.91	**0.87**	0.79	0.96	**1.0**	0.99	0.71	0.84	**0.95**	0.79	**0.98**	1.0

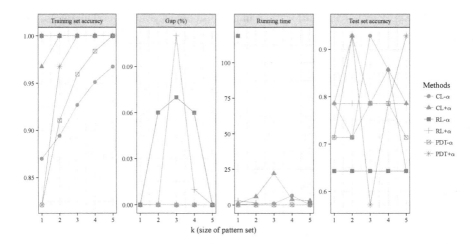

Fig. 5. Sensitivity to the parameter k of our approaches over Hepa dataset (time limit = 600 s).

9   Conclusion

Motivated and inspired by the high level modeling language offered by deep learning frameworks, this work proposed a unified modeling framework for various k-pattern set mining problems (concept learning, rule list learning, pattern-based decision trees, conceptual clustering and Boolean matrix factorization). The modeling language is independent from the optimization technology. We have shown that possible extensions of the language are possible and yield promising results, such as the soft gates. We have shown experimentally that despite the genericity of our framework, the performance of the approach is competitive to that of existing traditional learning approaches, and outperforms an earlier CP-based approach for pattern set mining.

Many future studies are possible. First, alternative optimization approaches are of interest, such as based on meta-heuristics (such as large neighborhood search) and gradient-based approaches, in particular those for learning *binarized neural networks* [15]. Furthermore, other links to deep learning can be explored further, for instance, in mixed networks that combine discrete and continuous components. In this work we did not restrict the form of Boolean circuit used; by adding restrictions on the form of the Boolean circuit, such as that the Boolean circuit is in decomposable negation normal form [7], it may be possible to build more optimized algorithms.

References

1. Angluin, D.: Queries and concept learning. Mach. Learn. **2**(4), 319–342 (1987)
2. Aoga, J.O.R., Guns, T., Nijssen, S., Schaus, P.: Finding probabilistic rule lists using the minimum description length principle. In: Soldatova, L., Vanschoren, J.,

Papadopoulos, G., Ceci, M. (eds.) DS 2018. LNCS (LNAI), vol. 11198, pp. 66–82. Springer, Cham (2018). https://doi.org/10.1007/978-3-030-01771-2_5

3. Aoga, J.O.R., Guns, T., Schaus, P.: Mining time-constrained sequential patterns with constraint programming. Constraints **22**(4), 548–570 (2017)

4. Bertsimas, D., Dunn, J.: Optimal classification trees. Mach. Learn. **106**(7), 1039–1082 (2017)

5. Clark, P., Niblett, T.: The CN2 induction algorithm. Mach. Learn. **3**, 261–283 (1989)

6. Coquery, E., Jabbour, S., Saïs, L., Salhi, Y.: A SAT-based approach for discovering frequent, closed and maximal patterns in a sequence. In: ECAI (2012)

7. Darwiche, A.: Compiling knowledge into decomposable negation normal form. In: Proceedings of the Sixteenth International Joint Conference on Artificial Intelligence, IJCAI 1999, pp. 284–289. Morgan Kaufmann Publishers Inc., San Francisco, CA, USA (1999)

8. De Raedt, L., Zimmermann, A.: Constraint-based pattern set mining. In: Proceedings of the Seventh SIAM International Conference on Data Mining, Minneapolis, Minnesota, USA, 26–28 April 2007, pp. 237–248 (2007)

9. Fisher, D.H., Langley, P.: Approaches to conceptual clustering. In: Joshi, A.K. (ed.) Proceedings of the 9th International Joint Conference on Artificial Intelligence, Los Angeles, CA, USA, August 1985, pp. 691–697. Morgan Kaufmann (1985)

10. Gay, D., Selmaoui, N., Boulicaut, J.: Pattern-based decision tree construction. In: Second IEEE International Conference on Digital Information Management (ICDIM), Lyon, France, Proceedings, 11–13 December 2007, pp. 291–296. IEEE (2007)

11. Guns, T., Dries, A., Tack, G., Nijssen, S., De Raedt, L.: MiningZinc: a modeling language for constraint-based mining. In: Twenty-Third International Joint Conference on Artificial Intelligence (2013)

12. Guns, T., Nijssen, S., De Raedt, L.: Evaluating pattern set mining strategies in a constraint programming framework. In: Huang, J.Z., Cao, L., Srivastava, J. (eds.) PAKDD 2011. LNCS (LNAI), vol. 6635, pp. 382–394. Springer, Heidelberg (2011). https://doi.org/10.1007/978-3-642-20847-8_32

13. Guns, T., Nijssen, S., De Raedt, L.: Itemset mining: a constraint programming perspective. Artif. Intell. **175**(12–13), 1951–1983 (2011)

14. Guns, T., Nijssen, S., De Raedt, L.: k-Pattern set mining under constraints. IEEE Trans. Knowl. Data Eng. **25**(2), 402–418 (2013)

15. Hubara, I., Courbariaux, M., Soudry, D., El-Yaniv, R., Bengio, Y.: Binarized neural networks. In: Lee, D.D., Sugiyama, M., Luxburg, U.V., Guyon, I., Garnett, R. (eds.) Advances in Neural Information Processing Systems 29, pp. 4107–4115. Curran Associates Inc., New York (2016). http://papers.nips.cc/paper/6573-binarized-neural-networks.pdf

16. Kemmar, A., Lebbah, Y., Loudni, S., Boizumault, P., Charnois, T.: Prefix-projection global constraint and top-k approach for sequential pattern mining. Constraints **22**(2), 265–306 (2017)

17. Lam, H.T., Pei, W., Prado, A., Jeudy, B., Fromont, É.: Mining top-k largest tiles in a data stream. In: Calders, T., Esposito, F., Hüllermeier, E., Meo, R. (eds.) ECML PKDD 2014. LNCS (LNAI), vol. 8725, pp. 82–97. Springer, Heidelberg (2014). https://doi.org/10.1007/978-3-662-44851-9_6

18. Lazaar, N., et al.: A global constraint for closed frequent pattern mining. In: Rueher, M. (ed.) CP 2016. LNCS, vol. 9892, pp. 333–349. Springer, Cham (2016). https://doi.org/10.1007/978-3-319-44953-1_22

19. Meo, R., Psaila, G., Ceri, S.: A new SQL-like operator for mining association rules. In: Vijayaraman, T.M., Buchmann, A.P., Mohan, C., Sarda, N.L. (eds.) VLDB 1996, Proceedings of 22th International Conference on Very Large Data Bases, Mumbai (Bombay), India, 3–6 September 1996, pp. 122–133. Morgan Kaufmann (1996)

20. Michalski, R.S.: On the quasi-minimal solution of the general covering problem (1969)

21. Negrevergne, B., Guns, T.: Constraint-based sequence mining using constraint programming. In: Michel, L. (ed.) CPAIOR 2015. LNCS, vol. 9075, pp. 288–305. Springer, Cham (2015). https://doi.org/10.1007/978-3-319-18008-3_20

22. Ouali, A., Loudni, S., Lebbah, Y., Boizumault, P., Zimmermann, A., Loukil, L.: Efficiently finding conceptual clustering models with integer linear programming. In: Kambhampati, S. (ed.) Proceedings of the Twenty-Fifth International Joint Conference on Artificial Intelligence, IJCAI 2016, New York, NY, USA, 9–15 July 2016, pp. 647–654. IJCAI/AAAI Press (2016)

23. Ouali, A., et al.: Integer linear programming for pattern set mining; with an application to tiling. In: Kim, J., Shim, K., Cao, L., Lee, J.-G., Lin, X., Moon, Y.-S. (eds.) PAKDD 2017. LNCS (LNAI), vol. 10235, pp. 286–299. Springer, Cham (2017). https://doi.org/10.1007/978-3-319-57529-2_23

24. Rudin, C., Ertekin, Ş.: Learning customized and optimized lists of rules with mathematical programming. Math. Program. Comput. **10**(4), 659–702 (2018)

25. Schaus, P., Aoga, J.O.R., Guns, T.: CoverSize: a global constraint for frequency-based itemset mining. In: Beck, J.C. (ed.) CP 2017. LNCS, vol. 10416, pp. 529–546. Springer, Cham (2017). https://doi.org/10.1007/978-3-319-66158-2_34

26. Sejnowski, T.J.: The Deep Learning Revolution. MIT Press, Cambridge (2018)

27. Verwer, S., Zhang, Y.: Learning optimal classification trees using a binary linear program formulation. In: 33rd AAAI Conference on Artificial Intelligence (2019)

Differential Privacy of Hierarchical Census Data: An Optimization Approach

Ferdinando Fioretto[(⊠)] and Pascal Van Hentenryck

Georgia Institute of Technology, Atlanta, GA, USA
fioretto@gatech.edu, pvh@isye.gatech.edu

Abstract. This paper is motivated by applications of a Census Bureau interested in releasing aggregate socio-economic data about a large population without revealing sensitive information. The released information can be the number of individuals living alone, the number of cars they own, or their salary brackets. Recent events have identified some of the privacy challenges faced by these organizations. To address them, this paper presents a novel differential-privacy mechanism for releasing hierarchical counts of individuals satisfying a given property. The counts are reported at multiple granularities (e.g., the national, state, and county levels) and must be consistent across levels. The core of the mechanism is an optimization model that redistributes the noise introduced to attain privacy in order to meet the consistency constraints between the hierarchical levels. The key technical contribution of the paper shows that *this optimization problem can be solved in polynomial time by exploiting the structure of its cost functions*. Experimental results on very large, real datasets show that the proposed mechanism provides improvements up to two orders of magnitude in terms of computational efficiency and accuracy with respect to other state-of-the-art techniques.

1 Introduction

The release of datasets containing sensitive information about a large number of individuals is central to a number of statistical analysis and machine learning tasks. For instance, the US Census Bureau publishes socio-economic information about individuals, which is then used as input to train classifiers/predictors and release important statistics about the US population.

One of the fundamental roles of a Census Bureau is to report *group size* queries, which are especially useful to study the skewness of a distribution. For instance, in 2010, the US Census Bureau released 33 datasets of such queries [24]. Group size queries partition a dataset in *groups* and evaluate the size of each group. For instance, a group may be the households that are families of four members, or the households owning three cars.

The challenge is to release these datasets without disclosing sensitive information about any individual in the dataset. Various techniques for limiting a-priori the disclosed information have been investigated in the past, including anonymization [22] and aggregations [25]. However, these techniques have been

© Springer Nature Switzerland AG 2019
T. Schiex and S. de Givry (Eds.): CP 2019, LNCS 11802, pp. 639–655, 2019.
https://doi.org/10.1007/978-3-030-30048-7_37

consistently shown ineffective in protecting sensitive data [14,22], For instance, the US Census Bureau confirmed [2] that the disclosure limitations used for the 2000 and 2010 censuses had serious vulnerabilities which were exposed by the Dinur and Nissim's reconstruction attack [6]. Additionally, the 2010 Census group sizes were truncated due to the lack of privacy methods for protecting these particular groups [4].

This paper addresses these limitations through the framework of *Differential Privacy* [7], that offers a formal approach to guarantee data privacy by bounding the disclosure risk of any individual participating in a dataset. Differential privacy is considered the de-facto standard for privacy protection and has been adopted by various corporations [9,23] and governmental agencies [1]. It works by injecting carefully calibrated noise to the data before release. However, while this process guarantees privacy, it also affects the fidelity of the released data. In particular, the injected noise often produces datasets that violate consistency constraints of the application domain. In particular, group size queries must be consistent in a geographical hierarchy, e.g., the national, state, and county levels. Unfortunately, the traditional injection of independent noise to the group sizes cannot ensure the consistency of hierarchical constraints.

To overcome this limitation, this paper casts the problem of privately releasing group size data as a *constraint optimization problem* that ensures consistency of the hierarchical dependencies. However, the optimization problem that redistributes noise optimally is intractable for real datasets involving hundreds of millions of individuals. In fact, even its convex relaxation, which does not guarantee consistency, is challenging computationally. This paper addresses these challenges by proposing mechanisms based on a dynamic programming scheme that leverages both the hierarchical nature of the problem and the structure of the objective function. The contributions of the paper are summarized as follows:

1. The paper introduces the *Privacy-preserving Group Size Release* (PGSR) problem, for releasing differentially private group sizes that preserves hierarchical consistency.
2. It proposes a differentially private mechanism that uses an optimization approach to release both accurate and consistent group sizes.
3. It shows that the differentially private mechanism can be implemented in polynomial time, using a dynamic program that exploits both the hierarchical nature of group size queries and the structure of the objective function.
4. Finally, it evaluates the mechanisms on very large datasets containing over 300,000,000 individuals. The results demonstrate the effectiveness and scalability of the proposed mechanisms that bring several orders of magnitude improvements over the state of the art.

2   Problem Specification

This paper is motivated by applications from the US Census Bureau, whose goal is to release socio-demographic features of the population grouped by census

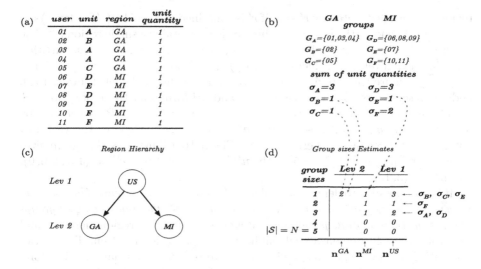

Fig. 1. A dataset D associating users to units and regions (a) and its associated groups and sum of units (b), region hierarchy (c), and the final hierarchical group-sizes (d).

blocks, counties, and states. For instance, the bureau is interested in releasing information such as the number of people in a household and how many cars they own. This section provides a generic formalization of this release problem.

Consider a dataset $D = \{(p_i, u_i, r_i, z_i)\}_{i=1}^{n}$ containing n tuples $(p_i, u_i, r_i, z_i) \in \mathcal{P} \times \mathcal{U} \times \mathcal{R} \times \mathcal{Z}$ denoting, respectively, a (randomly generated) identifier for user $i \in [n]$, its *unit identifier* (e.g., the home address where she lives), the *region* in which she lives, (e.g., a census block), and a *unit quantity* describing a socio-demographic feature, e.g., the number of cars she owns, or her salary bracket. The set of users sharing the same unit forms a *group* and $G_u = \{p_i \in \mathcal{P} \mid u_i = u\}$ denotes the group of unit u. The socio-demographic feature of interest is the sum of the *unit quantities* of a group G_u, i.e., $\sigma_u = \sum_{p_i \in G_u} z_i$. The set of all unit sizes, i.e., $\mathcal{S} \supseteq \{\sigma_u \mid u \in \mathcal{U}\}$, also plays an important role. Indeed, the bureau is interested in releasing, for every unit size $\sigma \in \mathcal{S}$, the quantity $n_\sigma = |\{G_u \mid u \in \mathcal{U}, \sigma_u = \sigma\}|$, i.e., the number of groups of size σ.

These concepts are illustrated in Fig. 1(a), which shows a dataset containing $n = 11$ users with their home addresses (units), their states (regions), and a 0/1 quantity denoting a feature of interest. In the running example, the feature is always 1, since the application is interested is the composition of the household, i.e., how many people live at the same address. Hence the *groups* identify households and the *sums of unit quantities* represent household sizes. For instance, G_A is the group of 3 users living in unit A and $\sigma_A = 3$ as shown in Fig. 1(b). The example also uses $\mathcal{S} = \{1, \ldots, 5\}$.

In addition to the dataset, the census bureau works with a *region hierarchy* that is formalized by a tree $\mathcal{T}_L$ of L levels. Each level $\ell \in [L]$ is associated with a set of regions $\mathcal{R}_\ell \subseteq \mathcal{R}$, forming a partition on D. Region r' is a subregion of region r, which is denoted by $r' \prec r$, if r' is contained in r and $\text{lev}(r') = \text{lev}(r)+1$, where $\text{lev}(r)$ denotes the level of r. The root level contains a single region $r^\top$. The children of r, i.e., $ch(r) = \{r' \in \mathcal{R} | r' \prec r\}$ is the set of regions that partition r in the next level of the hierarchy and $pa(r)$ denotes the parent of region r ($r \neq r^\top$). Figure 1(c) provides an illustration of a hierarchy of 2 levels. Each node represents a region. The regions GA and MI form a partition of region US.

The number of groups with size $\sigma \in \mathcal{S}$ and region $r \in \mathcal{R}$ is denoted by $n_\sigma^r = |\{u \in \mathcal{U} \mid \sigma_u = \sigma \land u \in r\}|$ and $\boldsymbol{n}^r = (n_1^r, \ldots, n_N^r)$ denotes the vector of *group sizes* for region r, where $N = |\mathcal{S}|$. Figure 1(d) illustrates the group sizes for each group size $s \in [N = 5]$ with: $\boldsymbol{n}^{GA} = (2,0,1,0,0)$, $\boldsymbol{n}^{MI} = (1,1,1,0,0)$, and $\boldsymbol{n}^{US} = (3,1,2,0,0)$.

It is now possible to define the problem of interest to the bureau: *The goal is to release, for every group size $s \in [N]$ and region $r \in \mathcal{R}$, the numbers n_s^r of groups of size s in region r, while preserving individual privacy.* The region hierarchy and the group sizes $\mathcal{S}$ are considered public *non-sensitive* information. The entries associating users with groups (see Fig. 1(a)) are *sensitive* information. Therefore, the paper focuses on protecting the privacy of such information. For simplicity, this paper assumes that the region hierarchy has exactly L levels and uses $\mathcal{T}$ as a shorthand for $\mathcal{T}_L$. The paper also focuses on the vastly common case when $z_i \in \{0,1\}, (i \in [n])$, but the results generalize to arbitrary z_i values.

3 Differential Privacy

This paper adopts the framework of differential privacy [7,8], which is the de-facto standard for privacy protection.

Definition 1 (Differential Privacy [7]). *A randomized algorithm $\mathcal{M} : \mathcal{D} \to \mathcal{R}$ with domain $\mathcal{D}$ and range $\mathcal{R}$ is ϵ-differentially private if*

$$\Pr[\mathcal{M}(D_1) \in O] \leq \exp(\epsilon) \Pr[\mathcal{M}(D_2) \in O], \tag{1}$$

for any output response $O \in \mathcal{R}$ and any two datasets $D_1, D_2 \in \mathcal{D}$ differing in at most one individual (called neighbors *and written $D_1 \sim D_2$).*

Parameter $\epsilon > 0$ is the *privacy budget* of the algorithm, with values close to 0 denoting strong privacy. Intuitively, the definition states that the probability of any event does not change much when a single individual data is added or removed to the dataset, limiting the amount of information that the output reveals about any individual.

This paper relies on the *global sensitivity method* [7]. The global sensitivity Δ_q of a function $q : \mathcal{D} \to \mathbb{R}^k$ (also called *query*) is defined as the maximum amount by which q changes when a single individual is added to, or removed from, a dataset:

$$\Delta_q = \max_{D_1 \sim D_2} \|q(D_1) - q(D_2)\|_1. \tag{2}$$

Queries in this paper concern the group size vectors $\boldsymbol{n}^r$ and neighboring datasets differ by the presence or absence of at most one record (see Fig. 1(a)).

The global sensitivity is used to calibrate the amount of noise to add to the query output to achieve differential privacy. There are several sensitivity-based mechanisms [7,20] and this paper uses the *Geometric* mechanism [13] for *integral* queries. It relies on a double-geometric distribution and has slightly less variance than the ubiquitous Laplace mechanism [7].

Definition 2 (Geometric Mechanism [13]**).** *Given a dataset D, a query $q \in \mathbb{R}^k$, and $\epsilon > 0$, the geometric mechanism adds independent noise to each dimension of the query output $q(D)$ using the distribution: $P(X = v) = \frac{1-e^{-\epsilon}}{1+e^{-\epsilon}} e^{(-\epsilon|v|/\Delta_q)}$.*

This distribution is also referred to as *double-geometric* with scale Δ_q/ϵ. In the following, $Geom(\lambda)^k$ denotes the i.i.d. double-geometric distribution over k dimensions with parameter λ. The Geometric Mechanism satisfies ϵ-differential privacy [13]. Differential privacy also satisfies several important properties [8].

Lemma 1 (Sequential Composition). *The composition of two ϵ-differentially private mechanisms ($\mathcal{M}_1$, $\mathcal{M}_2$) satisfies 2ϵ-differential privacy.*

Lemma 2 (Parallel Composition). *Let D_1 and D_2 be disjoint subsets of D and $\mathcal{M}$ be an ϵ-differential private algorithm. Computing $\mathcal{M}(D \cap D_1)$ and $\mathcal{M}(D \cap D_2)$ satisfies ϵ-differential privacy.*

Lemma 3 (Post-Processing Immunity). *Let $\mathcal{M}$ be an ϵ-differential private algorithm and g be an arbitrary mapping from the set of possible output sequences O to an arbitrary set. Then, $g \circ \mathcal{M}$ is ϵ-differential private.*

4   The Privacy-Preserving Group Size Release Problem

This section formalizes the Privacy-preserving Group Size Release (PGSR) problem. Consider a dataset D, a region hierarchy $\mathcal{T}$ for D, where each node a^r in $\mathcal{T}$ is associated with a vector $\boldsymbol{n}^r \in \mathbb{Z}_+^N$ describing the group sizes for region $r \in \mathcal{R}$, and let $G = \sum_{s \in [N]} n_s^\top$ be the total number of individual groups in D, which is public information (see Fig. 2(a) for an example). The PGSR problem consists in releasing a hierarchy of group sizes $\tilde{\mathcal{T}} = \langle \tilde{\boldsymbol{n}}^r \mid r \in \mathcal{R} \rangle^1$ that satisfies the following conditions:

1. *Privacy*: $\tilde{\mathcal{T}}$ is ϵ-differentially private.
2. *Consistency*: For each region $r \in \mathcal{R}$ and group size $s \in \mathcal{S}$, the group sizes in the subregions r' of r add up to those in region r: $\tilde{n}_s^r = \sum_{r' \in ch(r)} \tilde{n}_s^{r'}$.
3. *Validity*: The values $\tilde{n}_s^r$ are non-negative integers.
4. *Faithfulness*: The group sizes at each level ℓ of the hierarchy add up to the value G: $\sum_{r \in \mathcal{R}_\ell} \sum_{s \in [N]} \tilde{n}_s^r = G$.

These constraints ensure that the hierarchical group size estimates satisfy all publicly known properties of the original data.

---

[1] We abuse notation and use the angular parenthesis to denote a hierarchy.

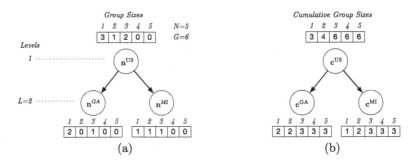

Fig. 2. Region hierarchies associated with the dataset of Fig. 1(a): *group size* hierarchy $\mathcal{T}_2$ (a), and *cumulative group size* hierarchy $\mathcal{T}_2^c$ (b).

5   The Direct Optimization-Based PGSR Mechanism

This section presents a two-step mechanism for the PGSR problem. The first step produces a noisy version of the group sizes, while the second step restores the feasibility of the PGSR constraints while staying as close as possible to the noisy counts. The first step produces a noisy hierarchy $\tilde{\mathcal{T}} = \{\tilde{\boldsymbol{n}}^r \mid r \in \mathcal{R}\}$ using the geometric mechanism with parameter $\lambda = \frac{2L}{\epsilon}$ on the vectors $\boldsymbol{n}^r$:

$$\tilde{\boldsymbol{n}}^r = \boldsymbol{n}^r + Geom\Big(\frac{2L}{\epsilon}\Big)^N. \qquad (3)$$

This step satisfies ϵ-differential privacy due to the following lemma.

Lemma 4. *The sensitivity Δ_n of the group estimate query is 2.*

The output of the first step satisfies Condition 1 of the PGSR problem but it will violate (with high probability) the other conditions. To restore feasibility, this paper uses a post-processing strategy similar to the one proposed in [10] for mobility applications. After generating $\tilde{\mathcal{T}}$ using Eq. (3), the mechanism post-processes the

$$\text{minimize}_{\{\hat{\boldsymbol{n}}^r\}_{r\in\mathcal{R}}} \quad \sum_{r\in\mathcal{R}} \|\hat{\boldsymbol{n}}^r - \tilde{\boldsymbol{n}}^r\|_2^2 \qquad (H1)$$

$$\text{s.t:} \quad \sum_{s\in[N]} \hat{n}_s^r = G \quad \forall r \in \mathcal{R} \qquad (H2)$$

$$\sum_{c\in ch(r)} \hat{n}_s^c = \hat{n}_s^r \quad \forall r \in \mathcal{R}, s \in [N] \qquad (H3)$$

$$\hat{n}_s^r \in D_s^r \quad \forall r \in \mathcal{R}, s \in [N] \qquad (H4)$$

Fig. 3. The $\mathcal{M}_H^{dp}$ bottom-up step.

values $\tilde{\boldsymbol{n}}^r$ of $\tilde{\mathcal{T}}$ through the *Quadratic Integer Program (QIP)* depicted in Fig. 3. Its goal is to find a new region hierarchy $\hat{\mathcal{T}}$, optimizing over the variables $\hat{\boldsymbol{n}}^r = (\hat{n}_1^r \ldots \hat{n}_N^r)$ for each $r \in \mathcal{R}$, so that their values stay close to the noisy counts of the first step, while satisfying faithfulness (Constraint (H2)), consistency (Constraint (H3)), and validity (Constraint (H4)). In the optimization model, D_s^r represents the domain (of integer, non-negative values) of $\hat{n}_s^r$. The resulting mechanism is called the *Hierarchical PGSR* and denoted by $\mathcal{M}_H$. It satisfies ϵ-differential privacy because of post-processing immunity of differential privacy (Lemma 3), since the post-processing step of $\mathcal{M}_H$ uses exclusively differentially private information ($\tilde{\mathcal{T}}$).

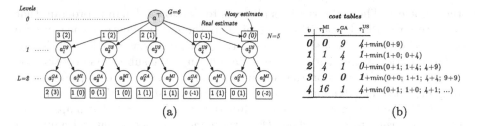

Fig. 4. (a): Region hierarchy $\mathcal{T}^{dp}$ associated with the dataset of Fig. 1(a). (b): Example of cost table computation for subtree associated to the group 1 estimates.

Solving this QIP is intractable for the datasets of interest to the census bureau. Therefore, the experimental results consider a version of $\mathcal{M}_H$ that *relaxes the integrability constraint* (H4) and rounds the solutions. The resulting optimization problem becomes convex but presents two limitations: *(i)* its final solution $\hat{T}$ may violate the PGSR *consistency* (2) and *faithfulness* (4) conditions, and *(ii)* the mechanism is still too slow for very large problems.

6   The Dynamic Programming PGSR Mechanism

This section proposes a dynamic-programming approach for the post-processing step to remedy the limitation of $\mathcal{M}_H$ and its convex relaxation. The resulting mechanism is called the *Dynamic Programming PGSR* mechanism and denoted by $\mathcal{M}_H^{dp}$. The dynamic program relies on a new hierarchy $\mathcal{T}^{dp}$ that modifies the original region hierarchy $\mathcal{T}$ as follows. It creates as many subtrees as the number N of groups in $\mathcal{S}$. The nodes of subtree s represent the groups of size s for the regions in $\mathcal{R}$. In other words, the root node a_s^r of subtree s–where r is level 1 in the region hierarchy (see Fig. 1(b))–is associated with the number n_s^r of groups of size s in region r. Its children $\{a_s^c\}_{c \in ch(r)}$ are associated with the numbers n_s^c, and so on. Finally, the new hierarchy has a root note $a^\top$ that represents the total number G of groups: It is associated with a *dummy* region $\top$ whose children are the N subtrees introduced above. The resulting region hierarchy is denoted $\mathcal{T}^{dp}$.

Example 1. The region hierarchy $\mathcal{T}^{dp}$ associated with the running example is shown in Fig. 4(a). The root node $a^\top$ is associated with the total number of groups in D, i.e., $G = 6$. Its children $a_1^{US}, \ldots, a_5^{US}$ represent the group sizes for the root of the region hierarchy for each group size $s \in [N = 5]$. Subtree 1, rooted at a_1^{US}, has two children: a_1^{GA} and a_1^{MI}, representing the number of groups of size 1: n_1^{GA} and n_1^{MI}. The figure illustrates the association of every node a_s^r with its *real* group size n_s^r (in red) and its noisy group size generated by the geometrical mechanism (in blue and parenthesis).

Note that *(i)* the value of a node equals to the sum of the values of its children, *(ii)* the sum of the group sizes at a given level add up to G, and *(iii)* the PGSR consistency conditions of the nodes in a subtree are independent of those of other subtrees. These observations allow us to develop a dynamic program that guarantees the PGSR conditions and exploits the independence of each subtree associated with groups of size s to solve the post-processing problem efficiently.

For notational simplicity, the presentation omits the subscripts denoting the group size s and focuses on the computation of a single subtree representing a group of size s. The dynamic program associates a *cost table* τ^r with each node a^r of $\mathcal{T}^{dp}$. The cost table represents a function $\tau^r : D^r \rightarrow \mathbb{R}_+$ that maps values (i.e., group sizes) to costs, where D^r is the *domain* (a set of natural numbers) of region r. Intuitively, $\tau^r(v)$ is the optimal cost for the post-processed group sizes in the subtree rooted at a^r when its post-processed group size is equal to v, i.e., $\hat{n}^r = v$. *The key insight of the dynamic program is the observation that the optimal cost for $\tau^r(v)$ can be computed from the cost tables τ^c of each of its children $c \in \mathrm{ch}(r)$ using the formula given in Figure* 5. In the formula, (d1) describes the cost for v of deviating from the noisy group size $\tilde{n}^r$. The function $\phi^r(v)$, defined in (d2), (d3), and (d4), uses the cost table of the children of r to find the combination of post-processed group sizes $\{x_c \in D^c\}_{c \in \mathrm{ch}(r)}$ of r's children that is consistent (d3) and minimizes the sum of their costs (d2).

The dynamic program exploits these concepts in two phases. The first phase is bottom-up and computes the cost tables for each node, starting from the leaves only, which are defined by (d1), and moving up, level by level, to the root. The cost table at the root is then used to retrieve the optimal cost of the problem. The second phase is top-down: Starting from the root, each node a^r receives its post-processed group size $\hat{n}^r$ and solves $\phi^r(\hat{n}^r)$ to

$$\tau^r(v) = \left(v - \tilde{n}^r\right)^2 + \qquad \text{(d1)}$$

$$\phi^r(v) = \min_{\{x_c\}_{c \in \mathrm{ch}(r)}} \sum_{c \in \mathrm{ch}(r)} \tau^c(x_c) \quad \text{(d2)}$$

$$\text{s.t.} \sum_{c \in \mathrm{ch}(r)} x_c = v \qquad \text{(d3)}$$

$$x_c \in D^c \ \ \forall c \in \mathrm{ch}(r) \quad \text{(d4)}$$

Fig. 5. The $\mathcal{M}_H^{dp}$ bottom-up step.

retrieve the optimal post-processed group sizes $\hat{n}^c = x_c$ for each child $c \in \mathrm{ch}(r)$.

An illustration of the process for the running example is illustrated in Fig. 4(b). It depicts the cost tables τ_1^{GA}, τ_1^{MI}, and τ_1^{US} related to the subtree rooted at a_1^{US} (groups of size 1) computed during the bottom-up phase. The values selected during the top-down phase are highlighted red.

In the implementation, the values $\phi^r(v)$ are computed using a constraint program where (d1) is implemented using a table constraint. The number of optimization problems in the dynamic program is given by the following theorem.

Theorem 1. *Constructing $\hat{\mathcal{T}}^{dp}$ requires solving $O(|\mathcal{R}|N\bar{D})$ optimization problems given in Fig. 5, where $\bar{D} = \max_{s,r} |D_s^r|$ for $r \in \mathcal{R}, s \in [N]$.*

7   A Polynomial-Time PGSR Mechanism

The dynamic program relies on solving an optimization problem for each region. This section shows that this optimization problem can be solved in polynomial time by exploiting the structure of the cost tables.

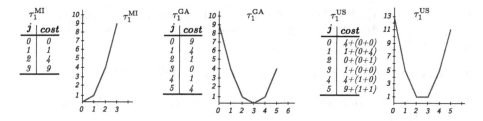

Fig. 6. Cost tables, extending Fig. 4(b), computed by the mechanism.

A cost table is a finite set of pairs (s, c) where s is a group size and c is a cost. When the pairs are ordered by increasing values of s and line segments are used to connect them as in Fig. 6, the resulting function is Piecewise Linear (PWL). For simplicity, we say that a cost table is PLW if its underlying function is PWL. Observe also that, at a leaf, the cost table is Convex PWL (CPWL), since the L2-Norm is convex (see Eq. (d1)).

The key insight behind the polynomial-time mechanism is the recognition that the function ϕ^r is CPWL whenever the cost tables of its children are CPWL. As a result, by induction, the cost table of every node a^r is CPWL.

Lemma 5. *The cost table τ_s^r of each node a_s^r of $\mathcal{T}^{dp}$ is CPWL.*

Lemma 5 makes it possible to design a polynomial-time algorithm to compute τ^r (subscripts omitted for succinctness) that replaces the constraint program used in the dynamic program. We give the intuition underlying the algorithm.

Given a node a^r, the first step of the algorithm is to select, for each node $c \in ch(r)$, the value v_c^0 with minimum cost, i.e., $v_c^0 = \operatorname{argmin}_v \tau^c(v)$. As a result, the value $V^0 = \sum_c v_c^0$ has minimal cost $\phi^r(V^0) = \sum_c \tau^c(v_c^0)$. Having constructed the minimum value in cost table ϕ^r, it remains to compute the costs of all values $V^0 + k$ for all integer $k \in [1, \max D^r - V^0]$ and all values $V^0 - k$ for all integer $k \in [1, V^0 - \min D^r]$. The presentation focuses on the values $V^0 + k$ since the two cases are similar. Let $\mathbf{v}^0 = \{v_c^0\}_{c \in ch(r)}$. The algorithm builds a sequence of vectors $\mathbf{v}^1, \mathbf{v}^2, \dots, \mathbf{v}^k, \dots$ that provides the optimal combinations of values for $\phi^r(V^0 + 1), \phi^r(V^0 + 2), \dots, \phi^r(V^0 + k), \dots$. Vector $\mathbf{v}^k$ is obtained from $\mathbf{v}^{k-1}$ by changing the value of a single child whose cost table has the smallest slope, i.e.,

$$v_c^k = \begin{cases} v_c^{k-1} + 1 & \text{if } c = \operatorname{argmin}_c \tau^c(v_c^{k-1} + 1) - \tau^c(v_c^{k-1}) \\ v_c^{k-1} & \text{otherwise.} \end{cases} \quad (4)$$

Once ϕ^r has been computed, cost table τ^r can be computed easily since both $(v - \tilde{n}^r)^2$ and ϕ^r are CPWL and the sum of two CPWL functions is CPWL.

Example 2. These concepts are illustrated in Fig. 6, where the values of ϕ^r are highlighted in blue, and in parenthesis, in the right table. The first step identifies that $v_{MI}^0 = 0$, $v_{GA}^0 = 3$, thus $V^0 = 3$ and $\phi^r(3) = \tau^{MI}(0) + \tau^{GA}(3) = 0 + 0 = 0$. In the example, $\mathbf{v}^0 = (v_{MI}^0, v_{GA}^0) = (0, 3)$ and $\mathbf{v}^1 = (v_{MI}^1, v_{GA}^1) = (1, 3)$, since $MI = \operatorname{argmin}\{\tau^{MI}(0+1) - \tau^{MI}(0), \tau^{GA}(3+1) - \tau^{GA}(3)\} = \operatorname{argmin}\{1 - 0, 1 - 0\}$.

Its associated cost, $\phi^{US}(\mathbf{v}^1) = 1 + 0 = 1$. $\mathbf{v}^2 = (1,4)$ since $GA = \operatorname{argmin}\{(\tau^{MI}(1+1) - \tau^{MI}(1)), (\tau^{GA}(3+1) - \tau^{GA}(3))\} = \operatorname{argmin}\{(4-1), (1-0)\}$, and its associated cost $\phi^{US}(\mathbf{v}^2) = 1 + 1 = 2$.

Theorem 2. *The cost table τ_s^r of each node a_s^r of $\mathcal{T}^{dp}$ is CPWL.*

Theorem 3. *The cost table τ_s^r for each region r and size s can be computed in time $O(\bar{D}\log\bar{D})$.*

8   Cumulative PGSR Mechanisms

In the mechanisms presented so far, each query has sensitivity $\Delta_n = 2$. This section exploits the structure of the group query to reduce the query sensitivity and thus the noise introduced by the geometric mechanism.

Define the operator $\oplus : \mathbb{Z}_+^N \to \mathbb{Z}_+^N$ that, given a vector $\mathbf{n} = (n_1, \ldots, n_N)$ of group sizes, returns its cumulative version $\mathbf{c} = (c_1, \ldots, c_N)$ where $c_s = \sum_{k=1}^{s} n_k$ is the cumulative sum of the first s elements of $\mathbf{n}$. This operator can be used to produce a hierarchy $\mathcal{T}^c = \{\mathbf{c}^r | r \in \mathcal{R}\}$ of cumulative group sizes. An example of such a hierarchy $\mathcal{T}^c$ is provided in Fig. 2(b).

Lemma 6. *The sensitivity Δ_c of the cumulative group estimate query is 1.*

The result follows from the fact that removing an element from a group in $\mathbf{c}$ only affects the group preceding it. This idea is from [15], where cumulative sizes are referred to as *unattributed histograms*.

To generate a privacy-preserving version $\tilde{\mathcal{T}}^c$ of $\mathcal{T}^c$, it suffices to apply the geometrical mechanism with parameter $\lambda = (L/\epsilon)^N$ on the vectors $\mathbf{c}^r$ associated with every node τ^r of the region hierarchy. Once the noisy sizes are computed, the noisy group sizes can be easily retrieved via an inverse mapping $\ominus : \mathbb{Z}_+^N \to \mathbb{Z}_+^N$ from the cumulative sums.

Note however that the resulting private versions $\tilde{c}^r$ of c^r may no longer be non-decreasing (or even non-negative) due to the added noise. Therefore, as in Sect. 5, a post-processing step is applied to restore consistency and to guarantee the PGSR conditions 2 to 4. The post-processing is illustrated in Fig. 7. It takes as input the noisy hierarchy of cumulative sizes $\tilde{\mathcal{T}}^c$ com-

$$\text{minimize}_{\{\hat{c}^r\}_{r \in \mathcal{R}}} \quad \sum_{r \in \mathcal{R}} \|\hat{c}^r - \tilde{c}^r\|_2^2 \tag{C1}$$

$$\text{s.t.:} \quad \hat{c}_N^\top = G \tag{C2}$$

$$\hat{c}_i^r \leq \hat{c}_{i+1}^r \quad \forall r \in \mathcal{R}, i \in [N-1] \tag{C3}$$

$$\sum_{r' \in ch(r)} \hat{c}_i^{r'} = \hat{c}_i^r \quad \forall r \in \mathcal{R}, i \in [N] \tag{C4}$$

$$\hat{c}_i^r \in \{0, 1, \ldots\} \quad \forall r \in \mathcal{R}, i \in [N] \tag{C5}$$

Fig. 7. The $\mathcal{M}_c$ post-processing step.

puted with the geometrical mechanism and optimizes over variables $\hat{c}^r = (\hat{c}_1^r, \ldots, \hat{c}_N^r)$ for $r \in \mathcal{R}$, minimizing the L2-norm wrt their noisy counterparts (Eq. ((C1)). Constraints (C2) guarantees that the sum of the sizes equals the public value G (PGSR condition 4), where $\top$ denotes the root region of the region hierarchy. Constraints (C3) guarantee consistency of the cumulative counts.

Finally, Constraints (C4) and (C5), respectively, guarantee the PGSR consistency (2) and validity (3) conditions.

Once the post-processed hierarchy $\hat{\mathcal{T}}^c$ is obtained, operator $\ominus$ is applied to obtain a post-processed version of the group size hierarchy $\hat{\mathcal{T}}$. It is easy to see that the above mechanism satisfies ϵ-differential privacy, due to post-processing immunity. This mechanism is called the cumulative PGSR and denoted by $\mathcal{M}_c$.

Like for $\mathcal{M}_H$, the structure of the PGSR problem can be exploited to create an efficient mechanism that satisfies the conditions of the PGSR problem using the cumulative group sizes. The resulting mechanism is called $\mathcal{M}_c^{dp}$ and operates in three steps:

1. $\mathcal{M}_c^{dp}$ creates a noisy hierarchy $\tilde{\mathcal{T}}^c$.
2. $\mathcal{M}_c^{dp}$ executes the post-processing step described in Algorithm 1.
3. $\mathcal{M}_c^{dp}$ runs the post-processing step of the polynomial-time PGSR mechanism (see Sect. 7).

Algorithm 1: $\mathcal{M}_c^{dp}.post\text{-}process$

input $: \tilde{\mathcal{T}}^c = \{\tilde{c}^r \mid r \in \mathcal{R}\}$

1 **foreach** $r \in \mathcal{R}$ **do**

2     $\hat{c}^r \leftarrow \text{argmin}_{\hat{c}} \|\hat{c} - \tilde{c}^r\|^2$
               s.t. $\hat{c}_i \leq \hat{c}_{i+1}$ $\forall i \in [N-1]$
                    $0 \leq \hat{c}_i \leq G$ $\forall i \in [N]$

3     $\bar{n} \leftarrow \ominus(round(\hat{c}^r))$

4     $\hat{\mathcal{T}} \leftarrow \bar{\mathcal{T}} \cup \{\bar{n}\}$

5 $\hat{\mathcal{T}} \leftarrow \mathcal{M}_{dp}.post\text{-}process(\hat{\mathcal{T}})$

The novelty is in step 2 which takes $\tilde{\mathcal{T}}^c$ as input and, for each node $\tilde{c}^r$, solves the *convex program* described in line 2 to create a new noisy hierarchy $\hat{c}^r$ that is non-decreasing and non-negative. The resulting cumulative vector $\hat{c}^r$ is then rounded and transformed to its corresponding group size vector through operator $\ominus(\cdot)$ (line 3). The resulting vector $\bar{n}^r$ is added to the region hierarchy $\hat{\mathcal{T}}$ (line 4). Observe that this post-processing step pays a polynomial-time penalty w.r.t. the runtime of the $\mathcal{M}_H^{dp}$ post-processing. The convex program of line (2) is executed in $O(\text{poly}(N))$ and the resulting post-processing step runtime is in $O(|\mathcal{R}|\text{poly}(N) + |\mathcal{R}|N\bar{D}\log\bar{D})$.

It is important to note that $\mathcal{M}_c^{dp}$ does not solve the same post-processing program as the cumulative PGSR mechanism specified by Eqs. (C1) to (C2), since it restores consistency of the cumulative counts locally. However, the experimental results show that it consistently reduces the final error: See Sect. 9 for detailed results.

9 Experimental Evaluation

This section evaluates the privacy-preserving mechanisms for the PGSR problem. The evaluation focuses on comparing runtime and accuracy. Consistent with the privacy literature, accuracy is measured in term of the L_1 difference between the privacy-preserving group sizes and the original ones, i.e., given the original group sizes $\mathcal{T} = \{n^r \mid r \in \mathcal{R}\}$, and their private counterparts $\hat{\mathcal{T}}$, the L_1-error is defined as $\sum_r \|n^r - \hat{n}^r\|_1$. Since the mechanisms are nondeterministic due to the noise added by the geometric mechanism, 30 instances are generated for each benchmark and the results report average values and standard deviations. Each mechanism is run on a single-core 2.1 GHz terminal with 24 GB of RAM and is implemented in Python 3 with Gurobi 8.0 for solving the convex quadratic optimization problems.

Mechanisms. The evaluation compares the PGSR mechanisms $\mathcal{M}_H$, its cumulative version $\mathcal{M}_c$, and their polynomial-time dynamic-programming (DP) counterparts $\mathcal{M}_H^{dp}$, and $\mathcal{M}_c^{dp}$. The former are referred to as OP-based methods and the latter as DP-based methods. In addition to $\mathcal{M}_H$ and $\mathcal{M}_c$, that solve the associated post-processing QIPs, the experiments evaluate the associated *relaxations*, $\mathcal{M}_H^r$ and $\mathcal{M}_c^r$, respectively, that relax the integrality constraints (H4) and (C5) and rounds the solutions. For completeness, the experiments also evaluate the performance of the optimization-based mechanism $\mathcal{M}_{dp}^{OP}$ that *does not* exploit the structure of the cost function to compute the cost tables.

Datasets. The mechanisms are evaluated on three datasets.

- **Census Dataset**: The first dataset has 117,630,445 groups, 7592 leaves, 305,276,358 individuals, 3 levels, and $N = 1,000$. Individuals live in facilities, i.e., households or dormitories, assisted living facilities, and correctional institutions. Due to privacy concerns and lack of available methods to protect group sizes during the 2010 Decennial Census release, group sizes were aggregated for any facility of size 8 or more (see Summary File 1 [24]). Therefore, following [17] and starting from the truncated group sizes Census dataset, the experiments augment the dataset with group sizes up to $N = 1,000$ that mimic the published statistics, but add a heavy tail to model group quarters (dormitories, correctional facilities, etc.). This was obtained by computing the ratio $r = n_7/n_6$ of household groups of sizes 7 and 6, subtracting from the aggregated groups n_{8+} M people according to the ratio r, and redistributed these M people in groups $k > 8$ so that the ratio between any two consecutive groups holds (in expectation). Finally, 50 outliers were added, chosen uniformly in the interval between 10 and $1,000$. The region hierarchy is composed by the National level, the State levels (50 states + Puerto Rico and District of Columbia), and the Counties levels (3143 in total).
- **NY Taxi Dataset**: The second dataset has 13,282 groups, 3,973 leaves, 24,489,743 individuals, 3 levels, and $N = 13,282$. The 2014 NY city Taxi dataset [3] describes trips (pickups and dropoffs) from geographical locations in NY city. The dataset views each taxi as a group and the size of the group is the number of pickups of the taxi. The region hierarchy has 3 levels: the entire NY city at level 1, the boroughs: *Bronx, Brooklyn, EWR, Manhattan, Queens*, and *Staten Island* at level 2, and a total of 263 zones at level 3.
- **Synthetic Dataset**: Finally, to test the runtime scalability, the experiments considered synthetic data from the NY Taxi dataset by limiting the number of group sizes N arbitrarily, i.e., removing group sizes greater than a certain threshold.

9.1 Scalability

The first results concern the scalability of the mechanisms, which are
evaluated on the synthetic datasets
for various numbers of group sizes.
Figure 8 illustrates the runtimes of
the algorithms at varying of the number of group sizes N from 5 to 50
for the synthetic dataset. The experiments have a timeout of 30 min
and the runtime is reported in log-
10 scale. The figure shows that the

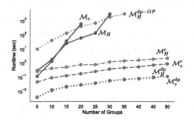

Fig. 8. The Runtime for the mechanisms:
Census data (left) and Taxi data (right).

exact OP-based approaches and $\mathcal{M}_{dp}^{OP}$ are not competitive, even for small groups
sizes. Therefore, these results rule out the following mechanisms: $\mathcal{M}_{dp}^{OP}$, $\mathcal{M}_H$,
and $\mathcal{M}_c$ and the remaining results focus on comparing the relaxed versions of
the OP-based mechanisms versus their proposed DP-counterparts.

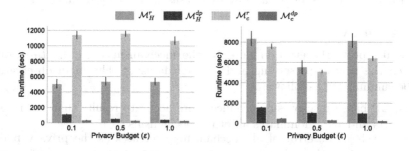

Fig. 9. Runtime (in seconds) at varying of the number of group size N.

9.2 Runtimes

Fig. 9 reports the runtime, in seconds, for the hierarchical mechanism $\mathcal{M}_H^r$ and
its DP-counterpart $\mathcal{M}_H^{dp}$, and the hierarchical cumulative mechanism $\mathcal{M}_c^r$ and
its dp-counterpart $\mathcal{M}_c^{dp}$. The left side of the figure illustrates the results for the
Census data and the right side those for the NY Taxi data. The main observations
can be summarized as follows:

1. Although the OP-based algorithms consider only a relaxation of the problem,
   the *exact* DP-versions are consistently faster. In particular, $\mathcal{M}_H^{dp}$ is up to one
   order of magnitude faster than its counterpart $\mathcal{M}^r$, and $\mathcal{M}_c^{dp}$ is up to two
   orders of magnitude faster than its counterpart $\mathcal{M}_c^r$.
2. $\mathcal{M}_c^r$ is consistently slower then $\mathcal{M}_H^r$. This is because, despite the fact that
   the two post-processing steps have the same number of variables, the $\mathcal{M}_c$
   post-processing step has many additional constraints of type (C3).

3. The runtime of the DP-based mechanisms decreases as the privacy budget increases, due to the sizes of the cost tables that depend on the noise variance.[2]
4. The cumulative version $\mathcal{M}_c^{dp}$ outperforms its $\mathcal{M}_H^{dp}$ counterpart. Once again, the reason is due to the domain sizes. In fact, due to reduced sensitivity, $\mathcal{M}_c^{dp}$ applies a smaller amount of noise than that required by $\mathcal{M}_H^{dp}$ to guarantee the same level of privacy and resulting in smaller domain sizes.

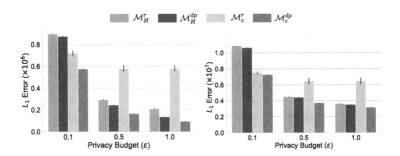

Fig. 10. The L_1 errors for the algorithms: Census data (left) and Taxi data (right).

9.3 Accuracy

Figure 10 reports the error induced by the mechanisms, i.e., the L_1-distance between the privacy-preserving and original datasets. The main observations can be summarized as follows:

1. The DP-based mechanisms produce more accurate results than their counterparts and $\mathcal{M}_{dp}^c$ dominates all other mechanisms.
2. As expected, the error of all mechanisms decreases as the privacy budget increases, since the noise decreases as privacy budget increases. The errors are larger in the NY Taxi dataset, which has a larger number of group sizes than the Census dataset.
3. Finally, the results show that the cumulative mechanisms tend to concentrate the errors on small group sizes. Unfortunately, these are also the most populated groups, and this is true for each subregion of the hierarchy. On the other hand, the DP-based version, that retains the integrality constraints, better redistributes the noise introduced by the geometrical mechanism and produce substantially more accurate results.

To shed further light on accuracy, Table 1 reports a breakdown of the average errors of each mechanism at each level of the hierarchies. Mechanism $\mathcal{M}_{dp}^c$ is clearly the most accurate. Note that the table reports the average number of *constraint violations* in the output datasets. A constraint violation is counted whenever a subtree of the hierarchy violates the PGRP *consistency* condition (2). Being exact, the DP-based methods report no violations. In contrast, both $\mathcal{M}_H^r$ and $\mathcal{M}_c^r$ report a substantial amount of constraint violations.

---

[2] The implementation uses $D_s^r = \{\tilde{n}_s^r - \delta \ldots \tilde{n}_s^r + \delta\} \cap \mathbb{Z}_+$, where $\delta = 3 \times \lceil 2\lambda^2 \rceil$, i.e., 3 times the variance associated with the double-geometrical distribution with parameter λ.

10   Related Work

The release of privacy-preserving datasets using differential privacy has been subject of extensive research [12,16,19]. However, these methods focus on creating *unattributed histograms* that count the number of individuals associated with each possible property in the dataset universe. Extensions to hierarchical problems were also explored. For instance, [15,21] study methods to answer count queries using a hierarchical structure to impose consistency of counts. Extensions for optimizing various count queries have also been proposed [5,18,21].

Table 1. L_1-errors and constraint violations (CV) for each level of the hierarchies.

		Taxi Data				Census Data			
		L_1 Errors ($\times 10^4$)			#CV	L_1 Errors ($\times 10^3$)			#CV
ϵ	Alg	Lev 1	Lev 2	Lev 3		Lev 1	Lev 2	Lev 3	
0.1	$\mathcal{M}_H^r$	25.4	158.7	904.4	18206	40.3	54.3	802.1	1966
	$\mathcal{M}_H^{dp}$	26.6	121.9	915.7	0	10.3	38.4	825.4	0
	$\mathcal{M}_c^r$	47.9	153.2	**551.6**	19460	23.1	64.5	632.2	1715
	$\mathcal{M}_c^{dp}$	**19.9**	**65.6**	644.3	0	**0.9**	**23.2**	**550.6**	0
0.5	$\mathcal{M}_H^r$	8.6	81.2	364.2	18591	39.4	37.9	216.3	1990
	$\mathcal{M}_H^{dp}$	5.5	31.0	408.9	0	2.4	9.4	230.8	0
	$\mathcal{M}_c^r$	46.7	153.5	450.7	19531	23.1	61.0	494.2	1718
	$\mathcal{M}_c^{dp}$	**4.0**	**16.4**	**352.9**	0	**0.2**	**5.8**	**159.1**	0
1.0	$\mathcal{M}_H^r$	7.7	77.2	**279.0**	18085	40.7	39.2	130.0	1989
	$\mathcal{M}_H^{dp}$	3.1	19.8	328.5	0	1.2	5.1	128.8	0
	$\mathcal{M}_c^r$	47.1	154.2	447.1	19706	24.1	63.0	494.5	1728
	$\mathcal{M}_c^{dp}$	**2.0**	**8.7**	307.8	0	**0.1**	**3.2**	**91.0**	0

These methods differ in two ways from the mechanisms proposed here: (1) They focus on histograms queries, rather than group queries; the latter generally have higher L_1-sensitivity and thus require more noise and (2) they ensure neither the consistency for integral counts nor the non-negativity of the release counts. They thus violate the requirements of group sizes (see Sect. 4).

Finally, [10] proposed a hierarchical-based solution based on minimizing the L2-distance between the noisy counts and their private counterparts. While this solution guarantees non-negativity of the counts, their mechanism, if formulated as a MIP/QIP, cannot cope with the scale of the census problems discussed here which compute privacy-preserving country-wise group sizes. If their solution is used as is, in its relaxed form, then it cannot guarantee the integrality of the counts. These mechanisms reduce to $\mathcal{M}_H^r$, which has been shown, in the previous section, to be strongly dominated by the DP-based mechanisms.

11   Conclusions

The release of datasets containing sensitive information concerning a large number of individuals is central to a number of statistical analysis and machine learning tasks. Of particular interest are hierarchical datasets, in which counts of individuals satisfying a given property need to be released at different granularities (e.g., the location of a household at a national, state, and county levels). The paper defined the *Privacy-perserving Group Release* (PGRP) problem and proposed an exact and efficient constrained-based approach to privately generate consistent counts across all levels of the hierarchy. This novel approach was evaluated on large, real datasets and results in speedups of up to two orders of magnitude, as well as significant improvements in terms of accuracy with respect to state-of-the-art techniques. Interesting avenues of future directions include exploiting different forms of parallelism to speed up the computations of the dynamic programming-based mechanisms even further, using, for instance, Graphical Processing Units as proposed in [11].

References

1. AAAS: New Privacy Protections Highlight the Value of Science Behind the 2020 census. https://www.aaas.org/news/new-privacy-protections-highlight-value-science-behind-2020-census. Accessed 23 Apr 2019
2. NBC News: Potential privacy lapse found in Americans' 2010 census data. https://www.nbcnews.com/news/us-news/potential-privacy-lapse-found-americans-2010-census-data-n972471. Accessed 23 Apr 2019
3. New York City Taxi Data. http://www.nyc.gov/html/tlc/html/about/trip_record_data.shtml. Accessed 20 Apr 2019
4. NY Times: To Reduce Privacy Risks, the Census Plans to Report Less AccurateData. https://www.nytimes.com/2018/12/05/upshot/to-reduce-privacy-risks-the-census-plans-to-report-less-accurate-data.html
5. Cormode, G., Procopiuc, C., Srivastava, D., Shen, E., Yu, T.: Differentially private spatial decompositions. In: 2012 IEEE 28th International Conference on Data Engineering, pp. 20–31. IEEE (2012)
6. Dinur, I., Nissim, K.: Revealing information while preserving privacy. In: Proceedings of the Twenty-Second ACM SIGMOD-SIGACT-SIGART Symposium on Principles of Database Systems, pp. 202–210. ACM (2003)
7. Dwork, C., McSherry, F., Nissim, K., Smith, A.: Calibrating noise to sensitivity in private data analysis. In: Halevi, S., Rabin, T. (eds.) TCC 2006. LNCS, vol. 3876, pp. 265–284. Springer, Heidelberg (2006). https://doi.org/10.1007/11681878_14
8. Dwork, C., Roth, A.: The algorithmic foundations of differential privacy. Theoret. Comput. Sci. **9**(3–4), 211–407 (2013)
9. Erlingsson, Ú., Pihur, V., Korolova, A.: Rappor: randomized aggregatable privacy-preserving ordinal response. In: Proceedings of the 2014 ACM SIGSAC Conference on Computer and Communications Security, pp. 1054–1067. ACM (2014)
10. Fioretto, F., Lee, C., Van Hentenryck, P.: Constrained-based differential privacy for private mobility. In: Proceedings of the International Joint Conference on Autonomous Agents and Multiagent Systems (AAMAS), pp. 1405–1413 (2018)

11. Fioretto, F., Pontelli, E., Yeoh, W., Dechter, R.: Accelerating exact and approximate inference for (distributed) discrete optimization with GPUs. Constraints **23**(1), 1–43 (2018)

12. Fioretto, F., Van Hentenryck, P.: Constrained-based differential privacy: releasing optimal power flow benchmarks privately. In: Proceedings of the International Conference on the Integration of Constraint Programming, Artificial Intelligence, and Operations Research (CPAIOR), pp. 215–231 (2018)

13. Ghosh, A., Roughgarden, T., Sundararajan, M.: Universally utility-maximizing privacy mechanisms. SIAM J. Comput. **41**(6), 1673–1693 (2012)

14. Golle, P.: Revisiting the uniqueness of simple demographics in the US population. In: Proceedings of the 5th ACM Workshop on Privacy in Electronic Society, pp. 77–80. ACM (2006)

15. Hay, M., Rastogi, V., Miklau, G., Suciu, D.: Boosting the accuracy of differentially private histograms through consistency. Proc. VLDB Endow. **3**(1–2), 1021–1032 (2010)

16. Huang, D., Han, S., Li, X., Yu, P.S.: Orthogonal mechanism for answering batch queries with differential privacy. In: Proceedings of the 27th International Conference on Scientific and Statistical Database Management, p. 24. ACM (2015)

17. Kuo, Y.H., Chiu, C.C., Kifer, D., Hay, M., Machanavajjhala, A.: Differentially private hierarchical group size estimation. arXiv preprint arXiv:1804.00370 (2018)

18. Li, C., Hay, M., Miklau, G., Wang, Y.: A data-and workload-aware algorithm for range queries under differential privacy. Proc. VLDB Endow. **7**(5), 341–352 (2014)

19. Li, T., Li, N.: On the tradeoff between privacy and utility in data publishing. In: Proceedings of the 15th ACM SIGKDD International Conference on Knowledge Discovery and Data Mining, pp. 517–526. ACM (2009)

20. McSherry, F., Talwar, K.: Mechanism design via differential privacy. In: 48th Annual IEEE Symposium on Foundations of Computer Science, 2007, FOCS 2007, pp. 94–103. IEEE (2007)

21. Qardaji, W., Yang, W., Li, N.: Understanding hierarchical methods for differentially private histograms. Proc. VLDB Endow. **6**(14), 1954–1965 (2013)

22. Sweeney, L.: k-anonymity: a model for protecting privacy. Int. J. Uncertainty, Fuzziness Knowl.-Based Syst. **10**(05), 557–570 (2002)

23. Team, A.D.P.: Learning with privacy at scale. Apple Mach. Learn. J. **1**(8), 1–25 (2017)

24. U.S. Census Bureau: 2010 census summary file 1: census of population and housing, technical documentation (2012). https://www.census.gov/prod/cen2010/doc/sf1.pdf

25. Winkler, W.: Single ranking micro-aggregation and re-identification. Statistical Research Division report RR 2002/8 (2002)

Generic Constraint-Based Block Modeling Using Constraint Programming

Alex Mattenet[1]([⊠]), Ian Davidson[2], Siegfried Nijssen[1], and Pierre Schaus[1]

[1] ICTEAM, UCLouvain, Ottignies-Louvain-la-Neuve, Belgium
{alex.mattenet,siegfried.nijssen,pierre.schaus}@uclouvain.be
[2] Computer Science Department, University of California, Davis, USA
davidson@cs.ucdavis.edu

Abstract. Block modeling has been used extensively in many domains including social science, spatial temporal data analysis and even medical imaging. Original formulations of the problem modeled the problem as a mixed integer programming problem, but were not scalable. Subsequent work relaxed the discrete optimization requirement, and showed that adding constraints is not straightforward in existing approaches. In this work, we present a new approach based on constraint programming, allowing discrete optimization of block modeling in a manner that is not only scalable, but also allows the easy incorporation of constraints. We introduce a new constraint filtering algorithm that outperforms earlier approaches, in both constrained and unconstrained settings. We show its use in the analysis of real datasets.

1 Introduction

Block modeling has a long history in the analysis of social networks [31]. The core problem is to take a graph and divide it into k clusters and interactions between those clusters described by a $k \times k$ image matrix. The purpose is to summarize a complex graph to be better understood by humans.

More formally, in its simplest formulation, the core problem is: given a graph $G(V, E)$ whose $n \times n$ adjacency matrix is X, simplify X into a symmetric trifactorization FMF^T. Here F is an $n \times k$ block allocation matrix with the blocks/clusters stacked column wise, and $F_{i,j} \in \{0, 1\}$. M is a $k \times k$ interaction (image) matrix showing the interaction between blocks. The objective function is to minimize the reconstruction error $||X - FMF^T||$.

This block modeling formulation has the advantage of identifying structural equivalence: if the reconstruction error is 0, any instance in cluster i must have the exact same neighbors in the graph. The reconstruction error ($||X - FMF^T||$) counts the number of edges that violate this property.

The original MIP formulations of block modeling were lacking in two directions. Firstly, they were not scalable; secondly, they often found results that were inconsistent with the expectations of domain experts. To solve both problems,

A. Mattenet—Supported by a FRIA grant.

T. Schiex and S. de Givry (Eds.): CP 2019, LNCS 11802, pp. 656–673, 2019.
https://doi.org/10.1007/978-3-030-30048-7_38

the use of constraints has been studied in the literature. For example: (i) Entry level constraints such as non-negativity [30], (ii) Incorporating simple composition constraints on the blocks such as spatial continuity or together/apart constraints [3,16], (iii) Constraints on the interaction/image matrix [15] and even (iv) simultaneous constraints on blocks and interaction/image matrices [2].

However, each of these studies yielded a different type of solver that is only scalable for one specific problem setting. For example, in [3] an update rule for the LM method is used, and in [2] a multiplicative update rule. Hence, it is impossible to use all of these constraints at the same time; the approaches are either not usable or not scalable without the predefined constraints.

In this paper we propose a novel approach to block modeling based on Constraint Programming. The advantage of CP is that it offers a generic and modular approach to solving constraint satisfaction and optimization problems by means of global constraints. Global constraints can be combined to solve problems involving multiple constraints. In this work, we introduce a global constraint for block modeling. This allows solving block modeling problems under additional constraints such as: (a) upper and lower bounds on the cluster size; (b) complex requirements in conjunctive normal form, such as that if vertex i is in the same cluster as j, then k and l must not be; (c) constraints on the structure of the image graph M, forcing it to be a tree, a ring graph, a star graph, ...; (d) connectivity constraints: we can require that the subgraph induced by the nodes in each cluster is connected; (e) bin packing constraints: given a weight for each vertex, limit the total weight of each cluster; and more.

Such constraints now allow combining strong semantic knowledge (the constraints) along with empirical evidence (the graph).

The focus of this work is primarily on how to build a filtering algorithm for block modeling that works well in practice. We will demonstrate this on a number of experiments on both datasets used in earlier studies and new problems that we propose in this work. We will show that our propagator is correct for the constraint that it implements and outperforms other methods by orders of magnitude.

2 Related Work

Block modeling in practice has two core computational challenges: (i) the problem needs to be solved as a discrete optimization problem to be truly interpretable. (ii) constraints are required to make results realistic in that they are consistent with human expectations.

Take for example the application of block modeling on Twitter data from the US elections. Each person/account should be allocated to a cluster, and we wish to efficiently find clusters consistent with our expectations (i.e. that Donald Trump will not be in the same cluster as Hillary Clinton).

There have been two lines of work to address both challenges, but no work attempts to address both. There are some MIP formulations of block modeling [9], but as we show in this paper (Table 3) their run time is extremely slow.

Instead, most work has focused on relaxing the problem to a continuous problem and adding constraints. There are a plethora of such constraints, some of which we outline in Table 1. Unfortunately, these methods cannot be combined to create a block modeling solver that uses all constraints as they use different underlying solving methods. Furthermore, these solvers do not yield exact solutions for the discrete allocation problem. All of these constraints and others mentioned in the introduction (i.e. cardinality constraints) can however easily be encoded in our exact CP model.

Table 1. A list of some complex constraints used to solve continuous optimization versions of block modeling. These methods cannot be combined as they use different underlying solvers, whereas our method can address all of these constraints.

Constraint	Description	Solver used
Spatial continuity [3]	A soft constraint based on a kernel	Additive update rule
Path [2]	All nodes in a block have a path to each other	Multiplicative update rule
Composition [16]	Must-link/cannot-link constraints	Gradient descent
Image structure [15]	Constraints on image matrix	Gradient descent

3   Problem Statement: Block Modeling for Structural Equivalence

The assumption underlying block modeling is that every vertex plays a role in the network, and the ties that this vertex will have with other vertices depend on their respective role. Vertices playing an equivalent role are grouped in clusters, and the structure of the graph is summarized with the graph of connections between the different clusters (the *image graph*).

Different definitions of equivalence between vertices have been proposed in the block modeling literature. The one most commonly used, called "structural equivalence", dictates that two vertices are equivalent if they are connected to exactly the same other vertices in the network [22]. Formally, given a graph $G = (V, E)$, vertices $u, v \in V$ are structurally equivalent $u \equiv v$ if and only if $\forall x \in V : (u, x) \in E \iff (v, x) \in E \land (x, u) \in E \iff (x, v) \in E$.

For example, consider the digraph, along with its adjacency matrix, in Fig. 1. Vertices 1 and 2 are structurally equivalent, since they are both connected to vertices 3 and 4 and nothing else. The equivalence classes according to $\equiv$ define a partition of the vertices into to three clusters, $V_1 = \{1, 2\}, V_2 = \{3, 4\}$ and $V_3 = \{5\}$. Observe that in the adjacency matrix, the rows and columns of equivalent vertices are identical. This gives rise to *blocks* in the matrix, delimited by lines in

Fig. 1. In this example, the vertices of the same block are numbered sequentially, but in practice the rows and columns have to be reordered to show the blocks in the matrix. Structural equivalence dictates that blocks be either *Null blocks* (containing only 0) or *Complete Blocks* (containing only 1) [4].

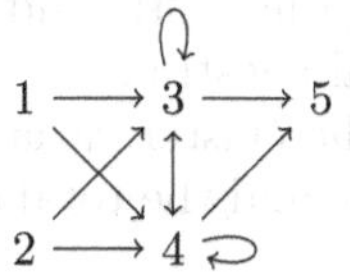

$$X = \left(\begin{array}{cc|cc|c} \cdot & \cdot & 1 & 1 & \cdot \\ \cdot & \cdot & 1 & 1 & \cdot \\ \hline \cdot & \cdot & 1 & 1 & 1 \\ \cdot & \cdot & 1 & 1 & 1 \\ \hline \cdot & \cdot & \cdot & \cdot & \cdot \end{array}\right)$$

Fig. 1. Small digraph along with its adjacency matrix X. According to structural equivalence, $1 \equiv 2$ and $3 \equiv 4$.

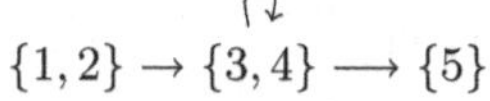

$$M = \begin{pmatrix} \cdot & 1 & \cdot \\ \cdot & 1 & 1 \\ \cdot & \cdot & \cdot \end{pmatrix}$$

Fig. 2. Image graph of Fig. 1 along with its image matrix M.

The image graph is shown in Fig. 2. It has one vertex for each cluster, and the edges are given by the blocks in X. We can reconstruct the adjacency matrix X from the image matrix M in the following way. Let F be a 5×3 matrix such that $F_{ik} = 1$ if vertex i is in cluster k, otherwise $F_{ik} = 0$. Then we have $X = FMF^T$, where F^T is F transposed.

Structural equivalence is a very strong requirement. In order to deal with the noise in real-world data, we will look for an F and M which approximate the base graph X with the least error, for a fixed model size k. We define the error (the *cost* of the solution) as the number of edges which must be added or deleted from our graph in order to fit the model perfectly: $||X - FMF^T|| = \sum_{i=1}^{n} \sum_{j=1}^{n} |X_{ij} - (FMF^T)_{ij}|$. Formally, the minimization problem BLOCKMODEL(X, k) that we are solving in the absence of other constraints is as follows: given $X \in \mathbb{B}^{n \times n}$ a binary adjacency matrix and number of clusters k, find F and M such that

$$\min_{F,M} ||X - FMF^T|| \tag{1}$$

$$\text{s.t.} \quad \sum_{c=1}^{k} F_{ic} = 1 \quad \forall i \in \{1..n\} \tag{2}$$

$$\sum_{i=1}^{n} F_{ic} \geq 1 \quad \forall c \in \{1..k\}. \tag{3}$$

$F \in \mathbb{B}^{n \times k}$ is the indicator matrix and $M \in \mathbb{B}^{k \times k}$ is the image matrix of our model. Equation (2) ensures that vertices are assigned to one cluster only, while Eq. (3) ensures that there are no empty clusters. To this model, additional constraints can be added.

4   CP Model for Block Modeling with a Global Constraint

The main contributions of this paper are (1) a CP model for the block modeling problem, (2) a global constraint used in this model, that we call `blockModelCost`, and (3) a tailored filtering algorithm for this constraint. We first describe the CP model, with its variables and constraints. Afterwards, we present the global constraint and its filtering algorithm. Finally, we present a heuristic and symmetry breaking scheme for the CP solver based on the global constraint.

There are four groups of variables in our model: the cluster variables C, the image matrix variables M, the block cost variables cost and the total cost of our solution totalCost. They are presented in this table:

Variable	Domain	Interpretation
C_i	$\{1..k\}$	$C_i = c$ if vertex i is in cluster c
M_{cd}	$\{0,1\}$	$M_{cd} = 0$ if the submatrix of rows in cluster c and columns in cluster d is a Null block, and $M_{cd} = 1$ if it is a Complete Block
cost_{cd}	$\{0..n^2\}$	Number of entries in the submatrix c, d which do not match M_{cd}
totalCost	$\{0..n^2\}$	The cost of the solution $\|X - FMF^T\|$

The variables are subject to the following constraints:

- `sum(cost, totalCost)`, which ensures that the total cost of the solution and the individual cost of every block stays consistent: $\sum_{c=1}^{k} \sum_{d=1}^{k} \text{cost}_{cd} = \text{totalCost}$. This constraint is already implemented in most CP systems.
- `atLeast`$(1, C, c), \forall c \in \{1..k\}$, which ensures that every value between 1 and k appears at least once in C—i.e. there are no empty clusters, as per Eq. 3.
- `blockModelCost`$(X, M, C, \text{cost}, \text{totalCost})$. This is the global constraint that we add to the solver, which filters the values of the different variables along the search. It ensures $\sum_{i=1}^{n} \sum_{j=1}^{n} |X_{ij} - M_{C_i C_j}| \leq \text{totalCost}$ and $\sum_{i=1}^{n} \sum_{j=1}^{n} (C_i = c) \cdot (C_j = d) \cdot |X_{ij} - M_{cd}| \leq \text{cost}_{cd} \ \forall c, d$.

Note that the constraint of Eq. (2) (vertices can only be in one cluster) is implicitly modeled by the variable C. Since in the final solution, all variables must be bound to a single value, no vertex is bound to more than one cluster.

The model can be extended with any set of existing additional constraints present in CP systems, on any of the variables, such as cardinality constraints or connectivity constraints.

5   A Global Constraint for Block Modeling

A global constraint [20] is a constraint that captures a relationship between a number of variables. Typically, a global constraint, as this one, can also be decomposed into several simpler constraints but considering it globally permits filtering more impossible values and is often also faster [7]. Global constraints

Table 2. Adjacency matrix with its columns and rows reordered to show the partial assignment of vertices into clusters.

Cluster	1				2	3		Unbound	
Vertex	1	2	7	9	5	3	4	6	8
1	·	·	·	1	·	1	1	1	·
2	·	·	·	·	1	1	1	1	·
7	·	1	·	·	·	1	·	·	·
9	·	·	·	·	·	·	1	1	·
5	·	·	·	·	1	1	1	1	1
3	1	1	·	·	·	·	·	·	·
4	·	1	1	1	·	·	1	·	·
6	1	1	·	1	·	1	·	·	·
8	·	·	·	·	1	1	1	1	·

are thus key to prune the search tree and solve complex problems efficiently with CP. The filtering algorithm of the global constraint is called every time the domain of one variable in its scope changes. This filtering does not need to be complete, although it needs to be able to check the feasibility when all the variables are bound and it must also guarantee that no valid values are removed.

In this subsection, we present `blockModelCost`, a global constraint for block modeling. We first give a concrete example to illustrate the filtering strategies. Then, we describe the pseudo code for the propagation method.

5.1   Illustration of the Different Filtering Strategies

To illustrate the filtering algorithm, let's consider the following partial assignment: $C = (\{1\}, \{1\}, \{3\}, \{3\}, \{2\}, \{1, 2, 3\}, \{1\}, \{1, 2, 3\}, \{1\})$, $\forall c, d : M_{cd} \in \{0, 1\}$, $cost_{cd} \in \{0..13\}$ and $totalCost = \{0..13\}$. In Table 2, we show the adjacency matrix X for this example with its rows and columns reordered to show the current partial assignment.

Filtering $cost_{cd}$. If we look at the submatrix defined by what is already assigned to the block $(1,1)$—i.e. the northwestern block in Table 2—we see that it contains fourteen 0s and two 1s. If $M_{1,1} = 0$, the block should be filled with 0s so the its cost will be at least 2, because of the two 1s. It could be more than 2 if other vertices are bound to cluster 1, but it can never be less than 2. If $M_{1,1} = 1$, the cost will be at least 14, because of the fourteen 0s. Thus, we can increase the lower bound of the domain of $cost_{1,1}$ to 2. Doing this for all blocks, we get

$$cost = \begin{pmatrix} \{2..13\} & \{1..13\} & \{2..13\} \\ \{0..13\} & \{0..13\} & \{0..13\} \\ \{3..13\} & \{0..13\} & \{1..13\} \end{pmatrix}$$

After propagating the sum constraint, we get totalCost $\in \{9..13\}$ and

$$\text{cost} = \begin{pmatrix} \{2..6\} \ \{1..5\} \ \{2..6\} \\ \{0..4\} \ \{0..4\} \ \{0..4\} \\ \{3..7\} \ \{0..4\} \ \{1..5\} \end{pmatrix}$$

Filtering M_{cd}. As observed previously, setting $M_{1,1}$ to 1 would bring the minimum cost of the block to 14. However, the value 14 is not in the domain of $\text{cost}_{1,1}$, so we can filter the value 1 from $M_{1,1}$, in effect binding it to $M_{1,1} = 0$.

Filtering C_i. If we were to assign vertex 6 to cluster 1, it would add six 1s to the $(1,1)$ block—three from the partial column representing edges from vertices in cluster 1 to vertex 6, and three more from the partial row representing edges from vertex 6 to vertices in cluster 1. Remember that $M_{1,1} = 0$, so each one would increase the cost of the block. The resulting cost (8) would exceed the maximum allowed value for $\text{cost}_{1,1}$, so we can remove 1 from the domain of C_6.

Tightening the Lower Bound on totalCost. In what has been described so far, the lower bound of totalCost is only the sum of the lower bounds of the individual cost variables. These take into account only the submatrix defined by the vertices already assigned to a specific cluster. We can improve the bound by also taking into account the unbound vertices (vertices 6 and 8 in our example). In Table 2, consider the horizontal rectangle in bold at row 6. It corresponds to the edges going from vertex 6 to vertices in cluster 1. Since $C_6 \in \{2,3\}$, we do not know yet in which block it will be, but those 4 values will stay together in the final assignment. If the 4 values end up in a Null block, their cost will be 3, and if they end up in a Complete block, their cost will be one, so we can at least increase the lower bound on totalCost by one. The same can be done for all other rectangles in the "unbound" part of Table 2 except for the southeastern corner (edges between unbound vertices). If we add all of these contributions, we get totalCost $\in \{12..13\}$.

5.2   Filtering Algorithm

The pseudocode for our propagation method is shown in Algorithm 1. In order to filter the domains of our CP variables efficiently, the number of zeroes and ones in the different "blocks" of our reordered matrix are computed. For efficiency reasons, those counters are stored on a trail [27], or more exactly inside reversible integers that are restored on backtracking. This permits an incremental update based on the changes since the last call to the filtering algorithm without having to worry about the restoration at backtracking. Specifically, these values are stored as reversible integers:

- nbOBlock, a $k \times k$ array reflecting the number of zeroes already assigned to each block: $\text{nbOBlock}_{cd} = \#\{X_{ij} \mid C_i = c, C_j = d, X_{ij} = 0\}$,
- nbORow, a $n \times b$ array where for all unbound vertices i:
  $\text{nbORow}_{ic} = \#\{X_{ij} \mid C_j = c, X_{ij} = 0\}$,

– nb0Col, a $b \times n$ array where for all unbound vertices i:
  $\text{nb0Col}_{ci} = \#\{X_{ji} \mid C_j = c, X_{ji} = 0\}$,

as well as their equivalent variables for the number of ones: nb1Block, nb1Row and nb1Row. The set of unbound vertices $\text{unboundVertices} = \{i \in \{1..n\} \mid 1 < |\text{dom}(C_i)|\}$ is maintained in a reversible sparse set [26].

A lower bound on the cost of the block c, d is:

$$\underline{cost}(c, d) = \begin{cases} \text{nb0Block}_{cd} & \text{if } \mathsf{M}_{cd} = \{1\} \\ \text{nb1Block}_{cd} & \text{if } \mathsf{M}_{cd} = \{0\} \\ \min(\text{nb0Block}_{cd}, \text{nb1Block}_{cd}) & \text{otherwise.} \end{cases}$$

We obtain a better bound by also maintaining $\underline{rowcost}$, using the method described in the earlier paragraph "Tightening the lower bound on totalCost", as follows:

$$\underline{rowcost}(c, i) = \begin{cases} \text{nb0Row}_{ic} & \text{if } \forall d : \mathsf{M}_{dc} = \{1\} \\ \text{nb1Row}_{ic} & \text{if } \forall d : \mathsf{M}_{dc} = \{0\} \\ \min(\text{nb0Row}_{ic}, \text{nb1Row}_{ic}) & \text{otherwise.} \end{cases}$$

Similarly we maintain $\underline{colcost}(c, i)$, defined equivalently from nb0Col and nb1Col. They put a lower bound on the cost incurred by rows and columns of vertices which have not been bound yet.

Finally, we can also calculate a lower bound on the added cost for block (c, d) if we put vertex i in cluster x, $\delta_{i \mapsto x}(c, d) = \underline{cost}_{i \mapsto x}(c, d) - \underline{cost}(c, d)$ where $\underline{cost}_{i \mapsto x}(c, d)$ is the value of $\underline{cost}(c, d)$ if vertex i is assigned to cluster x.

The first step in our algorithm is to process all the vertices that have been bound to a cluster since the propagation method was last called, and update the constraint's variables. Then, we filter the CP variables cost_{cd} with the new lower bounds $\underline{cost}(c, d)$, and filter totalCost further with $\underline{rowcost}$ and $\underline{colcost}$. Then we filter the values of M_{cd} by removing the values which lead to a cost higher than $\max(\text{cost}_{cd})$. Finally, we filter the values of C_i with the lower bounds of $\delta_{i \mapsto x}(c, d)$. This order of steps was chosen as we found it to perform well in practice.

5.3   Theoretical Properties of the Algorithm

The algorithm was designed with practical performance for block modeling in mind. For example, we only implemented filtering on the lower bound of the cost variables, since we are trying to minimize them. Of course, filtering their upper bound would also be possible, but was not implemented since it does not help solve the block modeling problem. Similarly, our algorithm does not have some of the theoretical guarantees ensured by well-known other global constraints, such as bound or arc consistency and idempotency. These would be very complex to implement for this problem, and would not improve the performance. We will nonetheless discuss them in this section.

Soundness, Completeness and Idempotency: The filtering is sound (any pruned value is inconsistent with respect to the objective) but it does not achieve any classical notion of consistency. Our focus was on practical performance rather than theoretical guarantees. We added fine-grained filtering only when it improved efficiency. Consequently, propagation is also not idempotent. For example, at the last step of Algorithm 1 (filter C) if a variable C_i is bound, we do not update the local counters and miss all further filtering arising from that if the propagation is not called again. A while loop in the propagator would

---

Algorithm 1. Propagation of our global constraint.

ΔC is a list of all the variables C_i which have been bound since the last propagation of this constraint.

1: /* *update local counters* */
2: **for all** $C_i = \{c\} \in \Delta C$ **do**
3:     `unboundVertices` $\leftarrow$ `unboundVertices` $\setminus \{i\}$
4:     **for all** j in `unboundVertices` **do**
5:         $\text{nb1Col}_{cj} \mathrel{+}= X_{ij}; \text{nb0Col}_{cj} \mathrel{+}= (1 - X_{ij})$
6:         $\text{nb1Row}_{jc} \mathrel{+}= X_{ji}; \text{nb0Row}_{jc} \mathrel{+}= (1 - X_{ji})$
7:     **end for**
8:     **for all** $d = 1$ to k **do**
9:         $\text{nb0Block}_{cd} \mathrel{+}= \text{nb0Row}_{id}; \text{nb1Block}_{cd} \mathrel{+}= \text{nb1Row}_{id}$
10:         $\text{nb0Block}_{dc} \mathrel{+}= \text{nb0Col}_{di}; \text{nb1Block}_{dc} \mathrel{+}= \text{nb1Col}_{di}$
11:     **end for**
12:     $\text{nb0Block}_{cc} \mathrel{+}= (1 - X_{ii}); \text{nb1Block}_{cc} \mathrel{+}= X_{ii}$
13: **end for**
14: /* *filter* cost *and* totalCost */
15: minCost $\leftarrow 0$
16: **for all** $c, d \in \{1..k\} \times \{1..k\}$ **do**
17:     update min of cost_{cd} to $\underline{cost}(c, d)$.
18:     minCost $\mathrel{+}= \underline{cost}(c, d)$.
19: **end for**
20: **for all** unbound vertex i, $c = 0$ to k **do**
21:     minCost $\mathrel{+}= \underline{colcost}(c, i) + \underline{rowcost}(c, i)$
22: **end for**
23: update min of totalCost to minCost
24: /* *filter* M */
25: **for all** $c, d \in \{1..k\} \times \{1..k\}$ if $M_{cd} = \{0, 1\}$ **do**
26:     **if** $\text{nb0Block}_{cd} > \max(\text{cost}_{cd})$ **then** $M_{cd} \leftarrow M_{cd} \setminus \{0\}$ **end if**
27:     **if** $\text{nb1Block}_{cd} > \max(\text{cost}_{cd})$ **then** $M_{cd} \leftarrow M_{cd} \setminus \{1\}$ **end if**
28: **end for**
29: /* *filter* C */
30: **for all** $i \in$ `unboundVertices`, $c \in C_i, d = 1$ to k **do**
31:     **if** $\underline{cost}(c, d) + \underline{\delta_{i \mapsto c}}(c, d) > \max(\text{cost}_{cd})$ **or** $\underline{cost}(d, c) + \underline{\delta_{i \mapsto c}}(d, c) > \max(\text{cost}_{dc})$
    then
32:         $C_i \leftarrow C_i \setminus \{c\}$
33:     **end if**
34: **end for**

solve this, but we found that such a loop reduces performance: intermediate propagation by other, lighter constraints helps in practice.

Time Complexity for One Execution: For practical block modeling applications, the complexity of one execution of Algorithm 1 is linear in terms of the number of unbound vertices. Let us define three variables: δ_C, the number of variables in C bound since the last call, u_C, the number of unbound variables in C, and k the number of clusters. The different steps of the algorithm have these complexities:

Step 1: updating local counters: $\mathcal{O}(\delta_C(u_C + k))$
Step 2: filtering cost and totalCost: $\mathcal{O}(k^2 + u_C k)$
Step 3: filtering M: $\mathcal{O}(k^2)$
Step 4: filtering C: $\mathcal{O}(u_C k^2)$

In total for one execution of the filtering algorithm this yields $\mathcal{O}(\delta_C(u_C + k) + k^2 + u_C k + k^2 + u_C k^2) = \mathcal{O}(\delta_C u_C + \delta_C k + u_C k^2)$. The value δ_C is assumed to be small between consecutive calls of the filtering algorithm, and the number of clusters k is typically small (10 at most) in block modeling applications, so we consider the complexity to be $\mathcal{O}(u_C)$.

Time Complexity Along a Branch: We will now consider the time complexity to reach the first solution from the root of the search tree. We consider the worst case, i.e. there is no additional constraint on the variables, and no constraint on the cost of the solution. We start from the root—all C and M variables unbound—and assign a value to the variables one by one.

Let's assign first the n variables in C, then the k^2 variables of M. For the first n variables, $\delta_C = 1$ and u_C decreases from $n - 1$ to 0, giving a complexity at each search node of $\mathcal{O}(nk^2)$. For the last k^2 nodes of the search tree, $\delta_C = 0 = u_C$, so the complexity is $\mathcal{O}(k^2)$. This gives a complexity along the branch of $\mathcal{O}(n^2k^2 + k^4)$.

6   Search Procedure for Block Modeling

In constraint programming, the formulation of the problem is kept separate from the search procedure. The search procedure is a branch and bound depth-first-search. Two important components of a search procedure are the variable and value ordering heuristics. These should permit discovering rapidly good incumbent solutions in order to prune the search tree. Since the problem also exhibits value symmetries, we use a dynamic symmetry breaking scheme during the search. When the search space becomes too large, and there is no hope to explore completely the search tree, LNS (Large Neighborhood Search) [28] can be used on top of CP to diversify the search and discover good solutions rapidly.

Value and Variable Ordering Heuristic. When arriving at a branching point in the search, the CP solver must decide which variable to branch on and what value to try first. These decisions are called *variable ordering* and *value ordering*. Selecting the right ordering for the problem can significantly improve the efficiency of the solver.

For the CP model presented here, there are two sets of variables we can branch on (C and M). Since the `blockModelCost` constraint filters mostly based on the vertices which have been bound, it is better to branch on those before branching on M variables. The ordering of the C variables can further be refined with modern first-fail learning heuristics [17,19,23]. A good value heuristic for the clusters can also be constructed from our global constraint. We calculate $\delta_{i \mapsto x}(c, d)$, a lower bound on the added cost of assigning vertex i to cluster x, so a good heuristic is to branch first on $C_i = \arg\min_x \sum_{c,d} \underline{\delta_{i \mapsto x}}(c, d)$, i.e., branch first on the value for which we expect the least increase in cost. Similarly for M, we branch first on $M_{cd} = 0$ if $\texttt{nb1Block}_{cd} < \texttt{nb0Block}_{cd}$, and $M_{cd} = 1$ otherwise.

Symmetry Breaking for the Block Modeling Problem. Symmetry breaking permits to drastically reduce the search. Symmetries can generally be avoided by adding constraints to the model. Unfortunately, this approach suffers from a bad interaction with the search as good solutions that were discovered early may become unfeasible because of the symmetry breaking constraints [29]. Therefore, a dynamic symmetry breaking during search strategy is generally more efficient. At every stage of the search, all-but one child nodes leading to symmetrical states are discarded.

The search space for this CP formulation of the block modeling problem has a number of symmetries. Firstly, it is clear that as long as the clusters stay the same, their labels can be changed—i.e. for any permutation $\sigma : \{0..k\} \to \{0..k\}$ and any state $S = (C, M)$, the permutated state $\sigma(S) = (\sigma(C_*), M_{\sigma(*)\sigma(*)})$ is symmetrical to S. If σ' is an automorphism of the graph X, then $S' = (C_{\sigma'(*)}, M_{**})$ is symmetrical to S. Finally, if σ'' is an automorphism of the graph M, then $S'' = (C_*, M_{\sigma''(*)\sigma''(*)})$ has the same error as S.

In our CP model, we are only concerned with the first kind of symmetries (permutations of the cluster labels); those are easier to break. The dynamic symmetry-breaking scheme is: when branching on a C_i variable, the solver explores branches $C_i = 1, C_i = 2, \ldots, C_i = m + 1$ where m is the largest value bound to a C variable $m = \max\{v \mid \exists i : C_i = \{v\}\}$.

Breaking the symmetries on the graph automorphisms of X and M is much more complicated and has not been considered for this paper. It is nonetheless an interesting direction for further work on this problem. For a related treatment of symmetry breaking of graph automorphisms, see [32].

7   Experiments

7.1   Comparison with MIP Model

The block modeling problem is often approximated using heuristic search. However, an approach to find the optimal solution is proposed in [9]. It builds on the work of [8], which defines a MIP model to find the optimal partition given a fixed image matrix M. We expand this approach to find the optimal solution by generating a minimal, representative set of image matrices of size k and running the MIP solver for each matrix in this set.

In this section, we compare the performance of the CP approach with this MIP approach. As both give exact solutions, the quality of the solutions are identical, and we only need to compare the running time. In order to evaluate the performance of our global constraint, we wrote three CP models. The first is used as a baseline. It follows the mathematical formulation of the problem, and uses our symmetry-breaking scheme. The second uses our global constraint for filtering with the same search procedure. The third uses our global constraint with the value ordering heuristic. The MIP model and the 3 CP approaches are compared on four small well-studied social networks, published and analyzed in depth in [12, Chapters 2, 6], namely: (a) the Transatlantic Industries little league baseball team network, [14], (b) the Sharpstone little league baseball team network, [14], (c) the political actor network (PA), and [11] (d) the Kansas search and air rescue (SAR) network [13].

The CP models were written and solved in OscaR [24]. The MIP model was written and solved in Java using Gurobi [18]. All experiments were run on a computer with Xeon Platinum 8160 24c/48t HyperThread processors. The results are shown in Table 3.

We clearly observe that the MIP approach does not scale and is inapplicable for non-trivial sizes. The effect of our global constraint and our value heuristic are also evident, making the search orders of magnitude faster.

7.2 Comparison with Local Search

The global constraint can also be used in local search by doing Large Neighborhood Search [28]. In this subsection, we compare the performance of the LNS approach with a local search algorithm for block modeling bundled in the popular graph processing software Pajek[1].

We generated synthetic graphs with 50, 100, 150 and 200 vertices—the classical block modeling algorithm included in Pajek [5] only supports graphs of less than 256 vertices—with a fixed block model structure of 5 clusters. We added 40% of noise to the data, then compared the evolution of the quality of the solution with time for both methods. The results are shown in Fig. 3. For all instances over 50 vertices, the LNS method outperformed Pajek's local search.

7.3 Scalability

We now show the scalability of the complete search and LNS method on larger instances. We once again generated synthetic graphs of different sizes n with a known block model structure and 20% of noise. In the first plot of Fig. 4, we report the runtime until proving optimality for different sizes n and number of

---

[1] http://mrvar.fdv.uni-lj.si/pajek/.

Table 3. Run time of the MIP approach compared to a baseline CP approach (CP(bsl)), a CP approach with our global constraint (CP(our)), and our constraint + our value heuristic (+heuris.) for different number of clusters k. "−" indicates a timeout after 2 h.

Dataset	n	k	CPU time (s)			
			MIP	CP(bsl)	CP(our)	+heuris
Transatlantic	13	2	1.73	0.80	0.45	0.28
		3	142.25	21.15	0.88	0.79
		4	−	386.20	2.94	2.07
Sharpstone	13	2	1.24	0.50	0.44	0.19
		3	62.46	13.57	1.17	0.85
		4	2952.13	221.41	2.78	1.82
		5	−	1102.68	2.31	1.30
Political actor	14	2	2.14	1.13	0.62	0.31
		3	155.90	60.15	1.32	0.89
		4	2178.42	1936.43	2.68	2.20
		5	−	−	2.93	2.25
Search and rescue	20	2	13.31	22.04	0.85	0.48
		3	−	−	6.01	5.18

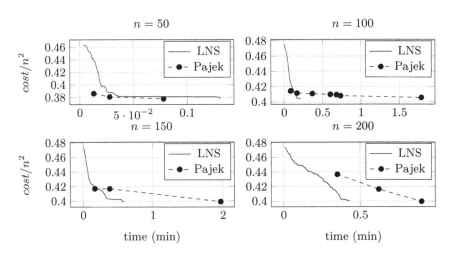

Fig. 3. Comparison of LNS search with the local search bundled in Pajek for synthetic dataset with 40% of noise. Each graph shows a different instance of the problem, for $n = 50$ to $n = 200$ by increments of 50.

clusters k. In the second plot of Fig. 4, we plot the convergence of Large Neighborhood Search over 10 min, with restarts every 1000 failed states and relaxation of 5% of the variables. We see that, while proving optimality is still prohibitively hard for large graphs, the LNS search converges quickly on a solution of optimal cost, even with thousands of vertices. Note that all of these graphs are too large for Pajek's method, but were solved by our LNS search in a handful of minutes.

7.4   Beyond Traditional Block Modeling

A real strength of a constraint programming formulation is the ability to add complex constraints on the clusters or the image graph, to combine multiple instances of the same constraint, and to optimize any of the variables. As an illustration, we explore the use of block modeling on migration data in Europe. In the first illustration, Fig. 5, we add the constraint that the clusters must be connected on the map—i.e. one can travel between any two countries of a cluster without leaving the cluster. This connectivity constraint is very complex to model in MIP but is an existing building block in CP [6, 10, 25]. In the second illustration, Fig. 6, we study a problem that involves multiple instances of our global constraint: we take the migration matrix at 5 different points in time. We build a blockmodel for each year, with the constraint that the clusters are the same in all models. We have five block models, so we minimize the sum of their costs. This is similar to the non-negative RESCAL setting [21].

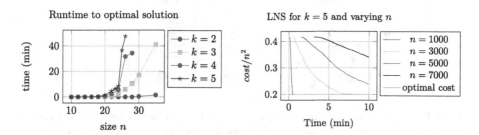

Fig. 4. Scalability on graphs with known block model and 20% of noise. The first graph shows runtime until the solution is proven optimal. The second shows the convergence of the solution with LNS.

The migration graphs were built from an open dataset provided by the World Bank [1]. An edge $X_{ab} = 1$ indicates that the number of migrants born in a living in b is more than 0.01% of the population of a. The dataset was limited to countries in continental Europe, excluding islands for the first illustration because of the connectivity constraint. The models were found after a Large Neighborhood Search of 10 min, with restarts after 1000 failed states relaxing 5% of the variables.

In Fig. 5, we clearly see the ex-Soviet block appear in cluster 4, with mostly internal migration and not much migration to Western Europe. Germany and

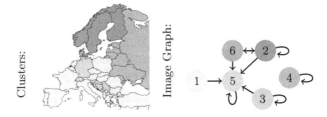

Fig. 5. A block model of migrant stocks in continental Europe in 2015, with geographically connected clusters.

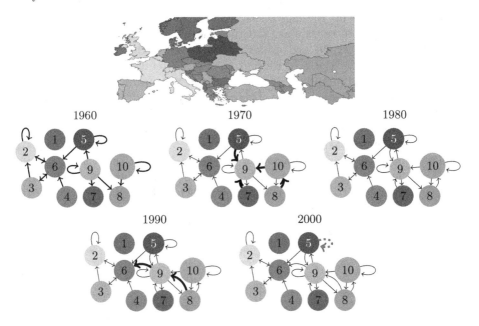

Fig. 6. RESCAL model for the evolution of migrant stocks in Europe. The edges which appeared in a decade are rendered in thick stroke, and those which disappeared are in dotted red stroke.

Switzerland appear as a core destination for migrants from most European countries. Denmark is the only member of its cluster, but it would have been in the same cluster as Fennoscandia if it did not violate the connectivity constraint. In Fig. 6, we observe for example the migration of people from Russia to Germany in the nineties (thick arrow between 9 and 6), which we can probably link to the fall of the Iron Curtain.

8   Conclusion and Further Work

We have introduced a CP approach to the block modeling problem, using a dedicated global constraint. It has the advantage of being able to easily incorporate any combination of additional constraints, contrary to previous works. Our experiments show that our approach is orders of magnitude faster than competing solutions to find optimal block models. Our CP formulation can also be used for heuristic search with Large Neighborhood Search.

This work could be further expanded with an equivalent global constraint for regular equivalence or generalized blockmodeling [12]. The search could be accelerated by breaking symmetries on the automorphisms of X and M and considering more advanced variable ordering schemes.

References

1. The world bank: Migration and remittances data. http://www.worldbank.org/en/topic/migrationremittancesdiasporaissues/brief/migration-remittances-dat
2. Bai, Z., Qian, B., Davidson, I.: Discovering models from structural and behavioral brain imaging data. In: Proceedings of the 24th ACM SIGKDD International Conference on Knowledge Discovery & Data Mining, pp. 1128–1137. ACM (2018)
3. Bai, Z., Walker, P., Tschiffely, A., Wang, F., Davidson, I.: Unsupervised network discovery for brain imaging data. In: Proceedings of the 23rd ACM SIGKDD International Conference on Knowledge Discovery and Data Mining, pp. 55–64. ACM (2017)
4. Batagelj, V.: Notes on blockmodeling. Soc. Netw. **19**(2), 143–155 (1997). https://doi.org/10.1016/S0378-8733(96)00297-3
5. Batagelj, V., Mrvar, A., Ferligoj, A., Doreian, P.: Generalized blockmodeling with pajek. Metodoloski zvezki **1**(2), 455 (2004)
6. Bessiere, C., Hebrard, E., Katsirelos, G., Walsh, T.: Reasoning about connectivity constraints. In: Twenty-Fourth International Joint Conference on Artificial Intelligence (2015)
7. Bessière, C., Van Hentenryck, P.: To be or not to be... a global constraint. In: Rossi, F. (ed.) CP 2003. LNCS, vol. 2833, pp. 789–794. Springer, Heidelberg (2003). https://doi.org/10.1007/978-3-540-45193-8_54
8. Brusco, M.J., Steinley, D.: Integer programs for one- and two-mode blockmodeling based on prespecified image matrices for structural and regular equivalence. J. Math. Psychol. **53**(6), 577–585 (2009). https://doi.org/10.1016/j.jmp.2009.08.003
9. Dabkowski, M., Fan, N., Breiger, R.: Exploratory blockmodeling for one-mode, unsigned, deterministic networks using integer programming and structural equivalence. Soc. Netw. **47**, 93–106 (2016). https://doi.org/10.1016/j.socnet.2016.05.005
10. Dooms, G.: The CP(Graph) computation domain in constraint programming. Ph.D. thesis, UCL - Université Catholique de Louvain (2006). https://dial.uclouvain.be/pr/boreal/object/boreal:107275
11. Doreian, P., Albert, L.H.: Partitioning political actor networks: some quantitative tools for analyzing qualitative networks. J. Quant. Anthropol. **1**, 279–291 (1989). https://www.ifip.com/PartitioningPoliticalActor.html
12. Doreian, P., Batagelj, V., Ferligoj, A.: Generalized Blockmodeling. Cambridge University Press, Cambridge (2005)
13. Drabek, T., Tamminga, H., Kilijanek, T., Adams, C.: Managing multi-organizational emergency responses. University of Colorado, Institute of Behavioral Science, BoulDder (1981)

14. Fine, G.A.: With the Boys: Little League Baseball and Preadolescent Culture. University of Chicago Press, Chicago (1987). google-Books-ID: 2qWgZPuNjEYC
15. Ganji, M., et al.: Image Constrained blockmodelling: a constraint programming approach. In: Proceedings of the 2018 SIAM International Conference on Data Mining, pp. 19–27. Proceedings, Society for Industrial and Applied Mathematics, May 2018. DOI: https://doi.org/10.1137/1.9781611975321.3
16. Ganji, M., et al.: Semi-supervised blockmodelling with pairwise guidance. In: Berlingerio, M., Bonchi, F., Gärtner, T., Hurley, N., Ifrim, G. (eds.) ECML PKDD 2018. LNCS (LNAI), vol. 11052, pp. 158–174. Springer, Cham (2019). https://doi.org/10.1007/978-3-030-10928-8_10
17. Gay, S., Hartert, R., Lecoutre, C., Schaus, P.: Conflict ordering search for scheduling problems. In: Pesant, G. (ed.) CP 2015. LNCS, vol. 9255, pp. 140–148. Springer, Cham (2015). https://doi.org/10.1007/978-3-319-23219-5_10
18. Gurobi Optimization, L.: Gurobi optimizer reference manual (2018). http://www.gurobi.com
19. Hebrard, E., Siala, M.: Explanation-based weighted degree. In: Integration of AI and OR Techniques in Constraint Programming - 14th International Conference, CPAIOR 2017, Padua, Italy, 5–8 June 2017, Proceedings, pp. 167–175 (2017). https://doi.org/10.1007/978-3-319-59776-8_13
20. van Hoeve, W.J., Katriel, I.: Global constraints. In: Rossi, F., Van Beek, P., Walsh, T. (eds.) Foundations of Artificial Intelligence, vol. 2, pp. 169–208. Elsevier, Amsterdam (2006)
21. Krompaß, D., Nickel, M., Jiang, X., Tresp, V.: Non-negative tensor factorization with RESCAL. In: Tensor Methods for Machine Learning, ECML Workshop (2013)
22. Lorrain, F., White, H.C.: Structural equivalence of individuals in social networks. J. Math. Sociol. 1(1), 49–80 (1971). https://doi.org/10.1080/0022250X.1971.9989788
23. Michel, L., Hentenryck, P.V.: Activity-based search for black-box constraint programming solvers. In: Proceedings of the 9th International Conference, CPAIOR 2012, Nantes, France, May 28–June 1 2012, pp. 228–243 (2012). https://doi.org/10.1007/978-3-642-29828-8_15
24. OscaR Team: OscaR: Scala in OR (2012). https://bitbucket.org/oscarlib/oscar
25. Prosser, P., Unsworth, C.: A connectivity constraint using bridges. In: ECAI, pp. 707–708 (2006)
26. le Clément de Saint-Marcq, V., Schaus, P., Solnon, C., Lecoutre, C.: Sparse-sets for domain implementation. In: The 19th International Conference on Principles and Practice of Constraint Programming, Uppsala, Sweden, 16–20 September 2013 (2013). https://dial.uclouvain.be/pr/boreal/object/boreal:135574
27. Schulte, C.: Comparing trailing and copying for constraint programming. In: Logic Programming: The 1999 International Conference, Las Cruces, New Mexico, USA, November 29–December 4, 1999, pp. 275–289 (1999)
28. Shaw, P.: Using constraint programming and local search methods to solve vehicle routing problems. In: Maher, M., Puget, J.-F. (eds.) CP 1998. LNCS, vol. 1520, pp. 417–431. Springer, Heidelberg (1998). https://doi.org/10.1007/3-540-49481-2_30
29. Van Hentenryck, P., Michel, L.: The steel mill slab design problem revisited. In: Perron, L., Trick, M.A. (eds.) CPAIOR 2008. LNCS, vol. 5015, pp. 377–381. Springer, Heidelberg (2008). https://doi.org/10.1007/978-3-540-68155-7_41
30. Wang, F., Li, T., Wang, X., Zhu, S., Ding, C.: Community discovery using nonnegative matrix factorization. Data Min. Knowl. Disc. 22(3), 493–521 (2011). https://doi.org/10.1007/s10618-010-0181-y

31. Wasserman, S., Faust, K.: Social Network Analysis: Methods and Applications, 4th edn. Cambridge University Press, Cambridge (2018). No. 8 in Structural analysis in the social sciences
32. Zampelli, S., Deville, Y., Dupont, P.: Symmetry breaking in subgraph pattern matching. In: Benhamou, F., Jussien, N., O'Sullivan, B. (eds.) Trends in Constraint Programming, pp. 203–218. Wiley, Hoboken (2006). (ISBN: 978-1-905209-97-2), https://dial.uclouvain.be/pr/boreal/object/boreal:85004

Reward Potentials for Planning with Learned Neural Network Transition Models

Buser Say[1,2(✉)], Scott Sanner[1,2], and Sylvie Thiébaux[3]

[1] University of Toronto, Toronto, Canada
{bsay,ssanner}@mie.utoronto.ca
[2] Vector Institute, Toronto, Canada
[3] Australian National University, Canberra, Australia
sylvie.thiebaux@anu.edu.au

Abstract. Optimal planning with respect to learned neural network (NN) models in continuous action and state spaces using mixed-integer linear programming (MILP) is a challenging task for branch-and-bound solvers due to the poor linear relaxation of the underlying MILP model. For a given set of features, potential heuristics provide an efficient framework for computing bounds on cost (reward) functions. In this paper, we model the problem of finding optimal potential bounds for learned NN models as a bilevel program, and solve it using a novel finite-time constraint generation algorithm. We then strengthen the linear relaxation of the underlying MILP model by introducing constraints to bound the reward function based on the precomputed reward potentials. Experimentally, we show that our algorithm efficiently computes reward potentials for learned NN models, and that the overhead of computing reward potentials is justified by the overall strengthening of the underlying MILP model for the task of planning over long horizons.

Keywords: Neural networks · Potential heuristics · Planning · Constraint generation

1 Introduction

Neural networks (NNs) have significantly improved the ability of autonomous systems to learn and make decisions for complex tasks such as image recognition [11], speech recognition [5], and natural language processing [4]. As a result of this success, formal methods based on representing the decision making problem with NNs as a mathematical programming model, such as verification of NNs [9,14] and optimal planning with respect to the learned NNs [18] have been studied.

In the area of learning and planning, Hybrid Deep MILP Planning [18] (HD-MILP-Plan) has introduced a two-stage data-driven framework that (i) learns

This work is done during author's visit to Australian National University.

T. Schiex and S. de Givry (Eds.): CP 2019, LNCS 11802, pp. 674–689, 2019.
https://doi.org/10.1007/978-3-030-30048-7_39

transitions models with continuous action and state spaces using NNs, and (ii) plans optimally with respect to the learned NNs using a mixed-integer linear programming (MILP) model. It has been experimentally shown that optimal planning with respect to the learned NNs [18] presents a challenging task for branch-and-bound (B&B) solvers [8] due to the poor linear relaxation of the underlying MILP model that has a large number of *big-M* constraints.

In this paper, we focus on the important problem of improving the efficiency of MILP models for decision making with learned NNs. In order to tackle this challenging problem, we build on potential heuristics [15,19], which provide an efficient framework for computing a lower bound on the cost of a given state as a function of its features. In this work, we describe the problem of finding optimal potential bounds for learned NN models with continuous inputs and outputs (i.e., continuous action and state spaces) as a bilevel program, and solve it using a novel finite-time constraint generation algorithm. Features of our linear potential heuristic are defined over the hidden units of the learned NN model, thus providing a rich and expressive candidate feature space. We use our constraint generation algorithm to compute the potential contribution (i.e., reward potential) of each hidden unit to the reward function of the HD-MILP-Plan problem. The precomputed reward potentials are then used to construct linear constraints that bound the reward function of HD-MILP-Plan, and provide a tighter linear relaxation for B&B optimization by exploring smaller number of nodes in the search tree. Experimentally, we show that our constraint generation algorithm efficiently computes reward potentials for learned NNs, and that the overhead computation is justified by the overall strengthening of the underlying MILP model for the task of planning over long horizons.

Overall this work bridges the gap between two seemingly distant literatures – research on planning heuristics for discrete spaces and decision making with learned NN models in continuous action and state spaces. Specifically, we show that data-driven NN models for planning can benefit from advances in heuristics and from their impact on the efficiency of search in B&B optimization.

2   Preliminaries

We review the HD-MILP-Plan framework for optimal planning [18] with learned NN models, potential heuristics [15] as well as bilevel programming [1].

2.1   Deterministic Factored Planning Problem Definition

A deterministic factored planning problem is a tuple $\Pi = \langle S, A, C, T, I, G, R \rangle$ where $S = \{s_1, \ldots, s_n\}$ and $A = \{a_1, \ldots, a_m\}$ are sets of state and action variables with continuous domains, $C : \mathbb{R}^{|S|} \times \mathbb{R}^{|A|} \to \{true, false\}$ is a function that returns true if action and state variables satisfy global constraints, $T : \mathbb{R}^{|S|} \times \mathbb{R}^{|A|} \to \mathbb{R}^{|S|}$ denotes the stationary transition function, and $R : \mathbb{R}^{|S|} \times \mathbb{R}^{|A|} \to \mathbb{R}$ is the reward function. Finally, $I : \mathbb{R}^{|S|} \to \{true, false\}$ represents the initial state constraints, and $G : \mathbb{R}^{|S|} \to \{true, false\}$ represents the goal constraints. For horizon H, a solution $\pi = \langle \bar{A}^1, \ldots, \bar{A}^H \rangle$ to problem Π (i.e. a plan for Π) is a value assignment to the action variables with values $\bar{A}^t = \langle \bar{a}_1^t, \ldots, \bar{a}_{|A|}^t \rangle \in \mathbb{R}^{|A|}$ for all time steps $t \in \{1, \ldots, H\}$ (and state variables with values $\bar{S}^t = \langle \bar{s}_1^t, \ldots, \bar{s}_{|S|}^t \rangle \in \mathbb{R}^{|S|}$ for all time steps $t \in \{1, \ldots, H+1\}$) such that $T(\langle \bar{s}_1^t, \ldots, \bar{s}_{|S|}^t, \bar{a}_1^t, \ldots, \bar{a}_{|A|}^t \rangle) = \bar{S}^{t+1}$ and $C(\langle \bar{s}_1^t, \ldots, \bar{s}_{|S|}^t, \bar{a}_1^t, \ldots, \bar{a}_{|A|}^t \rangle) = true$ for all time steps $t \in \{1, \ldots, H\}$, and the initial and goal state constraints are satisfied, i.e. $I(\bar{S}^1) = true$ and $G(\bar{S}^{H+1}) = true$, where $\bar{x}^t$ denotes the value of variable $x \in A \cup S$ at time step t. Similarly, an optimal solution to Π is a plan such that the total reward $\sum_{t=1}^{H} R(\langle \bar{s}_1^{t+1}, \ldots, \bar{s}_{|S|}^{t+1}, \bar{a}_1^t, \ldots, \bar{a}_{|A|}^t \rangle)$ is maximized. For notational simplicity, we denote the tuple of variables $\langle x_{d_1}, \ldots, x_{d_{|D|}} \rangle$ as $\langle x_d | d \in D \rangle$ given set D, and use the symbol $\frown$ for the concatenation of two tuples. Given the notations and the description of the planning problem, we next describe a data-driven planning framework using learned NNs.

2.2   Planning with Neural Network Learned Transition Models

Hybrid Deep MILP Planning [18] (HD-MILP-Plan) is a two-stage data-driven framework for learning and solving planning problems. Given samples of state transition data, the first stage of the HD-MILP-Plan process learns the transition function $\tilde{T}$ using a NN with Rectified Linear Units (ReLUs) [13] and linear activation units. In the second stage, the learned transition function $\tilde{T}$ is used to construct the learned planning problem $\tilde{\Pi} = \langle S, A, C, \tilde{T}, I, G, R \rangle$. As shown in Fig. 1, the learned transition function $\tilde{T}$ is sequentially chained over the horizon $t \in \{1, \ldots, H\}$, and compiled into a MILP. Next, we review the MILP compilation of HD-MILP-Plan.

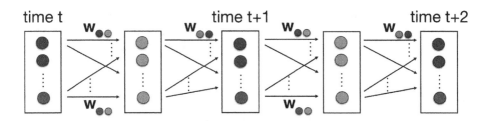

Fig. 1. Visualization of the learning and planning framework [18], where blue circles represent state variables S, red circles represent action variables A, gray circles represent ReLUs U and **w** represent the weights of a NN. During the learning stage, the weights **w** are learned from data. In the planning stage, the weights are fixed and the planner optimizes a given total (cumulative) reward function with respect to the set of free action variables A and state variables S. (Color figure online)

2.3   Mixed-Integer Linear Programming Compilation of HD-MILP-Plan

We begin with all notation necessary for HD-MILP-Plan.

Parameters

- U is the set of ReLUs in the neural network.
- O is the set of output units in the neural network.
- $w_{i,j}$ denotes the learned weight of the neural network between units i and j.
- $A(u)$ is the set of action variables connected as inputs to unit $u \in U \cup O$.
- $S(u)$ is the set of state variables connected as inputs to unit $u \in U \cup O$.
- $U(u)$ is the set of ReLUs connected as inputs to unit $u \in U \cup O$.
- $O(s)$ specifies the output unit that predicts the value of state variable $s \in S$.
- $B(u)$ is a constant representing the bias of unit $u \in U \cup O$.
- M is a large constant used in the big-M constraints.

Decision Variables

- $X_{a,t}$ is a decision variable with continuous domain denoting the value of action variable $a \in A$ at time step t.
- $Y_{s,t}$ is a decision variable with continuous domain denoting the value of state variable $s \in S$ at time step t.
- $P_{u,t}$ is a decision variable with continuous domain denoting the output of ReLU $u \in U$ at time step t.
- $P_{u,t}^b = 1$ if ReLU $u \in U$ is activated at time step t, 0 otherwise (i.e., $P_{u,t}^b$ is a Boolean decision variable).

MILP Compilation

$$\text{maximize } \sum_{t=1}^{H} R(\langle Y_{s,t+1} | s \in S \rangle ^\frown \langle X_{a,t} | a \in A \rangle) \tag{1}$$

subject to

$$I(\langle Y_{s,1} | s \in S \rangle) \tag{2}$$
$$C(\langle Y_{s,t} | s \in S \rangle ^\frown \langle X_{a,t} | a \in A \rangle) \tag{3}$$
$$G(\langle Y_{s,H+1} | s \in S \rangle) \tag{4}$$
$$P_{u,t} \leq M P_{u,t}^b \quad \forall u \in U \tag{5}$$
$$P_{u,t} \leq M(1 - P_{u,t}^b) + In(u,t) \quad \forall u \in U \tag{6}$$
$$P_{u,t} \geq In(u,t) \quad \forall u \in U \tag{7}$$
$$Y_{s,t+1} = In(u,t) \quad \forall u \in O(s), s \in S \tag{8}$$

for all time steps $t = 1, \ldots, H$ except for constraints (2)–(4). Expression $In(u,t)$ denotes the total weighted input of unit $u \in U \cup O$ at time step t, and is equivalent to $B(u) + \sum_{u' \in U(u)} w_{u',u} P_{u',t} + \sum_{s \in S(u)} w_{s,u} Y_{s,t} + \sum_{a \in A(u)} w_{a,u} X_{a,t}$.

In the above MILP, the objective function (1) maximizes the sum of rewards over a given horizon H. Constraints (2–4) ensure the initial state, global and goal state constraints are satisfied. Constraints (5–8) model the learned transition function $\tilde{T}$. Note that while constraints (5–7) are sufficient to encode the piecewise linear activation behaviour of ReLUs, the use of big-M constraints (5–6) can hinder the overall performance of the underlying B&B solvers that rely on the linear relaxation of the MILP. Therefore next, we turn to potential heuristics that will be used to strengthen the MILP compilation of HD-MILP-Plan.

2.4   Potential Heuristics

Potential heuristics [15,19] are a family of heuristics that map a set of features to their numerical potentials. In the context of cost-optimal classical planning, the heuristic value of a state is defined as the sum of potentials for all the features that are true in that state. Potential heuristics provide an efficient method for computing a lower bound on the cost of a given state.

In this paper, we introduce an alternative use of potential functions to tighten the linear relaxation of ReLU units in our HD-MILP-Plan compilation and improve the search efficiency of the underlying B&B solver. We define the features of the learned NN over its set of hidden units U (i.e., gray circles in Fig. 1), and compute the potential contribution (i.e., reward potential) of each hidden unit $u \in U$ to the reward function R for any time step t. These reward potentials are then used to introduce additional constraints on ReLU activations that help guide B&B search in HD-MILP-Plan. Specifically, we are interested in finding a set of reward potentials, denoted as v_u^{on} and v_u^{off} representing the activation (i.e., $P_{u,t}^b = 1$) and the deactivation (i.e., $P_{u,t}^b = 0$) of ReLUs $u \in U$, such that the relation $\sum_{u \in U} v_u^{on} P_{u,t}^b + v_u^{off}(1 - P_{u,t}^b) \geq R(\langle Y_{s,t+1}|s \in S\rangle \frown \langle X_{a,t}|a \in A\rangle)$ holds for all feasible values of $P_{u,t}^b$, $Y_{s,t+1}$ and $X_{a,t}$ at any time step t. Once values $\bar{v}_u^{on}$ and $\bar{v}_u^{off}$ are computed, we will add $\sum_{u \in U} \bar{v}_u^{on} P_{u,t}^b + \bar{v}_u^{off}(1 - P_{u,t}^b) \geq R(\langle Y_{s,t+1}|s \in S\rangle \frown \langle X_{a,t}|a \in A\rangle)$ as a linear constraint to strengthen HD-MILP-Plan. Next we describe bilevel programming that we use to model the problem of finding optimal reward potentials.

2.5   Bilevel Programming

Bilevel programming [1] is an optimization framework for modeling two-level asymmetrical decision making problems with a leader and a follower problem where the leader has complete knowledge of the follower, and the follower only observes the decisions of the leader to make an optimal decision. Therefore, the leader must incorporate the optimal decision of the follower to optimize its objective.

In this work, we use bilevel programming to compactly model the problem of finding the optimal reward potentials that has exponential number of constraints. In the bilevel programming description of the optimal reward potentials problem, the leader selects the optimal values $\bar{v}_u^{on}$ and $\bar{v}_u^{off}$ of reward potentials, and the follower selects the values of $P_{u,t}^b$, $Y_{s,t+1}$ and $X_{a,t}$ such that the expression $R(\langle Y_{s,t+1}|s \in S\rangle \frown \langle X_{a,t}|a \in A\rangle) - \sum_{u \in U} v_u^{on} P_{u,t}^b + v_u^{off}(1 - P_{u,t}^b)$ is maximized. That is, the follower tries to find values of $P_{u,t}^b$, $Y_{s,t+1}$ and $X_{a,t}$ that violate

the relation $\sum_{u \in U} v_u^{on} P_{u,t}^b + v_u^{off}(1 - P_{u,t}^b) \geq R(\langle Y_{s,t+1} | s \in S \rangle \frown \langle X_{a,t} | a \in A \rangle)$
as much as possible. Therefore the leader must select the values $\bar{v}_u^{on}$ and $\bar{v}_u^{off}$
of reward potentials by incorporating the optimal decision making model of the
follower. Next, we describe the reward potentials for learned NNs.

3   Reward Potentials for Learned Neural Networks

In this section, we present the optimal reward potentials problem and an efficient
constraint generation framework for finding reward potentials for learned NNs.

3.1   Optimal Reward Potentials Problem

The problem of finding the optimal reward potentials over a set of ReLUs U for
any time step t can be defined as the following bilevel optimization problem.

Leader Problem

$$\min_{v_u^{on}, v_u^{off}, Y_{s,t}, Y_{s,t+1}, X_{a,t}, P_{u,t}^b} \sum_{u \in U} v_u^{on} + v_u^{off} \tag{9}$$

subject to

$$\sum_{u \in U} v_u^{on} P_{u,t}^b + v_u^{off}(1 - P_{u,t}^b) \geq R(\langle Y_{s,t+1} | s \in S \rangle \frown \langle X_{a,t} | a \in A \rangle) \tag{10}$$

$$Y_{s,t}, Y_{s,t+1}, X_{a,t}, P_{u,t}^b \in \arg \text{Follower Problem}$$

Follower Problem

$$\max_{Y_{s,t}, Y_{s,t+1}, X_{a,t}, P_{u,t}^b} R(\langle Y_{s,t+1} | s \in S \rangle \frown \langle X_{a,t} | a \in A \rangle) - \sum_{u \in U} v_u^{on} P_{u,t}^b + v_u^{off}(1 - P_{u,t}^b)$$

$$\tag{11}$$

subject to
Constraints (3) and (5–8)
      In the above bilevel problem, the leader problem selects the values $\bar{v}_u^{on}$ and
$\bar{v}_u^{off}$ of the reward potentials such that their total sum is minimized (i.e., objective function (9)[1]), and their total weighted sum for all ReLU activations is an
upper bound to all values of the reward function R (i.e., constraint (10) and the
follower problem). Given the values $\bar{v}_u^{on}$ and $\bar{v}_u^{off}$ of the reward potentials, the
follower selects the values of decision variables $Y_{s,t}$, $Y_{s,t+1}$, $X_{a,t}$ and $P_{u,t}^b$ such
that the difference between the value of the reward function R and the sum of
reward potentials is maximized subject to constraints (3) and (5–8). Next, we
show the correctness of the optimal reward potentials problem as the bilevel program described by the leader (i.e., objective function (9) and constraint (10))
and the follower (i.e., objective function (11) and constraints (3) and (5–8))
problems.

---

[1] The objective function (9) is similar to the objective function of "All Syntactic
States" for potential heuristics used in classical planning [19].

Theorem 1 (Correctness of The Optimal Reward Potentials Problem). *Given constraints (3) and (5–8) are feasible, the optimal reward potentials problem finds the values $\bar{v}_u^{on}$ and $\bar{v}_u^{off}$ of reward potentials such that the relation $\sum_{u \in U} \bar{v}_u^{on} P_{u,t}^b + \bar{v}_u^{off}(1 - P_{u,t}^b) \geq R(\langle Y_{s,t+1}|s \in S\rangle^\frown \langle X_{a,t}|a \in A\rangle)$ holds for all values of $P_{u,t}^b$, $Y_{s,t+1}$ and $X_{a,t}$ at any time step t.*

Proof (by Contradiction). Let $\bar{v}_u^{on}$ and $\bar{v}_u^{off}$ denote the values of reward potentials selected by the leader problem that violate the relation $\sum_{u \in U} \bar{v}_u^{on} P_{u,t}^b + \bar{v}_u^{off}(1 - P_{u,t}^b) \geq R(\langle Y_{s,t+1}|s \in S\rangle^\frown \langle X_{a,t}|a \in A\rangle)$ for some values $\bar{Y}_{s,t+1}$, $\bar{X}_{a,t}$ and $\bar{P}_{u,t}^b$, implying $R(\langle \bar{Y}_{s,t+1}|s \in S\rangle^\frown \langle \bar{X}_{a,t}|a \in A\rangle) - \sum_{u \in U} \bar{v}_u^{on} \bar{P}_{u,t}^b + \bar{v}_u^{off}(1 - \bar{P}_{u,t}^b) > 0$. However, the feasibility of constraint (10) implies that the value of the objective function (11) must be non-positive (i.e., the follower problem is not solved to optimality), which yields the desired contradiction.

Note that we omit the case when constraints (3) and (5–8) are infeasible because it implies the infeasibility of the learned planning problem $\tilde{\varPi}$. Next, we describe a finite-time constraint generation algorithm for computing reward potentials.

3.2 Constraint Generation for Computing Reward Potentials

The optimal reward potentials problem can be solved efficiently through the following constraint generation framework that decomposes the problem into a master problem and a subproblem.[2] The master problem finds the values $\bar{v}_u^{on}$ and $\bar{v}_u^{off}$ of ReLU potential variables. The subproblem finds the values $\bar{P}_{u,t}^b$ of ReLU variables that violate constraint (10) the most for given values $\bar{v}_u^{on}$ and $\bar{v}_u^{off}$, and also finds the maximum value of reward function R for given $\bar{P}_{u,t}^b$ which is denoted as $R^*(\langle \bar{P}_{u,t}^b|u \in U\rangle)$. Intuitively, the master problem selects the values $\bar{v}_u^{on}$ and $\bar{v}_u^{off}$ of ReLU potentials that are checked by the subproblem for the validity of the relation $\sum_{u \in U} \bar{v}_u^{on} P_{u,t}^b + \bar{v}_u^{off}(1 - P_{u,t}^b) \geq R(\langle Y_{s,t+1}|s \in S\rangle^\frown \langle X_{a,t}|a \in A\rangle)$ for all feasible values of $P_{u,t}^b$, $Y_{s,t+1}$ and $X_{a,t}$ at any time step t. If a violation is found, a linear constraint corresponding to a given $\bar{P}_{u,t}^b$ and $R^*(\langle \bar{P}_{u,t}^b|u \in U\rangle)$ is added back to the master problem and the procedure is repeated until no violation is found by the subproblem.

Subproblem $\mathcal{S}$: For a complete value assignment $\bar{v}_u^{on}$ and $\bar{v}_u^{off}$ to ReLU potential variables, the subproblem optimizes the violation (i.e., objective function (11)) with respect to constraints (3) and (5–8) as follows.

---

[2] As noted by our reviewers, our constraint generation framework is related to Counterexample-guided Abstraction Refinement (CEGAR) [3]. The clear differences between the typical use of CEGAR and our work are: (i) problem formalizations (i.e., bilevel programming versus iterative model-checking) and (ii) purposes (i.e., obtaining valid bounds on planning reward function R versus verification of an abstract model). Naturally, what constitutes a violation is also different (i.e., error on reward estimation versus a spurious counterexample).

$$\max_{Y_{s,t}, Y_{s,t+1}, X_{a,t}, P^b_{u,t}} R(\langle Y_{s,t+1} | s \in S \rangle ^\frown \langle X_{a,t} | a \in A \rangle) - \sum_{u \in U} \bar{v}^{on}_u P^b_{u,t} + \bar{v}^{off}_u (1 - P^b_{u,t}) \tag{12}$$

subject to
Constraints (3) and (5–8)

We denote the optimal values of ReLU variables $P^b_{u,t}$, found by solving the subproblem as $\bar{P}^b_{u,t}$, and the value of the reward function R found by solving the subproblem as $R^*(\langle \bar{P}^b_{u,t} | u \in U \rangle)$. Further, we refer to subproblem as $\mathcal{S}$.

Master Problem $\mathcal{M}$: Given the set of complete value assignments K to ReLU variables with values $\bar{P}^{b,k}_{u,t}$ and optimal objective values $R^*(\langle \bar{P}^{b,k}_{u,t} | u \in U \rangle)$ for all $k \in K$, the master problem optimizes the regularized[3] sum of reward potentials (i.e., regularized objective function (9)) with respect to the modified version of constraint (10) as follows.

$$\min_{v^{on}_u, v^{off}_u} \sum_{u \in U} v^{on}_u + v^{off}_u + \lambda \sum_{u \in U} (v^{on}_u)^2 + (v^{off}_u)^2 \tag{13}$$

subject to

$$\sum_{u \in U} v^{on}_u \bar{P}^{b,k}_{u,t} + v^{off}_u (1 - \bar{P}^{b,k}_{u,t}) \geq R^*(\langle \bar{P}^{b,k}_{u,t} | u \in U \rangle) \quad \forall k \in K \tag{14}$$

We denote the optimal values of ReLU potential variables v^{on}_u and v^{off}_u, found by solving the master problem as $\bar{v}^{on}_u$ and $\bar{v}^{off}_u$, respectively. Further, we refer to master problem as $\mathcal{M}$.

Reward Potentials Algorithm. Given the definitions of the master problem $\mathcal{M}$ and the subproblem $\mathcal{S}$, the constraint generation algorithm for computing an optimal reward potential is outlined as follows.

Algorithm 1. Reward Potentials Algorithm

1: $k \leftarrow 1$, violation $\leftarrow \infty$, $\mathcal{M} \leftarrow$ objective function (13)
2: **while** violation > 0 **do**
3:     $\bar{v}^{on}_u, \bar{v}^{off}_u \leftarrow \mathcal{M}$
4:     $\bar{P}^{b,k}_{u,t}, \bar{Y}_{s,t+1}, \bar{X}_{a,t}, R^*(\langle \bar{P}^{b,k}_{u,t} | u \in U \rangle) \leftarrow \mathcal{S}(\bar{v}^{on}_u, \bar{v}^{off}_u)$
5:     violation $= R(\langle \bar{Y}_{s,t+1} | s \in S \rangle ^\frown \langle \bar{X}_{a,t} | a \in A \rangle) - \sum_{u \in U} \bar{v}^{on}_u \bar{P}^{b,k}_{u,t} + \bar{v}^{off}_u (1 - \bar{P}^{b,k}_{u,t})$
6:     $\mathcal{M} \leftarrow \mathcal{M} \cup \sum_{u \in U} v^{on}_u \bar{P}^{b,k}_{u,t} + v^{off}_u (1 - \bar{P}^{b,k}_{u,t}) \geq R^*(\langle \bar{P}^{b,k}_{u,t} | u \in U \rangle)$ (i.e., update constraint (14))
7:     $k \leftarrow k + 1$

---

[3] The squared terms penalize arbitrarily large values of potentials to avoid numerical issues. A similar numerical issue has been found in the computation of potential heuristics for cost-optimal classical planning problems with dead-ends [19].

Given constraints (3) and (5–8) are feasible, Algorithm 1 iteratively computes reward potentials v_u^{on} and v_u^{off} (i.e., line 3), and first checks if there exists an activation pattern, that is a complete value assignment $\bar{P}_{u,t}^{b,k}$ to ReLU variables, that violates constraint (10) (i.e., lines 4 and 5), and then returns the optimal reward value $R^*(\langle \bar{P}_{u,t}^{b,k}|u \in U\rangle)$ for the violating activation pattern. Given the optimal reward value $R^*(\langle \bar{P}_{u,t}^{b,k}|u \in U\rangle)$ for the violating activation pattern, constraint (14) is updated (i.e., lines 6–7). Since there are finite number of activation patterns and solving $\mathcal{S}$ gives the maximum value of $R^*(\langle \bar{P}_{u,t}^{b,k}|u \in U\rangle)$ for each pattern $k \in \{1, \ldots, 2^{|U|}\}$, the Reward Potentials Algorithm 1 terminates in at most $k \leq 2^{|U|}$ iterations with an optimal reward potential for the learned NN.

Increasing the Granularity of the Reward Potentials Algorithm. The feature space of Algorithm 1 can be enhanced to include information on each ReLUs input and/or output. Instead of computing reward potentials for only the activation $\bar{v}_u^{on}$ and deactivation $\bar{v}_u^{off}$ of ReLU $u \in U$, we (i) introduce an interval parameter N to split the output range of each ReLU u into N equal size intervals, (ii) introduce auxiliary Boolean decision variables $P'^b_{i,u,t}$ to represent the activation interval of ReLU u such that $P'^b_{i,u,t} = 1$ if and only if the output of ReLU u is within interval $i \in \{1, \ldots, N\}$, and $P'^b_{i,u,t} = 0$ otherwise, and (iii) compute reward potentials for each activation interval $\bar{v}_{u,1}^{on}, \ldots, \bar{v}_{u,N}^{on}$ and deactivation $\bar{v}_u^{off}$ of ReLU $u \in U$.

3.3   Strengthening HD-MILP-Plan

Given optimal reward potentials $\bar{v}_{u,1}^{on}, \ldots, \bar{v}_{u,N}^{on}$ and $\bar{v}_u^{off}$, the MILP compilation of HD-MILP-Plan is strengthened through the addition of following constraints:

$$\sum_{u \in U} \sum_{i=1}^{N} \bar{v}_{u,i}^{on} P'^b_{i,u,t} + \bar{v}_u^{off}(1 - x_u^t) \geq R(\langle Y_{s,t+1}|s \in S\rangle ^\frown \langle X_{a,t}|a \in A\rangle) \qquad (15)$$

$$\sum_{i=1}^{N} P'^b_{i,u,t} = P_{u,t}^b \qquad (16)$$

$$N_u \frac{(i-1)}{N} P'^b_{i,u,t} \leq P_{u,t} \leq N_u - (N_u - N_u \frac{i}{N}) P'^b_{i,u,t} \quad \forall i \in \{1, \ldots, N\}, u \in U \qquad (17)$$

for all time steps $t \in \{1, \ldots, H\}$ where N_u denotes the upperbound obtained from performing forward reachability on the output of each ReLU $u \in U$ in the learned NN. Briefly, constraint (15) provides the upperbound on the reward function R as a function of ReLU activation intervals and deactivations. Constraint (16) ensures that (i) at most one auxiliary variable $P'^b_{i,u,t}$ is selected, and (ii) at least one auxiliary variable $P'^b_{i,u,t}$ is selected if and only if ReLU u is activated. Constraint (17) ensures that the output of each ReLU is within its selected activation interval. Next, we present our experimental results to demonstrate the efficiency and the utility of computing reward potential and strengthening HD-MILP-Plan.

4 Experimental Results

In this section, we present computational results on (i) the convergence of Algorithm 1, and (ii) the overall strengthening of HD-MILP-Plan with the addition of constraints (15–17) for the task of planning over long horizons. First, we present results on the overall efficiency of Algorithm 1 and the strengthening of HD-MILP-Plan over multiple learned planning instances. Then, we focus on the most computationally expensive domain identified by our experiments to further investigate the convergence behaviour of Algorithm 1 and the overall strengthening of HD-MILP-Plan as a function of time.

4.1 Experimental Setup

The experiments were run on a MacBookPro with 2.8 GHz Intel Core i7 16 GB memory. All instances and the respective learned neural networks from the HD-MILP-Plan paper [18], namely *Navigation, Reservoir Control* and *HVAC* [18], were selected.[4] Both domain instance sizes and their respective learned NN sizes are detailed in Table 1 where columns from left to right denote the name of problem instances, the structures of the learned NNs where each number denotes the width of a layer and the values of the planning horizon H, respectively. The range bounds on action variables for Navigation domains were constrained to $[-0.1, 0.1]$. CPLEX 12.9.0 [8] solver was used to optimize both Algorithm 1, and HD-MILP-PLan, with 6000 s of total time limit per domain instance. In our experiments, we show results for the base model (i.e., objective (1) and constraints (2–8)) and the strengthened model with the addition of constraints (15–17) for the values of interval parameter $N = 2, 3$.[5] Finally in the master problem, we have chosen the regularizer constant λ in the objective function (9) to be $\frac{1}{\sqrt{M}}$ where M is the large constant used in the big-M constraints of HD-MILP-Plan (i.e., constraints (5–6)).

4.2 Overall Results

In this section, we present the experimental results on (i) the computation of the optimal reward potentials using Algorithm 1, (ii) and the performance of HD-MILP-Plan with the addition of constraints (15–17) over multiple learned planning instances over long horizons. Table 2 summarizes the computational results and highlights the best performing HD-MILP-Plan settings for each learned planning instance.

---

[4] https://github.com/saybuser/HD-MILP-Plan.

[5] The preliminary experimental results for interval parameter $N = 1$ have not shown significant improvements over the base encoding of HD-MILP-Plan.

Table 1. Domain and learned NN descriptions where columns from left to right denote the name of problem instances, the structures of NNs used to learn each transition model $\tilde{T}$ where each number denotes the width of a layer, and the values of the planning horizon H, respectively.

Domain instance	Network structure	Horizon
Navigation (8-by-8 maze)	4:32:32:2	100
Navigation (10-by-10 maze)	4:32:32:2	100
Reservoir control (3 reservoirs)	6:32:3	500
Reservoir control (4 reservoirs)	8:32:4	500
HVAC (3 rooms)	6:32:3	100
HVAC (6 rooms)	12:32:6	100

Table 2. Summary of experimental results on the computationally efficiency of Algorithm 1 and HD-MILP-Plan with the addition of constraint (15–17) over multiple learned planning instances with long horizons.

Domain setting	Algorithm 1	Cumul.	Primal	Dual	Open	Closed
Nav,8,100,Base	-	6000	-	−261.4408	16536	27622
Nav,8,100,N = 2	345	6000	-	**−267.1878**	6268	15214
Nav,8,100,N = 3	1150	6000	-	−267.056	6189	12225
Nav,10,100,Base	-	6000	-	−340.5974	17968	35176
Nav,10,100,N = 2	800	6000	-	**−340.6856**	14435	27651
Nav,10,100,N = 3	1700	6000	-	−339.8124	2593	7406
HVAC,3,100,Base	-	260.21	Opt. found	Opt. proved	0	289529
HVAC,3,100,N = 2	7	**88.21**	Opt. found	Opt. proved	0	2501
HVAC,3,100,N = 3	9	194.44	Opt. found	Opt. proved	0	10891
HVAC,6,100,Base	-	6000	−1214369.086	−1213152.304	618687	648207
HVAC,6,100,N = 2	8	6000	−1214365.427	**−1213199.787**	554158	567412
HVAC,6,100,N = 3	10	6000	**−1214364.704**	−1213025.189	1011348	1021637
Res,3,500,Base	-	**33.01**	Opt. found	Opt. proved	0	1
Res,3,500,N = 2	1	99.81	Opt. found	Opt. proved	0	714
Res,3,500,N = 3	2	90.27	Opt. found	Opt. proved	0	674
Res,4,500,Base	-	300.71	Opt. found	Opt. proved	0	1236
Res,4,500,N = 2	7	**109.66**	Opt. found	Opt. proved	0	1924
Res,4,500,N = 3	6	232.19	Opt. found	Opt. proved	0	1294

The first column of Table 2 identifies the domain setting of each row. The Z column denotes the runtime of Algorithm 1 in seconds. The third column (i.e., Cumul.) denotes the cumulative runtime of Algorithm 1 and HD-MILP-Plan in seconds. The remaining columns provide information on the performance of HD-MILP-Plan. Specifically, the fourth column (i.e., Primal) denotes the value of the incumbent plan found by HD-MILP-Plan, the fifth column (i.e., Dual) denotes the value of the duality bound found by HD-MILP-Plan, and the sixth and seventh columns (i.e., Open and Closed) denote the number of open and closed nodes in the B&B tree respectively. The bolded values indicate the best performing HD-MILP-Plan settings for each learned planning instance where the performance of each setting is evaluated first based on the runtime performance (i.e., Cumul. column), followed by the quality of incumbent plan (i.e., Primal column) and duality bound (i.e., Dual column) obtained by HD-MILP-Plan.

In total of five out of six instances, we observe that strengthened HD-MILP-Plan with interval parameter $N = 2$ performed the best. The pairwise comparison of the base HD-MILP-Plan and strengthened HD-MILP-Plan with interval parameter $N = 3$ shows that in almost all instances, the strengthened model performed better in comparison to the base model. The only instance in which the base model significantly outperformed the other two was the Reservoir Control domain with three reservoirs where the B&B solver was able to find an optimal plan in the root node. Overall, we found that especially in the instances where the optimality was hard to prove within the runtime limit of 6000 s (i.e., all Navigation instances and HVAC domain with 6 rooms), strengthened HD-MILP-Plan explored significantly less number of nodes in general while obtaining either higher quality incumbent plans or lower dual bounds. We observe that Algorithm 1 terminated with optimal reward potentials in less than 10 s in both Reservoir Control and HVAC domains, and took as much as 1700 s in Navigation domain – highlighting the effect of NN size and complexity (i.e., detailed in Table 1) on the runtime of Algorithm 1. As a result, next we focus on the most computationally expensive domain identified by our experiments, namely Navigation, to get a better understanding on the convergence behaviour of Algorithm 1 and the overall efficiency of HD-MILP-Plan as a function of time.

4.3   Detailed Convergence Results on Navigation Domain

In this section, we inspect the convergence of Algorithm 1 in the Navigation domain for computing an optimal reward potential for the learned NNs.

Figure 2 visualizes the violation of constraint (10) as a function of time over the computation of optimal reward potentials using the Reward Potentials Algorithm 1 for the learned NNs of both Navigation 8-by-8 (i.e., top) and Navigation 10-by-10 (i.e., bottom) planning instances. In both, we observe that the violation of constraint (10) decreases exponentially as a function of time, showcasing a long-tail runtime behaviour and terminates with optimal reward potentials.

4.4   Detailed Strengthening Results on Navigation Domain

Next, we inspect the overall strengthening of HD-MILP-Plan with respect to its underlying linear relaxation and search efficiency as a result of constraints (15–17), for the task of planning over long horizons in the Navigation domain.

Figures 3 and 4 visualize the overall effect of incorporating constraints (15–17) into HD-MILP-Plan as a function of time for the Navigation domain with (a) 8-by-8 and (b) 10-by-10 maze sizes. In both Figs. 3 and 4, linear relaxation (i.e. top), number of closed nodes (i.e., middle), and number open nodes (i.e., bottom), are displayed as a function of time. The inspection of both Figs. 3 and 4 show that once the reward potentials are computed, the addition of constraints (15–17) allows HD-MILP-Plan to obtain a tighter bound by exploring

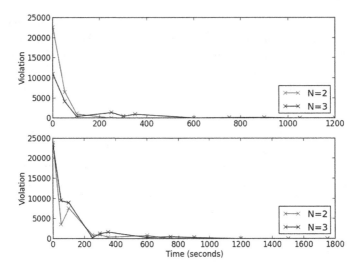

Fig. 2. Convergence of Algorithm 1 as a function of time for the learned NNs of both Navigation 8-by-8 (i.e., top) and Navigation 10-by-10 (i.e., bottom) planning instances. The violation of constraint (10) decreases exponentially as a function of time, showcasing a long-tail runtime behaviour and terminates with optimal reward potentials.

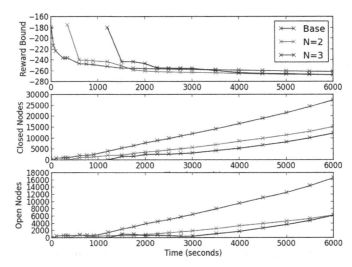

Fig. 3. Linear relaxation and search efficiency comparisons in Navigation domain with an 8-by-8 maze between the base and the strengthened HD-MILP-Plan using Algorithm 1 with interval parameter $N = 2, 3$. Overall, we observe that HD-MILP-Plan with constraints (15–17) outperforms the base HD-MILP-Plan by 1700 and 3300 s with interval parameter $N = 2, 3$, respectively.

significantly less number of nodes. In the 8-by-8 maze instance, we observe that HD-MILP-Plan with constraints (15–17) outperforms the base HD-MILP-Plan by 1700 and 3300 s with interval parameter $N = 2, 3$, respectively. In the 10-by-10 maze instance, we observe that HD-MILP-Plan with constraints (15–17) obtains a tighter bound compared to the base HD-MILP-Plan by 3750 s and almost reaches the same bound by the time limit (i.e., 6000 s) with interval parameter $N = 2, 3$, respectively.

The inspection of the top subfigures in Figs. 3 and 4 shows that increasing the value of the interval parameter N increases the computation time of Algorithm 1, but can also increase the search efficiency of the underlying B&B solver through increasing its exploration and pruning capabilities, as demonstrated by the middle and bottom subfigures in Figs. 3 and 4. Overall from both instances, we conclude that HD-MILP-Plan with constraints (15–17) obtains a linear relaxation that is at least as good as the base HD-MILP-Plan by exploring significantly less number of nodes in the B&B search tree.

5 Related Work

In this paper, we have focused on the important problem of improving the efficiency of B&B solvers for optimal planning with learned NN transition models in continuous action and state spaces. Parallel to this work, planning and decision making in discrete action and state spaces [12, 16, 17], verification of learned

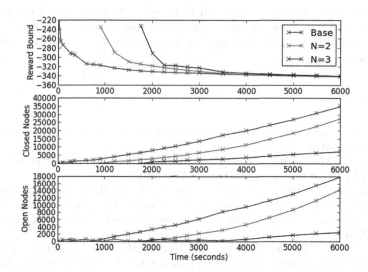

Fig. 4. Linear relaxation and search efficiency comparisons in Navigation domain with an 10-by-10 maze between the base and the strengthened HD-MILP-Plan using Algorithm 1 with interval parameter $N = 2, 3$. Overall, we observe that HD-MILP-Plan with constraints (15–17) obtains a tighter bound compared to the base HD-MILP-Plan by 3750 s and reaches almost the same bound by the time limit (i.e., 6000 s) with interval parameter $N = 2, 3$, respectively.

NNs [6,7,9,14], robustness evaluation of learned NNs [20] and defenses to adversarial attacks for learned NNs [10] have been studied with the focus of solving very similar decision making problems. For example, the verification problem solved by Reluplex [9][6] is very similar to the planning problem solved by HD-MILP-Plan [18] without the objective function and horizon $H = 1$. Interestingly, the verification problem can also be modeled as an optimization problem [2] and potentially benefit from the findings presented in this paper. For future work, we plan to explore how our findings in this work translate to solving other important tasks for learned neural networks.

6   Conclusion

In this paper, we have focused on the problem of improving the linear relaxation and the search efficiency of MILP models for decision making with learned NNs. In order to tackle this problem, we used bilevel programming to correctly model the optimal reward potentials problem. We then introduced a novel finite-time constraint generation algorithm for computing the potential contribution of each hidden unit to the reward function of the planning problem. Given the precomputed values of the reward potentials, we have introduced constraints to tighten the bound on the reward function of the planning problem. Experimentally, we have shown that our constraint generation algorithm efficiently computes reward potentials for learned NNs, and the overhead computation is justified by the overall strengthening of the underlying MILP model as demonstrated on the task of planning over long horizons. With this paper, we have shown the *potential* of bridging the gap between two seemingly distant literatures; the research on planning heuristics and decision making with learned NN models in continuous action and state spaces.

References

1. Bard, J.: Practical Bilevel Optimization: Algorithms And Applications. Springer, US (2000). https://doi.org/10.1007/978-1-4757-2836-1
2. Bunel, R., Turkaslan, I., Torr, P.H., Kohli, P., Kumar, M.P.: A unified view of piecewise linear neural network verification (2017)
3. Clarke, E., Grumberg, O., Jha, S., Lu, Y., Veith, H.: Counterexample-guided abstraction refinement. In: Emerson, E.A., Sistla, A.P. (eds.) CAV 2000. LNCS, vol. 1855, pp. 154–169. Springer, Heidelberg (2000). https://doi.org/10.1007/10722167_15
4. Collobert, R., Weston, J., Bottou, L., Karlen, M., Kavukcuoglu, K., Kuksa, P.: Natural language processing (almost) from scratch. J. Mach. Learn. Res. **12**, 2493–2537 (2011)
5. Deng, L., Hinton, G.E., Kingsbury, B.: New types of deep neural network learning for speech recognition and related applications: an overview. In: IEEE International Conference on Acoustics, Speech and Signal Processing, pp. 8599–8603 (2013)

---

[6] Reluplex [9] is a SMT-based learned NN verification software.

6. Ehlers, R.: Formal verification of piece-wise linear feed-forward neural networks. In: D'Souza, D., Narayan Kumar, K. (eds.) ATVA 2017. LNCS, vol. 10482, pp. 269–286. Springer, Cham (2017). https://doi.org/10.1007/978-3-319-68167-2_19

7. Huang, X., Kwiatkowska, M., Wang, S., Wu, M.: Safety verification of deep neural networks. In: Majumdar, R., Kunčak, V. (eds.) CAV 2017. LNCS, vol. 10426, pp. 3–29. Springer, Cham (2017). https://doi.org/10.1007/978-3-319-63387-9_1

8. IBM: IBM ILOG CPLEX Optimization Studio CPLEX User's Manual (2019)

9. Katz, G., Barrett, C., Dill, D., Julian, K., Kochenderfer, M.: Reluplex: an efficient SMT solver for verifying deep neural networks. In: Twenty-Ninth International Conference on Computer Aided Verification, CAV (2017)

10. Kolter Zico, J., Wong, E.: Provable defenses against adversarial examples via the convex outer adversarial polytope. In: Thirty-First Conference on Neural Information Processing Systems (2017)

11. Krizhevsky, A., Sutskever, I., Hinton, G.E.: Imagenet classification with deep convolutional neural networks. In: Twenty-Fifth Neural Information Processing Systems, pp. 1097–1105 (2012). http://dl.acm.org/citation.cfm?id=2999134.2999257

12. Lombardi, M., Gualandi, S.: A Lagrangian propagator for artificial neural networks in constraint programming. Constraints **21**, 435–462 (2016). https://doi.org/10.1007/s10601-015-9234-6

13. Nair, V., Hinton, G.E.: Rectified linear units improve restricted boltzmann machines. In: Twenty-Seventh International Conference on Machine Learning, pp. 807–814 (2010). http://www.icml2010.org/papers/432.pdf

14. Narodytska, N., Kasiviswanathan, S., Ryzhyk, L., Sagiv, M., Walsh, T.: Verifying properties of binarized deep neural networks. In: Thirty-Second AAAI Conference on Artificial Intelligence, pp. 6615–6624 (2018)

15. Pommerening, F., Helmert, M., Roger, G., Seipp, J.: From non-negative to general operator cost partitioning. In: Twenty-Ninth AAAI Conference on Artificial Intelligence, pp. 3335–3341 (2015)

16. Say, B., Sanner, S.: Compact and efficient encodings for planning in factored state and action spaces with learned binarized neural network transition models (2018)

17. Say, B., Sanner, S.: Planning in factored state and action spaces with learned binarized neural network transition models. In: Twenty-Seventh International Joint Conference on Artificial Intelligence, pp. 4815–4821 (2018). https://doi.org/10.24963/ijcai.2018/669

18. Say, B., Wu, G., Zhou, Y.Q., Sanner, S.: Nonlinear hybrid planning with deep net learned transition models and mixed-integer linear programming. In: Twenty-Sixth International Joint Conference on Artificial Intelligence, pp. 750–756 (2017). https://doi.org/10.24963/ijcai.2017/104

19. Seipp, J., Pommerening, F., Helmert, M.: New optimization functions for potential heuristics. In: Twenty-Fifth International Conference on Automated Planning and Scheduling, pp. 193–201 (2015)

20. Tjeng, V., Xiao, K., Tedrake, R.: Evaluating robustness of neural networks with mixed integer programming. In: Seventh International Conference on Learning Representations (2019)

Exploiting Counterfactuals for Scalable Stochastic Optimization

Stefan Kuhlemann[1,3], Meinolf Sellmann[2(✉)], and Kevin Tierney[1(✉)]

[1] Bielefeld University, Bielefeld, Germany
{stefan.kuhlemann,kevin.tierney}@uni-bielefeld.de
[2] GE Research, Niskayuna, USA
meinolf@ge.com
[3] Paderborn University, Paderborn, Germany

Abstract. We propose a new framework for decision making under uncertainty to overcome the main drawbacks of current technology: modeling complexity, scenario generation, and scaling limitations. We consider three NP-hard optimization problems: the Stochastic Knapsack Problem (SKP), the Stochastic Shortest Path Problem (SSPP), and the Resource Constrained Project Scheduling Problem (RCPSP) with uncertain job durations, all with recourse. We illustrate how an integration of constraint optimization and machine learning technology can overcome the main practical shortcomings of the current state of the art.

1 Introduction

Optimization relies on data. To solve a knapsack problem we need to know the profits and weights of the items, as well as the knapsack's capacity. To solve a shortest path or travelling salesperson problem, we need to know the lengths of the links in the network. To solve a revenue optimization problem, we need to know demand and how prices affect demand. In practice, we often lack perfect knowledge of the situation we ultimately needed to plan for. Profits, transition times, price sensitivity, and demands frequently have to be estimated.

One simple and still widely used approach is to optimize for point estimates of the data: We estimate demand, profits, transition times, etc, and optimize for the resulting optimization problem. The problem with using only one set of estimates, even if they represented the maximum likelihood scenario, is that the probability of exactly this scenario taking place is close to zero, and performance of the solution that is optimal for this one scenario may decline steeply across a range of scenarios that, together, would have a reasonable probability mass. In other words, a solution that is sub-optimal for all scenarios but works with good performance for a large number of potential futures will lead to much better *expected* performance than the solution that is provably optimal for the maximum likelihood scenario yet abysmal otherwise.

© Springer Nature Switzerland AG 2019
T. Schiex and S. de Givry (Eds.): CP 2019, LNCS 11802, pp. 690–708, 2019.
https://doi.org/10.1007/978-3-030-30048-7_40

1.1 Stochastic Optimization

The brittleness of solutions obtained by optimizing for one, point-estimated scenario only is well-studied in the field of stochastic optimization (SO). The objective of SO is to provide a solution that optimizes the expected returns over all possible futures.

This led to the idea of *two-stage stochastic optimization*: In the first stage, we need to take certain decisions based on uncertain data. After taking these decisions, the uncertainties are revealed and we can take the remaining decisions based on certain data. This allows us first to make up for certain inconsistencies our initial decisions might have created (note that the constraints are also based on estimates) and thus exercise certain *recourse actions* to regain feasibility, and second to optimize the second-stage decisions that can wait to be taken until we know the real data. An overview of two-stage stochastic integer programming problems can be found in [4], and [21] present a method to solve two-stage problems using the special form of these problems.

One crucial step in stochastic optimization is the generation of a representative set of potential futures (*scenarios*). Many methods exist to generate scenarios, and [13] points out that quality scenario generation is critical to the success using SO. [12] recommend that a number of different data sources should be used for scenario generation.

Obviously, solving SO problems to optimality gets harder the more scenarios are considered. Sample average approximation [14] has been developed to generate a small random sample of scenarios and approximate the expected value function. This technique has been be applied to a variety of problems (see, e.g., [16,20,22]) and can help the method scale a bit better. However, the fundamental problem remains that SO relies on a representative set of scenarios to be considered, and that it must make optimal first and second stage decisions for every scenario under consideration.

1.2 Multi-stage Stochastic Optimization

One practical aspect that we also need to take into account is that the execution of a planned solution is frequently disrupted by outside events: equipment or crew assumed to be available may suddenly go out of service, requiring adjustment of a plan during operations. Consequently, the plan may need to be adjusted multiple times.

This leads to the idea of *multi-stage stochastic optimization*. In multi-stage SO, uncertainty is revealed in multiple consecutive steps, and more decisions need to be taken at each stage. In these problems, random variables in later stages depend on the decisions taken in the earlier stages. Models and solutions to these problems are therefore structured in the form of a tree [7], with independent decisions at the root node, and dependencies between decisions modeled with parent-child relationships in the tree.

Due to their richer modeling power, these types of SO models are especially relevant for real-world decision making, but unfortunately explode in complexity very quickly, even when employing advanced decomposition techniques like presented by [3] who extend the work of [21]. Furthermore, the problem of scenario generation is even more daunting in this more realistic setting, as conditional scenarios need to be generated, and often the data needed for this purpose may not be available. An overview of scenario generation methods for multi-stage stochastic programs is provided in [7].

1.3   Simulation-Based Optimization

Modeling dynamic recourse and managing a meaningful number of scenarios in stochastic optimization is often cumbersome. An alternative is to employ a *simulator* that can evaluate a given plan on a number of scenarios, whereby the algorithm to generate recourse actions is built into the simulation. The recourse policy employed by a real-world organization may involve solving nested optimization problems on the go, as SO assumes, but oftentimes the real-world operational constraints may not allow for a full-fledged optimization, for example because the data needed is not readily available, or because re-optimization would be too time-consuming. A simulator can easily reflect the real recourse actions that would be taken, which are usually locally optimal only, or maybe just best-effort heuristics.

Simulation-based optimization is thus an alternative to stochastic optimization [1,9]. In this setting, a simulation is constructed to provide a stochastic evaluation of a provided solution. The search for good solutions can then be conducted by employing a meta-heuristic procedure. For example, [10] employs tabu search for this purpose.

An alternative to using a general local search heuristic is to apply bandit theory and to conduct a search based on Bayesian optimization [19]. In this method, the search space is traversed in a statistically principled way which balances exploitation and exploration by considering new solutions for simulation next which combine high expected performance with high uncertainty of this performance.

No matter which search method is employed, to compute an objective function value for an instance, we need to expose it to certain futures. In SO these were called scenarios, in simulation-based optimization the "scenario generation" is hidden in the simulator. However, both methods rely on an adequate representation of potential futures *of the world as it currently presents itself.*

2   Technology Gaps

In practice, many organizations do not take the uncertainty in their forecasts into account when devising their operational plans. In fact, this observation even holds for those organizations that would stand to benefit the most, because events disrupting their plans frequently ruin all operational success. Airlines are one prototypical example. Over decades, the airline industry has spent billions of dollars on optimization technology to improve their operational planning (e.g., in crew planning [17]). There is certainly no lack of affinity to optimization technology, nor a lack of understanding that their current optimized plans are very brittle. The question is, why then is decision-making under uncertainty not employed?

We believe there are three main factors that prevent current decision making under uncertainty technology from being applied in practice:

- Complexity of modeling the base problem
- Inability to generate meaningful future scenarios for the current situation
- Computational limitations preventing the scale-up to real-world numbers of primary and recourse decisions

Take the example of an airline again. Flights may be delayed due to weather or traffic. Gates may be occupied and have to be changed. Crew may be out of service because of sickness or because they are delayed and past their maximum allowed service time. Equipment may not be available because of technical issues or because other issues in the network prevented the plane from being at the airport where it was planned to be.

Modeling the operation of an airline is extremely complex to begin with, which is why airlines break down the original problem into network design, revenue management, fleet assignment, crew pairing, tail assignment, and crew scheduling problems. There are literally millions of decision variables to consider. Secondly, there are frequently no models available for assessing the probabilities of disruptions with any meaningful accuracy. This is especially true for the joint distributions of disruptions which are frequently correlated. And finally, the number of recourse decisions taken during operation is staggering: Airlines literally run their recovery solvers every minute to adjust their plans to ever new, thankfully usually minor, disruptions.

Stochastic optimization is not applicable, because computation times are prohibitively long, and the number of recourse decisions far exceeds efficient modeling capabilities. However, simulation-based optimization cannot handle the millions of decision variables or the complex constraints that govern whether solutions are even feasible.

This analysis is the starting point for our research. In the following, we propose a framework for decision making under uncertainty that overcomes the limitations of existing technology. In a nutshell, we propose a paradigm shift away from trying to anticipate the future and towards discovering structures in the solutions that correlate with historically good performance. In doing so, we trade dual bounds (i.e., a guarantee of the relative quality of the solution provided by SO) for scalability and easier modeling.

3   Learning from Counterfactuals

A key limitation of stochastic optimization is the need to model every decision. Not only does this put an enormous burden on the modeler and the optimization, it often also falsely assumes that we were able to optimize recourse decisions during operation. Another problem that both simulation-based optimization and stochastic optimization share is that they need to generate meaningful scenarios how the execution of a solution may unfold in the future. Finally, both methods scale to a hundreds, maybe a few thousand decision variables, before computation times become impractically long.

Our proposal is to combine both what disruptions are likely, as well as how well the initial solution is adjusted during execution, into one data-driven forecast.

Consider a model that, given two solutions to an optimization problem can provide a classification as to which of the two solutions will fare better when they are executed. Consider solving this problem in two stages. In the first stage, we solve the problem as if there were no uncertainty using the expected costs in the objective function, and generate multiple near-optimal solutions. The goal of the second stage is to determine which of the near-optimal solutions will likely lead to better results when executed. We train a model that compares these solutions on a pairwise basis and choose the solution that wins the most times against the other solutions. This general method alleviates many of the problems existing approaches encounter:

- The first-stage problem is as easy to model as the optimization problem without uncertainties.
- There is no need to generate future scenarios for the current data at hand.
- There is no complexity blow-up, no matter how many recourse actions are needed.

All of the complexity is off-loaded into the second stage model. The crucial question is, of course: How can we obtain a model that, given two solutions, can predict which one will fare better in operations?

Thesis: We can learn such a model from historical data.

We argue that all that is needed to learn such a model is to keep track of our estimates over time, and what eventually happened. Consider, e.g., the Stochastic Shortest Path Problem (SSPP), a problem that [5] argue is particularly difficult when there are no assumptions about the uncertain travel times. We can track how the arc transit time estimates for the entire network have evolved over time, and what they ended up being. Or, for the Stochastic Knapsack Problem (SKP), we can examine what our weight and profit estimates were before each decision for historical solutions and what they turned out to be in reality. That is to say: Historical data often enables us to compare *multiple* historical solutions, even though only one of them was actually executed in reality, whereas all others are essentially *counterfactuals*.

Please note a subtle but very important difference; the historical data is enough to compare two potential solutions for the optimization problem *as it presented itself in the past*. It would, however, not enable us to simulate two solutions for a new instance of our underlying optimization problem. Take the SSPP as an example. We may have a historic example where we needed to go from some node s to node t. We know how our estimated arc transition times evolved over time and the resulting values on all arcs in the network. With this, we can compare two paths P_1 and P_2 that connect s and t.

Now imagine we currently need a solution to go from node s to node t again. Our initial estimates are of course completely different from those in the historic example. Consequently, we cannot just simulate two paths Q_1 and Q_2 in the old scenario and assume that their relative performance would remain the same under the current conditions. In fact, if this were the case, we should forget our estimates altogether and just always go from s to t using the exact same route all the time.

However, if we could capture *estimate-dependent characteristics*, or *features*, of pairs of paths P_1 and P_2, and associate these characteristics with the *relative* performance of these paths, then, by repeating this exercise many times, we just might be able to *learn* to tell which of any given pairs of paths will probably execute better – albeit with no guarantees.

Through this framework, we have now decomposed the problem of making primary decisions based on uncertain forecasts and assumptions regarding estimate distributions and recourse policies into two tasks: We first need to model the primary optimization problem. Second, we need to use historical data to build a supervised set of examples of pairs of solutions, recording which one fared, or would have fared, better. Crucially, we need to devise a set of features to characterize the solutions *in the context of the problem instance they were generated for*.

We formalize our framework as follows. We are given a deterministic optimization problem P with decision variables $\mathbf{x}$. Let $f(\mathbf{x})$ be the objective function of the deterministic problem, and $f'(\mathbf{x}, \omega)$ be the objective function when the decisions are evaluated under scenario ω.

1. Training set generation: We first generate n solutions $\mathbf{x}_{i1}, \ldots, \mathbf{x}_{in}$ to the problem instance i in a set of training instances I, where all uncertain parameters take their expected value. The choice of such solutions is up to the user of this framework, but we recommend high quality solutions with some diversity. We associate a label $y_{ij} = \sum_{\omega \in \Omega_i} f'(\mathbf{x}_{ij}, \omega) / |\Omega_i|$ with each solution of each instance for a set of counterfactual scenarios Ω_i that are derived from the true scenario that unfolded for the historic problem instance i. Finally, we compute problem dependent feature vectors $\mathbf{u}_{ij} \in \mathbb{R}^f$ describing each solution j of instance i.

2. Learning a classifier: Next, we train a binary classifier M that, given two solutions j, k to a problem instance i, forecasts which of the two solutions will likely perform better when executed. The training input for this cost-sensitive learning task are triples $(u'_{ijk}, y_{ij}, y_{ik})$, where $u'_{ijk} \in \mathbb{R}^{3f}$ consists of a concatenation of feature vectors u_{ij}, u_{ik}, and $u_{ij} - u_{ik}$. We use the technique from [18] for this purpose.

3. Deployment: Given a problem instance i, we generate n solutions using the deterministic optimization model with expected values for uncertain parameters. We then compute the features u'_{ijk} for each pair of solutions j, k. Then, we query the model M for all such pairs and choose the solution that "wins" the most times.

In the following, we will exercise the above steps for three optimization problems: the SKP, the SSPP, and the Resource Constrained Project Scheduling Problem (RCPSP), each with recourse. The objective of this study is to investigate whether we can effectively learn which solution for a problem instance will perform better.

4 Stochastic Knapsack

The SKP is the stochastic variant of the well-known optimization problem: Given $n \in \mathbb{N}$ items $\{1, \ldots, n\}$ with profits $p_1, \ldots, p_n \in \mathbb{N}$, expected weights $w_1, \ldots, w_n \in \mathbb{N}$, and a capacity $C \in \mathbb{N}$, the objective is to find a subset of items $I \subseteq \{1, \ldots, n\}$ such that $\sum_{i \in I'} w'_i \leq C$ and $P = \sum_{i \in I'} p_i$ is maximized, where $w'_1, \ldots, w'_n$ are the actual weights incurred, and I' is the set of items we ultimately include in our knapsack.

4.1 Stochastic Environment

To complete the setup of our problem, we need to determine how the weights w'_i are derived from the expected weights w_i, and how I' derives from I during operations. This is precisely the task of determining the distributions of stochastic data, and the incorporation of recourse policies that we aim to avoid estimating and modeling when solving the stochastic variant of the underlying optimization problem. However, for the sake of experimentation, we obviously need to fix the stochastic environment.

We will assume that items have to be decided for inclusion or exclusion in sequence 1 to n.[1] That is, we first decide if we want to insert item 1 in the knapsack. If not, we can directly move on to the next item. If yes, then we add the actual weight w'_1 to our knapsack, the remaining capacity is reduced accordingly, and the profit p_1 is achieved. We consider all items in sequence. At stage i, we sample w'_i from a Pareto distribution with mean w_i (note: the nature of this distribution is not known to the optimization approach). In our variant of the problem, should the new item overload the knapsack, the item is automatically not inserted and we proceed as if we had never decided to include the item. However, if the item fits into the remaining capacity, we have to take it, even if the actual weight of the item is much larger than we had anticipated.

In terms of recourse, whenever during the sequential consideration of items our remaining capacity deviates from the anticipated capacity at that point in the sequence by more than a given percentage threshold p, we are allowed to reconsider our original plan and change the tail of our plan. However, if we

---

[1] This is in contrast to some theoretical results on the SKP that assume we can decide in what order we wish to consider the items [6]. We consider having this freedom less realistic.

include an item that was originally not planned to be included, we incur a profit penalty b (late buy penalty). Similarly, we incur a penalty s for items we do not include in our knapsack that we had originally committed to include (restocking fee). Finally, we cannot change the original plan for the next r items (minimum reaction time).

The recourse policy is to re-optimize the rest of the knapsack based on the profits adjusted for penalties, the originally estimated weights, and the remaining items and capacity. The selection of the next r items is fixed.

To support our introductory claim that existing technology is not feasible even for such a simple practical setting, we invite the reader to try to model this problem as an n-stage stochastic optimization problem or as a simulation-based optimization problem with n variables and an uncertain side constraint.

4.2   Winner Forecasting

In stage 1 of our approach, we consider the original knapsack problem with the given capacity, profits, and estimated weights. We solve the problem to optimality using dynamic programming and generate a desired number of solutions that are either optimal or as close to optimal as possible.

In stage 2, we need to characterize each solution with respect to the given problem instance. Before we list the features we introduce for this purpose, we define a number of quantities we can compute for any sequences of numbers.

For monotonically increasing (or decreasing) sequences, we define the following quantities (leading to $3q + 2$ quantities for q quantiles considered):

– The mean, and the mean of the second moment.
– The median and the median of the second moment.
– For a desired number q of quantiles over the range of the sequence, the percentage of numbers in the sequence before each quantile is first reached, depending on whether the sequence is increasing or decreasing (including the last quantile).
– For a desired number q of quantiles, the value of the sequence at each quantile of items in the sequence, and the corresponding values in the second moment (excluding the last quantile).

For general sequences, we define the following 8 quantities:

– The mean and the mean of the second moment.
– The median and the median of the second moment.
– The minimum and maximum, and the corresponding values of the second moment.

Now, to characterize a given solution to a knapsack instance, we consider the following five monotone sequences, and the six general sequences thereafter:

M 1: For each item i in the sequence, the total profits achieved so far, as a percentage of the maximum achievable profit (here and in the following per the given solution).

M 2: For each item i in the sequence, the remaining capacity as a percent of the total capacity.

M 3: For each item i in the sequence, the linear programming upper bound for the remaining items and the remaining capacity, as percentage of the optimum profit.

M 4: For each item i in the sequence, the linear programming upper bound for the remaining items and the total original capacity C, as percentage of the optimum profit achievable.

M 5: For each item i in the sequence, we compute the number d_i of items since the last item that was included in the solution. We aggregate and normalize these numbers by setting $D_i = \sum_{k \leq i} d_i/n$ and considering the monotone sequence $(D_i)_i$.

G 1–3: For each item selected in the given solution, its profit (as percentage of maximum profit), weight (as percentage of total capacity), and efficiency (the ratio of profit over weight).

G 4–6: The same three values as above, but over the items not selected in the solution.

We consider 5 quantiles, therefore the above yields $5(3 * 5 + 2) + 6 * 8 = 133$ features. We add two more by also computing the total efficiency of the solution, defined by the ratio of total profit divided by total capacity, and finally the LP/IP gap as percentage of maximum achievable profit. In total, for each solution we thus obtain 135 features. For a given pair of solutions, we concatenate the features of each solution, as well as the difference of the features of the two solutions. Our machine learning approach thus has access to $3 * 135 = 405$ features to decide which of the two solutions given is likely to perform better than the other.

To complete the data-driven part of our approach, we choose binary cost-sensitive classification to rank the solutions, in particular, the cost-sensitive hierarchical clustering approach from [18]. We use this technique in all following test cases.

4.3 Numerical Results

We generate knapsack instances with 1,000 items and (expected) weights drawn between 1 and 100 uniformly at random. The capacity is set to 10% of the total expected weights of all 1,000 items. Weakly correlated knapsack instances are generated by choosing the profit of item i with weight w_i in the interval $[w_i - 3, w_i + 3]$. Strongly correlated instances are generated by setting the profits to $w_i + 5$. Furthermore, almost strongly correlated instances are generated by choosing the profits in $[w_i + 4, w_i + 6]$ uniformly at random.

We build a simulation environment where the weight of an item i is drawn from a random variable following a Pareto distribution with mean w_i and minimum value $0.95w_i$. Note that the Pareto distribution is heavy-tailed: With the given parameters, there is only about a 20% chance of seeing a value larger than the mean, but a 1.5% chance to see a value of at least 1.5 times the mean, and about a 0.3% chance of encountering values of twice the mean or more.

We set the recourse threshold $p = 5\%$, the restocking fee and late buying fee $s = b = 10$, and the minimum reaction time $r = 5$. Whenever the remaining capacity in the knapsack deviates by more than $p = 5\%$, we solve a new knapsack problem (with adjusted profits to reflect the respective restocking and late buying fees) to determine our recourse action for the remaining items beyond the minimum reaction threshold.

Using this environment, we generate 100 instances of each knapsack type (weakly, strongly, and almost strongly correlated). To build our *test benchmarks*, we solve each knapsack to optimality using dynamic programming and choose ten near-optimal solutions. We then run each of these solutions through our simulation environment twenty times, so that each solution is exposed to the exact same twenty simulations. We then record the average performance for each near-optimal solution over the twenty simulations to grade them. In practice, there would only be one reality the selected solution would be exposed to, of course. We run each test solution through twenty potential futures to lower the possibility that we are just lucky with the scenario we encountered.

The task for our data-driven solution selector is to pick a solution from the set of ten that exhibits very good performance in the simulated environment. To train this assessor, we generate training data as follows: For each knapsack type, we generate 500 instances. For each instance, we generate twenty near-optimal solutions. Moreover, for each of these instances, we generate one, and only one, vector of weights for each item. Note that, in practice, we would equally have access to our originally expected weights w_i, and the actual weights w_i'.

Next, we need to counter-factually assess the performance of each solution. To lower the variance in these labels, we proceed as follows: First, we build twenty derived scenarios from each real scenario, by choosing weights $w_i'' \in [w_i' - \alpha, w_i' + \alpha]$, where $\alpha = \frac{|w_i - w_i'|}{2}$, uniformly at random. That is, we derive scenarios from the historical examples without any assumptions regarding, or knowledge of, any distributions. We merely consider the *actually encountered* deviations from our original estimates and derive scenarios by varying these deviations a little. Please note that these changes do not affect the *direction* of the deviations: A weight that was under-estimated, remains under-estimated in each derived scenario, and each weight that was over-estimated remains over-estimated.

Finally, we execute each of our twenty near-optimal solutions under each derived scenario (including the recourse actions we would have taken) and label each with the average performance observed. Note that all that is needed to conduct this counterfactual assessment of additional solutions is the knowledge about our original estimates and the real item weights that were encountered.

Test results on all three classes of knapsacks are shown in Table 1. In the first column we denote the parameters of the experiment: The number of training scenarios vs number of test scenarios (usually 500-100), late-buying and restocking fees on train vs test (usually 10-10 or 5-5), and the knapsack capacity on train vs test (usually 10% or 20% of the weight of all items for both).

Table 1. SKP results

Type			Max	Mean	Min	ML	GC
Weakly correlated							
500-100	10-10	10%-10%	15.29	10.67	5.79	9.10	32
500-100	5-5	10%-10%	14.30	9.93	5.47	8.95	22
500-100	10-10	20%-20%	9.75	6.69	3.82	5.77	32
100-100	10-10	10%-10%	15.29	10.67	5.79	9.39	26
500-100	5-10	10%-10%	15.29	10.67	5.79	9.38	26
Strongly correlated							
500-100	10-10	10%-10%	15.56	10.84	6.03	8.88	41
500-100	5-5	10%-10%	14.46	10.03	5.54	8.81	27
500-100	10-10	20%-20%	11.73	8.18	4.58	6.50	47
100-100	10-10	10%-10%	15.56	10.84	6.03	9.31	32
500-100	5-10	10%-10%	15.56	10.84	6.03	8.83	42
Almost strongly correlated							
500-100	10-10	10%-10%	15.63	11.12	5.86	9.42	32
500-100	5-5	10%-10%	14.21	10.03	5.44	8.93	24
500-100	10-10	20%-20%	11.82	8.31	4.43	7.31	26
100-100	10-10	10%-10%	15.63	11.12	5.86	9.68	27
500-100	5-10	10%-10%	15.63	11.12	5.86	9.59	29
Heterogeneous mix							
500-100	10-10	10%-10%	17.28	11.34	5.31	9.32	34
500-100	5-5	10%-10%	15.73	10.35	4.83	8.83	27
500-100	10-10	20%-20%	11.97	7.86	3.85	6.53	33
100-100	10-10	10%-10%	17.28	11.34	5.31	10.0	22
500-100	5-10	10%-10%	17.28	11.34	5.31	9.51	30

Next, we show the average of the worst of the ten solutions we generated for each test instance, the expected performance, and the performance if we chose the best of the ten solutions generated. Note that the latter is the maximum gain we can hope to achieve by selecting among the ten solutions generated. The numbers represent percentages above an imaginary best solution (since the ten we generated may obviously not include the optimum under uncertainty), which we set at three standard deviations below the average of the ten solutions, and whose performance itself we measure as percent above the best omniscient solution. In absolute terms, the numbers presented are thus percentages over percentages over the true profits.

Finally, we show the performance of counterfactual selection (ML), as well as the percent gap closed (GC) between the average performance and the best performance that is achievable by selecting among those select ten solutions for each instance.

Overall, we close between 22% and 47% of the gap between the average performance and the best solution available to us. That means that our forecasting models are certainly not optimal, but nevertheless effective at choosing solutions which are expected to perform better than the average near-optimal solution. This holds for varying knapsack types as well as different capacities and recourse penalties.

To assess how critical the amount of historical scenarios is, we lowered the training set to only 100 scenarios. On all knapsack types, this leads to a reduction in effectiveness, but the approach still works: We close 26%, 32% and 27% for the three knapsack types using only 20% of scenarios.

Encouragingly, we see that counterfactual forecasting can also be reasonably effective when the historical scenarios used were gathered under a different regime. For example, assume that, historically, the late-buy and restocking fees were 5, but now they are 10. Please note that what should be done when operational parameters change is to re-run the historical scenarios under the new penalties and to generate a new counterfactual training set this way. For experimental purposes only, we did not do that here so we can assess how robust our forecasting models are under varying parameters. Under [500-100 5–10 10%-10%] we see that we achieve 26%, 42% and 29% gap closed for weakly, strongly, and almost strongly correlated knapsacks, respectively.

Finally, we generated a benchmark which consists, in equal parts, of weakly, strongly, and almost strongly correlated knapsack instances, both for training and for testing. As the table shows, the counterfactuals-based predictive models work for heterogeneous mixes of different knapsack types as well.

Overall, we conclude that, for the SKP, we can learn an effective, though suboptimal, data-driven model to predict which near-optimal solution has greater chances of performing well in an uncertain future.

5   RCPSP with Uncertain Job Durations

The RCPSP with uncertain job durations involves the scheduling of a set of jobs J given a set of resources R and a set of time periods T. Each job j consumes u_{jr} units of resource r in each time period the job is running. Each resource has a maximum capacity k_r that may be consumed in each time period. A precedence graph $P = J \times J$ specifies an order in which jobs are executed, i.e., for $(i,j) \in P$ job i must be completed before j can start. In the deterministic case, each job j has a fixed duration d_j. We consider a version of the problem where the job duration is uncertain, and assume that, if a job takes longer than planned, it continues to consume u_{jr} resources in each additional time period. This version of the problem corresponds closely with real-world RCPSPs, such as construction or software projects in which delays are common, and resource consumption of jobs continues even if they take longer than expected.

The (deterministic) RCPSP can be modeled with the following constraint program [2], in which the start time of each job is given by S_j:

$$\min \max_{j \in J} S_j + d_j \tag{1}$$

$$\text{subject to } S_i + d_i \leq S_j \qquad\qquad \forall (i,j) \in P \tag{2}$$

$$\texttt{cumulative}(\mathbf{S}, \mathbf{d}, \mathbf{u}_{.r}, k_r) \qquad\qquad \forall r \in R \tag{3}$$

5.1   Stochastic Environment

We sample the job durations d'_j from a Pareto distribution with the expected value d_j. The simulation starts at time period 0 and iterates through each time period until the maximum time is reached or all jobs have been executed. For some time period t', all jobs ending in that period $(t' = S_j + d'_j)$ are ended and the resources they are consuming freed. We then start jobs that have a start time of the current time period, if their precedence constraints are satisfied and their resource consumption requirements can be met. If job j with $S_j = t'$ cannot start in t', $S_j \leftarrow S_j + 1$, i.e., we delay its start by one time period.

If significant delays occur, it may be appropriate to do recourse planning and find new start times for the remaining jobs based on the current forecast. The recourse planning involves simply fixing the start times of jobs that are finished, or running and updating the job durations with either the real duration for finished/running jobs or the current forecast for scheduled jobs. This deterministic problem can then be solved by any RCPSP algorithm, based on the CP model above.

We forecast job durations by assuming that, when a job i that must precede j is finished (i.e., $(i, j) \in P$), we know more about the duration of j than we did before i finished. We construct a graph with the same nodes and arcs as in the precedence graph, and assign the true duration d'_i to every arc (i, j). Let a_{ij} be the shortest path between all pairs of jobs on the newly constructed graph with Dijkstra's algorithm. We then compute the forecast as $f_{ij} := \text{round}(d_j + (1 - a_{ij}/\max_{k \in J}\{a_{ik}\})(d'_j - d_j))$, such that f_{ij} is the forecast for job j when job i is finished, assuming j is reachable from i in P. While simulating, when job i finishes we check f_{ij}, and if it is closer to the true duration of j, we update our expected duration.

5.2   Winner Forecasting

We propose the following groups of features to describe solutions to the RCPSP with uncertain job durations.

1. The expected makespan of the solution divided by the maximum time.
2. Let $B_{ij} := S_j - S_i$ for all $(i, j) \in P$ be the buffer between jobs with precedence relations. We compute the mean, median, standard deviation, skew, 25% quantile, and 75% quantile of the values in B.
3. Let $B^T_{ij} := B_{ij}(S_j/|T|)$. We compute the mean and skew of B^T.
4. We execute the solution with the expected durations and compute the percentage residual resource usage $\hat{k}_{tr}$ for each resource at each time period, and aggregate this into $\hat{k}_t := \sum_{r \in R} \hat{k}_{tr}/|R|$. We compute the mean, standard deviation, 25% quantile and 75% quantile over all values of $\hat{k}_t$.
5. Let m^1 and m^2 be the number of jobs *directly* affected (we do not examine network effects here) due to insufficient available resources if a job j starts 1 or 2 time periods later than planned, respectively.

Table 2. RCPSP results

Class	Max Dur.	Train	Test	Max	Mean	Min	ML	GC
j30	10	317	157	4.48	3.27	2.67	3.44	−28
	50	275	138	3.84	3.08	2.59	3.01	15
	100	260	135	3.70	3.05	2.63	3.02	7
j60	10	239	122	4.06	3.29	2.53	3.18	15
	50	225	109	3.48	2.88	2.27	2.70	30
	100	226	113	3.38	2.82	2.28	2.70	23

6. Let a *delay chain* be a path in the precedence graph which forms a sequence of jobs that are separated with buffer less than the 25% quantile of B. We compute the maximum length delay chain and divide it by the total number of jobs.

5.3   Numerical Results

We test our approach on the well-known instances from the PSPLIB [15]. We use the j30 and j60 categories, which have 30 and 60 jobs, respectively, and split each into 320 training instances and 160 testing instances. The maximum job duration in these categories is 10 time units, so we add two more instance categories containing randomly generated job durations with a maximum of 50 and 100 time units, respectively.

We solve the constraint programming model in (1) through (3) with Google OR Tools CP-SAT solver version 7.0 [11]. We first generate the optimal expected values solution. We note that, in the RCPSP, shifting the buffer of a few jobs results in a "new" solution, but this is not desirable for our approach, as the realized performance will be nearly the same. Therefore, to generate $k - 1$ solutions in addition to the optimal solution, we begin an iterative process. After a solution S' is found, we append the following constraints to require that a given percentage of the jobs have a different order than the previously found solution:

$$o_{ij} = 1 \Leftrightarrow S_i \circ S_j \qquad \forall S_i' \star S_j', \ (\circ, \star) \in \{(>,<),(<,>),(\neq,=)\} \qquad (4)$$

$$\sum_{i,j \in J, i < j} o_{ij} \geq h \qquad (5)$$

where the decision variable $o_{ij} \in \{0,1\}$ for $i,j \in J, i < j$ is 1 iff jobs i and j have a different order than in the previous solution. We require the number of job order changes in (5) to be greater than a threshold h, which we set to 5% of the unique job pairs ($|J||J - 1|$).

As for the SKP and SSPP, we test our approach on 20 simulations per instance, simulating training and testing instances the same way as the previous two problems, with training simulations being derived from only one real simulation without knowledge of the actual distributions of job delays, and test simulations running 20 real scenarios for proper evaluation. Table 2 shows the results in the same format as the SKP and SSPP, with the addition of columns indicating the number of training and testing instances.

We are able to achieve modest gains over using the expected value solution, except in the case of j30 with a maximum expected duration of 10. On this instance set, the learning algorithm failed to find a good way of identifying superior solutions. This may be due to our features, which focus closely on buffer, and this may not be sufficient when the durations are low.

On the j60 instances, we are able to close between 15% and 30% of the gap to the best available solution. Even though the absolute gain may seem small, as with many optimization under uncertainty problems, real-world RCPSP problems can involve expensive resources (specialized digging equipment, etc.), and even small absolute improvements often translate into significant cost savings, as well as time savings for the overall plan. Therefore, even though our method is heuristic in nature, it can be of high value in practice.

6   Stochastic Shortest Path Problem

In the SSPP, we are given a graph $G = (V, A)$ of nodes V and arcs A. Every arc $(i, j) \in A$ has an uncertain cost with an expected value of c_{ij}. The objective is to find a minimal cost path through the graph between a source node s and destination node t. The SSPP can model problems such as the routing of ships under the influence of weather, or routing a vehicle through a road network considering traffic delays.

6.1   Stochastic Environment

We base the stochastic environment for the SSPP on the one described for the SKP with a few problem-specific modifications. Given a solution to an SSPP instance, we first sample the realized costs c'_{ij} for each arc from a Pareto distribution with mean c_{ij} and the minimum at 90% of c_{ij}. We then begin executing the path given to us as one of the ten solutions, using c'_{ij} for each realized arc. If the accumulated delay exceeds 10% of the expected costs, we allow recourse planning every 5 nodes.

In the recourse planning, we adjust our forecast based on the current node. The assumption is that arcs close to this node have a more accurate forecast than those far away, since we would traverse these arcs in the nearer future. To assemble our forecast, let a_{ij} be the number of arcs between nodes i and j. Then, let the forecast cost for (i, j) be $f_{ij} := \text{round}(c_{ij} + (1 - a_{ij}/\Delta\})(c'_{ij} - c_{ij}))$ if $a_{ij} < 5$, and c_{ij} otherwise. We set Δ to 7 to keep the forecasts from becoming too accurate when we get close, but keep them inaccurate when we are far away.

6.2   Winner Forecasting

We introduce the following features to characterize an SSPP solution. For each feature set, we compute the minimum, maximum, mean, standard deviation,

Table 3. SSPP results

Graph Type	Nodes	Max	Mean	Min	ML	GC
G_{nm}	50k	207.38	91.68	43.00	68.63	47
Bottleneck	50k	165.55	97.06	54.27	71.80	59
Watts-Strogatz	25k	175.88	76.95	33.91	50.79	61
Mixed	50k/25k	182.81	92.98	43.50	61.17	64

skew and kurtosis of the array of values. For features using arc costs, we divide the costs by the average arc cost of the graph, and for features using node degrees, we use the average node degree of the entire graph.

1. Array of arc costs on the path
2. Array of arc costs over the set of arcs leaving nodes of the path going to nodes not on the path
3. Array of arc costs over the set of arcs leaving nodes that are connected to the path by a single node (excluding any arcs to nodes directly connected to the path or nodes on the path)
4. Array of node degrees in the path
5. Array of node degrees of nodes that are connected to nodes on the path
6. Array of node degrees over the set of nodes that are connected to the path by a single node (excluding any nodes directly connected to the path)

As in the case of the SKP, we concatenate the features for two given solutions with the difference between the features of both solutions, which are then used by the machine learning approach to determine the most promising solution.

6.3 Numerical Results

We build a dataset of SSPP instances consisting of graphs based on one of three graph types: G_{nm} [8], "bottleneck", and Watts-Strogatz small-world graphs. The bottleneck instances consist of five G_{nm} graphs of equal size connected sequentially with 5 links between each graph. We create 300 instances of each graph type and size and select random source and sink nodes for the path, splitting the instances into 200 train and 100 test. The expected arc costs c_{ij} are drawn uniformly random between 1 and 100. We further ensure that all graphs have no isolated components by adding arcs between such components and the rest of the graph.

For each SSPP graph, we generate the ten shortest paths for a given graph between s and t using Yen's algorithm [23]. We simulate using the same scenario structure as in the SKP. The "true" arc costs c'_{ij} are drawn from a Pareto distribution with an expected value c_{ij}, shifted so that the minimum is at $0.9c_{ij}$. Training instances are evaluated on 20 scenarios that are all variations of a single scenario (using the exact same scenario variation as in the SKP), and test instances are evaluated on 20 scenarios generated independently of each other. Every 5 nodes we check if the accumulated delay is more than 10% of the expected cost, and if it is, we run a recourse algorithm that tries to replan the shortest path to t from the current node.

Table 3 shows the results of our computational experiments. We compute the gap to the optimal shortest path considering c'_{ij} and average it over 20 scenarios as described above. Note that, for many graph instances, paths that were near optimal for the point-estimated scenario may perform much worse than the shortest path had we known the true arc distances beforehand. This leads to relatively high values in our table, but is really more a reflection of the inherent cost of uncertainty in this particular problem than the absolute performance of the particular algorithm used to optimize under the uncertainty. Looking closer at our data, we find that the expected path lengths of the solutions is usually about the same, with the bottleneck graphs exhibiting slightly higher variance than for the other graph types.

Despite the simplicity of our features (we just measure arc costs and node degrees), we are able to close the gap by around 50% in all graph types and 64% for the mixed setting. This provides further support that we can learn from historical data which solution features are favorable for later execution under stochastic disruption.

7   Conclusion

We have introduced a new methodology for modeling and heuristically solving stochastic optimization problems. The key idea is to move away from trying to accurately forecast the uncertainty in the problem instance at hand. Instead, we propose to use logs of historical estimates and the realities that followed for comparing various counterfactual solutions. Our thesis is that we can devise features that capture instance-dependent characteristics of the solutions that allow us to predict which solution from a solution pool will likely perform well for a new problem instance at hand.

The objective of this paper was to provide a proof of concept. We considered three stochastic optimization problems that would each be extremely hard to model and solve with existing approaches, even heuristically. For all three problems, we were able to quickly devise sets of features that were effective enough to choose solutions that were superior to picking an average solution from our pool of optimal (with respect to the underlying point-estimated optimization problem) or near-optimal solutions.

Note that we did not spend any time to optimize hyper-parameters of our learning approaches, or to engineer more effective features. Providing a general set of features and pairing it with off-the-shelf machine learning methods was enough to tackle each of the three optimization problems. We believe that the experimental results provided strongly support our thesis that we can learn from data which solutions will exhibit superior performance in an uncertain future. However, this is of course not to say that, in practice, one should not conduct feature engineering and hyper-parameter optimization to achieve even better results.

The ability to tackle complex stochastic optimization problems with thousands of recourse stages comes at a cost, though. The framework presented gives no guarantees regarding the quality of the solutions achieved, and dual bounds are not provided. Therefore, whenever traditional stochastic optimization is applicable and full online-reoptimization is feasible during real-world operations, we would recommend this approach. The framework introduced here is meant for situations when the traditional methods break down.

In the future, we intend to investigate if the models trained on historic counterfactuals can be mined to infer constraints to guide the search for less brittle solutions directly: solutions that are not only near-optimal for the "fair weather" data, but also have high probability of performing well under stochastic disruption. In this sense, the new framework opens the door for a comprehensive new research agenda for stochastic constraint optimization.

Acknowledgements. This work is partially supported by Deutsche Forschungsgemeinschaft (DFG) grant 346183302. We thank the Paderborn Center for Parallel Computation (PC^2) for the use of the OCuLUS cluster.

References

1. April, J., Glover, F., Kelly, J.P., Laguna, M.: Practical introduction to simulation optimization. In: Proceedings of the 35th Conference on Winter Simulation: Driving Innovation, pp. 71–78. Winter Simulation Conference (2003)
2. Berthold, T., Heinz, S., Lübbecke, M.E., Möhring, R.H., Schulz, J.: A constraint integer programming approach for resource-constrained project scheduling. In: Lodi, A., Milano, M., Toth, P. (eds.) CPAIOR 2010. LNCS, vol. 6140, pp. 313–317. Springer, Heidelberg (2010). https://doi.org/10.1007/978-3-642-13520-0_34
3. Birge, J.R.: Decomposition and partitioning methods for multistage stochastic linear programs. Oper. Res. **33**(5), 989–1007 (1985)
4. Birge, J.R., Louveaux, F.: Introduction to Stochastic Programming. Springer Science & Business Media, New York (2011). https://doi.org/10.1007/978-1-4614-0237-4
5. Cao, Z., Guo, H., Zhang, J., Niyato, D., Fastenrath, U.: Finding the shortest path in stochastic vehicle routing: a cardinality minimization approach. IEEE Trans. Intell. Transp. Syst. **17**(6), 1688–1702 (2015)
6. Dean, B.C., Goemans, M.X., Vondrák, J.: Approximating the stochastic knapsack problem: the benefit of adaptivity. Math. Oper. Res. **33**(4), 945–964 (2008)
7. Dupačová, J., Consigli, G., Wallace, S.W.: Scenarios for multistage stochastic programs. Ann. Oper. Res. **100**(1–4), 25–53 (2000)
8. Erdös, P., Rényi, A.: On random graphs. I. Publicationes Mathematicae (Debrecen) **6**, 290–297 (1959)
9. Fu, M.C., Glover, F.W., April, J.: Simulation optimization: a review, new developments, and applications. In: Proceedings of the Winter Simulation Conference, p. 13. IEEE (2005)
10. Glover, F., Kelly, J., Laguna, M.: New advances for wedding optimization and simulation. In: Winter Simulation Conference 1999 Proceedings, vol. 1, pp. 255–260. IEEE (1999)
11. Google: Google OR-Tools (2019). developers.google.com/optimization/

12. Hochreiter, R., Pflug, G.C.: Financial scenario generation for stochastic multi-stage decision processes as facility location problems. Ann. OR **152**(1), 257–272 (2007)
13. Kaut, M., Wallace, S.W.: Evaluation of scenario-generation methods for stochastic programming. Pac. J. Optim. **3**(2), 257–271 (2007)
14. Kleywegt, A.J., Shapiro, A., Homem-de Mello, T.: The sample average approximation method for stochastic discrete optimization. SIAM J. Opt. **12**(2), 479–502 (2002)
15. Kolisch, R., Sprecher, A.: PSPLIB-a project scheduling problem library. Eur. J. Oper. Res. **96**(1), 205–216 (1997)
16. Long, Y., Lee, L.H., Chew, E.P.: The sample average approximation method for empty container repositioning with uncertainties. Eur. J. Oper. Res. **222**(1), 65–75 (2012)
17. Luo, X., Dashora, Y., Shaw, T.: Airline crew augmentation: decades of improvements from sabre. INFORMS J. Appl. Anal. **45**(5), 409–424 (2015)
18. Malitsky, Y., Sabharwal, A., Samulowitz, H., Sellmann, M.: Algorithm portfolios based on cost-sensitive hierarchical clustering. In: Proceedings of the 23rd International Joint Conference on Artificial Intelligence (IJCAI), Beijing, China, 2013, pp. 608–614 (2013)
19. Pelikan, M., Goldberg, D.E., Cantú-Paz, E.: BOA: the Bayesian optimization algorithm. In: Proceedings of the 1st Annual Conference on Genetic and Evolutionary Computation-Volume 1, pp. 525–532. Morgan Kaufmann Publishers Inc. (1999)
20. Schütz, P., Tomasgard, A., Ahmed, S.: Supply chain design under uncertainty using sample average approximation and dual decomposition. Eur. J. Oper. Res. **199**(2), 409–419 (2009)
21. Van Slyke, R.M., Wets, R.: L-shaped linear programs with applications to optimal control and stochastic programming. SIAM J. Appl. Math. **17**(4), 638–663 (1969)
22. Verweij, B., Ahmed, S., Kleywegt, A.J., Nemhauser, G., Shapiro, A.: The sample average approximation method applied to stochastic routing problems: a computational study. Comput. Optim. Appl. **24**(2–3), 289–333 (2003)
23. Yen, J.Y.: An algorithm for finding shortest routes from all source nodes to a given destination in general networks. Q. Appl. Math. **27**(4), 526–530 (1970)

Structure-Driven Multiple Constraint Acquisition

Dimosthenis C. Tsouros[1]([⊠]), Kostas Stergiou[1], and Christian Bessiere[2]

[1] Department of Informatics and Telecommunications Engineering,
University of Western Macedonia, Kozani, Greece
{dtsouros,kstergiou}@uowm.gr
[2] CNRS, University of Montpellier, Montpellier, France
bessiere@lirmm.fr

Abstract. MQuAcq is an algorithm for active constraint acquisition that has been shown to outperform previous algorithms such as QuAcq and MultiAcq. In this paper, we exhibit two important drawbacks of MQuAcq. First, for each negative example, the number of recursive calls to the main procedure of MQuAcq can be non-linear, making it impractical for large problems. Second, MQuAcq, as well as QuAcq and Multi-Acq, does not take into account the structure of the learned problem. We propose MQuAcq-2, a new algorithm based on MQuAcq that integrates solutions to both these problems. MQuAcq-2 exploits the structure of the learned problem by focusing the queries it generates to quasi-cliques of constraints. When dealing with a negative query, it only requires a linear number of iterations. MQuAcq-2 outperforms MQuAcq, especially on large problems.

1 Introduction

Constraint acquisition learns the model of a constraint problem using a set of examples that are posted as queries to a human user or to a software system [1,2]. Constraint acquisition is an area where constraint programming meets machine learning, as the problem can be formulated as a concept learning task. In *passive* acquisition, examples of solutions and non-solutions are provided by the user. Based on these examples, the system learns a set of constraints that correctly classifies all the given examples [1,3–6]. A major limitation of passive acquisition is the requirement, from the user's part, to provide diverse examples of solutions and non-solution to the system. In contrast, *active* or *interactive* acquisition systems interact with the user while acquiring the constraint network. This is a special case of query-directed learning, also known as "exact learning" [7,8]. In such systems, the basic query is to ask the user to classify an example as solution or not solution. This "yes/no" type of question is called membership query [9], and this is the type of query that has received the most attention in active constraint acquisition [1,10,11]. The system can also ask the user to classify partial examples [12] or to provide a violated constraint when

© Springer Nature Switzerland AG 2019
T. Schiex and S. de Givry (Eds.): CP 2019, LNCS 11802, pp. 709–725, 2019.
https://doi.org/10.1007/978-3-030-30048-7_41

a proposed example is considered as incorrect [13]. Other types of queries, e.g. recommendation and generalization ones, have also been considered [14,15].

Quacq is a state-of-the-art interactive constraint acquisition algorithm that uses partial queries [12]. Given a negative example, QuAcq finds a constraint that is violated by repeatedly posting partial examples to the user. QuAcq needs a number of queries logarithmic in the size of the example to locate the scope of a violated constraint. Another relevant algorithm is MultiAcq [16]. This algorithm learns all the constraints that are violated by a negative example, but it needs a linear number of queries to learn each one. Recently, an algorithm called MQuAcq, that combines the strengths of QuAcq and MultiAcq and outperforms both of them, was proposed [17]. MQuAcq requires a logarithmic number of queries to locate the scope of each violated constraint, and discovers all the violated constraints from a negative example.

In this paper, we further enhance the efficiency of active constraint acquisition by identifying and addressing two important deficiencies of MQuAcq. We first show that there exist negative examples where the process of learning *all* the violated constraints can make $\Omega(|Y|^2)$ recursive calls, where $|Y|$ is the number of variables of the given example. This has important practical implications as MQuAcq becomes unacceptably slow when the size of the problems grows, and as a result it can be outperformed by its generally less efficient predecessor QuAcq. Another deficiency of MQuAcq (and also QuAcq and MultiAcq) is that although non-random problems usually display some structure/patterns in the way their constraints are interleaved, this is ignored by the acquisition process. By identifying and exploiting these patterns we could possibly speed up the process. Such patterns have for instance been exploited to detect types of variables suitable for generalization [18].

Aiming at addressing the above problems, we propose an algorithm called MQuAcq-2 that learns multiple constraints from a negative generated query, but not necessarily all of them as opposed to MQuAcq, and also exploits structure that may be present in the problem to better focus its queries. MQuAcq-2 blends together the following two ideas. First, MQuAcq-2 exploits the structure of the learned network to focus on some of the violated constraints instead of exhaustively searching in the generated example. In our implementation, we used the detection of quasi-cliques in the learned network and then focus on the missing constraints (i.e., the ones required to complete the cliques). Second, when trying to learn constraints from a negative example, the entire scope of a learned constraint is removed from the example as soon as the constraint is acquired. This means that the algorithm no longer guarantees to find all the violated constraints from a negative example, but nevertheless it may find several of them, and crucially, it only requires a linear number of iterations to achieve this. With the integration of these ideas we achieve the benefits of learning several constraints from each generated query and we also avoid the extensive search for scopes of MQuAcq. Experimental results with benchmark problems demonstrate that MQuAcq-2 offers significant improvements compared to MQuAcq, both in terms of time and number of queries, especially on large problems. Importantly, the new algorithm outperforms MQuAcq even in the absence of structure.

The rest of this paper is organized as follows. In Sect. 2 the necessary background on interactive constraint acquisition is presented. Section 3 reviews the basics of multiple constraint acquisition with MQuAcq. Section 4 presents the proposed methods. An experimental evaluation is presented in Sect. 5. Section 6 concludes the paper.

2   Background

The *vocabulary* (X, D) is a finite set of n variables $X = \{x_1, ..., x_n\}$ and a domain $D = \{D(x_1), ..., D(x_n)\}$, where $D(x_i) \subset \mathbb{Z}$ is the finite set of values for x_i. The vocabulary is the common knowledge shared by the user and the constraint acquisition system. A *constraint* c is a pair $(\mathrm{rel}(c), \mathrm{var}(c))$, where $\mathrm{var}(c) \subseteq X$ is the *scope* of the constraint and $\mathrm{rel}(c)$ is a relation between the variables in $\mathrm{var}(c)$ that specifies which of their assignments are allowed. $|\mathrm{var}(c)|$ is called the *arity* of the constraint. Two constraints $c1, c2$ are *overlapping* when $\mathrm{var}(c1) \cap \mathrm{var}(c2) \neq \emptyset$. A *constraint network* is a set C of constraints on the vocabulary (X, D). A constraint network that contains at most one constraint for each subset of variables (i.e., for each scope) is called a *normalized constraint network*. Following the literature, we will assume that the constraint network is normalized. Besides the vocabulary, the learner has a *language* Γ consisting of *bounded arity* constraints.

An example e_Y is an assignment on a set of variables $Y \subseteq X$. e_Y is rejected by a constraint c iff $\mathrm{var}(c) \subseteq Y$ and the projection $e_{var(c)}$ of e_Y on the variables in the scope $\mathrm{var}(c)$ of the constraint is not in $\mathrm{rel}(c)$. A complete assignment that is accepted by all the constraints in C is a solution to the problem. $sol(C)$ denotes the set of solutions of C. An assignment e_Y is called a partial solution iff it is accepted by all the constraints in C with a scope $S \subseteq Y$. Observe that partial solution is not necessarily part of a complete solution. An *implied* constraint c in C is a constraint such that, if removed from the constraint network, the set of solutions remains the same.

Using terminology from machine learning, concept learning can be defined as learning a Boolean function from examples. A *concept* is a Boolean function over D^X that assigns to each example $e \in D^X$ a value in $\{0, 1\}$, or in other words, classifies it as negative or positive. The target concept f_T is a concept that assigns 1 to e if e is a solution to the problem and 0 otherwise. In constraint acquisition, the target concept, also called target constraint network, is any constraint network C_T such that $sol(C_T) = \{e \in D^X \mid f_T(e) = 1\}$. The *constraint bias* B is a set of constraints on the vocabulary (X, D), built using the constraint language Γ. The bias is the set of all possible constraints from which the system can learn the target constraint network. $\kappa_B(e_Y)$ represents the set of constraints in B that reject e_Y.

In exact learning, the classification question asking the user to determine if an example e_X is a solution to the problem that the user has in mind is called a *membership query* $ASK(e)$. The answer to a membership query is positive if $f_T(e) = 1$ and negative otherwise. A *partial query* $ASK(e_Y)$, with $Y \subseteq X$, asks

---

Algorithm 1. The MQuAcq Algorithm

Input: B, X, D (B: the bias, X: the set of variables, D: the set of domains)
Output: C_L : a constraint network
 1: $C_L \leftarrow \emptyset$;
 2: **while** true **do**
 3:      $Scopes.clear()$;
 4:      **if** $sol(C_L) = \emptyset$ **then return** "collapse";
 5:      Generate e in D^X accepted by C_L and rejected by B;
 6:      **if** $e =$ nil **then return** "C_L converged";
 7:      **if** $\neg findAllCons(e, X, 0)$ **then return** "collapse";

---

the user to determine if e_Y, which is an assignment in D^Y, is a partial solution or not. Following the literature, we assume that all queries are answered correctly by the user.

The acquisition process has *converged* on the learned network $C_L \subseteq B$ iff C_L agrees with E and for every other network $C \subseteq B$ that agrees with E, we have $sol(C) = sol(C_L)$. If there does not exist a constraint network $C \subseteq B$ such that C agrees with E then the acquisition *collapses*. This happens when the target constraint network is not included in the bias, i.e. $C_T \nsubseteq B$.

3   Multiple Constraint Acquisition

We briefly describe the MQuAcq algorithm for multiple constraint acquisition [17], and we identify an important deficiency of this algorithm. MQuAcq (Algorithm 1) takes as input a bias B on a vocabulary (X, D), and returns a constraint network C_L equivalent to the target network C_T. It uses functions *FindScope-2* [17] and *FindC* [12].

MQuAcq starts by initializing the network C_L to the empty set (line 1) and then it enters the main loop (line 2). The array *Scopes*, which is initialized to be empty in line 3, is used within function *FindAllCons* as explained below. If C_L is unsatisfiable, the algorithm collapses (line 4). Otherwise, an assignment e is generated (line 5), satisfying C_L and violating at least one constraint in B. If such an example does not exist then the acquisition process has converged (line 6). Otherwise, it calls the function *FindAllCons* to find all the constraints that are violated by the example e and remove from B those that are surely not in C_T. If *findAllCons* return false then we have collapsed (line 7).

The recursive function *FindAllCons* (Algorithm 2) is used to find all the constraints from C_T that are violated by the generated negative example. It takes as parameters an example e, a set of variables Y, which defines the set of variables to search for the constraints, and an integer variable s, which is an identifier for the scopes. It returns false if collapse has occurred and true otherwise. *FindAllCons* adds to C_L all the constraints from C_T that are violated by the example e in Y. It uses the array *Scopes* to store all the scopes of the constraints that have been found from the current generated query.

Algorithm 2. findAllCons

Input: e, Y, s (e: the example, Y: set of variables, s: scopes identifier)
Output: $not\_collapsed$: returns false if collapsed, true otherwise
 1: **function** FindAllCons(e, Y, s)
 2:     **if** $\kappa_{B \backslash C_L}(e_Y) = \emptyset$ **then return** true;
 3:     **if** $s < |Scopes|$ **then**
 4:         **for** $x_i \in Scopes[s]$ **do**
 5:             **if** $\neg findAllCons(e, Y \backslash \{x_i\}, s+1)$ **then return** false;
 6:     **else**
 7:         **if** ASK(e_Y) = "yes" **then** $B \leftarrow B \backslash \kappa_B(e_Y)$;
 8:         **else**
 9:             $scope \leftarrow FindScope\text{-}2(e, \emptyset, Y, false)$;
10:             $c \leftarrow FindC(e, scope)$;
11:             **if** $c = $ nil **then return** false;
12:             **else** $C_L \leftarrow C_L \cup \{c\}$; $B \leftarrow B \backslash \{c\}$;
13:             $Scopes.push(scope)$;
14:             **if** $\neg findAllCons(e, Y, s)$ **then return** false;
15:     **return** true;

In any recursive call, *FindAllCons* starts by checking if there exists any violated constraint in B, not already in the learned network C_L. If not, it is implied that ASK(e_Y) = "yes" and the function returns true (line 2). After that, at line 3, *FindAllCons* checks if s is smaller than the size of *Scopes* (s acts an identifier of the scopes in which it has already branched). If $s < |Scopes|$, it means that the scope of a found violated constraint still exists in e_Y. Thus, *FindAllCons* is called recursively on each subset of Y created by removing one of the variables of the scope at position s of *Scopes* (lines 4–5), and increasing s by 1 to continue with the next scope in each recursive call.

If $s = |Scopes|$, branching has finished. The system asks the user to classify the partial example (line 7). If the answer is positive then the constraints in B that reject e are removed. Otherwise, function *FindScope-2* is called to find the scope of a violated constraint (line 9). *FindC* will then find a constraint from B with the discovered scope that is violated by e (lines 10–12). In lines 13–14, *FindAllCons* is called recursively to continue searching.

Functions *FindScope-2* and *FindC* are described in [17] and [12] respectively and are not included here due to space limitations.

MQuAcq models the query generation problem in line 5 of Algorithm 1 as an optimization problem that looks for a (partial) solution of C_L that maximizes the number of violated constraints in B. This heuristic is called max_B [17]. We will see in Sect. 5 that there are some cutoffs imposed.

Although MQuAcq offers improvements over its predecessors QuAcq and MultiAcq, both in terms of queries and cpu time, it still suffers from two weaknesses. We prove in Proposition 1 that the number of recursive calls to function *FindAllCons* to learn all the violated constraints from a negative example can be non-linear in the number of variables of the example given. This non-linear number of calls can significantly hinder the cpu time performance of MQuAcq.

Proposition 1. *Given a negative example e_Y, MQuAcq may require a number of recursive calls to function* FindAllCons *in $\Omega(|Y|^2)$ to learn all the constraints of C_T that are violated by e_Y.*

Proof. Consider a constraint network and a negative example e_Y with $|Y| = 2 \cdot p$. Assume that $\kappa_{C_T}(e_Y) = \{c_{1,2}, c_{2,3}, \ldots, c_{p-1,p}\}$, where $c_{i,j}$ denotes a constraint with scope $var(c) = \{i, j\}$. Assume that $\kappa_{B \setminus C_T}(e_Y) = \{c_{1,p+1}, c_{2,p+2}, \ldots, c_{p,2p}\}$. With this pattern, we have $|\kappa_{C_T}(e_Y)| = p - 1$ and $|\kappa_{B \setminus C_T}(e_Y)| = p$.

We know that the branching takes place for each one of the variables in the scope of each learned constraint, meaning that two recursive calls are made to function *FindAllCons* at each branching point. In addition, we know that the depth of the tree of the recursive calls to *FindAllCons* can be up to $|\kappa_{C_T}(e_Y)| + 1 = p$, as it branches once for each scope included in *Scopes* at line 5, whose size in the end will be equal to $|\kappa_{C_T}(e_Y)|$. The maximum depth is reached as all the constraints of $\kappa_{C_T}(e_Y)$ are learned in the first branch.

Due to the structure of $\kappa_{C_T}(e_Y)$ and $\kappa_{B \setminus C_T}(e_Y)$, in each level we will have one more branching point than the previous. This happens because every constraint in $\kappa_{C_T}(e_Y)$ has in common the first variable of its scope with one constraint in $\kappa_{B \setminus C_T}(e_Y)$. Thus, after the first variable removal in each level, one constraint from $\kappa_{B \setminus C_T}(e_Y)$ will not be violated by e'_Y and the algorithm will have to follow the right branch to remove it, adding a new branch to each level. Also, we know that the constraints from $\kappa_{B \setminus C_T}(e_Y)$ will not be removed by the function *FindScope-2*, as a constraint from $\kappa_{C_T}(e_Y)$ will be found first and returned.

Now, let us prove that this results in a total number of nodes $N = 1 + (\frac{Y}{2} - 1) \cdot \frac{Y}{2}$. As in each level l we have one more branching point than the previous, we know that each level l of the tree will have 2 more nodes than the level $l - 1$, without counting the first level with the root node. This results in $N = 1 + 2 + 4 + \ldots + 2 \cdot (p-1) = 1 + 2 \cdot \sum_{k=1}^{p-1} k = 1 + 2 \cdot \frac{(p-1) \cdot p}{2} = 1 + (\frac{Y}{2} - 1) \cdot \frac{Y}{2}$. Therefore, MQuAcq requires a number of recursive calls to function *FindAllCons* in $\Omega(|Y|^2)$ to learn all the constraints of C_T that are violated by e_Y. □

Another weakness of MQuAcq is that the extensive branching it makes to find all the constraints violated by a negative example yields a lot of (small) partial positive queries. It is better when positive queries violate a large number of constraints from B because we want to prune the bias as much as possible. It would thus be better to focus on asking small partial queries on specific constraints that have greater probability to be included in C_T instead of focusing on all the violated constraints.

4   MQuAcq-2

In this section we propose MQuAcq-2, a new algorithm that acquires multiple constraints from each negative generated example, but not necessarily all of them

---

Algorithm 3. MQuAcq-2: Quick Acquisition learning multiple scopes

---

Input: B, X, D (B: the bias, X: the set of variables, D: the set of domains)
Output: C_L : a constraint network
 1: $C_L \leftarrow \emptyset$;
 2: **while** true **do**
 3:     **if** $sol(C_L) = \emptyset$ **then return** "collapse";
 4:     Generate e in D^Y accepted by C_L and rejected by B;
 5:     **if** $e =$ nil **then return** "C_L converged";
 6:     $Y' \leftarrow Y$;
 7:     **do**
 8:         **if** ASK$(e_{Y'}) =$ "yes" **then** $B \leftarrow B \setminus \kappa_B(e_{Y'})$;
 9:         **else**
10:             $Scope \leftarrow FindScope\text{-}2(e_{Y'}, \emptyset, Y', false)$;
11:             $c \leftarrow FindC(e_{Y'}, Scope)$;
12:             **if** $c = nil$ **then return** "collapse";
13:             **else** $C_L \leftarrow C_L \cup \{c\}$; $B \leftarrow B \setminus \{c\}$;
14:             $NScopes \leftarrow Scope$;
15:             $NScopes \leftarrow NScopes \cup analyze\&Learn(e_Y)$;
16:             **for** $Scope \in NScopes$ **do**
17:                 $Y' \leftarrow Y' \setminus Scope$;
18:     **while** $\kappa_B(e_{Y'}) \neq \emptyset$

---

as opposed to MQuAcq. The intuition is to focus on constraints that are more likely to be included in C_T instead of exhaustively searching in the generated example, and thus to decrease the run time as well as the number of queries needed to learn the target network.

4.1 Algorithm Description

MQuAcq-2 (Algorithm 3) starts with an empty C_L and a bias B containing constraints that can be built using the constraint language Γ on the vocabulary X, D. MQuAcq-2 returns the learned constraint network C_L, equivalent to the target network C_T. MQuAcq-2 iteratively generates examples and posts them as queries to the user. If the answer from the user is negative (i.e., at least one constraint from C_T is violated from the query posted), it tries to learn multiple constraints. MQuAcq-2 achieves that with the two following steps:

- It exploits the structure of the learned network to focus on specific violated constraints from B,
- In case no more constraints can be learned by exploiting the structure of C_L, it tries to find some non-overlapping constraints of C_T. As we explain below, this allows us to alleviate the high run time that MQuAcq incurs when searching for all the violated constraints from each negative example.

MQuAcq-2 generates a (partial) example e satisfying C_L and rejecting at least one constraint from B (line 4). If it has not converged or collapsed, it tries to acquire multiple constraints of C_T violating e. At first it posts the example as a query to the user (line 8). In the case the answer of the user is positive then it removes from the bias the set $\kappa_B(e_{Y'})$ of all the constraints from B that reject $e_{Y'}$. If the answer is negative it tries to find a constraint by using the functions *FindScope-2* and *FindC* like in MQuAcq (lines 10–13).

The first novelty is that after a constraint is added to C_L, the system calls the function *analyze&Learn* (line 15) to analyze the structure of C_L and to ask partial queries on scopes of constraints violated by the initial example that seem to fit in that structure. The above steps are done repeatedly, removing from Y' the variables of the scope of each violated constraint it has already learned (lines 16–17) that are stored in the set $NScopes$. When no more constraint from B can be acquired by analyzing the structure of C_L, MQuAcq-2 tries to learn multiple non-overlapping constraints (lines 10–13). The iterative process ends when the example $e_{Y'}$ does not contain any violated constraint from the bias (line 18).

The second novelty is that MQuAcq-2 removes the entire scope of the acquired constraints at lines 16–17 to avoid the exhaustive branching that MQuAcq does by removing one variable in each call to *findAllCons*. We lose the guarantee to learn all the constraints violated by a generated example, as in MQuAcq, but on the other hand we achieve better performance in practice. The fact that MQuAcq-2 does not learn all the violated constraints from a generated example does not mean that it will not learn the entire network. The "missed" constraints will be learned at another example.

4.2   Using the Structure of the Problem to Learn Constraints

Function *analyze&Learn* (Algorithm 4) is used to analyze the structure of the learned network and then to focus on some of the violated constraints of the bias that fit in the structure, and thus are likely to be part of C_T. The type of structure that we have investigated so far is that of tightly connected groups of variables that form *quasi-cliques* that are hopefully extendable to complete cliques. Quasi-cliques are subgraphs with an edge density exceeding a threshold parameter [19,20]. More formally, given a threshold $\gamma \in [0,1]$, a (sub)graph $G = (V, E)$, with V the set of vertices and E the set of edges, is γ-dense if $|E(G)| \geq \gamma \cdot \frac{|V| * |V| - 1}{2}$. If in addition G is connected, it is a quasi-clique. We used quasi-clique detection to focus subsequent queries on the constraints that are still in B and could be included in C_L to possibly complete a detected quasi-clique to form a clique.

The algorithm we use for finding quasi-cliques is similar to the one used in [18]. It is based on the well-known Bron-Kerbosch's [21] algorithm for finding maximal cliques in a graph. It is a recursive backtracking function that searches for maximal quasi-cliques in the graph of constraints of C_L. We consider any type of constraint as an edge, as opposed to the algorithm used in [18], which considers only constraints with same relation.

Function *analyze&Learn* takes only the generated negative example e_Y as a parameter. It returns the set $NScopes$, which contains the scopes of the constraints learned. Function *analyze&Learn* starts by initializing the set $NScopes$

Algorithm 4. *analyze&Learn*

Input: e_Y : the example
Output: $NScopes$: the set of scopes of the constraints that have been learned
1: **function** *analyze&Learn*(e_Y)
2:     $NScopes \leftarrow \emptyset$;
3:     $QCliques \leftarrow FindQCliques(X, \emptyset, \emptyset)$;
4:     $C_Q \leftarrow \{c \mid c \in \kappa_B(e_Y) \setminus C_L \land \exists q \in QCliques \mid var(c) \subseteq q\}$;
5:     $PScopes \leftarrow \{Y' \mid c \in C_Q \land var(c) = Y'\}$;
6:     **for** $Y' \in PScopes$ **do**
7:         **if** $ASK(e_{Y'}) =$ "yes" **then** $B \leftarrow B \setminus \kappa_B(e_{Y'})$;
8:         **else**
9:             $Scope \leftarrow FindScope\text{-}2(e_{Y'}, \emptyset, Y', false)$;
10:            $c \leftarrow FindC(e_{Y'}, Scope)$;
11:            **if** $c = nil$ **then return** "collapse";
12:            **else** $C_L \leftarrow C_L \cup \{c\}$; $B \leftarrow B \setminus \{c\}$;
13:            $NScopes \leftarrow NScopes \cup Scope$;
14:        **if** $NScopes \neq \emptyset$ **then**
15:            $NScopes \leftarrow NScopes \cup analyze\&Learn(e_Y)$;
16:     **return** $NScopes$;

---

to the empty set (line 2). At line 3 it finds quasi-cliques in C_L via the function *FindQCliques*. A cutoff is imposed to this function, returning all the quasi-cliques found within this time limit. This is done to avoid the exponential time-complexity of finding all the quasi-cliques. $QCliques$ contains sets of variables where each set forms a quasi-clique in the graph of the already learned network. Using the quasi-cliques found, we fill the set C_Q with the predicted constraints, that is, the constraints of B that have not been already learned (i.e., not in C_L), have a scope that is included in a quasi-clique ($\exists q \in QCliques \mid var(c) \subseteq q$), and are violated by e_Y (are included in $\kappa_B(e_Y)$) (line 4). We only consider constraints violated by the current example to avoid the overhead of generating new examples to learn them. Next, we fill the set $PScopes$ with the scopes of these constraints (line 5). For each scope in $PScopes$ (line 6), the system posts a partial query to the user, focusing on the variables of the scope (line 7). If the answer is positive then the constraints that reject the example are removed from B. Otherwise, function *FindScope-2* is called to find the scope of the violated constraint (line 9). This is done to ensure that the violated constraint the user has in mind is not in a subscope. (As we use only binary problems in the experiments, that means it looks for unary constraints.) Next, *FindC* will select a constraint from B with the discovered scope that is violated by $e_{Y'}$ (line 10). If no constraint is found then the algorithm collapses (line 11). Otherwise, the constraint returned by *FindC* is added to C_L (line 12) and its scope is added to the set of found scopes (line 13). Finally, if any constraint is found (line 14), *analyze&Learn*(e_Y) is recursively called to check if new quasi-cliques have been formed (line 15). The scopes of the constraints learned by this call to *analyze&Learn* are added to $NScopes$, and $NScopes$ is returned (line 16).

Table 1. Execution of MQuAcq-2 at Example 1

Repetition	Y	e_Y	ASK	Constraint acquired
1	$x_1 - x_8$	$\{1, 2, 2, 2, 3, 3, 4, 4\}$	"no"	$\neq_{23}$
1.1	x_2, x_4	$\{-, 2, -, 2, -, -, -, -\}$	"no"	$\neq_{24}$
1.2	x_3, x_4	$\{-, -, 2, 2, -, -, -, -\}$	"no"	$\neq_{34}$
3	$x_1, x_5 - x_8$	$\{1, -, -, -, 3, 3, 4, 4\}$	"no"	$\neq_{56}$
3	x_1, x_7, x_8	$\{1, -, -, -, -, -, 4, 4\}$	"no"	$\neq_{78}$
3	x_1	$\{1, -, -, -, -, -, -, -\}$	-	-

We chose quasi-clique detection for the analysis of the structure of the network because cliques are a common structure in constraint networks. Function *analyze&Learn* could also look for other types of structures by simply replacing the search for quasi-cliques by any other type of structure (such as [22–25]). The problem of predicting which constraints of B are more likely to be included in C_T can also be seen as a link prediction problem. Any method which deals with this problem can be exploited (e.g., [14, 26–28]).

4.3 Example and Analysis of MQuAcq-2

Let us now illustrate the behavior of MQuAcq-2 on a simple example.

Example 1. The vocabulary is $X = \{x_1, ..., x_8\}$ and $D = \{D(x_1), ..., D(x_8)\}$ with $D(x_i) = \{1, ..., 8\}$, the target network C_T is $\{\neq_{12}, \neq_{13}, \neq_{14}, \neq_{23}, \neq_{24}, \neq_{34}, \neq_{56}, \neq_{78}\}$ and $B = \{\neq_{ij} \mid 1 <= i < 8 \wedge i < j <= 8\}$. Assume that the learned network so far is $C_L = \{\neq_{12}, \neq_{13}, \neq_{14}\}$ and $\gamma = 0.6$ in MQuAcq-2. Also, assume that the current example processed (generated at line 4 of MQuAcq-2) is $e = \{1, 2, 2, 2, 3, 3, 4, 4\}$.

The execution of MQuAcq-2 is presented in Table 1. The first column shows the iteration of the algorithm. In the second column the variables that are considered in Y are given, while in the third column the example e_Y is displayed. Column ASK shows the answer of the user to the query posted, if one is posted, − otherwise. Finally, the constraint learned is presented.

MQuAcq-2 will post the example to the user, and after receiving a negative answer it will find the constraint $\neq_{23}$ using functions *FindScope* and *FindC*. After learning this constraint, it detects a possible clique among variables x_1, x_2, x_3, x_4 as shown in Fig. 1. So the algorithm will now focus on constraints $\neq_{24}, \neq_{34}$ that are violated by e, and will learn them via the function *analyze&Learn* (iterations 1.1,1.2). As no other quasi-clique (with $\gamma = 0.6$) has been detected, the algorithm continues by removing the entire scope of the constraints learned from Y. In the next iteration, after the negative classification by the user, constraint $\neq_{56}$ will be learned and variables x_5, x_6 will be removed. In the same way, the constraint $\neq_{78}$ will be acquired next. As no other constraint from B rejects the example e_Y after removing the variables from the last constraint learned, MQuAcq-2 will return to the query generation step.

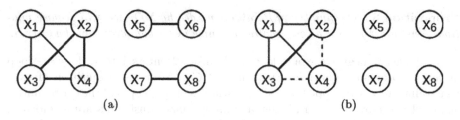

Fig. 1. (a) The target network of the problem. (b) The learned network so far and the predicted constraints

We now study the complexity of MQuAcq-2 in terms of the number of queries it needs to converge to the target network and in terms of the repetitions required to learn multiple violated constraints from a negative example.

Proposition 2. *Given a bias B built from a language Γ, with bounded arity constraints, and a target network C_T, MQuAcq-2 uses $O(|C_T| \cdot (\log |X| + |\Gamma|))$ queries to find the target network or to collapse and $O(|B|)$ queries to prove convergence.*

Proof. MQuAcq-2 learns each constraint from a negative example via the functions *FindScope* and *FindC* at lines 10–13 or via the function *analyze&Learn* using the same functions. We know that *FindScope* needs at most $|S| \cdot \log |Y|$ queries to locate a scope of a constraint from C_T, with $|S|$ being the arity of the scope and $|Y|$ the size of the example given to the function [12]. Since $Y \subseteq X$, *FindScope* needs in the worst case $|S| \cdot \log |X|$ queries to find a scope. In addition, we know that *FindC* needs at most $|\Gamma|$ queries to find a constraint from C_T in the scope it takes as parameter, if one exists [12]. In the case that none exists, the system collapses with the same bound. As a result, the number of queries necessary to find a constraint using the functions *FindScope* and *FindC* is $O(|S| \cdot \log |X| + |\Gamma|)$. Thus, the number of queries required for finding all the constraints in C_T or collapsing is at most $C_T \cdot (|S| \cdot \log |X| + |\Gamma|)$ which is $O(C_T \cdot (\log |X| + |\Gamma|))$ because $|S|$ is bounded. Concerning the convergence problem, it is proved when B is empty or contains only implied constraints. Constraints are removed from B when the answer from the user is "yes" in a query in the above cases. In the worst case, in which each positive query rejects only one constraint from B, it leads to at least one constraint removal in each query. This is because the example generated at line 4 of MQuAcq-2 violates at least one constraint from B and *analyze&Learn* does not ask a query $e_{Y'}$ when $\kappa_B(e_{Y'})$ is empty (lines 4–5 in *analyze&Learn*). This gives a total of $O(|B|)$ queries to converge. □

Therefore, MQuAcq-2 has a logarithmic complexity in terms of the number of queries needed to find the scope of a violated constraint, the same as QuAcq and MQuAcq. Now we turn out attention to the process of learning multiple constraints from a negative example e_Y.

Proposition 3. *The number of iterations needed by MQuAcq-2 to acquire multiple constraints from a given negative example e_Y is bounded above by $O(|Y|)$.*

Proof. Given a negative example e_Y, MQuAcq-2 enters into a do-while loop at line 7 to acquire multiple constraints of C_T. After the acquisition of each constraint, the entire scope (i.e., all the variables of the scope) of the constraint(s) acquired is removed from Y. Thus, assuming unary constraints are included in the target network of the problem, in the worst case only one variable will be removed from Y in each repetition. As a result, the worst case number of iterations made by MQuAcq-2 to acquire multiple constraints of C_T, is equal to $|Y|$. $\square$

Therefore, given a negative example, MQuAcq-2 learns multiple constraints of the target network in a complexity lower than MQuAcq. As we will see in the experiments, it significantly improves its time performance.

5 Experimental Evaluation

To evaluate our proposed algorithm, we ran experiments comparing MQuAcq-2 against MQuAcq. We also ran QuAcq as a reference point. Some more details about our experiments:

- All the experiments were conducted on a system carrying an Intel(R) Xeon(R) CPU E5-2667, 2.9 GHz clock speed, with 8 GB of RAM.
- The max_B heuristic [17] was used for the generation of the queries by all algorithms. max_B focuses on examples violating as many constraints as possible from B without necessarily building a complete variable assignment. *bdeg* was used for variable ordering, that is the variable with the most constraints in B is chosen. Random value ordering was used.
- For all the algorithms we set some cutoffs in the query generation step. The best (according to max_B) example found within 1 s is returned, even if not proved optimal. If after 5 s, not a single example is found, the system takes one by one each constraint c in B and tries to solve $C_L \cup \{\neg c\}$ with a additional cutoff of 5 s.
- We do not check for collapse before the generation of the queries, as it can be very time consuming, especially in large problems, with a lot of variables and a large C_T. We assume that the problem the user has in mind is solvable, the user's answers are correct and C_T is representable by B.
- In the function *FindQCliques*, γ was set to 0.8.
- As finding all the quasi-cliques is an NP-hard problem, we have added a cutoff of 1 s in the function *FindQCliques*, which then returns all the quasi-cliques found within this time limit.

We used the following benchmarks in our study:

Sudoku. The Sudoku puzzle is a $n^2 \times n^2$ grid. It must be completed in such a way that all the rows, all the columns and the n^2 non-overlapping $n \times n$ squares contain the numbers 1 to n. We use two variations of the problem with $n = 3$ and $n = 4$. This gives a *vocabulary* having 81 (256 respectively) variables and domains of size 9 (16 respectively). The target networks for the two problems are of size 810 and 4,992. The bias was initialized with 12,960 and 130,560 binary constraints respectively, using the language $\Gamma = \{=, \neq, >, <\}$.

Latin Square. The Latin square problem consists of an $n \times n$ table in which each element occurs once in every row and column. That means that the domain is of size n. We use two variations of the problem. The first one with 100 variables (i.e., $n = 10$) having a target network of 900 binary $\neq$ constraints on rows and columns, and the second with 225 variables (i.e., $n = 15$) and a target network of size 3,150. The language used was $\Gamma = \{=, \neq, >, <\}$, resulting in a bias of 19,800 binary constraints in the first problem and 100,800 constraints in the second.

Random. We used a problem with 100 variables and domains of size 5. We generated a random target network with 495 $\neq$ constraints. The bias was initialized with 19,800 constraints, using the language $\Gamma = \{=, \neq, >, <\}$.

Radio Link Frequency Assignment Problem. The RLFAP is the problem of providing communication channels from limited spectral resources [29]. We use a simplified version which consists of 50 variables with domains of size 40. The target network contains 125 binary distance constraints. We built the bias using a language of 2 basic distance constraints ($\{|x_i - x_j| > y, |x_i - x_j| = y\}$) with 5 different possible values for y. This led to a language of 10 different distance constraints. In total, B contains 12,250 constraints.

AllDiff. We used a problem with 50 variables and domains of size 50 with the condition that all variables must take different values. Thus, the target network contains a clique of 1,225 binary $\neq$ constraints. The bias was initialized with 4,900 constraints, using the language $\Gamma = \{=, \neq, >, <\}$.

To compare all the algorithms on the same simple scenario, all our experiments take the extreme case where we start from an empty constraint network. Even for the best algorithms, this scenario leads to an overall number of queries that can be considered as too large when the user is a human. In real applications, the user often has some background knowledge that give a frame of basic constraints. If not, the user may take other methods, such as ModelSeeker [6], to extract constraints from the structure of solutions of the problem. In this case, the interactive acquisition algorithm is only used to finalize the model.

5.1   Results

We measure the size of the learned network C_L, the average waiting time $\bar{T}$ (in seconds) for the user, the total number of queries $\#q$, the average size $\bar{q}$ of all queries, the number of complete queries $\#q_c$, and the total time needed to converge T_{total}. We present results of MQuAcq, QuAcq, MQuAcq-2 without *analyze&Learn* (denoted by MQuAcq-2 w/o A&L) and full MQuAcq-2. Each algorithm was run 5 times and the means are presented in Table 2.

Looking at the time performance of MQuAcq-2 without *analyze&Learn* we see that in all problems except AllDiff (which is relatively small) it decreases the total time of the acquisition process compared to MQuAcq. In the larger

Table 2. Results of MQuAcq-2

Benchmark	Algorithm	$|C_L|$	$\#q$	$\bar{q}$	$\#q_c$	$\bar{T}$	T_{total}
Latin 15×15	MQuAcq	3150	37176	101	27	0.18	6730.88
	QuAcq	3150	31426	117	2345	0.18	5808.12
	MQuAcq-2 w/o $A\&L$	3150	28364	70	42	0.04	1069.80
	MQuAcq-2	3150	25517	61	49	0.11	2757.66
Latin 10×10	MQuAcq	900	8457	46	16	0.02	155.84
	QuAcq	900	7997	53	784	0.13	1026.86
	MQuAcq-2 w/o $A\&L$	900	7265	33	31	0.02	139.58
	MQuAcq-2	900	6133	19	33	0.03	168.88
Sudoku 16×16	MQuAcq	4992	59953	90	18	0.49	29612.70
	QuAcq	4992	42648	52	120	0.33	14092.20
	MQuAcq-2 w/o $A\&L$	4992	43958	61	36	0.06	2771.94
	MQuAcq-2	4992	40905	51	35	0.15	6098.34
Sudoku 9×9	MQuAcq	810	6964	32	14	0.02	124.07
	QuAcq	810	6478	38	518	0.14	880.96
	MQuAcq-2 w/o $A\&L$	810	6136	24	31	0.02	94.12
	MQuAcq-2	810	4912	15	30	0.04	191.63
Random	MQuAcq	495	5959	45	10	0.03	168.52
	QuAcq	495	5500	53	472	0.10	570.96
	MQuAcq-2 w/o $A\&L$	495	4930	34	16	0.02	79.35
	MQuAcq-2	495	4962	34	16	0.02	79.05
RLFAP	MQuAcq	122	1520	22	26	0.14	222.72
	QuAcq	106	1168	25	71	0.23	274.01
	MQuAcq-2 w/o $A\&L$	124	1102	21	27	0.19	218.88
	MQuAcq-2	124	1113	21	23	0.19	237.27
AllDiff	MQuAcq	1225	3912	26	24	0.03	107.82
	QuAcq	1225	5082	34	1116	0.25	1280.27
	MQuAcq-2 w/o $A\&L$	1225	4153	23	51	0.03	135.24
	MQuAcq-2	1225	2774	14	37	0.12	345.65

problems the decrease in total time is quite significant: MQuAcq is 6.3 times slower on Latin 15×15 and 10.7 times slower on Sudoku 16×16. This confirms our intuition and complexity analysis. In terms of number of queries, we also have an important decrease compared to MQuAcq in all the problems except AllDiff (up to 26.7% in Sudoku 16×16). This decrease in the number of queries is mainly due to the fact that avoiding the extensive branching that MQuAcq makes, MQuAcq-2 also avoids posting a lot of small positive partial queries. As a result, the pruning of B is achieved with fewer queries. Another observation is that although the number of *complete* queries posted to the user is increased in all problems, the average size of the queries is reduced. This reduced size of the queries is mainly due to the smaller negative queries asked by MQuAcq-2 because, as opposed to MQuAcq, MQuAcq-2 removes the entire scope of the previous constraint learned (line 17) before proceeding with the search for constraints.

When MQuAcq-2 takes into account the structure of the problem by using *analyze&Learn*, we observe that the total time is larger than without using *analyze&Learn*, but it is still significantly faster than MQuAcq on the large problems. The decrease in time is 59% on Latin 15×15, 79.4% on Sudoku 16×16 and 46.9% on Random. In contrast, the run time is increased by 8.4% on Latin 10×10, 54.5% on Sudoku 9×9, 6.5% on RLFAP and 220.6% on AllDiff. The increase in time in these smaller problems is due to the overhead of *analyze&Learn*. The largest increase is in the AllDiff problem. This is not surprising as the AllDiff problem consists of a single clique of constraints. Focusing on the number of queries, by using *analyze&Learn* to exploit the structure of the learned network, MQuAcq-2 offers significant improvements compared to MQuAcq on all problems. The number of queries is decreased by 31.4% on Latin 15×15, 27.5% on Latin 10×10, 30.4% on Sudoku 16×16, 29.5% on Sudoku 9×9, 17.3% on Random 26.8% on RLFAP and 29.1% on AllDifferent. Interestingly, it seems that the larger the target network, the bigger the gain. We also observe that the more structured the problem is, the more queries full MQuAcq-2 saves compared to MQuAcq-2 without *analyze&Learn*. The average size of the queries is even smaller than with MQuAcq-2 without *analyze&Learn*. This is because *analyze&Learn* focuses on small scopes in the most promising parts of the network, avoiding the large negative queries that FindScope-2 would have asked to find these scopes.

Looking at the performance of full MQuAcq-2 on the Random problem, we see that both in terms of time and number of queries it dominates MQuAcq. Random is a pure random problem without any kind of structure. Thus, although MQuAcq-2 tries to exploit the structure of the problem to enhance the acquisition process, it performs quite well even in the absence of structure. The results with and without *analyze&Learn* are quite similar, confirming that in the absence of structure the overhead of *analyze&Learn* is relatively small. The same occurs on the RLFAP problem, which does not contain any cliques.

Finally, regarding QuAcq, we observe that this algorithm is better than MQuAcq both in run times and in number of queries on the largest problems that had not been considered before. However, it is inferior to MQuAcq-2 both in terms of time and number of queries on all problems.

6 Conclusion

Although the MQuAcq algorithm for constraint acquisition was shown to outperform previous algorithms such as QuAcq and MultiAcq, we have demonstrated that it suffers from two important drawbacks. First, the process of learning a maximum number of constraints from each negative generated example is time-consuming. This makes MQuAcq inefficient for large problems. Second, MQuAcq, as well as QuAcq and MultiAcq, does not take into account the structure revealed as constraints are learned. We have proposed a new algorithm, named MQuAcq-2, that integrates solutions to both of these problems. MQuAcq-2 exploits the structure of the learned problem by focusing the queries it generates to quasi-cliques of constraints that are being revealed. In addition, it alleviates the high cpu time requirements of MQuAcq by acquiring multiple constraints from each generated negative example, but not trying to learn all of them. Experiments with benchmark problems demonstrate that MQuAcq-2 outperforms MQuAcq both in terms of the number of queries and in the total time of the acquisition process, especially on large problems.

References

1. Bessiere, C., Koriche, F., Lazaar, N., O'Sullivan, B.: Constraint acquisition. Artif. Intell. **244**, 315–342 (2017)
2. Bessiere, C., et al.: New approaches to constraint acquisition. In: Bessiere, C., De Raedt, L., Kotthoff, L., Nijssen, S., O'Sullivan, B., Pedreschi, D. (eds.) Data Mining and Constraint Programming. LNCS (LNAI), vol. 10101, pp. 51–76. Springer, Cham (2016). https://doi.org/10.1007/978-3-319-50137-6_3
3. Bessiere, C., Coletta, R., Freuder, E.C., O'Sullivan, B.: Leveraging the learning power of examples in automated constraint acquisition. In: Wallace, M. (ed.) CP 2004. LNCS, vol. 3258, pp. 123–137. Springer, Heidelberg (2004). https://doi.org/10.1007/978-3-540-30201-8_12
4. Bessiere, C., Coletta, R., Koriche, F., O'Sullivan, B.: A SAT-based version space algorithm for acquiring constraint satisfaction problems. In: Gama, J., Camacho, R., Brazdil, P.B., Jorge, A.M., Torgo, L. (eds.) ECML 2005. LNCS (LNAI), vol. 3720, pp. 23–34. Springer, Heidelberg (2005). https://doi.org/10.1007/11564096_8
5. Lallouet, A., Lopez, M., Martin, L., Vrain, C.: On learning constraint problems. In: 22nd IEEE International Conference on Tools with Artificial Intelligence (ICTAI), vol. 1, pp. 45–52. IEEE (2010)
6. Beldiceanu, N., Simonis, H.: A model seeker: extracting global constraint models from positive examples. In: Milano, M. (ed.) CP 2012. LNCS, pp. 141–157. Springer, Heidelberg (2012). https://doi.org/10.1007/978-3-642-33558-7_13
7. Bshouty, N.: Exact learning boolean functions via the monotone theory. Inf. Comput. **123**(1), 146–153 (1995)
8. Bshouty, N.H.: Exact learning from an honest teacher that answers membership queries. Theor. Comput. Sci. **733**, 4–43 (2018)
9. Angluin, D.: Queries and concept learning. Mach. Learn. **2**(4), 319–342 (1988)
10. Bessiere, C., Coletta, R., O'Sullivan, B., Paulin, M., et al.: Query-driven constraint acquisition. In: International Joint Conference on Artificial Intelligence (IJCAI), vol. 7, pp. 50–55 (2007)
11. Shchekotykhin, K., Friedrich, G.: Argumentation based constraint acquisition. In: Ninth IEEE International Conference on Data Mining, pp. 476–482. IEEE (2009)
12. Bessiere, C., et al.: Constraint acquisition via partial queries. In: International Joint Conference on Artificial Intelligence (IJCAI), vol. 13, pp. 475–481 (2013)
13. Freuder, E.C., Wallace, R.J.: Suggestion strategies for constraint-based matchmaker agents. In: Maher, M., Puget, J.-F. (eds.) CP 1998. LNCS, vol. 1520, pp. 192–204. Springer, Heidelberg (1998). https://doi.org/10.1007/3-540-49481-2_15

14. Daoudi, A., Mechqrane, Y., Bessiere, C., Lazaar, N., Bouyakhf, E.H.: Constraint acquisition using recommendation queries. In: International Joint Conference on Artificial Intelligence (IJCAI), pp. 720–726 (2016)
15. Bessiere, C., Coletta, R., Daoudi, A., Lazaar, N., Mechqrane, Y., Bouyakhf, E.H.: Boosting constraint acquisition via generalization queries. In: European Conference on Artificial Intelligence (ECAI), pp. 99–104 (2014)
16. Arcangioli, R., Bessiere, C., Lazaar, N.: Multiple constraint aquisition. In: International Joint Conference on Artificial Intelligence (IJCAI), pp. 698–704 (2016)
17. Tsouros, D.C., Stergiou, K., Sarigiannidis, P.G.: Efficient methods for constraint acquisition. In: 24th International Conference on Principles and Practice of Constraint Programming (2018)
18. Daoudi, A., Lazaar, N., Mechqrane, Y., Bessiere, C., Bouyakhf, E.H.: Detecting types of variables for generalization in constraint acquisition. In: 2015 IEEE 27th International Conference on Tools with Artificial Intelligence (ICTAI), pp. 413–420. IEEE (2015)
19. Pardalos, J., Resende, M.: On maximum clique problems in very large graphs. DIMACS Ser. **50**, 119–130 (1999)
20. Abello, J., Resende, M.G.C., Sudarsky, S.: Massive quasi-clique detection. In: Rajsbaum, S. (ed.) LATIN 2002. LNCS, vol. 2286, pp. 598–612. Springer, Heidelberg (2002). https://doi.org/10.1007/3-540-45995-2_51
21. Bron, C., Kerbosch, J.: Algorithm 457: finding all cliques of an undirected graph. Commun. ACM **16**(9), 575–577 (1973)
22. Gibson, D., Kumar, R., Tomkins, A.: Discovering large dense subgraphs in massive graphs. In: Proceedings of the 31st International Conference on Very large data bases, VLDB Endowment, pp. 721–732 (2005)
23. Girvan, M., Newman, M.E.: Community structure in social and biological networks. Proc. Nat. Acad. Sci. **99**(12), 7821–7826 (2002)
24. Newman, M.E.: Modularity and community structure in networks. Proc. Nat. Acad. Sci. **103**(23), 8577–8582 (2006)
25. Papadopoulos, S., Kompatsiaris, Y., Vakali, A., Spyridonos, P.: Community detection in social media. Data Min. Knowl. Disc. **24**(3), 515–554 (2012)
26. Adamic, L.A., Adar, E.: Friends and neighbors on the web. Soc. Netw. **25**(3), 211–230 (2003)
27. Leicht, E.A., Holme, P., Newman, M.E.: Vertex similarity in networks. Phys. Rev. E **73**(2), 026120 (2006)
28. Liben-Nowell, D., Kleinberg, J.: The link-prediction problem for social networks. J. Am. Soc. Inform. Sci. Technol. **58**(7), 1019–1031 (2007)
29. Cabon, B., De Givry, S., Lobjois, L., Schiex, T., Warners, J.P.: Radio link frequency assignment. Constraints **4**(1), 79–89 (1999)

Computational Sustainability Track

Towards Robust Scenarios of Spatio-Temporal Renewable Energy Planning: A GIS-RO Approach

Nadeem Al-Kurdi[1], Benjamin Pillot[1], Carmen Gervet[2(✉)],
and Laurent Linguet[1]

[1] ESPACE-DEV, Univ Guyane, Univ Montpellier, IRD, Univ Réunion,
275 route de Montabo, BP 165, 97323 Cayenne Cedex, French Guiana
`nadeem.alkurdi@teledetection.fr`, `benjamin.pillot@ird.fr`,
`laurent.linguet@univ-guyane.fr`
[2] ESPACE DEV, Univ Montpellier, Univ Guyane, IRD, Univ Réunion,
500 rue Jean François Breton, 34090 Montpellier, France
`carmen.gervet@umontpellier.fr`

Abstract. Solar-based energy is an intermittent power resource whose potential pattern varies in space and time. Planning the penetration of such resource into a regional power network is a strategic problem that requires both to locate and bound candidate parcels subject to multiple geographical restrictions and to determine the subset of these and their size so that the solar energy production is maximized and the associated costs minimized. The problem is also permeated with uncertainty present in the estimated forecast energy demand, resource potential and technical costs. This paper presents a novel combination of Geographic Information Systems (GIS) and Robust Optimization (RO) to develop strategic planning scenarios of a collection of parcels that accounts for their spatio-temporal characteristics, and specifically their hourly radiation patterns that are location dependent, to best fit the network temporal demand and minimize technical costs.

The problem is formulated as a GIS spatial placement problem and a RO fractional knapsack problem to plan the effective power penetration and geographical suitability of new PV facilities. The combination GIS-RO generates an excellent decision support system that allows for the computation of optimized parcel scenarios (locations, sizes and power). The qualitative and quantitative effectiveness of the approach is demonstrated on real data on the French Guiana region. Results show that the proposed approach provides reliable fine grained planning that also accounts for the risk adversity of the decision maker towards forecast demand and solar potentials.

Keywords: Renewable energy planning ·
Geographic Information Systems · Robust optimization

1 Introduction and Related Work

Energy transition from a high-carbon regime born of fossil fuels to low-carbon solutions is a major challenge of current societies [10]. Most of the energy

T. Schiex and S. de Givry (Eds.): CP 2019, LNCS 11802, pp. 729–747, 2019.
https://doi.org/10.1007/978-3-030-30048-7_42

planning strategies therefore aim at enhancing the share of renewable energy (RE) sources within power networks [23]. When those energies are dispatchable, the integration remains pretty straightforward [16,36]. However, when it comes to volatile, or *intermittent* RE sources, this is no longer true. Their unstable and variability nature implies that the resulting aggregated power injected into the grid may threaten network's stability by not matching the power demand [23].

Thus, the full problem of integrating intermittent RE sources in power networks involves irregular spatio-temporal energy potential patterns, related to the location and dimensions of power facilities. It is a multi-criteria uncertain optimization and planning problem that combines the spatial placement of candidate parcels for installing power facilities subject to *geographical* constraints and *temporal* resource constraints, and the selection (location, size and capacity) of optimal plants such that the power into the network is increased at minimal costs and short-term unpredictability is limited to an acceptable level.

Taking into account spatio-temporal energy potential data, together with heterogeneous land, network, and technico-economic constraints for effective renewable energy planning remains a challenging multi-dimensional problem permeated with uncertain data [37]. The computational approaches are broadly divided into two research streams: (1) geographical information system (GIS) modeling with multi-criteria decision-making (MCDM) [3,6,13,34], and (2) bottom-up engineering approaches [7,23,35,37]. GIS models with MCDM focus on providing suitability maps based on static resource assessment and expert-based decision criteria. The maps depict areas with their respective weighted criteria values to be used in the MCDM model, such as economic, environmental or technical ones. These approaches do not aim at optimizing the actual parcel selection that would require taking into consideration the short-term temporal variation of both the resource and power demand, or their evolution in the long-term. On the other hand, bottom-up engineering approaches allow for time simulation and optimization of given energy system configurations. Their main objective is to guide energy policy road map often at a national scale and longer time horizon. This systemic approach gives a significant insight into the potential contribution of RE sources [11,12], but does not aim at identifying physical parcel locations. Similarly [22] addresses the resource management problem as a knapsack problem, that shows the suitability of linear programming to select among experts' given parcels, the ones with highest resource potential. It does not consider the hourly temporal patterns of the different sites and their projected uncertainty, the possibility to consider a fraction of a given parcel, nor the impact of geographical restrictions and distances, and the complexity of the associated technical costs.

In summary, to date we are not aware of computational approaches that tackle the spatio-temporal optimization problem consisting of identifying the best parcels that increase solar energy penetration into the network at minimal cost, while satisfying a region's specific constraints (terrain, resource, infrastructures, etc) and related costs. This paper addresses this problem by proposing a two-steps specification in terms of a spatial placement and a resource planning problem, and we propose an integrated computational approach. The approach contributes a novel framework based on GIS spatio-temporal data and constraint processing, connected to a Robust Optimization (RO) knapsack model to plan renewable energy scenarios. The combination GIS-RO generates an excellent decision support system that allows for the planning of parcel scenarios (locations and areas) that will best increase the RE power into the network at minimal cost, according to the decision maker risk adversity. A GIS can handle very large volumes of data, including remote sensing images for solar radiation indicators, land use maps, and various networks maps (electrical, roads, water). The application of global and multi-layers geographical constraints and various control parameters allow for an effective deterministic pruning of the region, to determine suitable candidate parcels, and their relevant properties without impairing the optimization problem.

The core contributions of this paper are: (1) the specification of a complex spatial placement and planning problem, (2) a computational approach that efficiently exploits GIS geographical constraints, and makes powerful use of large scale spatio-temporal environmental data, and (3) an integration of the spatial analysis with a robust optimization module through a comprehensive set of resource and contextual features. Through the use of Robust Optimization, data uncertainty present in the forecast figures for the planned horizon is tackled with a measure of robustness, allowing best and worst case scenarios to be studied according to various risk adversity positions of the decision maker.

The GIS-RO framework presented in this paper, is applied to a real world challenge of PV solar power plants planning in the region of French Guiana. It illustrates the qualitative and quantitative efficiency of the approach as a decision support system providing solution scenarios. The paper is organized as follows: Sect. 2 describes the problem and overall approach; Sect. 3 presents the GIS module; Sect. 4 the robust optimization model; Sect. 5 is the experimental section based on a real-world case study for robust spatial decision making from time series resources; and Sect. 6 concludes the paper.

2   Problem Description, Application and Approach

The problem is motivated by a renewable energy scenario planning problem from the 2015 Energy Transition Act. France's energy policy has the target for overseas regions, in particular French Guiana, of 50% of renewable energy in final consumption in 2020 and full energy self-sufficiency by 2030 [15]. The challenge is to identify suitable candidate parcels for RE parks and determine the ones, and their optimal size, that would maximize power network contribution at minimal costs. A candidate parcel must satisfy a number of geographical constraints, including topographic land use restrictions, type of ground surface, be at a maximum distance threshold from the electrical grid, and have a maximum surface with limited land slope. These constraints bound the areas for candidate parcels.

The scenarios for the best parcels selection deal with the resource potential and costs associated with each candidate parcel. The intermittent resource follows a temporal pattern specific to the geographic location. The costs are mainly technical costs (installation, maintenance, grid connection) that depend on the size of the parcel and its distance to the grid. These data have a degree of uncertainty in terms of their future value. The full problem can actually be defined as a spatial placement problem to identify candidate parcels, and a fractional uncertain knapsack problem with forecast time-series resource, to compute scenarios of optimal parcel selection and sizes. This paper aims at defining and showing the strengths of a combined GIS and RO computational approach to tackle complex spatial decision making problems with time-series resources, with application here to solar energy placement and planning.

Integrated Computational Work-flow. The work-flow depicted in Fig. 1 best describes the computational process and integration of the two modules. We first describe the main inputs, then each module. Geographical data layers and control parameters input the developed GIS GREECE module, which implements methods to determine the candidate parcels, and contributes a despatialization of the relevant features for each parcel (resource pattern, maximal size and costs), needed to enrich the RO model.

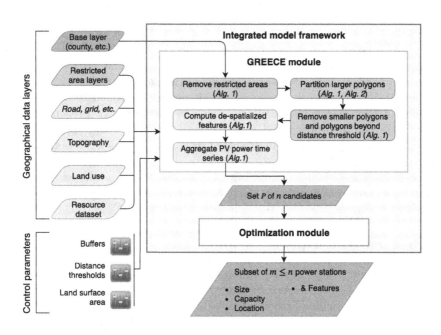

Fig. 1. GIS-RO integrated workflow

Data, Constraints and Control Parameters. In GIS terminology, the concept of layer corresponds to geographic datasets. When the dataset is an image, the term Raster is commonly used. The geometric objects are in vector mode and can be specified as polygons, lines or points. As depicted in Fig. 1, input data layers correspond to: (1) the study region or base layer, (2) the restricted area layers, (3) specific objects for which distance to resulting polygons must be computed (e.g. road, grid), (4) terrain features (land use and topography) and (5) the resource of interest (here solar radiation maps).

The restricted area layers stand for polygons where facilities cannot or should not be established. Typically, they include urban areas, ecological zones, watercourses, military sectors, cultural heritage, etc. They may also represent zones too far from specific objects (e.g. electrical grid). Geographic elements for which distance must be computed can be of any kind but are generally related to connection and accessibility costs such as road and grid networks. Topography and land use allow terrain to be characterized within each polygon. Finally, resource dataset is a set of raster or vector processed images, potentially with time series. Data layers and maps are retrieved from national and international geographic databases or remote sensing image processing.

The range control parameters are set by the user, and allow for different land management scenarios to be generated. Buffers surrounding geographic objects depend both on their type and on the kind of power station. Distance thresholds to given layers (e.g. road network, electrical grid) stand for the limit beyond which establishing a facility is not economically viable. Finally, land surface area specify a region's land management in terms of minimum and maximum allowed surface thresholds. Smaller or larger parcels might respectively be excluded from the study, or partitioned into suitable smaller parcels.

3   GIS Module: GREECE

We specify the Geographical REnewable Energy Candidate Extraction problem below, then describe our spatial partitioning and placement solution methods.

Given:

B_{layer}	Base layer (i.e. a set of polygons)
R_{layer}	Restricted area layers
D_{layer}	Distance threshold layers (i.e. sets of geometries)
h	Matrix of elevation values (DEM) associated to the base layer
R_{maps}	Set of resource raster maps
LU_{layer}	Land-use layer
NET_{layers}	Layers (e.g. grid, roads, ...) for which distance to each resulting polygon must be calculated
B_{uffers}	Control parameter: Set of buffer values, each associated with a given layer
$DistT$	Control parameter: Set of distance threshold values, defined for each D_{layer}
s_{min}, s_{max}	Control parameter: Surface thresholds

Find:

The set P of candidate parcels

The de-spatialized resource time series and geographical features for each parcel

Such that the following geographical restrictions hold:

A candidate parcel is geographically disjoint from all layers of restricted areas

The surface of a candidate parcel is within given bounds

A candidate parcel is within a bounded distance from the threshold layers

---

For space reasons, we give the main procedure in Algorithm 1, the spatial partitioning algorithm, and describe our spatial slicing and extraction methods, developed with Python GIS packages. The general procedure is decomposed in two main steps: (1) spatial placement and partitioning (Algorithm 1: line 2–10), (2) conversion of the resulting polygons defined by their geographical coordinates into *de-spatialized* items with relevant features (Algorithm 1: line 11–20). The set of relevant built-in Python GIS functions for topological, raster, set and graph operations is defined (packages used: geopandas, shapely, rtree, gdal [21], numpy [43], and networkx).

Topological geometric and raster images operators

DISTANCE(p,P)	Minimum euclidean distance between centroid of polygon p and all elements in set P
UNION($p_1, .., p_n$)	geometric union of the polygons p_i
RTREEIDX(P)	Compute spatial index idx of all elements in P
INTERSECTS(p,P,idx)	geometric intersection of p with elements in P
SHAPE(p)	Shape factor of polygon p (e.g. roundness)
SURFACE(p)	Surface of polygon p
SLOPE(h)	Slope raster from the Digital Elevation Model h
ASPECT(h)	Raster of slope orientation values from DEM h

Set and graph functions

HONEYCOMB(x,s_{hex})	Creates a honeycomb grid corresponding to polygon x of x with hexagonal elements having surface s_{hex}
PARTGRAPH(G, n, W_{part})	Partitions a graph G into n parts having weights W_{part}

Algorithm 1: GREECE main algorithm

```
 1  begin
 2  │   /* Compute candidate parcels specified spatially as polygons        */
 3  │   P ← MASK(B_layer, R_layers, B_uffers)
 4  │   for p ∈ P:
 5  │   │   if SURFACE(p) >= s_max:
 6  │   │   │   P.DELETE(p)
 7  │   │   └   P.INSERT(PARTITION(p, s_max, f_d))/* see algorithm 2            */
 8  │   for p ∈ P:
 9  │   │   if SURFACE(p) < s_min ∨ DISTANCE(p, i ∈ D_layer) > dt_i, dt_i ∈ DistT:
10  │   │   └   P.DELETE(p)
11  │   /* Compute, store de-spatialized features for each polygon p (Section
    │      3.2)                                                               */
12  │   for p ∈ P:
13  │   │   p.APPEND(DISTANCE(p, N ∈ NET_layers))
14  │   │   p.APPEND(SHAPE(p)), p.APPEND(SURFACE(p))
15  │   │   p.APPEND(s_k ∈ SURFACE(p ∩ INTERSECTS(p, LU_layer, RTREEIDX(LU_layer))))
16  │   │   /* Aggregate raster cell values, terrain features, within p      */
17  │   │   p.APPEND(μ_n ∈ ZONALSTAT({h, SLOPE(h), ASPECT(h)}, p, μ))
18  │   │   p.APPEND(σ_n ∈ ZONALSTAT({h, SLOPE(h), ASPECT(h)}, p, σ))
19  │   │   /* Mean energy resource values per raster time series            */
20  │   └   p.APPEND({μ_1, μ_2, ⋯, μ_t} ∈ ZONALSTAT(R_maps, p, μ))
```

3.1 Spatial Placement and Partitioning of Polygons

Slicing the Base Layer: Extracting Parcel Polygons. The slicing of the whole study area is divided into two main steps. The first consists in identifying and removing the restricted areas that intersect a base polygon (our study area) as well as zones beyond the distance threshold from given elements (electrical grid, road network, etc.). The procedure $\mathtt{Mask}(B_{layer}, R_{layers}, B_{uffers})$ applies set-based topological operators, that mask out portions of restricted layers and buffered zones, intersecting the base layer (Algorithm 1:line 3). It corresponds to a 2-dimensional difference operation. The result of this first step is a finite set P of new polygons, representing available land for potential power facilities, illustrated in Fig. 2(a).

The second step (Algorithm 1:lines 4–10), consists in filtering in a deterministic manner the polygons belonging to this set, based on their surface and distance to the grid. First we identify the parcels whose surface is beyond the allowed threshold. We developed a 2D space partitioning approach based on a k-way graph partitioning method, that partitions these parcels into smaller ones of suitable sizes (Algorithm 1: line7). Then, we prune further the resulting set of potential parcels according to the minimal surface threshold and distance to the grid, illustrated in Fig. 2(b). This last step is best handled using GIS geographical metric operators (Algorithm 1: lines 8–10). The extracted and computed parcels can all contribute to a solution, without inconsistent pruning of viable parcels from the standpoint of the threshold and land restriction constraints. Figure 2 illustrates the spatial layers masking process as well as the pruning of parcels below a surface threshold, and beyond a distance threshold to network layers (grid, roads, ...).

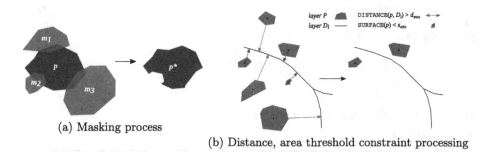

(a) Masking process

(b) Distance, area threshold constraint processing

Fig. 2. Pruning restricted areas and threshold layers

Spatial Partitioning Method. To partition a polygon into smaller plots of equal size, we propose a k-way graph partitioning approach, depicted in Algorithm 2. First, we specify the initial polygon as a honeycomb mesh, that is a set of connected hexagonal plots of given size (line 6). We then map the mesh to a graph, and apply a k-way graph partitioning (lines 7, 20–29). Each hexagon denotes a vertex connected to its concomitant neighbors by unweighted and undirected

---

Algorithm 2: Surface partitioning

---

1 **def** PARTITION(p, s_{max}, f_d): /* Partition polygon p into a set of
   polygons P of equal surface   */
    Input   : polygon p, maximal area per partition s_{max}, disaggregation
           factor f_d
    Output : a set P of polygons

2   /* Initialization   */
3     $k \leftarrow \lfloor \frac{\text{SURFACE}(p)}{s_{max}} \rfloor$ /* Number of targeted plots   */
4   **if** $k <= 1$ **and** SURFACE(p) $- s_{part} < s_{part}/f_d$:
5   **return** $\{p\}$
6     $H \leftarrow$ HONEYCOMB(p, **area**:s_{max}/f_d) /* create the mesh of hexagons   */
7     $G \leftarrow$ TOGRAPH(H, SURFACE($hexa \in H$))
8   **for** $i \in \{1,..,k\}$:
9   /* Set the weight for each sought plot   */
10   W_{plot}.APPEND(s_{max})
11   **if** SURFACE(p) mod s_{max} **not** $= 0$:
12   W_{plot}.APPEND(SURFACE(p) $- n \cdot s_{max}$)
13        $k \leftarrow k + 1$
14   /* k-way graph partitioning of G into a set of k clusters C */
15     $C \leftarrow$ PARTGRAPH(G, k, W_{plot})
16   /* Convert the clusters of vertices into a set polygons   */
17   **for** $set \in C$:
18   P.INSERT(UNION($p_i \mid i \in set$))
19   **return** P

20 **def** TOGRAPH(H, W_{hexa}): /* Convert the weighted mesh into a graph   */
    Input   : set of hexagons H, set of corresponding weights W_{hexa}
    Output : a graph G

21   /* Initialization   */
22     $G \leftarrow$ GRAPH()
23   **for** $(h,w) \in (H, W_{hexa})$:
24   /* set of polygons concomitant to h, frontiers   */
25        $F \leftarrow$ INTERSECTS(h, H, RTREEIDX(H))
26   **for** $f \in F$:
27   G.INSERT(EDGE(h, f))
28   G.INSERT(NODE(p, **weight**:w))
29   **return** G

---

edges. The k value is first initialized using a disaggregation factor, that sets the
size of each hexagon (lines 3–5). The bigger the factor value, the smaller each
hexagon and thus the more refined is the mesh. Each vertex is weighted with the
corresponding hexagon surface. The weight of the sought clusters (final plots)
is initialized (lines 8–13), to feed the k-way graph partitioning (line 15). The
procedure derives k clusters of vertices, to reach the surface threshold of each
plot. The algorithm minimizes the number of edge cuts and forces contiguous

partitions [29], so that the final aggregates present round-like geometries. The algorithm was developed using Python library networkx and the METIS package [30]. Finally the k-clusters are translated into k polygons (lines 17–18).

Example 1. The figure illustrate a 16-way graph partitioning of a 781 km^2 polygon into plots of 50 km^2, using an hexagonal honeycomb mesh. k is initialized to $\lfloor \frac{781}{50} \rfloor$. The algorithm derives 15 plots between 49 km^2 and 51 km^2 plus one (marked with an asterisk) which fills the leftover space ($\approx$ 32km^2). The output is a finite set of polygons of acceptable surface (below the maximum threshold).

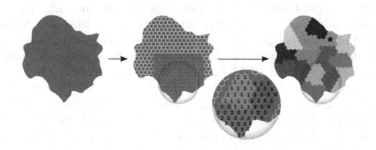

3.2 Contextual Data and Resource Time Series De-Spatialization

Geospatial data is an invaluable source of contextual information, to characterize in general, multiple land features and resources, but also their proximity with all sorts of infrastructure networks. The challenge when dealing with optimization problems evolving around resources, costs and constraints, is to seek a computational approach to specify and solve a problem as closely as possible to its real setting. A contribution of the proposed GIS-RO framework is to achieve this by using the qualitative and large spatio-temporal and contextual information analyzed and extracted through GREECE, with a de-spatilization process. The goal is to convert the polygons into items without their geo-referencing, and to associate to each the features relevant to the optimization module. Regarding energy planning, the features must capture many land properties, and resource time series, but also dimensions and distances that will allow a reliable conversion to technical costs for each candidate parcel.

The localization and maximal surface of a candidate parcel can provide numerous relevant information from the intersected data layers: (1) the distance to specific infrastructures such as substations, electrical grid or road network contributes to assessing energy losses as well as accessibility costs, (2) the maximal surface is linked to construction and maintenance costs [3,6,42]. In addition, terrain slope is a critical criterion for establishing power facilities [4,46]; in the same way, final cost might also be affected by land type and land use, or even the shape of the parcel. In the case of solar and wind energies, geo-referenced resource maps are available from satellite images or field studies [5,6,20]. By overlaying the previously sliced polygons with these raster images, it is therefore possible to aggregate the resource within each parcel. Essentially, the de-spatialization consists in translating the geography of each candidate into either static or dynamic quantifiable features. These features can then be integrated into energy models, converted into construction and operation costs, or used as constraints or in the objective functions of the optimization model. In the case of solar PV, GREECE extracts the following parameters from each candidate parcel: area, shape, distance to the grid, land use share, elevation, slope, aspect and solar GHI time series (Algorithm 1:lines 13–20). The whole implementation has made use of the python libraries referred earlier. SURFACE(p) provides information on the maximum power plant capacity that could eventually be set up within the parcel. SHAPE ranges from 0 to 1 and measures the roundness of a parcel, that is how spread out a solar PV plant might eventually be. DISTANCE to the grid is computed from the polygon centroid and is used to get both connection costs and transmission losses. *Land use share* can be correlated to construction costs, as well as *elevation* (from DEM h) using SLOPE and ASPECT. They may also be used as exclusion thresholds. Land use share is retrieved by computing the intersecting area between the given parcel and each attribute of a land use layer LU_{layer} (line 15). Elevation, slope and aspect are retrieved from the 3 arc second SRTM-based digital elevation model (DEM) [27] and by calculating average and standard deviation from raster cells overlapping with each candidate parcel (lines 17–18). Finally, *solar GHI time series* are obtained in the same way as with the DEM but for as many time steps as available solar radiation maps (line 20). As a result, each candidate parcel is now a compound item with geometric and terrain features, and time series of solar GHI values. In the case of solar PV, we have also converted GHI into power by adapting the pvlib library from the Sandia National Laboratory (SNL) [25].

4   Optimization Module - Fractional Knapsack Approach

The Optimal planning and sizing of PV plants (OPSPV) is an optimization problem permeated with uncertainty, rooted in projection estimates from current data relative to the growth of energy demand and the resource values. As shown in Fig. 1, it takes the candidate parcels with their de-spatialized features to select the ones and their optimal size such that the PV power penetration in the network is maximized at minimal cost. We propose a robust optimization approach based on the seminal works of [8,14], that specifies uncertainty using deterministic intervals. They denote the robust bounds within which the uncertain data is known to take its value. This modelling approach enables reliable best and worst case planning scenarios to guide the decision makers, and to assess the impact of his risk adversity impact on the output scenarios. The specification of the problem is given in Fig. 3.

Robust Constraint Optimization Model. The OPSPV, energy strategy planning problem can be modelled as a fractional knapsack problem with additional constraints. In this analogy, the knapsack corresponds to

Given:

Unit: Hour (per year) $t \in H = \{1, \cdots, 8760\}$

Unit: Candidate parcel $i \in N = \{1, \cdots, n\}$

Current hourly production from intermittent energy sources (KWh) $Eint_t$

Current hourly production from non-intermittent energy sources (KWh) Ep_t

Estimated global hourly power demand (KWh) $Dem_t = [\underline{Dem_t}, \overline{Dem_t}]$

Nominal power of a PV unit (to convert plant size into power) $Pnom$

Minimum and maximum *area* for each candidate parcel (m^2) $Smin, \quad Smax_i$

Estimated *hourly production* per PV unit (Wh/m^2) $Ppv_{t,i} = [\underline{Ppv_{t,i}}, \overline{Ppv_{t,i}}]$

Minimal distance from the grid to centroid of a candidate PV parcel (m) Dg_i

Transmission line unit cost (€/m) $Clan$

Substation unit power cost (€/KW) $Csta$

Find:

The set of parcels where power stations will be built

The surface to consider for each candidate parcel that is selected

Cost functions:

Sum of all costs of PV plant installation $C = \Sigma_i (Ccap_i + Cop_i + Ccon_i + Csta_i)$

Unit capital cost for installing a PV power plant (€) $Ccap_i$

Annual operational cost per new PV plant (€) Cop_i

Unit connection costs for each new PV plant, transmission lines (€) $Ccon_i$

Capital cost for new substation (€) $Csta_i$

Total added PV energy production $\Sigma_t PV_t$

Such that the following constraints hold:

PV newly added production plus existing production are below the hourly demand

PV existing and new production are less than 35% of the total energy demand per hour

PV parcel size cannot exceed a maximal set size

Fig. 3. OPSPV problem specification

the forecast energy demand to be provided by existing and new intermittent RE sources (in (KW/hr), and the items are the candidate parcels with their potential supply (KW/hr) plus their associated technical costs (installation relative to the size thus production $(€/KW)$, and connection to the grid and substation (€). The objective is to maximize hourly penetration of additional RE power in the network while minimizing global costs.

We first specify the problem and then describe the model developed in terms of variables, constraints and cost functions.

Variables. We consider two sets of variables that need to be linked to each other. Boolean variables relate to the selection or not of a candidate parcel, needed to determine whether the unit connection cost is applied or not $(Ccon_i)$. The area variables, ranging over a real interval, are involved in the energy production constraint and the installation, capital and operational costs, that depend on the size of a new PV plant.

$\forall\, i \in N, \mathbf{B_i} \in \{0, 1\}$ 1 if parcel is selected, 0 otherwise

$\forall\, i \in N, \mathbf{SA_i} \in [0.00..Smax_i]$ Area of a parcel

Knapsack Constraints. The first set of constraints relates to the forecast energy demand, using existing resources augmented with new PV production. It seeks to determine the capacity of new PV plants to contribute to the anticipated demand. Two scenarios are considered: (1) best case scenario (highest PV energy forecast and lowest forecast demand) and (2) worst case scenario (lowest PV energy forecast and highest forecast demand). This allows to study the impact of the decision maker risk adversity in planning the creation of new PV plants.

Best case scenario: $\forall\ t \in H, \quad \Sigma_i \mathbf{SA_i} \times \overline{Ppv_{t,i}} + Eint_t + Ep_t \leq \underline{Dem_t}$

Worst case scenario: $\forall\ t \in H, \quad \Sigma_i \mathbf{SA_i} \times \underline{Ppv_{t,i}} + Eint_t + Ep_t \leq \overline{Dem_t}$

Network Penetration Constraints. The second set of constraints states that the amount of intermittent energy resource into the network should be less than 35% of the total forecast energy demand per hour (upper bound) [17]. The time stamp is the hour. It is also set for the best and worst case scenarios:

Best case scenario: $\forall\ t \in H, \quad \Sigma_i \mathbf{SA_i} \times \overline{Ppv_{t,i}} + Eint_t \leq 0.35 \times \underline{Dem_t}$

Worst case scenario: $\forall\ t \in H, \quad \Sigma_i \mathbf{SA_i} \times \underline{Ppv_{t,i}} + Eint_t \leq 0.35 \times \overline{Dem_t}$

Connecting Parcel Selection and Size. The third set of constraints establishes a link between the Boolean and PV plant area variables. This relationship is needed to connect the energy production and the various costs. If a plant size is not null then the parcel is selected, and conversely if a parcel is not selected its size is forced to be null.

$$\forall i \in N, \ \mathbf{SA_i} \leq Smax_i \times \mathbf{B_i}, \ \mathbf{B_i} \times Smin \leq \mathbf{SA_i}$$

Objectives and Cost Functions. The OPSPV problem has two main objective functions: (1) to maximize the total hourly RE energy production over the year through new PV energy production, (2) to minimize the total technical costs. Since the functions are in different units, a single weighted function is not meaningful, instead we seek the pareto frontier, by optimizing PV production function while constraining the cost function with more restrictive values at each run.

Maximize PV Production: depending on the scenario considered, $Ppv_{t,i}$ will take its highest estimate ($\overline{Ppv_{t,i}}$ for best case) or lowest estimate ($\underline{Ppv_{t,i}}$ for the worst case). The cost function to maximize is:

$$\Sigma_i \Sigma_t \mathbf{SA_i} \times Ppv_{t,i}$$

Minimize Costs: Modelling Non-Linear Functions. Four cost functions are involved and relate to the installation and size of a PV plant as defined in Fig. 3. Typically, capital and operational costs are approximated as linear functions [9,19,24]. However, this approach is unrealistic as both the Cap_i and Cop_i costs are in fact non-linear, since they depend on the size of the plant (linked to the related number of PV panels) [31]. Basically the fewer the number of panels the highest the relative cost per panel. To get closer to reality, we thus consider an innovative approach using a piece-wise linear function such that a_i is the coefficient of the slope, and y_i the value of the coordinate where the new slope begins. It is illustrated below for Cap_i, and Cop_i follows a similar specification with different constants. Values have been set from [31]:

$$Cap_i = \begin{cases} a_1 \times Pnom \times \mathbf{SA_i} + y_1 & \text{if } 0_{MW} \leq \mathbf{SA_i} \times Pnom \leq 1_{MW} \\ a_2 \times Pnom \times \mathbf{SA_i} + y_2 & \text{if } 1_{MW} \leq \mathbf{SA_i} \times Pnom \leq 10_{MW} \\ a_3 \times Pnom \times \mathbf{SA_i} + y_3 & \text{if } 10_{MW} \leq \mathbf{SA_i} \times Pnom \end{cases}$$

On the other hand, the unit connection cost of a PV plant, and the capital cost of a new station for a plant, are both linear functions that depend respectively on the creation of the plant in a parcel with its Euclidian distance to the grid, and the unit cost of a substation proportional to the computed size of the plant. We have the following functions:

$$Ccon_i = Clan \times Dg_i \times \mathbf{B_i}, \quad Csta_i = Csta \times Pnom \times \mathbf{SA_i}$$

5   Experimental Study and Evaluation

The proposed GIS-RO framework seeks to make powerful use of multi-scale contextual information for spatial decision making and optimization. It is evaluated on the timely challenge in the region of French Guiana, where the objective is to reach the energy policy plan of 100% renewable by the year 2030, first by increasing PV, then biomass. The challenge lies in the strategic planning of solar PV scenarios using contextual real data characterized by spatio-temporal patterns and permeated with uncertainty (resource projections). In this section we present our results and analysis. Input data layers, retrieved from various national and world databases [26,28,32,33], and associated buffer values are depicted in Table 1. In addition, maximum distance to both power grid and road network has been set to 20 km, twice the value commonly used [6,39]. Land surface area minimum and maximum thresholds for establishing solar PV plants are respectively set to 1.5 ha [44] and 50 ha [17]. Finally, regarding plot resource, monthly solar GHI time series derived from satellite-based raster images [5,20] have been disaggregated at the hour using an updated version of a synthetic generation model [1,2,38].

Table 1. Land management scenario used in this study for restricted areas.

Layer	Protected areas	Forest	Urban areas	Flood savanna	Water bodies	Shore	Power grid	Road network	Wetland	Dune/ Sand	Rice/ Orchard
Buffer (m)	500	200	200	100	100	100	100	30	0	0	0
Refs	[45]	[6]	[3,45]	[41]	[40,41]	[40]	[3,39,41]	[3,39,41]	[41]		

The GREECE module was implemented in Python using GIS-Python libraries recalled in the paper. It provides de-spatialized data items to the OPSPV module, which was implemented using IBM ILOG OPL CPLEX Optimization studio, on a 2 processors (Intel(R) Xeon(R) CPU E5-2609 v4 @1.70GHz) of 32 Go RAM. The GREECE module led to the extraction of 133 candidate parcels with their relevant features. The execution CPU time for the GREECE module is about 460 s. (reading files: 30 s; mask: 380 s.; partitioning: 3 s.; feature + monthly resource extraction: 50 s.), handled as a one off spatial placement preprocessing. The optimization CPU time varies from 23 s to 47 s depending on the bound set on the constrained objective function (runtime differences come from handling piece-wise linear functions that depend on the park sizes).

Data Sets. Restricted area layers handled in MASK correspond to a total of 21088 geographical objects. Here, the base layer B_{layer} corresponds to Guiana's land use LU_{layer}, which gathers 2643 polygons. Road network and power grid used in distance threshold computation are made of 2247 lines. Monthly solar resource is represented by a raster set of $12 \times 3999 \times 3999$ cells. Global energy demand and existing production data are known from the sources and extractions from records of 2016 [18]. For the 2030 horizon we projected hourly energy demand values according to EDF estimations of worst case 5% annual growth and best case of 2% annual growth.

5.1 Results and Analysis

We analyzed three aspects of relevance to the decision maker and network manager (power plant energy investors in Guiana and EDF) that are made possible with the combination of geographical and temporal contextual information and optimization: (1) the impact of the spatio-temporal energy patterns on the geographic selection process (Figs. 4 and 6 (b)), (2) the study of the risk adversity comparing best and worst case scenarios (Fig. 5), (3) the identification of robust planning investment scenarios where the optimal plants show to be identical regardless of the degree of spatio-temporal uncertainty on the power resource (Fig. 6 (a)).

Figure 4 depicts the resulting spatial variation of the solar GHI patterns (derived by GREECE) on the produced power from the optimal solar PV plants (location and size) whose placement is visible in Fig. 6 (b). P1 and P4 sites (P2 and P3) have similar patterns for they are located in the same solar potential cluster zone. Essentially, it shows how our GIS-RO approach manages real site spatial arrangement so that the global output power is robust through time from the optimal PV plants: impact of their RE intermittency on future network power management is limited.

Figure 5 gives information about (a) the volume of solar PV plants one may install within the region without threatening the power grid in the long-term, and (b) its corresponding final share in the energy mix. As long as both Pareto lines remain together in Fig. 5 (a), the solution is robust, i.e. power generation

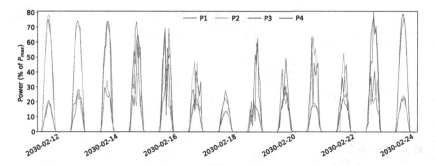

Fig. 4. Normalized output power from solar PV sites of Fig. 6.

over time from selected plants fills up *the same free energy volume* regardless of the scenario: corresponding PV sites can be explored safely in the limits of their maximum capacity estimated by the RO. In contrast, once the best case reaches its plateau , and so Pareto lines split in half (around 40 M€), power generation no longer fills up the same volume. At this point, the more the energy generated in the worst case scenario, the more it exceeds the energy generation limit in the best case: the risk grows as much as the gap between both lines.

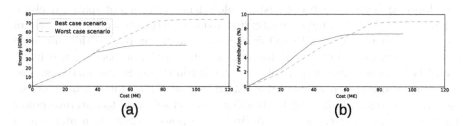

Fig. 5. Pareto charts for both scenarios with respect to (a) generated energy and (b) PV penetration.

Finally, Fig. 6 evaluates the robustness of the investment according to spatio-temporal uncertainty on the resource: (i) estimated GHI, energy potential (in blue), (ii) its mitigation by a random uniform noise per hour and per parcel, between 0 and 10% (in green) and (iii) between 0 and 20% (in red) respectively. The safety zone lies below a cost of 70 M€ (C70), meaning that the selected PV sites are the same for all resource time series projections. Those sites are sorted by ascending parcel size area from the smallest (P1) to the largest (P4). The investment costs grow naturally when the optimal PV plant size grows within its corresponding parcel area (fractional knapsack optimization).

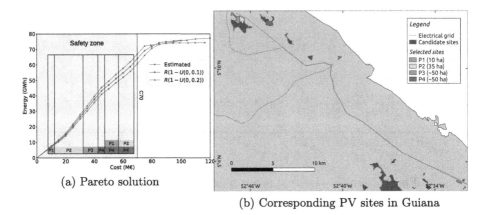

(a) Pareto solution

(b) Corresponding PV sites in Guiana

Fig. 6. Robustness of the worst-case scenario solution with decreasing resource (Color fig online)

6   Conclusion

In this paper we have proposed a two-step specification and novel computational approach of the spatio-temporal placement and energy planning problem. We addressed key challenges of sustainable science in terms of spatial decision making and usage of complex contextual data, constraints and time series of resources. The presented GIS-RO framework allows for real world and large scale applications to be solved through an integrated approach at the interface of GIS science, graph and robust optimization models and methods. The case study showed in particular the importance of taking into account actual resource patterns and allowing for fractional parcel selection to optimize power plants location and size, and thus optimize the power penetration at minimal cost. Current work includes its generalization to include the cost-effectiveness of energy storage considered not profitable to this date in Guiana, and the generalization to biomass resources that raises a complex temporal renewability issue of the resource.

References

1. Aguiar, R., Collares-Pereira, M.: TAG: a time-dependent, autoregressive, Gaussian model for generating synthetic hourly radiation. Solar Energy **49**(3), 167–174 (1992). https://doi.org/10.1016/0038-092X(92)90068-L
2. Aguiar, R.J., Collares-Pereira, M., Conde, J.P.: Simple procedure for generating sequences of daily radiation values using a library of Markov transition matrices. Solar Energy **40**(3), 269–279 (1988). https://doi.org/10.1016/0038-092X(88)90049-7
3. Al Garni, H.Z., Awasthi, A.: Solar PV power plant site selection using a GIS-AHP based approach with application in Saudi Arabia. Appl. Energy **206**, 1225–1240 (2017). https://doi.org/10.1016/j.apenergy.2017.10.024

4. Al Garni, H.Z., Awasthi, A.: Solar PV power plants site selection: a review. In: Yahyaoui, I. (ed.) Advances in Renewable Energies and Power Technologies, pp. 57–75. Elsevier, Amsterdam (2018). https://doi.org/10.1016/B978-0-12-812959-3. 00002-2

5. Albarelo, T., Marie-Joseph, I., Primerose, A., Seyler, F., Wald, L., Linguet, L.: Optimizing the heliosat-II method for surface solar irradiation estimation with GOES images. Can. J. Remote Sens. **41**(2), 86–100 (2015). https://doi.org/10. 1080/07038992.2015.1040876

6. Ali, S., Taweekun, J., Techato, K., Waewsak, J., Gyawali, S.: GIS based site suitability assessment for wind and solar farms in Songkhla. Thai. Renewable Energy **132**, 1360–1372 (2019). https://doi.org/10.1016/j.renene.2018.09.035

7. Barbosa, L., Bogdanov, D., Vainikka, P., Breyer, C.: Hydro, wind and solar power as a base for a 100% renewable energy supply for South and Central America. PLoS One **12**(3), 1–28 (2017). https://doi.org/10.1371/journal.pone.0173820

8. Ben-Tal, A., Nemirovski, A.: Robust solutions of uncertain linear programs. Oper. Res. Lett. **25**, 1–13 (1999)

9. Bogdanov, D., Breyer, C.: North-East Asian super grid for 100% renewable energy supply: optimal mix of energy technologies for electricity, gas and heat supply options. Energy Convers. Manage. **112**, 176–190 (2016). https://doi.org/10.1016/ j.enconman.2016.01.019

10. Bolwig, S., Bazbauers, G., Klitkou, A., Lund, P.D., Blumberga, A., Blumberga, D.: Review of modelling energy transitions pathways with application to energy system flexibility. Renew. Sustain. Energy Rev. **101**, 1–23 (2019). https://doi.org/ 10.1016/j.rser.2018.11.019

11. Breyer, C., et al.: Solar photovoltaics demand for the global energy transition in the power sector. Prog. Photovoltaics Res. Appl. **26**(8), 505–523 (2017). https:// doi.org/10.1002/pip.2950

12. Breyer, C., Bogdanov, D., Komoto, K., Ehara, T., Song, J., Enebish, N.: North-East Asian super grid renewable energy mix and economics. Jpn. J. Appl. Phys. **54**(8S1), 08KJ01 (2015)

13. Castro-Santos, L., Garcia, G.P., Simões, T., Estanqueiro, A.: Planning of the installation of offshore renewable energies: a GIS approach of the Portuguese roadmap. Renewable Energy **132**, 1251–1262 (2019). https://doi.org/10.1016/j.renene.2018. 09.031

14. Chinneck, J.W., Ramadan, K.: Linear programming with interval coefficients. J. Oper. Res. Soc. **51**(2), 209–220 (2000)

15. CTG: Programmation pluriannuelle de l'énergie (PPE) 2016–2018 et 2019–2023 de la Guyane. Tech. rep., Collectivité Territoriale de Guyane (2017)

16. Dotzauer, M., et al.: How to measure flexibility - performance indicators for demand driven power generation from biogas plants. Renewable Energy **134**, 135–146 (2019). https://doi.org/10.1016/j.renene.2018.10.021

17. EDF: Systèmes énergétiques insulaires Guyane - bilan prévisionnel de l'équilibre offre / demande d'électricité. Tech. rep., EDF - Direction des Systèmes Énergétiques Insulaires, Paris, France (2017)

18. EDF: Open Data EDF Guyane (2019). https://opendata-guyane.edf.fr/pages/ home/. Accessed 30 Apr 2019

19. Ferrer-Martí, L., Domenech, B., García-Villoria, A., Pastor, R.: A MILP model to design hybrid wind-photovoltaic isolated rural electrification projects in developing countries. Eur. J. Oper. Res. **226**(2), 293–300 (2013). https://doi.org/10.1016/j. ejor.2012.11.018

20. Fillol, E., Albarelo, T., Primerose, A., Wald, L., Linguet, L.: Spatiotemporal indicators of solar energy potential in the Guiana shield using GOES images. Renewable Energy **111**, 11–25 (2017). https://doi.org/10.1016/j.renene.2017.03.081

21. GDAL/OGR contributors: GDAL/OGR Geospatial Data Abstraction software Library. Open Source Geospatial Foundation (2019). https://gdal.org

22. Gervet, C., Atef, M.: Optimal allocation of renewable energy parks: a two-stage optimization model. RAIRO-Oper. Res. **47**, 125–150 (2013). https://doi.org/10.1051/ro/2013031

23. Hache, E., Palle, A.: Renewable energy source integration into power networks, research trends and policy implications: a bibliometric and research actorssurvey analysis. Energy Policy **124**, 23–35 (2019). https://doi.org/10.1016/j.enpol.2018.09.036

24. Heydari, A., Askarzadeh, A.: Optimization of a biomass-based photovoltaic power plant for an off-grid application subject to loss of power supply probability concept. Appl. Energy **165**, 601–611 (2016). https://doi.org/10.1016/j.apenergy.2015.12.095

25. Holmgren, W.F., Hansen, C.W., Mikofski, M.A.: Pvlib Python: a Python package for modeling solar energy systems. J. Open Source Softw. **3**(29), 884 (2018). https://doi.org/10.21105/joss.00884

26. IGN: BD TOPO® Version 2.2 - Descriptif de contenu. Institut Géographique National, Paris, France (2018). http://professionnels.ign.fr/doc/DC_BDTOPO_2-2.pdf

27. Jarvis, A., Reuter, H., Nelson, A., Guevara, E.: Hole-filled SRTM for the globe Version 4, available from the CGIAR-CSI SRTM 90m Database. Tech. rep. http://srtm.csi.cgiar.org (2008)

28. Juffe-Bignoli, D., Bingham, H., MacSharry, B., Deguignet, M., Milam, A., Kingston, N.: World database on protected areas - User manual 1.4. United Nations Environment Programme - World Conservation Monitoring Centre, 219 Huntingdon Road, Cambridge, UK (2016)

29. Karypis, G.: METIS: a software package for partitioning unstructured graphs, partitioning meshes, and computing fill-reducing orderings of sparse matrices. Department of Computer Science & Engineering, University of Minnesota, Minneapolis, MN 55455 (2013)

30. Karypis, G., Kumar, V.: A fast and highly quality multilevel scheme for partitioning irregular graphs. SIAM J. Sci. Comput. **20**(1), 359–392 (1999)

31. NREL: Distributed generation renewable energy estimate of costs. https://www.nrel.gov/analysis/tech-lcoe-re-cost-est.html (2016)

32. Guyane, O.N.F.: Programme régional de mise en valeur forestière pour la production de bois d'œuvre - période 2015–2019. Tech. rep., Direction Régionale ONF Guyane (2015)

33. ONF Guyane: Occupation du sol en 2015 sur la bande littorale de la Guyane et son évolution entre 2005 et 2015. Direction Régionale ONF Guyane (2017)

34. Ozdemir, S., Sahin, G.: Multi-criteria decision-making in the location selection for a solar PV power plant using AHP measurement. J. Int. Measur. Confederation **129**, 218–226 (2018). https://doi.org/10.1016/j.measurement.2018.07.020

35. Pfenninger, S., Hawkes, A., Keirstead, J.: Energy systems modeling for twenty-first century energy challenges. Renew. Sustain. Energy Rev. **33**, 74–86 (2014). https://doi.org/10.1016/j.rser.2014.02.003

36. Purkus, A., et al.: Contributions of flexible power generation from biomass to a secure and cost-effective electricity supply -a review of potentials, incentives and obstacles in Germany. Energy Sustain. Soc. **8**(1), 18 (2018). https://doi.org/10. 1186/s13705-018-0157-0

37. Ramirez Camargo, L., Stoeglehner, G.: Spatiotemporal modelling for integrated spatial and energy planning. Energy Sustain. Soc. **8**(1), 32 (2018). https://doi. org/10.1186/s13705-018-0174-z

38. Remund, J., Müller, S., Kunz, S., Huguenin-Landl, B., Studer, C., Cattin, R.: Meteonorm Handbook part II: Theory, Global Meteorological Database Version 7 Software and Data for Engineers, Planers and Education (2018), http://www. meteonorm.com

39. Sabo, M.L., Mariun, N., Hizam, H., Mohd Radzi, M.A., Zakaria, A.: Spatial matching of large-scale grid-connected photovoltaic power generation with utility demand in Peninsular Malaysia. Appl. Energy **191**, 663–688 (2017). https://doi.org/10. 1016/j.apenergy.2017.01.087

40. Siyal, S.H., Mörtberg, U., Mentis, D., Welsch, M., Babelon, I., Howells, M.: Wind energy assessment considering geographic and environmental restrictions in Sweden: a GIS-based approach. Energy **83**, 447–461 (2015). https://doi.org/10.1016/ j.energy.2015.02.044

41. Sultana, A., Kumar, A.: Optimal siting and size of bioenergy facilities using geographic information system. Appl. Energy **94**, 192–201 (2012). https://doi.org/10. 1016/j.apenergy.2012.01.052

42. Teixeira, T.R., et al.: Forest biomass power plant installation scenarios. Biomass Bioenergy **108**, 35–47 (2018). https://doi.org/10.1016/j.biombioe.2017.10.006

43. van der Walt, S., Colbert, S.C., Varoquaux, G.: The numpy array: a structure for efficient numerical computation. Comput. Sci. Eng. **13**(2), 22–30 (2011). https:// doi.org/10.1109/MCSE.2011.37

44. Wang, Q., M'Ikiugu, M., Kinoshita, I., Wang, Q., M'Ikiugu, M.M., Kinoshita, I.: A GIS-based approach in support of spatial planning for renewable energy: a case study of Fukushima. Japan. Sustain. **6**(4), 2087–2117 (2014). https://doi.org/10. 3390/su6042087

45. Watson, J.J., Hudson, M.D.: Regional scale wind farm and solar farm suitability assessment using GIS-assisted multi-criteria evaluation. Landscape Urban Plan. **138**, 20–31 (2015). https://doi.org/10.1016/j.landurbplan.2015.02.001

46. Woo, H., Acuna, M., Moroni, M., Taskhiri, M.S., Turner, P.: Optimizing the location of biomass energy facilities by integrating multi-criteria analysis (MCA) and geographical information systems (GIS). Forests **9**(10), 1–15 (2018). https://doi. org/10.3390/f9100585

Peak-Hour Rail Demand Shifting with Discrete Optimisation

John M. Betts, David L. Dowe, Daniel Guimarans$^{(\boxtimes)}$, Daniel D. Harabor, Heshan Kumarage, Peter J. Stuckey, and Michael Wybrow

Faculty of Information Technology, Monash University, Melbourne, Australia
{john.betts,david.dowe,daniel.guimarans,daniel.harabor,
heshan.kumarage,peter.stuckey,michael.wybrow}@monash.edu

Abstract. In this work we consider an information-based system to reduce metropolitan rail congestion in Melbourne, Australia. Existing approaches aim to reduce congestion by asking commuters to travel outside of peak times. We propose an alternative approach where congestion is reduced by enabling commuters to make an informed trade-off between travel time and ride comfort. Our approach exploits the differences in train frequency and stopping patterns between stations that results in trains, arriving within a short time of each other, to have markedly different levels of congestion, even during peak travel periods. We show that, in such cases, commuters can adjust their departure and arrival time by a small amount (typically under 10 min) in exchange for more comfortable travel. We show the potential benefit of making this trade-off with a discrete optimisation model which attempts to redistribute passenger demand across neighbouring services to improve passenger ride comfort overall. Computational results show that even at low to moderate levels of passenger take-up, our method of demand shifting has the potential to significantly reduce congestion across the rail corridor studied, with implications for the metropolitan network more generally.

1 Introduction

Home to more than 4.8 million residents, Melbourne is Australia's second-largest, and fastest growing city. Melbourne residents enjoy access to an extensive public transportation network which includes metropolitan rail, light rail and bus services. According to Public Transport Victoria (PTV)[1] there were 565 million trips on public transport in the Melbourne metropolitan area in the year from 1 July 2017 – 30 June 2018 [2]. Of these, the largest share belongs to rail, with 240.9 million trips recorded. One of the major challenges facing transport planners in Melbourne is that rail passengers often experience high levels of congestion, especially during morning and afternoon peak periods. Attempts to tackle Melbourne's congestion tend to focus on the addition of more infrastructure, that is, rail lines and trains. However, this approach is expensive, and

---

[1] PTV is the government agency responsible for providing and coordinating public transport for Melbourne and across the state of Victoria.

© Springer Nature Switzerland AG 2019
T. Schiex and S. de Givry (Eds.): CP 2019, LNCS 11802, pp. 748–763, 2019.
https://doi.org/10.1007/978-3-030-30048-7_43

sometimes impossible due to limitations on available space and other resources. Two alternative strategies, both widely studied in the research literature, are: (i) optimisation-based demand management and; (ii) demand management via incentives. We briefly discuss each.

Optimisation-based demand management, sometimes called passenger flow control, works by directing passengers to services based on their planned journey; e.g. [8,14]. This body of research shows that passenger load can be effectively moved downstream until the demand in a saturated system is resolved. Disadvantages of this type of approach include high planning overheads, as passenger flows need to be optimised in near real time, and a dependence on significant physical infrastructure, such as waiting areas, boarding areas and designated entries. This approach also presumes that passengers will tolerate it.

Incentive-based demand management, by comparison, exploits trade-offs that exist between passenger preferences for time, comfort and cost. The idea is to encourage passengers to travel during periods of reduced demand and to discourage travel during periods of peak-demand. Studies in this area often apply equilibrium modelling, seeking to quantify, under certain conditions, the dis-utility of travelling early or late against the cost of discomfort and the willingness to pay [12,17]. The main disadvantage of such in-principle economic models is that proposed fare structures are complicated and their actual effects on real schedules are usually not clear. When applied in practice, incentive-based systems employ more simplified structures. One example is PTV's *Early Bird train travel* [9] a scheme that allows Melbourne passengers to ride for free provided they arrive at their destination before 7:15am on weekdays. Another example is Singapore's INSINC [13] system, which rewards passengers who shift their travel away from periods of peak demand. These approaches report varying degrees of success but related studies [9,16] show that relatively high reductions in fare are sometimes necessary to overcome the reluctance of some passengers to avoid peak periods.

In this work we consider a different approach where we aim to shift demand *within* peak periods by encouraging commuters to make informed trade-offs. We are motivated by evidence from the literature which suggests that passengers are willing to incur some additional travel time in order to secure a more comfortable trip [3,5–7]. In the case of Melbourne, electronic noticeboards at rail stations, and also travel apps, show only the time of the next departure. However, many stations are serviced by multiple lines, including some with low levels of occupancy, even during peak periods. We posit that, if congestion information was made available, passengers could make informed trade-offs based on preferences and needs. For example, pregnant, elderly, or disabled passengers might prefer a longer seated trip to a shorter one that involves standing.

Working with our industrial partner, PTV, we undertake a capacity-based study to investigate the congestion-reducing benefits from such an information scheme including under varying degrees of passenger uptake. Our approach relies on travel-card data, from which we construct a detailed congestion model of two rail lines in the Melbourne network: Werribee and Williamstown. We combine this data with an optimisation-based model that measures the impact of

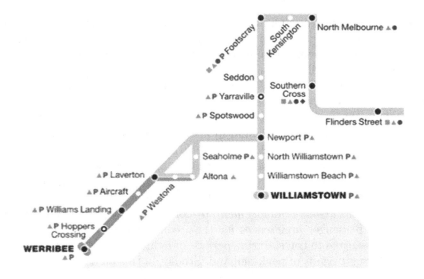

Fig. 1. The south-western region of the Melbourne Metropolitan Rail Network comprising the Werribee and Williamstown lines.

passenger demand shifting on congestion during times of peak demand. We show that with even modest levels of uptake (e.g., 20%) congestion measures can be almost halved. Meanwhile with an uptake of 60% we show that overcrowded trains can be almost entirely eliminated.

2   Background: The Melbourne Rail Network

The Melbourne Metropolitan Rail Network is a large hub-and-spoke system comprised of 217 stations, connected by 837 km of rail [1]. Figure 1 shows the south-western region of this network, the focus of our study.

The network consists of 16 lines which connect at a central terminus, *Flinders Street Station*. Trains in the network operate from 5am to midnight on weekdays and until 1am on weekends. Morning peak demand occurs between 7:00 am and 9:30 am, and afternoon peak between 3:30 pm and 7:00 pm week days.

The network is serviced by a fleet of more than 200 trains which are managed by Metro Trains, a privately-owned rail operator. The fleet is currently made up of 3 models [1]. These are: **Comeng**, having 536–556 seats, with a target capacity of 800 passengers; and **X' Trapolis** and **Siemens**, both having 528 seats and target capacity of 900 passengers. Any train which exceeds its target capacity is considered to be in *breach* of the service agreement between PTV and Metro Trains. Breach events are undesirable because they typically cause delays at stations and increase the risk of accidents when passengers are boarding and disembarking. Systematic breaching can result in penalties for the rail operator.

3 Modelling Assumptions

3.1 Trains and Rail Network

We focus our attention on the south-western section of the Melbourne Rail Network (Fig. 1). This network consists of two rail lines, with origins at Werribee and Williamstown, and three distinct types of train services:

- *Werribee Express* services, which originate at Werribee, have a frequency of approximately 10 min and run non-stop between the stations of Laverton and Newport and between Newport and Footscray.
- *Williamstown* services, which originate at Williamstown, have a frequency of approximately 20 min and stop at all stations.
- *Laverton* services, which originate at Laverton, have a frequency of approximately 20 min and stop at all stations, running through the so-called Altona loop.

We study demand shifting for the morning peak period. We consider all scheduled trains and we work with actual arrival times (cf. departure times) at each station, as measured by our industrial partner PTV. Additionally, owing to the configuration of the rail lines, overtaking is not possible. That means the relative order of arrival of trains at each station is fixed.

3.2 Measuring Congestion

In this section we discuss the region we study, the definitions of different levels of congestion, and the way we translate trip data into congestion measurements.

In the modelling that follows, we consider 5 levels of congestion. We use the capacity of Comeng trains as a reference, since these trains service the south-western rail network. We believe that this does not affect the generality of our conclusions. The congestion levels we use accord to PTV's own scale for congestion, and are broadly in line with those reported in [6].

1. **sparse** 0 to 264 passengers: no more than half seated capacity.
2. seated 265 to 528 passengers: fewer passengers than seats.
3. standing 529 to 662 passengers: more passengers than seats, but less than half standing capacity.
4. target 663 to 800 passengers: more passengers than seats, and more than half standing capacity.
5. breach 800+ passengers.

3.3 Data Collection and Train Occupancy Calculation

We calculate passenger counts on each train service from *smart card* data, which records the location and times at which a passenger entered and exited the rail system by *touch on* and *touch off*. The data used for this project comprised 3.6 m passenger touch on and 3.5 m touch off instances throughout Melbourne's rail network from January 30 to February 5 2017. We impute train occupancy from these data by assigning passenger trips to specific train services through the entry and exit stations during the corresponding time window using the following protocol. In each 10-min period, at each station:

1. Count the passengers arriving (touch on); identify each passenger's destination (touch off location).
2. Identify *eligible services* stopping at the current station (during the current period, or immediately after if no service is available during the current period) and stopping at the touch off station.
3. Remove passengers disembarking at their touch off station.
4. Assign waiting passengers equally to eligible services.
5. Record congestion level.

We assume that city-bound passengers exiting the train system outside the south-western region remain on board the train until the last hub. For the purpose of our study, this was Footscray. Figure 2a shows train occupancy during the Thursday, February 2, 2017 morning peak period calculated using this method. (Thursday has the busiest morning peak; 15,927 trips were identified over this period.) Passenger occupancy (shaded to show congestion) highlights the overcrowding that motivated this study.

4   Greedy Demand Shifting

To observe the potential reduction in congestion due to passenger demand shifting, we modify the passenger load calculation to simulate passengers choosing less congested adjacent services at Laverton and Newport. This includes trains arriving up to 10 min earlier and departing 10 min after the current time period. We use greedy shifting, moving as many boarding passengers as possible in order to keep the congestion score of the current train and alternative services to a minimum. Treating levels 4 and 5 as *congested*, we use the same protocol as shown above, modified as follows:

4. Check congestion level of the incoming train; Distribute boarding passengers equally to all eligible services; for congested services, reassign a proportion of these passengers *equally* to *all* the non-congested trains arriving during the interval from 10 min prior, to 10 min post current time window.

Figure 2b shows passenger occupancy and congestion during the February 2, 2017 morning peak after greedy demand shifting is applied at Laverton and Newport. We only present results up to 8:30 am to save space but note that congestion levels are low for all services departing any station in the south-western rail corridor after this time, with available seating in all cases. Comparing the two figures it is evident that the number of breach incidents decreases from 13 in the original case to 11 when greedy shifting is adopted.

Dep. Time	Werribee	Laverton	Altona	Williamstown	Newport	Footscray
7:04	435	530	-	-	707	691
7:13				171	242	238
7:16		108	256		499	487
7:19	532	662	-	-	900	877
7:29	620	749	-	-	971	933
7:35				202	450	433
7:38		159	358	-	578	554
7:39	648	785	-	-	1003	964
7:49	1001	1087	-	-	1233	1183
7:57				215	488	463
8:00		103	266	-	569	539
8:02	791	855		-	1091	1049
8:11	480	545	-	-	722	699
8:19				144	388	371
8:22		101	194	-	374	360
8:24	425	485	-	-	640	621
8:33	383	435	-	-	559	538

(a) Actual passenger count and congestion levels.

Dep. Time	Werribee	Laverton	Altona	Williamstown	Newport	Footscray
7:04	435	530	-	-	707	691
7:13				171	242	238
7:16		108	256	-	499	487
7:19	532	662	-	-	900	877
7:29	620	749	-	-	742	704
7:35				202	565	548
7:38		159	358	-	808	783
7:39	648	785	-	-	774	735
7:49	1001	961	-	-	945	895
7:57				215	684	659
8:00		229	392	-	858	828
8:02	791	762	-		750	709
8:11	480	592	-	-	851	828
8:19				144	471	453
8:22		148	241	-	420	407
8:24	425	485	-	-	640	621
8:33	383	435	-	-	559	538

(b) Passenger count and congestion levels with greedy demand shifting.

Fig. 2. Passenger counts and congestion levels during the morning peak for all services operating on the Werribee-Williamstown-Footscray network on February 2, 2017.

5 Reducing Congestion with Discrete Optimisation

Although the greedy approach shows some potential benefit from demand shifting, it does not reveal the greatest reduction in congestion that could be achieved if a whole network view was taken. In this section we describe a discrete optimisation model which assigns passengers to trains to reduce congestion globally. This enables us to determine the maximum reduction in congestion that could be achieved by demand shifting.

The principal data set used by the model contains trip information quantised into time blocks of 10 min. This shows, for each time block, the number of city-bound passengers travelling between each pair of stations during that period. From this, a flow network is constructed, which assigns the number of passengers embarking and disembarking for each service and each station. This enables the occupancy of each train to be calculated along its journey. Passengers are constrained to maintain a feasible trip across the network that is similar to their recorded trip touch on and touch off times. The objective is to minimise the congestion in the network. The optimal solution automatically reroutes passengers to reduce congestion. We recognise the solution given by this model represents an idealisation not achievable in practice, as it is based on perfect future knowledge, and assumes complete compliance by passengers. However, it does show the degree to which congestion could be reduced in a perfect situation.

The core sets of the model are given below, along with the name we shall use for indices that refer to elements in that set

ST, st	The set of stations considered
L, l	The set of lines considered
B, b	The set of time blocks considered
S, s	The set of train services considered

The core data of the problem model is given by

$mpax \in \mathbb{Z}$	Maximum passengers on any service
$seq_l \subseteq ST$	The sequence of stations for line l
$line_s \in L$	Which line is used by service s
$trip_{b,st_1,st_2} \in \mathbb{Z}$	The number of passengers commencing a journey at station st_1 in time block b to go to st_2
$comp_{b,st,s} \in \{true, false\}$	Is it possible for a passenger to enter service s at station st at time block b

The *compatibility*, $comp_{b,st,s}$, of a station st and time block b with a service s is determined as follows. Assuming we are allowing demand shifting be able to change passenger arrival times at their start station by no more than δ minutes from the time shown by recorded trip data, then s is compatible with b and st if the departure time for s from station st is no earlier than δ minutes before the time block commences, and no later than δ minutes after the time block ends. If we disallow forward shifting of passenger arrivals (i.e., assigning passengers to a train departing earlier than their touch on time), then only services that arrive between when the time block commences, and up to δ minutes later are compatible.

We assume sequence functions $first(Q)$ returning the first element q_1 of a sequence $Q = [q_1, q_2, \ldots, q_n]$, and $succ(Q)$ returning the set of adjacent pairs $\{(q_1, q_2), (q_2, q_3), \ldots, (q_{n-1}, q_n)\}$ of sequence Q. We compute auxiliary data

$on_{b,st} = \sum_{st' \in ST} trip_{b,st,st'}$	The number of passengers entering the system at station st at time b
$off_{b,st} = \sum_{st' \in ST} trip_{b,st',st}$	The number of passengers leaving the system at station st that entered at time b
$visits_{s,st} = st \in seq_{line_s}$	Whether service s visits the station st

The principle decisions *enter* and *exit* and auxiliary decision variables *pax* are defined below. They are all constrained to lie in the range $0..mpax$.

$enter_{b,st,s}$ The number of passengers entering service s at station st at time block b

$exit_{b,st_1,st_2,s}$ The number of passengers exiting service s at station st_2 that entered at station st_1 in time block b

$pax_{st,s}$ The number of passengers on service s when departing station st

We are now in a position to define the constraints of the problem.

$$enter_{b,st,s} = 0 \;\; \forall b \in B, st \in ST, s \in S, \neg visits_{s,st} \tag{1}$$

$$exit_{b,st_1,st_2,s} = 0 \;\; \forall b \in B, st_1, st_2 \in ST, s \in S, \neg visits_{s,st_1} \tag{2}$$

$$exit_{b,st,st,s} = 0 \;\; b \in B, st \in ST, s \in S \tag{3}$$

$$pax_{st,s} = 0 \;\; s \in S, \neg visits_{s,st} \tag{4}$$

Equation (1) ensures no passengers enter a service at a station it does not visit. Similarly, Eq. (2) ensures no passengers exiting a service commence at a station it does not visit. Equation (3) ensures no one enters and exits a service at the same station. Equation (4) ensures that the passengers for a station st not visited by a service s is 0.

$$\sum_{b \in B, s \in S} enter_{b,st,s} = \sum_{b \in B} on_{b,st} \;\; \forall st \in ST \tag{5}$$

$$\sum_{b \in B, st_1 \in ST, s \in S} exit_{b,st_1,st_2,s} = \sum_{b \in B} off_{b,st_2} \;\; \forall st_2 \in ST \tag{6}$$

$$\sum_{s \in S, comp_{b,st,s}} enter_{b,st,s} = on_{b,st} \;\; \forall b \in B, st \in ST \tag{7}$$

$$\sum_{s \in S} exit_{b,st_1,st_2,s} = trip_{b,st_1,st_2} \tag{8}$$

$$exit_{b,st_1,st_2,s} \leq enter_{b,st_1,s} \;\; \forall b \in B, st_1, st_2 \in ST, s \in S \tag{9}$$

$$pax_{st,s} = \sum_{b \in B} enter_{b,st,s} \;\; s \in S, st = first(seq_{line_s}) \tag{10}$$

$$pax_{st_2,s} = pax_{st_1,s} + \sum_{b \in B} enter_{b,st_2,s} - \sum_{st_3 \in ST, b \in B} exit_{b,st_3,st_2,s} \atop \forall s \in S, (st_1, st_2) \in succ(seq_{line_s}) \tag{11}$$

Equation (5) ensures that all passengers arriving in the system enter a train. Equation (6) similarly ensures that all passengers leaving the system at a station st_2 beginning at st_1 are exiting a service. Equation (7) ensures that every passenger entering the system at a station gets on a compatible service. Equation (8) ensures that the number of trips from station st_1 to st_2 commencing in block

b matches the travel data. Equation (9) ensures that no more passengers take a trip from st_1 to st_2 on service s commencing in block b than enter the service. Equation (10) ensures the passengers on the service s at its starting station are correct. Equation (11) ensures the passengers on the service s at later stations in the line are correct by adding in newly entering passengers and removing exiting passengers.

Finally we can specify the objective for optimising. We consider two objectives. The first is based on PTV's own congestion scale. It just counts the number of stations and services which reach each capacity level, and penalises each capacity level by a rapidly increasing amount. Let $sparse = 264$, $seated = 528$, $standing = 662$ and $target = 800$, then the first objective is simply:

$$\text{MINIMIZE} \quad \begin{aligned} &\sum_{st \in ST, s \in S, pax_{st,s} > sparse, pax_{st,s} \leq seated} 10 \\ + &\sum_{st \in ST, s \in S, pax_{st,s} > seated, pax_{st,s} \leq standing} 100 \\ + &\sum_{st \in ST, s \in S, pax_{st,s} > standing, pax_{st,s} \leq target} 1000 \\ + &\sum_{st \in ST, s \in S, pax_{st,s} > target} 10000 \end{aligned} \quad (12)$$

The objective above is deceptive: a train running at $seated + 1$ passengers is given the same objective cost as one running at $standing$. For example, this means that once a train needs more than $seated$ passengers the objective will try to fill it to $standing$.

An alternative objective builds a continuous piecewise linear function which defines a cost for each passenger load, which grows steadily with higher capacities increasing faster. The function we use is

$$cost(p) = \begin{cases} 0 & p \leq sparse \\ p - sparse & p > sparse, p \leq seated \\ 5 \times (p - seated) + cost(seated) & p > seated, p \leq standing \\ 10 \times (p - standing) + cost(standing) & p > standing, p \leq target \\ 100 \times (p - target) + cost(target) & p > target \end{cases}$$

where $cost(level)$ represents the cumulative costs per passenger before reaching the current congestion level. Then the objective is simply

$$\text{MINIMIZE} \quad \sum_{st \in ST, s \in S} cost(pax_{st,s}) \quad (13)$$

Restricting Passenger Movement. The model as defined above allows all passengers to be shifted from their original service. While this provides a strong lower bound on possible congestion, we are unlikely to be able to enforce this behaviour. We also consider cases where only some percentage p of the customers can be moved. This reflects an assumption that any take-up in advice will only ever be followed by at most $p\%$ of customers. Adding this to the model simply requires adding lower bounds to the $enter_{b,st,s}$ variables to be $(100 - p)\%$ of the baseline ridership on each service (as computed in Sect. 3.3).

6 Experiments and Results

6.1 Design of Experiments

In the experiments we compare the raw congestion values determined by the train occupancy calculation of Sect. 3.3, with the congestion values where we enact greedy policies that divert passengers away from congested services, as well as against the discrete optimisation model of Sect. 5.

The discrete optimisation model is written in MiniZinc [11] and solved with the Gurobi 8.1.0 mixed integer programming solver [4]. Note that the entire model of Sect. 5 is linear, except for the piece-wise linear objectives. We rely on MiniZinc's automatic linearisation to encode the objective for Gurobi.

We consider experiments where we are allowed to shift 100% of passengers, which gives us a lower bound on possible congestion. To be more realistic we also consider where at most some smaller percentage $p\%$ of customers can have their behaviour changed. We examine the cases where $p = 20, 40, 60$, and 80. With $p = 0$ there is no shifting possible, we just show the calculated congestion levels.

Forward shifting of passengers, which requires some way of informing passengers to arrive earlier at the station, is more complex than simply backward shifting, which just requires information available at the station. Our experiments consider allowing both backward and forward shifting of passengers, as well as disallowing forward shifting.

To reduce the computational complexity of the optimisation model we also simplify the network by merging passenger data for stations where there is no potential for demand shifting. Therefore, Werribee incorporates Hoppers Crossing, Williams Landing and Aircraft; Williamstown incorporates Williamstown Beach and North Williamstown; Newport incorporates Spotswood, Yarraville and Seddon. Passenger travel from Westona, Seaholme and Altona is low and we consolidate demand in the so-called Altona loop.

The resulting network consists of 7 stations, with an observed daily demand ranging between 15,000 and 17,000 passengers. The network is serviced by 34 trains running during the peak period: 16 Werribee Express services, and 9 trains for each Laverton and Williamstown lines. The resulting reduced-network model requires up to 15 s to be solved for most instances, with a few exceptions requiring up to a maximum of 3 h of execution.

6.2 Heatmaps

We re-examine the morning peak period for Thursday, February 2, 2017, now using the optimisation model. The resulting passenger counts and congestion levels are shown in Fig. 3a assuming forward shifting is not allowed, and Fig. 3b assuming forward shifting is allowed.

Clearly the optimisation-based solution drastically reduces congestion levels. With forward shifting, it is able to restrict congestion levels on all services to *standing*. Even without forward shifting, it is able to remove all breach events from the system.

Dep. Time	Werribee	Laverton	Altona	Williamstown	Newport	Footscray
7:04	505	528	-	-	662	648
7:13				171	264	261
7:16		115	264		662	652
7:19	510	528	-	-	662	647
7:29	614	625	-	-	739	662
7:35				202	662	654
7:38		329	528	-	662	648
7:39	662	662	-	-	662	662
7:49	762	744	-	-	759	662
7:57				215	662	649
8:00		374	528	-	798	783
8:02	542	528	-	-	662	596
8:11	516	528	-	-	662	651
8:19				144	528	521
8:22		220	338	-	528	520
8:24	454	528	-	-	525	510
8:33	452	440	-	-	528	506

(a) Forward shifting not allowed.

Dep. Time	Werribee	Laverton	Altona	Williamstown	Newport	Footscray
7:04	527	528	-	-	528	527
7:13				171	498	494
7:16		99	250	-	662	654
7:19	450	528	-	-	633	528
7:29	528	528	-	-	662	655
7:35				202	662	654
7:38		260	459	-	528	514
7:39	528	528	-	-	662	660
7:49	528	528	-	-	636	528
7:57				215	662	649
8:00		377	528	-	528	513
8:02	577	528	-	-	528	493
8:11	488	528	-	-	528	515
8:19				144	662	646
8:22		226	345	-	528	520
8:24	436	528	-	-	528	512
8:33	424	395	-	-	528	519

(b) Forward shifting allowed.

Fig. 3. Optimised passenger counts and congestion levels during the morning peak for all services operating on the Werribee-Williamstown-Footscray network on February 2, 2017.

6.3   Effect of Passenger Uptake

In the next experiment we vary passenger uptake levels, thus restricting the number of passengers that can be moved from their original service. We also consider the five different weekdays from Monday, January 30, 2017 until Friday, February 3, 2017.

Figure 4 shows on the left how the objective function of Eq. (12) changes for different levels of passenger uptake, for all five weekdays. We compare greedy demand shifting versus the optimisation-based demand shifting using the objective of Eq. (12), with and without forward shifting. Note that Monday has noticeably less passengers than the other days. Clearly, greedy demand

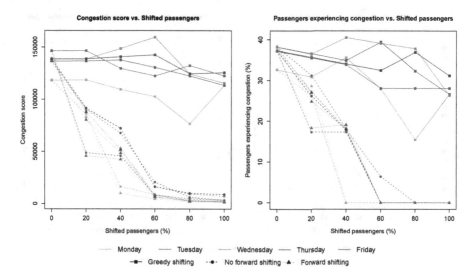

Fig. 4. Congestion levels (left) and percentage of passengers experiencing congestion at some point in their trip (right), using Eq. (12).

shifting only has a slightly beneficial effect on the network and can indeed *worsen* the congestion score, because it makes myopic decisions which end up leading to later congestion. In contrast, the optimisation-based approaches can drastically reduce congestion. With 100% take-up, congestion is always reduced to nearly zero when allowing forward shifting. Enabling forward shifting, while universally improving the results, does not appear to make that much difference, at least at the granularity visible in the plot. But the gains are substantial when considered in relative terms (c.f. Fig. 3).

Figure 4 on the right shows how many passengers experience congestion on their trip, that is, passengers that travel at least one segment in a breached train. Again we see the greedy demand shifting can worsen this measure, while the optimisation approaches can quickly find solutions where no passenger experiences congestion.

Figure 5 shows the results measured with the more fine-grained objective function of Eq. (13). In these experiments, the optimisation approach was run minimising this objective. On the left we see how the objective value changes as passenger uptake increases. Interestingly, using this measure we can see that greedy shifting is in fact reducing congestion per passenger, although not per service, and it does improve as passenger uptake increases. The optimisation solutions are again far superior, reducing the objective function to very low values smoothly as the uptake increases.

Figure 5 on the right shows the percentage of passengers experiencing congestion. The results for greedy shifting are unchanged. Again, the optimisation results show a significant reduction on the proportion of passengers travelling on breached services. The peak for Wednesday and 40% uptake can be explained by

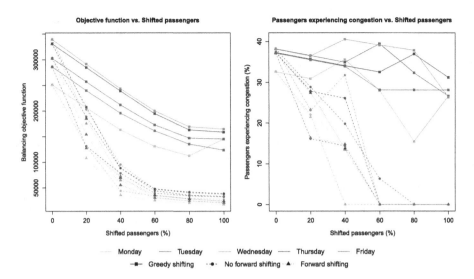

Fig. 5. Balancing objective function values (left) and percentage of passengers experiencing congestion at some point in their trip (right), using Eq. (13).

considering that our model is not directly minimising the percentage of passengers experiencing congestion. According to the recorded trips, several services ran well above the breach threshold on this day. Considering a small uptake (e.g., 20%) evens out the number of passengers on these services, but keeps most trains previously running at *target* on the same congestion level. At 40% uptake, the number of shifted passengers is not enough to bring the former services below *breach*, but some passengers can be re-allocated to different trains that were operating just under the breach threshold. Passengers on these services, who were not considered to be experiencing congestion before, are now travelling on breached trains. However, despite the percentage increasing, passenger numbers on board over-congested trains are lower and more balanced across services. Table 1 summarises the results obtained with the proposed fine-grained objective function aimed at balancing passengers between services, including the number of breach incidents and the percentage of passengers experiencing congestion for different levels of take-up.

6.4   Experimental Results and Discussion

If forward shifting is not allowed, the model yields results that will help alleviate congestion, but do not completely eliminate it. We still obtain some trains where utilisation is very close to the *breach* threshold. This increases the risk of over-congestion if passenger numbers keep rising. However, this kind of intervention is the easiest and most likely the cheapest, since it does not require a major change in passenger behaviour. Arrival and touch-on patterns remain the same, but passengers are given recommendations to board specific alternative trains

Table 1. Results obtained with our optimisation model using Eq. (13) for different percentages of shifted passengers showing number of breach incidents and percentage of passengers experiencing a breach service.

Day	Forward	0%		20%		40%		60%		80%		100%	
		Breach inc.	PAX (%)	Breach inc.	PAX (%)	Breach inc.	PAX (%)	Breach inc.	PAX (%)	Breach inc.	PAX (%)	Breach inc.	PAX (%)
Mon	✗	11	32.62	8	22.02	0	0.00	0	0.00	0	0.00	0	0.00
	✓	11	32.62	8	21.45	0	0.00	0	0.00	0	0.00	0	0.00
Tue	✗	13	38.21	10	28.86	7	19.79	1	6.30	0	0.00	0	0.00
	✓	13	38.21	9	27.84	4	13.53	0	0.00	0	0.00	0	0.00
Wed	✗	13	38.13	9	23.11	10	31.75	0	0.00	0	0.00	0	0.00
	✓	13	38.13	9	23.32	4	13.42	0	0.00	0	0.00	0	0.00
Thu	✗	13	37.29	10	27.83	7	26.11	0	0.00	0	0.00	0	0.00
	✓	13	37.29	10	27.42	4	13.76	0	0.00	0	0.00	0	0.00
Fri	✗	14	37.16	4	16.12	4	14.40	0	0.00	0	0.00	0	0.00
	✓	14	37.16	4	16.36	4	14.84	0	0.00	0	0.00	0	0.00

in order to avoid an uncomfortable trip on a congested service. This could be achieved by, e.g., providing additional on-screen information about alternative trains and congestion levels. One example of this is the smartphone app of NS, the Dutch national rail operator, which shows the expected congestion levels of arriving trains, from which passengers can make an informed choice of whether to wait for a less congested service or not. We observe the greatest reduction in congestion is achieved when forward shifting is allowed. For this to work, passengers might need to arrive at the station up to 10 min prior to their intended trip departure for a less congested ride. This would require a change in passenger behaviour, and might need to be implemented in conjunction with other incentive mechanisms, such as fare reduction. However, our results show that even a quite modest adoption of such a program (of the order of 20%) could provide a significant reduction on congestion levels during the morning peak.

Our optimisation results also show that demand shifting up the line (closer to where the service originates) can lead to a major reduction in congestion down the line—that is, closer to Melbourne. The results for simulated demand shifting show that myopic interventions lead at most to a minor alleviation of congestion for some services. The implications of this research for PTV is that better management of demand originating at Werribee, Hoppers Crossing, Williams Landing, or Aircraft, could reduce congestion at busy stations down the line. For example, many trains originating at Werribee currently depart Laverton and Newport at or above *target* levels, so that it is almost inevitable that these trains will become congested as they journey towards Melbourne. Results from our optimisation model show that reducing demand for these services when alternatives are available at origin could prevent congestion down the line. Our results, thus, give some guidance on where to trial interventions to reduce congestion in the morning peak.

7   Conclusions and Future Work

Much has been written on the various ways in which to optimise rail networks, but relatively little work that we are aware of exists on providing information for passengers to modify their behaviour in ways that improve the system for all users. Our pilot study shows how this might be achieved in the southwest Melbourne rail network. As Melbourne grows, the efficient usage of all public transport infrastructure will become more and more important, and modifying passenger behaviour is an attractive alternative to provisioning more services.

There are several ways in which the work presented here might be extended. The actual weights (or costs) in our objective functions from Sect. 5 are ad hoc. Analysis of the relative importance ascribed to comfort over the other competing concerns of passengers from the survey literature may, perhaps, more realistically weight the objective function in our model. We could consider the multi-objective problem measuring both trip time and comfort and explore the efficient (Pareto) set of possible solutions arising from this.

Deeper analysis of customer arrival times at stations could also be valuable. Patterns in customer arrival may arise from a mixture of behaviours—for example, we would expect to see *schedule-aware* customers whose arrival time is some function of when the next train is scheduled to depart, and *schedule-oblivious* customers whose arrival time is largely independent of the timetable. This leads to the possibility of using mixture modelling [10], possibly incorporating the Poisson distribution [15], to more accurately model customer behaviour. Further to such mixture modelling for different customer arrival times, re-visiting the discussion of [17], our greedy and optimisation models could be modified to treat customers of various behaviour types differently.

Acknowledgements. We acknowledge the Monash University Faculty of Information Technology for seed funding to begin this project. Daniel Harabor is funded by the Australian Research Council under the grant DE160100007.

References

1. Network Development Plan - Metropolitan Rail December 2012 (Updated 2016). Public Transport Victoria (2016)
2. Annual Report 2017–18. Public Transport Victoria, 140 p (2018)
3. Frappier, A., Morency, C., Trépanier, M.: A new method to measure the quality and diversity of transit trip alternatives. Technical report, October 2015. CIRRELT (Centre interuniversitaire de recherche sur les réseaux de l'interprise, la logistique et le transport) technical report CIRRELT-2015-51
4. Gurobi optimizer version 8.1 (2019). http://www.gurobi.com/products/gurobi-optimizer
5. Haywood, L., Koning, M.: The distribution of crowding costs in public transport: new evidence from Paris. Transp. Res. Part A: Policy Pract. **77**, 182–201 (2015)
6. Kroes, E., Kouwenhoven, M., Debrincat, L., Pauget, N.: On the value of crowding in public transport for Ile-de-France. international transport forum discussion papers, no. 2013/18 (2013)

7. Kroes, E., Kouwenhoven, M., Duchateau, H., Debrincat, L., Goldberg, J.: Value of punctuality on suburban trains to and from paris. Transp. Res. Rec. **2006**(1), 67–75 (2007)
8. Liu, R., Li, S., Yang, L.: Collaborative optimization for metro train scheduling and train connections combined with passenger flow control strategy. Omega (2018, in press)
9. Liu, Y., Charles, P.: Spreading peak demand for urban rail transit through differential fare policy: a review of empirical evidence. In: Australasian Transport Research Forum 2013. Queensland University of Technology, Brisbane, QLD (2013)
10. McLachlan, G.J., Peel, D.: Finite Mixture Models. Wiley, Hoboken (2000)
11. Nethercote, N., Stuckey, P.J., Becket, R., Brand, S., Duck, G.J., Tack, G.: MiniZinc: towards a standard CP modelling language. In: Bessière, C. (ed.) CP 2007. LNCS, vol. 4741, pp. 529–543. Springer, Heidelberg (2007). https://doi.org/10.1007/978-3-540-74970-7_38
12. de Palma, A., Lindsey, R., Monchambert, G.: The economics of crowding in rail transit. J. Urban Econ. **101**, 106–122 (2017)
13. Pluntke, C., Prabhakar, B.: INSINC: a platform for managing peak demand in public transit. J. Land Transp. Authority Acad. Singapore 31–39 (2013)
14. Shi, J., Yang, L., Yang, J., Gao, Z.: Service-oriented train timetabling with collaborative passenger flow control on an oversaturated metro line: an integer linear optimization approach. Transp. Res. Part B: Methodol. **110**, 26–59 (2018)
15. Wallace, C.S., Dowe, D.L.: MML clustering of multi-state, Poisson, von Mises circular and Gaussian distributions. Stat. Comput. **10**(1), 73–83 (2000)
16. Yen, B.T., Tseng, W.C., Chiou, Y.C., Lan, L.W., Mulley, C., Burke, M.: Effects of two fare policies on public transport travel behaviour: evidence from South East Queensland, Australia. J. Eastern Asia Soc. Transp. Stud. **11**, 425–443 (2015)
17. Zhang, J., Yang, H., Lindsey, R., Li, X.: Modeling and managing congested transit service with heterogeneous users under monopoly. Transp. Res. Part B: Methodol. (2019, in press)

CP and Life Sciences Track

Functional Significance Checking in Noisy Gene Regulatory Networks

S. Akshay[1(✉)], Sukanya Basu[1], Supratik Chakraborty[1],
Rangapriya Sundararajan[1,2], and Prasanna Venkatraman[2,3]

[1] Department of CSE, IIT Bombay, Mumbai, India
akshayss@cse.iitb.ac.in
[2] ACTREC, Kharghar, India
[3] Tata Memorial Hospital, HBNI, Mumbai, India

Abstract. Finding gene regulatory pathways that explain outcomes of
wet-lab experiments is one of the holy grails of systems biology. SAT-
solving techniques have been used in the past to find few small explana-
tory pathways assuming either zero or a few known perturbations in the
experimental observations. Unfortunately, these approaches do not work
when (i) there is noise in the experimental data or domain knowledge, as
opposed to known perturbations, and (ii) the number of possible pathways generated by repeatedly invoking a SAT-solver is too large to be
analyzed by enumeration. In such settings, determining if an actor plays
a functionally significant role towards explaining experimental observations is very difficult using existing SAT-based techniques.

In this paper, we formalize the problem of functional significance
checking in gene-regulatory pathways in the presence of a bounded
amount of noise. We show that this problem is Δ_2^P-hard and hence
cannot be efficiently encoded into SAT (unless the polynomial hierarchy collapses). We then propose an algorithm that uses a polynomial
number of SAT-oracle invocations to solve a practically useful version
of this problem. Finally, we present results on checking functional significance of suspect genes in real microarray data obtained from cancer
cell-line experiments, some of which are corroborated by subsequent wet-
lab knock-off experiments.

1 Introduction

A central problem in systems biology concerns finding gene regulatory pathways
that explain observed outcomes of wet-lab experiments. In a typical wet-lab
experiment, a pre-determined stimulus is given to specially prepared cells under
controlled conditions, and the expressions of various genes (i.e. concentrations
of corresponding gene products) measured at carefully timed instants. Practical
constraints (including cost, unknown time constants of biological processes etc.)
often limit the number of gene expression profiles that can be measured during the course of an experiment. In addition, measured gene expression profiles
almost inevitably have noise. As a consequence, it becomes difficult to infer if a

© Springer Nature Switzerland AG 2019
T. Schiex and S. de Givry (Eds.): CP 2019, LNCS 11802, pp. 767–785, 2019.
https://doi.org/10.1007/978-3-030-30048-7_44

suspected gene plays a functionally significant role in the outcome of the experiment. This motivates us to ask if we can computationally predict the functional significance of a gene even when a single noisy expression profile (in addition to a reference profile) is available, by taking into account domain knowledge about gene interactions from public-domain databases, and by bounding a quantitative metric of the admissible noise.

The gene expression profile (often measured using microarray [5] or RNA-sequencing [52]) is usually given as log fold changes relative to a reference profile corresponding to a normal (or wild-type) cell, and serves as a proxy for the activation level of a gene. An activated (resp. inhibited) gene in the experimental cells usually yields higher (resp. lower) concentrations of the corresponding gene product compared to a normal cell. The use of contextual gene interaction information from a public-domain database like KEGG [23] provides a reasonable encoding of domain knowledge. "Noise" in our setting can be along two dimensions: (a) some gene expression measurements can be erroneous, (b) interactions between gene pairs in the context of the experiment under study may differ from what is recorded in KEGG, giving rise to "noise" in gene interaction information. Given these noisy inputs, we wish to identify if a suspect gene plays a functionally significant role in the outcome of the wet-lab experiment. Informally, this happens if the presence of the gene makes it possible to "easily" explain the measured gene expression profile consistently with domain knowledge, while its absence makes it difficult to provide any such explanation. We quantify the "easiness" via a quantitative metric, which we formalize as the number of relaxations or changes that must be admitted in the input to obtain an explanation.

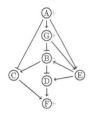

Fig. 1. Example gene interactions

An illustrative example: To better understand the computational aspects of the problem, consider a hypothetical wet-lab experiment in which cancer cells are treated with a drug known to activate gene A. Suppose we wish to determine how this affects the activation of another gene F in the cancerous cells. For simplicity, assume that only 7 genes named A, B, C, D, E, F and G potentially play any role in the outcome of the experiment. Let the gene expression profile obtained at an appropriate time instant be as follows: A, F, G over-expressed, B under-expressed, C, D, E did not show any significant difference in expressions relative to that of a normal (non-cancerous) cell. From this, we infer that A, F and G are activated and B is inhibited in the context of the experiment. Genes C, D and E in the cancerous cells could either be in their respective ground states (as in a normal cell), or could even be in mildly activated or mildly inhibited states (mild enough so that they do not express their effect overwhelmingly in the gene expression profile). Suppose we are also told that the domain knowledge about mutual interaction of genes A through G are as in the graph shown in Fig. 1 (sans the $\pm$ labelings). In this figure, a $\rightarrow$ denotes an activating interaction ($A \rightarrow E$ implies that if A is active, so must E be) and a $\dashv$ denotes an inhibiting edge ($G \dashv B$ implies that if G is active, B must be inactive). Since the ground state (in a normal cell) of a gene may itself be activated/inhibited, we must be careful in interpreting the $\rightarrow$ and $\dashv$ edges. For example, the edge $A \rightarrow E$ not only admits both A and E being activated, but also admits both being inhibited. To see why this makes sense, note that if E has an inhibited ground state, then an inhibited A cannot activate E through an $A \rightarrow E$ edge. Similarly, $G \dashv B$ not only admits G activated and B inhibited, but also vice versa, i.e. B has an activated ground state, and G being inhibited, cannot inhibit B.

Given domain knowledge encoded in a graph like Fig. 1, we represent activation levels of genes in the experiment under study by $\pm$ labelings of nodes, where activated genes are labeled "+" and inhibited genes are labeled "-". Our first goal is to determine if there exists a set of paths from A to F in Fig. 1, and a $\pm$ labeling of nodes along these paths, such that the labeling is consistent with both the observed gene expressions and the domain knowledge. Informally, such a set of paths "explains" the experimental observations consistently with domain knowledge. In this example, it is indeed possible to find such an explanation with three paths from A to F, namely: $A(+) \rightarrow G(+) \dashv B(-) \dashv D(+) \rightarrow F(+)$, $A(+) \rightarrow E(+) \rightarrow D(+) \rightarrow F(+)$ and $A(+) \rightarrow G(+) \rightarrow E(+) \rightarrow D(+) \rightarrow F(+)$ There are several points to note here: (i) although E was not differentially expressed in the observed profile, it is fine to assign label "+" to E in the explanation, since E could indeed have been in a mildly activated state that didn't result in a strong gene expression profile, (ii) although E and B are included in the explanation, the induced edge $E \rightarrow B$ is not included since the labelings of B, E are not consistent with $E \rightarrow B$, and (iii) the presence of a topological path from A to F through B doesn't necessarily imply that this path explains the experimental observations consistently with domain knowledge. For example, although there is a topological path $A(+) \rightarrow G(+) \dashv B(-) \rightarrow C(?) \rightarrow F(+)$ in Fig. 1, there is no way of assigning a label ("+" or "-") to C that is consistent with the interpretation of activating and inhibiting edges. Thus, finding explanations is significantly harder than finding topological paths or induced sub-graphs.

We now ask: *Does gene D play a functionally significant role in explaining the observed expressions consistently with domain knowledge?* While the precise notion of functional significance will be discussed later, informally, we ask if we can find a domain knowledge-consistent explanation of the observed gene expressions *even if node D is removed from Fig.* 1. It is easy to see from Fig. 1 that the answer is in the negative. In contrast, if node E or C (or both) is (are) removed from Fig. 1, the path $A(+) \rightarrow G(+) \dashv B(-) \dashv D(+) \rightarrow F(+)$ continues to explain the observed gene expressions. Therefore, if we assume that all gene expression measurements are noise-free, D is functionally significant, while E and C are not. However, if we admit that one gene expression measurement can be noisy, then functional significance of D warrants re-examination. Indeed, with D removed from Fig. 1, the paths $A(+) \rightarrow E(+) \rightarrow B(+) \rightarrow C(+) \rightarrow F(+)$ and $A(+) \rightarrow G(+) \rightarrow E(+) \rightarrow B(+) \rightarrow C(+) \rightarrow F(+)$ explain the observed gene expressions with the (noisy) label of B *changed* from "-" to "+". This shows that functional significance of a gene can vary depending on the admissible noise.

Generalizing from the above discussion, our objective is to study computational techniques that (i) work with a single gene expression profile (in addition to a reference profile), (ii) are tolerant to a bounded amount of noise in both gene expression measurements and in the encoding of domain knowledge as gene interactions, and (iii) allow us to check whether a suspect gene (provided as input) plays a functionally significant role in explaining observed gene expression levels. By bounded noise, we mean that the number of errors either in the gene expression measurements or difference wrt KEGG must be at most a fixed constant, which is typically small. Note that this does not mean we know the errors, just that their number is limited. This is a reasonable assumption since allowing an arbitrarily large amount of noise would invalidate the experiment and any inferences made from it entirely.

In this paper, we formalize the problem described above, show that it is Δ_2^P-hard and in Π_2^P as well as present an algorithm to solve a useful variant of the problem. We are not aware of any earlier proof of hardness of even the simplest problem of finding explanations in the absence of noise. We fill this gap and go much beyond to prove the Δ_2^P-hardness of functional significance checking with bounded weighted noise. This shows that functional significance checking cannot be reduced to propositional SAT-solving (unless the polynomial hierarchy collapses). Our treatment of noise is also more robust than that used in earlier work. Specifically, we allow different genes and gene interactions to contribute in a weighted manner to the overall noise metric. Additionally, we don't need the user to specify the exact set of gene expressions or gene interactions that may be noisy. Instead, we allow all combinations of noisy gene expressions and gene interactions subject to the weighted noise metric staying within specified bounds. This permits exploring a much larger space of possible explanations than that in earlier work (viz. [10]). Finally, our algorithm detects functional significance of a gene without actually enumerating the potentially explosively many explanations of the observed gene expressions while admitting bounded noise. This makes it possible to analyze much larger systems of gene interactions.

The entire work reported in this paper was done by a team of three computer scientists and two molecular biologists. As such, the biological relevance of modeling artifacts and predictions were discussed and validated at each step. However, this paper is focused more on the computational aspects.

Related Work. Biological phenomena have been modeled in various ways, viz. using Petri-nets, ODEs, sets of rules, Boolean networks, etc (see [20,51]). A popular way of representing biological networks, especially gene regulatory networks, is *influence graphs* [42], which are (partially) edge-labeled graphs, used to model incomplete data. The Sign Consistency Model (SCM) of [41] enhances this with a (partial) labeling of nodes such that the whole labeled graph is consistent with a set of constraints [19]. In [18], a SAT-solver and MAX-SAT solver are used to check for consistency, somewhat similar to our work. Answer set programming (ASP) is yet another technique for obtaining models to a set of logical constraints used in AI [43], for searching models of NP-hard problems [30] and to detect inconsistencies, repair and prediction in biological networks

[11,12]. The works in [33,46] model different notions of sign consistency and use an ILP solver to obtain a minimal set of nodes whose sign needs to be changed to be consistent. Such variants can also be encoded in our approach.

Several tools have been built over the years to analyze biological pathway networks [4,8,15,16,44,50]. Of these [15,16] apply statistical methods to correlate the network topology and gene expression data, which allows them to also identify some functional associations, assuming the availability of sufficient gene interaction data. The approaches in [6,9,31,38,53] try to find enriched pathways based only on gene information, whereas [2,7,29,32,48,54] use both gene-expression and topology information for selecting candidate enriched pathways. In spite of the apparent difficulty, some tools such as [13,21,35,45,49] have tried to exploit the full annotation on the interactions, while recently [24,27,28] have stringently analyzed relations between genes. In [14], SMT-solvers have been used to analyze robustness under mutations of gene regulatory networks.

While encoding the problem of finding an explanation from known pathways and expression data as a SAT problem has similarities with other work in literature [10,40], such an encoding often gives no explanation subgraphs or too many of them as solutions to the SAT problem. Crucially it doesn't solve functional significance checking when expression data and knowledge about pathways are noisy, unless we examine every explanation subgraph for all noisy inputs – an impractical task. The primary differentiator of our work vis-a-vis these earlier work is in the way we model noise and *implicitly* consider all possible noisy inputs subject to the weighted noise being bounded, while still requiring a polynomial number of SAT invocations.

Finally, identifying important actors in a network has been studied in multiple contexts, including the web, social media networks, gene regulatory and protein-protein interaction networks. Various graph theoretic metrics have been used to detect crosstalk and identify *hubs* and *bottlenecks* in large biological networks [37,55]. Our work can be used in tandem with these techniques by first obtaining potential candidates using graph theoretic techniques, and then checking their functional significance using our approach.

2   Problem Formulation

While we have used KEGG [23] to encode domain knowledge of gene interactions in our experiments, our abstract problem formulation is not KEGG specific. To keep the exposition simple, we assume that there are only two types of edges – activating ($\mathcal{A}$) and inhibiting ($\mathcal{I}$). The domain knowledge of gene interactions is given as an edge labeled graph $G_{dom} = (V, E, \mu)$, where V is the set of genes, $E \subseteq V \times V$ is the set of interactions (directed edges) between genes, and $\mu : E \longrightarrow L_e$ is a labeling of edges with $L_e = \{\mathcal{A}, \mathcal{I}\}$. The interpretation of activating and inhibiting edges is as follows: For an edge $e = (u, v)$, if $\mu(e) = \mathcal{A}$, then gene v must be activated whenever u is active. In addition, as discussed in Sect. 1, an activating edge (u, v) is consistent with both u and v being in inhibited states. Similarly, if $\mu(e) = \mathcal{I}$, v must be inhibited whenever u is active. In addition, an inhibiting edge (u, v) is consistent with u being inhibited and v being active.

In order to represent the gene expression profile, we decorate each node v in the graph G_{dom} with a label $\lambda(v)$ from the set $L_v = \{+, -, ?\}$. Here, $+$ denotes an over-expressed (and by implication, active) gene, $-$ denotes an under-expressed (and by implication, inhibited) gene, and $?$ denotes a gene that is not significantly differentially expressed with respect to the expression level of a normal cell. For clarity of exposition, we use $*$ to denote either $+$ or $-$, but not both.

The domain knowledge and gene expression profile can be represented together as a node- and edge-labeled graph $G = (V, E, \lambda, \mu)$, where $\lambda : V \to L_v$ and $\mu : E \to \{\mathcal{A}, \mathcal{I}\}$. We also assume that we are given three nodes $s, t, i \in V$ as follows: s represents a stimulus gene, the effect of whose activation we wish to study, t represents a target gene that is eventually activated (possibly after a long chain of interactions) due to activation of s, and i represents a suspect gene whose functional significance in the activation of t by s is the subject of our investigation. For simplicity, we fix $\lambda(s) = \lambda(t) = +$; other combinations of $\lambda(s)$ and $\lambda(t)$ are easily handled. To formally define the notion of functional significance, we first define an *explanation subgraph*. Informally, this is a subgraph of G that contains s-t paths along with a labeling of nodes that "explains" the observed gene expression profile while being consistent with the domain knowledge.

Definition 1 (Explanation subgraph). *Let $G = (V, E, \lambda, \mu)$ be as defined above, and let s and t be nodes in V s.t. $\lambda(s) = \lambda(t) = +$. An* explanation *subgraph of (G, s, t) is a node- and edge-labeled graph $G' = (V', E', \lambda', \mu')$ s.t.,*

1. **Subgraph containing s, t:** *We require $V' \subseteq V$, $E' \subseteq E \cap (V' \times V')$, μ' is the restriction of μ to E', and $s, t \in V'$.*
2. **Labels consistent with observed expressions:** $\lambda'(v) \in \{+, -\}$ *for all $v \in V'$, and $\lambda'(v) = \lambda(v)$ if $\lambda(v) \neq ?$.*
3. **No floating nodes:** *Every $v \in V'$ is reachable from s in G'.*
4. **Activity condition:** *Every s-t path of length > 1 in G' passes through some node $v \notin \{s, t\}$ with $\lambda(v) = +$, and every such node v in G' appears on some s-t path in G'. Effectively, for a pathway to credibly explain how s eventually activates t, it must be supported by at least one other active node along the pathway. Also, every node in G' that was originally active must contribute towards explaining how s activates t along some path in G'.*
5. **Compatible labeling:** *For every edge $e = (u, v)$ in E', if $\mu'(e) = \mathcal{A}$, then $\lambda'(u) = \lambda'(v)$, and if $\mu'(e) = \mathcal{I}$, then $\lambda'(u) \neq \lambda'(v)$. Moreover, every node other than s in G' must have at least one incoming compatible edge.*

For the example in Fig. 1, the path $A(+) \to E(+) \to D(+) \to F(+)$ doesn't constitute an explanation subgraph because the activity condition is violated. However, $A(+) \to G(+) \dashv B(-) \dashv D(+) \to F(+)$ is an explanation subgraph.

2.1   Graph Relaxation: Modeling Errors and Noise

A startling finding of our initial experiments with real micro-array data and KEGG pathways was that often no explanation subgraphs could be found at all. Delving deeper, we realized that there were two primary reasons for this:

(a) the pathway information in KEGG didn't relate to the context in which the experiments were performed (i.e., some edge attributes were incorrectly labeled), and (b) there was noise in the micro-array data (i.e., node attributes were incorrectly labeled). Thus we need to search for explanation subgraphs not on the original graphs, but graphs obtained by changing some (unknown) edges and nodes. To formalize this, we introduce the notion of *relaxations*. Specifically, we associate an integer relaxation weight to each node and edge, and allow node and edge labels to be changed when finding an explanation subgraph. The total *node noise* (resp. *edge noise*) introduced to obtain an explanation subgraph is simply the sum of relaxation weights of all nodes (resp. edges) whose labels had to be ignored or changed to obtain the explanation subgraph. We bound the admissible noise by specifying an upper bound (n, e) of node and edge noise respectively. For notational convenience, we refer to (n, e) as *relaxation bounds* in the subsequent discussion. Thus, these bounds provide a quantitative metric to deal with noise (caused by errors or inconsistencies in KEGG and microarray data), as mentioned in the introduction. Our definition of noise is driven by specific biological experiments and hypotheses as explained in Sect. 5. However, our techniques and encoding can also model other related notions considered in the literature, such as creation of new edges.

Formally, for a subgraph G' of G, a node in G is said to be *relaxed* in G' if one of the following hold: (i) it is labeled $+$ in G, but is absent in G', i.e. a node active in G is excluded from G', (ii) it is labeled $+$ (resp. $-$) in G, and is present but labeled $-$ (resp. $+$) in G'. If a node is inhibited in G but excluded from G', we do not treat it as relaxed. Similarly, edge $e = (u, v)$ in G is *relaxed* in G' if $u, v \in V', e \in E'$ and either $\mu(e) = \mathcal{I}$ and $\mu'(e) = \mathcal{A}$ or $\mu(e) = \mathcal{A}$ and $\mu'(e) = \mathcal{I}$.

Definition 2 (Relaxed explanation). *Given $G = (V, E, \lambda, \mu)$, source s, target t, a relaxation weight $R : V \cup E \to \mathbb{N}$ and $(n, e) \in \mathbb{N}^2$, we call H an (n, e)-relaxed explanation of (G, s, t) under R if (a) there exists a subgraph G' of G obtained by relaxing nodes and/or edges, (b) $\sum_{\{v \in V | v \text{ relaxed in } G'\}} R(v) \leq n$, (c) $\sum_{\{e \in E | e \text{ relaxed in } G'\}} R(e) \leq e$ (d) H is an explanation subgraph of (G', s, t).*

2.2   Pareto Optimality and Functional Significance

As mentioned earlier, often there are no explanation subgraphs with 0 node and edge relaxations. Interestingly, our experiments indicate that there is a large multiplicity (literally 1000s) of explanation subgraphs if we allow small node and/or edge relaxations. In this context, solutions obtained with very large values of node and/or edge relaxations may not be meaningful. Relaxing too many nodes allows activation status of many nodes to differ from the observed gene expression profile. Similarly, relaxing too many edges amounts to making significant modifications to a curated database of regulatory pathways. None of these are desirable. Indeed, if we allow all nodes or all edges to be relaxed, we can always find an explanation subgraph that may hardly relate to the wet-lab experiment under investigation. Hence it makes sense to ask for minimal node and edge relaxations that yield at least one explanation subgraph. Not surprisingly, increasing node relaxations reduces the requirement of edge relaxations, and vice versa. Therefore, we have a multi-objective optimization problem and obtain a set of minimal or Pareto-optimal (n, e) values. Further, since large node and edge relaxations are undesirable, we want explanations where node and edge relaxations are within given bounds. This motivates us to define a *window of relaxation* W as a pair of intervals, $\langle [n_l, n_u], [e_l, e_u] \rangle$, for node and edge relaxations respectively. We say that $(n, e) \in W$ iff $n \in [n_l, n_u]$ and $e \in [e_l, e_u]$.

Consider the partial order $\sqsubseteq$ on $\mathbb{N} \times \mathbb{N}$ defined by $(n', e') \sqsubseteq (n, e)$ iff $n' \le n$ and $e' \le e$. We say (n, e) *dominates* (n', e') if $(n', e') \sqsubseteq (n, e)$, and that (n, e) *strictly dominates* (n', e') if $(n', e') \sqsubseteq (n, e)$ but $(n', e') \ne (n, e)$. Given an input instance (G, s, t, W, R), where G, s, t, are as before, R is a relaxation weight function and W a relaxation window, let $Sol(G, s, t, W, R)$ denote the set of $(n, e) \in W$ such that there exists an (n, e)-relaxed explanation of (G, s, t) under R. If $(n, e) \in Sol(G, s, t, W, R)$ but both $(n - 1, e)$ and $(n, e - 1)$ are not in $Sol(G, s, t, W, R)$, we say (n, e) is on the *solution curve* of (G, s, t, W, R). The set of points on the solution curve forms a Pareto-optimal curve; any point in W that dominates a point on the curve is in $Sol(G, s, t, W, R)$ and any point in W that is strictly dominated by a point on the curve is not in $Sol(G, s, t, W, R)$.

We now make two reasonable, yet important, assumptions.

A1: The "golden truth" pathway for the wet-lab experiment under study, henceforth called *true explanation subgraph*, is present, modulo relaxations and inter-pathway crosstalk, in the input graph $G = (V, E, \lambda, \mu)$.

A2: The true explanation subgraph corresponds to a Pareto-optimal point $(n^\star, e^\star)$ in the relaxation window W of interest for the given relaxation weight function R. It is reasonable to expect $(n^\star, e^\star)$ to be a Pareto-optimal point, as otherwise, we'd have an alternative explanation of the microarray data with fewer relaxations than that required for the true explanation subgraph to provide a plausible explanation.

Definition 3. *Under assumptions A1 and A2, a node v is said to be functionally significant in (G, s, t, W, R) if its removal from G leaves no $(n^\star, e^\star)$-relaxed explanation subgraph. In other words, $Sol(G \setminus \{v\}, s, t, \langle [n^\star, n^\star], [e^\star, e^\star] \rangle, R) = \emptyset$.*

Unfortunately, Definition 3 does not yield a practical algorithm for checking functional significance of a node, due to two reasons. First, we do not know the values of $n^\star$ and $e^\star$ for a given experiment. Second, our studies show that there are literally thousands of explanation subgraphs at each Pareto-optimal point in the window of relaxation of interest. So, even if we knew $(n^\star, e^\star)$, it would be practically impossible to examine all $(n^\star, e^\star)$-relaxed explanation subgraphs and identify a common node. Thus, we must find a way to decide the functional significance of a node without knowing $(n^\star, e^\star)$ exactly, and without generating all explanation subgraphs corresponding to Pareto-optimal pairs. The following lemma provides a sufficient condition to surmount the above hurdles.

Lemma 1. *Suppose $Sol(G, s, t, W, R) \neq \emptyset$ and either $Sol(G \setminus \{i\}, s, t, W, R) = \emptyset$ or for every $(n, e) \in Sol(G \setminus \{i\}, s, t, W, R)$, there exists $(n', e') \in Sol(G, s, t, W, R)$ such that (n, e) strictly dominates (n', e'). Then i is functionally significant in (G, s, t, W, R) under assumptions A1 and A2.*

3   Complexity Results

Theorem 1. *Checking the existence of an explanation subgraph, even without relaxations, is NP-complete.*

Proof. It is easy to see that the problem is in NP, since we can guess the explanation subgraph, and check it in polynomial time. To prove NP-hardness, we reduce 3-SAT to our problem. Let φ be an instance of 3-SAT in CNF, with ℓ variables $x_1, \ldots, x_\ell$ and m clauses.

For each variable x_i, we first construct a gadget A_i of 6 nodes depicted in Fig. 2, three for variable x_i (which we call a_i, b_i, d_i) and three for $\neg x_i$ (which we denote na_i, nb_i, nd_i). We add activating edges from source s to a_i and na_i for all i. Also add 4 activating edges from a_i to b_i, b_i to d_i, na_i to nb_i and nb_i to d_i and 4 inhibiting edges from a_i to nb_i, na_i to b_i, b_i to nd_i and nb_i to nd_i. Finally, we add node label $+$ for d_i and nd_i and activating edges from both to target t. This gadget A_i ensures that x_i and its negation are not active at the same time.

Fig. 2. Gadget A_i

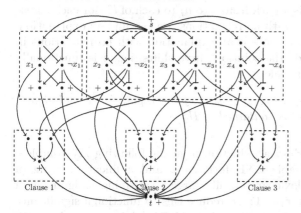

Fig. 3. Construction for reduction from 3-SAT

For each clause c, we construct gadget B_c with 4 nodes: one for each of the 3 literals that occurs in the clause, denoted l_i^c (i.e., $l_i^c = x_i$ or $\neg x_i$) and an additional node l_c^c. Now if $l_i^c = x_i$, (resp $= \neg x_i$) then we add an edge from b_i in gadget A_i to it, else we add an edge from nb_i. Further we add edges from each literal in a clause c to the additional node l_c^c. Each of these additional nodes l_c^c are labeled $+$, which is used to ensure that each clause does evaluate to true in a valid explanation. Each clause of the original instance is replicated by edges from a variable or its negation as appropriate, which finally converge at l_c^c emulating the disjunctions within each clause. Finally from each l_c^c node we add an edge to the target node t. Recall that the target is also labeled $+$.

We claim that an explanation subgraph from s to t exists iff the formula is satisfiable. In one direction, if there is an explanation from s to t, the additional node l_c^c at clause c for every clause must be active and each variable is assigned a unique value. Further, to make this active, by the compatibility condition, one of the literals in that clause gadget must be active. In turn to make that literal active, node corresponding to the literal should be active and this gives the satisfying assignment. Conversely, if the formula is satisfiable, then the satisfying assignment defines an explanation subgraph. This completes the proof of NP-hardness. An example is the formula $\varphi = (x_1 \vee \neg x_2 \vee x_3) \wedge (\neg x_3 \vee x_4 \vee \neg x_1) \wedge (x_2 \vee \neg x_3 \vee x_4)$, whose graph is shown in Fig. 3 with a source s and target t. □

Depending on whether the relaxation window W is fixed or part of the input to a decision procedure, we obtain the following results.

Theorem 2. *For every $(n, e) \in \mathbb{N}^2$, for every relaxation weight function R,*

1. *Checking for an (n, e)-relaxed explanation subgraph under R is NP-complete.*
2. *Checking functional significance of a node in (G, s, t, W, R), where $W = \langle [0, n], [0, e] \rangle$ is co-NP complete.*

Proof. Part 1. follows from proof of Theorem 1 with a simple modification: we replicate the gadget for each variable and clause $n + e + 1$ times, so that even if n nodes and e edges are relaxed, finding an explanation subgraph would require setting each variable in way that all clauses are satisfied. For Part 2., we modify the construction in Theorem 1 by adding a special node n_i, where i is the node whose significance we wish to check. We add an edge from the source s to n_i and from n_i to each node l_c^c for each clause c and to nodes d_i and nd_i for each x_i. In the resulting graph G', there is a path from s to n_i to each of l_c^c (for each clause c), d_i, nd_i and then to t. Thus, with no relaxations, we can find an explanation. However, if n_i is removed, then there is an explanation with no relaxations iff φ is satisfiable. In other words the solution curve shifts, i.e., i is functionally significant in G iff φ is unsatisfiable.

Theorem 3. *If the relaxation window is part of the input, functional significance checking is Δ_2^P-hard and is contained in Π_2^P.*

Proof. For the hardness, we show a reduction from the following Δ_2^P-complete problem [25]: Given a satisfiable CNF formula φ and a linear ordering of $x_1 \prec \ldots x_n$ in φ, does the lexicographically largest satisfying assignment of φ have its least significant bit $x_1 = 1$? To reduce this to functional significance checking, consider the construction in the proof of NP-hardness above, but with the following modification. The gadget in Fig. 2 is modified so that nodes a_i, na_i are removed and so are all edges coming in and out of them. We add an inhibiting edge from the source s to each nb_i, and an inhibiting edge from each nb_i to the corresponding node b_i. Let G be the resulting graph. Let R be the relaxation weight function that assigns weight 2^{i-1} to the inhibiting edge from s to nb_i and 2^n to all other edges. R also assigns weight 2^n to all nodes. We ask if b_1 is functionally significant in (G, s, t, W, R), where $W = \langle [0, 0], [0, 2^n - 1] \rangle$. The size of (G, s, t, W, R) is polynomial in $|\varphi|$. Also, the choice of W disallows relaxation of any node and any edge other than those from s to some nb_i. The lexicographically largest satisfying assignment of φ corresponds to an explanation graph with the smallest edge relaxation noise. If this explanation includes b_1, then removing b_1 from G disallows this explanation. This proves Δ_2^P hardness of functional significance checking.

Containment in Π_2^P is easy to see. We encode the problem as: for all (n', e')-relaxed solutions without the actor, there is an (n, e)-solution with the actor, where (n', e') strictly dominates (n, e) and both are within relaxation bounds. Since n, e, n', e' are integers within given relaxation bounds, the quantifier free part has a polynomial sized propositional encoding.

The problem of *counting explanation subgraphs* corresponds to #SAT, which is widely believed to be beyond the polynomial hierarchy (by Toda's theorem [47]). Thus, unless long-standing complexity-theory conjectures are falsified, checking functional significance (in Π_2^P) is easier than counting explanations.

4   SAT Encoding and Pareto-Curve Generation

Given a problem instance (G, s, t, W, R), and a path length bound Δ, we first extract a sub-graph $\hat{G} = (\hat{V}, \hat{E}, \hat{\lambda}, \hat{\mu})$ of G that contains every simple path of length $\leq \Delta$ from s to t in G. This can be done easily using a forward and backward bounded search. Once $\hat{V}$ is defined, $\hat{E}$, $\hat{\lambda}$ and $\hat{\mu}$ are obtained by restricting E, λ and μ respectively to $\hat{V}$ and $\hat{E}$. In practice, Δ is chosen based on domain expert inputs, such that all potentially important s-t paths are included. Henceforth, whenever we refer a labeled graph G, we mean the pruned graph $\hat{G}$ for a value of Δ that is assumed to be constant.

The problem of deciding whether an (n, e)-relaxed explanation subgraph exists was shown to be NP-complete in Sect. 3. A SAT encoding of the problem is rather straightforward. Given a labeled graph $G = (V, E, \lambda, \mu)$, nodes $s, t \in V$, a relaxation weight function R, and a relaxation window $W = \langle [0, n], [0, e] \rangle$, we construct a propositional formula $\varphi_{G,s,t,W,R}$ that is satisfiable iff there is an (n, e)-relaxed explanation subgraph of (G, s, t) under R. The formula $\varphi_{G,s,t,W,R}$ has seven sub-formulas: (i) φ_{conn} encoding topological connectivity between nodes in the explanation subgraph (this uses the fact that all paths are of length $\leq \Delta$), (ii) φ_{data} encoding the labeling of nodes obtained from microarray data, (iii) φ_{act} encoding the activity condition in Definition 1, (iv) φ_{comp} encoding the compatibility condition in Definition 1, (v) φ_{rel} encoding that total node relaxation is $\leq n$ and total edge relaxation is $\leq e$, and (vii) φ_{imp} encoding that every node is reachable from s by a path of length at most Δ. These sub-formulas use a set of variables as described below. For each $v \in V$, we use 3 boolean variables, p_v, a_v and r_v, that encode whether v is present, active and relaxed respectively, in the explanation subgraph. Similarly, for each edge $e \in E$, we use 3 boolean variables, p_e, r_e and f_e that encode whether e is present, relaxed and contributes to the activity condition in Definition 1 respectively, in the explanation subgraph. Finally, for each $v \in V$, we use $\log \Delta + 1$ propositional variables $d_{v,0}, \ldots d_{v,\log \Delta}$ to encode a measure of "distance" from source s to v in the explanation subgraph.

Once $\varphi_{G,s,t,W,R}$ is obtained, a SAT solver (Z3 [34] in our case) can be used to obtain an (n, e)-relaxed explanation subgraph. We exploit the observation that satisfiability of $\varphi_{G,s,t,W,R}$ implies satisfiability of $\varphi_{G,s,t,W',R}$ where $W' = \langle [0, n'], [0, e'] \rangle$ and $(n, e) \sqsubseteq (n', e')$. Therefore, given any set of (n, e) pairs linearly ordered w.r.t. $\sqsubseteq$, we can use binary search to determine the smallest (under $\sqsubseteq$) pair (n, e) for which $\varphi_{G,s,t,W,R}$ is satisfiable. This suggests the following simple algorithm for constructing the Pareto-optimal curve. We first use binary search along the (n_l, e_l) to (n_u, e_u) diagonal of the window $W = \langle [n_l, n_u], [e_l, e_u] \rangle$ to find the smallest (under $\sqsubseteq$) pair (n_d, e_d) for which $\varphi_{G,s,t,W_d,R}$ is satisfiable, where $W_d = \langle [n_l, n_d], [e_l, e_d] \rangle$. Note that (n_d, e_d) may not be a Pareto-optimal point. We then use binary search on (n, e) pairs in $\langle [n_d, e_l], [n_d, e_d] \rangle$ and $\langle [n_l, e_d], [n_d, e_d] \rangle$ to find the projections of (n_d, e_d) on the Pareto-optimal curve. Once a Pareto-optimal point (n_p, e_p) is obtained, the problem can be recursively decomposed into those of generating Pareto-optimal curves in the relaxation windows $\langle [n_p, n_u], [e_l, e_p] \rangle$ and $\langle [n_l, n_p], [e_p, e_u] \rangle$. This requires a total of $\mathcal{O}(k \log_2 k)$ invocations of a SAT solver, where $k = \max(n, e)$, and gives us the Pareto curves, from which we can determine functional significance.

Note that our methodology is not contingent on a specific choice of relaxation, but implicitly considers all relaxations within given bounds. However, our tool also has the functionality of printing a set of relaxations used to obtain explanation subgraphs, if the user so desires.

5   Experimental Results and a Case-Study

We began by constructing a database of existing pathways, by merging the 163 pathways from the KEGG database [22,23], giving a master network of 2498 nodes and 10497 edges. In discussion with molecular biologists, we then fixed the gene expression data from a specific microarray experiment, with the following features: (i) the source, target and the differentially expressed nodes were not merged with any other id, (ii) if a gene occurred more than once in the expression data, we took the average of the fold-change for more than one occurrence of a gene, (iii) after considering realistic lengths of regulatory chains in the biological context, the path bound (Δ in Sect. 4) was chosen to be 7. This resulted in a pruned subgraph with 297 nodes and 1858 edges. Of these nodes, 55 are up-regulated and 26 are down-regulated, as per the microarray data (see [1] for details). Finally, we also fixed an upper bound on number of relaxations that we allow among the nodes and edges in the worst case, i.e., the window size, denoted below as W to be at most 30×30. Note that this does not mean that we cannot have fewer perturbations, just that more than 30 errors (of either nodes or edges) were considered impractical. While we fix all the above parameters to be able to present results, we emphasize that these are easily tunable by the user. In our experiments, the relaxation weight function R assigned weight 1 to all nodes and edges. But the formulation allows generalizing to other weight functions, e.g., to not relax a node or edge, it suffices to assign a large weight to that node/edge.

With this setup, we encoded finding a relaxed explanation graph, as discussed in Sect. 4, and considered different source and target pairs, as well as different candidate actors which were checked for functional significance. We computed the Pareto optimal curves with and without the actor to check functional significance of the actor. All experiments were performed on an Intel(R)-Core(TM)-i7-3770 CPU. It had 8 cores with clock speed 3.40 GHz and total of 32 GB RAM. The code used C++ API of Z3 version 4.7.1 on Ubuntu 18.04.

One way to understand the explanations is to enumerate and exhaustively look at each solution. However, with window size 30×30, there are 900 points,

Table 1. Shift of Pareto curves

Source-target pair (Expt condition)	Func. Sign. Cand.	Pareto shift (Y/N)	# SAT Calls	Time (in hrs)
Synthetic1-5var-$W(5,5)$	x	Y	5	.035
Synthetic2-15var-$W(5,5)$	x	Y	6	.35
Synthetic3-45var-$W(0,0)$	x	Y	2	.004
TNFa-IkBa (Expr/Act merged)	None	-	62	5
TNFa-IkBa (Expr/Act merged)	p38	Y	72	5
TNFa-IkBa (Expr/Act merged)	ERK	N	62	2.6
TNFa-IkBa (Expr/Act merged)	PIK3CA	Y	71	1.5
TNFa-IkBa (Expr/Act merged)	AKT	Y	42	11
TNFa-IkBa (Expr only)	None	-	63	9
TNFa-IkBa (Expr only)	p38	Y	63	15
TNFa-IkBa (Expr only)	ERK	Y	63	15
TNFa-IkBa (Expr only)	PIK3CA	N	68	14
TNFa-IkBa (Expr only)	AKT	N	68	18.4
TNFa-IkBa (Act only)	None	-	64	15.6
TNFa-IkBa (Act only)	p38	Y	64	37
TNFa-IkBa (Act only)	ERK	N	64	25.8
TNFa-IkBa (Act only)	PIK3CA	Y	64	18.5
TNFa-IkBa (Act only)	AKT	Y	54	44
TNFa-A20	None	-	56	0.3
TNFa-A20	ERK	Y	57	0.7
TNFa-A20	AKT	N	52	0.3
TNFa-A20	p38	N	54	0.3

of which all points on or above the PO curve have multiple solutions. In our case, we found that for all such points there were at least >1000 solutions per point. And enumerating these, and printing the solutions for just 30 of them (for inspection), for a single PO curve took over 100 hours of computations time. Thus, examining all solutions even at each point on the Pareto-optimal curve (to identify key players in the solution) is already prohibitively expensive. This leads us to use the shift of the Pareto-optimality curves to identify key players in context of an experiment. In Table 1, we present the results for a few different source-target pairs, different candidate actors and whether a shift was observed in the Pareto-curves or not, along with the time taken to plot these curves. The Pareto-optimality curves themselves, along with further experiments with more source-target pairs including ITGB1-ACTB, ITGB1-STAT3 are in [1]. We also performed experiments on synthetically constructed benchmarks motivated by Proof of Theorem 2. The benchmarks were parametrized by number of variables (in the 3SAT problem), and node, edge relaxation upper bounds, and a special node x that was made functionally significant. A select few results are in Table 1, with more in [1]. Interestingly, almost the entire time taken by our tool went into SAT solving using a state-of-the-art solver (Z3). Our tool minimizes the number of SAT calls as described in Sect. 4. The scalability of our approach hence crucially depends on the performance of the SAT solver, and is expected to improve with further improvements in SAT solvers.

In Table 1, Act/Expr merged means we included both types of edges in our potential explanation. However, we also experimented by (i) asking for the target IkBa to be expressed, and not just activated (by required the solution to have at least one expression edge reaching the target) and (ii) asking target IkBa to be activated (by requiring the solution to have at least one activation edge reaching the target), which led to surprisingly different Pareto-shifts.

Case-Study: Role of ERK, A20 in PSMD9-Induced Inhibition of NFkB: We performed a detailed case-study on a mammalian cell line model system, created as part of a joint project with researchers from a Cancer research institute: these were the embryonic human kidney cell lines called HEK 293 cell that stably over express PSMD9 (an important gene associated with radio resistance in breast cancer and glioblastoma [26, 36].) and obtained differential gene expression data specific to PSMD9. Among the many signalling events that could possibly be modulated by PSMD9, we were interested in finding key players that regulated the expression of IkBa, for a very specific reason. IkBa is a potent inhibitor of NFkB a transcription factor induced upon chemo and radiation therapy in cancer treatment [3, 17].

One of the mechanisms by which PSMD9 may achieve this is by inducing NFkB activation [39]. However, besides the reported mechanism, there are a number of other ways in which the activity of this gene can be modulated and this can vary depending on the context. Several kinases and transcription factors are involved and inflammatory cytokines such as TNFa can modulate activity of these players. NFkB is also under a remarkable tight feed-back loop involving both positive and negative regulators that are transcribed by NFkB and other TFs. Therefore any attempt towards developing therapeutic mechanism to overcome therapy resistance associated with PSMD9 demands a comprehensive understand of the many mechanism leading to the expression of the target genes of NFkB including IkBa, the contribution of other TFs, the role of kinases and their crosstalk. Since this also involves feed-back loops and it can become challenging to identify the activation/repression status of the genes involved both for experimental verification and computational approaches. This provided us with a case study: we considered the gene expression data from above and took TNFa, a gene induced by PSMD9 overexpression as the stimulus and IkBa as the target to help uncover the key players involved in the expression of IkBa, the endogenous inhibition of NFkB. From the literature and using domain knowledge, 4 candidate key actors were chosen, namely p38, ERK, PIK3CA and AKT. The Pareto-optimality curves generated for TNFa to IkBa are shown in Fig. 4.

Biological Validation. Among nodes explored for functionality, we completed wetlab investigations at submission-time for ERK and AKT kinases, which showed PSMD9-induced phosphorylation. As mentioned earlier, we used a merged KEGGgraph combining activation and expression edges for simplicity. Since negative feedback loops involving both IkBa and A20 control NFkB activation and target gene expression, we also conducted experiments for both these targets after separating the composite graphs into activation and expression graphs (see [1]). Phosphorylation impacts (in)activation status of transcription factors (in-built in Response: KEGG) and hence must be integrated into gene expression studies. Indeed, excluding ERK from composite graphs did not induce Pareto shift, whereas separating into activation and expression graphs did. As can be

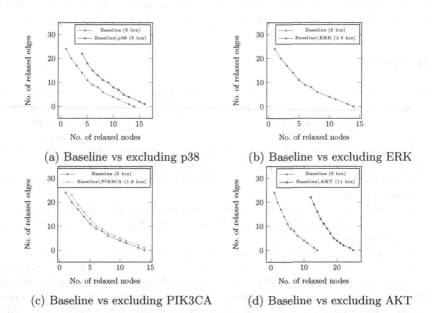

(a) Baseline vs excluding p38   (b) Baseline vs excluding ERK

(c) Baseline vs excluding PIK3CA   (d) Baseline vs excluding AKT

Fig. 4. Individual plots of the exclusion experiments for TNFa-IkBa

gleaned from Table 1 (and [1]) , ERK exclusion, but not AKT exclusion induced a Pareto shift indicating its requirement in both IkBa and A20 expression. Only AKT induced IkBa (in)activation Pareto shift (see [1]). We tested ERK's significance in IkBa and A20 gene expression using qPCR. A two-fold decrease in A20 mRNA was observed in PSMD9 overexpression cells upon ERK inhibition [p = 0.03] whereas AKT inhibition did not impact IkBa or A20 mRNA levels, a trend consistent even upon TNFa stimulation (t = 3hrs). The lack of impact of ERK inhibition on IkBa mRNA levels is likely due to as yet unexplored PSMD9-specific effects. The NFkB-dependence for IkBa or A20 expression was evident from lack of solutions upon its exclusion. The routinely-used PD98059 and LY294002 signaling inhibitors achieved ERK (~100%) and AKT (~90%) phosphorylation inhibition, respectively, at recommended IC50 values. They may have off-target effects. *Importantly, these inhibition-dependent mRNA level changes were PSMD9-specific, consistent with computational predictions.*

6   Conclusion

We presented a novel problem formulation to capture functional signficance of a node in an interaction pathway between a stimulus and a target observation, in a highly noisy environment with minimal experimental data and using publicly available pathway databases. Our definition comes closest to a computational simulation of a knockout experiment that is classically done to establish the functional significance of a node in wet-lab experiments. After showing theoretical hardness results, we design practical encodings using SAT, which we implemented and validated by some wet-lab experiments and domain knowledge.

References

1. Akshay, S., Basu, S., Chakraborty, S., Sundararajan, R., Venkatraman, P.: Constraint-based functional significance checking in biological networks. https://github.com/sukanyabasu2009/network_tool_CP19. supplementary material
2. Alcaraz, N., Kücük, H., Weile, J., Wipat, A., Baumbach, J.: Keypathwayminer: detecting case-specific biological pathways using expression data. Internet Math. **7**(4), 299–313 (2011). https://doi.org/10.1080/15427951.2011.604548
3. Bai, M., et al.: The accomplices of NF-kB lead to radioresistance. Curr. Protein. Peptide Sci. **16**(4), 279–294 (2015)
4. Beltrame, L., et al.: Using pathway signatures as means of identifying similarities among microarray experiments. PLOS One **4**(1), 1–11 (2009). https://doi.org/10.1371/journal.pone.0004128
5. Bolón-Canedo, V., Sánchez-Maroño, N., Alonso-Betanzos, A., Benítez, J.M., Herrera, F.: A review of microarray datasets and applied feature selection methods. Inf. Sci. **282**, 111–135 (2014). https://doi.org/10.1016/j.ins.2014.05.042
6. Cavalieri, D., et al.: Eu.Gene analyzer a tool for integrating gene expression data with pathway databases. Bioinformatics **23**(19), 2631–2632 (2007). https://doi.org/10.1093/bioinformatics/btm333

7. Chen, X., et al.: A sub-pathway-based approach for identifying drug response principal network. Bioinformatics **27**(5), 649–654 (2011). https://doi.org/10.1093/bioinformatics/btq714
8. Cline, M.S., et al.: Integration of biological networks and gene expression data using cytoscape. Nat. Protoc. **2**, 2366–82 (2007)
9. Drăghici, S., Khatri, P., Martins, R.P., Ostermeier, G., Krawetz, S.A.: Global functional profiling of gene expression. Genomics **81**(2), 98–104 (2003). http://www.sciencedirect.com/science/article/pii/S0888754302000216
10. Dunn, S.J., Martello, G., Yordanov, B., Emmott, S., Smith, A.G.: Defining an essential transcription factor program for naïve pluripotency. Science **344**(6188), 1156–1160 (2014). https://science.sciencemag.org/content/344/6188/1156
11. Gebser, M., Schaub, T., Thiele, S., Veber, P.: Detecting inconsistencies in large biological networks with answer set programming. CoRR abs/1007.0134 (2010). http://arxiv.org/abs/1007.0134
12. Gebser, M., Schaub, T., Thiele, S., Veber, P.: Detecting inconsistencies in large biological networks with answer set programming. TPLP **11**(2–3), 323–360 (2011). https://doi.org/10.1017/S1471068410000554
13. Geistlinger, L., Csaba, G., Küffner, R., Mulder, N., Zimmer, R.: From sets to graphs: towards a realistic enrichment analysis of transcriptomic systems. Bioinformatics **27**(13), i366–i373 (2011). https://doi.org/10.1093/bioinformatics/btr228
14. Giacobbe, M., Guet, C.C., Gupta, A., Henzinger, T.A., Paixão, T., Petrov, T.: Model checking gene regulatory networks. In: Proceedings of the Tools and Algorithms for the Construction and Analysis of Systems - 21st International Conference, TACAS 2015, Held as Part of the European Joint Conferences on Theory and Practice of Software, ETAPS 2015, London, UK, 11–18 April 2015, pp. 469–483 (2015)
15. Glaab, E., Baudot, A., Krasnogor, N., Schneider, R., Valencia, A.: Enrichnet: network-based gene set enrichment analysis. Bioinformatics **28**(18), i451–i457 (2012). https://doi.org/10.1093/bioinformatics/bts389
16. Glaab, E., Baudot, A., Krasnogor, N., Valencia, A.: Topogsa: network topological gene set analysis. Bioinformatics **26**(9), 1271–1272 (2010). https://doi.org/10.1093/bioinformatics/btq131
17. Godwin, P., Baird, A.M., Heavey, S., Barr, M., O'Byrne, K., Gately, K.: Targeting nuclear factor-kappa B to overcome resistance to chemotherapy. Front. Oncol. **3**, 120 (2013). https://www.frontiersin.org/article/10.3389/fonc.2013.00120
18. Guerra, J., Lynce, I.: Reasoning over biological networks using maximum satisfiability. In: Milano, M. (ed.) CP 2012. LNCS, pp. 941–956. Springer, Heidelberg (2012). https://doi.org/10.1007/978-3-642-33558-7_67
19. Guziolowski, C., Borgne, M.L., Radulescu, O.: Checking consistency between expression data and large scale regulatory networks: a case study (2007)
20. Jong, H.D.: Modeling and simulation of genetic regulatory systems: a literature review. J. Comput. Biol. **9**, 67–103 (2002)
21. Judeh, T., Johnson, C., Kumar, A., Zhu, D.: Teak: Topology enrichment analysis framework for detecting activated biological subpathways. Nucleic Acids Res. **41**(3), 1425–1437 (2013). https://doi.org/10.1093/nar/gks1299
22. Kanehisa, M., Goto, S.: KEGG: Kyoto encyclopedia of genes and genomes. Nucleic Acids Res. **28**, 27–30 (2000)
23. Kanehisa, M., Sato, Y., Kawashima, M., Furumichi, M., Tanabe, M.: KEGG as a reference resource for gene and protein annotation. Nucleic Acids Res. **44**, D457–D462 (2016)

24. Koumakis, L., et al.: Minepath: mining for phenotype differential sub-paths in molecular pathways. PLOS Comput. Biol. **12**(11), 1–40 (2016). https://doi.org/10.1371/journal.pcbi.1005187

25. Krentel, M.W.: The complexity of optimization problems. J. Comput. Syst. Sci. **36**(3), 490–509 (1988). https://doi.org/10.1016/0022-0000(88)90039-6

26. Langlands, F.E., et al.: PSMD9 expression predicts radiotherapy response in breast cancer. Mol. Cancer **13**(1), 73 (2014). https://doi.org/10.1186/1476-4598-13-73

27. Lee, H., Shin, M.: Mining pathway associations for disease-related pathway activity analysis based on gene expression and methylation data. BioData Min. **10**(1), 3 (2017). https://doi.org/10.1186/s13040-017-0127-7

28. Lee, S., Park, Y., Kim, S.: MIDAS: mining differentially activated subpaths of KEGG pathways from multi-class RNA-seq data. Methods **124**, 13–24 (2017). http://www.sciencedirect.com/science/article/pii/S1046202317300488. Integrative Analysis of Omics Data

29. Li, C., et al.: Subpathwayminer: a software package for flexible identification of pathways. Nucleic Acids Res. **37**(19), e131 (2009). https://doi.org/10.1093/nar/gkp667

30. Lifschitz, V.: What is answer set programming? In: Proceedings of the Twenty-Third AAAI Conference on Artificial Intelligence, AAAI 2008, Chicago, Illinois, USA, 13–17 July 2008, pp. 1594–1597 (2008). http://www.aaai.org/Library/AAAI/2008/aaai08-270.php

31. Ma, S., Kosorok, M.R.: Detection of gene pathways with predictive power for breast cancer prognosis. BMC Bioinformatics **11**(1), 1 (2010). https://doi.org/10.1186/1471-2105-11-1

32. Martini, P., Sales, G., Massa, M.S., Chiogna, M., Romualdi, C.: Along signal paths: an empirical gene set approach exploiting pathway topology. Nucleic Acids Res. **41**(1), e19 (2013). https://doi.org/10.1093/nar/gks866

33. Melas, I.N., Samaga, R., Alexopoulos, L.G., Klamt, S.: Detecting and removing inconsistencies between experimental data and signaling network topologies using integer linear programming on interaction graphs. PLOS Comput. Biol. **9**(9), 1–19 (2013)

34. de Moura, L., Bjørner, N.: Z3: An efficient SMT solver. In: Ramakrishnan, C.R., Rehof, J. (eds.) TACAS 2008. LNCS, vol. 4963, pp. 337–340. Springer, Heidelberg (2008). https://doi.org/10.1007/978-3-540-78800-3_24

35. Nam, S., et al.: Pathome: an algorithm for accurately detecting differentially expressed subpathways. Oncogene **33**(41), 4941–4951 (2014)

36. Rajendra, J., et al.: Enhanced proteasomal activity is essential for long term survival and recurrence of innately radiation resistant residual glioblastoma cells. Oncotarget **9**(25), 27667 (2018)

37. Ramadan, E., Alinsaif, S., Hassan, M.R.: Network topology measures for identifying disease-gene association in breast cancer. BMC Bioinformatics **17**(7), 274 (2016). https://doi.org/10.1186/s12859-016-1095-5

38. Rhodes, D.R., et al.: Oncomine 3.0: genes, pathways, and networks in a collection of 18,000 cancer gene expression profiles. Neoplasia **9**(2), 166–180 (2007). http://www.sciencedirect.com/science/article/pii/

39. Sahu, I., Sangith, N., Ramteke, M., Gadre, R., Venkatraman, P.: A novel role for the proteasomal chaperone PSMD9 and hnRNPA1 in enhancing IkBa degradation and NF-kB activation - functional relevance of predicted PDZ domain-motif interaction. FEBS Open Bio. **281**(11), 2688–2709 (2014)

40. Sharan, R., Karp, R.M.: Reconstructing boolean models of signaling. J. Comput. Biol. **20**(3), 249–257 (2013). https://doi.org/10.1089/cmb.2012.0241

41. Siegel, A., Radulescu, O., Borgne, M.L., Veber, P., Ouy, J., Lagarrigue, S.: Qualitative analysis of the relation between DNA microarray data and behavioral models of regulation networks. Biosystems **84**(2), 153–174 (2006). http://www.sciencedirect.com/science/article/pii/S0303264705001723. dynamical Modeling of Biological Regulatory Networks

42. Soule, C.: Mathematical approaches to differentiation and gene regulation. C. R. Biol. **329**(1), 13–20 (2006). http://www.sciencedirect.com/science/article/pii/S1631069105001800. modelisation de systemes complexes en agronomie et environnement

43. Steel, S., Alami, R. (eds.): ECP 1997. LNCS, vol. 1348. Springer, Heidelberg (1997). https://doi.org/10.1007/3-540-63912-8

44. Subramanian, A., et al.: Gene set enrichment analysis: a knowledge-based approach for interpreting genome-wide expression profiles. Proc. Nat. Acad. Sci. **102**(43), 15545–15550 (2005). http://www.pnas.org/content/102/43/15545.abstract

45. Tarca, A.L., et al.: A novel signaling pathway impact analysis. Bioinformatics **25**(1), 75–82 (2009). https://doi.org/10.1093/bioinformatics/btn577

46. Thiele, S., Cerone, L., Saez-Rodriguez, J., Siegel, A., Guziołowski, C., Klamt, S.: Extended notions of sign consistency to relate experimental data to signaling and regulatory network topologies. BMC Bioinformatics **16**(1), 345 (2015)

47. Toda, S.: PP is as hard as the polynomial-time hierarchy. SIAM J. Comput. **20**(5), 865–877 (1991)

48. Ulitsky, I., Krishnamurthy, A., Karp, R.M., Shamir, R.: Degas: de novo discovery of dysregulated pathways in human diseases. PLOS One **5**(10), 1–14 (2010). https://doi.org/10.1371/journal.pone.0013367

49. Vaske, C.J., et al.: Inference of patient-specific pathway activities from multidimensional cancer genomics data using paradigm. Bioinformatics **26**(12), i237–i245 (2010). https://doi.org/10.1093/bioinformatics/btq182

50. Wang, L., Zhang, B., Wolfinger, R.D., Chen, X.: An integrated approach for the analysis of biological pathways using mixed models. PLOS Genet. **4**(7), 1–9 (2008). https://doi.org/10.1371/journal.pgen.1000115

51. Wang, R.S., Saadatpour, A., Albert, R.: Boolean modeling in systems biology: an overview of methodology and applications. Phys. Biol. **9**(5), 055001 (2012). http://stacks.iop.org/1478-3975/9/i=5/a=055001

52. Wang, Z., Gerstein, M., Snyder, M.: RNA-Seq: a revolutionary tool for transcriptomics. Nat. Rev. Genet. **10**(1), 57–63 (2009). https://doi.org/10.1038/nrg2484

53. Warde-Farley, D., et al.: The genemania prediction server: biological network integration for gene prioritization and predicting gene function. Nucleic Acids Res. **38**(suppl 2), W214–W220 (2010). https://doi.org/10.1093/nar/gkq537

54. Xia, J., Wishart, D.S.: MetPA: a web-based metabolomics tool for pathway analysis and visualization. Bioinformatics **26**(18), 2342–2344 (2010). https://doi.org/10.1093/bioinformatics/btq418

55. Yu, H., Kim, M.P., Sprecher, E., Trifonov, V., Gerstein, M.: The importance of bottlenecks in protein networks correlation with gene essentiality and expression dynamics. PLoS Comput. Biol. **3**(4), e59 (2007). https://doi.org/10.1371/journal.pcbi.0030059

Correction to: Dual Hashing-Based Algorithms for Discrete Integration

Alexis de Colnet and Kuldeep S. Meel

Correction to:
Chapter "Dual Hashing-Based Algorithms for Discrete Integration" in: T. Schiex and S. de Givry (Eds.):
Principles and Practice of Constraint Programming,
LNCS 11802, https://doi.org/10.1007/978-3-030-30048-7_10

In the version of this paper that was originally published, reference was made to an incorrect grant number. This has now been corrected.

The updated version of this chapter can be found at
https://doi.org/10.1007/978-3-030-30048-7_10

T. Schiex and S. de Givry (Eds.): CP 2019, LNCS 11802, p. C1, 2019.
https://doi.org/10.1007/978-3-030-30048-7_45

Author Index

Printed in the United States
By Bookmasters